Sustainable Environmental Waste Management Strategies

Sustainable waste management is an absolute imperative in a world that is already dealing with the effects of climate change. This book series seeks to publish an extensive collection of scientific books targeting researchers, scientists, professionals, academics, students, and anyone else involved in the wide, diverse field of sustainable waste management. The series accepts monographs, edited volumes, textbooks, and conference proceedings. Topics covered include but are not limited to environmental pollution and remediation, sources of environmental waste, environmental and human health, sustainable waste management, circular economy, nature-inspired environmental solutions, pollution prevention, and energy-efficient technologies for waste management. The series puts a particular emphasis on developing countries/the majority world but accepts proposals globally as long as the topic is in scope.

Ipsita Dipamitra Behera · Alok Prasad Das ·
Manisha Bal
Editors

Emerging Micropollutants

Status in the Environment and Management through Bioremediation

Springer

Editors
Ipsita Dipamitra Behera
Department of Chemical Engineering
Indira Gandhi Institute of Technology, Sarang
Dhenkanal, Odisha, India

Manisha Bal
Department of Chemical Engineering
NIT Durgapur
Durgapur, West Bengal, India

Alok Prasad Das
Department of Life Sciences
Rama Devi Women's University
Bhubaneswar, Odisha, India

ISSN 3005-1622 ISSN 3005-1630 (electronic)
Sustainable Environmental Waste Management Strategies
ISBN 978-3-032-03529-5 ISBN 978-3-032-03530-1 (eBook)
https://doi.org/10.1007/978-3-032-03530-1

This Springer imprint is published by the registered company Springer Nature Switzerland AG
The registered company address is: Gewerbestrasse 11, 6330 Cham, Switzerland

Contents

Table 1.1 (continued)

Contaminant Category	Examples	Description	Sources	Effects on Human Health	Effects on Environment
Persistent Organic Pollutants	Poly-chlorinated biphenyls, dioxins, furans, aldrin	Chemicals that retain in the ecosystems for long periods, bio magnify	Industries, incinerators, agricultural-runoff	Cancer, neurological and immune system damage, reproductive harm	Organic pollutants accumulate in the food chain, harming predators, especially in the Arctic, and disrupting biodiversity
Radioactive Contaminants	Radon, uranium, cesium-137, strontium-90	Radioactive materials that emit harmful radiation affecting human tissues and biological processes	Mining, nuclear waste, medical waste, natural sources	Lung cancer. kidney damage, radiation, poisoning and cancer	Radioactive contamination can persist for long periods, causing soil, water, and air contamination, harming ecosystems and wildlife

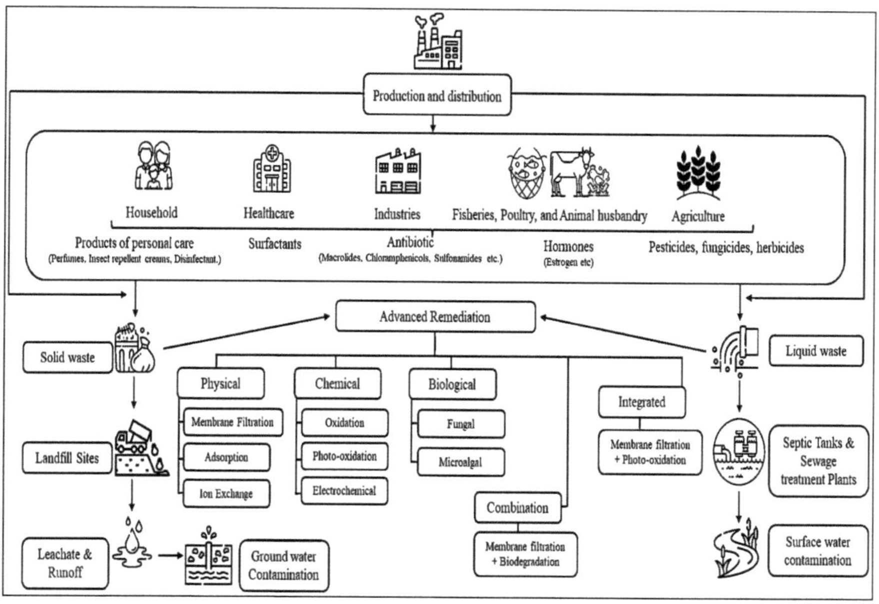

Fig. 1.2 Remediation methods for pharmaceuticals and other metabolites entering to aquatic systems (Gupta et al., 2024)

Environmental Impacts of Micro Pollutants

Soil Contamination

The term "soil contamination" describes the existence of hazardous chemicals, heavy metals, or other contaminants in the soil that impair its quality and hinder its capacity to support microbial and plant life. Usually, human activities like mining, farming, industrial waste disposal, and poor waste management are the cause of this contamination. Heavy metals (such as lead, mercury, and cadmium), insecticides, fungicides, surfactants and solvents are examples of micro pollutants. The presence of pollutants in soil can change its chemical makeup, making it harder for plants to get the vital nutrients they need to develop. The pH of the soil can also be affected by toxic compounds, making it too acidic or too alkaline for most crops. Reduced crop yields and decreased agricultural pollinators follow from the resulting drop in soil fertility, use of pesticides and rise in pest infestation that are resistant to these chemicals. This presents a serious risk to food security, particularly in countries where agriculture is the main source of income. The decomposition of organic matter, soil structure maintenance and nitrogen cycling are all greatly aided by soil microbes (Das et al., 2024; Das et al., 2024a, 2024b, 2024c; Dey et al., 2024a, 2024b; Dey et al., 2024a, 2024b; Mishra et al., 2023a, 2023b). On the other hand, microbial activity is disrupted when harmful substances are introduced, herbicides and pesticides, for instance, can lower soil microbiome biodiversity, which compromises soil fertility and health. The health of plants may be further impacted by the growth of pathogenic bacteria. Through decreased seed germination rates, stunted plant growth, or crop death, soil pollution can have a direct impact on agricultural growth. Long-term exposure to contaminated soil causes the quality of the soil to decline, requiring costly chemicals and fertilizers to restore productivity, which worsens environmental issues (Du et al., 2016; Ferreira et al., 2020; Fu et al., 2023).

Water Pollution

Water pollution arises hazardous materials like chemicals, poisons, and waste products pollute rivers, lakes, seas, and groundwater. These contaminants majorly impact aquatic ecosystems and human health in addition to deteriorating water quality. Aquatic ecosystems have a very delicate balance which can be easily disrupted by micropollutants (Das & Ghosh, 2023; Dey et al., 2023; Jagatee et al., 2024; Priyadarshini & Das, 2024). Agricultural runoff, industrial chemicals, heavy metals, and municipal waste are some of the examples of chemical pollutants that can seriously deteriorate aquatic life and reduce local biodiversity. Elevated nutrient levels, such as organic matter and inorganic minerals, cause eutrophication, a phenomenon that leads to algal blooms. Agricultural runoff frequently introduces micropollutants like nitrogen and phosphorus, intensifying the process. These blooms cause "dead zones" in the water by declining its oxygen content, during which it is impossible for any marine life to survive. Furthermore, harmful elements such as lead, cadmium and mercury build up in aquatic specie's tissues, rendering them unfit for human consumption. Human health can be directly threatened by polluted water seeping into surface or groundwater. Heavy metals, herbicides, and microbial infestations are examples of few pollutants that can lead to a variety of health issues, including neurological damage, cancers and gastrointestinal disorders. In developing nations, where access to clean water may be restricted, drinking water contamination is especially concerning (Adegoke et al., 2022; Glassmeyer et al., 2005; Handford et al., 2015; Jiang et al., 2013).

Air Pollution

The persistent presence of hazardous compounds in the atmosphere that might deteriorate air quality and endanger human health. These micro pollutants originate from a number of sources, including natural processes like wildfires, volcanic eruptions and through daily activities such as exhaust from automobiles and other home appliances, industrial emissions, and agricultural practices. oxides of nitrogen and sulphur, insecticides, and persistent organic pollutants are examples of micro-pollutants that can travel great distances through the atmosphere. Wind currents can carry these pollutants and deposit them far from their original source (Bala et al., 2024a, 2024b; Bala et al., 2024a, 2024b; Mohanty et al., 2024a, 2024b, 2024c). For instance, industrial-related mercury can transit continents and settle in far-flung locations, affecting human populations and ecosystems at great distances from the source. Extended exposure to air pollution also contributes to respiratory problems in humans and can exacerbate conditions like asthma and bronchitis. Once airborne pollutants settle onto the ground through processes like rain or dry deposition, they can contaminate soil, water bodies, and vegetation. Acid rain, which is caused by atmospheric sulphur dioxide and nitrogen oxide emissions, can harm crops, forests, and aquatic ecosystems. (Ahmed et al., 2019; Ramírez et al., 2012; Shen et al., 2014).

Biodiversity Loss

A number of factors, such as climate change, pollution, habitat destruction, and overexploitation, contribute to biodiversity loss, which is the decline or extinction of species within an ecosystem. Pollution, particularly in the soil, air and water, can directly harm organisms, making it more difficult to live and procreate. The reproductive health of species can be significantly impacted by pollution, for instance, endocrine disruptors and pesticides have an impact on wildlife's reproductive systems, resulting in decreased fertility and altered behaviour. Deforestation and wetland draining are two examples of habitat degradation that increases the danger of extinction for numerous species. Exposure to contaminants such as insecticides, aerosols and micro plastic can diminish the success of reproduction and induce developmental abnormalities in offspring in aquatic organisms. In a similar vein, exposure to these chemicals in the environment makes reproduction difficult for terrestrial species as well. In order to sustain ecological processes like pollination, water purification, and nutrient cycling, biodiversity is essential. Ecosystems become less robust when species disappear because these processes are disrupted. For instance, crop productivity and food security are directly impacted by the reduction of pollinator species, such as bees (Dulsat-Masvidal et al., 2023; Su et al., 2021; Vörösmarty et al., 2010).

Bioaccumulation and Biomagnification

Bioaccumulation is a process in which accumulation of hazardous substances in an organism over a period of time. Usually non-biodegradable, these compounds can remain in the tissues of an organism for a long time, frequently at higher concentrations greater than those in the neighbouring environment. Conversely, biomagnification describes how these harmful compounds become more prevalent as they ascend the food chain in a particular ecosystem. Pollutants build up in organisms, including pesticides, heavy metals (such cadmium, lead and arsenic), and organic pollutants. Producers such as microbes, fungi, and plants absorb these toxins from soil or water, as do primary consumers like plankton, insects, fish, and small mammals at the base of the food chain. Predator's bodies get more concentrated with toxins as they eat their prey species. When tertiary creatures of food web, including humans, ingest tainted fish, meat, or plants, it might cause health issues. Micro Pollutants buildup at higher trophic levels can cause serious health problems, such as immune system weakness, developmental abnormalities, and infertility (Das & Singh, 2011; Paul et al., 2023, 2024; Raj & Das, 2023). Mercury poisoning, for example, can cause brain damage and even death in top predators like humans, dolphins, and eagles. Apex predators experience toxic concentrations of contaminants as a result of the biomagnification process, which has a hazardous knock-on effect on ecosystems. (Adegoke et al., 2022; Du et al., 2016; Gupta et al., 2024; Vadiraj et al., 2021).

Toxicological Effects and Consequences for Human Health

Chronic exposure to micro-pollutants, which include substances has been confederated with a number of serious health effects. Cancers are one of the main concern, as research shows that carcinogenic compounds like dioxins, benzene, and persistent organic pollutants can build up in the tissues and organs over time and raise the risk of developing several types of cancer both in humans and wildlife. Endocrine disruptors from micropollutants may interfere with the organism's ability to regulate hormones, developmental delays, thyroid issues, and infertility are some of the issues linked to these disturbances. Immune system suppression, liver damage, and neurotoxicity are further possible health effects. In the long run, human concentrations of microscopic pollutants may rise as a result of their accumulation in the environment and bioaccumulation in food chains. Emerging pollutants are especially hazardous because of their propensity to linger in the environment, posing health concerns to the general public years after their initial exposure. Controlling the impact of micropollutants on the environment and human health requires an understanding of their toxicological consequences (Vadiraj et al., 2021; Wu et al., 2019). Micro-pollutants in the environment pose significant risks to public health through various channels, including soil, water, food, and air. Contamination of drinking water with micro-pollutants such as pharmaceuticals, industrial chemicals, and heavy metals can lead to major health issues, including kidney disease, neurological disorders, and cancers. Similarly, agricultural runoff containing pesticides or pharmaceutical residues can contaminate food sources, which then enter the human food web. Long-term consumption of these tainted food products can result in a build-up of toxic chemicals in the body, leading to adverse health effects. Airborne pollutants, including aerogels, particulate matter and organic compounds, can cause neurological and cardiovascular diseases. People living near industrial areas or in regions with high levels of automobile emissions are at a heightened risk. Gradually, micropollutants can build up in the environment and cause widespread contamination that impacts public health systems as well as individual health, increasing healthcare expenses and burdening society. To reduce these dangers, micropollutants must be effectively monitored and regulated (Devi et al., 2017; Du et al., 2019; Tiwari et al., 2021). Children, the elderly, and marginalized communities are among the groups most vulnerable to the negative health effects of micropollutants. Due to their developing specie's increased sensitivity to environmental contaminants, children are more vulnerable. Immune systems, growth, and cognitive development of internship can all be hampered by exposure to endocrine-disrupting substances. The long-term impacts of micro-pollutants, such as respiratory and cardiovascular disorders, are more likely to affect the elderly, whose immune systems and organ functioning may already be compromised. Additionally, because they live close to contaminated sources like factories, landfills, and busy roads, marginalized communities, especially those in low-income or industrial areas are frequently disproportionately exposed to these contaminants. This community's vulnerability is increased since they lack the means to relocate to cleaner areas or obtain treatment. These groups are especially susceptible to the toxicological effects of micro-pollutants due to the cumulative effect of multiple exposures to pollutants and social determinants of health. This underscores the necessity of environmental justice and focused public health interventions (Du et al., 2016; Enyoh et al., 2020).

Global and Regional Trends

The impacts of micropollutants vary depending on urban, agricultural, and industrial activity. Pharmaceuticals, pesticides, heavy metals, plastics, and other micropollutants can be found in elevated concentrations in the air, water, soil, and food around the world. These pollutants are frequently the result of human endeavours such as industrial manufacturing, poor waste disposal, and agricultural practices. Even in the most remote places, these pollutants can contribute to environmental contamination because they are persistent, can build up in ecosystems, and can travel across areas through natural processes like air and water currents. The lack of efficient international regulations makes this broad dispersion even worse by enabling pollutants to traverse national boundaries and affect far-off areas. Micro-pollutants have become a significant global concern that cuts across national boundaries and need international cooperation since they are especially vulnerable to developing nations with weaker environmental regulations, infrastructure, or resources for pollution control. Due to heavy industrial and agricultural operations, some regions have exceptionally high concentrations of micro-pollutants, making them regional hotspots of specific pollutants. These hotspots are frequently located close to urban centers, agricultural regions, and heavily industrialized areas. In some parts of Asia, for example, large concentrations of chemical pollutants, such as pesticides, heavy metals, and industrial waste products, contaminate the air and water due to the region's extensive manufacturing industries and intensive agricultural runoff. Comparable problems crop up in Europe, where pollutants including nitrogen and sulphur compounds, particulate matter, and microplastics are frequently found due to its extensive transportation networks and dense network of chemical production businesses. Micropollutants frequently impact agricultural land, water bodies, and air quality in these regional hotspots, leading to environmental deterioration and human health problems like respiratory, cardiovascular, endocrinal and neurological disorders, waterborne infections, and food web contamination. These areas face disproportionate risks from emerging pollutants due to regulatory inadequacies, limited access to clean water, and restricted healthcare availability. As such, they are important targets for intervention and sustainable management techniques.

Emerging contaminants present additional environmental and public health risks in addition to the ones that are already present. Pollutants that have only lately been recognized as dangerous but whose long-term effects on the environment and human wellbeing are still not entirely known are known as emerging contaminants. Pharmaceuticals, personal care items, and endocrine-disrupting chemical compounds like phthalates and bisphenol A (BPA), which have been found in a variety of environmental media like land, water and air, are a few examples of these emerging pollutants. Due to their extensive presence in ecosystems and capacity to endure for decades, microplastics (minute plastic particles) are another newly discovered contaminant that has grown to be a serious ecological concern. These contaminants may build up in the tissues of an organism over time and magnify contamination. Since they are often resistant to conventional waste management procedures, these new pollutants are especially concerning because it is unclear how they may affect ecosystems, wildlife, biodiversity and human health when exposed to various emerging micropollutants for longer durations (Fig. 1.3).

Environmental rules do not control many of these emerging compounds, and research is currently ongoing to determine whether they could damage wildlife, disturb ecosystems, or endanger human health. Concerns regarding the effects of these pollutants on biodiversity and human health, particularly in vulnerable communities, are further raised by their persistence and bioaccumulation in the food web. Understanding the behaviour, sources, and the consequences of emerging contaminants is crucial for creating management plans and policies that better safeguard the environment and human health (Kanungo et al., 2023a, 2023b) (Das et al., 2023, 2025). For this reason, research on these pollutants is critical. Given the rapid development of new chemicals and products, addressing emerging contaminants requires adaptive environmental governance and monitoring systems that can quickly identify and mitigate potential risks before they become widespread. (Akdogan & Guven, 2019a; Hamscher & Hartung, n.d.; Juliano & Magrini, 2017; Sola et al., 2014; Tolls et al., 2010).

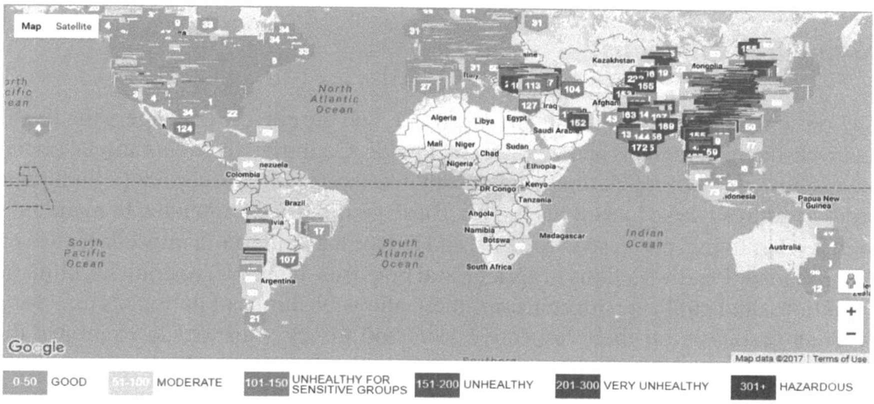

Fig. 1.3 Interactive 3D Earth showing most polluted places (Interactive 3D Earth Shows Most Polluted Places in the World, n.d.)

Regulatory and Policy Responses

International Agreements and Frameworks

UNEP or the United Nations Environment Programme, is very much required in order to handle global environmental issues, such as pollution. It gives nations a forum to work along together on issues including ecosystem preservation, sustainable development, and the reduction of pollution. The United Nations Environment Programme's priorities include supporting global environmental governance and minimizing the release of harmful compounds like micro-pollutants. The Stockholm Convention, a worldwide agreement to eradicate or subside the usage and production of persistent organic pollutants, a significant category of micropollutants. Various nations signed it to protect ecosystems, wildlife, and public health from persistent substances that bioaccumulate in living organisms and remain in the environment. The deal targets substances that can cause reproductive harm, endocrine disruption, and cancer, among other long-term implication on the living organisms (Lemm et al., 2021; Wang, 2023).

National Regulations

The presence of micropollutants in soil, water, and air gravely endangers the health and well-being of humans and wildlife worldwide. Countries and international bodies have set of rules & regulations and acceptable thresholds for certain pollutants in order to address these hazardous issues. These guidelines are intended to maintain a high level of living, protect ecosystems, and reduce harmful health impacts. However, because to regional variations in agricultural practises, industrial operations, climatic conditions, and public health concerns, these allowable limits of pollutants may vary. With an emphasis on their respective laws and effects, this article examines the standards for acceptable pollution levels in soil, water, and air in developing nations. Because pollutants in water can have detrimental effects on aquatic life, ecosystems, and human health, water pollution is a serious problem.

The World Health Organization (WHO), along with regional bodies like the European Union (EU) and the U.S. Environmental Protection Agency (EPA), has established standards for acceptable pollutant levels in drinking water and water bodies. With the goal of achieving "Good Status" in all water bodies, the EU Water Framework Directive (WFD) establishes the benchmark for water quality in all of its member states. The directive restricts pollutants like pesticides, nitrates, and heavy metals. Because airborne pollutants directly affect public health and cause environmental harm, air quality is another area where allowable pollutant levels are regularly evaluated. Particulate matter ($PM_{2.5}$ and PM_{10}), nitrogen dioxide (NO_2), carbon dioxide (CO_2), carbon monoxide (CO), sulphur dioxide (SO_2), ozone (O_3), and volatile organic compounds (VOCs) are among the air contaminants for which the WHO, EPA, and EU have established particular standards (Backhaus & Faust, 2012; Karthikeyan & Subagunasekar, 2023; Regulations.Gov n.d.). According to WHO guidelines, the average yearly PM2.5 concentration should not exceed 5 $\mu g/m^3$, and the average daily concentration should not exceed 15 $\mu g/m^3$. The recommended daily exposure to PM_{10} is 45 $\mu g/m^3$, and the recommended annual exposure is 15 $\mu g/m^3$. These pollutants have comparable yearly limits in the EU, with $PM_{2.5}$ having a limit of 25 $\mu g/m^3$ and PM_{10} having a limit of 40 $\mu g/m^3$. Nevertheless, different European cities frequently report levels beyond the advised limits, posing a risk to health, particularly cardiovascular and neurological conditions, and enforcement differs among member states of the union. The Clean Air Act requires the EPA to establish National Ambient Air Quality Standards (NAAQS) in the United States for six primary pollutants: $PM_{2.5}$, PM_{10}, CO, NO_2, O_3, and SO_2. $PM_{2.5}$ and NO_2 have annual averages of 12 $\mu g/m^3$ and 53 ppb, respectively.

However, areas of the United States like Los Angeles that have a lot of industrial activity or heavy automobile activity frequently have air quality problems that go beyond these limitations, which can be harmful to health when exposed for longer periods. Some of the worst air quality in the world is found in Asian nations like China and India. Cities like Delhi, Dhaka and Beijing record $PM_{2.5}$ concentrations well above WHO recommendations, indicating that the allowable threshold are frequently surpassed in metropolitan areas. China has a norm of 35 $\mu g/m^3$, which is still more than the WHO's recommended level, whereas India has a tolerable maximum of 40 $\mu g/m^3$ per year. Different degrees of industrialization, urbanization, seasonal changes and regulatory enforcement are reflected in the disparities in standards (Altuwayjiri et al., 2021; Dallarosa et al., 2005; Krishnan et al., 1993). Soil and water pollution, caused by natural disasters, industrial processes, agricultural practices, and municipal waste disposal, negatively impacts soil fertility, agricultural yields, and ecosystems. To guarantee safe usage for agricultural purposes and to avoid contaminating food supplies, the allowable amounts of harmful contaminants in soil are regulated. The kind of micropollutant and their origins affect these requirements. Soil Protection Strategy in the country addresses soil contamination and establishes acceptable limits for a number of contaminants, including insecticides, heavy metals, and persistent organic pollutants. 1.4 mg/kg of cadmium and 50 mg/kg of lead are acceptable concentrations in soils used for agriculture purposes. In order to prevent soil pollution from surpassing acceptable levels that could harm crops and groundwater, the European union also imposes pesticide limitations. In the United States, the EPA regulates soil pollution through various programs, including the Superfund program, which deals with the cleanup of hazardous waste sites. While specific permissible limits for soil contaminants like heavy metals changes depending on the use of the land (e.g., residential, agricultural, industrial), the EPA sets standards for certain substances, such as a limit of 0.2 mg/kg for lead in residential soils (Central Pollution Control Board, n.d.; Interactive 3D Earth Shows Most Polluted Places in the World, n.d.; Regulations.Gov n.d.; Wuijts et al., 2021). Permissible levels of heavy metals like lead and mercury may not be universally enforced in developing countries due to the absence of thorough soil pollution legislation. Similar to this, China has experienced severe contamination due to soil pollution from industrial processes and agricultural pesticides (Behera et al., 2020, 2021, 2022). Although the government has established guidelines for acceptable concentrations of contaminants such as arsenic, lead and mercury, implementation of these regulations is uneven in rural parts. The allowable levels of soil contamination may be higher or less regulated in many poorer nations, which is a reflection of resource constraints and poor governance. (Akdogan & Guven, 2019b; Gani et al., 2016; Rutgersson et al., 2014; Sola et al., 2014).

Challenges in Regulation

International organizations such as the WHO and UNEP establish global frameworks for controlling pollutants. However, significant differences remain in enforcement and the actual permissible levels of pollutants in soil, water, and air across various regions. Because they have greater resources, technology, and public awareness, developed regions like Europe, North America, and Japan typically have stronger rules, whereas developing nations frequently find it difficult to enact and enforce such standards. This discrepancy is further exacerbated by elements like urbanization, industrial expansion, and insufficient environmental monitoring. For example, fast industrialization, deforestation, and a lack of waste treatment facilities cause significant air and water pollution in many developing nations, including sections of Africa and Southeast Asia. Establishing standards and working towards to their application are both global challenges. Regional variations highlight the need for specialized policies and international collaboration to address the unequal distribution of pollution and resources, even if acceptable levels of pollutants in ecosystems are crucial for safeguarding human health and the environment. To guarantee that pollution levels are successfully managed worldwide, international agreements, funding for environmental initiatives, and capacity building in underdeveloped nations are crucial measures (Central Pollution Control Board, n.d.; Akdogan & Guven, 2019b; Gani et al., 2016; Rutgersson et al., 2014; Sola et al., 2014).

Conclusion

Micropollutants seep into the ecosystems through different human activities. The disintegration of larger plastic products, such as packaging, clothing's synthetic fibres, and plastic waste from industrial processes, produces microplastics. Human waste and inappropriate disposal methods of medications allow pharmaceuticals, including antiperspirant, antibiotics and other medications, to enter water bodies. Modernized agriculture practices that include the usage of pesticides, herbicides, chemical fertilizers and insecticides also contaminate water and soil, while heavy metals from automobile emissions and industrial discharges are long-term hazards to human health and the ecosystem. Micropollutants have wide-ranging and intricate effects. These materials have the potential to make their way up the food chain, especially in aquatic environments where fish and other aquatic creatures consume microplastics, heavy metals and other toxins. Because they alter growth, interfere with reproductive processes, and even induce mortality, these pollutants can harm consumers, with prolonged chronic effects. For instance, when consumed, microplastics can physically injure aquatic life by obstructing digestion or causing internal injuries. Furthermore, ecosystems are constantly contaminated due to the persistence of micropollutants in the environment, which threatens biodiversity. Exposure to polluted air and the consumption of tainted food or water can cause a number of health problems in humans. Recent trends in research shows that, harmful chemicals and microplastics may have negative health impacts on organisms, such as immune system damage, hormone imbalances, and an increased risk of cancer. Antibiotic resistant bacteria are a major hazard to public health worldwide and are a result of antibiotic usage in poultry and dairy, disposal of pharmaceuticals, particularly antibiotics, in the water supply. Lead and mercury are two examples of heavy metals that can build up in the body and cause long-term health problems such kidney damage, neurological illnesses, and developmental problems in children. Even while more people are becoming aware of micro pollution, most study on the topic is still in its early phases. To completely comprehend the long-term effects of these contaminants on the ecosystem and human health, rigorous research is essential. Scientists are studying the most dangerous types of micropollutants, their ecological pathways, and their impacts on both humans and wildlife. Additionally, research is required to ascertain how well the present methods of waste management and treatment are reducing the amount of micropollutants entering in the environment. Public understanding of the origins and effects of micropollutants must be increased as science advances. To lessen the impact and usage of these contaminants, educational initiatives, stronger waste disposal laws, and improved product chemical management are required. It is essential for communities of various sectors to understand the importance of reducing plastic waste, properly disposing of pharmaceuticals, and adopting sustainable agricultural practices to mitigate pesticide runoff.

An integrated strategy is required to address the growing problem of micro pollution. This entails including a variety of stakeholders in concerted efforts, such as governments, businesses, scientists, environmental organizations, and the general people. Stricter laws governing industrial and vehicle emissions, chemical use, and waste management should be passed and enforced by governments. The introduction of micropollutants into the environment can be significantly reduced by policies that encourage the creation of biodegradable substitutes, usage of greener chemicals and the secure disposal of day-to-day chemicals. To prevent micropollutants from getting into soil and rivers, industries can implement better waste management techniques, use less plastic packaging, and switch to cleaner production methods at the industrial level. The spread of micropollutants can also be lessened by research and development in new technology for removing or filtering them from the air and wastewater. Consumers and sellers can also contribute by choosing to use less plastic, dispose of chemicals properly, and support environmentally friendly, industrial and agricultural methods. Addressing micro pollution also requires initiatives in national and international cooperation. Since air and water currents can carry pollution over national boundaries, international cooperation is required to establish universal norms and laws. Effective methods for lowering micro pollution will be developed with the assistance of international research and the best practice exchanges.

References

Adegoke, K. A., Olagunju, A. O., Alagbada, T. C., Alao, O. C., Adesina, M. O., Afolabi, I. C., Adegoke, R. O., & Bello, O. S. (2022). Adsorptive removal of endocrine-disrupting chemicals from aqueous solutions: A review. *Water, Air, & Soil Pollution, 233*(2), 38. https://doi.org/10.1007/s11270-021-05405-8

Ahel, M., & Jeličić, I. (2001). *Phenazone analgesics in soil and groundwater below a municipal solid waste landfill* (pp. 100–115). https://doi.org/10.1021/bk-2001-0791.ch006

Ahmed, F., Hossain, S., Hossain, S., Fakhruddin, A. N. M., Abdullah, A. T. M., Chowdhury, M. A. Z., & Gan, S. H. (2019). Impact of household air pollution on human health: Source identification and systematic management approach. *SN Applied Sciences, 1*(5), 418. https://doi.org/10.1007/s42452-019-0405-8

Akdogan, Z., & Guven, B. (2019). Microplastics in the environment: A critical review of current understanding and identification of future research needs. *Environmental Pollution, 254*, Article 113011. https://doi.org/10.1016/j.envpol.2019.113011

Altuwayjiri, A., Pirhadi, M., Taghvaee, S., & Sioutas, C. (2021). Long-term trends in the contribution of PM 2.5 sources to organic carbon (OC) in the Los Angeles basin and the effect of PM emission regulations. *Faraday Discussions, 226*, 74–99. https://doi.org/10.1039/D0FD00074D

Arnold, K. E., Boxall, A. B. A., Brown, A. R., Cuthbert, R. J., Gaw, S., Hutchinson, T. H., Jobling, S., Madden, J. C., Metcalfe, C. D., Naidoo, V., Shore, R. F., Smits, J. E., Taggart, M. A., & Thompson, H. M. (2013). Assessing the exposure risk and impacts of pharmaceuticals in the environment on individuals and ecosystems. *Biology Letters, 9*(4), 20130492. https://doi.org/10.1098/rsbl.2013.0492

Backhaus, T., & Faust, M. (2012). Predictive environmental risk assessment of chemical mixtures: A conceptual framework. *Environmental Science & Technology, 46*(5), 2564–2573. https://doi.org/10.1021/es2034125

Bala, K., Anshul, Verma, S., Dogra, A., Das, A. P., Kumar, R., & Nandini. (2024). Sustainable management for urban plastic waste generation. *Sustainable Management of Urban Plastic Waste Through Circular Economic Approaches*, 211–236. https://doi.org/10.1201/9781003465447-13

Bala, K., Dogra, A., Lata, P., & Das, A. P. (2024). Discussing Sources and Biological-Chemical Remediation Approaches for Microplastic Pollution. *Sustainable Microbial Technology for Synthetic and Cellulosic Microfiber Bioremediation*, 37–59. https://doi.org/10.1007/978-3-031-62600-5_3

Barnes, K. K., Christenson, S. C., Kolpin, D. W., Focazio, M. J., Furlong, E. T., Zaugg, S. D., Meyer, M. T., & Barber, L. B. (2004). Pharmaceuticals and other organic waste water contaminants within a leachate plume downgradient of a municipal landfill. *Groundwater Monitoring & Remediation, 24*(2), 119–126. https://doi.org/10.1111/j.1745-6592.2004.tb00720.x

Barzman, M., Bàrberi, P., Birch, A. N. E., Boonekamp, P., Dachbrodt-Saaydeh, S., Graf, B., Hommel, B., Jensen, J. E., Kiss, J., Kudsk, P., Lamichhane, J. R., Messéan, A., Moonen, A.-C., Ratnadass, A., Ricci, P., Sarah, J.-L., & Sattin, M. (2015). Eight principles of integrated pest management. *Agronomy for Sustainable Development, 35*(4), 1199–1215. https://doi.org/10.1007/s13593-015-0327-9

Beckers, L.-M., Busch, W., Krauss, M., Schulze, T., & Brack, W. (2018). Characterization and risk assessment of seasonal and weather dynamics in organic pollutant mixtures from discharge of a separate sewer system. *Water Research, 135*, 122–133. https://doi.org/10.1016/j.watres.2018.02.002

Behera I. D., Nayak M., Biswas S., Meikap B. C, Sen, R. (2021). Enhanced biodegradation of total petroleum hydrocarbons by implementing a novel two-step bioaugmentation strategy using indigenous bacterial consortium. *Journal of Environmental Management, 292*, 112746. ISSN 0301-4797. https://doi.org/10.1016/j.jenvman.2021.112746

Behera, I. D., Nayak, M., Mishra, A., Meikap, B. C., Sen, R. (2022). Strategic implementation of integrated bioaugmentation and biostimulation for efficient mitigation of petroleum hydrocarbon pollutants from terrestrial and aquatic environment. *Marine Pollution Bulletin, 177*, 113492. ISSN 0025-326X. https://doi.org/10.1016/j.marpolbul.2022.113492

Behera, I. D., Basak, G., Kumar, R. R., Sen, R., & Meikap, B. C. (2020). Treatment of petroleum refinery sludge by petroleum degrading bacterium *Stenotrophomonas pavanii* IRB19 as an efficient novel technology. *Journal of Environmental Science and Health, Part A, 56*(2), 226–239. https://doi.org/10.1080/10934529.2020.1866924

Central Pollution Control Board. (n.d.). Retrieved February 5, 2025 from http://www.cpcb.nic.in

Dallarosa, J. B., Teixeira, E. C., Pires, M., & Fachel, J. (2005). Study of the profile of polycyclic aromatic hydrocarbons in atmospheric particles (PM10) using multivariate methods. *Atmospheric Environment, 39*(35), 6587–6596. https://doi.org/10.1016/j.atmosenv.2005.07.034

Das, A. P., & Ghosh, S. (2023). Metagenomic insights into unculturable microbial diversity. *Manganese Mining Microorganisms*, 63–76. https://doi.org/10.1016/B978-0-12-822147-1.00002-9

Das, A. P., & Singh, S. (2011). Occupational health assessment of chromite toxicity among Indian miners. *Indian Journal of Occupational and Environmental Medicine, 15*(1), 6. https://doi.org/10.4103/0019-5278.82998

Das, A. P., Behera, I. D., & Das, N. P. (2024). *Renewable energy generation and value addition from environmental microfiber pollution through advanced greener solution* (p. 278).

Das, A., Dey, S., & Das, A. P. (2024). Ecotoxicological impacts of synthetic microfiber pollutants and development of sustainable mitigation strategies. *Environmental Chemistry and Ecotoxicology*. https://doi.org/10.1016/J.ENCECO.2024.12.003

Das, A., Dey, S., Bala, K., Gnanasekaran, R., & Das, A. P. (2024). Microbial bioremediation technology for sustainable treatment and management of synthetic microfiber waste. In *Sustainable microbial technology for synthetic and cellulosic microfiber bioremediation* (pp. 221–238). https://doi.org/10.1007/978-3-031-62600-5_12

Das, P., & Bal, M. (2025). Modeling and optimization study for microplastic removal from aquatic medium using Chroococcidiopsis species. *Chemical Engineering Science, 304*, Article 121085. https://doi.org/10.1016/j.ces.2024.121085

Das, P., Halder, G., & Bal, M. (2023). A critical review on remediation of microplastics using microalgae from aqueous system. *Science of the Total Environment, 898*, Article 166425. https://doi.org/10.1016/j.scitotenv.2023.166425

Devi, P. I., Thomas, J., & Raju, R. K. (2017). Pesticide consumption in India: A spatiotemporal analysis. *Agricultural Economics Research Review, 30*(1), 163. https://doi.org/10.5958/0974-0279.2017.00015.5

Dey, S., Das, A., Mallick, K., Sahu, A., & Das, A. P. (2023). *Environmental petroleum waste: Pollution, toxicity, sustainable remediation* (pp. 159–175). https://doi.org/10.1007/978-3-031-48220-5_7

Dey, S., Dash, A., & Das, A. P. (2024). Bioremediation as a sustainable and circular solution for mining waste management and bioenergy generation. In *Sustainable management of mining waste and tailings: A circular economy approach* (pp. 107–119). https://doi.org/10.1201/9781003442455-6

Dey, S., Dash, N., Ray, S. K., Priyadarsini, S., Chhotaray, C., & Das, A. P. (2024). Occurrence and quantification of tire wear plastics pollutants from Highway City Dust of Bhubaneswar

City of Odisha. In *Sustainable microbial technology for synthetic and cellulosic microfiber bioremediation* (pp. 77–89). https://doi.org/10.1007/978-3-031-62600-5_5

Du, B., Haddad, S. P., Luek, A., Scott, W. C., Saari, G. N., Burket, S. R., Breed, C. S., Kelly, M., Broach, L., Rasmussen, J. B., Chambliss, C. K., & Brooks, B. W. (2016). Bioaccumulation of human pharmaceuticals in fish across habitats of a tidally influenced urban bayou. *Environmental Toxicology and Chemistry, 35*(4), 966–974. https://doi.org/10.1002/etc.3221

Du, J., Li, H., Xu, S., Zhou, Q., Jin, M., & Tang, J. (2019). A review of organophosphorus flame retardants (OPFRs): Occurrence, bioaccumulation, toxicity, and organism exposure. *Environmental Science and Pollution Research, 26*(22), 22126–22136. https://doi.org/10.1007/s11356-019-05669-y

Dulsat-Masvidal, M., Ciudad, C., Infante, O., Mateo, R., & Lacorte, S. (2023). Water pollution threats in important bird and biodiversity areas from Spain. *Journal of Hazardous Materials, 448*, Article 130938. https://doi.org/10.1016/j.jhazmat.2023.130938

Enyoh, C. E., Verla, A. W., Qingyue, W., Ohiagu, F. O., Chowdhury, A. H., Enyoh, E. C., Chowdhury, T., Verla, E. N., & Chinwendu, U. P. (2020). An overview of emerging pollutants in air: Method of analysis and potential public health concern from human environmental exposure. *Trends in Environmental Analytical Chemistry, 28*, Article e00107. https://doi.org/10.1016/j.teac.2020.e00107

Ferreira, A. R., Guedes, P., Mateus, E. P., Ribeiro, A. B., & Couto, N. (2020). Emerging organic contaminants in soil irrigated with effluent: Electrochemical technology as a remediation strategy. *Science of the Total Environment, 743*, Article 140544. https://doi.org/10.1016/j.scitotenv.2020.140544

Fu, Y., Zhu, Y., Dong, H., Li, J., Zhang, W., Shao, Y., & Shao, Y. (2023). Effects of heavy metals and antibiotics on antibiotic resistance genes and microbial communities in soil. *Process Safety and Environmental Protection, 169*, 418–427. https://doi.org/10.1016/j.psep.2022.11.020

Gani, K. M., Rajpal, A., & Kazmi, A. A. (2016). Contamination level of four priority phthalates in North Indian wastewater treatment plants and their fate in sequencing batch reactor systems. *Environmental Science: Processes & Impacts, 18*(3), 406–416. https://doi.org/10.1039/C5EM00583C

Glassmeyer, S. T., Furlong, E. T., Kolpin, D. W., Cahill, J. D., Zaugg, S. D., Werner, S. L., Meyer, M. T., & Kryak, D. D. (2005). Transport of chemical and microbial compounds from known wastewater discharges: Potential for use as indicators of human fecal contamination. *Environmental Science & Technology, 39*(14), 5157–5169. https://doi.org/10.1021/es048120k

Gupta, A., Kumar, S., Bajpai, Y., Chaturvedi, K., Johri, P., Tiwari, R. K., Vivekanand, V., & Trivedi, M. (2024). Pharmaceutically active micropollutants: Origin, hazards and removal. *Frontiers in Microbiology, 15*. https://doi.org/10.3389/fmicb.2024.1339469

Hamscher, G., & Hartung, J. (n.d.). Veterinary antibiotics in dust: Sources, environmental concentrations, and possible health hazards. In *Pharmaceuticals in the environment* (pp. 95–102). Springer. https://doi.org/10.1007/978-3-540-74664-5_7

Handford, C. E., Elliott, C. T., & Campbell, K. (2015). A review of the global pesticide legislation and the scale of challenge in reaching the global harmonization of food safety standards. *Integrated Environmental Assessment and Management, 11*(4), 525–536. https://doi.org/10.1002/ieam.1635

Heberer, T., & Feldmann, D. (2005). Contribution of effluents from hospitals and private households to the total loads of diclofenac and carbamazepine in municipal sewage effluents—Modeling versus measurements. *Journal of Hazardous Materials, 122*(3), 211–218. https://doi.org/10.1016/j.jhazmat.2005.03.007

Interactive 3D Earth shows most polluted places in the world. (n.d.). Retrieved February 5, 2025, from https://geospatialworld.net/blogs/3d-earth-shows-most-polluted-places/

Jagatee, S., Priyadarshini, S., Rath, C. C., & Das, A. P. (2024). Synthetic microfiber: An enduring environmental problem linked to sustainable development. *Environmental Science and Engineering, Part F2333*, 93–112. https://doi.org/10.1007/978-3-031-51792-1_6

Jiang, J.-Q., Zhou, Z., & Sharma, V. K. (2013). Occurrence, transportation, monitoring and treatment of emerging micro-pollutants in waste water—A review from global views. *Microchemical Journal, 110*, 292–300. https://doi.org/10.1016/j.microc.2013.04.014

John, A., & Das, A. P. (2024). Government strategies for achieving circular economy for plastic waste management in Southeast Asia. In *Sustainable management of urban plastic waste through circular economic approaches* (pp. 72–82). https://doi.org/10.1201/9781003465447-6

Juliano, C., & Magrini, G. (2017). Cosmetic ingredients as emerging pollutants of environmental and health concern: A mini-review. *Cosmetics, 4*(2), 11. https://doi.org/10.3390/cosmetics4020011

Kanungo, J., Sahoo, T., Bal, M., & Behera, I. D. (2023a). Performance of bioremediation strategy in waste lubricating oil pollutants: A review. *Geomicrobiology Journal, 41*(4), 360–373. https://doi.org/10.1080/01490451.2023.2245395

Kanungo, J., Sahoo, T., Swain, L.P., & Behera, I.D. (2023). Toxicity of persistent hydrocarbon pollutants, sources and sustainable remediation process. In: Behera, I. D., & Das, A. P. (eds.). *Impact of petroleum waste on environmental pollution and its sustainable management through circular economy. environmental science and engineering*. Springer. https://doi.org/10.1007/978-3-031-48220-5_2

Karthikeyan, P., & Subagunasekar, M. (2023). Microplastics pollution studies in India: A recent review of sources, abundances and research perspectives. *Regional Studies in Marine Science, 61*, Article 102863. https://doi.org/10.1016/j.rsma.2023.102863

Krishnan, A. V., Stathis, P., Permuth, S. F., Tokes, L., & Feldman, D. (1993). Bisphenol-A: An estrogenic substance is released from polycarbonate flasks during autoclaving. *Endocrinology, 132*(6), 2279–2286. https://doi.org/10.1210/endo.132.6.8504731

Lemm, J. U., Venohr, M., Globevnik, L., Stefanidis, K., Panagopoulos, Y., van Gils, J., Posthuma, L., Kristensen, P., Feld, C. K., Mahnkopf, J., Hering, D., & Birk, S. (2021). Multiple stressors determine river ecological status at the European scale: Towards an integrated understanding of river status deterioration. *Global Change Biology, 27*(9), 1962–1975. https://doi.org/10.1111/gcb.15504

Lishman, L., Smyth, S. A., Sarafin, K., Kleywegt, S., Toito, J., Peart, T., Lee, B., Servos, M., Beland, M., & Seto, P. (2006). Occurrence and reductions of pharmaceuticals and personal care products and estrogens by municipal wastewater treatment plants in Ontario, Canada. *Science of the Total Environment, 367*(2–3), 544–558. https://doi.org/10.1016/j.scitotenv.2006.03.021

Liu, N., Jin, X., Yan, Z., Luo, Y., Feng, C., Fu, Z., Tang, Z., Wu, F., & Giesy, J. P. (2020). Occurrence and multiple-level ecological risk assessment of pharmaceuticals and personal care products (PPCPs) in two shallow lakes of China. *Environmental Sciences Europe, 32*(1), 69. https://doi.org/10.1186/s12302-020-00346-1

Mishra, R. K., Mentha, S. S., Misra, Y., & Dwivedi, N. (2023a). Emerging pollutants of severe environmental concern in water and wastewater: A comprehensive review on current developments and future research. *Water-Energy Nexus, 6*, 74–95. https://doi.org/10.1016/j.wen.2023.08.002

Mishra, S., Dash, D., & Das, A. P. (2023). Aquatic microbial diversity on plastisphere: Colonization and potential role in microplastic biodegradation. *Geomicrobiology Journal*, 1–12. https://doi.org/10.1080/01490451.2023.2209750

Mishra, S., Swain, S., Sahoo, M., Mishra, S., & Das, A. P. (2021). Microbial colonization and degradation of microplastics in aquatic ecosystem: A review. *Geomicrobiology Journal*. https://doi.org/10.1080/01490451.2021.1983670

Mohanty, J., Dey, S., Bala, K., & Das, A. P. (2024). Recent advances in microplastic pollution for its sustainable management. In: *Sustainable management of urban plastic waste through circular economic approaches* (pp. 27–48). https://doi.org/10.1201/9781003465447-3

Mohanty, M., Tripathy, B., Koteswara Reddy, G., & Das, A. P. (2024a). Greener technologies for establishing restraint over microplastic pollution. In *Sustainable microbial technology for synthetic and cellulosic microfiber bioremediation* (pp. 239–258). https://doi.org/10.1007/978-3-031-62600-5_13

Morin-Crini, N., Lichtfouse, E., Liu, G., Balaram, V., Ribeiro, A. R. L., Lu, Z., Stock, F., Carmona, E., Teixeira, M. R., Picos-Corrales, L. A., Moreno-Piraján, J. C., Giraldo, L., Li, C., Pandey, A., Hocquet, D., Torri, G., & Crini, G. (2022). Worldwide cases of water pollution by emerging contaminants: A review. *Environmental Chemistry Letters, 20*(4), 2311–2338. https://doi.org/10.1007/s10311-022-01447-4

Paul, A., Dey, S., Ram, D. K., & Das, A. P. (2023). Hexavalent chromium pollution and its sustainable management through bioremediation. *Geomicrobiology Journal*, 1–11. https://doi.org/10.1080/01490451.2023.2218377

Paul, A., Priyadarsini, S., Karakuş, S., Behera, J. K., Gnanasekaran, R., & Das, A. P. (2024). Fate and accumulation of synthetic microfibers in the environment. In *Sustainable microbial technology for synthetic and cellulosic microfiber bioremediation* (pp. 61–76). https://doi.org/10.1007/978-3-031-62600-5_4

Priyadarshini, B., & Das, A. P. (2024). The adverse impact of synthetic microplastics released from disposable face masks pollutants on aquatic microbiota, aquatic species and humans: A review. *Environmental Science and Engineering, Part, F2827*, 67–81. https://doi.org/10.1007/978-3-031-50840-0_5

Priyadarshini, B., Santhosh Kumar, M., & Das, A. P. (2024). Microbial degradation of synthetic plastics released from COVID-19 generated single use disposable face masks. *Environmental Science and Engineering, Part, F2827*, 1–16. https://doi.org/10.1007/978-3-031-50840-0_1

Raj, K., & Das, A. P. (2023). Lead pollution: Impact on environment and human health and approach for a sustainable solution. *Environmental Chemistry and Ecotoxicology, 5*, 79–85. https://doi.org/10.1016/J.ENCECO.2023.02.001

Ramírez, N., Cuadras, A., Rovira, E., Borrull, F., & Marcé, R. M. (2012). Chronic risk assessment of exposure to volatile organic compounds in the atmosphere near the largest Mediterranean industrial site. *Environment International, 39*(1), 200–209. https://doi.org/10.1016/j.envint.2011.11.002

Regulations.gov. (n.d.). Retrieved February 5, 2025, from https://www.regulations.gov/

Rudel, R. A., Melly, S. J., Geno, P. W., Sun, G., & Brody, J. G. (1998). Identification of alkylphenols and other estrogenic phenolic compounds in wastewater, septage, and groundwater on Cape Cod, Massachusetts. *Environmental Science & Technology, 32*(7), 861–869. https://doi.org/10.1021/es970723r

Rutgersson, C., Fick, J., Marathe, N., Kristiansson, E., Janzon, A., Angelin, M., Johansson, A., Shouche, Y., Flach, C.-F., & Larsson, D. G. J. (2014). Fluoroquinolones and *qnr* genes in sediment, water, soil, and human fecal flora in an environment polluted by manufacturing discharges. *Environmental Science & Technology, 48*(14), 7825–7832. https://doi.org/10.1021/es501452a

Sahu, A., Dey, S., Mallick, K., Priyadarsini, S., Behera, I. D., & Das, A. P. (2024). A review on advanced technology for sustainable management of synthetic microplastic waste. In *Sustainable management of urban plastic waste through circular economic approaches* (pp. 262–279). https://doi.org/10.1201/9781003465447-15

Sahu, S., & Das, A. P. (2024). *Microplastics degradation mechanism by aquatic plastiomes: A review* (pp. 213–226). https://doi.org/10.1007/978-3-031-50840-0_12

Shen, G. F., Yuan, S. Y., Xie, Y. N., Xia, S. J., Li, L., Yao, Y. K., Qiao, Y. Z., Zhang, J., Zhao, Q. Y., Ding, A. J., Li, B., & Wu, H. S. (2014). Ambient levels and temporal variations of PM 2.5 and PM 10 at a residential site in the mega-city, Nanjing, in the western Yangtze River Delta, China. *Journal of Environmental Science and Health, Part A, 49*(2), 171–178. https://doi.org/10.1080/10934529.2013.838851

Sola, P., Mvumi, B. M., Ogendo, J. O., Mponda, O., Kamanula, J. F., Nyirenda, S. P., Belmain, S. R., & Stevenson, P. C. (2014). Botanical pesticide production, trade and regulatory mechanisms in sub-Saharan Africa: Making a case for plant-based pesticidal products. *Food Security, 6*(3), 369–384. https://doi.org/10.1007/s12571-014-0343-7

Sondhi, S., Sharma, R., Matta, S., & Singh, C. (2023). Biotechnology-based strategies for removal of emerging contaminants. In *Management and mitigation of emerging pollutants* (pp. 95–124). https://doi.org/10.1007/978-3-031-41005-5_4

Su, G., Logez, M., Xu, J., Tao, S., Villéger, S., & Brosse, S. (2021). Human impacts on global freshwater fish biodiversity. *Science, 371*(6531), 835–838. https://doi.org/10.1126/science.abd3369

Tiwari, A., Dhanker, R., Saxena, A., Goyal, S., Melchor-Martínez, E. M., Iqbal, H. M. N., & Parra-Saldívar, R. (2021). Toxicity evaluation of personal care and household products as silent killers on the survival of Daphnia magna. *Case Studies in Chemical and Environmental Engineering, 4*, Article 100124. https://doi.org/10.1016/j.cscee.2021.100124

Tolls, J., Berger, H., Klenk, A., Meyberg, M., Beiersdorf, A., Müller, R., Rettinger, K., & Steber, J. (2010). Environmental safety aspects of personal care products—A European perspective. *Environmental Toxicology and Chemistry, 28*(12), 2485–2489. https://doi.org/10.1897/09-104.1

Tripathy, B., Dash, A., & Das, A. P. (2022). Detection of Environmental microfiber pollutants through vibrational spectroscopic techniques: Recent advances of environmental monitoring and future prospects. In *Critical reviews of analytical chemistry* (pp. 1–11). https://doi.org/10.1080/10408347.2022.2144994

Tripathy, B., Dey, S., Dubey, N. K., Lo, W.-C., & Das, A. P. (2024). *An omics-based approach for managing the risk of COVID-19-generated wastes* (pp. 51–66). https://doi.org/10.1007/978-3-031-50840-0_4

Tripathy, B., Sahoo, P. P., Sundaray, H., & Das, A. P. (2024b). In-vitro biodegradation of discarded marine microplastics across the eastern coast of the Bay of Bengal, India using Exiguobacterium sp. *Environmental Chemistry and Ecotoxicology, 6*, 236–247. https://doi.org/10.1016/J.ENCECO.2024.07.003

Vadiraj, K. T, Achar, R. R., & Siriger, S. (2021). A review on emerging micropollutants: sources, environmental concentration and toxicity. *Bionatura, 6*(4), 2305–2325. https://doi.org/10.21931/RB/2021.06.04.29

Vörösmarty, C. J., McIntyre, P. B., Gessner, M. O., Dudgeon, D., Prusevich, A., Green, P., Glidden, S., Bunn, S. E., Sullivan, C. A., Liermann, C. R., & Davies, P. M. (2010). Global threats to human water security and river biodiversity. *Nature, 467*(7315), 555–561. https://doi.org/10.1038/nature09440

Wang, S. (2023). International law-making process of combating plastic pollution: Status Quo, debates and prospects. *Marine Policy, 147*, Article 105376. https://doi.org/10.1016/j.marpol.2022.105376

Wu, L., Luo, X.-S., Li, H., Cang, L., Yang, J., Yang, J., Zhao, Z., & Tang, M. (2019). Seasonal levels, sources, and health risks of heavy metals in atmospheric PM2.5 from four functional areas of Nanjing City, Eastern China. *Atmosphere, 10*(7), 419. https://doi.org/10.3390/atmos10070419

Wuijts, S., Claessens, J., Farrow, L., Doody, D. G., Klages, S., Christophoridis, C., Cvejić, R., Glavan, M., Nesheim, I., Platjouw, F., Wright, I., Rowbottom, J., Graversgaard, M., van den Brink, C., Leitão, I., Ferreira, A., & Boekhold, S. (2021). Protection of drinking water resources from agricultural pressures: Effectiveness of EU regulations in the context of local realities. *Journal of Environmental Management, 287*, Article 112270. https://doi.org/10.1016/j.jenvman.2021.112270

Chapter 2
Assessment and Fate of Micropollutants in the Environment

Neha Kharadi, B. G. Chudasma, Mohit Bamaniya, Mayur Bhadarka, and Hardik Sikotariya

Introduction

Micropollutants, often referred to as emerging contaminants, encompass a wide and growing range of substances derived from both human activities and natural sources (Luo et al., 2014). Micropollutants are significant environmental contaminants originating from a variety of sources, such as pharmaceuticals and personal care products, industrial discharges, and agricultural runoff. These substances, frequently detected in wastewater, can disrupt aquatic ecosystems and pose health risks to humans due to their extensive use and improper disposal practices (Sharma et al., 2024a, 2024b). Industrial operations release chemicals like benzene into water bodies, highlighting the need for enhanced regulatory oversight and better waste management to reduce their negative effects (Çimen Mesutoğlu & Gök, 2024). Additionally, runoff from agriculture, which includes pesticides and fertilizers, contributes to water contamination, jeopardizing water quality and posing serious threats to aquatic organisms and drinking water sources (Almasi, 2024). In recent years, a variety of treatment methods and technologies have been developed to eliminate micropollutants from water, often focusing on specific micropollutants as target contaminants to assess their removal effectiveness (Yang et al., 2021). Micropollutants have a profound impact on aquatic ecosystems, contributing to biodiversity loss and degradation of these

N. Kharadi (✉) · B. G. Chudasma · M. Bamaniya
Department of Fish Processing Technology, College of Fisheries Science, Kamdhenu University, Veraval, Gujarat, India
e-mail: neha05022002@gmail.com

M. Bhadarka
Department of Aquatic Environment Management, College of Fisheries Science, Kamdhenu University, Veraval, Gujarat, India

H. Sikotariya
Department of Fisheries Resource Management, College of Fisheries Science, Kamdhenu University, Veraval, Gujarat, India

I. D. Behera et al. (eds.), *Emerging Micropollutants*, Sustainable Environmental Waste Management Strategies, https://doi.org/10.1007/978-3-032-03530-1_2

environments (Bhatt et al., 2022). These substances can disrupt the reproductive and developmental processes of aquatic species, leading to population declines and jeopardizing the overall stability of ecosystems as they accumulate in the food web (Dey et al., 2024a, 2024b; Ray et al., 2024). Moreover, human exposure to micropollutants poses serious health threats, including endocrine disruption and the development of antibiotic resistance (Servien et al., 2023). The presence of these contaminants in drinking water and food supplies raises concerns about potential long-term health consequences, especially for vulnerable populations such as children and pregnant women. Additionally, the rise of antibiotic-resistant bacteria linked to micropollutants represents a significant public health challenge, complicating infection treatment and highlighting the urgent need for intervention. Advanced detection of these micropollutants is achieved through sophisticated analytical techniques, including hybrid methods that combine chromatography and spectroscopy, allowing for the identification of trace levels within complex environmental samples (Sharma et al., 2024a). Despite substantial efforts to tackle the issue of micropollutants, there are still hurdles in comprehensively understanding their long-term effects and formulating effective removal strategies. Additional research is necessary to clarify the intricate interactions between micropollutants and different environmental factors, as these interactions can influence their behavior and impact. Moreover, the development of efficient removal technologies that are both cost-effective and capable of large-scale application remains a significant challenge for water treatment facilities and policymakers.

Categories of Micropollutants

Organic Micropollutants

Organic micropollutants (OMPs) are a diverse group of contaminants that significantly threaten environmental health and human well-being. Originating primarily from anthropogenic activities, these compounds are highly resistant to natural degradation and tend to accumulate in living organisms over time.

Types of Organic Micropollutants

Pharmaceuticals and Personal Care Products (PPCPs): These pollutants encompass medications, beauty products, and other personal care items that are often introduced into water systems via household wastewater. Many of these compounds are biologically active, potentially disrupting the natural behavior and physiological processes of aquatic organisms (Kaur et al., 2024).

Pesticides: Commonly employed in agricultural practices for pest control, pesticides frequently seep into surrounding soil and water systems during rainfall events.

Their exposure to non-target species can lead to adverse physiological, reproductive, and behavioral effects, which may cascade through ecosystems (Kumar et al., 2024).

Polychlorinated Biphenyls (PCBs): PCBs are synthetic chemicals historically used in various industrial applications, including electrical equipment. Although banned in many regions, their persistence in the environment makes them a lingering pollutant that bioaccumulates in the food chain, posing long-term ecological and health risks (Gupta, 2024).

Polycyclic Aromatic Hydrocarbons (PAHs): These hazardous substances are byproducts of incomplete combustion processes involving coal, oil, or wood. They are frequently detected in industrial discharges and urban runoff, where they settle into sediments and remain a source of prolonged contamination (Gupta, 2024).

Perfluoroalkyl and Polyfluoroalkyl Substances (PFAs): PFAs are man-made chemicals widely used in products like non-stick cookware, firefighting foams, and stain-resistant fabrics. Their exceptional stability and resistance to natural degradation make them pervasive pollutants that contaminate water supplies and persist in the environment for decades (Kaur et al., 2024).

Sources of Organic Micropollutants

Industrial Discharges: Factories are a primary contributor to OMP pollution, releasing untreated or partially treated wastewater containing these contaminants into nearby rivers and lakes. This not only pollutes water bodies but also affects downstream ecosystems and human populations (Kaur, 2024).

Agricultural Runoff: Excess fertilizers, pesticides, and herbicides from agricultural fields are carried into water systems during rainstorms. This runoff introduces harmful chemicals into aquatic ecosystems, disrupting ecological balance and water quality (Kumar et al., 2024).

Wastewater Treatment Plants: Many conventional wastewater treatment facilities are not equipped to adequately remove OMPs, allowing these substances to escape into natural water systems. Once released, they persist and accumulate, impacting aquatic and terrestrial ecosystems (Włodarczyk-Makuła, 2024).

Environmental and Human Health Impacts

Ecosystem Disruption: OMPs negatively affect aquatic life by interfering with growth, reproduction, and survival. Extended exposure to these contaminants can cause a decline in sensitive species, reduce biodiversity, and alter the structure and function of ecosystems (Włodarczyk-Makuła, 2024).

Human Health Risks: Long-term exposure to OMPs through contaminated water, food, or air is associated with various health problems, including hormone disruption, reproductive issues, and an increased risk of cancer (Bala et al., 2024a, 2024b; Paul et al., 2024). Many OMPs act as endocrine disruptors, mimicking or blocking natural hormones and causing adverse health effects (Kumar et al., 2024).

Inorganic Micropollutants

Inorganic micropollutants present a significant environmental challenge, primarily stemming from industrial, agricultural, and urban activities. These pollutants include heavy metals, inorganic pesticides, and long-lasting chemical compounds, all of which pose serious threats to ecosystems and human health (Das et al., 2024; Das et al., 2023a, 2023b; Dey, Tripathy, et al., 2023a, 2023b). The following sections explore the major categories of these pollutants and their associated effects.

Types of Inorganic Micropollutants

Heavy Metals: Toxic metals such as lead, mercury, and cadmium are widely found in air, water, and soil due to activities like industrial operations, mining, and agricultural runoff. These metals accumulate within living organisms, leading to severe health consequences such as neurological disorders, kidney damage, and cancer. Their persistence in the environment makes their removal and remediation highly complex, intensifying their long-term ecological impacts (Pasqual et al., 2024).

Pesticides: Inorganic pesticides, extensively used in agriculture, often enter soil and water systems through runoff and seepage. These chemical residues disrupt endocrine systems, leading to developmental and reproductive harm in both humans and wildlife. Furthermore, their long-lasting presence in the environment poses chronic exposure risks, particularly for aquatic organisms and other unintended targets (Gooty et al., 2022; Sharma et al., 2024a).

Persistent Organic Pollutants (POPs): Persistent organic pollutants, including compounds such as polychlorinated biphenyls (PCBs) and dioxins, are highly resistant to degradation. These substances accumulate in the food chain, causing long-term health effects such as hormone disruption, immune system damage, and increased cancer risk. Their extensive presence in ecosystems underscores their role as a global environmental challenge (Abbasi et al., 2022).

Sources of Inorganic Micropollutants

Industrial Discharges: Heavy metals such as lead, mercury, and cadmium are commonly introduced into the environment through industrial activities like mining, manufacturing, and energy production. These metals persist in water, soil, and air, accumulating over time and posing significant risks to ecosystems and human health (Pasqual et al., 2024).

Agricultural Runoff: The excessive application of fertilizers and pesticides in agriculture is a major contributor to the contamination of water systems with inorganic micropollutants. During rainfall, these substances are washed into nearby rivers, lakes, and groundwater, leading to the degradation of aquatic ecosystems and reduced water quality (Sharma et al., 2024a).

Wastewater Treatment: Inefficient or incomplete wastewater treatment processes often allow inorganic micropollutants to escape into the environment. These pollutants subsequently disperse through aquatic systems, causing far-reaching ecological damage and health concerns (Servien et al., 2023).

Environmental and Human Health Impacts

Ecosystem Disruption: Inorganic micropollutants significantly affect biodiversity in aquatic environments by impairing the survival, reproduction, and behavior of various species. These disruptions can destabilize food chains and lead to cascading ecological effects, ultimately threatening ecosystem resilience and functionality.

Soil Contamination: The accumulation of heavy metals in soil disrupts nutrient absorption in plants, leading to reduced agricultural productivity and crop toxicity. This contamination not only endangers plant health but also compromises food security and overall ecosystem balance (Pasqual et al., 2024).

Toxicity and Disease: Prolonged exposure to heavy metals is linked to a wide range of severe health problems, including organ damage, neurological issues, and an elevated risk of cancer. These pollutants can enter the human body via contaminated air, water, and food, resulting in chronic health conditions (Pasqual et al., 2024).

Microplastics and Nanoplastics

Microplastics and nanoplastics have emerged as major environmental contaminants, with a wide range of types and sources driving their presence in ecosystems (Ghosh et al., 2023; Pradhan et al., 2024; Priyadarshini & Das, 2024). Defined by their size microplastics measuring between 1 μm and 5 mm, and nanoplastics being smaller than 1 μm these particles originate from the breakdown of larger plastic materials or are purposefully produced for specific applications (Arif et al., 2024; Cherian et al., 2024).

Types of Microplastics and Nanoplastics

Primary Microplastics: These are deliberately manufactured small plastic particles found in items like facial cleansers, toothpaste, and cleaning products. They frequently enter water systems via wastewater, evading conventional treatment methods and contributing significantly to global pollution.

Secondary Microplastics: These particles result from the degradation of larger plastic waste, such as bottles, bags, and fishing nets, due to factors like sunlight exposure, physical abrasion, and wave action. Their prevalence in the environment is difficult to control as they continually break down into smaller particles over time (Arif et al., 2024).

Common Polymers: The materials commonly forming microplastics and nanoplastics include polyethylene, polystyrene, polypropylene, and polyvinyl chloride. These polymers are extensively used in industries like packaging, construction, and household goods, making their presence widespread in both land-based and aquatic ecosystems (Manikandan et al., 2024).

Sources of Microplastics and Nanoplastics

Land-Based Sources: Ineffective waste management and urban runoff are significant contributors to marine plastic pollution, facilitating the movement of microplastics and nanoplastics from terrestrial environments into oceans. Furthermore, stormwater runoff can transport debris from streets and industrial sites, exacerbating the issue of marine contamination (Kushwaha et al., 2024).

Agricultural Practices: The incorporation of plastic materials in agriculture, such as plastic mulch and greenhouse covers, leads to the introduction of microplastics and nanoplastics into soil ecosystems. Additionally, the use of sewage sludge as fertilizer can result in the accumulation of these particles in agricultural lands, potentially transferring them to crops (Nath et al., 2024).

Food Contamination: Microplastics have been found in a variety of food products, including seafood, salt, and certain fruits and vegetables. This contamination can occur throughout the food chain, presenting potential health risks to consumers, as these particles may transport harmful substances or disrupt normal biological functions (Carvalho et al., 2024).

Environmental and Human Health Impacts

Toxicological Effects: Microplastics and nanoplastics can enter the human body via the ingestion of contaminated food and drinks, as well as through the inhalation of airborne particles (Das et al., 2024a, 2024b, 2024c, 2024d; Dey et al., 2023a, 2023b; Singh et al., 2020). Their presence raises concerns regarding respiratory issues, cardiovascular diseases, and possible cancer risks, prompting ongoing studies into their long-term health effects (Khanna et al., 2024).

Bioaccumulation: Research has indicated that microplastics and nanoplastics can accumulate in human tissues, with findings showing their presence in blood, organs, and breast milk. This accumulation raises significant health concerns, as these particles may disrupt biological functions and contribute to chronic health issues (Carvalho et al., 2024).

Ecosystem Disruption: The presence of microplastics and nanoplastics can have a detrimental impact on soil health by affecting water retention, nutrient availability, and overall soil fertility. Additionally, these particles can disrupt microbial communities and food webs, resulting in decreased ecosystem resilience and reduced biodiversity (Nath et al., 2024).

Chemical Pollution: Microplastics and nanoplastics have the ability to adsorb toxic pollutants from their environment, including heavy metals and persistent organic pollutants. This process intensifies chemical contamination in ecosystems, posing risks to aquatic organisms and potentially entering the food chain (Kushwaha et al., 2024).

Fate of Micropollutants in the Environment

Bioaccumulation and Biomagnification

The persistence and potential toxic effects of micropollutants in the environment, especially regarding bioaccumulation, are significant concerns for ecosystems and human health. Emerging micropollutants, such as pharmaceuticals, microplastics, and organic compounds, can accumulate in aquatic organisms, leading to serious ecological and health consequences. These micropollutants frequently enter aquatic systems due to inadequate removal during sewage treatment processes, causing their buildup in water bodies (Eregowda & Mohapatra, 2020). Higher trophic level organisms, like fish, act as bioindicators because of their ability to accumulate these pollutants, which can affect human health when consumed (Basu et al., 2023). The bioaccumulation of organic micropollutants may result in chronic effects in aquatic species, including endocrine disruption and increased vulnerability to diseases (Eregowda & Mohapatra, 2020; Gautam & Anbumani, 2020). Prolonged exposure to these pollutants has been linked to various health issues in humans, including hormonal imbalances and a greater risk of cancer (Gautam & Anbumani, 2020). Furthermore, microplastics serve as vectors for other pollutants, increasing their potential for bioaccumulation in marine organisms (Pelamatti et al., 2022). The combination of microplastics with adsorbed contaminants poses additional risks, highlighting the need for further research into their ecotoxicological effects (Pelamatti et al., 2022).

The fate of micropollutants in the environment, particularly regarding biomagnification, represents an important area of research due to the growing presence of these contaminants in aquatic ecosystems (Bala et al., 2024a, 2024b; Prasad et al., 2024). Micropollutants, such as pharmaceuticals and heavy metals, frequently evade conventional wastewater treatment processes, leading to their accumulation in water bodies and subsequent absorption by aquatic organisms. This accumulation can trigger biomagnification, whereby pollutant concentrations escalate at higher trophic levels, posing risks to both wildlife and human health. Hydrophobic organic chemicals (HOCs) are especially prone to biomagnification as they tend to build up in the fatty tissues of living organisms (Jonker, 2012). This process can result in elevated concentrations within predator species, ultimately interfering with their physiological functions and reproductive health. The biomagnification factor (BMF) measures the extent of this accumulation, demonstrating that pollutant levels rise with each successive trophic level (Jonker, 2012). As a result, apex predators may be exposed to significantly higher concentrations than their prey, potentially leading to serious health problems and declines in population numbers. Moreover, biomagnification can disrupt aquatic food webs considerably, causing decreases in biodiversity and altering the dynamics of communities. The extinction of sensitive species can weaken ecosystems, making them less resilient and less capable of recovering from environmental challenges. Additionally, elevated levels of biomagnified pollutants in human food sources, particularly fish, present significant health threats, including neurological damage and reproductive issues. Long-term exposure to these contaminants can lead to chronic health conditions, highlighting the need for public health strategies to mitigate the risks associated with dietary exposure.

Degradation and Transformation

The degradation mechanisms of micropollutants, such as pharmaceuticals and personal care products, are vital areas of research in environmental science. Biodegradation involves the breakdown of these contaminants by microbial action in aquatic environments, where specific bacteria and fungi play crucial roles in metabolizing these compounds (Yusuf et al., 2024). The efficiency of biodegradation can be influenced by several factors, including temperature, nutrient availability, and the presence of other organic matter (John & Das, 2024; Mishra et al., 2024a; Mohanty et al., 2024a; Tripathy et al., 2024). Photodegradation utilizes reactive radicals generated by ultraviolet light to effectively degrade micropollutants, with efficiency varying according to the complexity of the water matrix (Huang et al., 2023). Environmental conditions, such as the presence of dissolved organic matter and the specific wavelengths of light available, can also impact this process. Advanced oxidation processes, such as gamma irradiation and ozonation, have demonstrated over 90% degradation efficiency for certain micropollutants, highlighting their potential for treating contaminated water (Jang et al., 2024; Singh et al., 2024). These methods generate highly reactive species capable of breaking down even recalcitrant compounds; however, their applicability may vary depending on the type of micropollutant and treatment conditions. The degradation of micropollutants often results in the formation of transformation products (TPs), which may have different toxicity profiles compared to their parent compounds (Behera et al., 2020, 2021, 2022). For instance, gamma irradiation has been found to produce TPs that present lower environmental risks, emphasizing the need for thorough assessments of these byproducts (Jang et al., 2024). Understanding the nature and toxicity of TPs is essential, as they may still pose risks to aquatic ecosystems and human health despite being degradation products.

Persistence and Transport

Micropollutants (MPs) are recognized for their significant toxicity and resistance to degradation, allowing them to persist in the environment for extended periods (Kumar et al., 2024). This durability raises alarms regarding their long-term effects on ecosystems and human health, as accumulating concentrations may result in increasingly severe biological consequences over time. Organic micropollutants, including organochlorine pesticides and polycyclic aromatic hydrocarbons, can build up in living organisms, disrupting essential biological functions and causing harmful health effects in both wildlife and humans (Gupta, 2024), (Kanungo et al., 2023a, 2023b) (Das et al., 2023a, 2023b, 2025). Such accumulation can lead to bioaccumulation and biomagnification, posing considerable risks to higher trophic levels and potentially entering the human food chain (Das et al., 2024a, 2024b, 2024c, 2024d; Mishra et al., 2024c; Mohanty et al., 2024b; Sahu et al., 2024). Additionally, micropollutants can traverse through water bodies and soil, influenced by factors such as solubility, adsorption properties, and environmental conditions (Dawen & Nabi, 2024). Their movement can result in extensive contamination across various ecosystems, complicating remediation efforts and impacting overall ecosystem health. Furthermore, the transport of MPs can lead to the redistribution of pollutants, forming contamination hotspots in both terrestrial and aquatic environments (Oliveira et al., 2024). This widespread effect can disrupt food webs and ecosystem services, emphasizing the pressing need for effective management strategies to address the risks posed by micropollutant transport.

Assessment Techniques for Micropollutant

Detection and Measurement

Assessment techniques for micropollutants are primarily aimed at detecting and quantifying their presence across diverse environments, with a particular focus on agricultural and urban areas (Dey et al., 2024a, 2024b; Moharana et al., 2024). Recent research underscores the importance of both targeted and non-targeted screening methods in identifying a broad spectrum of organic micropollutants (OMPs) that pose significant risks to ecosystems and human health.

Non-targeted Screening

Comprehensive Identification: Non-targeted screening has uncovered more than 400 micropollutants, including pesticides and pharmaceuticals, within agricultural environments through advanced mass spectrometry (Qiu et al., 2024). High-resolution tools are employed to detect trace-level compounds in a variety of samples.

Ongoing improvements in methodology enhance both the precision and reliability of these assessments in real-world applications. Methods like suspect and non-target screening (SNTS) have been utilized to uncover a variety of micropollutants in urban stormwater runoff, identifying 65 different substances during storm events (Kang et al., 2024). This technique is instrumental in detecting emerging contaminants that conventional methods may miss, offering a more thorough insight into water quality challenges.

Detecting Emerging Contaminants: This approach is invaluable for identifying new contaminants that conventional techniques might miss, enabling a more holistic environmental evaluation (Kang et al., 2024). It also aids in recognizing novel chemical structures with potential risks. Such findings can shape regulatory policies and improve risk evaluation strategies.

Targeted Analysis

Fluorescence-Based Detection: Fluorescence spectroscopy offers a cost-efficient solution for the quick identification of fluorescent OMPs in water, with detection limits ranging from 1.0 to 3.3 $\mu g\ L^{-1}$ (Paradina-Fernández et al., 2023). Its portable design allows for effective on-site testing in isolated or resource-constrained locations. Advances in sensor technology have further broadened its range of applications and improved its detection capabilities.

Integrated Techniques: The combination of chromatography and spectroscopy enhances the ability to detect micropollutants, particularly in complex matrices such as wastewater (Sharma et al., 2024a, 2024b). This integration provides better resolution for analyzing intricate compound mixtures. Optimized targeted analysis methods, including gas chromatography and liquid chromatography, are highly effective for identifying specific micropollutants in sewage sludge and agricultural environments (Rede et al., 2023; Qiu et al., 2024). For example, research has detected 42 organic compounds in sewage sludge with detection limits ranging from 0.00700 to 0.271 $\mu g\ g^{-1}$, highlighting the precision and efficacy of these approaches in quantifying micropollutants (Rede et al., 2023). Additionally, these hybrid techniques are versatile and can be adapted for use in other environmental samples like soil and sediments.

Source-Controlled Approaches

Proactive Measures: Source-controlled strategies, such as advanced oxidation processes and membrane filtration systems, effectively reduce the release of micropollutants into the environment (Sharma et al., 2024a, 2024b). These methods are particularly beneficial for industries with significant pollutant discharges. Their integration into treatment systems ensures compliance with environmental regulations and helps mitigate pollution risks.

Hybrid Techniques: The integration of chromatography and spectroscopy is commonly used for detecting micropollutants, such as pharmaceuticals and pesticides (Sharma et al., 2024a, 2024b). This combined methodology enhances both the resolution and accuracy of pollutant analyses, especially in intricate environmental samples like wastewater. Furthermore, it allows for the simultaneous separation and characterization of several compounds within a single analytical process, which conserves time and resources.

Advanced Instruments: Techniques such as triple quadrupole liquid chromatography-mass spectrometry and nuclear magnetic resonance are utilized for accurately identifying and quantifying micropollutants (Ahmed et al., 2020). These sophisticated methods offer high sensitivity and specificity, enabling the detection of contaminants at trace levels. Moreover, recent improvements in instrumentation have increased their throughput, making them well-suited for extensive environmental monitoring initiatives.

Health and Environmental Implications

The buildup of micropollutants in aquatic organisms, such as fish, presents significant health hazards, emphasizing the need for comprehensive monitoring and assessment methods (Nazar et al., 2024). These pollutants can negatively impact aquatic biodiversity by disrupting physiological and metabolic functions in marine life. Studies show that micropollutants can enter the food chain, creating serious concerns for food safety and public health (Nazar et al., 2024). This underscores the pressing need for enhanced regulatory frameworks and advanced treatment technologies to reduce environmental pollution.

Toxicity Testing

Assessment techniques for micropollutants, especially in toxicity testing, utilize diverse methodologies to evaluate the effects of chemical contaminants on ecosystems and human health. These methods play a vital role in identifying the risks posed by micropollutants in water systems and shaping effective regulatory frameworks.

Bioassays

The use of bioassays in assessing micropollutants (MPs) is a vital approach to understanding their ecological effects and evaluating the effectiveness of wastewater treatment processes. Bioassays offer a biological perspective on the toxicity of water samples, serving as a complement to traditional chemical analysis. This method is particularly effective in identifying the impacts of low-concentration MPs that may evade removal during conventional wastewater treatments (Escher et al., 2014; Toušová et al., 2019).

Types of Bioassays

Cell-Based Bioassays: These assays focus on health-related biological endpoints, measuring toxicity pathways such as xenobiotic metabolism and hormone-mediated activities (Escher et al., 2014). They are especially useful in detecting endocrine-disrupting compounds and other bioactive substances that might escape chemical detection. Their sensitivity to molecular changes makes them a powerful tool for early-stage risk assessment.

Mode-of-Action Test Batteries: These test systems combine multiple bioassays, including bioluminescence inhibition and growth rate inhibition tests, to evaluate various toxicological modes of action (Escher et al., 2008). By encompassing a wide range of toxic responses, they provide a holistic understanding of the ecological risks associated with MPs. Additionally, they help identify specific stressors within wastewater, aiding in targeted treatment improvements.

Efficacy in Wastewater Treatment

Challenges in MP Removal: Research shows that while traditional wastewater treatments significantly reduce MP levels, certain groups, such as herbicides and insecticides, remain difficult to fully eliminate (Ma et al., 2020). This underscores the need for integrating advanced treatment methods, like membrane filtration or advanced oxidation processes, to enhance removal efficiency. Furthermore, persistent MPs can accumulate in aquatic environments, amplifying long-term ecological risks.

Toxicity of Treated Effluents: Bioassays reveal that wastewater treatment plants often retain residual toxicity in their effluents, indicating the need for more sophisticated technologies to address this issue (Abbasi et al., 2022; Escher et al., 2008). Advanced bioassay applications can guide the development of treatment strategies tailored to mitigate specific toxic effects. These insights are critical for ensuring that treated water meets stringent ecological and health standards.

Biosensors

Biosensors have become a transformative technology for monitoring micropollutants, capitalizing on breakthroughs in nanotechnology and biological engineering. These devices combine high sensitivity, specificity, and real-time monitoring capabilities, making them indispensable for environmental applications (Das, 2016).

Types of Biosensors

Nanobiosensors: These sensors use nanomaterials to achieve superior detection performance, enabling real-time analysis with minimal sample preparation (Kuswandi, 2018; Quazi et al., 2023). The integration of nanotechnology enhances signal transduction, improving both the accuracy and speed of pollutant detection. Additionally, their compact design makes them adaptable for in-field applications in remote areas.

Microbial Biosensors: Genetically engineered microorganisms in these biosensors detect pollutants via specific biochemical reactions, offering both cost-efficiency and portability (Bala et al., 2022). They are particularly effective for identifying a broad range of organic micropollutants. Advances in genetic engineering have further improved their selectivity and durability in varying environmental conditions.

Optical Biosensors: Systems such as surface plasmon resonance (SPR) provide ultrasensitive detection of micropollutants, achieving detection limits in the nanogram-per-liter range (Tan et al., 2023). These biosensors excel in high-precision applications, offering unparalleled accuracy for complex water samples. Recent developments in optical technologies have also increased their robustness and usability in diverse environmental settings.

Advantages of Biosensors

High Sensitivity and Specificity: Biosensors can detect trace amounts of micropollutants, which is essential for early warning systems and environmental monitoring (Dhir, 2023). Their ability to distinguish between similar compounds reduces false positives and ensures more reliable data for analysis. This capability is vital for maintaining ecosystem health and public safety.

Real-Time Monitoring: Biosensors enable continuous monitoring of micropollutants, allowing for timely identification and intervention (Bala et al., 2022). This real-time feedback is critical for dynamic environments, such as industrial discharge points and stormwater systems. Moreover, it enhances decision-making for environmental management.

Ease of Use: Many biosensors are designed for straightforward operation, making them accessible for various applications, including on-site monitoring by non-specialists (Dhir, 2023). The user-friendly interfaces of modern biosensors minimize training requirements, expanding their usability in both developed and resource-limited regions. This ease of use also facilitates large-scale deployment for environmental protection efforts.

Acute and Chronic Toxicity Assessment

The assessment of acute and chronic toxicity of micropollutants (MPs) is essential for comprehending their environmental effects. Various techniques have been developed to evaluate these toxicities across different trophic levels, focusing on both immediate and long-term consequences for aquatic ecosystems.

Acute Toxicity Assessment Techniques

Organism Selection: Common test organisms include Daphnia magna, Danio rerio, and marine algae, each demonstrating varying sensitivities to different micropollutants (Yilmaz et al., 2022). The choice of organisms is critical, as their responses can provide insights into the potential ecological impacts of specific contaminants. Additionally, using a diverse range of test organisms helps to build a more comprehensive understanding of toxicity across trophic levels.

Testing Methods: Standardized bioassays are used to determine the effective concentration (EC50) of pollutants, such as 1-Chloro-2,4 dinitrobenzene, which exhibited significant toxicity (Yilmaz et al., 2022). These assays follow rigorous protocols to ensure consistency and reproducibility of results, making them valuable for regulatory assessments. Moreover, the EC50 values obtained can inform risk assessments and help establish safe concentration limits for various micropollutants.

Data Utilization: Acute toxicity data can be aggregated to derive predicted no-effect concentrations (PNECs) for various active pharmaceutical ingredients (APIs) (Vestel et al., 2016). This process aids in developing guidelines for environmental safety and public health. Furthermore, the establishment of PNECs enables stakeholders to prioritize monitoring efforts for the most concerning pollutants based on their toxicity profiles.

Chronic Toxicity Assessment Techniques

Long-term Exposure Studies: Chronic toxicity assessments examine the effects of prolonged exposure to MPs, which may remain undetected in acute tests (Arslan et al., 2017). These studies are essential for understanding the cumulative effects of low-level exposures on aquatic organisms. Additionally, long-term assessments can reveal potential population-level impacts, such as reproductive success and growth rates, that are critical for ecosystem sustainability.

Bioaccumulation Studies: Research highlights the role of aquatic and terrestrial plants in the bioaccumulation of organic micropollutants, emphasizing potential risks to ecosystems (Arslan et al., 2017). These studies are crucial for understanding how contaminants move through food webs and impact higher trophic levels. Furthermore, bioaccumulation data can inform risk assessments by identifying species that are particularly vulnerable to toxic effects.

Molecular Techniques: Advanced methods, such as real-time PCR, are employed to investigate the molecular effects of MPs, including oxidative stress responses in organisms like Hydra (Vimalkumar et al., 2022). These molecular approaches provide deeper insights into the cellular mechanisms affected by exposure to micropollutants. Additionally, they enable researchers to identify specific biomarkers of toxicity that can be used for early detection of environmental stressors.

Modeling and Predictive Tools

Modeling and predictive assessments of micropollutants play a critical role in understanding their behavior within wastewater treatment processes and their effects on the environment. A range of methodologies has been developed to enhance the accuracy of predictions regarding the removal of micropollutants, focusing on various treatment technologies and the factors that influence their degradation and rejection.

Wastewater Treatment Modeling

The Simple Treatment model has been utilized to assess the removal of micropollutants in conventional activated sludge systems, revealing that the biodegradation constant is a significant factor affecting prediction accuracy (Di Marcantonio et al., 2024). This model aids in identifying optimal conditions for treatment processes, thereby increasing the removal efficiency of micropollutants (Mishra & Das, 2021; Mishra et al., 2024b). Furthermore, integrating site-specific data can improve the model's reliability and applicability across diverse treatment facilities (Di Marcantonio et al., 2024). A three-year monitoring initiative demonstrated that while the initial predictions from the model were often inadequate, modifications based on comprehensive full-scale data led to better outcomes. This iterative process emphasizes the necessity for ongoing validation and enhancement of predictive models to align them more closely with actual performance. The study also highlights the importance of updating models regularly as new data emerges, thereby improving their accuracy over time (Di Marcantonio et al., 2024).

Nanofiltration and Membrane Technology

The effectiveness of nanofiltration membranes in removing micropollutants was assessed, identifying crucial factors such as membrane pore size and operating pressure that are vital for determining rejection rates (Xu et al., 2024). By optimizing these parameters, substantial improvements can be achieved in the removal of various contaminants. Additionally, gaining insight into the interactions between these factors facilitates improved design and operation of nanofiltration systems for different applications. A machine learning model indicated a low error rate in predicting micropollutant rejection, highlighting the potential for refining membrane design and operational strategies (Xu et al., 2024). Incorporating machine learning techniques enables researchers to swiftly adapt to variations in influent water quality and operational conditions, thereby enhancing system performance. Moreover, the integration of predictive models into real-time monitoring frameworks allows for proactive modifications to treatment processes (Xu et al., 2024).

Advanced Oxidation Processes

Modeling of UV-based advanced oxidation processes (UV-AOPs) was conducted to predict the degradation of micropollutants, highlighting the importance of the water matrix's complexity on degradation rates (Huang et al., 2023). This modeling strategy aids in identifying the best conditions for treatment, ensuring optimal degradation of specific micropollutants. Understanding how various water constituents influence degradation also informs treatment design under different water quality conditions (Huang et al., 2023). The research established a facilitated prediction method that combines simulation with portable measurements, optimizing the treatment process. This novel approach enables real-time evaluations and adjustments during treatment, resulting in more efficient degradation of contaminants. Incorporating portable measurements into the modeling framework also aids in validating predictions in situ, thereby increasing confidence in the outcomes (Huang et al., 2023).

Urban Water Cycle Insights

Investigations into the urban water cycle have revealed the persistence and seasonal dynamics of various micropollutants, emphasizing the need for ongoing monitoring (Arlos et al., 2023). This continuous assessment is vital for comprehending how micropollutant concentrations vary throughout the year and their potential impacts on water quality. Such monitoring efforts can also guide policymakers in implementing strategies to mitigate pollution during high-concentration periods (Arlos et al., 2023). The research identified certain substances as effective tracers for wastewater contamination, which can assist in future monitoring efforts. By utilizing these tracers, researchers can effectively track the movement and fate of micropollutants in urban water systems. This information can further support targeted remediation initiatives and enhance overall urban water management practices (Arlos et al., 2023).

Adsorption Mechanisms

The application of zeolites for micropollutant adsorption was modeled using quantitative structure–activity relationships (QSAR), showcasing strong predictive capabilities for various micropollutants (Cho et al., 2024). This modeling approach helps pinpoint the most effective adsorbents for specific contaminants, improving the efficiency of treatment processes. Additionally, the predictive power of QSAR models can aid in designing customized adsorbents for various environmental applications. The results indicate that both empirical data and in silico descriptors can accurately predict adsorption affinities. This combined approach enables researchers to utilize existing data while also incorporating computational methods to forecast adsorption behavior (Cho et al., 2024).

Challenges and Limitations

Detection and Analysis

The detection and analysis of micropollutants in the environment present several significant challenges. These stem from the complex nature of these contaminants, their presence at extremely low concentrations, and the shortcomings of current detection methods. Overcoming these obstacles is essential for implementing effective strategies to reduce their environmental impact.

Complexity of Micropollutants: Micropollutants comprise a wide range of substances, including pharmaceuticals, pesticides, and industrial chemicals, each with distinct chemical characteristics and environmental behaviors (Ganguly et al., 2024; Sharma et al., 2024a, 2024b). Their diverse origins and behaviors make detection difficult, as specific analytical techniques are often required for each category (Sharma et al., 2024a, 2024b). Furthermore, their interactions with environmental matrices can lead to complex mixtures, complicating accurate identification and measurement. Traditional detection methods frequently lack the precision and sensitivity needed to identify micropollutants at trace levels, which can result in underestimating their presence (Madhesh & Mohamed, 2024).

Complexity of Co-Exposure: Research often emphasizes individual pollutants, overlooking the combined or interactive effects that arise from simultaneous exposures to multiple contaminants (Eick & Hüls, 2022). This gap in understanding does not reflect real-world conditions, where organisms encounter pollutant mixtures, potentially underestimating associated risks.

Detection Challenges: A common issue in studies is the inability to detect certain pollutants at very low concentrations, which hampers statistical analysis and risk assessment efforts (Eick & Hüls, 2022). These detection limitations often lead to incomplete datasets, diminishing the reliability of predictive models and hindering the development of effective mitigation measures. Although advanced technologies such as hybrid chromatography-spectroscopy are emerging as solutions, they are limited by their high costs, the need for specialized expertise, and restricted accessibility, particularly in less developed regions (Sharma et al., 2024a, 2024b). These constraints hinder their broader application in regular monitoring practices.

Gaps in Regulation and Monitoring: The lack of cohesive and comprehensive regulatory systems poses a major barrier to effectively monitoring and managing micropollutants in the environment. Additionally, existing databases often fail to cover the full spectrum of micropollutants, limiting their usefulness for thorough risk assessments. Many newly identified contaminants remain unaccounted for, further delaying necessary mitigation measures and increasing the potential risks they pose (Ganguly et al., 2024).

Understanding Fate and Transport

Understanding the behavior and movement of micropollutants in the environment poses numerous challenges and limitations. These difficulties arise from the complex interplay of environmental factors and the distinct properties of micropollutants.

Variability of Pollutants: Micropollutants, including substances like pharmaceuticals and pesticides, possess diverse chemical characteristics that complicate their interactions within different environmental systems (Kumar et al., 2023). Their interactions with organic matter and fluctuating environmental conditions further increase the unpredictability of their behavior. Additionally, the mixed and inconsistent composition of waste in landfills produces highly variable leachates, making effective monitoring and treatment a significant challenge (Kumar et al., 2023).

Challenges in Experimental Studies: Laboratory column experiments, though valuable for studying transport mechanisms, often fail to replicate real-world conditions due to unique field boundaries and non-equilibrium dynamics (Banzhaf & Hebig, 2016). Conducting field-scale studies is resource-intensive and often impractical, which limits their use in validating experimental findings. Furthermore, differences in experimental setups can result in inconsistent outcomes, making it difficult to compare results across different studies (Banzhaf & Hebig, 2016).

Interactions with Environmental Factors: Micropollutants distribute unevenly across water, sediment, and living organisms, requiring a multi-compartment approach to accurately assess risks (Pulgarin-Zapata et al., 2024). This distribution is also influenced by environmental factors such as temperature and pH, which can vary significantly across ecosystems. Local biogeochemical processes, including adsorption and degradation, add further complexity to understanding the behavior and fate of micropollutants in natural systems (Pachepsky et al., 2021).

Data Gaps and Uncertainty in Environmental Impact

Lack of Comprehensive Data: There is a significant lack of historical data on the toxicological effects of micropollutants, especially those involving engineered nanoparticles and pharmaceuticals (Williams et al., 2022; Ahkola et al., 2024). This data gap hinders the development of effective regulatory frameworks and limits our ability to predict long-term environmental and health consequences. Moreover, the rapid emergence of novel pollutants like nanoplastics exacerbates this challenge by introducing uncertainties into current monitoring practices.

Uncertain Health Effects: The health effects of micropollutants, including their potential to cause genetic damage and increase cancer risks, remain poorly understood (Ganguly et al., 2024; Joshi & Sarma, 2021). The lack of comprehensive epidemiological studies further limits our understanding of chronic exposure effects. Additionally, the cumulative and low-dose impacts of these pollutants on vulnerable populations are often overlooked in current research.

Inconsistent Risk Assessments: Environmental risk assessments for pharmaceuticals frequently lack consistency, creating significant uncertainties in predicted no-effect concentrations (Ahkola et al., 2024). These inconsistencies stem from variations in methodologies and incomplete datasets used in evaluations. Additionally, the absence of standardized testing protocols across regions undermines the reliability and comparability of risk predictions.

Future Trends in Micropollutant in Environment

Future trends in micropollutant management are increasingly centered on advanced treatment technologies and preventive strategies to reduce their environmental impact. Pharmaceuticals and personal care products, in particular, remain major threats to ecosystems and human health, making the development of improved detection and remediation techniques critical.

Emerging Treatment Technologies

Advanced Oxidation Processes (AOPs): Methods like ozonation and UV irradiation are becoming increasingly popular due to their ability to effectively degrade micropollutants in wastewater (Derco et al., 2024). These techniques produce highly reactive radicals capable of breaking down even the most persistent pollutants into safer byproducts.

Nitrification–Denitrification (ND): This eco-friendly approach has demonstrated potential in transforming micropollutants through biological processes by optimizing parameters such as pH and dissolved oxygen levels (Muhammad et al., 2024). Furthermore, integrating ND with bioaugmentation has shown promise in improving the efficiency of micropollutant removal.

Source-Controlled Approaches

Preventive Measures: Industries are increasingly adopting proactive strategies to reduce the release of micropollutants into the environment (Sharma et al., 2024a, 2024b). Examples include initiatives like pharmaceutical collection programs and the development of environmentally friendly product alternatives.

Legislative Support: Recent policies are requiring the monitoring and removal of targeted micropollutants, fostering collaboration among various stakeholders (Derco et al., 2024). These regulations are also accelerating the creation of standardized testing protocols and sophisticated analytical methods to ensure compliance.

Conclusion

Micropollutants pose a serious threat to environmental sustainability due to their varied origins, persistence in ecosystems, and potential for long-term adverse effects on both ecological and human health. Tackling this issue requires a holistic strategy that involves identifying their types, studying their behavior in the environment, and improving methods for their detection and removal. Both organic and inorganic micropollutants, along with newer concerns such as microplastics and nanoplastics, demand detailed investigations into their bioaccumulation, transformation, and movement within ecological systems. Despite advancements in detection technologies, toxicity assessments, and predictive modeling, significant challenges remain, including data shortages and the complexity of environmental interactions. Moving forward, adopting innovative solutions such as advanced oxidation processes and predictive modeling, coupled with effective source control practices, will be essential. Collaboration among scientists, policymakers, and industries, combined with heightened public awareness and robust regulatory policies, holds promise for reducing the impact of micropollutants and fostering a sustainable future.

References

Abbasi, N. A., Shahid, S. U., Majid, M., & Tahir, A. (2022). Ecotoxicological risk assessment of environmental micropollutants. In *Environmental micropollutants* (pp. 331–337). Elsevier.

Ahkola, H., Äystö, L., Sikanen, T., Riikonen, S., Pihlaja, T., & Kauppi, S. (2024). Current uncertainties and challenges of publicly available pharmaceutical environmental risk assessment data. *European Journal of Pharmaceutical Sciences*, 106769.

Ahmed, M. B., Johir, M. A. H., Ngo, H. H., Guo, W., Zhou, J. L., Belhaj, D., & Moni, M. A. (2020). Methods for the analysis of micro-pollutants. In *Current developments in biotechnology and bioengineering* (pp. 63–86). Elsevier.

Almasi, H. (2024). Environmental impacts of emerging micropollutants in the environment: Chemical properties, behavior, toxicology, health effects and fate: A. Journal ISSN, 2766, 2276.

Arif, Y., Mir, A. R., Zieliński, P., Hayat, S., & Bajguz, A. (2024). Microplastics and nanoplastics: Source, behavior, remediation, and multi-level environmental impact. *Journal of Environmental Management, 356*, Article 120618.

Arlos, M. J., Arnold, V. I., Bumagat, J. S., Zhou, J., Cereno, K. M., Deas, A., Munkittrick, K. R., et al. (2023). Combining chemical, bioanalytical and predictive tools to assess persistence, seasonality, and sporadic releases of organic micropollutants within the urban water cycle. *Water Research, 244*, Article 120454.

Arslan, M., Ullah, I., Müller, J. A., Shahid, N., & Afzal, M. (2017). Organic micropollutants in the environment: Ecotoxicity potential and methods for remediation. In *Enhancing cleanup of environmental pollutants* (Vol. 1, pp. 65–99). Biological Approaches.

Bala, K., Anshul, Verma, S., Dogra, A., Das, A. P., Kumar, R., & Nandini. (2024). Sustainable management for urban plastic waste generation. In *Sustainable management of urban plastic waste through circular economic approaches* (pp. 211–236). https://doi.org/10.1201/9781003465447-13

Bala, K., Dogra, A., Lata, P., & Das, A. P. (2024). Discussing sources and biological-chemical remediation approaches for microplastic pollution. In *Sustainable microbial technology for synthetic and cellulosic microfiber bioremediation* (pp. 37–59). https://doi.org/10.1007/978-3-031-62600-5_3

Bala, R., Kumar, M., Somal, M. K., Karnwal, A., & Sachan, R. S. K. (2022). Microbial biosensors for environmental monitoring. In *Environmental microbiology: Advanced research and multidisciplinary applications* (pp. 105–136). Bentham Science Publishers.

Banzhaf, S., & Hebig, K. H. (2016). Use of column experiments to investigate the fate of organic micropollutants—A review. *Hydrology and Earth System Sciences, 20*(9), 3719–3737.

Basu, A., Moulik, S., & Karmakar, S. (2023). Bioaccumulation of different organic micropollutants in fishes and its toxicological and stress impacts: A review. *International Journal for Research in Applied Science & Engineering Technology (IJRASET), 11*(II), 1549–1564.

Bhatt, P., Bhandari, G., & Bilal, M. (2022). Occurrence, toxicity impacts and mitigation of emerging micropollutants in the aquatic environments: Recent tendencies and perspectives. *Journal of Environmental Chemical Engineering, 10*(3), Article 107598.

Behera I. D., Nayak M., Biswas S., Meikap, B. C, Sen R.(2021) Enhanced biodegradation of total petroleum hydrocarbons by implementing a novel two-step bioaugmentation strategy using indigenous bacterial consortium. *Journal of Environmental Management, 292*, 112746. ISSN 0301-4797. https://doi.org/10.1016/j.jenvman.2021.112746

Behera, I. D., Basak, G., Kumar, R. R., Sen, R., & Meikap, B. C. (2020). Treatment of petroleum refinery sludge by petroleum degrading bacterium *Stenotrophomonas pavanii* IRB19 as an efficient novel technology. *Journal of Environmental Science and Health, Part A, 56*(2), 226–239. https://doi.org/10.1080/10934529.2020.1866924

Behera I. D., Nayak M., Mishra A., Meikap B. C., Sen R., (2022). Strategic implementation of integrated bioaugmentation and biostimulation for efficient mitigation of petroleum hydrocarbon pollutants from terrestrial and aquatic environment. *Marine Pollution Bulletin, 177*, 113492. ISSN 0025-326X. https://doi.org/10.1016/j.marpolbul.2022.113492

Das, P., & Bal, M. (2025). Modeling and optimization study for microplastic removal from aquatic medium using Chroococcidiopsis species. *Chemical Engineering Science, 304*, Article 121085. https://doi.org/10.1016/j.ces.2024.121085

Das, P., Halder, G., & Bal, M. (2023a). A critical review on remediation of microplastics using microalgae from aqueous system. *Science of the Total Environment, 898*, Article 166425. https://doi.org/10.1016/j.scitotenv.2023.166425

Carvalho, J. G. D., Augusto, H. C., Ferraz, R., Delerue-Matos, C., & Fernandes, V. C. (2024). Micro (nano) plastic and related chemicals: Emerging contaminants in environment, food and health impacts. *Toxics, 12*(10), 762.

Cherian, E., Mohan, L., & Manoj, S. E. (2024). Microplastics and nanoplastics and related chemicals: The physical–chemical interactions. In *Toxic effects of micro-and nanoplastics: Environment, food and human health* (pp. 73–86).

Cho, B. G., Lee, K. Y., Mun, S. B., Lim, C. R., Yun, Y. S., & Cho, C. W. (2024). Adsorptive removal of micropollutants by natural and faujasite zeolites: Structural effect of micropollutants on adsorption. *Ecotoxicology and Environmental Safety, 270*, Article 115869.

Çimen Mesutoğlu, Ö., & Gök, O. (2024). Assessment and modeling of benzene micropollutant in surface waters proximal to coal-fired thermal power plants. *International Journal of Coal Preparation and Utilization, 44*(6), 740–755.

Das, A. P. (2016). Biosensors: The future of diagnostics. *International Journal of Sensor Networks and Data Communications, 01*(s1). https://doi.org/10.4172/2090-4886.S1-e001

Das, A. P., Mallavarapu, M., & Ghosh, S. (2023b). Ecotoxicity of mining pollutants on the environment and their remediation. *Environmental Chemistry and Ecotoxicology, 5*, 165–167. https://doi.org/10.1016/J.ENCECO.2023.08.002

Das, A. P., Behera, I. D., & Das, N. P. (2024). *Renewable energy generation and value addition from environmental microfiber pollution through advanced greener solution* (p. 278).

Das, A., Dey, S., & Das, A. P. (2024). Ecotoxicological impacts of synthetic microfiber pollutants and development of sustainable mitigation strategies. *Environmental Chemistry and Ecotoxicology*. https://doi.org/10.1016/J.ENCECO.2024.12.003

Das, A., Dey, S., Bala, K., Gnanasekaran, R., & Das, A. P. (2024). Microbial bioremediation technology for sustainable treatment and management of synthetic microfiber waste. In *Sustainable microbial technology for synthetic and cellulosic microfiber bioremediation* (pp. 221–238). https://doi.org/10.1007/978-3-031-62600-5_12

Das, A., Ray, S. K., Mohanty, M., Mohanty, J., Dey, S., & Das, A. P. (2024d). Ecotoxicity of microplastic wastes and their sustainable management: A review. *Environmental Chemistry and Ecotoxicology, 6*, 144–152. https://doi.org/10.1016/J.ENCECO.2024.05.003

Dawen, G., & Nabi, M. (2024). Fate and transport of emerging pollutants in aquatic environment. In *Novel approaches towards wastewater treatment: Effective strategies and techniques* (pp. 315–416). Springer Nature Switzerland.

Derco, J., Žgajnar Gotvajn, A., Guľašová, P., Šoltýsová, N., & Kassai, A. (2024). Selected micropollutant removal from municipal wastewater. *Processes, 12*(5), 888.

Dey, S., Bissoyi, A., Koteswara Reddy, G., & Das, A. P. (2024). Synthetic microfibers: An urban emerging pollutant demanding for sustainable environmental solutions. In *Sustainable microbial technology for synthetic and cellulosic microfiber bioremediation* (pp. 107–121). https://doi.org/10.1007/978-3-031-62600-5_7

Dey, S., Das, A., Mallick, K., Sahu, A., & Das, A. P. (2023). *Environmental petroleum waste: Pollution, toxicity, sustainable remediation* (pp. 159–175). https://doi.org/10.1007/978-3-031-48220-5_7

Dey, S., Dash, N., Ray, S. K., Priyadarsini, S., Chhotaray, C., & Das, A. P. (2024). Occurrence and quantification of tire wear plastics pollutants from Highway City Dust of Bhubaneswar City of Odisha. In *Sustainable microbial technology for synthetic and cellulosic microfiber bioremediation* (pp. 77–89). https://doi.org/10.1007/978-3-031-62600-5_5

Dey, S., Tripathy, B., Kumar, M. S., & Das, A. P. (2023). Ecotoxicological consequences of manganese mining pollutants and their biological remediation. *Environmental Chemistry and Ecotoxicology*. https://doi.org/10.1016/J.ENCECO.2023.01.001

Dhir, B. (2023). Techniques in prevention, detection and monitoring of environmental contaminants. In *Technology for a sustainable environment* (pp. 271–292). Bentham Science Publishers.

Di Marcantonio, C., Chiavola, A., Spagnoli, D., Meşe, B., Margarita, F., Gioia, V., Boni, M. R., et al. (2024). Linking conventional activated sludge treatment plant performances for micropollutants removal to environmental risk and SimpleTreat model assessment. *International Journal of Environmental Science and Technology*, 1–14.

Eick, S. M., & Hüls, A. (2022). Invited perspective: Challenges and opportunities for missing data in the context of environmental mixture methods. *Environmental Health Perspectives, 130*(11), Article 111305.

Eregowda, T., & Mohapatra, S. (2020). Fate of micropollutants in engineered and natural environment. In *Resilience, response, and risk in water systems: Shifting management and natural forcings paradigms* (pp. 283–301).

Escher, B. I., Allinson, M., Altenburger, R., Bain, P. A., Balaguer, P., Busch, W., Leusch, F. D., et al. (2014). Benchmarking organic micropollutants in wastewater, recycled water and drinking water with in vitro bioassays. *Environmental Science & Technology, 48*(3), 1940–1956.

Ganguly, A., Nanda, S., Das, K., Ghanty, S., Biswas, G., Mandi, M., & Rajak, P., et al. (2024). *Micropollutants in environment: Sources, ecotoxicity, and strategies for remediation. Ecosystem management: Climate change and sustainability* (pp. 453–492).

Gautam, K., & Anbumani, S. (2020). Ecotoxicological effects of organic micro-pollutants on the environment. In *Current developments in biotechnology and bioengineering* (pp. 481–501). Elsevier.

Ghosh, S., Sinha, J. K., Ghosh, S., Vashisth, K., Han, S., & Bhaskar, R. (2023). Microplastics as an emerging threat to the global environment and human health. *Sustainability 15*(14), 10821. https://doi.org/10.3390/SU151410821

Gooty, J. M., Srinivasulu, M., Mosquera, J. A. N., & Llaguno, S. N. S. (2022). Occurrence and fate of micropollutants in surface waters. In *Environmental micropollutants* (pp. 233–269). Elsevier.

Gupta, S. (2024). Organic micropollutants in the environment: Ecotoxicity potential and bioremediation approaches. In *Organic micropollutants in aquatic and terrestrial environments* (pp. 249–263). Springer Nature Switzerland.

Jang, H., Kim, T. H., & Jeon, J. (2024). Degradation of micropollutants by gamma irradiation: Insight of transformation product formation explored by suspect and non-target screening using LC-HRMS. *Journal of Environmental Chemical Engineering, 12*(1), Article 111659.

John, A., & Das, A. P. (2024). Government strategies for achieving circular economy for plastic waste management in Southeast Asia. In *Sustainable management of urban plastic waste through circular economic approaches* (pp. 72–82). https://doi.org/10.1201/9781003465447-6/

Jonker, M. T. O. (2012). What is causing biomagnification of persistent hydrophobic organic chemicals in the aquatic environment? *Environmental Science & Technology, 46*(1), 110–111. https://doi.org/10.1021/es204161k

Joshi, S. J., & Sarma, H. (2021). Micropollutants in the environment: Challenges and bioremediation strategies. *Open Biotechnology Journal, 15*(Special issue 1), 68–70.

Kang, D., Yun, D., Cho, K. H., Baek, S. S., & Jeon, J. (2024). Profiling emerging micropollutants in urban stormwater runoff using suspect and non-target screening via high-resolution mass spectrometry. *Chemosphere, 352*, Article 141402.

Kanungo, J., Sahoo, T., Bal, M., & Behera, I. D. (2023a). Performance of bioremediation strategy in waste lubricating oil pollutants: A review. *Geomicrobiology Journal, 41*(4), 360–373. https://doi.org/10.1080/01490451.2023.2245395

Kanungo, J., Sahoo, T., Swain, L. P., Behera, I. D. (2023). Toxicity of persistent hydrocarbon pollutants, sources and sustainable remediation process. In Behera, I. D., Das, A. P. (eds.), *Impact of petroleum waste on environmental pollution and its sustainable management through circular economy*. Environmental Science and Engineering. Springer. https://doi.org/10.1007/978-3-031-48220-5_2

Kaur, J. (2024). Assessment, obstacles, and risk communication for organic micropollutants in the urban water. In *Organic micropollutants in aquatic and terrestrial environments* (pp. 181–200). Springer Nature Switzerland.

Kaur, R., Kaur, H., Singh, S., Jagota, N., Sharma, A., & Sharma, A. (2024). Organic micropollutants in the urban soils: Technological advances and management concerns. In *Organic micropollutants in aquatic and terrestrial environments* (pp. 47–63).

Khanna, R., Chandra, A., Sen, S., Konyukhov, Y., Fuentes, E., Burmistrov, I., & Kravchenko, M. (2024). Microplastics and nanoplastics as environmental contaminants of emerging concern: Potential hazards for human health. *Sustainability, 16*(19), 8704.

Kumar, A., Deepika, Tyagi, D., Tarkeshwar, & Kapinder. (2024). Organic micropollutants and their effects on the environment and human health. In *Organic micropollutants in aquatic and terrestrial environments* (pp. 87–105). Springer Nature Switzerland.

Kumar, V., Sharma, N., Umesh, M., Chakraborty, P., Kaur, K., Duhan, L., Maitra, S. S., et al. (2023). Micropollutants characteristics, fate, and sustainable removal technologies for landfill leachate: A technical perspective. *Journal of Water Process Engineering, 53*, Article 103649.

Kushwaha, M., Shankar, S., Goel, D., Singh, S., Rahul, J., Rachna, K., & Singh, J. (2024). Microplastics pollution in the marine environment: A review of sources, impacts and mitigation. *Marine Pollution Bulletin, 209*, Article 117109.

Kuswandi, B. (2018). Nanobiosensors for detection of micropollutants. *Environmental Nanotechnology, 1*, 125–158.

Luo, Y., Guo, W., Ngo, H. H., Nghiem, L. D., Hai, F. I., Zhang, J., Wang, X. C., et al. (2014). A review on the occurrence of micropollutants in the aquatic environment and their fate and removal during wastewater treatment. *Science of the Total Environment, 473*, 619–641.

Ma, X. Y., Dong, K., Tang, L., Wang, Y., Wang, X. C., Ngo, H. H., Wang, N., et al. (2020). Investigation and assessment of micropollutants and associated biological effects in wastewater treatment processes. *Journal of Environmental Sciences, 94*, 119–127.

Madhesh, J. C., & Mohamed, S. (2024). Impact of micropollutants on the environment and detection by biomarker-based approach. In *Futuristic trends in biotechnology* (Vol. 3, Book 20, pp. 173–182). IIP Series. https://doi.org/10.58532/V3BJBT20P4CH1

Manikandan, S., Preethi, B., Deena, S. R., Vijayan, D. S., Subbaiya, R., Vickram, S., Govarthanan, M., et al. (2024). Systematic assessment of mechanisms, developments, innovative solutions, and future perspectives of microplastics and ecotoxicity—A review. *Advanced Sustainable Systems, 8*(11), 2400294.

Mishra, S., & Das, A. P. (2021). Current treatment technologies for removal of microplastic and microfiber pollutants from wastewater. In *Wastewater treatment: Cutting-edge molecular tools, techniques and applied aspects* (pp. 237–251). https://doi.org/10.1016/B978-0-12-821881-5.00011-8

Mishra, S., Dash, D., & Das, A. P. (2024a). Aquatic microbial diversity on plastisphere: Colonization and potential role in microplastic biodegradation. *Geomicrobiology Journal, 41*(4), 312–323. https://doi.org/10.1080/01490451.2023.2209750

Mishra, S., Dash, D., & Das, A. P. (2024b). Isolation and molecular characterization of polyester-degrading bacterial strains from municipal wastewater and their application in microplastic degradation in contaminated wastewater. *Geomicrobiology Journal*. https://doi.org/10.1080/01490451.2024.2376101

Mishra, S., Dash, D., & Das, A. P. (2024c). Microbial degradation of polyester microfibers using indigenously isolated bacterial strain Exiguobacterium Sp. *CLEAN – Soil, Air, Water, 52*(12), e202300343. https://doi.org/10.1002/CLEN.202300343

Mohanty, M., Tripathy, B., Koteswara Reddy, G., & Das, A. P. (2024a). Greener technologies for establishing restraint over microplastic pollution. In *Sustainable microbial technology for synthetic and cellulosic microfiber bioremediation* (pp. 239–258). https://doi.org/10.1007/978-3-031-62600-5_13

Moharana, S., Dey, S., Priyadarsini, S., Kumar, M. S., & Das, A. P. (2024). Identification and characterization of microplastic pollutants from the marine sediments of Paradeep Coast of Bay of Bengal, India for their sustainable management. *Environmental Science and Engineering, Part, F2333*, 57–73. https://doi.org/10.1007/978-3-031-51792-1_4

Mohanta, S., Pradhan, B., & Behera, I. D. (2023). Impact and remediation of petroleum hydrocarbon pollutants on agricultural land: A review. *Geomicrobiology Journal, 41*(4), 345–359. https://doi.org/10.1080/01490451.2023.2243925

Muhammad, H. A., Masyab, H. M., Othman, B. A., & Mahmood, Y. N. (2024). Micropollutant control in wastewater treatment. *ARO-THE Scientific Journal of Koya University, 12*(2), 130–138.

Nath, S., Enerijiofi, K. E., Astapati, A. D., & Guha, A. (2024). Microplastics and nanoplastics in soil: Sources, impacts, and solutions for soil health and environmental sustainability. *Journal of Environmental Quality, 53*(6), 1048–1072.

Nazar, N., Athira, A. S., Krishna, R. D., Panda, S. K., Banerjee, K., & Chatterjee, N. S. (2024). Large scale screening and quantification of micropollutants in fish from the coastal waters of Cochin, India: Analytical method development and health risk assessment. *Science of the Total Environment, 954*, Article 176515.

Oliveira, G. H. D., Lovato, G., Rodrigues, J. A. D., & Zaiat, M. (2024). Fate of micropollutants. In *Anaerobic treatment of domestic wastewater: Present status and potentialities.*

Pachepsky, Y., Anderson, R., Harter, T., Jacques, D., Jamieson, R., Jeong, J., Zhang, W., et al. (2021). Fate and transport in environmental quality. *Journal of Environmental Quality, 50*(6), 1282–1289.

Paradina-Fernández, L., Wünsch, U., Bro, R., & Murphy, K. (2023). Direct measurement of organic micropollutants in water and wastewater using fluorescence spectroscopy. *ACS ES&T Water, 3*(12), 3905–3915.

Pasqual, O., Rathnayake, A., Thiripuranathar, G., & Ekanayake, S. (2024). Main inorganic pollutants and their risk to living beings. In *Nanotechnology-based sensors for detection of environmental pollution* (pp. 11–38). Elsevier.

Paul, A., Priyadarsini, S., Karakuş, S., Behera, J. K., Gnanasekaran, R., & Das, A. P. (2024). Fate and accumulation of synthetic microfibers in the environment. In *Sustainable microbial technology for synthetic and cellulosic microfiber bioremediation* (pp. 61–76). https://doi.org/10.1007/978-3-031-62600-5_4

Pelamatti, T., Cardelli, L. R., & Rios-Mendoza, L. M. (2022). The role of microplastics in bioaccumulation of pollutants. *Handbook of microplastics in the environment* (pp. 667–696). Springer International Publishing.

Pradhan, G., Dey, S., Priyadarsini, S., Kumar, M. S., & Das, A. P. (2024). Characterization and quantification of microplastics pollutants in sediment samples from Daya River of Odisha State in India for their appropriate management. *Environmental Science and Engineering, Part, F2333*, 75–92. https://doi.org/10.1007/978-3-031-51792-1_5

Priyadarshini, B., & Das, A. P. (2024). The adverse impact of synthetic microplastics released from disposable face masks pollutants on aquatic microbiota, aquatic species and humans: A review. *Environmental Science and Engineering, Part, F2827*, 67–81. https://doi.org/10.1007/978-3-031-50840-0_5

Pulgarin-Zapata, D., Bragg, L. M., Cardenas-Soraca, D. M., Marjan, P., Munkittrick, K. R., Servos, M. R., Arlos, M. J., et al. (2024). Multicompartment examination of micropollutant partitioning in replicate artificial streams highlights the limitations of assessing water matrices alone. *ACS ES&T Water, 4*(9), 4165–4174.

Qiu, Y., Liu, L., Xu, C., Zhao, B., Lin, H., Liu, H., Yang, X., et al. (2024). Farmland's silent threat: Comprehensive multimedia assessment of micropollutants through non-targeted screening and targeted analysis in agricultural systems. *Journal of Hazardous Materials, 476*, Article 135064.

Quazi, S., Malik, J. A., Prakash, A., & Tiwari, P. (2023). Nano (bio) sensors in detection of micropollutants. In *Implications of nanoecotoxicology on environmental sustainability* (pp. 76–101). IGI Global.

Ray, S. K., Dey, S., Das, A. P., & John, A. (2024). Microplastic waste management through greener technologies: A review. In *Sustainable management of urban plastic waste through circular economic approaches* (pp. 237–261). https://doi.org/10.1201/9781003465447-14

Sahu, A., Dey, S., Mallick, K., Priyadarsini, S., Behera, I. D., & Das, A. P. (2024). A review on advanced technology for sustainable management of synthetic microplastic waste. In *Sustainable management of urban plastic waste through circular economic approaches* (pp. 262–279). https://doi.org/10.1201/9781003465447-15

Servien, R., Bonnot, K., Latrille, E., Hélias, A., & Patureau, D. (2023). *Insights on impact estimation of unmeasured micropollutants from WWTP release.*

Sharma, A., Jadeja, F., Gosai, H. G., & Jain, S. (2024a). Organic micropollutants in the freshwater ecosystem: Environmental effects, potential treatments, and limitations. In *Organic micropollutants in aquatic and terrestrial environments* (pp. 203–224). Springer Nature Switzerland.

Sharma, M., Bains, A., Sridhar, K., Chawla, P., & Sharma, M. (2024). Emerging micropollutants—Environmental impact and source-controlled approaches: Current state and future prospects. In *Food and chemical toxicology* (p. 115038).

Singh, A., Majumder, A., Saidulu, D., Bhattacharya, A., Bhatnagar, A., & Gupta, A. K. (2024). Oxidative treatment of micropollutants present in wastewater: A special emphasis on transformation products, their toxicity, detection, and field-scale investigations. *Journal of Environmental Management, 354*, Article 120339.

Singh, R. P., Mishra, S., & Das, A. P. (2020). Synthetic microfibers: Pollution toxicity and remediation. *Chemosphere, 257*, Article 127199. https://doi.org/10.1016/j.chemosphere.2020.127199

Tan, J., Dai, Z., Zhou, K., Zhang, L., He, M., Tan, Y., & Zhou, X. (2023). An ultrasensitive and universal surface plasmonic biosensor for detection of micropollutants in aquatic environments. *Environmental Science & Technology, 57*(22), 8313–8322.

Toušová, Z., Vrana, B., Smutná, M., Novák, J., Klučárová, V., Grabic, R., Hilscherová, K., et al. (2019). Analytical and bioanalytical assessments of organic micropollutants in the Bosna River using a combination of passive sampling, bioassays and multi-residue analysis. *Science of the Total Environment, 650*, 1599–1612.

Tripathy, B., Sahoo, P. P., Sundaray, H., & Das, A. P. (2024). In-vitro biodegradation of discarded marine microplastics across the eastern coast of the Bay of Bengal, India using Exiguobacterium sp. *Environmental Chemistry and Ecotoxicology, 6*, 236–247. https://doi.org/10.1016/J.ENCECO.2024.07.003

Vestel, J., Caldwell, D. J., Constantine, L., D'Aco, V. J., Davidson, T., Dolan, D. G., Wilson, P., et al. (2016). Use of acute and chronic ecotoxicity data in environmental risk assessment of pharmaceuticals. *Environmental Toxicology and Chemistry, 35*(5), 1201–1212.

Vimalkumar, K., Sangeetha, S., Felix, L., Kay, P., & Pugazhendhi, A. (2022). A systematic review on toxicity assessment of persistent emerging pollutants (EPs) and associated microplastics (MPs) in the environment using the Hydra animal model. *Comparative Biochemistry and Physiology Part c: Toxicology & Pharmacology, 256*, Article 109320.

Williams, M. A., & Bannon, D. I. (2022). Knowledge gaps, future directions, and the emergence of nanoplastics as an environmental threat pollutant. In *Toxicology of nanoparticles and nanomaterials in human, terrestrial and aquatic systems* (pp. 423–451).

Włodarczyk-Makuła, M. (2024). Selected organic micropollutants in the aquatic environment. *Desalination and Water Treatment, 317*, Article 100061.

Xu, R., Zhang, Z., Deng, C., Nie, C., Wang, L., Shi, W., Yang, Q., et al. (2024). Micropollutant rejection by nanofiltration membranes: A mini review dedicated to the critical factors and modelling prediction. *Environmental Research, 244*, Article 117935.

Yang, Y., Zhang, X., Jiang, J., Han, J., Li, W., Li, X., Alvarez, P. J., et al. (2021). Which micropollutants in water environments deserve more attention globally? *Environmental Science & Technology, 56*(1), 13–29.

Yılmaz, H., Avaz, G., Yetiş, Ü., & Özkan, M. (2022). Toxicity of environmentally important micropollutants on three trophic levels. *Aquatic Research, 5*(1), 20–28.

Yusuf, A., Ajibade, F. O., Ajibade, T. F., Ifeoluwa, O. B., Lasisi, K. H., Nwogwu, N. A., & Akinbile, C. O. (2024). Fate of organic micropollutants in aquatic environment: policies and regulatory measures. In *Organic micropollutants in aquatic and terrestrial environments* (pp. 331–357). Springer Nature Switzerland.

Chapter 3
Assessment and Fate of Micropollutants in Aquatic Environments

Abhishek Kumar and Deeptimayee Pal

Introduction

The prevalence of micropollutants in aquatic ecosystems and their potential to threaten human health and ecological well-being have rendered them a significant environmental problem. Heavy metals, pesticides, industrial chemicals, pharmaceuticals, personal care products, microplastics, and other pollutants are present in small concentrations and are frequently neglected by conventional monitoring techniques. Contrary to conventional pollutants, micropollutants can persist in the environment after wastewater treatment, which might cause them to accumulate in water bodies and possibly make a pathway into food chains. Their profound but subtle effects affect biological systems even at very low doses (Trapido et al., 2014). The significance of micropollutants is derived from the variety of biological impacts they may induce. Pesticides and industrial chemicals provide hazards, including genotoxicity and bioaccumulation, that impact whole ecosystems, whereas pharmaceuticals, for example, can disturb microbial communities and endocrine systems (Behera et al., 2020, 2021, 2022). Microplastics worsen the ecological effect of other pollutants by acting as carriers for them. Managing water quality has become more complex by these compounds' interactions in aquatic ecosystems, resulting in advanced evaluation techniques and a better understanding of their behavior. With a particular emphasis on their occurrence, distribution, transformation, and impact, this

A. Kumar (✉) · D. Pal
Department of Chemical Engineering, National Institute of Technology (NIT) Rourkela, Rourkela 769008, Odisha, India
e-mail: abhisheknitrkl@outlook.com; ak170192@gmail.com

D. Pal
e-mail: deeptimayeenitrkl@outlook.com

A. Kumar
Information Management & Dissemination Centre (IMDC), CSIR–National Metallurgical Laboratory (CSIR–NML), Jamshedpur 831007, Jharkhand, India

I. D. Behera et al. (eds.), *Emerging Micropollutants*, Sustainable Environmental Waste Management Strategies, https://doi.org/10.1007/978-3-032-03530-1_3

chapter examines the assessment and environmental fate of micropollutants in aquatic environments (Dey et al., 2024a, 2024b, 2024c; Moharana et al., 2024; Pradhan et al., 2024). It is essential to comprehend how micropollutants enter aquatic systems from various sources, such as industrial effluents, agricultural discharge, and urban runoff. The chapter additionally addresses advanced analytical methods for detecting and quantifying micropollutants, including gas chromatography-mass spectrometry (GC–MS) and liquid chromatography-mass spectrometry (LC–MS), as well as advancements in monitoring technologies like suspect and non-target screening. These methods can thoroughly assess micropollutants, and their prevalence in various aquatic systems can be better understood comprehensively (Siegrist & Joss, 2012).

A complex interaction between physical, chemical, and biological elements affects the fate of micropollutants in aquatic environments. While partitioning behaviors affect their distribution across dissolved phases, sediments, and biota, transport mechanisms, including advection and diffusion, govern their movement through water bodies. Both biotic and abiotic transformation mechanisms have implications in determining the persistence of harmful micropollutants (Mohanty et al., 2024a, 2024b; Sahu et al., 2024). Light availability, pH, and other environmental factors affect abiotic transformation processes like photolysis and hydrolysis. The breakdown or biotransformation of micropollutants can be caused by biotic activities driven mainly by microbial populations. These processes often produce metabolites with distinctive features and toxicities. Subsequently, a multidisciplinary study integrates ecological modeling, environmental chemistry, and microbiology to comprehend these processes. In addition to environmental factors like temperature and competing pollutants, external variables that impact micropollutant fate include wastewater treatment conditions, residence time, treatment technology, and operational parameters (Bala et al., 2024; Das et al., 2024a, 2024b, 2024c; Das et al., 2024a, 2024b, 2024c). Internal elements that further complicate the behavior and interactions of micropollutants include their physicochemical characteristics, such as hydrophobicity, biodegradability, and volatility. Predicting the environmental pathways and impacts of micropollutants requires thorough models that include these factors (Yang et al., 2022).

A comprehensive understanding of the environmental fate of micropollutants and strong assessment frameworks are necessary to address this serious and complicated challenge to aquatic habitats. Although useful for focused investigation, conventional analytical techniques have limitations and often fail to capture the whole spectrum of micropollutants. High-resolution mass spectrometry (HRMS) and multi-omics methods are two emerging tools transforming scientific research by allowing for the thorough detection and integrated assessment of biological consequences. The complex relationships between micropollutants and their surroundings must be addressed by these methods, which have the potential to improve the understanding of their behavior and impacts significantly. By emphasizing the complex interrelationships among micropollutants, environmental factors, and biological systems, this chapter advances the general objective of protecting ecosystem health and water quality against increasing human-induced stressors. A better understanding of these elements is essential to reduce the environmental and human health hazards posed by micropollutants and to guide sustainable water quality management methods (Barbosa et al., 2016).

Classification and Sources of Micropollutants

Micropollutants can be categorized based on their chemical composition, source, and environmental behavior. Organic micropollutants are diversified, including pharmaceuticals, personal care products, insecticides, surfactants, and industrial chemicals (Kanungo et al., 2023a, 2023b) (Das et al., 2023a, 2023b, 2025). Pharmaceuticals, such as antibiotics and analgesics, are formulated to interact with biological systems, hence posing hazards of endocrine disruption and microbial resistance (Pal & Sen, 2024b). Pesticides, including insecticides and herbicides, are inherently harmful and can bioaccumulate, impacting non-target organisms. Surfactants, used in detergents and cleansers, can exhibit toxicity to aquatic organisms even at low concentrations. Industrial chemicals, including polycyclic aromatic hydrocarbons (PAHs) and polychlorinated biphenyls (PCBs), are persistent and often hydrophobic, facilitating accumulation in sediments and biological tissues. Inorganic micropollutants include heavy metals, metalloids, and nanomaterials. Heavy metals like lead, mercury, and cadmium are non-biodegradable and exhibit acute and chronic toxicity (Pal & Sen, 2024a). These metals can bioaccumulate and biomagnify within aquatic food systems, leading to considerable ecological and human health hazards (Das & Ghosh, 2018, 2023; Dey, Tripathy et al., 2023a, 2023b). Metalloids, including arsenic, pose significant challenges owing to their toxicity and endurance. Engineered nanomaterials, such as titanium dioxide and silver nanoparticles, constitute a growing class of micropollutants characterized by distinctive features, including increased surface area and reactivity, potentially resulting in unforeseen environmental interactions (Kumar & Gangawane, 2022b; Kumar & Gangawane, 2022a). Microplastics constitute a specific yet widely identified class of micropollutants. These tiny plastic particles, often measuring less than 5 mm, originate from primary and secondary sources. Primary microplastics, exemplified as microbeads in personal care products, are deliberately produced at minute scales. Secondary microplastics arise from the disintegration of more oversized plastic materials due to environmental degradation. Microplastics present not just physical hazards, such as ingestion by aquatic species, but also serve as vectors for other pollutants, including hydrophobic organic compounds and heavy metals (Dey, Dash, & Das, 2024a, 2024b, 2024c; J. Mohanty et al., 2024a, 2024b; Priyadarsini & Das, 2024). Their prevalence in marine ecosystems and potential for bioaccumulation provide considerable ecological and health issues.

Micropollutants originate from several sources, which can be classified into urban, agricultural, and industrial categories, as illustrated in Fig. 3.1. Urban environments significantly contribute to pollution through untreated or insufficiently treated municipal wastewater. This effluent comprises a complex amalgamation of micropollutants, including pharmaceuticals, personal care products, and microplastics, originating from domestic activities. Wastewater treatment plants (WWTPs) are crucial for reducing pollution, yet they often do not eliminate micropollutants, leading to their persistence in effluents and eventual release into surface waters (Mishra & Das, 2021, 2023; Mishra et al., 2024). Urban stormwater runoff transports contaminants from roadways, structures, and urban environments into aquatic ecosystems. Agricultural practices are a notable source of micropollutants (Kumar et al., 2022). The widespread use of pesticides, herbicides, and fertilizers results in the leaching of these compounds into groundwater or their flow into surface waterways. Veterinary drugs, such as antibiotics and hormones used in animal husbandry, increase the accumulation of micropollutants in agricultural runoff. Moreover, agricultural plastic mulching contributes microplastics to soils, which could be carried to aquatic ecosystems through erosion and runoff. The industrial sector contributes a significant source of micropollutants, with the characteristics and extent of emissions differing throughout industries. The chemical production sector emits many organic substances, such as solvents, dyes, and intermediates (Bhattacharjee et al., 2021). The mining and metallurgical sectors release substantial amounts of heavy metals and metalloids into aquatic ecosystems (Das & Ghosh, 2022; Das et al., 2023a, 2023b; Pradhan et al., 2023). The textile and plastic industries release synthetic dyes, finishing chemicals, and microplastics, while the electronics sector contributes nanomaterials and rare earth elements. Untreated or poorly managed industrial effluents act as concentrated point sources of pollutants, requiring rigorous regulatory supervision (Sánchez-Avila et al., 2012).

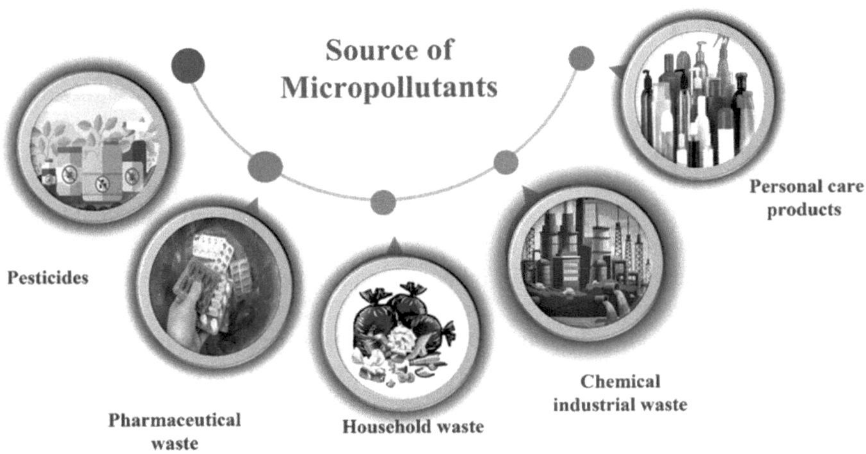

Fig. 3.1 Sources of micropollutants (Balu et al., 2024)

Assessment of Micropollutants

Advanced analytical methods that can detect and quantify these compounds at trace levels are required to identify micropollutants in aquatic environments precisely. The two widely used techniques among these are GC–MS and LC–MS/MS. For the analysis of polar and thermally volatile compounds, including pharmaceuticals and personal care products, LC–MS/MS can be highly effective. Combining mass spectrometry for detection and liquid chromatography for separation results in its outstanding sensitivity and selectivity (Rosen, 2007). Multiple steps of ion fragmentation are made possible by tandem mass spectrometry, which improves selectivity and is crucial for detecting complex micropollutant combinations. However, GC–MS is a preferable option for volatile and semi-volatile organic substances, such as industrial chemicals and insecticides. While mass spectrometry allows structural clarification and quantification, gas chromatography ensures effective chemical separation. The capabilities of HRMS have been further enhanced by developments like time-of-flight (TOF) and Orbitrap technology. Non-target screening, enabled by HRMS, makes it accessible to detect micropollutants and their transformation products that were previously unidentified. Combining hybrid methods like GCxGC-MS with isotopic tagging provides improved separation and identification of complex mixtures for thorough environmental evaluations.

Even with advancements in detection methods, measuring micropollutants at low concentrations may be difficult. Ultra-sensitive analytical techniques are required because these compounds frequently exist in parts-per-billion (ppb) or parts-per-trillion (ppt) concentrations. The identification of target analytes may be hindered by matrix effects, which evolve from the co-extraction of natural organic materials and other components. For example, ion enhancement or suppression in mass spectrometry might result in inaccurate quantification. Sample preparation methods, including liquid–liquid extraction (LLE) and solid-phase extraction (SPE), are used to remove interfering chemicals, concentrate analytes, and resolve these challenges. The efficiency and selectivity of these procedures have increased because of automated extraction devices and advanced sorbents like molecularly imprinted polymers (MIPs). The complexity of environmental matrices, including numerous chemical elements, poses another difficulty. Conventional-focused analytical techniques can overlook novel or unidentified micropollutants since they concentrate on pre-selected substances. Suspect and non-target screening techniques have become complementary strategies that use HRMS to provide thorough chemical fingerprints of environmental products. Nevertheless, these techniques need much data processing and interpretation, sometimes requiring massive spectrum databases and complex software. The absence of established procedures and reference materials makes the measurement and inter-laboratory comparability of numerous micropollutants more difficult (Schmidt, 2018).

Because aquatic ecosystems are dynamic, it is necessary to continuously monitor them to record changes in micropollutant concentrations across time and space. While effective for snapshot evaluations, conventional grab sampling techniques have limitations concerning providing real-time data. Passive sampling tools have been designed to navigate this restriction, including semipermeable membrane devices (SPMDs) and polar organic chemical integrative samplers (POCIS). By accumulating micropollutants over time, these sensors improve the detection of episodic pollution incidents and provide time-weighted average concentrations. A significant development in monitoring technology, real-time sensors allow for the in situ detection and measurement of micropollutants (Schmidt, 2018). For instance, electrochemical sensors employ specific electrodes to identify ionic pollutants and trace metals. Organic substances are monitored via optical sensors, such as fluorescence and absorbance-based systems. Pharmaceuticals and endocrine-disrupting substances are now included in the range of real-time detection owing to recent advancements in biosensors that include biological recognition components like enzymes or antibodies. Rapid reaction times, mobility, and the ability to be used in distant or resource-constrained environments are characteristics of these sensors. Smart monitoring systems have been made possible by combining Internet of Things (IoT) technology with real-time sensors. Real-time data collection, transmission, and analysis are made possible by IoT-enabled systems, which enable timely reactions to pollution incidents (Houthuijs et al., 2023).

Complex datasets produced by these systems are increasingly interpreted using advanced data analytics, such as ML techniques. Proactive water quality management is supported by ML algorithms that can recognize patterns and forecast changes in micropollutant concentrations.

It is anticipated that the development of portable and miniature analytical tools will completely transform the evaluation of micropollutants. Rapid, on-site analysis may be possible using lab-on-a-chip systems, which combine many analytical processes onto a single microfluidic substrate. These devices are perfect for outdoor applications because they minimize the sample volume and the analysis time. Lab-on-a-chip systems have the potential to significantly increase the accessibility and effectiveness of micropollutant monitoring when combined with developments in nanotechnology, such as the use of nanomaterials for improved sensitivity and selectivity. The use of multiomics techniques, which integrate proteomics, metabolomics, and genomes to evaluate the biological effects of micropollutants, is another promising direction. These methods supplement conventional chemical analysis by offering insights into the molecular processes driving pollutant toxicity. For instance, metabolomic profiling may detect exposure-related metabolic changes, whereas transcriptome investigations can demonstrate changes in gene expression in aquatic organisms exposed to micropollutants. Advanced analytical techniques combined with multiomics data will provide a more comprehensive knowledge of the relationships between marine ecosystems and micropollutants (Paradina-Fernández et al., 2023). The advantages and disadvantages of various analytical methods for micropollutant assessment have been listed in Table 3.1.

Table 3.1 Analytical techniques for micropollutant assessment (Ahmed et al., 2020; Houthuijs et al., 2023; Paradina-Fernández et al., 2023)

Analytical technique	Principle	Target compound	Advantages	Disadvantages
GCMS	Volatile compound analysis	POPs	Effective for organic micropollutants	Limited to volatile or semi-volatile compounds
LC–MS/MS	Separation through liquid chromatography and detection through mass spectrometry	Pharmaceuticals, pesticides	High selectivity and high sensitivity	It is expensive and requires skilled operation
HRMS	High-resolution mass spectrometry	Emerging contaminants	Accurate identification of unknown compounds	Expensive and data-intensive
AAS	Absorption of light by atoms	Lead, mercury, cadmium	Cost-effective for heavy metals	Limited detection range
FTIR	Infrared absorption for functional group analysis	Microplastics, polymers	Non-destructive	Less sensitive compared to mass spectrometry

Occurrence and Distribution of Micropollutants

Micropollutants are prevalent in many water systems, such as rivers, lakes, estuaries, and seas, and are frequently encountered in aquatic habitats worldwide. The distribution of micropollutants is disparate because of regional industrial practices, population density, anthropogenic activities, and wastewater treatment systems. Large-scale urbanization and industrialization in developed areas have led to high levels of industrial chemicals, personal care products, and pharmaceuticals in waterways. For example, research conducted in North America and Europe has proven that riverine and estuary systems include significant concentrations of hormones, antibiotics, and synthetic surfactants. These results are explained by the extensive usage of these substances and the fact that wastewater treatment only partially removes them. On the other hand, distinct patterns of micropollutant pollution are frequently observed in emerging countries. Significant pollution in aquatic systems is caused by agricultural runoff, rapid urbanization, and insufficient wastewater treatment plants (Kumar & Gangawane, 2023b). For instance, Southeast Asia and Sub-Saharan Africa are heavily contaminated with veterinary medications and pesticides because of intense agricultural practices. Furthermore, due to untreated domestic sewage, household-derived micropollutants, including detergents and personal care items, are more common in highly populated metropolitan areas. The regional differences in micropollutant profiles highlight the need for localized monitoring and mitigation methods, which reflect varying socioeconomic situations and environmental management techniques. Micropollutant pollution may occur in marine environments (Khatun et al., 2023; Mallick et al., 2024a; Mishra et al., 2019). Numerous micropollutants, such as microplastics and persistent organic pollutants (POPs), have been found in open seas and coastal waterways. Rivers and atmospheric deposition from terrestrial sources often transport these pollutants. Despite their distance, micropollutants have been stated to exist in the Arctic and Antarctic, demonstrating the persistence and worldwide movement of these substances (Félix-Cañedo et al., 2013). Micropollutants are ubiquitous, as shown by how they spread to even the most pristine habitats via long-distance air transport and marine currents. The various sources and pathways of micropollutants are illustrated in Fig. 3.2.

Anthropogenic inputs, environmental factors, and the physicochemical characteristics of the pollutants interact extensively to regulate the concentration of micropollutants in aquatic ecosystems. The primary source of micropollutants in aquatic systems is human activity. Point sources mainly influence local micropollutant concentrations, including industrial discharges, landfill leachates, and wastewater treatment plant (WWTP) effluents. For example, pharmaceuticals, personal care items, and synthetic organic compounds are often found in WWTP effluents. These facilities usually discharge micropollutants into receiving waters because they cannot entirely remove them. Similarly, industrial effluents may introduce a wide range of chemicals, such as solvents, dyes, and heavy metals depending on the industry and

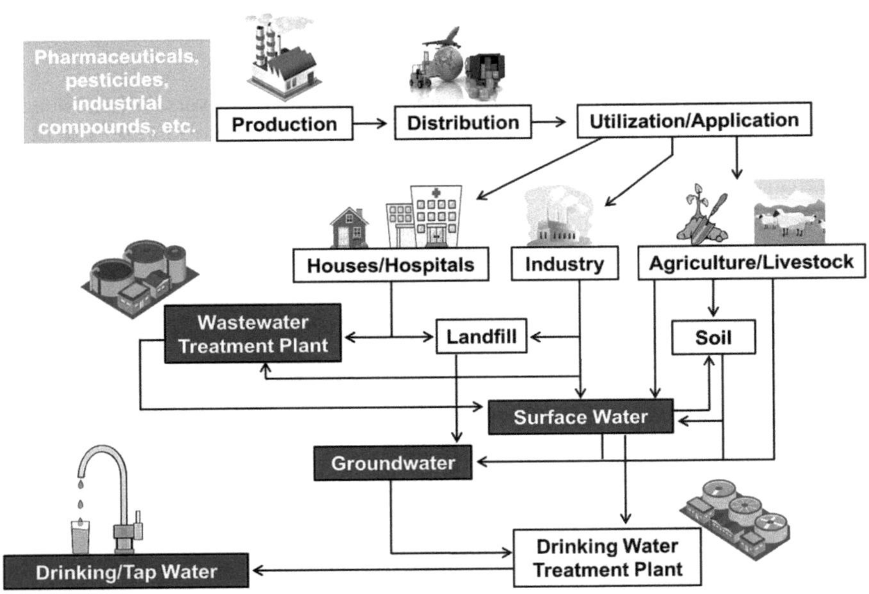

Fig. 3.2 Sources and pathways of micropollutants (Barbosa et al., 2016). Reproduced with permission from (Barbosa et al., 2016), Copyright Water Research 2016

operations. Non-point sources, such as urban stormwater and agricultural runoff, mainly cause diffuse pollution. Surface and groundwater systems attain pesticides, fertilizers, and veterinary medications from agricultural operations. The magnitude of these inputs depends on crop type, rainfall patterns, and the intensity of agricultural operations. Various micropollutants, such as metals, hydrocarbons, and microplastics, are carried into aquatic ecosystems by urban runoff from roads, roofs, and urban landscapes, especially during periods of intense rainfall (Jiang et al., 2013).

Environmental conditions, including land use, temperature, and hydrology, greatly influence the geographical distribution and concentration of micropollutants. The dilution and dispersion of pollutants are determined by hydrological circumstances, such as river flow rates, water residence durations, and mixing patterns. While stagnant or slowly flowing streams may allow micropollutant concentrations to accumulate, high flow rates in rivers may dilute them. Seasonal changes in precipitation and temperature also impact micropollutant levels. For example, greater temperatures can accelerate the breakdown of certain chemicals, while more precipitation during rainy seasons might improve runoff and pollution loading into aquatic systems. Urbanization, agriculture, and industrial growth are examples of land use patterns that influence the input and transportation paths of micropollutants (Das et al., 2024; Dey et al., 2024a, 2024b, 2024c; Dey et al., 2023a, 2023b; Paul et al., 2023). While agricultural regions contribute a significant amount of fertilizer and pesticides, urban areas are hotspots for household-derived pollution and microplastics. The release of heavy metals and specialized chemicals is a characteristic of industrial zones. The interaction between hydrological connectivity and land use determines the geographical distribution of micropollutants across landscapes (Musolff et al., 2009).

The intrinsic characteristics of micropollutants, such as their solubility, hydrophobicity, volatility, and biodegradability, significantly impact how they behave in the environment. Because they prefer to remain dissolved in water, hydrophilic substances like certain medications make long-distance transportation easier. On the other hand, hydrophobic micropollutants, such as polychlorinated biphenyls (PCBs) and polycyclic aromatic hydrocarbons (PAHs), have a significant attraction for organic matter and sediments (Pal & Sen, 2023a). These substances are often trapped in sedimentary layers, which could persist long. Micropollutant dispersion between aquatic and atmospheric compartments is influenced by volatility. Certain insecticides and other volatile substances may disperse locally and globally by evaporating off water surfaces and traveling into the atmosphere. Their biodegradability determines the persistence of micropollutants in aquatic environments. Microbial activity has a greater chance of transforming readily biodegradable chemicals, which lowers their quantities. Nevertheless, their breakdown products could still be harmful. Therefore, transformation paths need to be carefully considered.

In aquatic environments, micropollutants undergo several transformation processes that affect their toxicity and concentrations. Micropollutants may break down into simpler compounds via abiotic processes, including photolysis and hydrolysis. Sunlight-driven photolysis works especially well for substances that are exposed to surface waters. Temperature and pH impact hydrolysis processes, which aid in the disintegration of fragile compounds. Another important pathway is biotic transformations, which are mediated by microbial populations. Complete mineralization or the generation of intermediate metabolites with unique characteristics and toxicities may result from microbial breakdown. Sediments are secondary sources of micropollutants and serve as sinks for them. PCBs and PAHs are examples of hydrophobic chemicals likely to sorb onto soil particles, decreasing their bioavailability in the water column (Pal & Sen, 2023b). However, these pollutants may be reintroduced into the water column due to environmental changes, such as sediment resuspension during storms. Micropollutants interact with biota, which includes aquatic plants and organisms, via mechanisms including bioaccumulation and biomagnification. Consumption of seafood may pose health concerns to humans because lipophilic chemicals are more likely to bioaccumulate in aquatic organisms, resulting in larger quantities in higher trophic levels. Hotspots for micropollutant pollution have been found in several locations worldwide (Sousa et al., 2020). The Rhine River Basin in Europe is a well-known example, with high industrial chemicals, pesticides, and pharmaceuticals. Comprehensive monitoring initiatives have yielded significant details on the occurrence and transport of micropollutants in the basin. Industrial discharges, urban runoff, and air deposition heavily contaminate the Great Lakes area of North America. The difficulties in controlling legacy pollutants and emerging pollutants are introduced by the enduring presence of POPs and microplastics in these water bodies. Two well-known examples of highly contaminated aquatic systems in Asia are the Yangtze River in China and the Ganges River in India. The varied micropollutant profiles observed in these rivers result from high population density, widespread agricultural practices, and rapid industrialization. Due to agricultural activities, herbicides and veterinary medications have emerged as important pollutants in Southeast Asia's Mekong River Basin, posing serious issues (Yang et al., 2017).

Fate of Micropollutants

A complex interaction of physical, chemical, and biological processes governs the fate of these pollutants in aquatic ecosystems. Each process substantially impacts their persistence, transformation, and bioavailability, thus affecting their ecological and human health concerns.

Physical Processes

Micropollutants are transported in aquatic habitats by processes including advection, diffusion, and dispersion. Advection denotes the transport of micropollutants alongside the bulk flow of water, driven by hydrodynamic forces such as river currents, tidal movements, and reservoir water circulation. Diffusion, propelled by concentration gradients, facilitates the transfer of micropollutants from areas of high concentration to those of low concentration. Turbulent dispersion, conversely, results in the dissemination of pollutants owing to water turbulence and mixing. The transportation of micropollutants determines their geographical distribution and exposure routes, affecting their interaction with different environmental compartments. Adsorption is an essential mechanism in which micropollutants adhere to the surfaces of sediments, suspended particles, or organic materials (Kumar & Gangawane, 2023a). The extent of adsorption is dependent upon the physicochemical attributes of the micropollutant, including hydrophobicity and charge, in addition to the features of the adsorbing substance. Hydrophobic chemicals, such as polycyclic aromatic hydrocarbons (PAHs), have a great affinity for organic carbon-rich sediments. Adsorption can lower the bioavailability of micropollutants in the water column, although it may also function as a temporary reservoir, with desorption resulting in their reintroduction under fluctuating environmental circumstances. Sedimentation involves the deposition of particle-bound micropollutants at the bottom of aquatic environments, leading to their accumulation in sediments (Paul et al., 2024). Variables like particle size, density, and water velocity affect sedimentation rates. Although sedimentation may efficiently eliminate micropollutants from the water column, these contaminants may persist in sediments in a bioavailable form or may be remobilized owing to physical disturbances (e.g., dredging, storms) or chemical alterations (e.g., redox conditions) (Kim & Zoh, 2016).

Chemical Processes

Photolysis refers to the degradation of micropollutants resulting from exposure to sunlight, especially ultraviolet (UV) radiation. Direct photolysis transpires when the pollutant absorbs photons, resulting in molecular breakdown. Indirect photolysis involves reactive species, such as hydroxyl radicals ($OH^{\cdot}$), generated by photochemical processes that oxidize micropollutants. Factors affecting photolysis encompass light intensity, water depth, turbidity, and the existence of photosensitizers such as dissolved organic materials. Although photolysis may result in full mineralization, it often produces transformation products with distinctive, and at times higher, toxicities. Hydrolysis is the chemical interaction between micropollutants and water, breaking them down. This technique is especially pertinent for molecules containing hydrolyzable functional groups, like esters and amides. The rates of hydrolysis are affected by variables like pH, temperature, and the presence of catalytic ions. Acidic or alkaline environments often accelerate hydrolysis. Hydrolysis may diminish the persistence of certain micropollutants, while the resultant byproducts can exhibit differing degrees of toxicity and environmental mobility. Oxidation entails the interaction of micropollutants with environmental oxidizing agents, such as oxygen, ozone, or reactive oxygen species (ROS). This process modifies the chemical composition of micropollutants, perhaps diminishing their toxicity or converting them into more reactive intermediates. Advanced oxidation processes (AOPs), which use hydroxyl radicals, are very efficient in degrading persistent pollutants such as pharmaceuticals and pesticides. In natural environments, oxidation is affected by aspects like redox conditions, sunshine, and the presence of catalysts such as metal oxides (Cho et al., 2014).

Biological Processes

Microbial degradation is an essential process in the natural attenuation of micropollutants. It entails the enzymatic degradation of pollutants by microorganisms, such as bacteria, fungi, and algae. Microbial degradation may transpire in aerobic or anaerobic environments, with its pathway and route determined by the composition of the microbial population, the availability of electron donors and acceptors, and environmental factors like temperature and nutrient availability. Aerobic degradation is more effective for several organic micropollutants, but anaerobic processes are essential for chemicals that resist oxygen-dependent degradation, like chlorinated solvents. Besides degradation, bacteria may convert micropollutants into metabolites with modified chemical properties. These biotransformation products may possess varying toxicological profiles, sometimes demonstrating heightened toxicity or persistence. The microbial transformation of pharmaceuticals such as diclofenac often results in intermediate molecules that exhibit resistance to further degradation. Comprehending microbial transformation processes is essential for forecasting the environmental fate of micropollutants and evaluating the dangers linked to their metabolites (Ternes, 2007).

The partitioning behavior influences the distribution of micropollutants across water, sediment, and biota. The process is affected by the compound's physicochemical characteristics, including the octanol–water partition coefficient (Kow), which signifies hydrophobicity. Hydrophobic micropollutants are often acquired in sediments or biological tissues, while hydrophilic chemicals are dissolved in water. The dynamics of partitioning influence the bioavailability and durability of micropollutants, affecting their ecological and toxicological effects. Sediments function as both reservoirs and secondary sources of micropollutants. The sorption of pollutants to sediment particles diminishes their mobility in the water column but may result in prolonged accumulation. Under certain circumstances, such as changes in pH, temperature, or redox potential, micropollutants may desorb from sediments and re-enter the aquatic environment, presenting additional risks. Micropollutants that partition into biota can undergo bioaccumulation, resulting in increased concentrations inside an organism over time and biomagnification, leading to amplified concentrations across trophic levels in the food chain. Lipophilic substances, like PCBs and mercury, are especially susceptible to these reactions. Bioaccumulation and biomagnification have considerable ecological consequences, resulting in detrimental impacts on aquatic organisms and possible hazards to human health by ingesting contaminated seafood (Nam et al., 2014).

The interaction of physical, chemical, and biological processes in influencing the fate of micropollutants emphasizes the complexity of their behavior in aquatic ecosystems. These mechanisms are interdependent and interact dynamically, influenced by environmental circumstances and the unique characteristics of pollutants. Understanding these pathways is crucial for formulating prediction models and efficient mitigation techniques. Future research must concentrate on assessing the relative contributions of various processes, identifying critical transformation products, and evaluating their ecological and toxicological relevance. Progress in analytical methodologies, including high-resolution mass spectrometry and molecular biology instruments, will be essential in understanding the environmental behavior of micropollutants.

Adverse Effects of Micropollutants

Micropollutants in aquatic habitats provide considerable toxicological hazards, disturbing the delicate equilibrium of these environments. Although considerable focus has been given on deadly concentrations of pollutants, sub-lethal impacts, which arise at lower, ecologically relevant concentrations, are progressively acknowledged as significant factors in ecological degradation. These consequences may result in physiological, behavioral, and reproductive alterations in aquatic organisms, often cascading throughout the ecosystem. The synergistic effects of micropollutant combinations aggravate these consequences, posing a complicated challenge for environmental health and biodiversity conservation.

Sub-lethal toxicity implies detrimental effects that do not cause immediate death but considerably compromise the health, behavior, and functioning of organisms. These impacts are especially pernicious since they may go unnoticed in standard monitoring but nonetheless lead to population decreases and environmental instability. Aquatic flora and algae serve as primary producers in aquatic habitats, constituting the foundation of the food web. Sub-lethal exposure to micropollutants impairs metabolic and reproductive functions. Herbicides, including atrazine, inhibit the photosynthetic processes of aquatic flora and algae, diminishing their capacity to generate oxygen and organic matter (Van Der Kraak et al., 2014). Prolonged exposure results in decreased growth and production, causing further impacts on higher trophic levels. Aquatic flora such as duckweed and water hyacinth accumulate heavy metals and micropollutants, disrupting their nutrient cycle functions and rendering them carriers of pollution for herbivores. Pharmaceuticals, such as antibiotics, affect the composition of algal populations by selectively suppressing certain species. This results in alterations to community structure, often promoting detrimental algal blooms. Aquatic organisms, including invertebrates, fish, and amphibians, have various sub-lethal reactions to micropollutants that endanger their life and ecological functions. Exposure to neurotoxic micropollutants, including specific pesticides and medications, impairs typical activities such as predator avoidance, foraging, and mating. For example, anti-anxiety pharmaceuticals in wastewater effluents have been shown to increase fish aggression or diminish their responsiveness to predators. Sub-lethal quantities of heavy metals, including cadmium and lead, disrupt normal development patterns in fish and amphibians, resulting in malformations and low fitness (Baby et al., 2011).

Micropollutants, especially endocrine-disrupting chemicals (EDCs), impair immunological function, rendering aquatic organisms more vulnerable to illnesses and infections. EDCs are a group of micropollutants that disrupt hormonal signaling pathways. Their impacts are significant since they interfere with essential physiological processes, including growth, reproduction, and metabolism. EDCs, such as certain pesticides, plasticizers (e.g., bisphenol A), and pharmaceuticals, either imitate or obstruct natural hormones, resulting in endocrine system impairment. Endocrine-disrupting chemicals may result in distorted sex ratios within aquatic species. Exposure to estrogenic substances, including synthetic hormones in contraceptives, causes feminization in male fish, as demonstrated by the synthesis of egg yolk proteins and decreased sperm production. Specific industrial pollutants, including polychlorinated biphenyls (PCBs), interfere with thyroid hormone signaling, hindering growth and development in amphibians and fish. Seasonal reproductive cycles, typically governed by hormones, are altered by EDCs, leading to asynchronous spawning and diminished reproductive success (Lauretta et al., 2019). Micropollutants cause genetic modifications in aquatic species, resulting in immediate and long-term consequences. Polycyclic aromatic hydrocarbons (PAHs) and heavy metals induce DNA damage, resulting in mutations that impair fitness and increase cancer risks. Recent data indicates that some micropollutants may modify gene expression without changing the DNA sequence, possibly impacting several generations. Prolonged exposure to pharmaceuticals, including antibiotics, promotes the emergence of resistant bacterial strains. This phenomenon mostly influences microbial populations, but it also affects higher organisms by modifying their symbiotic interactions with bacteria. The reproductive health of aquatic organisms is especially susceptible to micropollutant exposure. Pharmaceuticals such as antidepressants and contraceptives decrease fertility in fish and amphibians by modifying hormone levels and gamete quality. Progeny exposed to micropollutants during embryonic development often displays developmental anomalies, such as deformed fins, diminished size, and compromised swimming proficiency. The aggregate effect of reproductive failure results in diminished population numbers, endangering the viability of animals with prolonged reproductive cycles (Pal & Sen, 2025).

Aquatic species in natural habitats are rarely subjected to isolated contaminants. Instead, they confront intricate combinations of micropollutants, often exhibiting unforeseen and intensified consequences. The interaction of numerous micropollutants often results in a combined toxicity that surpasses the aggregate of their individual effects. Mixtures of pesticides and heavy metals demonstrate increased toxicity in aquatic invertebrates, concurrently affecting many physiological processes. The intricate nature of micropollutant combinations complicates the prediction of their effects. The combination of pharmaceuticals and EDCs may generate whole new forms of toxicity, increasing dangers to aquatic life. Organisms subjected to prolonged exposure to low quantities of micropollutants often collect contaminants, resulting in delayed harmful consequences. The enduring presence of some micropollutants in sediments and biota extends exposure, even after a reduction in external inputs, hence hindering ecosystem recovery. Sensitive species are often substituted by more resilient counterparts, modifying community dynamics and diminishing biodiversity. Synergistic toxicity at lower trophic levels disseminates the food chain, amplifying effects on higher organisms, including predators (Kumar et al., 2020).

The toxicological effects of micropollutants have extensive consequences for aquatic ecosystems. The combined impacts of sub-lethal toxicity, reproductive failure, and population decrease lead to the extinction of vulnerable species. Decreased biodiversity weakens ecological resilience, rendering aquatic systems more susceptible to other stresses, like climate change and habitat degradation. Micropollutants interfere with essential ecosystem functions, such as nutrient cycling, primary production, and decomposition. The suppression of microbial populations by antibiotics diminishes organic matter decomposition, hence impacting nutrition availability for higher trophic levels. The bioaccumulation and biomagnification of micropollutants in consumable species, including fish and shellfish, provide substantial health hazards to people and wildlife. Prolonged exposure to pollutants, including mercury and PCBs, via food consumption, is associated with neurological and developmental abnormalities. The bioavailability and bioaccumulation of micropollutants are crucial for comprehending their ecological effects, toxicological importance, and enduring presence in aquatic ecosystems. (Stamm et al., 2016). Bioavailability signifies the fraction of a micropollutant available for absorption and biological activity by an organism. It assesses the short-term and long-term impacts of various chemicals on aquatic ecosystems. A complex interaction of chemical, environmental, and biological variables determines the bioavailability of micropollutants. These interrelated conditions affect the interactions of micropollutants with species, their transmission across food webs, and their contribution to wider ecological concerns. Bioaccumulation is the gradual accumulation of micropollutants inside an organism over time by direct absorption from water, sediment, or food sources. In contrast to bioavailability, bioaccumulation considers the exposure time and the pathways involved.

Challenges in the Assessment of Micropollutants

The accurate assessment and monitoring of micropollutants in aquatic ecosystems are essential for comprehending their distribution, fate, and effects. Nonetheless, additional challenges hinder this process, including analytical constraints, the lack of standardized protocols, and uncertainties in predicting long-term behavior. Addressing these challenges requires progress in methodologies, international collaboration, and a cohesive strategy that connects environmental chemistry, toxicity, and ecological modeling.

A primary challenge in assessing micropollutants is their low concentration in environmental matrices. Numerous micropollutants exist at parts-per-billion (ppb) or parts-per-trillion (ppt) concentrations, necessitating extremely sensitive analytical techniques for their detection. Although advanced techniques such as LC–MS/MS and GC–MS may identify trace levels of pollutants, their sensitivity often relies on the matrix composition. Complex environmental substances, such sediments or wastewater, provide interferences that diminish detection performance. High-resolution equipment are expensive and need skilled personnel for operation and maintenance, hence restricting their accessibility in resource-limited areas (Shimabuku et al., 2017). Emerging contaminants, such as transformation products and unidentified micropollutants, may evade detection in conventional targeted analyses that concentrate on known components. The processes of sample collection, storage, and preparation substantially affect the accuracy of micropollutant evaluations. Environmental samples, including water, sediment, and biota, have a diverse array of organic and inorganic elements that inhibit micropollutant identification. Sample pretreatment procedures, including filtration, extraction, and cleaning, are labor-intensive and susceptible to variation. A multitude of micropollutants, including pharmaceuticals and pesticides, are prone to degradation during sample storage. Ensuring sample integrity requires strict compliance with storage conditions, including temperature and preservative application, which is not always practicable. The variability of micropollutant concentrations across geographical and temporal dimensions challenges the acquisition of representative samples. Seasonal variations, including higher pesticide runoff during precipitation, require regular and methodical sampling initiatives. Accurately quantifying micropollutants and identifying their transformation products continue to pose considerable challenges. A multitude of micropollutants undergoes chemical or biological transformation in the environment, resulting in byproducts that may vary in toxicity and persistence. Analytical techniques often fail to detect these products owing to the lack of reference standards. The interference or amplification of signals due to the presence of concurrent chemicals in environmental matrices results in quantification errors. The concurrent existence of many micropollutants in environmental samples generates overlapping signals in chromatographic and spectrometric investigations, hence limiting identification and quantification (Ahmed et al., 2020).

The lack of standardized protocols for addressing the variety of emerging pollutants obstructs consistent monitoring initiatives. A multitude of new pollutants lacks regulatory standards for acceptable concentrations, resulting in inconsistencies in monitoring priorities and methodology. The lack of comprehensive and centralized databases for emerging contaminants and their characteristics, including partition coefficients and degradation rates, limits the establishment of standardized methods. Reference standards for emerging contaminants, especially transformation products, are often lacking, limiting their accurate identification and quantification (Mallick et al., 2024b; Ray et al., 2024). Variations in equipment, techniques, and calibration processes across laboratories lead to inconsistent data, limiting the comparability of monitoring findings. Although new techniques like HRMS exhibit potential, their extensive use is impeded by the lack of standardized protocols for validation and quality assurance. Global surveillance of micropollutants is crucial because of their transboundary characteristics; yet, the lack of standardized methods hinders multinational initiatives. Variations in regulatory frameworks across countries result in differences in monitoring priorities and methodologies. Developing countries often lack the infrastructure and skills necessary for advanced monitoring, leading to data deficiencies that distort global evaluations. Collaborative initiatives, such as those under the United Nations or regional accords, have logistical and technological obstacles in standardizing methodologies across diverse regions (Pan et al., 2020).

Micropollutants undergo many physical, chemical, and biological processes in aquatic environments, resulting in uncertainties over their long-term fate. Variables like temperature, pH, light intensity, and microbial activity vary geographically and temporally, affecting the rates of micropollutant degradation and transformation. The distribution of micropollutants among water, sediment, and biota is dependent upon their hydrophobicity and environmental conditions, which provide significant challenges. Abiotic and biotic transformation processes can generate a range of byproducts with unique characteristics and toxicities. Predicting the formation and behavior of these byproducts requires thorough kinetic and mechanistic investigations. Numerous environmental fate models depend on simplified assumptions, such as steady-state conditions or homogeneous pollutant distribution, which fail to represent real-world complexity. The lack of empirical data regarding emerging pollutants, such as degradation rates and partition coefficients, limits the accuracy of predictive models. The interaction of transport, degradation, and bioaccumulation processes generates feedback loops that are challenging to incorporate into models (Schwarzenbach et al., 2006).

Seasonal variations in temperature and hydrology influence the behavior of micropollutants. Increased photolysis in summer and less microbial activity in winter alter degradation pathways. Point sources, like industrial effluents and wastewater treatment facilities, generate localized concentrations of micropollutants that differ from regional norms. Acquiring these hotspots requires high-resolution spatial data. Variations in temperature, precipitation patterns, and severe weather phenomena affect the movement and transformation of micropollutants, hence developing more uncertainty. Increased precipitation may augment the flow of pesticides, pharmaceuticals, and nanomaterials, surpassing the capacity of monitoring programs to adjust, and resulting in shortcomings in comprehending their long-term behavior. The advancement of increasingly sensitive and selective equipment, like TOF–MS and nuclear magnetic resonance (NMR), may improve detection and identification capabilities. Progress in sensor technologies, such as biosensors and electrochemical sensors, enables in situ, real-time monitoring of micropollutants (Mishra et al., 2022; Tripathy et al., 2022). Formulating worldwide standards for the surveillance of micropollutants through entities such as ISO or WHO ensures uniformity and comparability. Establishing centralized, open-access databases for micropollutant characteristics and monitoring data may enhance cooperation and minimize duplication. Integrating models that include molecular, ecological, and global dynamics may enhance predictions of micropollutant behavior. Utilizing machine learning (MI) and artificial intelligence (AI) for pattern detection and data analysis may improve the prediction accuracy of environmental models. Improving technical proficiency in resource-limited areas via training programs and seminars may augment global monitoring capacities. Augmenting financial support for research and infrastructure development might mitigate resource inequities and promote innovation in micropollutant assessment (Brogat et al., 2013).

Conclusions and Future Perspectives

Micropollutants, including various synthetic organic and inorganic compounds, pose a substantial threat to worldwide aquatic ecosystems. Their persistence, bioaccumulative potential, and toxicity pose complex environmental and ecological hazards. The increasing concern over micropollutants in aquatic ecosystems necessitates innovative strategies for their detection, monitoring, and control. Future directions include technology advancements, improved comprehension of transformation processes, and incorporating predictive modeling methodologies. Future developments in analytical chemistry will prioritize enhancing the sensitivity, specificity, and efficiency of micropollutant detection. Techniques like TOF–MS and Orbitrap mass spectrometry are becoming essential for untargeted analysis. These approaches facilitate the identification of unidentified pollutants and their transformation products, hence increasing the range of micropollutant detection. Integrating chromatographic techniques, such as liquid and gas chromatography, with spectroscopic instruments, including Fourier-transform infrared spectroscopy and nuclear magnetic resonance, improves chemical separation and structural determination.

Integrating remote sensing technologies with the IoT offers promising opportunities for extensive monitoring. These platforms can assess water quality metrics, including turbidity and chlorophyll-a concentrations, which are associated with micropollutant levels. Networks of IoT-enabled sensors can provide continuous, real-time data on water quality, enabling proactive control of pollution hotspots. The integration of remote sensing with in-situ measurements augments geographical and temporal coverage, hence enhancing the accuracy of monitoring initiatives. Miniaturized gadgets, referred to as lab-on-a-chip systems, are transforming environmental monitoring. These devices integrate analytical processes, including sampling, separation, and detection, onto a single chip, facilitating fast and economical analysis. Lab-on-a-chip devices are very effective for identifying micropollutants in distant locations or during emergency pollution incidents.

Advanced methodologies, including metagenomics and proteomics, may reveal microbial genes and enzymes implicated in the breakdown of micropollutants. Transformation products (TPs) often exhibit variations in toxicity and durability compared to their parent chemicals. Formulating methodologies to discover and characterize transformation products, including non-targeted screening techniques. Assessing the ecotoxicological effects of TPs is essential for comprehending their environmental hazards. Examining TPs in ecologically relevant circumstances, such as fluctuating solar exposure or microbial populations, will improve practical relevance.

Future modeling methodologies will include incidents happening at molecular, local, and global levels. Quantum chemical computations and molecular dynamics simulations can forecast the reaction kinetics and thermodynamics of micropollutant transformations. Integrating hydrodynamic, biogeochemical, and ecological models can replicate the distribution and fate of micropollutants in aquatic systems. ML and AI are becoming potent instruments for analyzing intricate environmental data. ML algorithms can discern trends in previous monitoring data to forecast future micropollutant activity. AI can amalgamate varied information, including chemical characteristics, environmental factors, and toxicity profiles, to enhance risk evaluations. The amalgamation of real-time monitoring data with prediction models provides substantial benefits. Integrating real-time data into models enables adaptive simulations that react to fluctuating environmental circumstances. Predictive models using continuous monitoring may activate alarms for impending pollution incidents, facilitating prompt actions.

Mitigating the issues posed by micropollutants necessitates interdisciplinary cooperation. Integrating chemical analysis and toxicological investigations will provide a more thorough comprehension of the effects of micropollutants. Comprehending the socio-economic factors influencing micropollutant pollution may guide policy formulation and public awareness initiatives. International collaboration is crucial for tackling transboundary micropollutant contamination. Formulating standardized protocols for the detection, analysis, and risk evaluation of micropollutants. Establishing open-access platforms for the dissemination of monitoring data and research outcomes. Assisting developing countries with resources and training for monitoring and controlling micropollutants. Designing environmentally friendly compounds with lower toxicity and better degradability may lessen micropollutant environmental inputs. Formulating pharmaceuticals that decompose swiftly post-excretion without generating detrimental byproducts. Formulating biodegradable insecticides with less non-target impacts. Future wastewater treatment methods must prioritize the removal of a wider array of micropollutants. Developing economical and energy-efficient advanced oxidation processes for the degradation of micropollutants. Developing advanced membranes with enhanced selectivity and resistance to fouling. Enhancing public awareness and fortifying regulatory frameworks are essential for lowering micropollutant emissions. Advocating for the proper use and disposal of pharmaceuticals and personal care items. Implementing stricter rules on industrial effluents and agricultural methodologies.

Regulations and policies are pivotal in the management of micropollutants. Establishing global standards for the monitoring and evaluating micropollutants may enhance international collaboration and data exchange. Policies aimed at source reduction, including the restriction of high-risk chemicals and the promotion of environmentally friendly alternatives, are crucial for sustainable solutions. Educating stakeholders, such as industries, governments, and the general populace, may promote responsible behaviors and minimize pollutant contributions. Research on micropollutants necessitates collaborative endeavors across scientific fields and geopolitical borders. Integrating knowledge from chemistry, biology, toxicology, engineering, and social sciences may provide comprehensive answers to micropollutant issues. Establishing collaborative networks for information sharing and capacity development, especially in developing areas, can improve global initiatives to combat micropollutant pollution.

References

Ahmed, M. B., Johir, M. A. H., Ngo, H. H., Guo, W., Zhou, J. L., Belhaj, D., & Moni, M. A. (2020). Methods for the analysis of micro-pollutants. In *Current developments in biotechnology and bioengineering* (pp. 63–86). Elsevier. https://doi.org/10.1016/B978-0-12-819594-9.00004-8

Baby, J., Raj, J., Biby, E., Sankarganesh, P., Jeevitha, M., Ajisha, S., & Rajan, S. (2011). Toxic effect of heavy metals on aquatic environment. *International Journal of Biological and Chemical Sciences, 4*(4), 939–952. https://doi.org/10.4314/ijbcs.v4i4.62976

Bala, K., Anshul, Verma, S., Dogra, A., Das, A. P., Kumar, R., & Nandini. (2024). Sustainable management for urban plastic waste generation. In *Sustainable management of urban plastic waste through circular economic approaches* (pp. 211–236). https://doi.org/10.1201/9781003465447-13

Balu, S., Ganapathy, D., Arya, S., Atchudan, R., & Sundramoorthy, A. K. (2024). Advanced photocatalytic materials based degradation of micropollutants and their use in hydrogen production—A review. *RSC Advances, 14*(20), 14392–14424. https://doi.org/10.1039/D4RA01307G

Barbosa, M. O., Moreira, N. F. F., Ribeiro, A. R., Pereira, M. F. R., & Silva, A. M. T. (2016). Occurrence and removal of organic micropollutants: An overview of the watch list of EU decision 2015/495. *Water Research, 94*, 257–279. https://doi.org/10.1016/j.watres.2016.02.047

Bhattacharjee, J., Mishra, S., & Das, A. P. (2021). Recent advances in sensor-based detection of toxic dyes for bioremediation application: A review. *Applied Biochemistry and Biotechnology, 2021*, 1–20. https://doi.org/10.1007/S12010-021-03767-7

Brogat, M., Cadiere, A., Sellier, A., Thomas, O., Baures, E., & Roig, B. (2013). MSPE/UV for field detection of micropollutants in water. *Microchemical Journal, 108*, 215–223. https://doi.org/10.1016/j.microc.2012.10.025

Behera I. D., Nayak, M., Biswas S., Meikap B. C., Sen, R. (2021). Enhanced biodegradation of total petroleum hydrocarbons by implementing a novel two-step bioaugmentation strategy using indigenous bacterial consortium. *Journal of Environmental Management, 292*, 112746. ISSN 0301-4797. https://doi.org/10.1016/j.jenvman.2021.112746.

Behera, I. D., Basak, G., Kumar, R. R., Sen, R., & Meikap, B. C. (2020). Treatment of petroleum refinery sludge by petroleum degrading bacterium *Stenotrophomonas pavanii* IRB19 as an efficient novel technology. *Journal of Environmental Science and Health, Part A, 56*(2), 226–239. https://doi.org/10.1080/10934529.2020.1866924

Behera I. D., Nayak, M., Mishra, A., Meikap, B. C., Sen, R. (2022). Strategic implementation of integrated bioaugmentation and biostimulation for efficient mitigation of petroleum hydrocarbon pollutants from terrestrial and aquatic environment. *Marine Pollution Bulletin, 177*, 113492. ISSN 0025-326X. https://doi.org/10.1016/j.marpolbul.2022.113492

Das, P., & Bal, M. (2025). Modeling and optimization study for microplastic removal from aquatic medium using Chroococcidiopsis species. *Chemical Engineering Science, 304*, Article 121085. https://doi.org/10.1016/j.ces.2024.121085

Das, P., Halder, G., & Bal, M. (2023a). A critical review on remediation of microplastics using microalgae from aqueous system. *Science of the Total Environment, 898*, Article 166425. https://doi.org/10.1016/j.scitotenv.2023.166425

Cho, E., Khim, J., Chung, S., Seo, D., & Son, Y. (2014). Occurrence of micropollutants in four major rivers in Korea. *Science of the Total Environment, 491–492*, 138–147. https://doi.org/10.1016/j.scitotenv.2014.03.025

Das, A., Dey, S., Bala, K., Gnanasekaran, R., & Das, A. P. (2024). Microbial bioremediation technology for sustainable treatment and management of synthetic microfiber waste. In *Sustainable microbial technology for synthetic and cellulosic microfiber bioremediation* (pp. 221–238). https://doi.org/10.1007/978-3-031-62600-5_12

Das, A., Dey, S., & Das, A. P. (2024). Ecotoxicological impacts of synthetic microfiber pollutants and development of sustainable mitigation strategies. In *Environmental chemistry and ecotoxicology*. https://doi.org/10.1016/J.ENCECO.2024.12.003

Das, A., & Ghosh, S. (2022). Role of microorganisms in extenuation of mining and industrial wastes. *Geomicrobiology Journal, 39*, 173–175. https://doi.org/10.1080/01490451.2022.2038953

Das, A. P., & Ghosh, S. (2018). Bioleaching of Manganese from mining waste materials. *Materials Today: Proceedings, 5*(1), 2381–2390. https://doi.org/10.1016/J.MATPR.2017.11.459

Das, A. P., & Ghosh, S. (2023). Manganese and its application. In *Manganese mining microorganisms* (pp. 29–40). https://doi.org/10.1016/B978-0-12-822147-1.00007-8

Das, A. P., Mallavarapu, M., & Ghosh, S. (2023b). Ecotoxicity of mining pollutants on the environment and their remediation. *Environmental Chemistry and Ecotoxicology, 5*, 165–167. https://doi.org/10.1016/J.ENCECO.2023.08.002

Das, A. P., Behera, I. D., & Das, N. P. (2024). *Renewable energy generation and value addition from environmental microfiber pollution through advanced greener solution* (p. 278).

Dey, S., Bissoyi, A., Koteswara Reddy, G., & Das, A. P. (2024). Synthetic microfibers: An urban emerging pollutant demanding for sustainable environmental solutions. In *Sustainable microbial technology for synthetic and cellulosic microfiber bioremediation* (pp. 107–121). https://doi.org/10.1007/978-3-031-62600-5_7

Dey, S., Das, A., Mallick, K., Sahu, A., & Das, A. P. (2023). *Environmental petroleum waste: Pollution, toxicity, sustainable remediation* (pp. 159–175). https://doi.org/10.1007/978-3-031-48220-5_7

Dey, S., Dash, A., & Das, A. P. (2024). Bioremediation as a sustainable and circular solution for mining waste management and bioenergy generation. In *Sustainable management of mining waste and tailings: a circular economy approach* (pp. 107–119). https://doi.org/10.1201/9781003442455-6

Dey, S., Dash, N., Ray, S. K., Priyadarsini, S., Chhotaray, C., & Das, A. P. (2024). Occurrence and quantification of tire wear plastics pollutants from Highway City Dust of Bhubaneswar City of Odisha. In *Sustainable microbial technology for synthetic and cellulosic microfiber bioremediation* (pp. 77–89). https://doi.org/10.1007/978-3-031-62600-5_5

Dey, S., Tripathy, B., Kumar, M. S., & Das, A. P. (2023). Ecotoxicological consequences of manganese mining pollutants and their biological remediation. In *Environmental chemistry and ecotoxicology*. https://doi.org/10.1016/J.ENCECO.2023.01.001

Félix-Cañedo, T. E., Durán-Álvarez, J. C., & Jiménez-Cisneros, B. (2013). The occurrence and distribution of a group of organic micropollutants in Mexico City's water sources. *Science of the Total Environment, 454–455*, 109–118. https://doi.org/10.1016/j.scitotenv.2013.02.088

Houthuijs, K. J., Horn, M., Vughs, D., Martens, J., Brunner, A. M., Oomens, J., & Berden, G. (2023). Identification of organic micro-pollutants in surface water using MS-based infrared ion spectroscopy. *Chemosphere, 341*, Article 140046. https://doi.org/10.1016/j.chemosphere.2023.140046

Jiang, J.-Q., Zhou, Z., & Sharma, V. K. (2013). Occurrence, transportation, monitoring and treatment of emerging micro-pollutants in waste water—A review from global views. *Microchemical Journal, 110*, 292–300. https://doi.org/10.1016/j.microc.2013.04.014

Kanungo, J., Sahoo, T., Bal, M., & Behera, I. D. (2023a). Performance of Bioremediation strategy in waste lubricating oil pollutants: A review. *Geomicrobiology Journal, 41*(4), 360–373. https://doi.org/10.1080/01490451.2023.2245395

Kanungo, J., Sahoo, T., Swain, L. P., & Behera, I. D. (2023). Toxicity of persistent hydrocarbon pollutants, sources and sustainable remediation process. In: I. D. Behera, & A. P. Das (eds.), *Impact of petroleum waste on environmental pollution and its sustainable management through circular economy*. Environmental Science and Engineering. Springer. https://doi.org/10.1007/978-3-031-48220-5_2

Khatun, R., Singh, S., Dubey, N. K., & Prasad Das, A. (2023). A review on marine-based phytochemicals and their application in biomedical research. In *Recent frontiers of phytochemicals: Applications in food, pharmacy, cosmetics, and biotechnology* (pp. 383–395). https://doi.org/10.1016/B978-0-443-19143-5.00013-X

Kim, M.-K., & Zoh, K.-D. (2016). Occurrence and removals of micropollutants in water environment. *Environmental Engineering Research, 21*(4), 319–332. https://doi.org/10.4491/eer.2016.115

Kumar, A., & Gangawane, K. M. (2022a). Effect of precipitating agents on the magnetic and structural properties of the synthesized ferrimagnetic nanoparticles by co-precipitation method. *Powder Technology, 401*, Article 117298. https://doi.org/10.1016/j.powtec.2022.117298

Kumar, A., & Gangawane, K. M. (2022b). Synthesis and effect on the surface morphology & magnetic properties of ferrimagnetic nanoparticles by different wet chemical synthesis methods. *Powder Technology, 410*, Article 117867. https://doi.org/10.1016/j.powtec.2022.117867

Kumar, A., & Gangawane, K. M. (2023a). Advanced computational approaches for water treatment. In *Advanced computational approaches for water treatment* (pp. 1–12). CRC Press. https://doi.org/10.1201/9781003325147-1

Kumar, A., & Gangawane, K. M. (2023b). Molecular dynamics for the membrane process. In *Advanced computational approaches for water treatment* (pp. 13–38). CRC Press. https://doi.org/10.1201/9781003325147-2

Kumar, A., Gangawane, K. M., & Ramanjaneyulu, B. (2022). Ferrofluids for waste-water treatment. In *Advances in chemical, bio and environmental engineering* (pp. 723–744). Springer. https://doi.org/10.1007/978-3-030-96554-9_48

Kumar, N. M., Sudha, M. C., Damodharam, T., & Varjani, S. (2020). Micro-pollutants in surface water: Impacts on the aquatic environment and treatment technologies. In *Current developments in biotechnology and bioengineering* (pp. 41–62). Elsevier. https://doi.org/10.1016/B978-0-12-819594-9.00003-6

Lauretta, R., Sansone, A., Sansone, M., Romanelli, F., & Appetecchia, M. (2019). Endocrine disrupting chemicals: Effects on endocrine glands. *Frontiers in Endocrinology, 10*, 1–7. https://doi.org/10.3389/fendo.2019.00178

Mallick, K., Sahu, S., Sahu, A., Dey, S., & Das, A. P. (2024). A critical review of marine microfiber pollution routes, toxicity, and its sustainable remediation. *Environmental Science and Engineering, Part, F2333*, 189–211. https://doi.org/10.1007/978-3-031-51792-1_10

Mishra, S., & Das, A. P. (2021). Current treatment technologies for removal of microplastic and microfiber pollutants from wastewater. In *Wastewater treatment: Cutting-edge molecular tools, techniques and applied aspects* (pp. 237–251). https://doi.org/10.1016/B978-0-12-821881-5.00011-8

Mishra, S., & Das, A. P. (2023). Microbial remediation of synthetic microfiber contaminated wastewater. In M. P. Shah (Ed.), *Microbial technologies in industrial wastewater treatment* (pp. 337–350). Springer. https://doi.org/10.1007/978-981-99-2435-6_17

Mishra, S., Dash, D., & Das, A. P. (2022). Detection, characterization and possible biofragmentation of synthetic microfibers released from domestic laundering wastewater as an emerging source of marine pollution. *Marine Pollution Bulletin, 185*, Article 114254. https://doi.org/10.1016/J.MARPOLBUL.2022.114254

Mishra, S., Dash, D., & Das, A. P. (2024). Isolation and molecular characterization of polyester-degrading bacterial strains from municipal wastewater and their application in microplastic degradation in contaminated wastewater. *Geomicrobiology Journal*. https://doi.org/10.1080/01490451.2024.2376101

Mishra, S., Rath, C., Charan, & Das, A. P. (2019). Marine microfiber pollution: A review on present status and future challenges. *Marine Pollution Bulletin, 140*, 188–197. https://doi.org/10.1016/J.MARPOLBUL.2019.01.039

Mohanty, J., Dey, S., Bala, K., & Das, A. P. (2024). Recent advances in microplastic pollution for its sustainable management. In *Sustainable management of urban plastic waste through circular economic approaches* (pp. 27–48). https://doi.org/10.1201/9781003465447-3

Mohanty, M., Tripathy, B., Koteswara Reddy, G., & Das, A. P. (2024). Greener technologies for establishing restraint over microplastic pollution. In *Sustainable microbial technology for synthetic and cellulosic microfiber bioremediation* (pp. 239–258). https://doi.org/10.1007/978-3-031-62600-5_13

Moharana, S., Dey, S., Priyadarsini, S., Kumar, M. S., & Das, A. P. (2024). Identification and characterization of microplastic pollutants from the marine sediments of Paradeep Coast of Bay of Bengal, India for their sustainable management. *Environmental Science and Engineering, Part, F2333*, 57–73. https://doi.org/10.1007/978-3-031-51792-1_4

Musolff, A., Leschik, S., Möder, M., Strauch, G., Reinstorf, F., & Schirmer, M. (2009). Temporal and spatial patterns of micropollutants in urban receiving waters. *Environmental Pollution, 157*(11), 3069–3077. https://doi.org/10.1016/j.envpol.2009.05.037

Nam, S.-W., Jo, B.-I., Yoon, Y., & Zoh, K.-D. (2014). Occurrence and removal of selected micropollutants in a water treatment plant. *Chemosphere, 95*, 156–165. https://doi.org/10.1016/j.chemosphere.2013.08.055

Pal, D., & Sen, S. (2023a). Emerging petroleum pollutants and their adverse effects on the environment. In *Impact of petroleum waste on environmental pollution and its sustainable management through circular economy* (pp. 103–137). Springer. https://doi.org/10.1007/978-3-031-48220-5_5

Pal, D., & Sen, S. (2023b). In-depth coverage of petroleum waste sources, characteristics, environmental impact, and sustainable remediation process. In *Impact of petroleum waste on environmental pollution and its sustainable management through circular economy* (pp. 1–38). Springer. https://doi.org/10.1007/978-3-031-48220-5_1

Pal, D., & Sen, S. (2024a). Optimal synthesis of dolochar derived faujasite zeolite X for highly effective Cd(II) removal. *Environmental Research, 240*, Article 117494. https://doi.org/10.1016/j.envres.2023.117494

Pal, D., & Sen, S. (2024b). Valorization of sponge iron industrial waste into iron-modified zeolite X for ciprofloxacin removal: A multi-parameter optimization study. *Environmental Science and Pollution Research, 31*(40), 53381–53398. https://doi.org/10.1007/s11356-024-34730-8

Pal, D., & Sen, S. (2025). Toxicity and health impacts of emerging pollutants. In *Biotechnological interventions in removal of emerging pollutants*. Springer. https://doi.org/10.1007/978-981-97-9922-0_4

Pan, B., Feng, M., McDonald, T. J., Manoli, K., Wang, C., Huang, C.-H., & Sharma, V. K. (2020). Enhanced ferrate(VI) oxidation of micropollutants in water by carbonaceous materials: Elucidating surface functionality. *Chemical Engineering Journal, 398*(Vi), 125607. https://doi.org/10.1016/j.cej.2020.125607

Paradina-Fernández, L., Wünsch, U., Bro, R., & Murphy, K. (2023). Direct measurement of organic micropollutants in water and wastewater using fluorescence spectroscopy. *ACS ES&T Water, 3*(12), 3905–3915. https://doi.org/10.1021/acsestwater.3c00323

Paul, A., Dey, S., Ram, D. K., & Das, A. P. (2023). Hexavalent chromium pollution and its sustainable management through bioremediation. *Geomicrobiology Journal*, 1–11. https://doi.org/10.1080/01490451.2023.2218377

Paul, A., Priyadarsini, S., Karakuş, S., Behera, J. K., Gnanasekaran, R., & Das, A. P. (2024). Fate and accumulation of synthetic microfibers in the environment. In *Sustainable microbial technology for synthetic and cellulosic microfiber bioremediation* (pp. 61–76). https://doi.org/10.1007/978-3-031-62600-5_4

Pradhan, G., Dey, S., Priyadarsini, S., Kumar, M. S., & Das, A. P. (2024). Characterization and quantification of microplastics pollutants in sediment samples from Daya River of Odisha State in India for their appropriate management. *Environmental Science and Engineering, Part, F2333*, 75–92. https://doi.org/10.1007/978-3-031-51792-1_5

Pradhan, G., Tripathy, B., Ram, D. K., Digal, A. K., & Das, A. P. (2023). Bauxite mining waste pollution and its sustainable management through bioremediation. *Geomicrobiology Journal*, 1–10. https://doi.org/10.1080/01490451.2023.2235353

Priyadarsini, S., & Das, A. P. (2024). Analytical and structural characterization of waste lithium-ion batteries for their effective recycling strategy. *Environmental Chemistry and Ecotoxicology*. https://doi.org/10.1016/j.enceco.2024.12.004

Ray, S. K., Dey, S., Das, A. P., & John, A. (2024). Microplastic waste management through greener technologies: A review. In *Sustainable management of urban plastic waste through circular economic approaches* (pp. 237–261). https://doi.org/10.1201/9781003465447-14

Rosen, R. (2007). Mass spectrometry for monitoring micropollutants in water. *Current Opinion in Biotechnology, 18*(3), 246–251. https://doi.org/10.1016/j.copbio.2007.03.005

Sahu, A., Dey, S., Mallick, K., Priyadarsini, S., Behera, I. D., & Das, A. P. (2024). A review on advanced technology for sustainable management of synthetic microplastic waste. In *Sustainable management of urban plastic waste through circular economic approaches* (pp. 262–279). https://doi.org/10.1201/9781003465447-15

Sánchez-Avila, J., Tauler, R., & Lacorte, S. (2012). Organic micropollutants in coastal waters from NW Mediterranean Sea: Sources distribution and potential risk. *Environment International, 46*, 50–62. https://doi.org/10.1016/j.envint.2012.04.013

Schmidt, T. C. (2018). Recent trends in water analysis triggering future monitoring of organic micropollutants. *Analytical and Bioanalytical Chemistry, 410*(17), 3933–3941. https://doi.org/10.1007/s00216-018-1015-9

Schwarzenbach, R. P., Escher, B. I., Fenner, K., Hofstetter, T. B., Johnson, C. A., von Gunten, U., & Wehrli, B. (2006). The challenge of micropollutants in aquatic systems. *Science, 313*(5790), 1072–1077. https://doi.org/10.1126/science.1127291

Shimabuku, K. K., Kennedy, A. M., Mulhern, R. E., & Summers, R. S. (2017). Evaluating activated carbon adsorption of dissolved organic matter and micropollutants using fluorescence spectroscopy. *Environmental Science & Technology, 51*(5), 2676–2684. https://doi.org/10.1021/acs.est.6b04911

Siegrist, H., & Joss, A. (2012). Review on the fate of organic micropollutants in wastewater treatment and water reuse with membranes. *Water Science and Technology, 66*(6), 1369–1376. https://doi.org/10.2166/wst.2012.285

Sousa, J. C. G., Barbosa, M. O., Ribeiro, A. R. L., Ratola, N., Pereira, M. F. R., & Silva, A. M. T. (2020). Distribution of micropollutants in estuarine and sea water along the Portuguese coast. *Marine Pollution Bulletin, 154*, 111120. https://doi.org/10.1016/j.marpolbul.2020.111120

Stamm, C., Räsänen, K., Burdon, F. J., Altermatt, F., Jokela, J., Joss, A., Ackermann, M., & Eggen, R. I. L. (2016). Unravelling the impacts of micropollutants in aquatic ecosystems. In *Advances in ecological research* (Vol. 55, pp. 183–223). https://doi.org/10.1016/bs.aecr.2016.07.002

Ternes, T. (2007). The occurrence of micropollutants in the aquatic environment: A new challenge for water management. *Water Science and Technology, 55*(12), 327–332. https://doi.org/10.2166/wst.2007.428

Trapido, M., Epold, I., Bolobajev, J., & Dulova, N. (2014). Emerging micropollutants in water/wastewater: Growing demand on removal technologies. *Environmental Science and Pollution Research, 21*(21), 12217–12222. https://doi.org/10.1007/s11356-014-3020-7

Tripathy, B., Dash, A., & Das, A. P. (2022). Detection of environmental microfiber pollutants through vibrational spectroscopic techniques: Recent advances of environmental monitoring and future prospects. In *Critical reviews of analytical chemistry* (pp. 1–11). https://doi.org/10.1080/10408347.2022.2144994

Van Der Kraak, G. J., Hosmer, A. J., Hanson, M. L., Kloas, W., & Solomon, K. R. (2014). Effects of Atrazine in fish, amphibians, and reptiles: An analysis based on quantitative weight of evidence. *Critical Reviews in Toxicology, 44*(sup5), 1–66. https://doi.org/10.3109/10408444.2014.967836

Yang, Y., Zhang, X., Jiang, J., Han, J., Li, W., Li, X., Yee Leung, K. M., Snyder, S. A., & Alvarez, P. J. J. (2022). Which micropollutants in water environments deserve more attention globally? *Environmental Science & Technology, 56*(1), 13–29. https://doi.org/10.1021/acs.est.1c04250
Yang, Y.-Y., Toor, G. S., Wilson, P. C., & Williams, C. F. (2017). Micropollutants in groundwater from septic systems: Transformations, transport mechanisms, and human health risk assessment. *Water Research, 123*, 258–267. https://doi.org/10.1016/j.watres.2017.06.054

Chapter 4
Toxicity Assessment of Micropollutants in the Environment

Reshmi John, Usha K. Aravind, and C. T. Aravindakumar

Introduction

Micropollutants are widespread organic and inorganic chemical compounds that enhance human life standards. Due to their widespread use, they became a significant contaminant in soil, water and the broader environment (Kim & Zoh 2016). They are classified as emerging contaminants or prospective pollutants due to their persistence, dynamic biological characteristics and continuous release to the environment (Arslan et al., 2017). Depending on the physicochemical properties, they are categorized into organic substances such as polychlorinated biphenyls, polycyclic aromatic hydrocarbons (PAH), pesticides, endocrine disruptors, pharmaceuticals and personal care products as well as inorganic substances like heavy metals (Arman et al., 2021). Even at low concentrations ranging from micrograms to nanograms and picograms, they can have detrimental effects on living beings in the environment (Gul et al., 2022). Micropollutants entered the environment through various sources, including agricultural practices, the domestic use of chemicals, and industrial activities (Gul et al., 2022). With the progression of time, they accumulate in living tissues and in the environment, leading to chronic or long-term negative health effects on living beings. Chronic exposure to these micropollutants mainly affects the biotic characteristics in the environment (Aemig et al., 2021). However, the health effects in humans associated with exposure to micropollutants are numerous and this is an extensively studied

R. John · C. T. Aravindakumar (✉)
Inter University Instrumentation Centre (IUIC), Mahatma Gandhi University, Kottayam, Kerala, India
e-mail: cta@mgu.ac.in

U. K. Aravind
School of Environmental Studies, Cochin University of Science and Technology (CUSAT), Kochi, Kerala, India

C. T. Aravindakumar
School of Environmental Sciences, Mahatma Gandhi University, Kottayam, Kerala, India

I. D. Behera et al. (eds.), *Emerging Micropollutants*, Sustainable Environmental Waste Management Strategies, https://doi.org/10.1007/978-3-032-03530-1_4

area. Further various classes of these micropollutants are specifically identified as harmful compounds (Bhatt et al., 2022). For example, the chemical compounds used to control pests and weeds in agriculture are called pesticides. Its exposure in humans results in numerous health issues, such as neurological disorders, cancer and reproductive issues (Aemig et al., 2021). Whereas hormonal imbalances and antibiotic resistance are health issues linked to antibiotic exposure (Ambaye et al., 2023). A similar hormonal imbalance is associated with exposure to another class of micropollutants, the endocrine disruptors (Das et al., 2024b; Dey et al., 2023b). As the endocrine disruptors can mimic, interfere and block natural hormones in living beings (Ambaye et al., 2023). Thus, the deleterious impact of micropollutants on the health of humans is numerous and complex.

Once the micropollutants enter the environment, they undergo various pathways such as dissolution in water, (bio) chemical transformation, and binding to solids. Only a small fraction of the micropollutants in the environment undergo degradation, while the remaining part is unaffected (Mishra et al., 2021, 2024; Sahu & Das, 2024). The physicochemical properties of micropollutants such as stability, water solubility, water partitioning, vapor pressure, the environmental compartment in which it is released, and the metabolic activity of microbes on the micropollutants control its fate in the environment (Kim & Zoh 2016; Gautam & Anbumani 2020). Consequently, micropollutants and their degradation products marked their presence in various environmental matrices. Therefore, to reduce exposure to micropollutants, increased public awareness, stringent regulations on its use and disposal, and more effective remediation methods must be adopted (Arslan et al., 2017). Also, controlling the entry of micropollutants into the environment is challenging because they are by-products of our daily lives. While the production of these chemicals to meet human demands is on the rise. In this context, the possibility of exposure to these chemicals increases, highlighting the importance of assessing the toxicity of these substances in aquatic organisms and humans (Das et al. 2024c; Das et al., 2023a; Dey et al., 2023b).

Types of Micropollutants and Their Sources

Most of the micropollutants are xenobiotic chemicals, including pesticides, herbicides, insect repellants, pharmaceuticals, antibiotics, pain killers, anti-inflammatory medications, plasticizers, ultraviolet filters, preservatives, artificial sweeteners and personal care products (Tijani et al., 2016). They enter the environment through domestic use, industrial activities, agricultural practices, human and animal excretion, wastewater treatment plants (WWTPs) effluents, and hospital discharges (Sanchez-Huerta et al., 2024). The different types of micropollutants and their source of entry into the environment are listed in Table 4.1.

Table 4.1 Different types of micropollutants and their source of entry in the environment

Sl. no.	Category	Important classes	Major sources	Reference
1	Pesticides	Herbicides Fungicides Insecticides, Rodenticides	Agricultural waste residues, livestock practices and leachate from agricultural lands	Narwal et al. (2023)
2	Pharmaceuticals	Antibiotics, Antidepressants Anticonvulsants, Antifungals, Antidiabetics, β-blockers Lipid regulators Non-steroidal anti-inflammatory drugs (NSAIDs)	Medical waste, pharmaceutical effluents, veterinary use, industrial wastewater, hospital effluents	Ashiwaju et al. (2024)
3	Personal care products	Disinfectants, fragrances, insect repellants, preservatives, UV filters, antiseptics, sunscreens and nutritional supplements	Individual households, manufacturing sites, dumping sites, large farms, Sewage effluents from wastewater treatment plants,	Chopra and Kumar (2018)
4	Industrial chemicals	Toxic metals, aromatic and nonaromatic hydrocarbons, chlorinated solvents, plastics, plasticizers, cyanides and sulfides	Solid, liquid and slurry waste from industries, chemical manufacturing plants	Brusseau and Artiola (2019), Dsikowitzky and Schwarzbauer (2014)
5	Endocrine disruptors	Phthalates, organohalogen compounds, per and poly-fluoroalkyl compounds, Alkyl phenols, organotins, and natural toxins	Incinerators, cement kilns, discharges from wastewater treatment plants, hospital waste, chemical leachates from industrial and household items	Metcalfe et al. (2022), Pironti et al. (2021)
6	Atmospheric micropollutants	Polychlorinated biphenyls (PCBs), polycyclic aromatic hydrocarbons (PAHs), dioxins, furans	Incinerators, industrial processes, coal gasification plants, scraping of materials, emissions from vehicles, power plants and aluminum smelters	Thakur and Ganguly (2022)

Transformation Pathways and Environmental Fate of Micropollutants

The identification and detection of micropollutants in various environmental matrices have raised significant concerns about their transport and fate in the environment (Priyadarsini & Das, 2024; Raj & Das, 2023; Tripathy et al., 2024b). In the environment, these micropollutants undergo a complex transformation process, leading to the formation of more toxic or persistent by-products (La Farre et al., 2008). In the environment, the transformation pathways and environmental fate of micropollutants are influenced by various physical, chemical and biological processes and are described below (Tong et al., 2022).

(a) **Physical processes**

Through adsorption, micropollutants are adsorbed to the surface of organic matter and suspended sediments. Then, the adsorbed micropollutants get released into the aquatic environments by desorption. Thus, the mobility, persistence and bioavailability of micropollutants depend on adsorption and desorption (Glaser et al., 2019; Tong et al., 2022). Whereas in soil also the micropollutants are transformed through the sorption/desorption mechanism without a change in its structure (Nováková et al., 2024). In aquatic environments, dilution, and dispersion are other important factors that influence the concentration and distribution of micropollutants (Kashyap et al., 2023). When micropollutants are introduced into large aquatic matrices, their concentration decreases due to dilution. Whereas dispersion helps in the diffusion of micropollutants in aquatic environments through natural processes such as water currents and turbulent mixing. In addition, the hydrodynamic conditions (advection, diffusion and vertical mixing) significantly affect the transport of micropollutants in aquatic environments (Kashyap et al., 2023). In flowing water, advection plays a crucial role in the long-range transport of micropollutants. While diffusion facilitates the movement of micropollutants from areas of lower concentration to those of higher concentration. In addition, vertical mixing driven by water pressure and gravitational changes aids in the vertical transport of micropollutants within aquatic environments (Dey et al., 2023a; Ghosh et al., 2023; Jagatee et al., 2024; Mishra et al., 2022; Paul et al., 2024). Various hydrodynamic parameters, including topography, weather patterns, wind currents, aquatic plants and the roughness of the water bed, also influence the transportation of micropollutants in aquatic environments (Kashyap et al., 2023).

(b) **Chemical processes**

Hydrolysis and Photolysis are two chemical processes that transform micropollutants into less toxic and more biodegradable products (Guo et al., 2023; Margot et al., 2015). In hydrolysis, the chemical bonds in micropollutants get cleaved followed by water addition, leading to less toxic products (Margot et al., 2015). While during photolysis, micropollutants get degraded by the action of sunlight, which form new products (Guo et al., 2023). In contrast to hydrolysis, photolysis controls the fate and transformation of several micropollutants, which are susceptible to direct or indirect photochemical reactions. This process is a significant transformation pathway for several emerging micropollutants, including endocrine disruptors and UV filters. The quality of light available is another key factor that influences the degradation of micropollutants (Guo et al., 2023).

(c) **Biological processes**

Biodegradation, bioaccumulation and biomagnification are the key biological processes that influence the transformation of many micropollutants (Kanaujiya et al., 2019; Zenker et al., 2014). Biodegradation is the microorganism mediated degradation of micropollutants. Algae, fungi and bacteria are the microorganisms involved in biodegradation (Kanaujiya et al., 2019). For biologically labile molecules, biodegradation is the most suitable degradation method. Also, it is one of the suitable degradation methods for various personal care products and pharmaceuticals in wastewater treatment plants (Kanaujiya et al., 2019). However, for soil environments, transformation and fate of micropollutants are entirely dependent on microbial consortia (Nováková et al., 2024). In addition, using biofilms where a group of microorganisms exists facilitates the biodegradation of micropollutants aided by a wide range of enzymatic reactions and pathways by microorganisms. Another factor that aids in effective biodegradation is the environmental context, including aerobic and anaerobic conditions where the biodegradation takes place (Kanaujiya et al., 2019).

Meanwhile, the intake and retention of micropollutants by living beings in the environment are known as bioaccumulation. This process leads to an increase in the concentration of micropollutants at higher trophic levels within a food web, resulting in biomagnification. Both bioaccumulation and biomagnification of micropollutants can negatively affect ecology and pose health risks to humans (Zenker et al., 2014).

Other factors that influence transformation pathways and environmental fate are detailed in Table 4.2.

Table 4.2 The key elements influencing the transformation and fate of micropollutants in the environment

Elements	Components	Reference
Environmental factors	pH, organic matter, temperature and redox conditions	Zhong et al., (2019)
Microbial community structure	Diversity and abundance of microbial community, nature of bacterial strain and composition of bacterial community	Kanaujiya et al. (2019)
Presence of other contaminants	Humic substances, heavy metals	Zhong et al. (2019)

Health Effects Associated with Exposure to Micropollutants

The occurrence, distribution, and fate of micropollutants raise significant concerns because of the toxicity associated with their exposure by living beings in the environment. Even small quantities of micropollutants can harm human health and the environment (Das et al., 2024b). This concern arises mainly from the continuous, multigenerational exposure to these substances, which are persistently produced and released into aquatic ecosystems (Yang et al., 2021). Among the micropollutants, organic micropollutants significantly affect the biodiversity of freshwater ecosystems (Arslan et al., 2017). These micropollutants can harm microbial diversity, resulting in genetic changes in the microbial ecology, soil quality, and disrupt various soil biogeochemical processes (Kumar et al., 2024). Furthermore, the emerging micropollutants pose a potential threat to the energy flow, elemental cycles and biogeochemical cycles within ecosystems (Ahmed et al., 2022). Acute environmental changes such as climate changes, land degradation, damage to vegetation, water scarcity and food shortages are also caused by these micropollutants (Ahmed et al., 2022; Kanungo et al., 2023a, b; Das et al., 2023a, b; Das and Bal 2025). Also, micropollutants can cause human health issues such as oxidative stress, cytotoxicity, cancers, endocrine disruption, miscarriages, neuropathy, hypertension, immune system impairment and encephalopathy (Ahmed et al., 2022) (Table 4.3). As a result, the increasing presence of micropollutants in various environmental matrices and their exposure are reducing human lifespan. Therefore, it is crucial to address these issues to mitigate risks to human health.

Challenges in Toxicity Assessment of Micropollutants

Numerous analytical techniques have been developed for the identification and quantification of micropollutants. The continuous emergence of new micropollutants, their extremely low concentrations in the environment, the complex nature of environmental samples, and the presence of micropollutants in the mixture present significant analytical challenges for their detection (Shao et al., 2019). However, regarding toxicity assessment traditional toxicity assessment methods for micropollutants are limited by complexities such as unknown micropollutants, the current micropollutant level in the environment, low concentrations, mixture effects (synergistic effects) of micropollutants and the different types of species being exposed (Sanganyado et al., 2021). Usually, the toxicological impact of a particular micropollutant is analyzed using in vivo and in vitro techniques (Rehberger et al., 2018). In vivo studies use captured animals in a laboratory setting under controlled conditions, which can complicate the extrapolation of results to free-ranging populations. Additionally, regulatory and ethical constraints regarding the use of animals present challenges during the toxicity assessment of micropollutants (Sanganyado et al., 2021). While in vitro studies examine the impact of a particular micropollutant at the

Table 4.3 Multiple effects of different classes of micropollutants on human health

Sl. no.	Micropollutants	Multiple effects	Reference
1	Pesticides	Birth defects, reduced birth weight, fetal death, nervous system disruptors, neurodegenerative disorders, cancer, respiratory diseases, diabetes, leukemia, reproductive disorders	Kim et al. (2017)
2	Pharmaceuticals	Endocrine disorders, carcinogenic, mutagenic, attention deficit hyperactivity disorder, neurological disorders	Vaudin et al. (2022)
3	Personal care products	Increased risk of preterm birth, pulmonary toxicity, skin allergies, dermatitis, hypersensitivity	Khalid and Abdollahi (2021)
4	Heavy metals	Cardiovascular disorders, reproductive disorders, neurological disorders, risk of cancer and diabetes, renal injuries, vitamin D deficiency, premature birth	Rehman et al. (2018)
5	Industrial chemicals	Developmental toxicity, neurotoxicity, reproductive toxicity, pulmonary toxicity, endocrine disruption, carcinogenic	Li and Suh (2019)
6	Endocrine disruptors	Disrupt the body's normal function, interferes with hormone action, Oestrogenic effect, breast cancer, birth defects, biocidal effects and contributes to bacterial-resistant dermatitis, skin irritation, hepatotoxicity, peripheral and central neurotoxicity	Encarnação et al. (2019)
7	Atmospheric micropollutants	Respiratory problems, lung cancer, bronchiolitis, dysfunctions of the central nervous system and cardiovascular problems	Manisalidis et al. (2020)

cellular and subcellular levels (Rehberger et al., 2018). This type of study appears unusual, resulting in limited interest. Compounding this issue, some micropollutants are transformed in the environment and these transformation products are often more abundant than the parent compounds (Escher and Fenner 2011). Whereas a vast array of transformation products has yet to be identified. Additionally, it is important to consider factors such as whether the transformation products are formed in high yields, whether they are more persistent or labile compared to the parent compounds, and whether they exhibit greater toxicity than the parent compound. These concerns surrounding transformation products further complicate the process of chemical risk assessment (Escher and Fenner 2011).

Toxicity Testing Methods of Micropollutants

1. In vitro testing

In in vitro studies, the target cells were treated with a micropollutant under study, and the cellular changes were analyzed using different assays. It gives repeatable and meaningful results within a short period, making it more helpful for analyzing the toxicities using many samples (Verma et al., 2021). Compared to in vivo experiments, in vitro studies have advantages like lower cost, faster and minimum ethical concerns (Kumar et al., 2017). Also, the results obtained from in vitro experiments are not the same as those obtained from in vivo experiments; only an indication regarding the toxicity of the micropollutant under study is obtained. It also helps to decide whether to proceed with in vivo studies after the in vitro studies (Verma et al., 2021). In vitro toxicity assessment tests can be grouped into the following categories proliferation, apoptosis, necrosis, oxidative stress and DNA damage assays (Kumar et al., 2017).

(i) Proliferation assays

It is generally used to analyze the cellular metabolism in metabolically active cells using the reagent 3—(4, 5-Dimethyl-thiazol-2-yl)-2, 5-diphenyltetrazolium bromide (MTT). MTT is one of the most used tetrazolium salts for in vitro toxicity assessments. The positive charge and lipophilic nature of the side groups in MTT make it penetrate the cell membrane of the viable cells and get reduced by the cell plasma or mitochondrial enzymes such as oxidases, peroxidases, oxidoreductases and dehydrogenases in the presence of co substrates like NADPH, NADP, succinate and pyruvate. The reduced form of the tetrazolium salts, which is colored, is determined by basic spectroscopic methods (Präbst et al., 2017). Other tetrazolium salts used for the cell viability studies include dimethylthiazol-carboxymethoxyphenyl-sulfophenyl-tetrazolium (MTS), methoxynitrosulfophenyl-tetrazolium carboxanilide (XTT), 2-(2-methoxy-4-nirophenyl)-3-(4-nitrophenyl)-5-(2,4-disulfophenyl)-2H tetrazolium (WSTs) and resazurin (Stoddart 2011).

In the MTT assay, tetrazolium salt is reduced to an insoluble formazan with a purple color. The crystal spikes of the solid crystalline formazan destroy cell integrity, leading to cell death. After cell death, formazan formation was stopped, and the reactions endpoint was used to evaluate the cell culture viability. Therefore, this kind of assay is referred to as endpoint determination assay (Präbst et al., 2017). These types of assays are rapid and simple. The unavoidable cell death and for recording the absorbance of formazan, it must be dissolved, i.e., an additional dissolving step is necessary are the disadvantages associated with MTT assays (Präbst et al., 2017). In WST-based assays, soluble formazan is formed; therefore, an additional solvation step is not required (Kari et al., 2022). Viability assays using MTT and WSTs follow pseudo first-order kinetics, and the metabolic activities was evaluated from the reaction rates. Thus, it makes consistent reaction conditions essential for proliferation assays. Factors such as temperature, pH and even small changes in incubation time can alter the reaction rate (Präbst et al., 2017). While the cell viability assays using resazurin follow a pseudo-first-order reaction kinetics, a fluorescent product is formed here. The formation of this fluorescent product enhances the sensitivity and range of measurements, specifically for small cell concentrations. Compared with MTT and WST assays, resazurin based assays have more pitfalls (Präbst. et al., 2017).

(ii) **Apoptosis assay**

Apoptosis is referred to as regulated or programmed cell death, which helps to maintain tissue homeostasis. Excessive generation of free radicals is the cause of apoptosis. Usually, in apoptosis, unwanted cells are removed during normal development. It is a process that also occurs in the adult stage, where the cells are removed at the termination of their lifespan. It also occurs when a cell is infected with a virus, damaged beyond repair, or when subjected to stress. Two or more assays are used to detect apoptosis accurately because apoptosis and necrosis share some common features. Necrosis is another process in cells where cell death occurs due to infection, toxins, or trauma (Butler et al., 2014; Darzynkiewicz et al., 1994; Kari et al., 2022; Kumar et al., 2017). Detecting apoptosis involves measuring a proportion of cells going through early, mid and late stages of apoptosis. Each stage of apoptosis was detected from the stage-specific structures and proteins labeled by fluorescent probes. Combining all these assays will give an idea regarding the mechanism of apoptosis induction after specific stimuli (Butler et al., 2014).

In the early stages of apoptosis, the cell loses its membrane phospholipid asymmetry. As a result, the phosphatidylserine (PS) gets exposed to the outer membrane surface rather than the inner membrane surface of the cell. The PS molecule that is exposed is stained with annexin V molecules, which are conjugated to a fluorescein isothiocyanate (FITC) or R-phycoerythrin (PE) fluorescence dyes, which will help to distinguish the early stages of apoptosis. Further, the early apoptotic cells can be distinguished from the necrotic cells by staining the annexin V-stained cells using propidium iodide (PI) or 7-aminoactinomycin D (7-AAD) (Butler et al., 2014; Kumar et al., 2017).

Another method by which apoptosis is detected by fluorescently labeling the caspase protein using an antibody or an inhibitor of multiple caspases. The caspase proteins are involved in the apoptotic cascade, causing cellular breakdown and processing cytokines. The fluorescent labeling for detecting apoptosis uses an antibody against a fragment of the specific caspase. At the same time, the inhibitors of multiple caspases used for detecting apoptosis include sulforhodamine-valyl-alanyl-aspartylfluoromethyl-ketone (SR-VAD-FMK) (Butler et al., 2014).

The biochemical assay, DNA gel electrophoresis, detects the late stage of apoptosis. DNA gel electrophoresis, detects DNA laddering occurring at the late stage of apoptosis. It is detected by the fluorescent labeling of the 3′ hydroxyl ends of the fragmented DNA using fluorescence groups such as fluorescein, digoxigenin, biotin, and uridine. This reaction is accelerated by the terminal deoxynucleotidyl transferase (TdT) or DNA polymerase enzymes. This biochemical assay is also referred to as the TUNNEL assay (Butler et al., 2014; Kumar et al., 2017).

(iii) **Necrosis assay**

Necrosis is another form of cell death that occurs at normal and disease pathologies. It is morphologically characterized by the inflammation and rupture of organelles within a cell, followed by the breakdown of plasma membrane. The breakdown of the plasma membrane causes the leakage of intracellular contents of the necrotic cells to the extracellular milieu. Thus, necrotic cell death stimulates inflammatory responses, closely associated with inflammatory diseases (Butler et al., 2014; Chan et al., 2013; Darzynkiewicz et al., 1994). The cytotoxic assays for detecting necrosis can be grouped into two categories: one which shows differences in the uptake of dyes that bind to DNA (e.g., propidium iodide) that inhibits the crossing of the plasma membrane in living cells and the other based on the leakage of molecules present within a cell through the plasma membrane of the cell (Butler et al., 2014; Darzynkiewicz et al., 1994). The cytoplasmic enzyme Lactate dehydrogenase (LDH) is present in nearly all cells. When the plasma membrane is damaged LDH gets released into the extracellular space. To detect the leakage of LDH into the cell culture medium, tetrazolium salt is commonly preferred. When LDH catalyzes lactate oxidation to pyruvate, reduced nicotinamide adenine dinucleotide (NADH) is produced. The NADH formed and an electron acceptor present in the medium will convert the tetrazolium salt to a colored formazan product. The colored formazan formed is detected and quantified by standard spectroscopic techniques. Because of the linearity associated with this assay, the percentage of necrotic cells in a sample can also be identified (Butler et al., 2014; Chan et al., 2013).

(iv) **Oxidative stress assay**

Oxidative stress is produced because of the generation of short-lived oxidants and radicals. It is difficult to detect these short-lived species and to directly detect oxidative stress. Therefore, sentinel agents are often appropriate to detect and quantify the level of oxidative stress in cellular systems. It is because, in response to reactive oxygen species (ROS), sentinel agents show color changes or fluorescence. The oxidative stress in living cells is also measured by the oxidative damage of protein, DNA and lipids within a cell (Awashra & Młynarz 2023). The most common assays used to measure cellular oxidative stress include DCF fluorescence, CellROX, and GSH/GSSG assay (Pereira et al., 2012; Mitov et al., 2016; Camera & Picardo 2002).

In the DCF fluorescence assay, the cells are loaded with a nonfluorescent 2′,7′—Dichlorodihydrofluorescein (DCFH) sensitive probe using a cell-permeable nonfluorescent fluorescein derivative 2′,7′—Dichlorodihydrofluorescein diacetate (H_2DCFDA). The DCFH-sensitive probe accumulates within the cellular cytosol by losing the acetate groups by intracellular esterases. ROS causes the oxidation of the DCFH molecule to form the fluorescent 2′,7′-dichlorofluorescein molecule (DCF). The fluorescent DCF molecule is said to have a 485 nm excitation and 535 nm emission maximum wavelengths, respectively. It is one of the widely used assays, but it possesses some drawbacks related to the markers and mechanism that form DCF (Pereira et al., 2012).

CellROX dyes are used in CellROX assays, which are derivatives of the molecules in the xanthene family that work similarly to that of the DCF fluorescence assay. The oxidation products formed are stable and fluorescent and suffer less photobleaching.

CellROX deep red, CellROX orange and CellROX green are the most used dyes for this purpose. Each CellROX dye listed is characterized by an excitation and emission wavelength of 640 and 665 nm, 545 and 565 nm, 485 and 520 nm, respectively (Mitov et al., 2016).

Monitoring GSH status is another way to detect cellular oxidative stress. GSH Ɣ-l-glutamyl-l-cysteinyl-glycine tripeptide, known as glutathione, is a low molecular weight antioxidant synthesized in cells. As a result of oxidative events inside the cells, the GSH is converted to GSSG its disulfide form. Therefore, GSH to GSSH ratio measures the level of oxidative stress within the cells. The GSH to GSSH ratios can be determined directly using HPLC or fluorescent dyes. Using orthophthaldehyde, Ellman's reagent, and ThiolTracker Violet dye, GSH concentrations can be determined. Further, GSSG concentrations can be determined by the enzymatic reduction of GSSG to GSH using the GSH reductase enzyme. The formed GSH concentrations are then determined the same way as stated above (Camera & Picardo 2002).

Besides the in vitro based oxidative stress assays, the oxidative damage in cells can be measured through protein oxidation, lipid peroxidation and DNA oxidation (Awashra & Młynarz 2023). A vast number of methods are available for the detection of protein, lipid and DNA oxidation, which are summarized in the Table 4.4.

(v) **DNA damage assay**

The single-cell gel electrophoresis or comet assay is a microgel electrophoresis method commonly used to study DNA repair and damage. In this assay, the cells were dispersed in thin agarose gel, and displayed on a glass slide. It is then subjected to lysis and electrophoresis, and stained with a fluorescent dye that binds to DNA. DNA damage in cells causes the migration of chromosomal DNA from the nucleus to the anode, which has a comet shape. The amount of DNA migrated to the anode corresponds to DNA damage in the cell (Speit and Hartmann 1999). However, the alkaline version of the comet assay detected single-strand breaks and alkali labile sites in DNA. This alkaline version of the assay is most used nowadays and has a high sensitivity for detecting induced DNA damage (Mohanty et al., 2024b; Priyadarshini et al., 2024; Sahu et al., 2024). In 1988, Singh and his coworkers introduced the alkaline type of the comet assay. Here, the electrophoresis was carried out at a pH > 13.0. Other modifications in this assay involve the lysis in alkaline conditions and thereafter electrophoresis in either mild or neutral alkaline conditions (pH ~ 12.1). This modification detects either double-strand or single strand breaks in DNA, respectively (Speit and Hartmann 1999). Further, the comet assay results are categorized into type 0, 1, 3, and 4. This categorization is made based on the level of DNA damage. When no DNA damage occurs, it is grouped into the type 0 category. Type 1, 3, and 4 correspond to minor DNA damage, extensive DNA damage, and complete DNA destruction, respectively (Jiang et al., 2023). The advantages of these techniques are the cost of performing this assay is extremely low, its rapid and simple performance, the data analysis at the level of single cells, its sensitivity for detecting DNA damage, the use of extremely small samples, and its application can be extended to a eukaryote cell population (Collins et al., 2023). Some disadvantages of the comet assay are the use of small cell samples, which can result in sample bias, single-cell

Table 4.4 Biomarkers for oxidative damage of biomacromolecules protein, lipids and DNA, its detection methods

Biomacromolecule	Biomarkers for oxidation	Analytical methods/ assays for detection	Reference
Protein	o-tyrosine (o-Tyr), o,o′-dityrosine (DiY), 3-chlorotyrosine (3-CIY) and 3-nitrotyrosine (3-NT),	Enzyme-linked immunosorbent assay (ELISA), HPLC-PDA, LC-MS/MS, electrochemical sensors, fluorescent sensors and surface plasmon resonance sensors	Kehm et al. (2021), Mateos and Bravo (2007)
Lipids	A series of aldehydes, isoprostanes (IsoPs), Malondialdehyde (MDA), 4-hydroxynonenal (HNE), 8-isoprostaglandin F2α (8-isoPGF2α)	ELISA, TBARS assay (Thiobarbituric acid), HPLC-UV, HPLC-FLD, GC-MS and LC-MS	Mateos and Bravo (2007), Demirci-Cekic et al. (2022)
DNA	8-oxo-7, 8-guanosine (8-oxoGuo), 8-oxo-7, 8-dihydro-20-deoxyguanosine (8-oxodGuo, 8-oxodG and 8-OHdG), 8-oxo-7, 8-dihydroguanine (8-oxoGua), and 8-oxo-7, 8-dihydro-20-deoxyadenosine (8-oxo-dA)	ELISA, SCGE (single cell gel electrophoresis assay i.e. comet assay), chromatographic methods HPLC-ECD, LC-MS/MS, DNA damage sequencing	Mateos and Bravo (2007), Chao et al. (2021)

data, technical variability, and interpretation (Jiang et al., 2023). The comet assays also find wide application in DNA damage and repair studies, genotoxicity testing, human population monitoring, and environmental biomonitoring (Collins 2004).

2. **In vivo testing**

Besides the advantages of several in vitro based assays, they also have many limitations. In vitro studies provide information about the first insights into the in vivo conditions. However, it lacks information about the intact physiology present in an organism. Also, invitro results underestimate the toxicological properties of xenobiotics, leading to unauthorized restrictions and safety attestations of the xenobiotics under study. In contrast to in vitro studies, in vivo studies provide information from an organism, such as the complex nature of interacting metabolism, signaling mechanisms, and homeostasis. Moreover, in vitro experiments must require in vivo validation to be useful (Lanzerstorfer et al., 2021).

In vivo studies mainly use animal models for various toxicological bioassays. Rats and mice belonging to the category of rodents are the animal models frequently used for studies. The advantage of using mouse models is that they are small, portable and can be raised in a laboratory. In addition, mice have relatively short gestation periods, and many offspring can be obtained for in vivo studies. Therefore, mice are the suitable animal models for in vivo studies among rodents. Meanwhile, guinea pigs, rabbits and hamsters are less frequently used. Invertebrate model organisms like fruit fly (Drosophila melanogaster) and nematode worm (Caenorhabditis elegans) are also used to assess many biological activities in an organism (Benítez-Chao et al., 2021). Finally, Zebrafish is the widely used and rapidly developing vertebrate model for various toxicity studies (Zon and Peterson 2005).

Additionally, to conduct toxicity studies using animal models or tissues/cell lines from animals, one should get approval from the institute's animal ethics committee (IAEC). Otherwise, the protocol for the study should meet the guidelines put forward by the local governing body. The rules and the regulatory provisions for conducting animal models for in vivo studies differ from place to place. While in India, the committee's guidelines for control and supervision of experiments on animals (CPCSEA) are strictly followed for keeping the animals used for experiments (Parasuraman 2011).

In vivo studies usually use an animal model to monitor a particular xenobiotic's acute and chronic toxicity. In acute toxicity testing, the impact of a single dose of the micropollutant on a specific animal model are studied. One rodent and one non rodent animal species are considered for carrying out acute toxicity studies. The animal species under study were administered with different dose levels of the investigational product and the effects were monitored for 14 days. During the experimental period, the mortalities caused by the investigational product were recorded, followed by the dead animal's morphological, pathological, histological, and biochemical changes. Further acute toxicity testing helps to determine the 50% lethal dose (LD_{50}) of the investigational product. To determine LD_{50}, many animals have to be used for the experiments, and a high mortality ratio was observed. Because of these constraints associated with LD_{50} determination, modified methods like fixed dose procedure (FDP), acute toxic category method (ATC) and up and down method (UDP) were developed. FDP method determined the nonlethal dose rather than the lethal dose for toxicity (Parasuraman 2011; Stallard & Whitehead, 1995). In ATC, four predefined starting doses were utilized, and the selection of the test doses was based on the globally accepted classification system. Compared to FDP, ATC involves a stepwise procedure utilizing three animals of the same sex in each step (Parasuraman 2011; Diener et al., 1994). A similar staircase design is observed in the UDP approach, which is regulatory agencies most recommended toxicological testing approach (Parasuraman 2011; Zhang et al., 2022). This method minimizes the number of vertebrate animals for the studies, and female rodents are preferable. Here, the animals were exposed to a dose less than the LD_{50} value and monitored for about 48 hrs. If the animal survives, it is administered twice the dose that was initially administered. If the animal dies, testing is conducted using a lower dose than what was initially administered to another animal of the same sex as the original animal.

On the other hand, chronic toxicity testing gives insight into the long-term toxic effects of investigational products on animals, allowing the results to be extrapolated to assess human safety (Saganuwan et al., 2017). Here also, one rodent and nonrodent animal species are taken. The animals were administered the test doses for more than 90 days, and observed for a specific period. During the study period, the animals were noticed for their normal physiological functions, biochemical parameters and behavioral changes. When the study concluded, histological analysis was conducted on tissues collected from different animal parts.

Similarly, a test compounds mutagenicity, carcinogenicity, cardiotoxicity, neurotoxicity, reproductive toxicity, developmental toxicity, teratogenic and genetic toxicity can be studied using in vivo methods (Dusinska et al., 2017).

3. Ecotoxicological methods

Ecotoxicological assessment involves the evaluation of potential toxicity caused by the micropollutants on the ecosystem by combining exposure and effect data. It involves exposure assessment, dose-response assessment, hazard identification and risk characterization (Tamis et al., 2021). Further, using the approaches risk quotients (RQs), species sensitivity distributions (SSD), and community and population level models, the ecological risks caused by the micropollutants are analyzed (Raimondo & Forbes 2022; Pereira et al., 2018; Mentzel et al., 2021; Clements & Rohr 2009; Albarano et al., 2020; Kramer et al., 2011). The risk quotient approach correlates with exposure to environmental concentrations of micropollutants and reported ecological effects on sentinel species. Even though many countries have been using this approach for some regulatory purposes, this deterministic approach has limitations, such as its conservative assumptions and overestimated risks (Raimondo & Forbes 2022).

Meanwhile, in SSD, the variations in sensitivity of a species to environmental stress are monitored. SSD is created by fitting a cumulative distribution function to species toxicity data, ranking them by their centiles. From this cumulative distribution, the hazardous concentration (HC) is calculated. The percentage of species expected to be affected is estimated from these HC values. An HC5 value in SSD indicates the probability of 5% of species being affected by the toxicity of micropollutants under consideration (Pereira et al., 2018). Later, the SSD was refined to form the probabilistic risk assessment technique, which integrates the effect data with the distributions of environmental exposure to toxicants. The probabilistic risk assessment gives a better picture of the probability of exposure to micropollutants and their associated effects (Mentzel et al., 2021). However, the community-level model of ecotoxicological assessment offers important insight into the mechanism of micropollutant effects at lower levels to higher levels of biological organization. In the hierarchy of biological organization, communities represent the midpoints between the population and the ecosystem (Clements & Rohr 2009). The experimental system M-cosms (micro to mesocosms) helps to evaluate how organisms or communities respond to environmental changes and exposure to micropollutants. M-cosms are a bridge connecting the real environment and laboratory, instrumental in evaluating the biological effects of micropollutants (Albarano et al., 2020). Like the community-level model, the population-level ecological risk assessment model integrates information from lower levels of biological organization and transfers the data to higher-level landscape and ecosystem models if needed (Kramer et al., 2011). The data from these approaches will help the decision-makers to formulate monitoring programs, environmental regulations, and remediation strategies.

Apart from these approaches, selecting suitable bioindicators based on their role in the ecosystem and their importance in the ecosystem functioning and services will help to identify the threat caused by the micropollutants in specific ecological compartments. The bioindicators are a species or group of species that indicate the changes in their environment due to the presence of micropollutants that affect the biodiversity of the environment. The responses from the bioindicators can be grouped into the following categories: (1) ecological change—change in population diversity and species diversity; (2) physiological change—bioaccumulation and microbial activity; (3) behavioral changes—changes in feeding and movement. The best examples of bioindicators for the ecological risk assessment include plankton, macrophytes, benthic invertebrates, mollusks and fish (Kumari & Paul 2020; Bartell, 2006).

4. **Omics based approaches**

Omics is one of the toxicological approaches used to monitor an organism's responses to an external substance through its interaction study with biomolecules like DNA, RNA, proteins, and lipids (Li et al., 2024a, b). The basic principle of this approach is that the toxic impact of a particular toxicant is reflected first at the cellular level through its effect on gene and protein expression, as well as the production of minor metabolites (Ouedraogo et al., 2012; Tripathy et al., 2024a). This approach is beneficial in elucidating the molecular mechanism resulting in toxicity. Genomics, transcriptomics, epigenomics, proteomics, metabolomics and lipidomics are the different omics approaches used for toxicological studies (Li et al., 2024a, b).

(i) **Genomics**

The essential layout of all living organisms lies in the genome sets, and minute changes in the genome will result in drastic changes in all levels of biological organization, from cells to population (Martins et al., 2019). Genomics studies focus on the expression, regulation, and function of genes. The changes observed in the expression of genes in response to toxicants were drastic and rapid and are the cause for several toxic effects. Thus, it provides a thorough understanding of molecular mechanism and mode of action of toxicants, which are challenging to obtain by conventional toxicological methods (Ma et al., 2021). Toxicological studies using genomics are often associated with some challenges. They are the complexity of gene-environment interactions and environmental factors such as diet, lifestyle, and exposure to other toxicants make it difficult to analyze the toxicological impact of a particular toxicant and to understand the specific contribution of genetic factors. Further ethical considerations are required to carry out human genomics studies in the context of toxicology (Patil et al., 2023).

(ii) **Transcriptomics**

Transcriptomics studies mRNA expression from all genes to understand toxicants intoxication and detoxification process (Nam et al., 2023). Specifically, transcriptomics helps to identify genes that are differently expressed in response to a toxicant (Patil et al., 2023). It also helps to examine short noncoding RNAs, which are considered biomarkers due to their dysregulation in diseases. This field has advanced much with the advent of reverse transcription quantitative real-time polymerase chain reaction (RT-qPCR), DNA microarrays and next-generation sequencing (NGS) based RNA sequencing (RNA seq) (Li et al., 2024a, b). The complexity of gene regulation is the challenge associated with integrating transcriptomics in toxicology studies. The gene regulation is controlled by several factors such as epigenetic modifications, transcription factors and noncoding RNAs. Also, there may be differences in gene expression among individuals due to genetic and environmental factors (Patil et al., 2023).

(iii) **Epigenomics (emerging toxicological indicator)**

Epigenomics deals with the study of reversible modifications in DNA or DNA-associated proteins, known as epigenetic modification. The epigenetic modifications include DNA methylation and histone acetylation, which regulate gene transcription (Li et al., 2024a, b). These modifications occur as a result of external stimuli like toxic chemical exposure. This DNA methylation is responsible for downregulating gene transcription, thereby preventing transcription machinery from binding. In addition, the progression and malignancy in tumors arise because of the hypomethylation of DNA. Likewise, histone proteins are responsible for the structural organization of DNA in eukaryotic cells. To proceed with transcription, the histones must detach from DNA, which is catalyzed by the enzyme histone deacetylases. The increased activity of the enzyme histone deacetylase is also involved in cancer progression. These epigenetic changes cause mutations and transformations in cells, serving as indicators of genotoxicity (Fröhlich, 2017). DNA methylation was assessed using the modified genome sequencing approach Methylome-seq, polymerase chain reaction, pyrosequencing, HPLC and ELISA. The histone modification was investigated using Western blotting and the Chromatin-immuno precipitation method followed by ChIP sequencing (Canzler et al., 2020). The complex interplay between genetic and environmental factors controls epigenetic modifications (Li et al., 2024a, b). However, intensive infrastructure and bioinformatics skills are necessary to analyze the data in epigenomics approach (Nguyen et al., 2022).

(iv) **Proteomics**

The final products of gene expression are the proteins. The study of proteins helps to understand the various physiological processes that are taking place in an organism. In addition, proteomics helps to point out the expression and modification of all the proteins in cells, tissues and organs. Also, the biological signaling pathways can be qualitatively and quantitatively analyzed in response to a toxicant using proteomics. It will thus help to evaluate the toxic effects of micropollutants (Ma et al., 2021). Different methods are employed to identify and quantify proteins. The traditional methods of Western blotting and ELISA detects protein expression using particular antibodies that binds with target proteins. But, some drawbacks of these methods are that previous knowledge of protein characteristics and the availability of specific antibodies are necessary. Also, only a limited number of proteins can be detected in a single experiment (Nguyen et al., 2022). Other widely used analytical techniques for protein analysis are two-dimensional gel electrophoresis (separates proteins by mass and pI) and mass spectrometry (MS). In addition, gel free techniques helped to quantify the proteins using MS (Ma et al., 2021; Min et al., 2021). The MS techniques for protein analysis are of two types global and targeted proteomics. In Global proteomics, all proteins in a sample are identified and quantified and the differences in proteome are compared between samples without prior assumptions. By contrast, there is a limit in the number of proteins to be detected in targeted proteomics, which emphasizes the need to optimize methods for the highest sensitivity (Nguyen et al., 2022). Further, based on MS analysis procedures, the analysis of protein can be done using peptide based (shotgun proteomics) and protein-based approaches. The peptide-based approach analyses the MS of peptides to identify proteins after subjecting the proteins to enzymatic digestion. It is a bottom-up strategy for protein analysis. At the same time, the protein-based approach is a recently developed top-down strategy that analyzes all proteins. The worse ionization, fractionation and fragmentation of proteins in the gaseous phase make the top-down strategy a less universal than the bottom-up approach. Therefore, depending on the research purpose, suitable MS methods and analysis procedures must be selected (Fröhlich, 2017; Nguyen et al., 2022). The advantages of proteomics analysis are that it can be done quickly and comprehensively. The proteomics study can gather other information such as biomarkers of poisoning and detoxification, biological and molecular mechanisms of the toxic action of micropollutants (Ma et al., 2021). To date, it is difficult to obtain all the proteins present within a cell using proteomics. Furthermore, the requirement of large sized samples, difficulty in detecting low abundance proteins, not user friendly and not transparent nature of pipeline analysis of proteomics, sample contamination, the variation in protein expression in response to circardian cycle, age, sex, diseases, the lower sensitivity of MS compared to other detection techniques, only proteins that are water soluble and fall within a limited range of mass and isoelectric point can be analyzed are some of the draw backs associated with this technique (Fröhlich, 2017; Nguyen et al., 2022).

(v) **Metabolomics**

Metabolites in a biological system (organs, blood and urine) are metabolomes. Analyzing all these chemically heterogeneous small biological molecules is collectively called metabolomics (Karkossa et al., 2020). Separation, enrichment and quantification of metabolites are the steps involved in metabolomics analysis. The quantification of the metabolites is done with the help of nuclear magnetic resonance (NMR) and MS approaches. MS approach is usually coupled with GC or LC to separate metabolites before identification and quantification. Moreover, the MS approach is more advantageous than NMR due to its high sensitivity and ability to quantify large number of metabolites. Also, the MS techniques are costly, requires complex sample preparation and provides average reproducibility of results (Li et al., 2024a, b).

In contrast, NMR can be used to detect different types of metabolites simultaneously and complex sample preparation is not required. NMR can also be used to analyze nonionizable metabolites and metabolites that have to be derivatized for MS analysis. Further, it is used to analyze metabolites in biological samples such as saliva, urine, and solid tissue samples. The samples are not destroyed in NMR and can be reused for further analysis. Thus, NMR and MS techniques have advantages and disadvantages and are suitable for different purposes (Nguyen et al., 2022). The metabolomics study can also be done using targeted and untargeted methods and both methods can be used for toxicity related investigations. In the targeted method using stable isotope labeled reference standard, the samples were quantified either relative to each other or absolutely and were detected using MS/MS. In comparison the untargeted method follows a label free technique to quantify as many endogenous small metabolites as possible. Not even a single approach is available for detecting all the metabolites in a given biological sample. This is due to the high variability in metabolite properties and the complex nature of the metabolome (Karkossa et al., 2020). Due to the wide variety of metabolites, the metabolomic analysis faces several challenges, such as difficulty in covering a wide range of concentrations during screening and the metabolites may have different chemical structures; therefore, for its identification, different techniques may be adopted (Li et al., 2024a, b).

(vi) **Lipidomics**

It is a branch of metabolomics in which mediator lipidomics and membrane lipidomics were considered. Mediator lipidomics focuses on quantifying low abundant bioactive lipid species. Meanwhile the lipid constituents in the membrane and its structural aspects were considered in membrane lipidomics. Ultra-high performance liquid chromatography mass spectrometry (UHPLC-MS) is used for lipid profiling. It helps in identifying disease related alterations and biomarkers that result from exposure to toxicants (Li et al., 2024a, 2024b).

Biomarkers for Toxicity Assessment

A biomarker is a change in biological response in living beings in response to exposure to toxicants (Lomartire et al., 2021). The biomarkers are measured as parent compounds or metabolites from the tissues or body fluids. A biomarker should be selected based on the following criteria: it should be persistent and its concentration in tissues or body fluids must allow for their detection with adequate specificity and precision. Biomarkers can be grouped into three categories: biomarkers of exposure, biomarkers of response/ biomarkers of effect and biomarkers of susceptibility. Biomarkers of exposure indicates that exposure to a particular toxicant has taken place. It also provides information such as route, pathway and source of exposure. DNA damage is an example of a biomarker of exposure. It is further subdivided into markers of internal dose and effective dose. Markers of internal dose reflects the occurrence and extent of exposure that an organism has experienced. The concentration of a chemical in blood measured following its exposure is an example of biomarkers of internal dose. While markers of effective dose accurately indicate the extent of toxicant exposure to target molecules, tissue or cells (Das & Ghosh, 2023; Satapathy et al., 2024). In contrast, biomarkers of effect show the biological response on exposure to the toxicant. It includes biochemical, behavioral and physiological changes in the tissues of the organism that occur as a result of exposure (Behera et al., 2020, 2021, 2022). It helps to assess toxicological impact, monitoring disease progression and prognosis and the risk of adverse health effects in an organism. Genotoxicity biomarkers are examples of biomarkers of effect. Biomarkers of early disease are a subclass of biomarkers of effect that indicates an early reversible clinical response. Biomarkers of susceptibility indicates individual's intrinsic sensitivity to the toxicants (Owolabi et al., 2024; Lionetto et al., 2019; Ladeira and Viegas 2016). The enzymes that help to transform and metabolize toxicants are examples of susceptibility biomarkers. Thus, the study of biomarkers helps to understand the effects of toxicants on living beings and the environment (Özgenç et al., 2023).

Ecological or Environmental Risk Assessment (ERA) Framework

ERA framework aims to evaluate the damage to the ecosystem or environment caused by exposure to stressors (risk factors). ERA framework is a prerequisite in the decision-making process. It is integrated into environmental policies globally to manage the authorization and marketing of new chemicals (prospective ERA) and to assess potentially contaminated sites (retrospective ERA) (Pereira et al., 2018). The risk assessment process can be divided into scientific risk analysis and political risk management. The key elements involved in risk analysis comprise exposure assessment, effect assessment, hazard identification and risk characterization. Risk management is concerned with regulatory measures according to information gathered from risk analysis. Both risk analysis and risk management are interconnected processes. Risk analysis determines the risk of a particular situation, while risk management explores solutions to that problem (Van der Oost et al., 2003). ERA is primarily concerned with dealing with chemicals. There are eight steps involved in the overall risk assessment process.

1. Hazard identification—It involves recognizing the harmful effects that may result from exposure to chemicals (Van der Oost et al., 2003).
2. Effect assessment—Evaluates the relationship between chemical exposure levels and adverse effects. Here, the experiments are carried out to determine a no-effect level (NEL), which is then converted to a predicted no-effect level (PNEL) or predicted no-effect concentration (PNEC) for other species (Van der Oost et al., 2003). In the case of the risk analysis performed for the toxicants present in the

aquatic environments PNEC is calculated by dividing the value of acute (short-term) or chronic (long-term) toxicity by an assessment factor (AF). The median lethal concentration (LC50) or mean effective concentration (EC50) values were used to assess acute toxicity, where the AF value is 1000. Whereas chronic toxicity is assessed from the no observable effect concentration (NOEC) values, here AF values can be 100, 50 and 10 depending on the trophic levels (Vasilachi et al., 2021).

3. Exposure assessment—Estimates the concentration or dose to which a human population or different environmental compartments are exposed or may be exposed. It involves emission assessment and behavior and the fate of toxicant in the environment. The existing chemical exposure can be assessed by measuring its concentration in the environment, but for newer chemicals, a predicted environmental concentration (PEC) needs to be estimated (Van der Oost et al., 2003).
4. Risk characterization—It involves integrating the initial three steps of the risk assessment process, which assesses the severity and adverse effects likely to occur due to exposure to the actual or predicted exposure to chemicals. Risk quotient (RQ) is often measured for newly developed chemicals and is the ratio of PEC/PNEC. The RQ measurements help to link the risk analysis to risk management by quantifying the hazards and risks for specific situations. If RQ $\geq$ 1, it is necessary to propose a risk mitigation measure for reducing the risk to an acceptable level. The environmental risk persists if RQ remains $\geq$ 1 (Cizmas et al., 2015; Van der Oost et al., 2003).
5. Risk classification—Evaluate risks to decide whether a risk reduction is required. The classification is done at two levels: the upper limit represents the maximum permissible level, while the lower limit denotes the negligible level (NL) (Van der Oost et al., 2003).
6. Risk benefit analysis involves creating a balance sheet that weighs the risks against the benefits of a proposed risk reduction action (Van der Oost et al., 2003).
7. Risk reduction—It involves measures to protect people and the environment from identified risks. It can be implemented by establishing safety standards, such as acceptable daily intake (ADI) (Van der Oost et al., 2003).
8. Monitoring—An essential stage in the ERA process, carried out according to a prearranged schedule over time and space using comparable and standardized methods. Usually, chemical or biological elements are monitored during risk analysis, which will help in exposure characterization. Risk management monitoring will support verifying control strategy results and checking of control functions (Van der Oost et al., 2003).

Thus, the ERA process will help to manage micropollutants in a risk-based approach and help to choose the right remediation option regarding risk reduction, cost effectiveness and efficiency (Cizmas et al., 2015). Also, a single method is not enough for toxicity assessment; a combined method is needed. Moreover, it is the need of the hour to develop more innovative and integrated strategies for the risk assessment associated with micropollutants. The new methodologies will ultimately help in the efforts to support environmental protection and public health (Mofijur et al., 2024).

Measures for the Mitigation of Micropollutants, Policy Recommendations, Research Priorities and Future Challenges

1. **Preventing the entry of micropollutants right from their source**

This approach aims to reduce the entry of micropollutants into the environment right from their source. This aim can be achieved by the proper disposal of micropollutants, the use of less harmful alternative chemicals in industrial processes, carrying out best management practices to prevent the runoff of pesticides and nutrients to different environmental matrices (Bala et al., 2024; Das et al. 2024a; Mohanty et al., 2024a; Mohanty et al., 2024c; Ray et al., 2024). Also, practicing circular economy concepts for waste reduction, resource efficiency and product life extension will help to minimize the entry and the impact of micropollutants on the environment and human health (Kosek et al., 2020).

2. **Monitoring and surveillance programs to improve our understanding of micropollutants in the environment**

Here, the occurrence, fate and effects of micropollutants are monitored for risk assessment, remediation efforts, and for formulating policy developments. Standardized sampling and analytical methods are necessary for the continuous monitoring and surveillance programs of micropollutants. Also, it is a prerequisite to expand monitoring efforts and incorporate new techniques like passive sampling and biomonitoring to improve our understanding of the types of micropollutants present in the environment (Kosek et al., 2020).

Regulations and Guidelines for Emerging Micropollutants to Manage Their Potential Risks to Human Health and Environment

It is becoming a big task for environmental managers and the scientific community to prioritize chemical contaminants for pollution prevention and control. While, for most of the emerging micropollutants, right from its observation, proper risk assessment and risk ranking are lacking due to knowledge gaps. Therefore, there is a need to develop an efficient, rational and holistic prioritization approach to analyze the knowledge gaps relating to emerging micropollutants systematically.

Furthermore, efficient analytical methods still need to be developed to investigate individual micropollutants of emerging concern and their transformation products. In addition, it is important to know about their occurrence and how they get released or formed in the environment. Also, it is essential to know the potential risks they pose to the environment and human health (Khan et al., 2022; Li et al., 2024a).

3. **Public awareness and education**

Public awareness and education will inform the public of the source, risk and mitigation measures for micropollutants. It will thus help to develop sustainable consumption patterns, better waste management practices and the use of environment friendly products. Initiatives like public information campaigns, community engagement programs and school curricula will help the public to develop environmentally responsible behaviors (Kosek et al., 2020).

4. **Research priorities and future challenges**

Future research aims to identify new micropollutants and their transformation products using advanced analytical techniques. Further, the study will evaluate the long term ecological and human health effects caused by these micropollutants. Apart from all these cost effective, innovative and sustainable remediation technologies are required for the proper disposal of the micropollutants. In addition, already existing and emerging remediation measures must be assessed. Interdisciplinary collaboration among policy makers, scientists, industry and other stake holders is required to take necessary measures for mitigating the risks associated with micropollutants (Li et al., 2024a).

Conclusion

There is an alarming rise in micropollutants in different environmental compartments. Due to their low concentrations, they always remain unnoticed. There is a possibility that the transformations of these micropollutants are there. Due to the lack of proper sophisticated analytical instrument facilities, the identification and quantification of these micropollutants and their transformation products have now become a big task. Most of them are found to be persistent and can remain in different environmental matrices for years. In addition, exposure to these micropollutants and transformation products creates potential risks to the health of living beings and the environment. Thus, a toxicity assessment of micropollutants is necessary for the development of a risk-based assessment and screening approach to prioritize the micropollutants present in different environmental matrices. Also, there are drawbacks associated with the conventional toxicity assessment methods. The toxicity assessment test we are carrying out in laboratories often fail to replicate real world conditions, thus resulting in a gap between experimental findings and actual ecological and human health risks. Therefore, a modification in the toxicity assessment is a prerequisite to incorporate real world complexities like environmental variability, species interaction and multiple stressors.

References

Aemig, Q., Hélias, A., & Patureau, D. (2021). Impact assessment of a large panel of organic and inorganic micropollutants released by wastewater treatment plants at the scale of France. *Water Research, 188*, 116524.

Ahmed, T., & Hashmi, M. Z. (Eds.). (2022). *Hazardous Environmental Micro-Pollutants, Health Impacts and Allied Treatment Technologies*. Springer Nature.

Albarano, L., Costantini, M., Zupo, V., Lofrano, G., Guida, M., & Libralato, G. (2020). Marine sediment toxicity: A focus on micro-and mesocosms towards remediation. *Science of the Total Environment, 708*, 134837.

Ambaye, A. D., Fito, J., Nxumalo, E., & Nkambule, T. T. I. (2023). The challenges of micropollutants and their environmental detection through biochar-based nanostructured electrochemical sensors: A review. *International Journal of Environmental Science and Technology, 20*(10), 11741–11762.

Arman, N. Z., Salmiati, S., Aris, A., Salim, M. R., Nazifa, T. H., Muhamad, M. S., & Marpongahtun, M. (2021). A review on emerging pollutants in the water environment: Existences, health effects and treatment processes. *Water, 13*(22), 3258.

Arslan, M., Ullah, I., Müller, J. A., Shahid, N., & Afzal, M. (2017). Organic micropollutants in the environment: ecotoxicity potential and methods for remediation. In *Enhancing Cleanup of Environmental Pollutants: Volume 1: Biological Approaches* (pp. 65–99).

Ashiwaju, B. I., Uzougbo, C. G., & Orikpete, O. F. (2024). Environmental impact of pharmaceuticals: A comprehensive review. *Matrix Science Pharma, 7*(3), 85–94.

Awashra, M., & Młynarz, P. (2023). The toxicity of nanoparticles and their interaction with cells: An in vitro metabolomic perspective. *Nanoscale Advances, 5*(10), 2674–2723.

Bala, K., Anshul, Verma, S., Dogra, A., Das, A. P., Kumar, R., & Nandini. (2024). Sustainable Management for Urban Plastic Waste Generation. *Sustainable Management of Urban Plastic Waste Through Circular Economic Approaches*, 211–236. https://doi.org/10.1201/9781003465447-13

Bartell, S. M. (2006). Biomarkers, bioindicators, and ecological risk assessment—a brief review and evaluation. *Environmental Bioindicators, 1*(1), 60–73.

Behera, I. D., Basak, G., Kumar, R. R., Sen, R., & Meikap, B. C. (2020). Treatment of petroleum refinery sludge by petroleum degrading bacterium *Stenotrophomonas pavanii* IRB19 as an efficient novel technology. *Journal of Environmental Science and Health, Part A, 56*(2), 226–239. https://doi.org/10.1080/10934529.2020.1866924

Behera I. D., Nayak M., Biswas S., Meikap B. C, Sen R. (2021) Enhanced biodegradation of total petroleum hydrocarbons by implementing a novel two-step bioaugmentation strategy using indigenous bacterial consortium, *Journal of Environmental Management, 292*, 112746, ISSN 0301-4797. https://doi.org/10.1016/j.jenvman.2021.112746

Behera I. D., Nayak M., Mishra A., Meikap B. C., Sen R. (2022). Strategic implementation of integrated bioaugmentation and biostimulation for efficient mitigation of petroleum hydrocarbon pollutants from terrestrial and aquatic environment. *Marine Pollution Bulletin, 177*, 113492, ISSN 0025-326X. https://doi.org/10.1016/j.marpolbul.2022.113492

Benítez-Chao, D. F., León-Buitimea, A., Lerma-Escalera, J. A., & Morones-Ramírez, J. R. (2021). Bacteriocins: An overview of antimicrobial, toxicity, and biosafety assessment by in vivo models. *Frontiers in Microbiology, 12*, 630695.

Bhatt, P., Bhandari, G., & Bilal, M. (2022). Occurrence, toxicity impacts and mitigation of emerging micropollutants in the aquatic environments: Recent tendencies and perspectives. *Journal of Environmental Chemical Engineering, 10*(3), 107598.

Brusseau, M. L., & Artiola, J. F. (2019). Chemical contaminants. In *Environmental and Pollution Science* (pp. 175–190). Academic Press.

Butler, M., Spearman, M., & Braasch, K. (2014). Monitoring cell growth, viability, and apoptosis. In *Animal Cell Biotechnology: Methods and Protocols* (pp. 169–192).

Camera, E., & Picardo, M. (2002). Analytical methods to investigate glutathione and related compounds in biological and pathological processes. *Journal of Chromatography B, 781*(1–2), 181–206.

Canzler, S., Schor, J., Busch, W., Schubert, K., Rolle-Kampczyk, U. E., Seitz, H., Kamp, H., Von Bergen, M., Buesen, R. & Hackermüller, J., (2020). Prospects and challenges of multi-omics data integration in toxicology. *Archives of Toxicology, 94*, 371–388

Chan, F. K. M., Moriwaki, K., & De Rosa, M. J. (2013). Detection of necrosis by release of lactate dehydrogenase activity. *Immune Homeostasis: Methods and Protocols*, 65–70.

Chao, M. R., Evans, M. D., Hu, C. W., Ji, Y., Møller, P., Rossner, P., & Cooke, M. S. (2021). Biomarkers of nucleic acid oxidation—a summary state-of-the-art. *Redox Biology, 42*, 101872.

Chopra, S., & Kumar, D. (2018). Pharmaceuticals and personal care products (PPCPs) as emerging environmental pollutants: toxicity and risk assessment. In *Advances in Animal Biotechnology and its Applications* (pp. 337–353).

Cizmas, L., Sharma, V. K., Gray, C. M., & McDonald, T. J. (2015). Pharmaceuticals and personal care products in waters: Occurrence, toxicity, and risk. *Environmental Chemistry Letters, 13*, 381–394.

Clements, W. H., & Rohr, J. R. (2009). Community responses to contaminants: Using basic ecological principles to predict ecotoxicological effects. *Environmental Toxicology and Chemistry: An International Journal, 28*(9), 1789–1800.

Collins, A. R. (2004). The comet assay for DNA damage and repair: Principles, applications, and limitations. *Molecular Biotechnology, 26*(3), 249–261.

Collins, A., Møller, P., Gajski, G., Vodenková, S., Abdulwahed, A., Anderson, D., Bankoglu, E. E., Bonassi, S., Boutet-Robinet, E., Brunborg, G., & Chao, C. (2023). Measuring DNA modifications with the comet assay: a compendium of protocols. *Nature Protocols, 18*(3), 929–989.

Darzynkiewicz, Z., Li, X., & Gong, J. (1994). Assays of cell viability: Discrimination of cells dying by apoptosis. *Methods in Cell Biology, 41*, 15–38.

Das, P., & Bal, M. (2025). Modeling and optimization study for microplastic removal from aquatic medium using Chroococcidiopsis species. *Chemical Engineering Science, 304*, 121085. https://doi.org/10.1016/j.ces.2024.121085

Das, A. P., Mallavarapu, M., & Ghosh, S. (2023a). Ecotoxicity of mining pollutants on the environment and their remediation. *Environmental Chemistry and Ecotoxicology, 5*, 165–167. https://doi.org/10.1016/J.ENCECO.2023.08.002

Das, P., Halder, G., & Bal, M. (2023b). A critical review on remediation of microplastics using microalgae from aqueous system. *Science of the Total Environment, 898*, 166425. https://doi.org/10.1016/j.scitotenv.2023.166425

Das, A., Ray, S. K., Mohanty, M., Mohanty, J., Dey, S., & Das, A. P. (2024c). Ecotoxicity of microplastic wastes and their sustainable management: A review. *Environmental Chemistry and Ecotoxicology, 6*, 144–152. https://doi.org/10.1016/J.ENCECO.2024.05.003

Das, A. P., & Ghosh, S. (2023). Microorganisms in manganese biomining processes. *Manganese Mining Microorganisms*, 55–61. https://doi.org/10.1016/B978-0-12-822147-1.00013-3

Das, A. P., Behera, I. D., & Das, N. P. (2024a). *Renewable Energy Generation and Value Addition From Environmental Microfiber Pollution Through Advanced Greener Solution* (p. 278)

Das, A., Dey, S., & Das, A. P. (2024b). Ecotoxicological impacts of synthetic microfiber pollutants and development of sustainable mitigation strategies. *Environmental Chemistry and Ecotoxicology*. https://doi.org/10.1016/J.ENCECO.2024.12.003

Demirci-Cekic, S., Özkan, G., Avan, A. N., Uzunboy, S., Çapanoğlu, E., & Apak, R. (2022). Biomarkers of oxidative stress and antioxidant defense. *Journal of Pharmaceutical and Biomedical Analysis, 209*, 114477.

Dey, S., Das, A., Mallick, K., Sahu, A., & Das, A. P. (2023a). *Environmental Petroleum Waste: Pollution, Toxicity, Sustainable Remediation* (pp. 159–175). https://doi.org/10.1007/978-3-031-48220-5_7

Dey, S., Tripathy, B., Kumar, M. S., & Das, A. P. (2023b). Ecotoxicological consequences of manganese mining pollutants and their biological remediation. *Environmental Chemistry and Ecotoxicology*. https://doi.org/10.1016/J.ENCECO.2023.01.001

Diener, W., Siccha, L., Mischke, U., Kayser, D., & Schiede, E. (1994). The biometry evaluation of the acute-toxic-class method (oral). *Archives of Toxicology, 68*, 599–610.

Dsikowitzky, L., & Schwarzbauer, J. (2014). Industrial organic contaminants: Identification, toxicity and fate in the environment. *Environmental Chemistry Letters, 12*, 371–386.

Dusinska, M., Rundén-Pran, E., Schnekenburger, J., & Kanno, J. (2017). Toxicity tests: In vitro and in vivo. In *Adverse Effects of Engineered Nanomaterials* (pp. 51–82). Academic Press.

Encarnação, T., Pais, A. A., Campos, M. G., & Burrows, H. D. (2019). Endocrine disrupting chemicals: Impact on human health, wildlife and the environment. *Science Progress, 102*(1), 3–42.

Escher, B. I., & Fenner, K. (2011). Recent advances in environmental risk assessment of transformation products. *Environmental Science & Technology, 45*(9), 3835–3847.

Fröhlich, E. (2017). Role of omics techniques in the toxicity testing of nanoparticles. *Journal of Nanobiotechnology, 15*, 1–22.

Gautam, K., & Anbumani, S. (2020). Ecotoxicological effects of organic micro-pollutants on the environment. In *Current Developments in Biotechnology and Bioengineering* (pp. 481–501). Elsevier.

Ghosh, S., Sinha, J. K., Ghosh, S., Vashisth, K., Han, S., & Bhaskar, R. (2023). Microplastics as an emerging threat to the global environment and human health. *Sustainability, 15*(14), 10821. https://doi.org/10.3390/SU151410821

Glaser, C., Schwientek, M., & Zarfl, C. (2019). Designing field-based investigations of organic micropollutant fate in rivers. *Environmental Science and Pollution Research, 26*, 28633–28649.

Gul, B., Naseem, M. K., Malik, W. U. N., Gurmani, A. R., Mehmood, A., & Rafique, M. (2022). Environmental micropollutants and their impact on human health with special focus on agriculture. *Hazardous Environmental Micro-pollutants, Health Impacts and Allied Treatment Technologies* (pp. 1–19). Springer International Publishing.

Guo, Z., Kodikara, D., Albi, L. S., Hatano, Y., Chen, G., Yoshimura, C., & Wang, J. (2023). Photodegradation of organic micropollutants in aquatic environment: Importance, factors and processes. *Water Research, 231*, 118236.

Jagatee, S., Priyadarshini, S., Rath, C. C., & Das, A. P. (2024). Synthetic microfiber: An enduring environmental problem linked to sustainable development. *Environmental Science and Engineering, Part, F2333*, 93–112. https://doi.org/10.1007/978-3-031-51792-1_6

Jiang, N., Naz, S., Ma, Y., Ullah, Q., Khan, M. Z., Wang, J., Lu, X., Luosang, D. Z., Tabassum, S., Chatha, A. M. M., & Basang, W. D. (2023). An overview of comet assay application for detecting DNA damage in aquatic animals. *Agriculture, 13*(3), 623.

Kanaujiya, D. K., Paul, T., Sinharoy, A., & Pakshirajan, K. (2019). Biological treatment processes for the removal of organic micropollutants from wastewater: A review. *Current Pollution Reports, 5*, 112–128.

Kanungo, J., Sahoo, T., Bal, M., & Behera, I. D. (2023a). Performance of bioremediation strategy in waste lubricating oil pollutants: A review. *Geomicrobiology Journal, 41*(4), 360–373. https://doi.org/10.1080/01490451.2023.2245395

Kanungo, J., Sahoo, T., Swain, L.P., Behera, I.D. (2023). Toxicity of persistent hydrocarbon pollutants, sources and sustainable remediation process. In I. D Behera & A. P. Das (Eds.) *Impact of Petroleum Waste on Environmental Pollution and its Sustainable Management Through Circular*

Economy. Environmental Science and Engineering. Springer, Cham. https://doi.org/10.1007/978-3-031-48220-5_2

Kari, S., Subramanian, K., Altomonte, I. A., Murugesan, A., Yli-Harja, O., & Kandhavelu, M. (2022). Programmed cell death detection methods: A systematic review and a categorical comparison. *Apoptosis, 27*(7), 482–508.

Karkossa, I., Raps, S., von Bergen, M., & Schubert, K. (2020). Systematic review of multi-omics approaches to investigate toxicological effects in macrophages. *International Journal of Molecular Sciences, 21*(24), 9371.

Kashyap, A., Nishil, B., & Thatikonda, S. (2023). Experimental and numerical elucidation of the fate and transport of antibiotics in aquatic environment: A review. *Environmental Monitoring and Assessment, 195*(8), 942.

Kehm, R., Baldensperger, T., Raupbach, J., & Höhn, A. (2021). Protein oxidation-formation mechanisms, detection and relevance as biomarkers in human diseases. *Redox Biology, 42*, 101901.

Khalid, M., & Abdollahi, M. (2021). Environmental distribution of personal care products and their effects on human health. *Iranian Journal of Pharmaceutical Research: IJPR, 20*(1), 216.

Khan, S., Naushad, M., Govarthanan, M., Iqbal, J., & Alfadul, S. M. (2022). Emerging contaminants of high concern for the environment: Current trends and future research. *Environmental Research, 207*, 112609.

Kim, M. K., & Zoh, K. D. (2016). Occurrence and removals of micropollutants in water environment. *Environmental Engineering Research, 21*(4), 319–332.

Kim, K. H., Kabir, E., & Jahan, S. A. (2017). Exposure to pesticides and the associated human health effects. *Science of the Total Environment, 575*, 525–535.

Kosek, K., Luczkiewicz, A., Fudala-Książek, S., Jankowska, K., Szopińska, M., Svahn, O., Tränckner, J., Kaiser, A., Langas, V., & Björklund, E. (2020). Implementation of advanced micropollutants removal technologies in wastewater treatment plants (WWTPs)-examples and challenges based on selected EU countries. *Environmental science & policy, 112*, 213–226.

Kramer, V. J., Etterson, M. A., Hecker, M., Murphy, C. A., Roesijadi, G., Spade, D. J., Spromberg, J. A., Wang, M., & Ankley, G. T. (2011). Adverse outcome pathways and ecological risk assessment: Bridging to population-level effects. *Environmental Toxicology and Chemistry, 30*(1), 64–76.

Kumar, V., Sharma, N., & Maitra, S. S. (2017). In vitro and in vivo toxicity assessment of nanoparticles. *International Nano Letters, 7*(4), 243–256.

Kumar, A., Deepika, Tyagi, D., Tarkeshwar, & Kapinder. (2024). Organic micropollutants and their effects on the environment and human health. In *Organic Micropollutants in Aquatic and Terrestrial Environments* (pp. 87–105). Cham: Springer Nature Switzerland.

Kumari, D., & Paul, D. K. (2020). Assessing the role of bioindicators in freshwater ecosystem.

La Farre, M., Pérez, S., Kantiani, L., & Barceló, D. (2008). Fate and toxicity of emerging pollutants, their metabolites and transformation products in the aquatic environment. *TrAC Trends in Analytical Chemistry, 27*(11), 991–1007.

Ladeira, C., & Viegas, S. (2016). Human biomonitoring: An overview on biomarkers and their application in occupational and environmental health. *Biomonitoring, 3*(1), 15–24.

Lanzerstorfer, P., Sandner, G., Pitsch, J., Mascher, B., Aumiller, T., & Weghuber, J. (2021). Acute, reproductive, and developmental toxicity of essential oils assessed with alternative in vitro and in vivo systems. *Archives of Toxicology, 95*, 673–691.

Li, D., & Suh, S. (2019). Health risks of chemicals in consumer products: A review. *Environment International, 123*, 580–587.

Li, X., Shen, X., Jiang, W., Xi, Y., & Li, S. (2024a). Comprehensive review of emerging contaminants: Detection technologies, environmental impact, and management strategies. *Ecotoxicology and Environmental Safety, 278*, 116420.

Li, Y., Vulpe, C., Lammers, T., & Pallares, R. M. (2024). Assessing inorganic nanoparticle toxicity through omics approaches. *Nanoscale.*

Lionetto, M. G., Caricato, R., & Giordano, M. E. (2019). Pollution biomarkers in environmental and human biomonitoring. *The Open Biomarkers Journal, 9*(1).
Lomartire, S., Marques, J. C., & Gonçalves, A. M. (2021). Biomarkers based tools to assess environmental and chemical stressors in aquatic systems. *Ecological Indicators, 122*, 107207.
Ma, Y., Liu, H., Du, X., Shi, Z., Liu, X., Wang, R., Zhang, S., Tian, Z., Shi, L., Guo, H., & Zhang, H. (2021). Advances in the toxicology research of microcystins based on omics approaches. *Environmental International, 154.*
Manisalidis, I., Stavropoulou, E., Stavropoulos, A., & Bezirtzoglou, E. (2020). Environmental and health impacts of air pollution: A review. *Frontiers in Public Health, 8*, 14.
Margot, J., Rossi, L., Barry, D. A., & Holliger, C. (2015). A review of the fate of micropollutants in wastewater treatment plants. *Wiley Interdisciplinary Reviews: Water, 2*(5), 457–487.
Martins, C., Dreij, K., & Costa, P. M. (2019). The state-of-the art of environmental toxicogenomics: Challenges and perspectives of "omics" approaches directed to toxicant mixtures. *International Journal of Environmental Research and Public Health, 16*(23), 4718.
Mateos, R., & Bravo, L. (2007). Chromatographic and electrophoretic methods for the analysis of biomarkers of oxidative damage to macromolecules (DNA, lipids, and proteins). *Journal of Separation Science, 30*(2), 175–191.
Mentzel, S., Grung, M., Tollefsen, K. E., Stenrød, M., Petersen, K., & Moe, S. J. (2021). Development of a Bayesian network for probabilistic risk assessment of pesticides. *Integrated Environmental Assessment and Management, 18*(4), 1072–1087.
Metcalfe, C. D., Bayen, S., Desrosiers, M., Muñoz, G., Sauvé, S., & Yargeau, V. (2022). An introduction to the sources, fate, occurrence and effects of endocrine disrupting chemicals released into the environment. *Environmental Research, 207*, 112658.
Min, E. K., Lee, A. N., Lee, J. Y., Shim, I., Kim, P., Kim, T. Y., Kim, K.T & Lee, S. (2021). Advantages of omics technology for evaluating cadmium toxicity in zebrafish. *Toxicological Research*, 1–9.
Mishra, S., Swain, S., Sahoo, M., Mishra, S., & Das, A. P. (2021). Microbial colonization and degradation of microplastics in aquatic ecosystem: A review. *Geomicrobiology Journal.* https://doi.org/10.1080/01490451.2021.1983670
Mishra, S., Dash, D., Al-Tawaha, A. R. M. S., & Das, A. P. (2022). A Review on heavy metal ion adsorption on synthetic microfiber surface in aquatic environments. *Applied Biochemistry and Biotechnology, 194*(10), 4639–4654. https://doi.org/10.1007/S12010-022-04029-W/METRICS
Mishra, S., Dash, D., & Das, A. P. (2024). Microbial degradation of polyester microfibers using indigenously isolated bacterial strain Exiguobacterium Sp. *CLEAN—Soil, Air, Water, 52*(12), e202300343. https://doi.org/10.1002/CLEN.202300343
Mitov, M. I., Patil, V. S., Alstott, M. C., Dziubla, T., & Butterfield, D. A. (2016). In vitro cellular assays for oxidative stress and biomaterial response. In *Oxidative Stress and Biomaterials* (pp. 145–186). Academic Press.
Mofijur, M., Hasan, M. M., Ahmed, S. F., Djavanroodi, F., Fattah, I. M. R., Silitonga, A. S., Kalam, M. A., Zhou, J. L., & Khan, T. Y. (2024). Advances in identifying and managing emerging contaminants in aquatic ecosystems: Analytical approaches, toxicity assessment, transformation pathways, environmental fate, and remediation strategies. *Environmental Pollution, 341*, 122889.
Mohanty, J., Dey, S., Bala, K., & Das, A. P. (2024a). Recent advances in microplastic pollution for its sustainable management. *Sustainable Management of Urban Plastic Waste Through Circular Economic Approaches*, 27–48. https://doi.org/10.1201/9781003465447-3
Mohanty, M., Mohanty, J., Dey, S., Dutta, K., Shah, M. P., & Das, A. P. (2024b). The face mask: A tale from protection to pollution and demanding sustainable solution. *Emerging Contaminants, 10*(2), 100298. https://doi.org/10.1016/J.EMCON.2023.100298
Mohanty, M., Tripathy, B., Koteswara Reddy, G., & Das, A. P. (2024c). Greener technologies for establishing restraint over microplastic pollution. *Sustainable Microbial Technology for Synthetic and Cellulosic Microfiber Bioremediation*, 239–258. https://doi.org/10.1007/978-3-031-62600-5_13

Nam, S. E., Bae, D. Y., Ki, J. S., Ahn, C. Y., & Rhee, J. S. (2023). The importance of multi-omics approaches for the health assessment of freshwater ecosystems. *Molecular & Cellular Toxicology, 19*(1), 3–11.

Narwal, N., Katyal, D., Kataria, N., Rose, P. K., Warkar, S. G., Pugazhendhi, A., Ghotekar, S., & Khoo, K. S. (2023). Emerging micropollutants in aquatic ecosystems and nanotechnology-based removal alternatives: a review. *Chemosphere*, 139945.

Nguyen, N., Jennen, D., & Kleinjans, J. (2022). Omics technologies to understand drug toxicity mechanisms. *Drug Discovery Today, 27*(11), 103348.

Nováková, P., Kodešová, R., Fedorova, G., Bořík, A., Sadchenko, A., & Grabic, R. (2024). Identifying organic micropollutants' transformation products from the soil dissipation experiment by non-targeted high-resolution mass spectrometry approach: Can we gain more than transformation product identity? *Environmental Pollution, 351*, 124038.

Ouedraogo, M., Baudoux, T., Stévigny, C., Nortier, J., Colet, J. M., Efferth, T., Qu, F., Zhou, J., Chan, K., Shaw, D., & Duez, P. (2012). Review of current and "omics" methods for assessing the toxicity (genotoxicity, teratogenicity and nephrotoxicity) of herbal medicines and mushrooms. *Journal of Ethnopharmacology, 140*(3), 492–512.

Owolabi, I. O., Siwarak, K., Greer, B., Rajkovic, A., Dall'asta, C., Karoonuthaisiri, N., Uawisetwathana, U., Elliott, C. T., & Petchkongkaew, A. (2024). Applications of mycotoxin biomarkers in human biomonitoring for exposome-health studies: Past, present, and future. *Exposure and Health, 16*(3), 837–859.

Özgenç, E., Keleş, E., & Töre, G. Y. (2023). Assessment of biomarker-based ecotoxic effects in combating microplastic pollution-a review. *Global NEST Journal, 26*(1), 05398.

Parasuraman, S. (2011). Toxicological screening. *Journal of Pharmacology & Pharmacotherapeutics, 2*(2), 74.

Patil, R., Satpute, R., & Nalage, D. (2023). The application of omics technologies to toxicology. *Toxicology Advances, 5*(6).

Paul, A., Priyadarsini, S., Karakuş, S., Behera, J. K., Gnanasekaran, R., & Das, A. P. (2024). Fate and accumulation of synthetic microfibers in the environment. *Sustainable Microbial Technology for Synthetic and Cellulosic Microfiber Bioremediation*, 61–76. https://doi.org/10.1007/978-3-031-62600-5_4

Pereira, C. V., Nadanaciva, S., Oliveira, P. J., & Will, Y. (2012). The contribution of oxidative stress to drug-induced organ toxicity and its detection in vitro and in vivo. *Expert Opinion on Drug Metabolism & Toxicology, 8*(2), 219–237.

Pereira, R., Cachada, A., Sousa, J. P., Niemeyer, J., Markwiese, J., & Andersen, C. P. (2018). Ecotoxicological effects and risk assessment of pollutants. In *Soil Pollution* (pp. 191–216). Academic Press.

Pironti, C., Ricciardi, M., Proto, A., Bianco, P. M., Montano, L., & Motta, O. (2021). Endocrine-disrupting compounds: An overview on their occurrence in the aquatic environment and human exposure. *Water, 13*(10), 1347.

Präbst, K., Engelhardt, H., Ringgeler, S., & Hübner, H. (2017). Basic colorimetric proliferation assays: MTT, WST, and resazurin. In *Cell Viability Assays: Methods and Protocols* (pp. 1–17).

Priyadarshini, B., Santhosh Kumar, M., & Das, A. P. (2024). Microbial degradation of synthetic plastics released from COVID-19 generated single use disposable face masks. *Environmental Science and Engineering, Part, F2827*, 1–16. https://doi.org/10.1007/978-3-031-50840-0_1

Priyadarsini, S., & Das, A. P. (2024). Analytical and structural characterization of waste lithium-ion batteries for their effective recycling strategy. *Environmental Chemistry and Ecotoxicology*. https://doi.org/10.1016/j.enceco.2024.12.004

Raimondo, S., & Forbes, V. E. (2022). Moving beyond risk quotients: Advancing ecological risk assessment to reflect better, more robust and relevant methods. *Ecologies, 3*(2), 145–160.

Raj, K., & Das, A. P. (2023). Lead pollution: Impact on environment and human health and approach for a sustainable solution. *Environmental Chemistry and Ecotoxicology, 5*, 79–85. https://doi.org/10.1016/J.ENCECO.2023.02.001

Ray, S. K., Dey, S., Das, A. P., & John, A. (2024). Microplastic waste management through greener technologies : a review. *Sustainable Management of Urban Plastic Waste Through Circular Economic Approaches*, 237–261. https://doi.org/10.1201/9781003465447-14

Rehberger, K., Kropf, C., & Segner, H. (2018). In vitro or not in vitro: A short journey through a long history. *Environmental Sciences Europe, 30*(1), 23.

Rehman, K., Fatima, F., Waheed, I., & Akash, M. S. H. (2018). Prevalence of exposure of heavy metals and their impact on health consequences. *Journal of Cellular Biochemistry, 119*(1), 157–184.

Saganuwan, S. A. (2017). Toxicity studies of drugs and chemicals in animals: an overview. *Bulgarian Journal of Veterinary Medicine, 20*(4).

Sahu, S., & Das, A. P. (2024). Microplastics degradation mechanism by aquatic Plastiomes: A review. *Environmental Science and Engineering, Part, F2827*, 213–226. https://doi.org/10.1007/978-3-031-50840-0_12

Sahu, A., Mallick, K., & Das, A. P. (2024). Environmental impact of single-use synthetic face mask and its recycling: a sustainable approach (pp. 197–212). https://doi.org/10.1007/978-3-031-50840-0_11

Sanchez-Huerta, C., Baasher, F., Wang, C., Zhang, S., & Hong, P. Y. (2024). Global occurrence of organic micropollutants in surface and ground water: Highlighting the importance of wastewater sanitation to tackle organic micropollutants. *Journal of Environmental Chemical Engineering, 12*(3), 112861.

Sanganyado, E., Bi, R., Teta, C., Buruaem Moreira, L., Yu, X., Yajing, S., Dalu, T., Rajput, I. R., & Liu, W. (2021). Toward an integrated framework for assessing micropollutants in marine mammals: Challenges, progress, and opportunities. *Critical Reviews in Environmental Science and Technology, 51*(23), 2824–2871.

Satapathy, S., Kar, D., Priyadarsini, S., Das, A. P., & Kuanar, A. (2024). Application of microorganisms in emerging waste management strategies and remediation of synthetic and cellulosic microfibers. *Sustainable Microbial Technology for Synthetic and Cellulosic Microfiber Bioremediation*, 183–206. https://doi.org/10.1007/978-3-031-62600-5_10

Shao, Y., Chen, Z., Hollert, H., Zhou, S., Deutschmann, B., & Seiler, T. B. (2019). Toxicity of 10 organic micropollutants and their mixture: Implications for aquatic risk assessment. *Science of the Total Environment, 666*, 1273–1282.

Speit, G., & Hartmann, A. (1999). The comet assay (single-cell gel test) a sensitive genotoxicity test for the detection of DNA damage and repair. In *DNA Repair Protocols: Eukaryotic Systems* (pp. 203–212).

Stallard, N., & Whitehead, A. (1995). Reducing animal numbers in the fixed-dose procedure. *Human & Experimental Toxicology, 14*(4), 315–323.

Stoddart, M. J. (2011). Cell viability assays: introduction. In *Mammalian Cell Viability: Methods and Protocols* (pp. 1–6).

Tamis, J. E., Koelmans, A. A., Dröge, R., Kaag, N. H., Keur, M. C., Tromp, P. C., & Jongbloed, R. H. (2021). Environmental risks of car tire microplastic particles and other road runoff pollutants. *Microplastics and Nanoplastics, 1*, 1–17.

Thakur, D., & Ganguly, R. (2022). Occurrence and fate of micropollutants in air. In *Environmental Micropollutants* (pp. 305–313). Elsevier.

Tijani, J. O., Fatoba, O. O., Babajide, O. O., & Petrik, L. F. (2016). Pharmaceuticals, endocrine disruptors, personal care products, nanomaterials and perfluorinated pollutants: A review. *Environmental Chemistry Letters, 14*, 27–49.

Tong, X., Mohapatra, S., Zhang, J., Tran, N. H., You, L., He, Y., & Gin, K. Y. H. (2022). Source, fate, transport and modelling of selected emerging contaminants in the aquatic environment: Current status and future perspectives. *Water Research, 217*, 118418.

Tripathy, B., Sahoo, P. P., Sundaray, H., & Das, A. P. (2024b). In-vitro biodegradation of discarded marine microplastics across the eastern coast of the Bay of Bengal, India using Exiguobacterium sp. *Environmental Chemistry and Ecotoxicology, 6*, 236–247. https://doi.org/10.1016/J.ENCECO.2024.07.003

Tripathy, B., Dey, S., Dubey, N. K., Lo, W.-C., & Das, A. P. (2024). An omics-based approach for managing the risk of COVID-19-generated wastes. 51–66. https://doi.org/10.1007/978-3-031-50840-0_4

Van der Oost, R., Beyer, J., & Vermeulen, N. P. (2003). Fish bioaccumulation and biomarkers in environmental risk assessment: A review. *Environmental Toxicology and Pharmacology, 13*(2), 57–149.

Vasilachi, I. C., Asiminicesei, D. M., Fertu, D. I., & Gavrilescu, M. (2021). Occurrence and fate of emerging pollutants in water environment and options for their removal. *Water, 13*(2), 181.

Vaudin, P., Augé, C., Just, N., Mhaouty-Kodja, S., Mortaud, S., & Pillon, D. (2022). When pharmaceutical drugs become environmental pollutants: Potential neural effects and underlying mechanisms. *Environmental Research, 205*, 112495.

Verma, R., Pathak, S., Srivastava, A. K., Prawer, S., & Tomljenovic-Hanic, S. (2021). ZnO nanomaterials: Green synthesis, toxicity evaluation and new insights in biomedical applications. *Journal of Alloys and Compounds, 876*, 160175.

Yang, Y., Zhang, X., Jiang, J., Han, J., Li, W., Li, X., Yee Leung, K. M., Snyder, S. A., & Alvarez, P. J. (2021). Which micropollutants in water environments deserve more attention globally? *Environmental Science & Technology, 56*(1), 13–29.

Zenker, A., Cicero, M. R., Prestinaci, F., Bottoni, P., & Carere, M. (2014). Bioaccumulation and biomagnification potential of pharmaceuticals with a focus to the aquatic environment. *Journal of Environmental Management, 133*, 378–387.

Zhang, Y. Y., Huang, Y. F., Liang, J., & Zhou, H. (2022). Improved up-and-down procedure for acute toxicity measurement with reliable LD50 verified by typical toxic alkaloids and modified Karber method. *BMC Pharmacology and Toxicology, 23*(1), 3.

Zhong, C., Zhao, H., Cao, H., & Huang, Q. (2019). Polymerization of micropollutants in natural aquatic environments: A review. *Science of the Total Environment, 693*, 133751.

Zon, L. I., & Peterson, R. T. (2005). In vivo drug discovery in the zebrafish. *Nature Reviews Drug Discovery, 4*(1), 35–44.

Chapter 5
Assessment of the Impact of Micropollutants on Environment

T. V. Karthik and Debashis Dey

Introduction

Environmental science and research have become growing in significance as concern over the effects of micropollutants on the environment intensifies. Trace pollutants known as micropollutants may have major and enduring impacts on ecosystems, nature, and human wellness even at very low quantities. These chemicals, which include insecticides, heavy metals, cosmetics, endocrine-disrupting chemicals, personal care items, and industrial chemicals, are set free into the environment through a medley of channels by many sources, including atmospheric absorption, natural calamities, sewage disposal, agricultural runoff, and industrial byproducts (Barnes et al., 2004; Jyotsna & Tandon, 2018). Since these pollutants are used so extensively used, they are frequently detected in soil, water, air, and biota, which makes monitoring and controlling their effects on the ecosystem challenging. Even at lower concentrations, micropollutants have the ability to cause detrimental impacts, which makes evaluating their effects on public well-being and environment crucial. Since many of these materials are difficult to break down, they build up in natural entities which includes soil, water, and food chain (Mishra et al., 2024; Sahu et al., 2024a). Chronic exposure to ecosystems and creatures, including humans, due to the persistence of micropollutants can cause minor but substantial disturbances in metabolic processes, reproduction, and biodiversity (Ajmal & Khan, 1985; Morin-Crini et al., 2022). Furthermore, micropollutants can interact synergistically with other pollutants, magnifying their effects and creating more complicated environmental concerns. For instance, pesticides can disturb ecosystems by killing non-target organisms like pollinators and lady bugs, changing biodiversity, while medications can alter the behaviour, growth, and fertility of aquatic animals (Das et al., 2023; Kanungo et al., 2023a, b). Monitoring, assessing, and detecting the presence and

T. V. Karthik (✉) · D. Dey
Department of Chemical Engineering, IIT Guwahati, Guwahati, Assam, India
e-mail: t.karthik@iitg.ac.in

I. D. Behera et al. (eds.), *Emerging Micropollutants*, Sustainable Environmental Waste Management Strategies, https://doi.org/10.1007/978-3-032-03530-1_5

consequences of these pollutants necessitates an integrated approach that integrates various evaluation techniques (Priyadarsini & Das, 2024, 2025). These techniques include sophisticated monitoring and analytical methods that enable the identification and quantification of micropollutants within environmental samples, such as spectrometry, chromatography, and biosensors (Ahmadi Doabi et al., 2019a, b; Ajmal & Khan, 1985). Ecotoxicological analysis sheds light on the biological effects on various species, while Environmental Risk Assessment (ERA) approaches aid in speculating the possible negative impacts of these micro pollutants on ecosystems and human health (Das et al. 2024b; Dey et al., 2024a; Sahu et al., 2024a). This chapter lines up to provide a comprehensive overview of the key methods used to assess the environmental impression of micropollutants, highlighting their advantages, challenges, and limitations, additionally, by estimating the long-term environmental behaviour of micropollutants (Ali et al., 2019a, b; Arguello-Pérez et al., 2019).

Methods of Assessing the Impact of Micro Pollutants

Monitoring and Detection Techniques

Determining the prevalence and spread of micropollutants requires both monitoring and detection. Sampling is the first step in this process and can be done in a variety of environmental media, including water, air, soil and biota (living organisms). Water sampling is frequently focused on groundwater, surface waters, and waste water runoff, while air and dirt samples help detect effluents from industries and automobiles that may be accumulated in a specific region or plumes transferred by the atmosphere. The ability of living organisms, which includes both terrestrial and aquatic species, to accumulate these micropollutants can reveal patterns of bioaccumulation. Advanced analytical and detection methods are used to examine the samples that were gathered from various sources (Mohanty et al., 2024c; Tripathy et al., 2022, 2024). Chromatography is frequently used to separate and quantify micropollutants in complicated samples. This includes gas chromatography (GC) and liquid chromatography (LC). Spectrometric methods such as mass spectrometry (MS) are paired with chromatography for highly sensitive analysis and specific identification of elements and compounds. Biosensors, an emerging technique, offer a real-time, cost-effective alternative for detecting micropollutants in the environment. They use bio-based components to detect specific contaminants, offering rapid and specific monitoring capabilities. The procedure of monitoring micropollutants usually starts before sampling and pre concentration steps. Filed sampling collects a single sample at a certain location and time period, whereas composite sampling averages pollutant concentrations over a range of time through several sample collections. Passive sampling, on the other hand, uses instruments that gather contaminants over time, resulting in integrated data interpretation. Pre-concentration procedures, such as solid-phase extraction and liquid phase extraction, are frequently used to improve detection sensitivity by separating micropollutants from complicated matrices (Arguello-Pérez et al., 2019; Barhoum et al., 2023; Cabrera-Peralta & Peña-Alvarez, 2018a, b; Archana et al., 2016).

Quantification of Micro Pollutants

Understanding the abundance of micropollutants in the environment and their possible effects requires precise quantification. As they present in very low concentrations, these pollutants require extremely sensitive instruments to measure their concentrations. This makes analysis and detection difficult to perform, particularly in intricate environmental factors. Certain micropollutants may be hard to detect because they are trace elements, and their concentration levels may shift over time or in response to climatic and seasonal variations. Furthermore, measurements may become more difficult due to matrix effects, which include interference from other active materials in soil, water, or biota (Behera et al., 2020, 2021, 2022). The detection of these pollutants presents an additional challenge, since many micropollutants share molecular and structural similarities, it takes sophisticated methods to distinguish between them and precisely measure them (Mohanty et al. 2024c; Pradhan et al., 2023). To get precise data on contaminant levels, thorough sampling techniques and high-resolution analytical technologies are therefore required (Ahmed et al., 2017; Archer et al., 2017; Barhoum et al., 2023; Basem et al., 2024).

Environmental Risk Assessment (ERA)

An organized method for assessing the possible negative environmental impacts of micropollutants is environmental risk assessment, or ERA. It usually entails a number of procedures, such as risk characterization, exposure analysis, dose–response assessment, and hazard identification. While dose–response evaluations look at how the intensity of effects varies with pollutant concentration, threat identification identifies the kinds of negative effects that could arise from exposure to a specific micropollutant. While risk characterization synthesizes the data to evaluate the overall risk to ecosystems and human health, exposure assessment calculates the concentration of contaminants in various natural compartments (such as water, soil, and air). In order to comprehend the possible impacts of micropollutants at varying concentrations, toxicological investigations and laboratory evaluations are essential (Fig. 5.1).

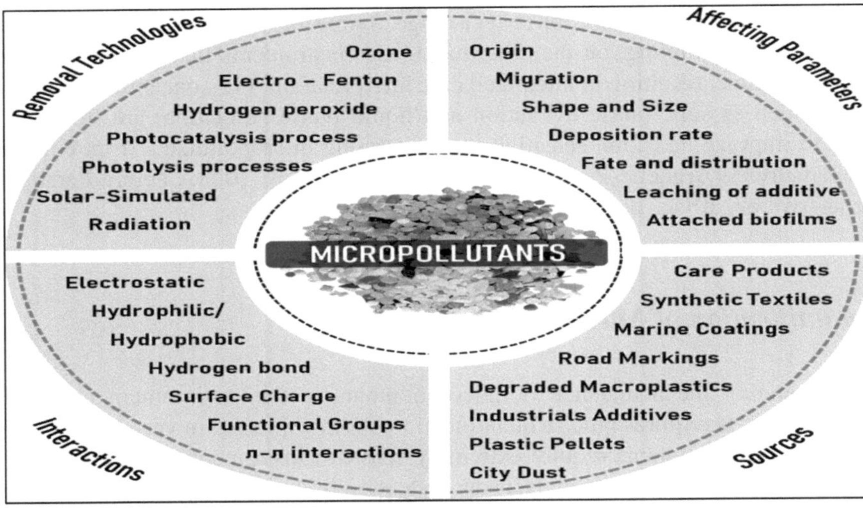

Fig. 5.1 Advanced technologies in monitoring and detecting micropollutants (Ratchnashree et al., 2023)

Ecotoxicology Testing, Modelling and Predictive Tools

Research and progress in ecotoxicology yields important insights into the biological impacts of micropollutants. Usually, these tests evaluate both acute and long-term toxicity. While chronic exposure to this toxicity leads to consequences like altered growth, reproduction, and survival rates, acute toxicity testing assesses a pollutant's immediate effects on tissues of organisms, typically over a brief exposure period. To evaluate the impact of micropollutants on ecosystem health, bioassays using indicator species such as fish, algae, and other invertebrates are frequently employed in such practises. These species can be utilized to identify pollution thresholds for ecosystem conservation and act as early indicators of environmental stress. Predictive and modelling methods are crucial for comprehending the behaviour and destiny of environmental micropollutants. It is possible to forecast the degradation, distribution, and migration of pollutants via various environmental media, including soil, water, and air, with the use of environmental fate modelling (Dey et al., 2024b; Paul et al., 2024; Priyadarshini et al., 2024). The potential dangers of micropollutants under various environmental circumstances can be ascertained by using these models to simulate different situations. When evaluating the long-term effects of micropollutants, predictive models for bioaccumulation are another essential tool to assess the risk. These models forecast the accumulation of contaminants in species over time, which is crucial for comprehending the dynamics of food chains and evaluating possible hazards to the health of humans and animals. When these models are combined with analytical and empirical data, a more thorough knowledge of the long-term biological and environmental effects of micropollutants is made possible. A multifaceted approach that combines advanced detection, quantification, risk assessment, ecotoxicology testing, and modelling is necessary to fully understand the impact of micropollutants on the environment (Ali et al., 2019b; Arguello-Pérez et al., 2019; Escher et al., 2008; Lv & Wang, 2019a, b; Yang et al., 2019a, c).

Impact of Micro Pollutants on Ecosystems

The fate of micropollutants in the environment is determined by their sources, transport mechanisms, transformation processes, and persistence. Once introduced, micropollutants travel through water, air, soil, and sediments, frequently dissolving, adsorbing to particles. Their transformation includes mechanisms such as photodegradation by sunshine, microbial biodegradation, hydrolysis, and redox reactions, however some defy breakdown and persist in the environment. Persistent micropollutants can build in sediments and organisms, causing bioaccumulation and possible entry into food chains, with long-distance transit harming isolated ecosystems (Bala et al., 2024b; Das et al., 2024a; Mohanty et al., 2024a).

Water and Marshy Systems

Due to direct contact to micropollutants from wastewater discharges and runoff from residences, companies, and farms, freshwater and marine water systems are particularly susceptible to the infiltration of micropollutants. Water quality, nitrogen cycling, biodiversity, and aquatic ecosystems have all been significantly and permanently impacted by micropollutants, even at trace amounts (Fig. 5.2).

Since water bodies are interconnected worldwide, the impacts of micropollutants in a single region can have a domino effect on larger ecosystems, making it a serious ecological concern. Micropollutants in freshwater and marine habitats can negatively impact these ecosystem's ability to operate in various ways. In freshwater systems, contaminants like pharmaceuticals, heavy metals, and pesticides can alter the chemical composition of water, leading to toxic conditions for aquatic life (Bala et al., 2024a; Mohanty et al., 2024b; Priyadarshini & Das, 2024; Sahu et al., 2024b). Marine ecosystems are also at risk, especially corals in coastal areas that receive high levels of runoff or effluents from land-based activities.

The viability of marine life, including fish, shellfish, and corals, can be impacted by micropollutants in marine habitats, which may cause disruptions to food webs and ecosystem functions. The identification of micropollutants in water systems has rapidly increased globally in recent years, and new emerging contaminants are being found in aquatic settings. More than 80% of water samples from rivers, lakes, and groundwater sources contained micropollutants. The extensive pollution of surface waterways with inorganic metals, pesticides, and personal care products all of which have been demonstrated to have an impact on aquatic ecosystems has also been

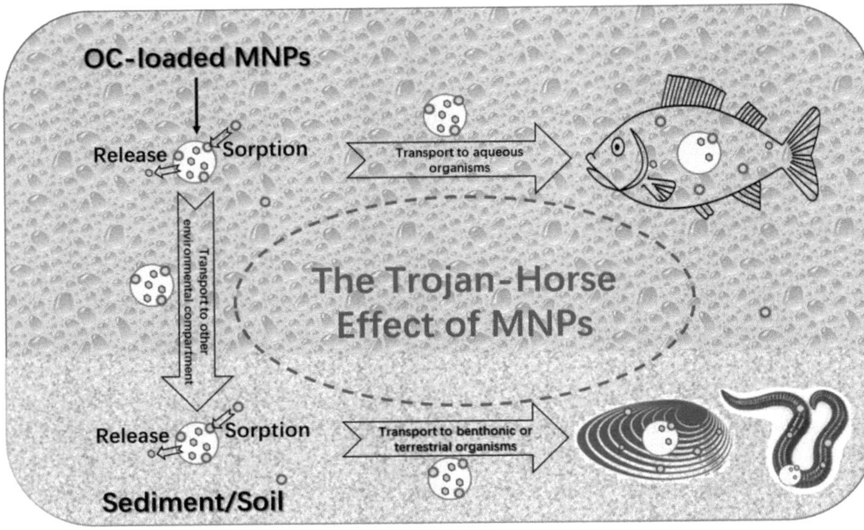

Fig. 5.2 Trojan-Horse effect for organic contaminants (Zhang & Xu, 2022)

brought to light by studies conducted across the globe. In particular, the contamination of water bodies by various micropollutants has been linked to decreases in the diversity of freshwater species of fish. Pollutant buildup is affecting coral reefs in many tropical and coastal areas, compromising coral health, impeding reproduction, and facilitating the spread of chronic disease. Aquatic invertebrates, the foundation of many aquatic food webs, can be extremely poisoned by pesticides and heavy metals like lead and mercury (Raj & Das, 2023). The whole aquatic food chain may be upset if invertebrates exposed to these contaminants suffer from delayed growth, poor reproduction, and get terminated. Higher trophic levels like secondary and tertiary consumers that depend on these species for food, such bigger fish, birds, and mammals are affected in a cascade manner by their loss. Additionally, as pollutants bioaccumulate through the food chain, they can raise the levels of toxins in predatory animals, such as humans and other wildlife, who eat polluted fish and shellfish. Global trends show that aquatic biodiversity is widely declining, and this reduction is frequently attributed to pollution, especially micropollutants.

In freshwater and marine environments, micropollutants can also affect the cycling of nutrients and the general quality of the water. Eutrophication, which results in algae blooms in water causes oxygen depletion in water bodies, can be caused by the presence of specific chemicals, such as nitrogen, sulphur and phosphorus compounds from fertilizers and wastewater. Although nutrient pollutants are more frequently linked to this occurrence, certain micropollutants, such as medications and personal care items, can also cause it by upsetting microbial communities involved in the cycling of nutrients. Pollutant-induced disruption of microbial communities can hinder organic matter's normal breakdown pattern in water, resulting in decreased water quality and altered nutrient dynamics. For instance, research has indicated that certain antibiotics present in water can interfere with the nitrogen-fixing bacteria and may hinder or boost its functionality. Nutrient imbalances can therefore arise, impacting the development of aquatic plants and algae and resulting in rapid algal blooms that can deteriorate water quality, reduce oxygen levels, and release chemicals that are detrimental to aquatic life. Micropollutants can intensify the effects of eutrophication in freshwater and marine habitats (Anumol et al., 2016; Baronti et al., 2000; Bartelt-Hunt et al., 2011; Wightwick et al., 2012; Zhang & Xu, 2022).

Land, Soil Bed, Reef and Swarm Ecosystems

Micropollutants are having a greater effect on soil and sediment ecosystems, which are essential to the biodiversity, food production, and to have control of the water and nutrient cycles. Numerous processes, such as industrial runoff, municipal water irrigation, modern agricultural practices, and atmospheric contaminants deposition, can introduce these micro pollutants which include pesticides, heavy metals, industrial chemicals, and pharmaceuticals into terrestrial systems. Micropollutants have the ability to linger in soil for extended periods of time, which may result in long-term alterations to biodiversity, soil health, and the food chain. Heavy metals including lead, cadmium, arsenic and mercury as well as organic pollutants like herbicides, medications, and industrial chemicals are frequently found in soil and sediment, making micropollutant contamination a global problem. Since they don't break down easily, heavy metals tend to build up in soils, especially around mining operations, industrial districts, and agricultural areas with high usage of synthetic chemicals and fertilizers. Similarly, organic pollutants can attach to soil particles and build up over time, causing long-term environmental contamination. This is especially true of persistent organic pollutants, which include industrial chemicals and ed banner pesticides. Global developments in micropollutant pollution of soil are a reflection of intensifying agricultural practices, urbanization, and industrialization.

The use of tainted water for irrigation purpose and the unregulated disposal of e-waste in some emerging areas worsen soil pollution. The wildlife and general health of the soil are significantly impacted by the existence of micropollutants in the soil. A complex community of microbes, invertebrates, and small vertebrates live in soil, and they are all essential to the breakdown of organic matter, the cycling of nutrients, and the preservation of soil structure. However, by damaging soil organisms, decreasing biodiversity, and lowering soil fertility, micro pollution can interfere with these processes. For earthen organisms, other insects and their larvae, heavy metals including lead, arsenic, and cadmium are chronic. The survival, development, and breeding of soil invertebrates such as nematodes and earthworms that are vital to soil aeration and cycling of nutrients may be adversely impacted by these pollutants. High levels of heavy metal exposure can occasionally cause the total extinction of soil biodiversity, which lowers ecosystem benefits. Furthermore, some organic pollutants, such as pesticides, can disrupt the microbial populations in the soil, especially those that fix nitrogen, which is essential for preserving soil fertility. Micropollutant's effects on the condition of soil can have an impact on agricultural yield as well. The presence of pesticides and other organic pollutants can affect the microbial activity required for nutrient mineralization, further hindering plant growth.

Another significant concern is the uptake of micro pollutants by plants, which might allow chronic substances to get into the food chain as plants are one of the primary producers of food web. Through their roots, crops cultivated in polluted soils can absorb organic contaminants and heavy metals, which can then build up in the plant's edible sections, including leaves, roots, fruits, and seeds. This presents immediate health hazards to people since eating tainted crops can cause hazardous compounds to build up in the tissues of body over time, which can have negative health impacts like cancer, renal damage, and neurological issues. It is well known that crops like wheat, maize, rice, and vegetables can easily absorb heavy metals like cadmium and arsenic, contaminating the food supply. Similar to this, crops cultivated in pesticide-contaminated soils, such as spinach, lettuce, and tomatoes, can absorb these chemicals and may retain them in the plant's edible portions long after harvest and processing. An increasing global concern is the bioaccumulation of these contaminants in crops, particularly in areas where industrial agriculture is prevalent and soil pollution regulations are not rigorously enforced. Globally, there are trends in crop uptake of micropollutants and soil pollution, for instance, allegations of heavy metal contamination in agricultural soils are growing worldwide, especially in nations with significant industrial activity (Satapathy et al., 2024). In developing countries, the unregulated use of wastewater for irrigation often leads to high levels of pollutants, including pharmaceuticals and pesticides, being absorbed by crops. These trends underscore the need for improved monitoring and regulation of soil and sediment contamination, particularly in agricultural areas, to protect food security and public health. With considerable regional heterogeneity in the sources and types of pollutants, micropollutant contamination is a serious global concern. Developing efficient remediation techniques and conducting thorough monitoring are necessary to address the problems caused by micropollutants in soil. These could involve the application of bioremediation, which uses microorganisms to break down contaminants, and phytoremediation, which uses plants to absorb toxins. (Ali et al., 2019b; Chirinos-Peinado & Castro-Bedriñana, 2020; Parizanganeh et al., 2010; Waterhouse et al., 2012; Zhang & Xu, 2022) (Fig. 5.3).

Air and Atmospheric Impact

The ability of micropollutants to travel great distances away from their source of emission is one of their main features in the atmosphere. Numerous pollutants continue to seep into the atmosphere as the result of human activities. These contaminants, which can affect areas distant from their origin, can travel great distances and include both organic and inorganic micropollutants. Once in the air, they can deteriorate, interact with other pollutants of the atmosphere, or return to Earth in other forms, which can

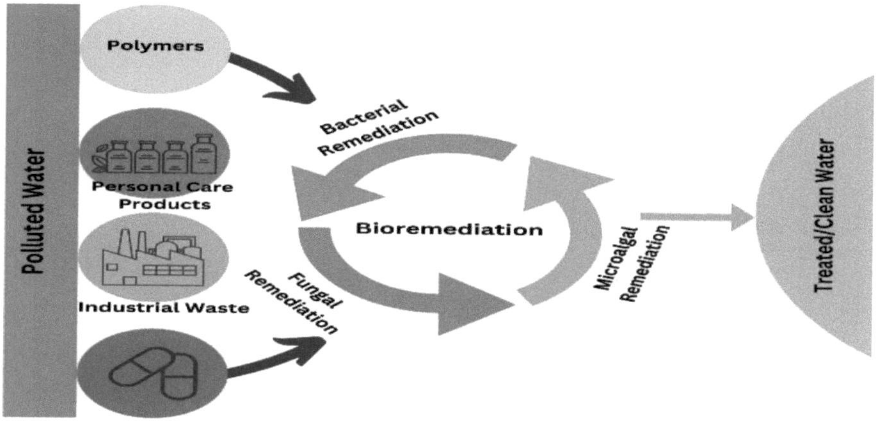

Fig. 5.3 Schematic presentation of bioremediation of emerging pollutants (Mishra et al., 2023a)

contaminate the land and water. The significance of maintaining and comprehending the behaviour of micropollutants in the atmosphere is underscored by their migration through atmospheric processes and subsequent impacts on ecological and human health. Agricultural insecticides, trash incineration, vehicle exhaust, and industrial emissions are some of the operations that can emit effluents into the atmosphere. The trade winds have the ability to carry these contaminants across great distances once they are in the atmosphere. Pollutants discharged in agricultural or industrial zones, for instance, have the potential to travel to far-off places, such as isolated or pristine ecosystems.

Significant global repercussions may result from this long-distance transportation. Often discovered in polar regions, far from their original origins, persistent organic pollutants can move across continents and include harmful chemicals and pesticides. In places where industrial or agricultural operations do not immediately impact the environment, these pollutants build up in soils, sediments, and biota. Studies have found elevated levels of micropollutants in remote areas such as the arctic and antarctica, where pollutants have travelled through atmospheric currents. Once airborne micropollutants reach the ground, they undergo deposition in the form of wet (rain, snow, fog) or dry deposition (direct settling onto surfaces). This deposition can affect both terrestrial and aquatic ecosystems, leading to contamination of ecosystems that were not directly exposed to the original emissions. Contaminants like polycyclic aromatic hydrocarbons from industrial emissions or heavy metals from vehicle exhaust can be deposited in agricultural fields, affecting crops and soil health.

Once released into the atmosphere, micropollutants may undergo a variety of physical and chemical transformations. Atmospheric degradation occurs through processes such as photolysis (breakdown by sunlight), oxidation, and reactions with other atmospheric compounds, which can alter the chemical structure of the pollutants. Some organic micropollutants may break down into smaller, more reactive

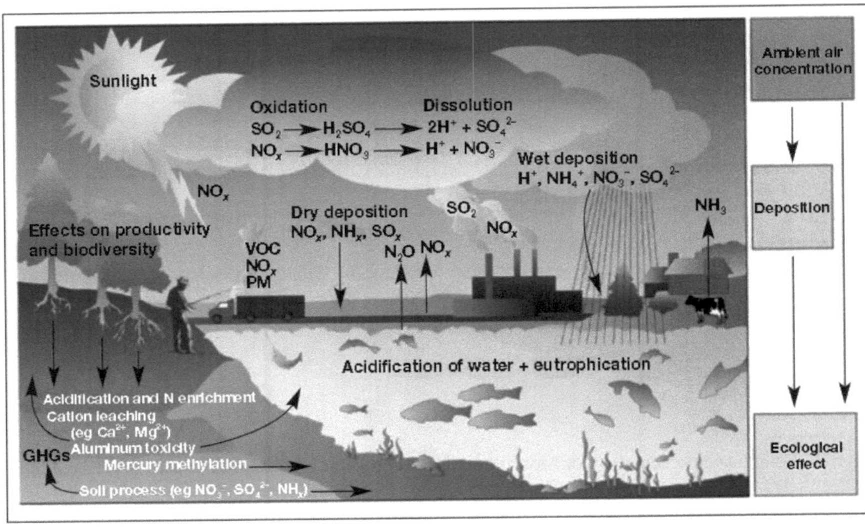

Fig. 5.4 Simplified diagram of the ecological effects caused by air pollutants (Greaver et al., 2012)

molecules when exposed to sunlight. This process can either reduce or increase the toxicity of the original pollutant, depending on the nature of the transformation in addition to the degradation of the original pollutants, atmospheric processes can lead to the formation of secondary pollutants. Secondary pollutants are new compounds that are formed through chemical reactions in the atmosphere, often as a result of interactions between primary pollutants and other components of the atmosphere such as ozone, nitrogen oxides, and volatile organic compounds. Another important atmospheric process is the formation of ground-level ozone, a secondary pollutant formed when nitrogen oxides from vehicle emissions and industrial sources react with volatile organic compounds in the presence of sunlight (Fig. 5.4).

Ground-level ozone is a potent air pollutant that can harm human respiratory health and reduce crop yields. While ozone itself is not a micropollutant, the interactions of volatile organic compounds and micropollutants like pesticides in the atmosphere can contribute to the formation of this harmful compound. Additionally, the degradation of micropollutants in the atmosphere may lead to the formation of more persistent or toxic compounds. Some chemical reactions may result in the creation of more reactive intermediates or toxic byproducts, which can have adverse effects on both human health and ecosystems when they are inhaled or deposited onto land and water. Concern over the long-distance movement of airborne micropollutants is developing worldwide, especially in areas far from the contamination source (Das et al., 2024b; John & Das, 2024; Mishra et al., 2023b). Micropollutants are widely dispersed across continents in industrialized countries due to emissions from manufacturing, transportation, and agriculture. Concerns regarding their possible effects on ecosystems and human health are raised by the increased detection of new contaminants in the atmosphere, such as medications and personal care items. As an illustration of how pharmaceutical residues travel through atmospheric processes, investigations conducted have found them in both ambient air and precipitation. It contributes to the deposition of heavy metals, and other micropollutants into both terrestrial and aquatic habitats. In certain areas, the effects of atmospheric deposition can cause respiratory disorders and other health issues, as well as jeopardize biodiversity and agricultural sustainability. Another trend that needs to be addressed is the increasing prevalence of secondary pollutants created by air degradation and interactions. Numerous health issues, including as lung cancer, asthma, and cardiovascular disorders, have been connected to secondary pollutants like ozone, particulate matter, and secondary organic aerosols. (Filonchyk & Peterson, 2023; Ghanem et al., 2023; Greaver et al., 2012; Lapworth et al., 2012; Sun et al., 2017; Yesu Ramya & Kumawat, 2024).

Advanced Pollutant Treatment Process and Challenges

Advanced waste and pollutant treatment processes are designed to enhance the removal of micropollutants from wastewater, industrial effluents and vehicle emissions, addressing limitations of conventional treatment methods. Common techniques include advanced oxidation processes, which use reactive species like hydroxyl radicals to break down pollutants, novel membrane filtration systems, such as reverse osmosis and nanofiltration, which physically separate contaminants, and adsorption using activated carbon to trap organic micropollutants. Ozonation and Ultra violet based treatments are effective at degrading persistent pollutants, whereas constructing wetlands and biofiltration use natural processes to remediate these pollutants. These technologies can considerably reduce micropollutant concentrations, resulting in cleaner effluents for discharge or reuse. The high operational and capital expenses make widespread adoption of these processes challenging, particularly for small and underfunded establishments. Energy intensive technologies like advanced oxidation processes and membrane filtration increase the carbon footprint of wastewater treatment. Secondary waste, such as concentrated brine from membranes or results of oxidation, presents further disposal issues. Furthermore, these procedures may not completely eradicate all micropollutants, as some chemicals can withstand degradation or produce potentially dangerous transformation products. Maintaining consistent performance across various effluent compositions remains a technical challenge. (Aghababai Beni & Jabbari, 2022; Barhoum et al., 2023; Basque et al., 2023; Dasenaki et al., 2015; Escher et al., 2008; Greaver et al., 2012) (Table 5.1).

Table 5.1 Advanced treatment process, their advantages and limitations. (Dasenaki et al., 2015; Greaver et al., 2012)

Treatment process	Advantages	Limitations
Conventional biological activated carbon	• Removes various contaminants from wastewater • Removes leftover disinfection and oxidation products • Generates non-toxic active compounds	• High operational and maintenance costs • Sludge regeneration and disposal issues • Sludge processing can raise total costs by 50–60%
Microalgae reactor	• Recovers algal biomass for fertilizer • Produces high-quality effluent with no acute toxicity risk	• Winter season affects removal efficiency • Microplastics are not properly decomposed
Activated sludge	• Lower capital and operational expenditures compared to oxidation processes • More environmentally friendly than chlorination	• Insufficient efficiency for medications • Large volume of sludge with emerging pollutants • Not appropriate for oxygen levels beyond 4000 mg L^{-1}
Non-conventional constructed wetland	• Low energy usage and maintenance expenses • Effective removal of oestrogens, personal products, pesticides, and viruses	• Clogging, solids entrapment, and sediment formation • Biofilm growth, chemical precipitation, and seasonal variation • Requires a vast land area and extended retention period
Chemical process coagulation	• Minimised clarity caused by stranded inorganic and biological particles • Increased deposition rate due to suspended granular particle production	• Inadequate micropollutant expulsion • Excessive sludge generation and incorporation of coagulant slats in the aqueous phase
Ozonation	• Hydrogen peroxide has a strong sensitivity for contaminants and acts as a selective oxidant, promoting disinfection and sanitation	• High usage of energy, resulting in oxidised byproducts • Interference by radical contaminants
Photocatalysis	• Catalyses the degradation of persistent organic molecules at high rates • TiO_2 catalysts are inexpensive, chemically stable, and easy to recover	• Difficulty in treating big volumes of wastewater • High cost of artificial UV lights and electricity • Separating and reusing photocatalytic particles from slurry suspension

(continued)

Table 5.1 (continued)

Treatment process	Advantages	Limitations
Physical process micro- or ultra-filtration	• Can remove pathogens • Applicable for heavy metal removal	• Inadequate removal due to pore diameters ranging from 100 to 1000 times bigger than micropollutants • High operating costs and maintenance is difficult
Nanofiltration	• Effective in neutralising saline water and wastewater treatment influents • Can remove dyes and insecticides	• High energy requirement, membrane fouling, and disposal concerns • Limited applicability in pharmaceutical removal
Reverse osmosis	• Suitable for cleaning saline water and sewage processing plant influents • Capable of removing personal care products and medications	• High energy consumption, membrane fouling, and disposal issues • Finished water is corrosive and has a reduced medication removal rate

Challenges in Assessing the Impact of Micro Pollutants

Complexity of Micro Pollutants

The complexity of micropollutants presents a significant challenge in understanding their impacts on environmental aspects. These elements encompass a diverse range, including pharmaceuticals, personal care products, pesticides, microplastics, and heavy metals, each with unique physical and chemical properties. Interactions between different types of micropollutants further complicate assessments, as combined exposure can result in synergistic or antagonistic effects. Without harmonized methods for analysing, comparing results across studies or accurately predicting ecological outcomes remains a significant hurdle in micropollutant assessments. Heavy metals and microplastics, may combine to increase toxicity to aquatic creatures, but secondary pollutants produced by chemical changes such as photodegradation or microbial activity might have unknown repercussions. Furthermore, micropollutant behaviour varies significantly depending on potential, dissolution, and availability. Persistent organic pollutants, accumulate over time in the tissues, whereas water-soluble substances can dissipate fast yet still cause significant environmental damage. The lack of standardised analytical procedures adds to this difficulty. Current evaluation frameworks frequently fail to account for all types of contaminant interactions or variations in concentrations, which can vary seasonally or geographically. Seasonal agricultural methods and industrial activities exacerbate these variations, making it harder to estimate long-term effects. (Ahmadi Doabi et al., 2019a, b; Ali et al., 2019b; Archana et al., 2016; Arguello-Pérez et al., 2019; Khan et al., 2022; Mishra et al., 2023a).

Lack of Long-Term Data

The lack of long-term data is a key impediment to understanding the full impact of micropollutants on ecosystems, biodiversity, and public health. Unlike well-documented contaminants like carbon dioxide or nitrogen-based substances, micropollutants such as few pharmaceuticals, personal care products, insecticides, and microplastics have only recently received attention as detection technology have improved significantly I recent years. As a result, previous data is often unavailable, leaving researchers with inadequate fundamental understanding to examine patterns, forecast upcoming hazards or devise effective mitigation techniques. Many micropollutants exhibit low concentrations in the environment, leading to acute rather than chronic effects. Subtle impacts, such as endocrine disruption, weakened immune responses, or reduced reproductive success, may take years or decades to manifest. Endocrine disruptors like bisphenol A or synthetic hormones can impair reproduction and development in aquatic organisms over generations, leading to population declines that may not be immediately visible and Persistent Organic Pollutants accumulate in food chains, causing long-term effects such as immune suppression or carcinogenesis in apex predators like carnivores and omnivores. These impacts often become evident only after populations reach critical thresholds of extinction.

Persistent micropollutants build in organisms and spread through trophic levels, with the impacts intensifying every level. However, data on long-term bioaccumulation dynamics are limited. Heavy metals, such as mercury and cadmium, accumulate in aquatic food webs, affecting predators such as lager fish, migratory birds, and marine mammals. The delayed onset of symptoms, such as neurological diseases in mammals or eggshell thinning in reptiles and birds, makes it difficult to link these consequences to pollution exposure without conducting extensive observational studies.

Lack of data is due in part to technological and logistical obstacles. Traditional environmental monitoring strategies frequently focus on well-known pollutants, leaving little capacity to study upcoming emergent pollutants. Micropollutants such as medicines and Nano plastics necessitate specialised analytical procedures like high-resolution mass spectrometry, which are expensive and not commonly available. Micropollutant behaviour and impact are influenced by environmental changes such as natural disasters, anthropogenic activities, climate change, habitat loss, and nutrient loading. Rising temperatures can enhance the solubility and bioavailability of certain pollutants, exacerbating their toxic effects. Habitat degradation like wetland loss reduces natural filtration systems, increasing the persistence and spread of micropollutants.

Previous studies are crucial to understand how these combined stressors influence pollutant distribution and their ecological consequences. Which includes (a) Establishing Global Monitoring Networks, coordinated efforts across nations can provide consistent and comparable long-term datasets (b) Investing in Advanced Technologies, Tools like molecular biomarkers, remote sensing, and big data analytics can enhance detection and monitoring capabilities. (c) Incorporating public Science

Engaging local like "The Lake Health Index" is a citizen science program that aims to bring Lake Health to the forefront by emphasising the importance of indicators that affect lake health in terms of water, ecology, and biodiversity. Maintaining consistent performance across various effluent compositions remains a technical challenge. (d) Developing Predictive Models, Artificial intelligence and Machine learning and simulation models can estimate long-term impacts based on existing short-term data and known ecological trends. (Biazar et al., 2006; Dasenaki et al., 2015; Goulart et al., 2017; Ma et al., 2020; Mishra et al., 2023a; Nicolaou & Checkley, 2021; Pinder et al., 2019; Vilcassim & Thurston, 2023).

Detecting Low Concentrations

Detecting low concentrations of micropollutants in the environment is one of the most challenging aspects of addressing their impact. Micropollutants, such as pharmaceuticals, pesticides, and microplastics, are often present in trace amounts ranging from nanograms to micrograms per Liter in water or parts per billion in soil and sediments. These concentrations, while seemingly negligible, can still exert significant biological effects, especially over long periods. However, existing detection methods are frequently insufficient for precisely analysing and quantifying such low levels, especially in complex environment matrices such as seawater, sediments, and living organisms (Das et al., 2024c; Ray et al., 2024). Modern methods such as high-resolution mass spectroscopy, gas chromatography- mass spectrometry, and liquid chromatography-mass spectrometry tandem have increased sensitivity, but their high cost and specialised skills limit their use. Environmental samples frequently contain combinations of contaminants that can interfere with identification, necessitating advanced sample preparation techniques to separate target molecules. Emerging contaminants, such as nanomaterials or novel medicinal residues, provide further hurdles because detection necessitates the development of whole new analytical procedures. As a result, many micropollutants are undetected or inadequately quantified, leaving considerable gaps in our comprehension of their environmental behaviour and effects. (Abedi et al., 2024; Castiglioni et al., 2005; Pinder et al., 2019; Subedi et al., 2017; Vilcassim & Thurston, 2023; Zavala-Yoe et al., 2017).

Technological and Methodological Advances in Impact Assessment

Advances in Detection Technology

Recent advances in sensor technology have significantly enhanced the ability to detect micropollutants in the environment, resulting in unparalleled sensitivity and selectivity. Electrochemical sensors, have developed as an effective instrumentation for detecting contaminants based on changes in electrical impulses induced by chemical reactions. The incorporation of nanomaterials such as graphene, nanotubes, and quantum dots has dramatically increased the sensitivity of these sensors. Graphene-based electrodes can detect heavy metals such as mercury, lead, and cadmium at parts-per-trillion levels in water, making them ideal for real-time monitoring of aquatic ecosystems. In contrast, optical biosensors use light-based detection techniques that are sometimes combined with organic recognition components such as enzymes or antibodies. Surface plasmon resonance technology, a notable advancement, allows these sensors to detect endocrine disrupting compounds such as bisphenol A, with high precision. Molecularly Imprinted Polymers, another cutting edge innovation, mimic natural receptors to bind specific pollutants. Nanostructured Molecularly Imprinted Polymers, which offer enhanced binding capacity and selectivity, are being utilized to detect persistent organic pollutants and pharmaceuticals in wastewater. Remote sensing technologies have transformed wide-scale environmental monitoring, allowing for the detection of micropollutants across large areas without requiring direct sampling. Satellite imaging, using hyperspectral and multispectral sensors, has proved critical in detecting chemical signatures associated with fertiliser runoff, algal blooms, and surface water pollution. For example, the European Space Agency's Sentinel-2 satellites provide high-resolution photos that highlight the spread of pollution in oceans and smaller water bodies. Drones and unmanned aerial vehicles outfitted with spectrometers and multispectral cameras have broadened the scope of surveillance by reaching remote and hazardous areas. These Unmanned air vehicles have been used to monitor microplastic pollution along shoreline and oil spills in maritime areas. These technologies not only provide comprehensive data but also enhance our understanding of the dynamic behaviour of micropollutants across different ecosystems. The portability and ease of use of these technologies have enabled broader environmental monitoring efforts, particularly in areas where traditional laboratory facilities are unavailable or infeasible.

Portable analytical technologies have transformed field monitoring, allowing for on-site analysis with minimal delay. Handheld gas chromatography–mass spectrometry instruments provide accurate identification of organic contaminants such as polycyclic aromatic hydrocarbons in the field itself samples to a laboratory. Microfluidic technologies, such as paper-based testing platforms, are small and inexpensive methods for quickly detecting contaminants like insecticides in agricultural runoff with small sample quantities. Similarly, portable Raman spectrometers are increasingly being employed for non-invasive monitoring of microplastics in sediments and water. These devices use Raman scattering to determine the chemical makeup of particles, even those in complex combinations and no need for any sample preparations as well. The incorporation of Internet of Things technologies has improved real-time micropollutant detection, monitoring, revolutionising data collection and analysis. This enabled water sensors installed in rivers, lakes, and wastewater treatment plants can continually monitor crucial factors including heavy metal concentrations, nutrient levels, and pH. The Libelium Smart Water IoT platform employs a network of sensors to detect nitrates, ammonia, and arsenic in real time, giving essential information for water quality management. Wearable air quality monitors, like Plume Labs' Flow, allow users to evaluate their personal exposure to urban air contaminants including NO_2 and PM2.5. These devices, combined with cloud-based data analytics, allow for real-time visualization and tracking of pollution events on regional and global scales. By leveraging IoT, researchers and policymakers can access high-resolution datasets that inform immediate responses to pollution incidents, while also contributing to long-term strategies for environmental sustainability. (Benjamin & Júnior, 2022; Dasenaki et al., 2015; Escher et al., 2008, 2011; Li et al., 2016; Plume Labs: Be Empowered against Air Pollution, n.d.; Water Management in Green Areas—Libelium, n.d.)

Integrated Risk Assessment Models

Integrated Risk Assessment (IRA) models have become essential tools for evaluating the multifaceted risks posed by micropollutants to both ecosystems and humans. These models integrate data from various disciplines, such as ecotoxicology, environmental chemistry, ecology, and socioeconomics, to provide a comprehensive framework for assessing the cumulative impacts of micropollutants. Risk assessment models take several factors into account, their interconnections, and their overall impacts on ecosystems and communities, in contrast to traditional risk evaluations that concentrate only on individual contaminants or particular exposure pathways. This all-encompassing method is especially helpful in comprehending the intricate and frequently cooperative impacts of environmental micropollutants. The ability of these models to integrate data from several temporal and spatial scales is one of its primary characteristics. For example, dynamic models take into consideration hydrological processes, chemical degradation, and bioaccumulation when simulating the dispersion of pollutants over time. These models are able to forecast the long-term effects on aquatic ecosystems of persistent micropollutants like heavy metals or medications. Unlike traditional risk assessments that focus on isolated pollutants or specific exposure pathways, IRA models consider multiple stressors, their interactions, and their combined effects on ecosystems and populations. This holistic approach is particularly valuable in understanding the complex and often synergistic effects of micropollutants in the environment.

These models are able to predict the long-term effects on aquatic ecosystems of persistent micropollutants like heavy metals or medications. Similar to this, spatially explicit models map pollution hotspots and evaluate the dangers to particular areas or ecosystems by integrating geographic information system (GIS) data. These models offer a more accurate representation of pollutant behaviour and its ecological effects by fusing temporal and geographical components. The incorporation of biological and molecular data into conventional frameworks is another important development in IRA models. Higher-level consequences like decreased reproduction or loss of population are linked to molecular-level effects like enzymatic malfunctions or changes in gene expression through the use of tools like Adverse Outcome Pathways (AOPs). For example, AOP-based models have been used to assess the impacts of endocrine-disrupting chemicals on fish populations by linking hormonal changes to population-level effects. This molecular-to-ecosystem integration allows for the identification of critical thresholds and early warning indicators, enhancing the predictive power of risk assessments. (Biazar et al., 2006; Goulart et al., 2017; Jiang et al., 2013; Peng et al., 2023; Yang et al., 2019b).

Innovative Ecotoxicological Methods

Understanding of the effects of micropollutants on ecosystems has significantly increased thanks to the development of novel ecotoxicological techniques. Researchers may now evaluate the impacts of contaminants at several biological levels, ranging from molecular interactions to ecosystem-wide alterations, by combining molecular biology tools and sophisticated biomonitoring techniques. These techniques offer a more dynamic and in-depth viewpoint, facilitating more precise environmental risk assessments and focused responses. Molecular Biology Tools in Ecotoxicology, the integration of genomic, proteomic, and metabolomic approaches has transformed ecotoxicological studies, offering unprecedented insights into how organisms respond to micropollutant exposure.

- Genomics: Through the use of genomic technologies like transcriptomics and whole-genome sequencing, scientists may detect alterations in gene expression brought on by micropollutants. Transcriptome study of fish exposed to pharmaceutical residues, for instance, can identify changes in genes related to reproductive health and endocrine disturbance. Early detection of ecotoxicological stress is improved by the discovery of biomarkers for particular pollutants made possible by high-throughput sequencing technologies.
- Proteomics: Proteomic methods examine how pollutants affect the expression and function of proteins. Through protein profiling, researchers can identify biological pathways altered by exposure to micropollutants such as pesticides, heavy metals, or pharmaceutical residues. For example, alterations in antioxidant enzymes or stress proteins in invertebrates exposed to agricultural runoff reveal information about sub-lethal toxicity processes.
- Metabolomics: Analysing the minuscule molecules (metabolites) created during biological processes is the main goal of metabolomics, which provides a glimpse

into the organism's metabolic condition under stress. Certain metabolites, such as those linked to energy metabolism or oxidative stress, may be present in response to exposure to industrial chemicals or contaminants like polycyclic aromatic hydrocarbons (PAHs). This method is especially useful for comprehending the long-term, less-than-lethal impacts of contaminants at low concentrations.

- Advanced Biomonitoring Techniques: Advanced biomonitoring techniques have enhanced our ability to detect and quantify the effects of micropollutants in real-world settings, bridging the gap between laboratory studies and environmental applications.
- Biomarker-Based Monitoring: For assessing how life forms react to contaminants, biomarkers including oxidative stress indicators, hormone levels, and enzyme activity are frequently employed. For instance, detecting neurotoxic chemicals in freshwater systems has been made possible by monitoring the activity of acetylcholinesterase in aquatic organisms. Actual time, field-based biomarker assessment is now possible because of developments in biosensors and bioassays, which eliminates the need for lab space.
- Passive Sampling Devices:Monitoring the bioactive fraction of microscopic pollutants in water and sediments has been transformed by passive sampling techniques including diffusive gradients in thin films (DGTs) and polar organic chemical integrative samplers (POCIS). By simulating biological uptake, these tools provide ecologically meaningful exposure data that is not possible with conventional grab sample techniques.
- Bioassays with Model Organisms: Imaging and automated tracking technologies are being used to improve bioassays that use model organisms such as water fleas (Daphnia magna) or zebrafish (Danio rerio). Video analysis behavioural assays, for example, can identify minute alterations in feeding, movement, or predator avoidance brought on by exposure to sub-lethal pollutants. These tests provide quick, high-throughput screening for possible environmental hazards.
- Integrative Ecosystem Monitoring: In order provide an in-depth understanding of the effects of pollutants, new methods combine physical, chemical, and biological monitoring. For example, the disruption of food webs and trophic relationships by micropollutants can be evaluated using ecological network analysis in conjunction with sophisticated biomonitoring methods. Large-scale pollution evaluations are improved when remote sensing data is combined with biomarker-based research to provide spatial perspectives on ecosystem health. (Avnery et al., 2011; Martín et al., 2021; Martínez-Morcillo et al., 2020; Mwakalapa et al., 2018; Peng et al., 2023; Schultz et al., 2015; Verlicchi et al., 2010; Xu et al., 2021)

Conclusion

Micropollutant research has drawn a lot of attention because of their prevalence and possible threats to both environmental and human health. Effective detection and monitoring methods are required to determine the extent of their occurrence and distribution in different ecosystems. The quantification of these contaminants gives crucial data for their evaluation, whereas environmental risk assessments and ecotoxicology investigations provide insight into their potential to upset ecological equilibrium. These evaluations highlight the importance of standardised methodologies for precisely measuring the hazards presented by micropollutants. Micro-pollutants have a profound impact on diverse ecosystems, affecting water bodies, marshy systems, soil beds, reefs, and even the atmosphere. In aquatic and marshy environments, they disrupt habitats and jeopardize aquatic species, while in soil and terrestrial systems, they alter nutrient cycles and reduce fertility. Furthermore, atmospheric contamination by micropollutants jeopardises air quality and contributes to climate change concerns. The widespread and bio accumulative character of these contaminants highlights the urgent need for comprehensive mitigation techniques designed to protect these essential ecosystems from long-term damage. Despite advancements in pollution treatment technologies, significant obstacles remain in tackling the danger of micropollutants. Their complicated chemical makeup, combined with a paucity of long-term environmental data, makes it difficult to forecast their cumulative impacts over time. Furthermore, because micropollutants exist at extremely low concentrations, detecting and quantifying them might be difficult. These problems underscore the need for more advanced research and the development of cutting-edge techniques that can increase detection accuracy, boost risk assessments, and guide appropriate contamination management policies. Advances in detection technologies, such as high-resolution analytical methods, have enabled more precise identification of micro-pollutants, while integrated risk assessments provide a holistic understanding of their impacts. To address these concerns, technological and methodological innovations are imperative. Furthermore, innovative ecotoxicological approaches are paving the way for more sustainable solutions to manage and mitigate their effects. A multidisciplinary effort, combining advancements in science, technology, and policy, is crucial to tackle the challenges posed by micro-pollutants and ensure a balanced and sustainable coexistence between human activities and natural ecosystems.

References

Abedi, G., Moghadam, A. J., Montazeri, A., Hajyani, Z., Nezarat, A., & Aghababai Beni, A. (2024). Synthesis of green ceramic adsorbent for the treatment of tire factory effluent containing lead, zinc, aluminum, cobalt, iron and manganese ions. *Results in Engineering, 23*, 102591. https://doi.org/10.1016/j.rineng.2024.102591

Aghababai Beni, A., & Jabbari, H. (2022). Nanomaterials for environmental applications. *Results in Engineering, 15*, 100467. https://doi.org/10.1016/j.rineng.2022.100467

Ahmadi Doabi, S., Karami, M., & Majid, A. (2019a). Heavy metal pollution assessment in agricultural soils of Kermanshah province Iran. *Environmental Earth Sciences, 78*, 58–101. https://doi.org/10.1007/s12665-019-8093-7

Ahmadi Doabi, S., Karami, M., & Afyuni, M. (2019b). Heavy metal pollution assessment in agricultural soils of Kermanshah province Iran. *Environmental Earth Sciences, 78*(3), 70. https://doi.org/10.1007/s12665-019-8093-7

Ahmed, M. B., Zhou, J. L., Ngo, H. H., Guo, W., Thomaidis, N. S., & Xu, J. (2017). Progress in the biological and chemical treatment technologies for emerging contaminant removal from wastewater: A critical review. *Journal of Hazardous Materials, 323*, 274–298. https://doi.org/10.1016/j.jhazmat.2016.04.045

Ajmal, M., & Khan, A. U. (1985). Effects of a textile factory effluent on soil and crop plants. *Environmental Pollution Series a, Ecological and Biological, 37*(2), 131–148. https://doi.org/10.1016/0143-1471(85)90004-2

Ali, H., Khan, E., & Ilahi, I. (2019a). Environmental chemistry and ecotoxicology of hazardous heavy metals: Environmental persistence, toxicity, and bioaccumulation. *Journal of Chemistry, 2019*, e6730305. https://doi.org/10.1155/2019/6730305

Ali, H., Khan, E., & Ilahi, I. (2019b). Environmental chemistry and ecotoxicology of hazardous heavy metals: environmental persistence, toxicity, and bioaccumulation. *Journal of Chemistry, 2019*, 1–14. https://doi.org/10.1155/2019/6730305

Anumol, T., Vijayanandan, A., Park, M., Philip, L., & Snyder, S. A. (2016). Occurrence and fate of emerging trace organic chemicals in wastewater plants in Chennai, India. *Environment International, 92–93*, 33–42. https://doi.org/10.1016/j.envint.2016.03.022

Archana, G., Dhodapkar, R., & Kumar, A. (2016). Offline solid-phase extraction for preconcentration of pharmaceuticals and personal care products in environmental water and their simultaneous determination using the reversed phase high-performance liquid chromatography method. *Environmental Monitoring and Assessment, 188*(9), 512. https://doi.org/10.1007/s10661-016-5510-1

Archer, E., Petrie, B., Kasprzyk-Hordern, B., & Wolfaardt, G. M. (2017). The fate of pharmaceuticals and personal care products (PPCPs), endocrine disrupting contaminants (EDCs), metabolites and illicit drugs in a WWTW and environmental waters. *Chemosphere, 174*, 437–446. https://doi.org/10.1016/j.chemosphere.2017.01.101

Arguello-Pérez, M. A., Mendoza-Pérez, J. A., Tintos-Gómez, A., RamírezAyala, E., Godínez-Domínguez, E., & Silva-Bátiz, F. A. (2019). Ecotoxicological analysis of emerging contaminants from wastewater discharges in the coastal zone of Cihuatlán (Jalisco, Mexico). *Water, 11*(7), 1386. https://doi.org/10.3390/w11071386

Avnery, S., Mauzerall, D. L., Liu, J., & Horowitz, L. W. (2011). Global crop yield reductions due to surface ozone exposure: 1. Year 2000 crop production losses and economic damage. *Atmospheric Environment, 45*(13), 2284–2296. https://doi.org/10.1016/j.atmosenv.2010.11.045

Bala, K., Anshul, Verma, S., Dogra, A., Das, A. P., Kumar, R., & Nandini. (2024a). Sustainable management for urban plastic waste generation. *Sustainable Management of Urban Plastic Waste Through Circular Economic Approaches*, 211–236. https://doi.org/10.1201/9781003465447-13

Bala, K., Dogra, A., Lata, P., & Das, A. P. (2024b). Discussing sources and biological-chemical remediation approaches for microplastic pollution. *Sustainable Microbial Technology for Synthetic and Cellulosic Microfiber Bioremediation*, 37–59. https://doi.org/10.1007/978-3-031-62600-5_3

Barhoum, A., Hamimed, S., Slimi, H., Othmani, A., Abdel-Haleem, F. M., & Bechelany, M. (2023). Modern designs of electrochemical sensor platforms for environmental analyses: Principles, nanofabrication opportunities, and challenges. *Trends in Environmental Analytical Chemistry, 38*, e00199. https://doi.org/10.1016/j.teac.2023.e00199

Barnes, K. K., Christenson, S. C., Kolpin, D. W., Focazio, M. J., Furlong, E. T., Zaugg, S. D., Meyer, M. T., & Barber, L. B. (2004). Pharmaceuticals and other organic waste water contaminants within a leachate plume downgradient of a municipal landfill. *Groundwater Monitoring & Remediation, 24*(2), 119–126. https://doi.org/10.1111/j.1745-6592.2004.tb00720.x

Baronti, C., Curini, R., D'Ascenzo, G., Di Corcia, A., Gentili, A., & Samperi, R. (2000). Monitoring natural and synthetic estrogens at activated sludge sewage treatment plants and in a receiving river water. *Environmental Science & Technology, 34*(24), 5059–5066. https://doi.org/10.1021/es001359q

Bartelt-Hunt, S., Snow, D. D., Damon-Powell, T., & Miesbach, D. (2011). Occurrence of steroid hormones and antibiotics in shallow groundwater impacted by livestock waste control facilities. *Journal of Contaminant Hydrology, 123*(3–4), 94–103. https://doi.org/10.1016/j.jconhyd.2010.12.010

Basem, A., Jasim, D. J., Majdi, H. S., Mohammed, R. M., Ahmed, M., Al-Rubaye, A. H., & Kianfar, E. (2024a). Adsorption of heavy metals from wastewater by chitosan: A review. *Results in Engineering, 23*, 102404. https://doi.org/10.1016/j.rineng.2024.102404

Basque, J., Lavoie, J., Reynier, N., & Larivière, D. (2023). Quantitative separation of thorium from rare earth elements and uranium in a rare earth element sulfuric acid leachate using cloud point extraction. *Microchemical Journal, 190*, 108724. https://doi.org/10.1016/j.microc.2023.108724

Behera, I. D., Basak, G., Kumar, R. R., Sen, R., & Meikap, B. C. (2020). Treatment of petroleum refinery sludge by petroleum degrading bacterium *Stenotrophomonas pavanii* IRB19 as an efficient novel technology. *Journal of Environmental Science and Health, Part A, 56*(2), 226–239. https://doi.org/10.1080/10934529.2020.1866924

Behera I. D., Nayak M., Biswas S., Meikap B. C, Sen R., (2021). Enhanced biodegradation of total petroleum hydrocarbons by implementing a novel two-step bioaugmentation strategy using indigenous bacterial consortium. *Journal of Environmental Management*, *292*, 112746, ISSN 0301-4797. https://doi.org/10.1016/j.jenvman.2021.112746

Behera I. D., Nayak M., Mishra A., Meikap B. C., Sen R., (2022). Strategic implementation of integrated bioaugmentation and biostimulation for efficient mitigation of petroleum hydrocarbon pollutants from terrestrial and aquatic environment. *Marine Pollution Bulletin, 177*, 113492, ISSN 0025-326X. https://doi.org/10.1016/j.marpolbul.2022.113492

Benjamin, S. R., & Júnior, E. J. M. R. (2022). Graphene-Based electrochemical sensors for detection of environmental pollutants. *Current Opinion in Environmental Science & Health, 29*, 100381. https://doi.org/10.1016/j.coesh.2022.100381

Biazar, J., Farrokhi, L., & Islam, M. R. (2006). Modeling the pollution of a system of lakes. *Applied Mathematics and Computation, 178*(2), 423–430. https://doi.org/10.1016/j.amc.2005.11.056

Cabrera-Peralta, J., & Peña-Alvarez, A. (2018). Simple method for the determination of personal care product ingredients in lettuce by ultrasound-assisted extraction combined with solid-phase microextraction followed by GC–MS. *Journal of Separation Science, 41*(10), 2253–2260. https://doi.org/10.1002/jssc.201701244

Cabrera-Peralta, J., & Peña-Alvarez, A. (2018b). Simple method for the determination of personal care product ingredients in lettuce by ultrasound-assisted extraction combined with solid-phase microextraction followed by GC–MS. *Journal of Separation Science, 41*(10), 2253–2260. https://doi.org/10.1002/jssc.201701244

Castiglioni, S., Bagnati, R., Calamari, D., Fanelli, R., & Zuccato, E. (2005). A multiresidue analytical method using solid-phase extraction and high-pressure liquid chromatography tandem mass spectrometry to measure pharmaceuticals of different therapeutic classes in urban wastewaters. *Journal of Chromatography A, 1092*(2), 206–215. https://doi.org/10.1016/j.chroma.2005.07.012

Chirinos-Peinado, D. M., & Castro-Bedriñana, J. I. (2020). Lead and cadmium blood levels and transfer to milk in cattle reared in a mining area. *Heliyon, 6*(3), e03579. https://doi.org/10.1016/j.heliyon.2020.e03579

Das, P., Halder, G., & Bal, M. (2023). A critical review on remediation of microplastics using microalgae from aqueous system. *Science of the Total Environment, 898*, 166425. https://doi.org/10.1016/j.scitotenv.2023.166425

Das, A., Dey, S., Bala, K., Gnanasekaran, R., & Das, A. P. (2024a). Microbial bioremediation technology for sustainable treatment and management of synthetic microfiber waste. *Sustainable Microbial Technology for Synthetic and Cellulosic Microfiber Bioremediation*, 221–238. https://doi.org/10.1007/978-3-031-62600-5_12

Das, A., Dey, S., & Das, A. P. (2024b). Ecotoxicological impacts of synthetic microfiber pollutants and development of sustainable mitigation strategies. *Environmental Chemistry and Ecotoxicology*. https://doi.org/10.1016/J.ENCECO.2024.12.003

Das, A., Ray, S. K., Mohanty, M., Mohanty, J., Dey, S., & Das, A. P. (2024c). Ecotoxicity of microplastic wastes and their sustainable management: A review. *Environmental Chemistry and Ecotoxicology, 6*, 144–152. https://doi.org/10.1016/J.ENCECO.2024.05.003

Dasenaki, M. E., Bletsou, A. A., Koulis, G. A., & Thomaidis, N. S. (2015). Qualitative multiresidue screening method for 143 veterinary drugs and pharmaceuticals in milk and fish tissue using liquid chromatography quadrupole-time-of-flight mass spectrometry. *Journal of Agricultural and Food Chemistry, 63*(18), 4493–4508. https://doi.org/10.1021/acs.jafc.5b00962

Dey, S., Dash, A., & Das, A. P. (2024a). Bioremediation as a sustainable and circular solution for mining waste management and bioenergy generation. *Sustainable Management of Mining Waste and Tailings: A Circular Economy Approach*, 107–119. https://doi.org/10.1201/9781003442455-6

Dey, S., Dash, N., Ray, S. K., Priyadarsini, S., Chhotaray, C., & Das, A. P. (2024b). Occurrence and quantification of tire wear plastics pollutants from highway city Dust of Bhubaneswar city of Odisha. *Sustainable Microbial Technology for Synthetic and Cellulosic Microfiber Bioremediation*, 77–89. https://doi.org/10.1007/978-3-031-62600-5_5

Escher, B. I., Bramaz, N., Mueller, J. F., Quayle, P., Rutishauser, S., & Vermeirssen, E. L. M. (2008). Toxic equivalent concentrations (TEQs) for baseline toxicity and specific modes of action as a tool to improve interpretation of ecotoxicity testing of environmental samples. *Journal of Environmental Monitoring, 10*(5), 612. https://doi.org/10.1039/b800949j

Escher, B. I., Lawrence, M., Macova, M., Mueller, J. F., Poussade, Y., Robillot, C., Roux, A., & Gernjak, W. (2011). Evaluation of contaminant removal of reverse osmosis and advanced oxidation in full-scale operation by combining passive sampling with chemical analysis and bioanalytical tools. *Environmental Science & Technology, 45*(12), 5387–5394. https://doi.org/10.1021/es201153k

Filonchyk, M., & Peterson, M. P. (2023). An integrated analysis of air pollution from US coal-fired power plants. *Geoscience Frontiers, 14*(2), 101498. https://doi.org/10.1016/j.gsf.2022.101498

Ghanem, S., Ferrini, S., & Di Maria, C. (2023). Air pollution and willingness to pay for health risk reductions in Egypt: A contingent valuation survey of Greater Cairo and Alexandria households. *World Development, 172*, 106373. https://doi.org/10.1016/j.worlddev.2023.106373

Goulart, A. G. O., Lazo, M. J., Suarez, J. M. S., & Moreira, D. M. (2017). Fractional derivative models for atmospheric dispersion of pollutants. *Physica a: Statistical Mechanics and Its Applications, 477*, 9–19. https://doi.org/10.1016/j.physa.2017.02.022

Greaver, T. L., Sullivan, T. J., Herrick, J. D., Barber, M. C., Baron, J. S., Cosby, B. J., Deerhake, M. E., Dennis, R. L., Dubois, J. B., Goodale, C. L., Herlihy, A. T., Lawrence, G. B., Liu, L., Lynch, J. A., & Novak, K. J. (2012a). Ecological effects of nitrogen and sulfur air pollution in the US: What do we know? *Frontiers in Ecology and the Environment, 10*(7), 365–372. https://doi.org/10.1890/110049

https://plumelabs.com/en/

https://www.libelium.com/iot-solutions/smart-water/

Jiang, J.-Q., Zhou, Z., & Sharma, V. K. (2013). Occurrence, transportation, monitoring and treatment of emerging micro-pollutants in waste water—a review from global views. *Microchemical Journal, 110*, 292–300. https://doi.org/10.1016/j.microc.2013.04.014

John, A., & Das, A. P. (2024). Government strategies for achieving circular economy for plastic waste management in Southeast Asia. *Sustainable Management of Urban Plastic Waste Through Circular Economic Approaches*, 72–82. https://doi.org/10.1201/9781003465447-6

Jyotsna, K., & Tandon, A. (2018). A nonlinear mathematical model investigating the sustainability of an urban system in the presence of haphazard urban development and excessive pollution. *Natural Resource Modeling, 31*(2). https://doi.org/10.1111/nrm.12163

Kanungo, J., Sahoo, T., Bal, M., & Behera, I. D. (2023a). Performance of bioremediation strategy in waste lubricating oil pollutants: a review. *Geomicrobiology Journal, 41*(4), 360–373. https://doi.org/10.1080/01490451.2023.2245395

Kanungo, J., Sahoo, T., Swain, L. P., Behera, I. D. (2023). Toxicity of persistent hydrocarbon pollutants, sources and sustainable remediation process. In: I. D Behera D & A. P. Das (Eds.)

Impact of Petroleum Waste on Environmental Pollution and its Sustainable Management Through Circular Economy. Environmental Science and Engineering. Springer, Cham. https://doi.org/10.1007/978-3-031-48220-5_2

Khan, J., Singh, R., Upreti, P., & Yadav, R. K. (2022). Geo-statistical assessment of soil quality and identification of Heavy metal contamination using Integrated GIS and multivariate statistical analysis in industrial region of Western India. *Environmental Technology & Innovation, 28*, 102646. https://doi.org/10.1016/j.eti.2022.102646

Lapworth, D., Baran, N., Stuart, M., & Ward, R. (2012a). Emerging organic contaminants in groundwater: A review of sources, fate and occurrence. *Environmental Pollution, 163*, 287–303. https://doi.org/10.1016/j.envpol.2011.12.034

Lv, J., & Wang, Y. (2019a). PMF receptor models and sequential Gaussian simulation to determine the quantitative sources and hazardous areas of potentially toxic elements in soils. *Geoderma, 353*, 347–358. https://doi.org/10.1016/j.geoderma.2019.07.020

Lv, J., & Wang, Y. (2019b). PMF receptor models and sequential Gaussian simulation to determine the quantitative sources and hazardous areas of potentially toxic elements in soils. *Geoderma, 353*, 347–358. https://doi.org/10.1016/j.geoderma.2019.07.020

Ma, Q., Qi, Y., Shan, Q., Liu, S., & He, H. (2020). Understanding the knowledge gaps between air pollution controls and health impacts including pathogen epidemic. *Environmental Research, 189*, 109949. https://doi.org/10.1016/j.envres.2020.109949

Martín, C., Fajardo, C., Costa, G., Sánchez-Fortún, S., Andrés, M. D. S., González, F., Nande, M., Mengs, G., & Martín, M. (2021). Bioassays to assess the ecotoxicological impact of polyethylene microplastics and two organic pollutants, simazine and ibuprofen. *Chemosphere, 274*, 129704. https://doi.org/10.1016/j.chemosphere.2021.129704

Martínez-Morcillo, S., Rodríguez-Gil, J. L., Fernández-Rubio, J., Rodríguez-Mozaz, S., Míguez-Santiyán, M. P., Valdes, M. E., Barceló, D., & Valcárcel, Y. (2020). Presence of pharmaceutical compounds, levels of biochemical biomarkers in seafood tissues and risk assessment for human health: Results from a case study in North-Western Spain. *International Journal of Hygiene and Environmental Health, 223*(1), 10–21. https://doi.org/10.1016/j.ijheh.2019.10.011

Mishra, R. K., Mentha, S. S., Misra, Y., & Dwivedi, N. (2023a). Emerging pollutants of severe environmental concern in water and wastewater: A comprehensive review on current developments and future research. *Water-Energy Nexus, 6*, 74–95. https://doi.org/10.1016/j.wen.2023.08.002

Mishra, S., Dash, D., & Das, A. P. (2023b). Aquatic microbial diversity on plastisphere: Colonization and potential role in microplastic biodegradation. *Geomicrobiology Journal*, 1–12. https://doi.org/10.1080/01490451.2023.2209750

Mishra, S., Dash, D., & Das, A. P. (2024). Microbial degradation of polyester microfibers using indigenously isolated bacterial strain Exiguobacterium Sp. *CLEAN—Soil, Air, Water, 52*(12), e202300343. https://doi.org/10.1002/CLEN.202300343

Mohanty, J., Dey, S., Bala, K., & Das, A. P. (2024). Recent advances in microplastic pollution for its sustainable management. *Sustainable Management of Urban Plastic Waste Through Circular Economic Approaches*, 27–48. https://doi.org/10.1201/9781003465447-3

Mohanty, M., Mohanty, J., Dey, S., Dutta, K., Shah, M. P., & Das, A. P. (2024b). The face mask: A tale from protection to pollution and demanding sustainable solution. *Emerging Contaminants, 10*(2), 100298. https://doi.org/10.1016/J.EMCON.2023.100298

Mohanty, M., Tripathy, B., Koteswara Reddy, G., & Das, A. P. (2024c). Greener technologies for establishing restraint over microplastic pollution. *Sustainable Microbial Technology for Synthetic and Cellulosic Microfiber Bioremediation*, 239–258. https://doi.org/10.1007/978-3-031-62600-5_13

Morin-Crini, N., Lichtfouse, E., Liu, G., Balaram, V., Ribeiro, A. R. L., Lu, Z., Stock, F., Carmona, E., Teixeira, M. R., Picos-Corrales, L. A., Moreno-Piraján, J. C., Giraldo, L., Li, C., Pandey, A., Hocquet, D., Torri, G., & Crini, G. (2022). Worldwide cases of water pollution by emerging contaminants: A review. *Environmental Chemistry Letters, 20*(4), 2311–2338. https://doi.org/10.1007/s10311-022-01447-4

Mwakalapa, E. B., Mmochi, A. J., Müller, M. H. B., Mdegela, R. H., Lyche, J. L., & Polder, A. (2018). Occurrence and levels of persistent organic pollutants (POPs) in farmed and wild marine fish from Tanzania. A pilot study. *Chemosphere, 191*, 438–449. https://doi.org/10.1016/j.chemosphere.2017.09.121

Nicolaou, L., & Checkley, W. (2021). Inequities in air pollution exposure and gaps in air quality monitoring. *Journal of Allergy and Clinical Immunology, 148*(1), 64–66. https://doi.org/10.1016/j.jaci.2021.04.014

Parizanganeh, A., Hajisoltani, P., & Zamani, A. (2010). Concentration, distribution and comparison of total and bioavailable metals in top soils and plants accumulation in Zanjan Zinc industrial town-Iran. *Procedia Environmental Sciences, 2*, 167–174. https://doi.org/10.1016/j.proenv.2010.10.020

Paul, A., Priyadarsini, S., Karakuş, S., Behera, J. K., Gnanasekaran, R., & Das, A. P. (2024). Fate and accumulation of synthetic microfibers in the environment. *Sustainable Microbial Technology for Synthetic and Cellulosic Microfiber Bioremediation*, 61–76. https://doi.org/10.1007/978-3-031-62600-5_4

Peng, X., Mousa, S., Sarfraz, M., A, N. A., & Haffar, M. (2023). Improving mineral resource management by accurate financial management: Studying through artificial intelligence tools. *Resources Policy, 81*, 103323. https://doi.org/10.1016/j.resourpol.2023.103323

Pinder, R. W., Klopp, J. M., Kleiman, G., Hagler, G. S., Awe, Y., & Terry, S. (2019). Opportunities and challenges for filling the air quality data gap in low-and middle-income countries. *Atmospheric Environment, 215*, 116794. https://doi.org/10.1016/j.atmosenv.2019.06.032

Pradhan, G., Tripathy, B., Ram, D. K., Digal, A. K., & Das, A. P. (2023). Bauxite mining waste pollution and its sustainable management through bioremediation. *Geomicrobiology Journal*, 1–10. https://doi.org/10.1080/01490451.2023.2235353

Priyadarshini, B., & Das, A. P. (2024). The adverse impact of synthetic microplastics released from disposable face masks pollutants on aquatic microbiota, aquatic species and humans: A review. *Environmental Science and Engineering, Part, F2827*, 67–81. https://doi.org/10.1007/978-3-031-50840-0_5

Priyadarshini, B., Santhosh Kumar, M., & Das, A. P. (2024). Microbial degradation of synthetic plastics released from COVID-19 generated single use disposable face masks. *Environmental Science and Engineering, Part, F2827*, 1–16. https://doi.org/10.1007/978-3-031-50840-0_1

Priyadarsini, S., & Das, A. P. (2024). Analytical and structural characterization of waste lithium-ion batteries for their effective recycling strategy. *Environmental Chemistry and Ecotoxicology*. https://doi.org/10.1016/j.enceco.2024.12.004

Priyadarsini, S., & Das, A. P. (2025). Lithium bioleaching: A review on microbial-assisted sustainable technology for lithium bio-circularity. *Journal of Water Process Engineering, 69*, 106744. https://doi.org/10.1016/j.jwpe.2024.106744

Raj, K., & Das, A. P. (2023). Lead pollution: Impact on environment and human health and approach for a sustainable solution. *Environmental Chemistry and Ecotoxicology, 5*, 79–85. https://doi.org/10.1016/J.ENCECO.2023.02.001

Ratchnashree, S., Karmegam, N., Selvam, M., Manikandan, S., Deena, S. R., Subbaiya, R., Vickram, A., Kim, W., & Govarthanan, M. (2023). Advanced technologies for the determination of quantitative structure-activity relationships and degradation efficiency of micropollutants and their removal in water—a review. *The Science of the Total Environment, 904*, 166563. https://doi.org/10.1016/j.scitotenv.2023.166563

Ray, S. K., Dey, S., Das, A. P., & John, A. (2024). Microplastic waste management through greener technologies : a review. *Sustainable Management of Urban Plastic Waste Through Circular Economic Approaches*, 237–261. https://doi.org/10.1201/9781003465447-14

Sahu, A., Dey, S., Mallick, K., Priyadarsini, S., Behera, I. D., & Das, A. P. (2024a). A review on advanced technology for sustainable management of synthetic microplastic waste. *Sustainable Management of Urban Plastic Waste Through Circular Economic Approaches*, 262–279. https://doi.org/10.1201/9781003465447-15

Sahu, A., Mallick, K., & Das, A. P. (2024b). Environmental impact of single-use synthetic face mask and its recycling: A sustainable approach (pp. 197–212). https://doi.org/10.1007/978-3-031-50840-0_11

Satapathy, S., Kar, D., Priyadarsini, S., Das, A. P., & Kuanar, A. (2024). Application of microorganisms in emerging waste management strategies and remediation of synthetic and cellulosic microfibers. *Sustainable Microbial Technology for Synthetic and Cellulosic Microfiber Bioremediation*, 183–206. https://doi.org/10.1007/978-3-031-62600-5_10

Schultz, L., Zurich, M., Culot, M., Da Costa, A., Landry, C., Bellwon, P., Kristl, T., Hörmann, K., Ruzek, S., Aiche, S., Reinert, K., Bielow, C., Gosselet, F., Cecchelli, R., Huber, C. G., Schroeder, O. H., Gramowski-Voss, A., Weiss, D. G., & Bal-Price, A. (2015). Evaluation of drug-induced neurotoxicity based on metabolomics, proteomics and electrical activity measurements in complementary CNS in vitro models. *Toxicology in Vitro, 30*(1), 138–165. https://doi.org/10.1016/j.tiv.2015.05.016

Subedi, B., Balakrishna, K., Joshua, D. I., & Kannan, K. (2017). Mass loading and removal of pharmaceuticals and personal care products including psychoactives, antihypertensives, and antibiotics in two sewage treatment plants in southern India. *Chemosphere, 167*, 429–437. https://doi.org/10.1016/j.chemosphere.2016.10.026

Sun, F., Dai, Y., & Yu, X. (2017). Air pollution, food production and food security: A review from the perspective of food system. *Journal of Integrative Agriculture, 16*(12), 2945–2962. https://doi.org/10.1016/S2095-3119(17)61814-8

Tripathy, B., Dash, A., & Das, A. P. (2022). Detection of environmental microfiber pollutants through vibrational spectroscopic techniques: Recent advances of environmental monitoring and future prospects. *Critical Reviews of Analytical Chemistry*, 1–11. https://doi.org/10.1080/10408347.2022.2144994

Tripathy, B., Sahoo, P. P., Sundaray, H., & Das, A. P. (2024). In-vitro biodegradation of discarded marine microplastics across the eastern coast of the Bay of Bengal, India using Exiguobacterium Sp. *Environmental Chemistry and Ecotoxicology, 6*, 236–247. https://doi.org/10.1016/J.ENCECO.2024.07.003

Verlicchi, P., Galletti, A., Petrovic, M., & Barceló, D. (2010). Hospital effluents as a source of emerging pollutants: An overview of micropollutants and sustainable treatment options. *Journal of Hydrology, 389*(3–4), 416–428. https://doi.org/10.1016/j.jhydrol.2010.06.005

Vilcassim, R., & Thurston, G. D. (2023). Gaps and future directions in research on health effects of air pollution. *eBioMedicine, 93*, 104668. https://doi.org/10.1016/j.ebiom.2023.104668

Waterhouse, J., Brodie, J., Lewis, S., & Mitchell, A. (2012). Quantifying the sources of pollutants in the Great Barrier Reef catchments and the relative risk to reef ecosystems. *Marine Pollution Bulletin, 65*(4–9), 394–406. https://doi.org/10.1016/j.marpolbul.2011.09.031

Wightwick, A. M., Bui, A. D., Zhang, P., Rose, G., Allinson, M., Myers, J. H., Reichman, S. M., Menzies, N. W., Pettigrove, V., & Allinson, G. (2012). Environmental fate of fungicides in surface waters of a horticultural-production catchment in Southeastern Australia. *Archives of Environmental Contamination and Toxicology, 62*(3), 380–390. https://doi.org/10.1007/s00244-011-9710-y

Xu, L., Li, J., Najeeb, U., Li, X., Pan, J., Huang, Q., Zhou, W., & Liang, Z. (2021). Synergistic effects of EDDS and ALA on phytoextraction of cadmium as revealed by biochemical and ultrastructural changes in sunflower (*Helianthus annuus* L.) tissues. *Journal of Hazardous Materials, 407*, 124764. https://doi.org/10.1016/j.jhazmat.2020.124764

Yang, S., Zhao, J., Chang, S. X., Collins, C., Xu, J., & Liu, X. (2019a). Status assessment and probabilistic health risk modeling of metals accumulation in agriculture soils across China: A synthesis. *Environment International, 128*, 165–174. https://doi.org/10.1016/j.envint.2019.04.044

Yang, S., He, M., Zhi, Y., Chang, S. X., Gu, B., Liu, X., & Xu, J. (2019b). An integrated analysis on source-exposure risk of heavy metals in agricultural soils near intense electronic waste recycling activities. *Environment International, 133*, 105239. https://doi.org/10.1016/j.envint.2019.105239

Yang, S., Zhao, J., Chang, S. X., Collins, C., Xu, J., & Liu, X. (2019dc). Status assessment and probabilistic health risk modeling of metals accumulation in agriculture soils across China: A synthesis. *Environment International, 128*, 165–174. https://doi.org/10.1016/j.envint.2019.04.044

Yesu Ramya, K., & Kumawat, A. S. (2024). A mini review on the catalytic technology development for the mitigation of various air pollutants. *Materials Today: Proceedings, 102*, 297–304. https://doi.org/10.1016/j.matpr.2023.05.322

Zavala-Yoe, R., Ramírez-Mendoza, R. A., & García-Lara, S. (2017). A 3-SPS-1S parallel robot-based laser sensing for applications in precision agriculture. *Soft Computing, 21*(3), 641–650. https://doi.org/10.1007/s00500-016-2457-7

Zhang, M., & Xu, L. (2022). Transport of micro- and nanoplastics in the environment: Trojan-Horse effect for organic contaminants. *Critical Reviews in Environmental Science and Technology, 52*(5), 810–846. https://doi.org/10.1080/10643389.2020.1845531

Chapter 6
Detailed Examination of the Adverse Effects of Micropollutants on Ecosystems and Human Health

G. Sree Pooja, S. Mounika, B. B. Nayak, M. Raviteja, and Ch. Navya

Introduction

Micropollutants represent a varied collection of synthetic, non-natural contaminants and organic chemical compounds that are introduced into the environment, typically at very low concentrations, ranging from nanograms to micrograms per liter. This category encompasses a range of substances, including pharmaceuticals, agrochemicals (such as pesticides and herbicides), and microplastics. These organic chemical compounds enter the environment through numerous anthropogenic activities, including those related to households, industries, and agriculture. Although present in minimal concentrations, micropollutants can remain in the environment and accumulate within biological organisms, thereby posing considerable risks to ecosystems and human health (El Hammoudani & Dimane, 2021a, 2021b). Pharmaceuticals, including antibiotics and analgesics, often find their way into water systems

G. Sree Pooja
Department of Agricultural Extension Education, Professor Jayashankar Telangana Agricultural University, Hyderabad, India

S. Mounika
Department of Agronomy, Professor Jayashankar Telangana Agricultural University, Hyderabad, India

B. B. Nayak
School of Sciences, Woxsen University, Kamkole, Telangana, India

M. Raviteja (✉)
School of Natural Resource Management, ICAR-Indian Agricultural Research Institute, Dhemaji, Assam, India
e-mail: raviteja.machanuru@icar.gov.in

Ch. Navya
Lab Technician, Woxsen University Telangana, Kamkole, India

I. D. Behera et al. (eds.), *Emerging Micropollutants*, Sustainable Environmental Waste Management Strategies, https://doi.org/10.1007/978-3-032-03530-1_6

via human excretion and improper disposal methods, resulting in ecological disturbances and the emergence of antibiotic-resistant bacteria. Agrochemicals, which include fertilizers, pesticides, and herbicides, are extensively utilized in agriculture to boost crop production and safeguard against pests. However, their runoff can lead to the contamination of water bodies, endangering aquatic life and potentially impacting human health through the food chain. Herbicides such as glyphosate and atrazine are particularly alarming due to their persistence in the environment and their potential to disrupt endocrine functions. Furthermore, microplastics—tiny plastic particles that arise from the degradation of larger plastic products—are increasingly detected in soil and water, raising concerns regarding their long-term effects on ecosystems and human health. The presence of these micropollutants presents significant challenges for environmental management, as conventional wastewater treatment methods frequently do not adequately eliminate them from effluents before they re-enter natural water systems (Wasser 3.0, 2024; Envirotech Online, 2022; BAFU, 2023; Betram et al., 2022).

Types of Micropollutants

Micropollutants can be classified into various categories based on their chemical composition and origins:

Pharmaceuticals and Personal Care Products (PPCPs): This group encompasses a diverse array of substances, including antibiotics, pain relievers, hormones, and other therapeutic agents. Their occurrence in aquatic ecosystems is frequently attributed to inadequate removal during wastewater treatment operations.

Pesticides: Widely utilized in agricultural practices, pesticides may enter water bodies through runoff during precipitation or irrigation, resulting in the contamination of both surface and groundwater sources.

Heavy Metals: Elements such as lead, mercury, and cadmium can infiltrate water systems via industrial effluents and urban runoff. These metals are recognized for their toxic properties and potential to bioaccumulate within living organisms.

Microplastics: Characterized as plastic fragments smaller than 5 mm, microplastics can arise from the degradation of larger plastic products or be intentionally manufactured for use in cosmetics and other applications. They represent a widespread pollutant that can adversely impact marine ecosystems and infiltrate the food chain (Das et al., 2023; Das & Bal, 2025).

Endocrine Disruptors: These substances can disrupt hormonal functions in both wildlife and humans. Notable examples include bisphenol A (BPA) and phthalates, which are commonly found in plastic materials.

Polychlorinated Biphenyls (PCBs): Previously prevalent in electrical equipment, PCBs are persistent organic pollutants that can accumulate in the environment and within living organisms. They are linked to various health hazards, including cancer and immune system dysfunction.

Polycyclic Aromatic Hydrocarbons (PAHs): PAHs are organic compounds generated during the incomplete combustion of organic matter. They can contaminate soil and water through industrial emissions or vehicular exhaust and are recognized for their carcinogenic effects (Behera et al., 2020, 2021, 2022; Kanungo et al., 2023a, 2023b).

Persistent Organic Pollutants (POPs): These organic compounds resist environmental degradation through chemical, biological, and photolytic processes. Examples include certain pesticides and industrial chemicals that can accumulate in the environment.

Industrial Chemicals: Various chemicals used in manufacturing processes can also be classified as micropollutants. These may include solvents, surfactants, and other additives that do not easily degrade in the environment.

Micropollutants in Different Ecosystems

Micropollutants, such as pharmaceuticals and pesticides, represent considerable dangers to diverse ecosystems. In aquatic environments, these substances can interfere with reproductive functions in fish and amphibians, resulting in issues like the feminization of male fish and diminished fertility among aquatic species (WHO). In terrestrial environments, the use of pesticides and herbicides leads to soil contamination, which negatively impacts soil health and agricultural yield. Moreover, the presence of microplastics in soil can modify its physical characteristics, thereby reducing its ability to retain water. Freshwater ecosystems are also affected; the occurrence of pharmaceuticals and polycyclic aromatic hydrocarbons (PAHs) can increase toxicity levels for aquatic organisms, leading to long-term consequences for biodiversity and ecosystem integrity (Borreca et al., 2024). Marine ecosystems are threatened by persistent organic pollutants, such as polychlorinated biphenyls (PCBs), which accumulate within marine food webs and pose health risks to both marine organisms and humans through the consumption of seafood. Additionally, heavy metals can adversely affect marine life, resulting in population declines. In soil ecosystems, heavy metals from industrial processes accumulate, diminishing microbial diversity and hindering plant growth, while microplastics further exacerbate soil health issues by disrupting nutrient cycling (Rubin & Ines, 2022). Collectively, these effects highlight the critical necessity for comprehensive strategies to address micropollutants across all ecosystems (Table 6.1).

Micropollutants in Agriculture

Micropollutants in agriculture pose significant environmental and health risks due to their extensive application and persistence within ecosystems. The use of pesticides leads to soil contamination and a decline in biodiversity, often resulting in

Table 6.1 Micropollutants in different ecosystemts

Ecosystem type	Common micropollutants	Sources	Effects on ecosystem	References
Aquatic ecosystems	Pharmaceuticals, pesticides, heavy metals	Agricultural runoff, wastewater discharge	Disruption of aquatic life (e.g., fish reproductive health); Bioaccumulation in aquatic organisms; Altered microbial communities	Antakyali et al. (2015), Yang et al. (2021)
Terrestrial ecosystems	Pesticides, herbicides, microplastics	Agricultural practices, plastic waste	Soil degradation; Reduced soil fertility; Impacts on terrestrial flora and fauna	EEA (2008), Launay et al. (2013)
Freshwater Systems	PPCPs (e.g., sulfamethoxazole), PAHs	Urban runoff, industrial discharge	Increased toxicity to aquatic organisms; Chronic effects on biodiversity	Narwal et al. (2023), Luo et al. (2014)
Marine ecosystems	PCBs, heavy metals, microplastics	Industrial effluents, shipping activities	Accumulation in marine food chains; Toxic effects on marine life (e.g., fish, mollusks)	EEA (2008), Launay et al. (2013)
Soil ecosystems	Heavy metals, microplastics	Landfills, agricultural runoff	Altered soil structure; Reduced microbial diversity; Impaired plant growth	Musolff (2009), Antakyali et al. (2015)

their accumulation in agricultural runoff, adversely affecting non-target species and infiltrating the food chain. Research indicates that herbicides such as atrazine and metolachlor can pollute surface waters, with atrazine detected at average concentrations of 1.82 mg/L in rivers, raising alarms regarding endocrine disruption and potential carcinogenic risks (WHO). The presence of pharmaceuticals in agricultural environments is increasing, primarily due to runoff from wastewater treatment facilities, which contributes to the emergence of antibiotic-resistant bacteria and hormonal imbalances in both wildlife and humans (Eggen et al., 2014). Additionally, heavy metals originating from industrial processes can accumulate in soils, threatening plant health and resulting in toxic effects when these plants are ingested by humans or livestock (Naeem et al., 2023). Furthermore, microplastics introduced through agricultural plastic materials can disrupt soil structure, diminish water retention capacity, and may be absorbed by crops. The presence of organic micropollutants, including polycyclic aromatic hydrocarbons (PAHs) and polychlorinated biphenyls (PCBs), in wastewater utilized for irrigation further jeopardizes crop safety and human health through contaminated produce (Borreca et al., 2024). Lastly, the overuse of nitrogen fertilizers leads to nutrient runoff, resulting in eutrophication of adjacent water bodies, which depletes oxygen levels and endangers aquatic life (Liu et al., 2023). Collectively, these challenges highlight the pressing need for sustainable agricultural practices that reduce the reliance on harmful chemicals and safeguard both ecosystems and human health (Table 6.2).

As of 2020, over 350,000 chemicals were documented for global production, with a considerable number categorized as micropollutants (Borreca et al., 2024; Wang et al., 2020). Research conducted by Yang et al. (2021) identified and prioritized 53 micropollutants, which included 36 pharmaceuticals (notably sulfamethoxazole and ibuprofen), 7 pesticides, and 10 industrial chemicals that are vital for effective risk management and remediation strategies. The application of synthetic pesticides, herbicides, and fertilizers in agricultural practices raises serious concerns due to their detrimental impacts on soil health, water quality, and the overall stability of food chains. Although these agrochemicals play a crucial role in enhancing crop yields and controlling pests, they are linked to numerous environmental challenges. The use of synthetic pesticides and fertilizers can result in soil degradation by disrupting the natural microbiome that is essential for nutrient cycling and maintaining soil fertility. While pesticides are formulated to target specific pests, they frequently harm non-target organisms, including beneficial soil microorganisms. Research indicates that extended exposure to pesticides can diminish microbial diversity and disrupt biochemical processes vital for sustaining healthy soil ecosystems (Gangola et al., 2023; Janarthanan & Murugan, 2017). For example, the use of certain insecticides has been shown to interfere with nitrogen fixation in leguminous plants, ultimately leading to decreased crop yields. This degradation not only compromises soil quality but also impairs its capacity to retain moisture, which is particularly important during periods of drought.

The runoff from agricultural fields that utilize synthetic pesticides and fertilizers plays a substantial role in water pollution. Research shows that more than 60% of the world's agricultural land is susceptible to pesticide contamination from various active ingredients. When these chemicals infiltrate water systems, they can produce harmful effects on aquatic ecosystems. For instance, studies have revealed that herbicides such as glyphosate can interfere with food chains and negatively impact fish populations (Environmental Protection Agency (EPA, 2023). Furthermore, fertilizers can induce eutrophication in aquatic environments, resulting in algal blooms that reduce oxygen levels and endanger aquatic life (Food and Agriculture Organization (FAO, 2023).

The presence of synthetic pesticides in the environment presents significant risks to both ecosystems and human health via the food chain. These pesticides can accumulate in living organisms, resulting in elevated concentrations in apex predators. Such bioaccumulation may lead to long-term health complications for individuals who ingest contaminated food items. The Food and Agriculture Organization (FAO) has indicated that certain pesticide residues are associated with a range of health issues, including endocrine disruption and an increased likelihood of cancer (FAO, 2023). Additionally, the combined effects of pesticide mixtures can enhance toxicity levels beyond those observed with single substances (EEA, 2023).

Table 6.2 Micropollutants in Agriculture and its health risks

Micropollutant type	Source	Common examples	Impact on agriculture	Detection levels	Health risks	References
Pharmaceuticals	Agricultural runoff, wastewater	Antibiotics (e.g., tetracycline), hormones (e.g., estradiol)	Accumulation in crops can affect human health through consumption; potential development of antibiotic resistance bacteria	ng/L to µg/L in irrigation water	Hormonal disruption, antibiotic resistance	Eggen et al. (2014), Gul et al. (2022), Gupta et al. (2024)
Pesticides	Agricultural application	Glyphosate, atrazine, chlorpyrifos	Contamination of soil and water; potential uptake by crops; impacts on non-target species	Varies widely; often detected in the low µg/L range	Toxicity to humans and wildlife; potential carcinogenic effects	Eggen et al. (2014), Gul et al. (2022), Bruzzoniti et al. (2024)
Herbicides	Agricultural application, runoff	Atrazine, metolachlor, simazine	Soil degradation; water contamination; affects biodiversity and crop health	Average concentration of atrazine found at 1.82 mg/L in rivers	Endocrine disruption; potential carcinogenic effects	Bruzzoniti et al. (2024)
Heavy metals	Industrial runoff, sewage sludge	Lead, cadmium, mercury	Accumulation in soil and crops; can affect plant growth and soil health	Typically measured in mg/kg in soil	Neurological damage; kidney damage; cancer risk	Eggen et al. (2014), Gul et al. (2022)
Microplastics	Breakdown of plastic products, agricultural films	Polyethylene, polypropylene particles	Potential uptake by plants; impacts on soil health and food safety	Detected in soil and water at varying concentrations	Long-term health effects; potential endocrine disruption	Gul et al. (2022), Eggen et al. (2014), Gupta et al. (2024)

(continued)

Table 6.2 (continued)

Micropollutant type	Source	Common examples	Impact on agriculture	Detection levels	Health risks	References
Polycyclic aromatic hydrocarbons (PAHs)	Combustion processes, industrial activities	Benzo[a]pyrene, phenanthrene	Can contaminate crops through irrigation with polluted water	MQLs around 3.9 ng/L for regulated compounds like benzo[a]pyrene	Carcinogenic risks associated with consumption of contaminated crops	Bruzzoniti et al. (2024)
Persistent organic pollutants (POPs)	Agricultural runoff, industrial discharge	PCBs, dioxins	Bioaccumulation in the food chain; affects crop safety and human health	Detected at low concentrations in agricultural soils	Long-term health risks including cancer and reproductive issues	Eggen et al. (2014), Gul et al. (2022), Bruzzoniti et al. (2024)
Nutrients (excess nitrogen)	Fertilizer application	Nitrate, ammonium	Eutrophication of water bodies; soil acidification; reduced crop yield over time	Varies widely based on application rates and soil conditions	Groundwater contamination leading to methemoglobinemia ("blue baby syndrome")	FAO (2021)

Adverse Effects of Micropollutants

1. **Ecosystem Impacts**

Micropollutants pose a significant threat to aquatic ecosystems, particularly by disrupting microbial communities and food webs. Increased concentrations of micropollutants, including pharmaceuticals and pesticides, have been demonstrated to affect planktonic food webs, with algal communities playing a crucial role in mediating cascading effects on higher trophic levels. This disruption can lead to altered community structures and a decline in biodiversity within urban rivers (Adyari et al., 2023). Additionally, these contaminants can affect the diversity and composition of prokaryotic communities in aquatic settings; for example, exposure to herbicides such as metolachlor and terbutryn has been linked to changes that may diminish ecosystem functionality (Borreca et al., 2024).

Research demonstrates that micropollutants can interfere with trophic interactions within ecosystems. For instance, pesticides have been shown to influence zooplankton that graze on algae, subsequently affecting nutrient cycling and primary productivity (Betram et al., 2022). The presence of these pollutants has detrimental effects on aquatic organisms, as evidence suggests that pharmaceuticals can disrupt reproductive processes in fish and amphibians, resulting in declines in their populations (Betram et al., 2022). Furthermore, micropollutants compromise soil quality by altering microbial diversity and biogeochemical processes that are vital for nutrient cycling. Heavy metals such as cadmium and lead also accumulate in agricultural soils, adversely affecting soil fertility and crop yields (Gul et al., 2022; Naeem et al., 2023). Together, these findings highlight the urgent necessity for effective management strategies to alleviate the impacts of micropollutants across diverse ecosystems.

2. **Agricultural Impacts**

The occurrence of micropollutants in agricultural environments can result in considerable soil degradation and contamination of crops. Agricultural runoff frequently transports micropollutants, which accumulate in the soil and negatively impact soil health and fertility, thereby impeding plant growth and diminishing agricultural output (Narwal et al., 2023). The prolonged use of contaminated irrigation water intensifies this problem, as it contributes to the buildup of micropollutants that disrupt soil structure and nutrient availability; for instance, microplastics have been demonstrated to decrease the soil's capacity to retain water (Gil, 2019; Naeem et al., 2023).

Crops have the ability to absorb micropollutants, which may pose risks to human health when consumed. Substantial levels of herbicides, such as atrazine, have been identified in crops that are irrigated with contaminated water (Narwal et al., 2023). Furthermore, micropollutants, including pharmaceuticals and pesticides, can accumulate in crops through the use of contaminated irrigation water or fertilizers that contain these chemicals. Research indicates that antibiotics can interfere with plant growth processes, resulting in diminished crop yields and quality (Moghadam et al.,

2023; Naeem et al., 2023). Additionally, agricultural practices frequently involve the application of pesticides, which can wash into adjacent water bodies during rainfall or irrigation, thereby contaminating aquatic ecosystems and posing threats to non-target species.

3. **Human Health Risks**

Micropollutants present considerable health hazards through contaminated water sources and food supplies. The detection of emerging micropollutants in drinking water has heightened public health concerns due to their possible long-term impacts, which include hormonal disruptions and an increased risk of cancer linked to toxic substances from agricultural runoff and wastewater discharges (Narwal et al., 2023; Betram et al., 2022; Moghadam et al., 2023). Moreover, the presence of pharmaceuticals in the environment fosters the emergence of antibiotic-resistant bacteria, which poses a significant threat to human health. This resistance often develops as a result of the selective pressure from low concentrations of antibiotics present in water systems (Betram et al., 2022; Stamm et al., 2016; Moghadam et al., 2023; Gul et al., 2022). Additionally, the accumulation of micropollutants in crops raises concerns regarding food safety, as consumers may inadvertently consume harmful substances through contaminated food products. Such exposure can result in potential health issues, including endocrine disruption and developmental complications (Naeem et al., 2023; Betram et al., 2022).

Synergistic Effects of Micropollutants

The relationships among micropollutants, especially pesticides and heavy metals, demonstrate intricate dynamics that can profoundly affect both ecosystems and human health. A comprehensive review indicated that although genuine synergistic interactions among chemicals are uncommon, they may arise at elevated concentrations, particularly with specific mixtures of pesticides, including cholinesterase inhibitors and azole fungicides. These combinations have been found to increase toxicity beyond predictions derived from individual evaluations, resulting in more significant ecological consequences (Cedergreen, 2014).

The degradation of pharmaceuticals in the environment can lead to the formation of various by-products, each with unique toxicological characteristics. Studies have shown that the photodegradation products of sulcotrione, a widely utilized herbicide, demonstrate higher levels of cytotoxicity and genotoxicity compared to the original compound, thereby complicating risk assessments and heightening health risks (Ledoigt et al., 2015). In aquatic ecosystems, laboratory investigations have revealed that the combined effects of metformin (an antidiabetic medication), metolachlor (an herbicide), and terbutryn (a biocide) can produce both synergistic and antagonistic effects on prokaryotic communities within sediment–water microcosms. This shift in community composition and diversity highlights the significant impact that interactions among contaminants can have on ecological outcomes (Borreca et al., 2024).

Microplastics have been identified as potential carriers of micropollutants, which aids in their distribution and increases their toxicity upon ingestion by various organisms. Research indicates that microplastics can enhance the bioavailability of these associated contaminants, resulting in increased toxicity for both human cells and aquatic life (Rubin & Ines, 2022). The interplay between natural toxins, such as those generated by harmful algal blooms, and human-made micropollutants further intensifies neurotoxic impacts within aquatic environments. For example, prymnesins produced by specific algal species demonstrate heightened neurotoxicity when in the presence of significant amounts of organic micropollutants, revealing intricate interactions that may obscure the true risks to aquatic organisms (Borreca et al., 2024). These insights underscore the necessity for thorough risk assessments that take into account not only individual pollutants but also their potential synergistic interactions. A comprehensive understanding of these relationships is essential for formulating effective environmental management strategies aimed at reducing the risks associated with micropollutants in various ecosystems.

Bioremediation

Bioremediation represents a cutting-edge and eco-friendly strategy that employs living organisms, mainly microorganisms and plants, to break down or stabilize pollutants found in soil, water, and sediments. This method provides a viable solution for mitigating pollution resulting from hazardous materials, especially heavy metals, organic contaminants, and various other environmental pollutants (Fig. 6.1).

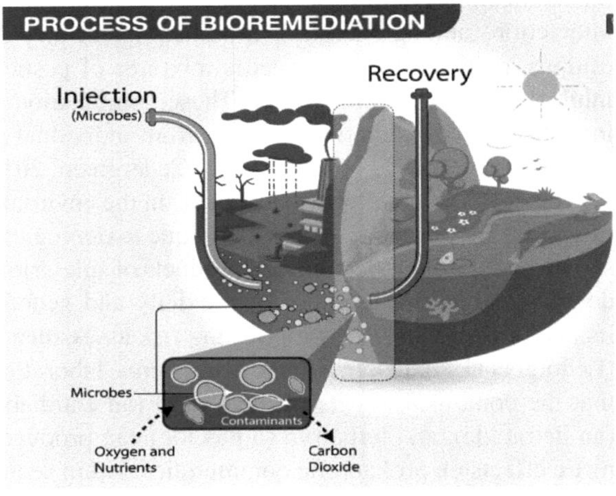

Fig. 6.1 Process of bioremediation

Mechanisms of Bioremediation

Microbial Bioremediation: Microorganisms, including bacteria, fungi, and algae, are essential in the bioremediation process as they utilize contaminants for energy. These microbes can convert harmful substances into less toxic or harmless forms through various biochemical mechanisms, such as redox reactions. For instance, specific bacteria are capable of degrading petroleum hydrocarbons via aerobic respiration, resulting in the production of carbon dioxide and water (Nester et al., 2001). Furthermore, microbial techniques such as biosorption, bioaccumulation, and biotransformation are utilized to eliminate heavy metals from polluted environments (Saha et al., 2021).

Phytoremediation: This approach employs plants to extract, stabilize, or decompose pollutants present in the environment. Plants can take up heavy metals through their root systems, either storing them in their tissues or converting them into less harmful forms. The success of phytoremediation is influenced by various factors, including the species of plants used, the bioavailability of the contaminants, and the conditions of the soil (Food and Agriculture Organization (FAO, 2023). For example, hyperaccumulator plants are known for their ability to withstand high levels of heavy metals and are particularly effective in the remediation of contaminated soils (Ojuederie & Babalola, 2017).

Microbe-Plant Interactions: The synergistic application of microbes and plants significantly improves the effectiveness of bioremediation efforts. Plants can foster a conducive environment for microbial proliferation by releasing root exudates that supply essential nutrients to beneficial soil microbes. In turn, these microbes can enhance the plants' nutrient absorption and aid in the breakdown of pollutants (Raklami et al., 2022). This collaborative relationship is crucial for optimizing bioremediation techniques.

Challenges of Bioremediation

Bioremediation presents numerous advantages in comparison to conventional physicochemical remediation techniques, contributing to its growing preference for tackling environmental pollution. A key advantage is its cost-effectiveness, as bioremediation methods typically demand less energy and fewer resources, leveraging natural processes that can lead to substantial reductions in operational expenses (Raklami et al., 2022). Moreover, bioremediation is noted for its environmentally friendly nature, as it minimizes ecological disruption and often produces fewer toxic by-products than chemical alternatives (Raklami et al., 2022). This approach facilitates the natural breakdown of contaminants, aiding in the restoration of polluted sites while simultaneously bolstering ecosystem resilience. Additionally, bioremediation is highly adaptable, making it applicable to a diverse array of contaminated locations, ranging from localized small-scale interventions to extensive industrial cleanups

(Ojuederie & Babalola, 2017). These benefits underscore the promise of bioremediation as a viable and sustainable strategy for alleviating pollution and restoring ecological integrity.

Bioremediation, while beneficial, encounters various challenges that may restrict its efficacy. A primary concern is the complexity of contaminants; the presence of a wide range of intricate pollutants can obstruct the processes of microbial degradation. Certain contaminants may exhibit resistance to microbial breakdown or may even suppress microbial functions, complicating the achievement of effective bioremediation results (Raklami et al., 2022). Furthermore, environmental factors are critical to the success of bioremediation initiatives; elements such as temperature, pH, moisture levels, and nutrient availability can greatly affect microbial performance and the overall effectiveness of remediation efforts (Ojuederie & Babalola, 2017). For example, extreme temperatures or unsuitable pH conditions can impede or halt the metabolic activities of the microorganisms engaged in bioremediation. These challenges underscore the importance of thoroughly assessing site-specific conditions and the characteristics of contaminants when formulating bioremediation approaches.

Global Efforts to Mitigate the Hazards Posed by Micropollutants

A significant initiative is the Financing Agrochemical Reduction and Management Programme (FARM), which has been introduced by the United Nations Environment Programme (UNEP). This program, with a budget of $379 million and a duration of five years, is designed to tackle pollution caused by pesticides and plastics in agriculture across various nations, including Ecuador, India, Kenya, Laos, the Philippines, Uruguay, and Vietnam. The FARM initiative aims to realign financial incentives to motivate farmers to transition to low- and non-chemical alternatives to harmful agrochemicals. It is anticipated that this initiative will prevent more than 51,000 tons of hazardous pesticides and 20,000 tons of plastic waste from contaminating the environment, while also protecting over 3 million hectares of land from degradation (UNEP, 2024).

A significant area of concern is the reduction of nitrogen pollution originating from agricultural lands worldwide. A comprehensive analysis of more than 1500 field observations has revealed essential strategies that can decrease nitrogen losses by 30–70%, while simultaneously improving crop yield and nitrogen use efficiency (NUE) by 10–30%. These strategies encompass the use of enhanced-efficiency fertilizers, organic amendments such as manure and straw, as well as agricultural practices like crop rotation and the establishment of buffer zones. The adoption of these approaches not only mitigates nitrogen pollution but also boosts agricultural productivity (Gu et al., 2023).

The implementation of best management practices (BMPs) in agriculture is crucial for decreasing dependence on synthetic fertilizers and lessening environmental repercussions. The 4R nutrient stewardship framework—appropriate fertilizer type, correct quantity, suitable placement, and optimal timing—has been highlighted as an effective strategy to enhance fertilizer utilization while alleviating its negative impacts on ecosystems (FAO, 2023). Nevertheless, challenges such as elevated implementation costs and insufficient knowledge among farmers frequently obstruct broader adoption. Additionally, innovative agricultural technologies present opportunities for addressing micropollutant risks. Methods that promote carbon sequestration in soils and enhance crop management can substantially lower greenhouse gas emissions while fostering soil health. Approaches such as bioenergy production, management of grazing lands, and the restoration of degraded areas contribute to a more sustainable agricultural framework (Uprety et al., 2012).

Conclusion

The widespread concern regarding micropollutants, especially within agricultural settings, presents considerable risks to both ecosystems and human health. This chapter has underscored the harmful impacts of various micropollutants, such as pesticides, pharmaceuticals, heavy metals, and microplastics, on soil quality, water purity, and biodiversity. The extensive application of synthetic agrochemicals has resulted in soil degradation, the pollution of freshwater resources, and disturbances in food webs, which carry serious consequences for environmental health and food safety. Programs like the Financing Agrochemical Reduction and Management Programme (FARM) exemplify a proactive strategy aimed at alleviating these issues by advocating for sustainable agricultural methods and decreasing dependence on hazardous chemicals. Nevertheless, tackling the intricacies of micropollutants necessitates coordinated global initiatives that engage policymakers, researchers, and farmers. In our efforts to safeguard ecosystems and promote human health, it is crucial to embrace innovative approaches such as bioremediation and to adopt best management practices that emphasize environmental sustainability. Through these collaborative endeavors, we can aspire to create a healthier planet for future generations while preserving the essential resources required for sustainable agriculture.

References

Adyari, B., Zhang, L., Nascimento, F. J., Zhang, Y., Cao, M., Wang, J., & Hu, A. (2023). Strong cascading impacts of micropollutants on planktonic food web in urban river. *bioRxiv, 12*.

Antakyali, D., Morgenschweis, C., De Kort, T., Sasse, R., Schulz, J., & Herbst, H. (2015). Micropollutants in the aquatic environment and their removal in wastewater treatment works. In *9th European waste water management conference, Manchester UK*.

BAFU. (2023). Micropollutants in watercourses. https://www.bafu.admin.ch/bafu/en/home/topics/water/info-specialists/state-of-waterbodies/state-of-watercourses/water-quality-in-watercourses/micropollutants-in-watercourses.html
Behera I. D., Nayak, M., Biswas, S., Meikap, B. C., & Sen, R. (2021). Enhanced biodegradation of total petroleum hydrocarbons by implementing a novel two-step bioaugmentation strategy using indigenous bacterial consortium. *Journal of Environmental Management, 292*, 112746, ISSN 0301-4797. https://doi.org/10.1016/j.jenvman.2021.112746
Behera, I. D., Nayak, M., Mishra, A., Meikap, B. C., Sen, R., (2022). Strategic implementation of integrated bioaugmentation and biostimulation for efficient mitigation of petroleum hydrocarbon pollutants from terrestrial and aquatic environment. *Marine Pollution Bulletin, 177*, 113492, ISSN 0025-326X. https://doi.org/10.1016/j.marpolbul.2022.113492
Behera, I. D., Basak, G., Kumar, R. R., Sen, R., & Meikap, B. C. (2020). Treatment of petroleum refinery sludge by petroleum degrading bacterium *Stenotrophomonas pavanii* IRB19 as an efficient novel technology. *Journal of Environmental Science and Health, Part A, 56*(2), 226–239. https://doi.org/10.1080/10934529.2020.1866924
Bertram, M. G., Martin, J. M., Wong, B. B. M., & Brodin, T. (2022). Micropollutants. *Current Biology, 32*(1), 17–19. https://doi.org/10.1016/j.cub.2021.11.038
Borreca, A., Vuilleumier, S., & Imfeld, G. (2024). Combined effects of micropollutants and their degradation on prokaryotic communities at the sediment–water interface. *Science and Reports, 14*, 16840. https://doi.org/10.1038/s41598-024-67308-y
Bruzzoniti, M. C., Del Bubba, M., Giordani, E., Fibbi, D., Beldean-Galea, M. S., Piesik, D., & Rivoira, L. (2024). Organic micropollutants in the agricultural chain of production of strawberries by irrigation with treated wastewater and assessment of human health implications. *Water, 16*(6), 830.
Cedergreen, N. (2014). Quantifying synergy: A systematic review of mixture toxicity studies within environmental toxicology. *PLoS ONE, 9*(5), Article e96580.
Das, P., & Bal, M. (2025). Modeling and optimization study for microplastic removal from aquatic medium using Chroococcidiopsis species. *Chemical Engineering Science, 304*, Article 121085. https://doi.org/10.1016/j.ces.2024.121085
Das, P., Halder, G., & Bal, M. (2023). A critical review on remediation of microplastics using microalgae from aqueous system. *Science of the Total Environment, 898*, Article 166425. https://doi.org/10.1016/j.scitotenv.2023.166425
Eggen, R. I., Hollender, J., Joss, A., Schärer, M., & Stamm, C. (2014). Reducing the discharge of micropollutants in the aquatic environment: the benefits of upgrading wastewater treatment plants. *Environmental Science and Technology*, 7683–7689.
El Hammoudani, Y., & Dimane, F. (2021a). Assessing behavior and fate of micropollutants during wastewater treatment: Statistical analysis. *Environmental Engineering Research, 26*(5), 35–41.
El Hammoudani, Y., & Dimane, F. (2021b). Assessing behavior and fate of micropollutants during wastewater treatment: Statistical analysis. *Environmental Engineering Research, 26*(5).
Environmental Protection Agency (EPA). (2023). Pesticide effects on aquatic life. https://www.epa.gov/pesticides
European Environment Agency. (2008). Organic micropollutants. https://www.eea.europa.eu/publications/92-9167-001-4/page025.html
European Environment Agency (EEA). (2023). How pesticides impact human health and ecosystems in Europe. https://www.eea.europa.eu/publications/how-pesticides-impact-human-health
FAO. (2021). Sources of soil pollution and major contaminants in agricultural areas. FAO Open Knowledge.
FAO. (2023). *Mitigating nitrogen pollution*. FAO Open Knowledge.
Food and Agriculture Organization (FAO). (2023). Pesticides: Environmental impact. FAO Open Knowledge.
Gangola, S., Joshi, S., Bhandari, G., Pant, G., Sharma, A., Perveen, K., & Rani, R. (2023). Exploring microbial diversity responses in agricultural fields: A comparative analysis under pesticide stress and non-stress conditions. *Frontiers in Microbiology, 14*, 1271129.

Gil, J. L. (2019). Micropollutants in wastewater irrigation systems: Impacts and perspectives. *Annals of Advances in Chemistry, 3*(1), 011–013.

Gu, B., Zhang, X., Lam, S. K., Yu, Y., Van Grinsven, H. J., Zhang, S., & Chen, D. (2023). Cost-effective mitigation of nitrogen pollution from global croplands. *Nature, 613*(7942), 77–84.

Gul, B., Naseem, M. K., Malik, W. U. N., Gurmani, A. R., Mehmood, A., & Rafique, M. (2022). Environmental micropollutants and their impact on human health with special focus on agriculture. *Hazardous environmental micro-pollutants, health impacts and allied treatment technologies* (pp. 1–19). Springer International Publishing.

Gupta, A., Kumar, S., Bajpai, Y., Chaturvedi, K., Johri, P., Tiwari, R. K., & Trivedi, M. (2024). Pharmaceutically active micropollutants: Origin, hazards and removal. *Frontiers in Microbiology, 15*, 1339469.

Janarthanan, R., & Murugan, A. (2017). Effect of applied pesticide on the microbial system of agriculture soil. *Journal of Pure and Applied Microbiology, 11*(1), 349–353.

Kanungo, J., Sahoo, T., Bal, M., & Behera, I. D. (2023a). Performance of bioremediation strategy in waste lubricating oil pollutants: A review. *Geomicrobiology Journal, 41*(4), 360–373. https://doi.org/10.1080/01490451.2023.2245395

Kanungo, J., Sahoo, T., Swain, L. P., & Behera, I. D. (2023b). Toxicity of persistent hydrocarbon pollutants, sources and sustainable remediation process. In I. D. Behera, A.P. Das (Eds.), *Impact of petroleum waste on environmental pollution and its sustainable management through circular economy*. Environmental Science and Engineering. Springer. https://doi.org/10.1007/978-3-031-48220-5_2

Launay, M., Dittmer, U., Steinmetz, H., & Kuch, B. (2013). Temporal variations of organic pollutants levels during storm events in an urban receiving water body. In *Novatech 2013–8ème Conférence internationale sur les techniques et stratégies durables pour la gestion des eaux urbaines par temps de pluie/8th International conference on planning and technologies for sustainable management of water in the city*.

Ledoigt, G., Sta, C., Goujon, E., Souguir, D., & El Ferjani, E. (2015). Synergistic health effects between chemical pollutants and electromagnetic fields. *Reviews on Environmental Health, 30*(4), 305–309.

Liu, P., Wu, Q., Hu, W., Tian, K., Huang, B., & Zhao, Y. (2023). Effects of atmospheric deposition on heavy metals accumulation in agricultural soils: Evidence from field monitoring and Pb isotope analysis. *Environmental Pollution, 330*, Article 121740.

Luo, Y., Guo, W., Ngo, H. H., Nghiem, L. D., Hai, F. I., Zhang, J., & Wang, X. C. (2014). A review on the occurrence of micropollutants in the aquatic environment and their fate and removal during wastewater treatment. *Science of the Total Environment, 473*, 619–641.

Moghadam, A. A., Shuai, W., & Hartmann, E. M. (2023). Anthropogenic antimicrobial micropollutants and their implications for agriculture. *Current Opinion in Biotechnology, 80*, Article 102902.

Mohanta, S., Pradhan, B., & Behera, I. D. (2023). Impact and remediation of petroleum hydrocarbon pollutants on agricultural land: A review. *Geomicrobiology Journal, 41*(4), 345–359. https://doi.org/10.1080/01490451.2023.2243925

Musolff, A. (2009). Micropollutants: Challenges in hydrogeology. *Hydrogeology Journal, 17*(4), 763–766.

Naeem, M., Gill, R., Gill, S. S., Singh, K., Sofo, A., & Tuteja, N. (2023). Emerging contaminants and their effect on agricultural crops. *Frontiers in Plant Science, 14*, 1296252.

Narwal, N., Katyal, D., Kataria, N., Rose, P. K., Warkar, S. G., Pugazhendhi, A., Ghotekar S., & Khoo, K. S. (2023). Emerging micropollutants in aquatic ecosystems and nanotechnology-based removal alternatives: a review. *Chemosphere*, 139945.

Nester, E. W., Anderson, D. G., Roberts Jr, C. E., & Nester, M. T. (2001). *Microbiology: A human perspective.*

Ojuederie, O. B., & Babalola, O. O. (2017). Microbial and plant-assisted bioremediation of heavy metal polluted environments: A review. *International Journal of Environmental Research and Public Health, 14*(12), 1504.

Envirotech Online. (2022). A simple guide to micropollutants—types, sources and removal. https://www.envirotech-online.com/news/wastewater-analysis/155/breaking-news/a-simple-guide-to-micropollutants-types-sources-removal/58108
Raklami, A., Meddich, A., Oufdou, K., & Baslam, M. (2022). Plants—Microorganisms-based bioremediation for heavy metal cleanup: Recent developments, phytoremediation techniques, regulation mechanisms, and molecular responses. *International Journal of Molecular Sciences, 23*(9), 5031.
Rubin, A., & Zucker, I. (2022). *Synergistic toxicity of micro-plastic and micro-pollutants in human cells.* https://doi.org/10.13140/RG.2.2.27026.73928
Saha, L., Tiwari, J., Bauddh, K., & Ma, Y. (2021). Recent developments in microbe–plant-based bioremediation for tackling heavy metal-polluted soils. *Frontiers in Microbiology, 12*, Article 731723.
Stamm, C., Räsänen, K., Burdon, F. J., Altermatt, F., Jokela, J., Joss, A., & Eggen, R. I. (2016). Unravelling the impacts of micropollutants in aquatic ecosystems: interdisciplinary studies at the interface of large-scale ecology. In *Advances in ecological research* (Vol. 55, pp. 183–223).
UNEP. (2024). New initiative aims to curb toxic impacts of agriculture. https://www.unep.org/news-and-stories/press-release/new-initiative-aims-curb-toxic-impacts-agriculture
Uprety, D. C., Dhar, S., Hongmin, D., Kimball, B. A., Garg, A., & Upadhyay, J. (2012). Technologies for climate change mitigation—agriculture sector. UNEP Risø Centre on Energy, Climate and Sustainable Development.
Wang, Z., Walker, G. W., Muir, D. C. G., & Nagatani-Yoshida, K. (2020). Toward a global understanding of chemical pollution: A first comprehensive analysis of national and regional chemical inventories. *Environmental Science and Technology, 54*, 2575–2584.
Wasser 3.0. (2024). What are micropollutants? https://wasserdreinull.de/en/knowledge/micropollutants/
Yang, Y., Zhang, X., Jiang, J., Han, J., Li, W., Li, X., & Alvarez, P. J. (2021). Which micropollutants in water environments deserve more attention globally? *Environmental Science & Technology, 56*(1), 13–29.

Chapter 7
Comprehensive Analysis of Micropollutants and Their Impact on Aquatic Ecosystems and Biodiversity

Sunil Kumar Singh, Aniruddha Deb, B. Anil Kumar Naik, and Awadh Kishor Kumar

Introduction

Defining Micropollutants and Their Importance

Micropollutants, often called emerging contaminants, include various chemical substances in the environment at trace levels, typically measured in nanograms to micrograms per liter (ng/L to μg/L) (Yang et al., 2021). These compounds, derived from anthropogenic and natural sources, are characterized by their persistence, bioactivity, and potential to disrupt ecosystems, even at low concentrations. Examples include pharmaceuticals, personal care products, industrial chemicals, pesticides, and hormones. Micropollutants are not a new phenomenon; however, their significance as an environmental challenge has gained global attention due to advancements in analytical techniques that can detect these compounds at ultra-trace levels (Vaye et al., 2022) (Fig. 7.1).

The ubiquity of micropollutants in aquatic ecosystems is increasingly alarming. Unlike conventional pollutants such as nutrients or heavy metals, micropollutants represent a diverse chemical spectrum with varying physicochemical properties and environmental behaviors (Grandclément et al., 2017). Their presence threatens the balance of aquatic ecosystems by disrupting biological processes, reducing biodiversity, and endangering the health of organisms across trophic levels. In addition,

S. K. Singh
Department of Chemical Engineering, BIT Sindri, Dhanbad, Jharkhand, India

A. Deb
Department of Chemical Engineering, BIT Mesra, Ranchi, Jharkhand, India

B. Anil Kumar Naik · A. K. Kumar (✉)
Department of Chemical Science and Technology, NIT Patna, Patna, India
e-mail: awadh.ch@nitp.ac.in

I. D. Behera et al. (eds.), *Emerging Micropollutants*, Sustainable Environmental Waste Management Strategies, https://doi.org/10.1007/978-3-032-03530-1_7

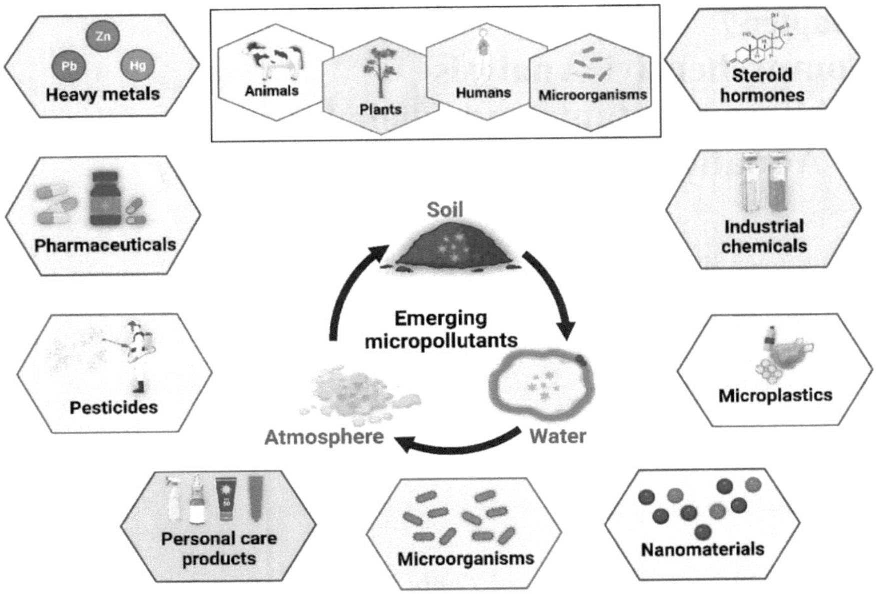

Fig. 7.1 Different types of emerging micropollutants (Narwal et al., 2023)

their potential to bioaccumulate and biomagnify in food chains amplifies risks to apex predators, including humans (Bashir et al., 2020).

Sources and Pathways into Aquatic Ecosystems

The sources of micropollutants are diverse and often diffuse, complicating efforts to regulate and mitigate their impacts. Common origins include effluents from wastewater treatment plants (WWTPs), industrial discharges, agricultural runoff, and leachate from landfills. Urbanization, combined with the proliferation of pharmaceuticals and personal care products, has significantly contributed to their release into the environment (Mohanty et al., 2024a, 2024b, 2024c; Ray et al., 2024; Sahu et al., 2024a). The pathways through which these pollutants enter aquatic systems are equally varied, ranging from direct discharges into water bodies to atmospheric deposition and stormwater runoff (Noorani et al., 2024).

The Challenge of Micropollutants

What makes micropollutants particularly challenging is their persistence and ability to evade conventional wastewater treatment processes. Most WWTPs are designed to remove bulk organic matter, nutrients, and pathogens but cannot effectively degrade or eliminate trace organic compounds (Zheng et al., 2013). As a result, treated effluent becomes a significant vector for their entry into natural water systems. Compounding this challenge is the synergistic or antagonistic effects micropollutants can exhibit when present in complex mixtures, further complicating risk assessments and ecological predictions.

Why Biodiversity Matters in the Context of Micropollutants

Biodiversity underpins ecosystem resilience and functionality. Aquatic ecosystems rely on a delicate balance of organisms ranging from microbial communities to fish and aquatic plants. Micropollutants pose a significant threat to this balance. Studies suggest that diverse microbial and phytoplankton communities are critical for these contaminants' natural attenuation and transformation. However, exposure to these compounds often results in losing sensitive species, reducing the system's ability to recover and function effectively (Stravs et al., 2019). For example, pharmaceuticals such as antidepressants have been shown to alter predator–prey dynamics in aquatic environments by affecting the behavior and reproductive success of fish (Das et al., 2024b; Dey et al., 2023b; Dey et al., 2024a; 2024b). Similarly, pesticides like neonicotinoids are linked to declines in aquatic insect populations, which serve as a food source for higher trophic levels (Yamamuro et al., 2019). The degradation of biodiversity reduces ecosystem services and hampers natural remediation processes, creating a vicious cycle of ecological harm.

Objectives and Structure of the Chapter

This chapter aims to comprehensively examine micropollutants and their impacts on aquatic ecosystems and biodiversity. It is structured as follows:

Sources and Pathways of Micropollutants: Exploring the primary sources and environmental transport mechanisms.

Ecological and Biological Impacts: A detailed analysis of how micropollutants affect biodiversity, from microbial communities to apex predators.

Biodiversity as a Mediator of Pollutant Transformation: A discussion on the role of biological diversity in enhancing natural detoxification processes.

Nanotechnology and Bioremediation for Micropollutant Removal: Insights into advanced technologies and biological strategies for mitigating micropollutant impacts.

Challenges and Future Directions: Addressing regulatory, technical, and scientific hurdles.

This chapter highlights the urgency of addressing micropollutants and the innovative pathways forward by bridging the gap between emerging scientific insights and practical applications.

Sources and Pathways of Micropollutants

Major Categories of Micropollutants and Their Origins

Micropollutants encompass various chemical compounds, each with distinct origins and environmental behaviors. Their classification into major categories provides insight into their sources and pathways into aquatic ecosystems (Fig. 7.2).

Pharmaceuticals and Personal Care Products: Pharmaceuticals, including antibiotics, analgesics, and hormonal treatments, are primarily derived from domestic

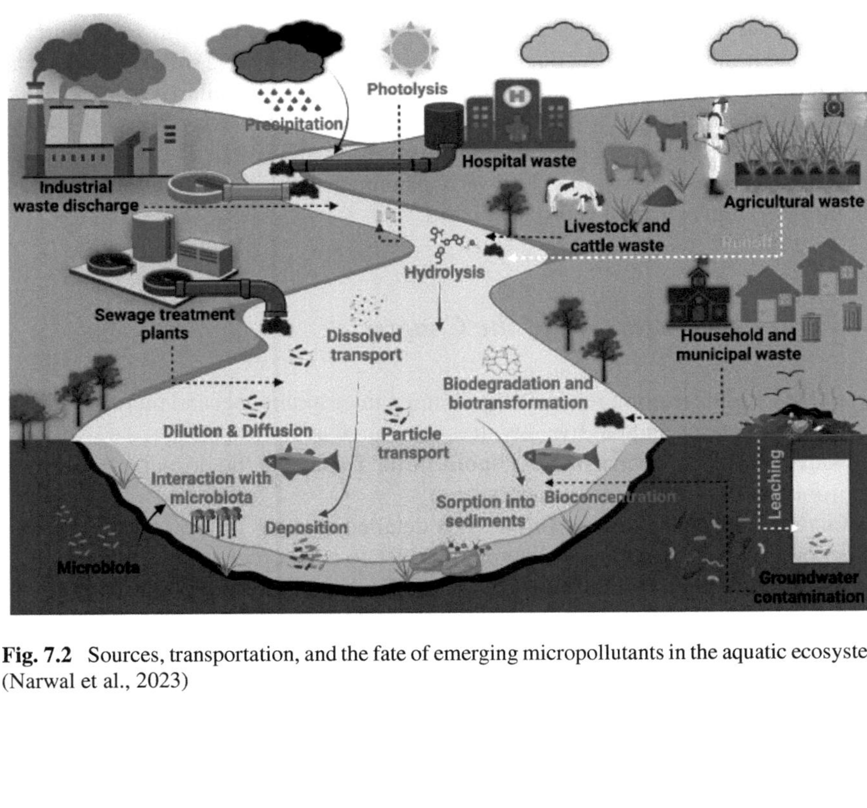

Fig. 7.2 Sources, transportation, and the fate of emerging micropollutants in the aquatic ecosystem (Narwal et al., 2023)

wastewater and hospital effluents. Improper disposal of unused medications further exacerbates their presence in water systems. Personal care products, such as shampoos, lotions, and sunscreens, contribute UV filters, fragrances, and antimicrobials like triclosan to aquatic pollution (Srinivasulu et al., 2022).

Pesticides and Agrochemicals: Pesticides are a significant source of micropollutants in agricultural runoff. Compounds such as glyphosate, atrazine, and neonicotinoids persist in soil and are transported to water bodies during irrigation or rainfall. Their use is linked to biodiversity declines, particularly in aquatic insects and amphibians (Brühl & Zaller, 2021).

Industrial Chemicals and Flame Retardants: Industrial chemicals like plasticizers (e.g., bisphenol A), surfactants, and flame retardants (e.g., polybrominated diphenyl ethers) are often released during manufacturing processes or through improper disposal of industrial waste. Their persistence and toxicity are especially concerning due to their widespread use in consumer goods (Maddela et al., 2023).

Nanomaterials and Microplastics: Emerging pollutants like engineered nanomaterials and microplastics are gaining attention due to their unique physicochemical properties. Nanomaterials, used in cosmetics and electronics products, can adsorb other micropollutants, creating complex mixtures with unpredictable environmental behaviors. Microplastics, resulting from the degradation of larger plastic debris, serve as carriers for persistent organic pollutants (POPs) (Fawcett et al., 2023; McGlade et al., 2021).

Environmental Pathways of Micropollutants

The journey of micropollutants from their sources to aquatic systems is complex and multifaceted. These contaminants traverse various pathways and mechanisms, contributing to their widespread presence and persistence in the environment. Their mobility is influenced by environmental factors such as water flow, soil permeability, atmospheric conditions, and human activities (Luo et al., 2014). Understanding these pathways is essential for designing effective strategies to mitigate their impacts.

Point Sources

Point sources, such as municipal wastewater treatment plants (WWTPs), industrial discharge outlets, and healthcare facilities, represent the most direct pathways through which micropollutants enter aquatic systems (Kosek et al., 2020).

Municipal WWTPs: WWTPs are designed to treat conventional pollutants such as organic matter, nutrients, and pathogens. However, their processes often fail to remove micropollutants, including pharmaceuticals, personal care products, and industrial chemicals, due to the trace concentrations and chemical stability of these compounds. Studies have detected carbamazepine and diclofenac in treated effluents at concentrations sufficient to affect aquatic life, demonstrating the inefficiency of conventional treatments (Capodaglio et al., 2018). Advanced treatment methods like ozonation and activated carbon adsorption are being developed but are not yet widely implemented due to cost constraints (Fig. 7.3).

Industrial Discharges: Industrial facilities release effluents containing persistent organic pollutants (POPs), heavy metals, and solvents. These effluents often have higher toxicity and complexity than municipal wastewater (Das et al., 2024a; Dey et al., 2023a; Mallick et al., 2024). Textile industries in South Asia discharge azo dyes into rivers, which resist degradation and accumulate in sediments, impacting aquatic biodiversity (Jahan & Singh, 2023).

Healthcare Facilities: Hospitals and laboratories are emerging contributors of micropollutants, releasing pharmaceuticals, radiological agents, and disinfectants into sewage systems. Antibiotics in these effluents contribute to developing antimicrobial resistance, posing risks to ecosystems and human health (Ajala et al., 2022).

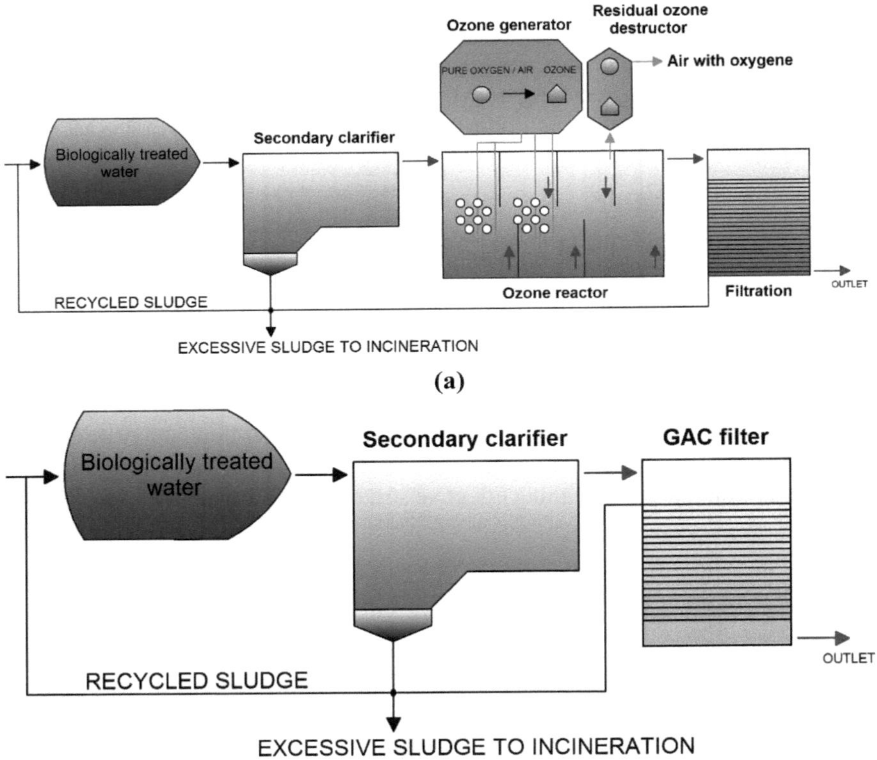

Fig. 7.3 Examples of **a** ozonation and **b** activated carbon process (Kosek et al., 2020)

Non-point Sources

Non-point sources are diffuse and more challenging to regulate, as they indirectly involve contaminants entering water bodies through surface runoff, leaching, or atmospheric deposition (Singh et al., 2024).

Agricultural Runoff: Fertilizers and pesticides applied to agricultural fields are washed into nearby water bodies during rainfall or irrigation. Excessive use of pesticides such as atrazine and glyphosate has led to contamination of groundwater and rivers. Studies in North America have shown high concentrations of nitrates and pesticides in agricultural watersheds, leading to eutrophication and toxic effects on aquatic organisms (Brain & Prosser, 2022).

Atmospheric Deposition: Micropollutants can volatilize into the atmosphere, travel long distances, and deposit into aquatic environments through wet or dry deposition. Industrial emissions and vehicular exhaust are major contributors. Polycyclic aromatic hydrocarbons (PAHs) and heavy metals like mercury have been detected in remote lakes, far from their original emission sources (Du & Jing, 2018).

Leachate from Landfills

Landfills are reservoirs of improperly disposed consumer products, including electronics, pharmaceuticals, and personal care items. Over time, rainwater infiltrates the landfill layers, dissolving contaminants and creating leachate (Elbeshbishy & Okoye, 2019) (Fig. 7.4).

Leachate Composition: Leachate contains a mixture of organic and inorganic pollutants, including bisphenol A (BPA), phthalates, and heavy metals. These contaminants can persist in the environment for decades, seeping into groundwater or being transported to nearby water bodies (Pisharody et al., 2022).

Environmental Impact: Leachate often bypasses natural filtration systems due to its high pollutant concentration. Once it enters aquatic systems, it can significantly alter water chemistry, harm aquatic organisms, and compromise drinking water sources. Studies in Europe have highlighted high levels of pharmaceuticals and plasticizers in leachate samples from older landfills, indicating long-term pollution risks (Kotowska et al., 2020).

Transformational Processes in the Environment

Once introduced into aquatic ecosystems, micropollutants undergo a series of physical, chemical, and biological transformations that influence their environmental persistence, mobility, and ecological impact. These processes determine whether a micropollutant remains bioavailable, degrades into less harmful compounds, or transforms into byproducts that may retain or amplify toxicity (Tiwari et al., 2017).

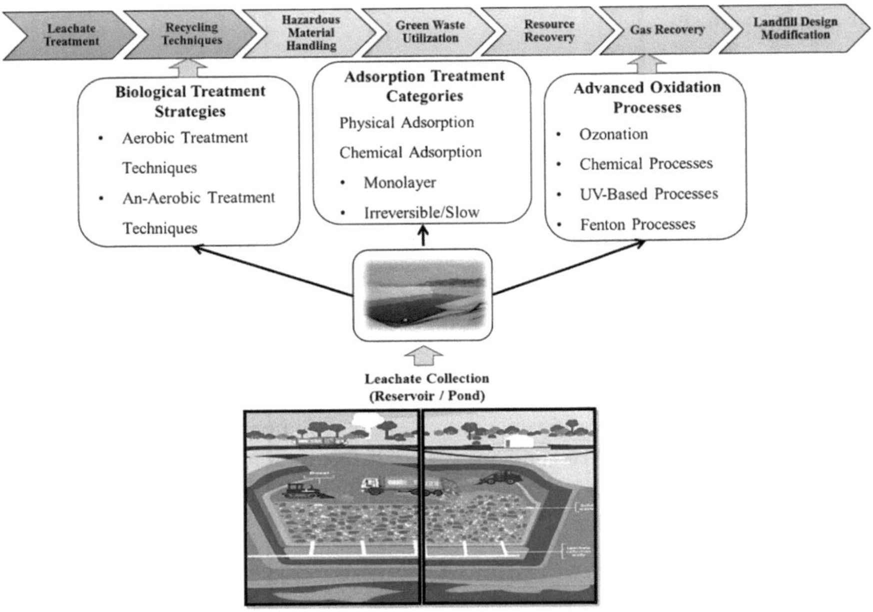

Fig. 7.4 Leachate from landfills (Siddiqi et al., 2022)

Sorption: Sorption is one of the initial and critical processes affecting micropollutants in aquatic environments. Hydrophobic compounds, such as pesticides, polycyclic aromatic hydrocarbons (PAHs), and industrial chemicals, strongly tend to adsorb onto sediments and organic matter (Saeedi et al., 2020). This sorption reduces the immediate bioavailability of these pollutants, lowering their interaction with aquatic organisms. However, sorption also increases their environmental persistence, as contaminants bound to sediments can remain sequestered for decades (Sarkar et al., 2021). Under certain conditions, such as sediment resuspension caused by dredging or flooding, these bound micropollutants can re-enter the water column, leading to secondary pollution. For example, the pesticide DDT, although banned in many countries, persists in sediments, periodically releasing trace amounts into aquatic systems (Herrero et al., 2018).

Photodegradation: Photodegradation, driven by sunlight, is a natural attenuation process particularly effective for micropollutants that absorb ultraviolet (UV) radiation. Compounds like triclosan, atrazine, and certain pharmaceuticals can undergo photolytic breakdown when exposed to sunlight in shallow water bodies or during wastewater treatment (Bavumiragira & Yin, 2022). This degradation leads to simpler molecular structures, reducing their persistence. However, photodegradation can produce intermediate byproducts, some of which retain significant toxicity or exhibit new harmful properties (Zada et al., 2021). For instance, the breakdown of triclosan can result in the formation of dioxins under specific environmental conditions, complicating its ecological impact. Additionally, photodegradation efficiency is influenced by factors such as water turbidity, depth, and natural organic matter, which can either facilitate or inhibit the process (Bora & Mewada, 2017).

Biodegradation: Biodegradation is another key transformation pathway where microbial communities metabolize micropollutants into less harmful or simpler compounds (Garcia-Becerra & Ortiz, 2018). Aerobic biodegradation occurs in oxygen-rich environments, breaking down compounds such as non-steroidal anti-inflammatory drugs (NSAIDs) and certain pesticides (Shanmuganathan et al., 2023). Anaerobic biodegradation, often occurring in sediments and oxygen-deprived zones, is particularly effective for pollutants like chlorinated hydrocarbons (Elyamine et al., 2021). However, the rate and efficiency of biodegradation depend on several factors, including the chemical structure of the pollutant, microbial diversity, environmental pH, temperature, and nutrient availability (Fouad et al., 2022). Persistent organic pollutants (POPs), such as polychlorinated biphenyls (PCBs) and certain flame retardants, resist biodegradation due to their stable molecular structures (Akhtar et al., 2021). Additionally, introducing novel synthetic chemicals may outpace microbial evolution, leaving some pollutants recalcitrant in the environment.

Bioaccumulation and Biomagnification: Bioaccumulation and Biomagnification are significant ecological processes that amplify the risks posed by micropollutants (Zenker et al., 2014). Lipophilic compounds, such as mercury, PCBs, and DDT, readily dissolve in the fatty tissues of organisms. These pollutants are absorbed faster than they are metabolized or excreted, leading to their accumulation in individual organisms. As contaminated prey are consumed by predators, the concentration of these pollutants magnifies through the food web (Mohanty et al., 2024b; Moharana et al., 2024). Apex predators, including large fish, birds, and mammals, are particularly vulnerable due to prolonged exposure and high trophic positions (Ritchie & Johnson, 2009). For example, mercury accumulation in fish affects aquatic ecosystems and poses health risks to humans who consume contaminated seafood. The long-term impacts of biomagnified micropollutants include reproductive failure, neurological disorders, and population declines in wildlife.

Case Studies Highlighting Pathways and Impacts

Case Study 1: Pharmaceuticals in Urban Water Systems

Research conducted in European cities has revealed that wastewater treatment plant (WWTP) effluents are a major source of pharmaceutical contaminants in urban rivers. These effluents often contain a cocktail of drugs, including antidepressants, analgesics, antibiotics, and hormonal compounds, which persist even after conventional treatment processes. Studies have shown that antidepressants, such as fluoxetine, and analgesics, including ibuprofen and diclofenac, are among the most frequently detected compounds in these waters (Bala et al., 2024; Das et al., 2023a). The concentrations of these pharmaceuticals, although typically in the nanogram to microgram per liter range, are sufficient to cause significant ecological disruptions. For instance, fluoxetine exposure has been linked to altered predator–prey dynamics in fish, as it affects their swimming patterns and feeding behaviors, making them more vulnerable to predators. On the other hand, Diclofenac has been observed to cause tissue damage in fish, particularly in their gills, kidneys, and livers, thereby impairing their physiological functions. Moreover, hormonal compounds like ethinylestradiol from contraceptive pills have induced feminization in male fish, leading to reduced fertility and population declines in some species (Das et al., 2024d). These impacts collectively disrupt aquatic food webs and biodiversity, emphasizing the critical need for advanced treatment technologies and regulatory measures to mitigate the release of such contaminants into urban water systems (Marcal et al., 2021; Tahar et al., 2018).

Case Study 2: Agricultural Runoff and Pesticides

In Spain, agricultural runoff containing pesticides, particularly neonicotinoids, has significantly impacted aquatic ecosystems, especially in biodiversity-rich areas such as the Ebro River Delta and Doñana National Park. Neonicotinoids, widely used for pest control in crops like olives and citrus fruits, are highly soluble in water and persist in the environment, accumulating in nearby streams, rivers, and wetlands through surface runoff and leaching. Studies have found concentrations of neonicotinoids in these water bodies exceeding safety thresholds, particularly during peak agricultural seasons, causing severe disruptions to local ecosystems.

Aquatic insects, vital components of freshwater food webs, suffer acute toxicity and sublethal effects from pesticide exposure, leading to population declines and disrupting nutrient cycling. Amphibians, which rely on these insects as a food source, face indirect effects, such as reduced survival and reproduction, alongside direct toxicity from pesticide-contaminated habitats (Das et al., 2024c; Mohanty et al., 2024c; Sahu et al., 2024b). These cascading impacts threaten the overall health and resilience of aquatic ecosystems. Efforts to mitigate these effects, including partial bans on neonicotinoids in the European Union and promoting sustainable farming practices, have been initiated but require stricter enforcement and broader adoption to curb further ecosystem degradation (Calvo et al., 2021; Herrero et al., 2018).

Case Study 3: Microplastics as Pollutant Vectors

Studies in estuarine ecosystems have shown that microplastics, due to their small size and large surface area, act as effective carriers for various pollutants, particularly persistent organic pollutants (POPs). These pollutants include hazardous compounds such as polycyclic aromatic hydrocarbons (PAHs), widely known for their environmental persistence and toxicity. The hydrophobic nature of microplastics allows them to adsorb hydrophobic organic compounds, including PAHs, from the surrounding water, significantly enhancing the bioavailability of these pollutants. In estuarine ecosystems, microplastics often accumulate contaminants from runoff, wastewater, and industrial discharge, resulting in high concentrations of these chemicals on their surfaces. This accumulation makes microplastics a focal point for contaminant transfer. When aquatic organisms such as fish, mollusks, and crustaceans ingest microplastics, they inadvertently consume these adsorbed pollutants. This exposes organisms to the harmful effects of PAHs and facilitates the biomagnification of these toxic substances throughout the food chain (Bakir et al., 2014; Okoye et al., 2022).

Furthermore, microplastics have been found to interact with a wide range of environmental toxins, including heavy metals, pesticides, and pharmaceuticals. Their role as pollutant vectors not only raises concerns about the health of marine life but also underscores the potential risk to human health through seafood consumption. Studies have demonstrated that the adsorption of PAHs onto microplastics in estuarine environments can lead to higher-than-expected concentrations of these compounds in organisms and an increase in toxicity due to the combined effect of both the microplastics and the pollutants they carry. The long-term implications of this phenomenon are still being explored, but it is evident that microplastics significantly influence the transport and bioavailability of toxic pollutants, exacerbating their impacts on marine and freshwater ecosystems (Stravs et al., 2019).

Ecological and Biological Impacts

Micropollutants exert significant ecological and biological effects on aquatic ecosystems due to their persistence, bioactivity, and potential for bioaccumulation. These compounds disrupt ecosystem dynamics by altering the physiology, behavior, and reproduction of aquatic organisms and degrading the structural integrity of microbial communities, which are essential for nutrient cycling and pollutant degradation (Zenker et al., 2014). Persistent exposure to micropollutants, such as pharmaceuticals and endocrine-disrupting compounds (EDCs), has been shown to cause reproductive abnormalities, hormonal imbalances, and reduced fitness in aquatic species, leading to population declines and disrupted food web interactions. Additionally, the bioaccumulative nature of many micropollutants amplifies their impacts at higher trophic levels, where predators face magnified concentrations of toxic substances. For example, lipophilic contaminants, including PCBs and DDT, accumulate in the fatty tissues of fish and aquatic mammals, causing chronic health effects such as immunosuppression, neurological damage, and reduced reproductive success. Micropollutants also alter microbial diversity and functionality, undermining the ecosystem's ability to degrade organic matter and other pollutants naturally (Jagatee et al., 2024; Mishra & Das, 2021). These disruptions can lead to long-term ecological imbalances, diminished biodiversity, and compromised ecosystem services, such as water purification and habitat stability, vital for aquatic life and human populations (Arslan et al., 2017).

Effects on Aquatic Biodiversity

Impacts on Primary Producers

Primary producers such as phytoplankton and aquatic plants form the foundation of aquatic food webs, playing a critical role in oxygen production, carbon sequestration, and nutrient cycling. Micropollutants like herbicides directly inhibit photosynthetic activity, reducing biomass and altering community composition. For example, glyphosate, a widely used herbicide, has been shown to decrease algal diversity and promote the dominance of resistant species, disrupting natural nutrient cycling and impairing ecosystem stability. These changes not only affect the productivity of primary producers but also have cascading effects on higher trophic levels, as reduced algal biomass limits the availability of food for zooplankton and other herbivores (Lobus & Kulikovskiy, 2023).

UV filters from personal care products, another category of micropollutants, can impair light penetration in water bodies, reducing photosynthesis and growth rates in submerged vegetation. This phenomenon is particularly concerning in shallow aquatic systems, with limited light availability. Furthermore, some UV filters have been found to accumulate in the tissues of aquatic plants, leading to physiological stress and further inhibiting growth (Mishra et al., 2019, 2020). Over time, these impacts can shift the balance of aquatic ecosystems, favoring opportunistic or invasive species that are more tolerant of micropollutant exposure, while native and sensitive species decline. Such alterations undermine the resilience and biodiversity of aquatic ecosystems, with long-term consequences for their ecological function and services (Polińska et al., 2021).

Effects on Invertebrates

Invertebrates, including zooplankton and benthic organisms, are highly susceptible to micropollutant exposure, given their roles as primary consumers and detritivores in aquatic ecosystems. Pesticides, such as neonicotinoids, have been linked to significant declines in aquatic insect populations, particularly mayflies, caddisflies, and dragonflies. These declines disrupt critical ecosystem services, such as nutrient recycling, sediment stabilization, and the maintenance of water quality. Reduced populations of zooplankton, which are vital grazers of algae, can lead to algal blooms, further destabilizing aquatic ecosystems and exacerbating oxygen depletion in affected waters (Bellasi et al., 2020).

Bioaccumulation of heavy metals and persistent organic pollutants (POPs) in mollusks and crustaceans disrupts metabolic processes, including enzyme functions and energy production, reducing reproduction, growth rates, and survival (Camacho-Jiménez et al., 2023). For instance, cadmium exposure in bivalves has been shown to impair shell formation and increase mortality during larval stages. In crustaceans like shrimps and crabs, POPs can interfere with molting cycles and neural functions, reducing their ability to forage and evade predators. The vulnerability of invertebrates to micropollutants not only compromises their populations but also affects higher trophic levels that rely on them as a food source, triggering cascading effects throughout aquatic food webs. Furthermore, the loss of benthic invertebrates, which play a crucial role in breaking down organic matter, can lead to the accumulation of detritus and a decline in sediment health, further diminishing ecosystem resilience (Hauer et al., 2018).

Impacts on Fish and Amphibians

Pharmaceuticals and endocrine-disrupting chemicals (EDCs) pose severe risks to fish and amphibians, as their bioactive properties directly interfere with critical physiological and behavioral functions. Antidepressants like fluoxetine, frequently detected in aquatic environments, have been shown to impair predation efficiency, alter mating behaviors, and disrupt social hierarchies in fish. These changes weaken population dynamics by reducing survival and reproductive success. Similarly, EDCs such as bisphenol A (BPA), a common industrial compound, mimic natural hormones and bind to hormone receptors, causing reproductive abnormalities, including intersex conditions where male fish develop female reproductive traits. Over time, these disruptions can lead to population imbalances, skewed sex ratios, and declines in species abundance (Kasonga et al., 2021).

Amphibians, known for their permeable skin and dependence on aquatic habitats for reproduction, are particularly vulnerable to micropollutants. Pesticide exposure such as atrazine has been linked to delayed metamorphosis, reduced growth rates, and impaired limb development in amphibian larvae. Pharmaceuticals in water can also suppress immune responses in amphibians, making them more susceptible to infections and diseases like chytridiomycosis, which has decimated populations worldwide. In addition to direct toxicity, sublethal exposure to contaminants can reduce their ability to evade predators and locate breeding sites, further threatening their survival. The compounded impacts on fish and amphibians disrupt entire aquatic ecosystems, as these organisms play essential roles in regulating prey populations, maintaining trophic balance, and supporting nutrient cycling. Without targeted mitigation efforts, the long-term persistence of micropollutants will continue to jeopardize the health of these keystone species and their habitats (Sharma et al., 2023).

Impact on Microbial Communities

Microbial communities are integral to ecosystem health, decomposers and mediators of nutrient cycling and pollutant degradation. They play a vital role in maintaining water quality and ecosystem stability. However, micropollutants significantly affect these communities, altering their composition, functionality, and resilience to environmental stressors.

Reduction in Diversity and Functionality: Exposure to pharmaceuticals, such as antibiotics, disrupts microbial diversity by selectively targeting susceptible species. This reduces the abundance of beneficial microbes and creates ecological niches that are rapidly filled by antibiotic-resistant bacteria. The proliferation of resistant strains decreases the functional redundancy of microbial communities, making ecosystems more vulnerable to additional stressors. For example, the overrepresentation of resistant bacteria can hinder the natural degradation of organic matter and pollutants, thereby impairing self-purification processes in aquatic systems. Furthermore, horizontal gene transfer among microbial populations enhances the spread of resistance genes, complicating efforts to mitigate the environmental and public health risks associated with antimicrobial resistance (Chaturvedi et al., 2021).

Disruption of Biogeochemical Cycles: Pesticides and industrial chemicals interfere with microbial enzymes critical for nutrient transformation and decomposition processes. For instance, herbicides like glyphosate inhibit the activity of nitrifying bacteria, reducing the conversion of ammonia to nitrate and disrupting the nitrogen cycle. Similarly, industrial solvents can inhibit carbon-degrading microbes, slowing the breakdown of organic matter and reducing carbon turnover rates. These disruptions affect the availability of essential nutrients for primary producers and destabilize the entire ecosystem. Over time, altered nutrient dynamics can lead to phenomena such as eutrophication, where excessive nutrient buildup promotes harmful algal blooms that further degrade water quality (Bose et al., 2021; Parte et al., 2017).

Formation of Toxic Byproducts: Incomplete microbial degradation of micropollutants often leads to the formation of intermediate compounds, which can exhibit higher toxicity or persistence than the parent compounds. For instance, during the microbial breakdown of pharmaceuticals like diclofenac, reactive metabolites are formed that can further inhibit enzymatic activity within microbial communities. These byproducts may also affect other organisms, amplifying their environmental impact. Additionally, the accumulation of toxic intermediates can alter microbial metabolic pathways, forcing communities to expend more energy on stress responses rather than on ecological functions like decomposition and nutrient cycling. This shift reduces the overall efficiency of pollutant degradation and compromises ecosystem recovery (Garcia-Becerra & Ortiz, 2018).

Bioaccumulation and Biomagnification

The lipophilic nature of many micropollutants enables their accumulation in the fatty tissues of aquatic organisms, a process known as bioaccumulation. Over time, these contaminants magnify through the food web, resulting in higher concentrations at each successive trophic level—a phenomenon called biomagnification. Apex predators, such as birds, mammals, and humans, face the greatest risks due to their long lifespans and high trophic positions, which lead to prolonged exposure to these toxic compounds. This can result in chronic health issues, including reproductive failure, immunosuppression, and neurological damage, threatening both wildlife populations and human health (Borgå et al., 2022).

Bioaccumulation in Aquatic Fauna

Studies on aquatic fauna, particularly fish, have demonstrated significant bioaccumulation of pollutants like mercury, polychlorinated biphenyls (PCBs), and brominated flame retardants. These compounds concentrate in liver and muscle tissues, disrupting endocrine systems, weakening immune responses, and altering natural behaviors such as migration and spawning. In mollusks and filter feeders, bioaccumulation is particularly severe as they continuously ingest pollutants from the water column and sediments during filter feeding. For example, oysters exposed to heavy metals exhibit impaired shell formation and reduced reproductive success, while bivalves in polluted waters show increased mortality during larval stages. The persistence of these pollutants in sediments exacerbates their long-term bioavailability, creating ongoing risks for benthic species and the broader aquatic food web (Maia et al., 2020).

Trophic Transfer of Micropollutants

Micropollutants, especially persistent organic pollutants (POPs), move through the food web via trophic transfer, increasing in concentration as they ascend trophic levels. For example, insecticides ingested by zooplankton are transferred to small fish, which are then consumed by piscivorous birds and mammals. This results in biomagnified concentrations in top predators, often reaching levels that are lethal or cause severe sublethal effects (Das et al., 2023b; Mallick et al., 2023; Mishra et al., 2024). In birds, exposure to organochlorine insecticides has been linked to eggshell thinning, reduced clutch sizes, and reproductive failures, which have significantly impacted populations of predatory bird species like ospreys and eagles. Similarly, marine mammals such as seals and whales accumulate high concentrations of POPs, leading to compromised immunity and increased susceptibility to infectious diseases. The widespread nature of trophic transfer highlights the interconnectedness of aquatic and terrestrial ecosystems, as pollutants in aquatic environments ultimately affect land-based species, including humans (Fu et al., 2022).

Cascading Effects on Ecosystem Services

The ecological disruptions caused by micropollutants have cascading effects on vital ecosystem services, which include primary production, nutrient cycling, and water purification. Reduced photosynthetic efficiency in primary producers and declining populations of key species disrupt the balance of aquatic ecosystems, undermining their capacity to support biodiversity and human needs. For instance, the degradation of microbial communities and the loss of invertebrates affect nutrient cycling, leading to an accumulation of organic matter and reduced ecosystem productivity (Fu et al., 2022).

Water purification, a critical ecosystem service, is also severely impacted as the natural breakdown of pollutants by microbial communities becomes less effective. This leads to higher pollutant loads in water bodies, necessitating expensive treatment processes to ensure water quality for human consumption and industrial use. Moreover, the loss of fisheries due to bioaccumulation and declining aquatic biodiversity has significant economic consequences for communities that rely on these resources. The combined ecological and economic impacts emphasize the need for proactive measures to mitigate micropollutant contamination (Griebler & Avramov, 2015).

Case Studies Highlighting Ecological Impacts

Case Study 1: Decline in Amphibian Populations

The widespread use of pesticides, such as atrazine, has been directly linked to amphibian population declines in North America. Atrazine, a herbicide commonly used in agricultural practices, disrupts hormonal pathways in amphibians, causing reproductive deformities such as feminization in male frogs. Furthermore, atrazine weakens the immune systems of amphibians, increasing their susceptibility to diseases like chytridiomycosis, a fungal infection responsible for significant amphibian mortality worldwide. The loss of amphibians, which serve as both predators and prey in aquatic ecosystems, destabilizes food webs and reduces ecosystem resilience (Hayes et al., 2010).

Case Study 2: Pharmaceutical-Induced Behavioral Changes

A study conducted in European rivers revealed that antidepressants such as fluoxetine in water bodies reduced the predatory efficiency of fish. This behavioral disruption altered predator–prey dynamics, leading to an overabundance of certain invertebrates, which subsequently affected algal growth and nutrient cycling. Such imbalances disrupt aquatic ecosystems and reduce the overall stability and biodiversity of these environments (Moreira et al., 2022).

Case Study 3: Coral Reef Degradation

In coastal ecosystems, UV filters from sunscreens have been identified as a major contributor to coral bleaching and reduced resilience to environmental stressors. These chemicals interfere with the photosynthetic symbiosis between corals and algae, essential for coral health and reef survival. Prolonged exposure to UV filters has also been linked to structural weakening of coral skeletons, threatening the long-term viability of reef ecosystems and the diverse marine species they support (Moeller et al., 2021).

Biodiversity as a Mediator of Pollutant Transformation

Biodiversity, encompassing the variety of life forms within an ecosystem, plays a pivotal role in regulating ecological processes, including the attenuation and transformation of micropollutants. A diverse array of microbial, plant, and animal communities contributes to ecosystem resilience by enhancing the ability to cope with environmental stressors, such as contamination from micropollutants. This section explores how biodiversity facilitates the degradation, sequestration, and detoxification of these pollutants, emphasizing the need to preserve and leverage biodiversity for effective environmental management.

The Role of Microbial Diversity

Microbial communities are the primary drivers of micropollutant degradation due to their remarkable metabolic versatility. These communities, composed of bacteria, fungi, and archaea, break down a wide range of organic and inorganic pollutants, converting them into less harmful or benign compounds. High microbial diversity amplifies this functionality, as complementary metabolic pathways among species enable ecosystems to process complex pollutant mixtures more efficiently.

Metabolic Pathways in Degradation

Microbial species employ a range of metabolic pathways to degrade micropollutants, depending on environmental conditions such as oxygen availability:

Aerobic Metabolism: In oxygen-rich environments, microbes oxidize pollutants like antibiotics, pesticides, and hydrocarbons, breaking them into less toxic or more biodegradable intermediates. For example, bacteria in wastewater treatment plants efficiently metabolize ibuprofen and diclofenac under aerobic conditions (Gupta et al., 2017).

Anaerobic Degradation: In oxygen-depleted environments, such as sediments, anaerobic microbes degrade pollutants like chlorinated hydrocarbons and nitroaromatic compounds. These processes often result in slower degradation rates but significantly reduce the environmental persistence of these compounds (Kumari & Das, 2023).

Synergistic Effects of Microbial Consortia

Microbial consortia, or communities of interacting microbial species, exhibit synergistic interactions that enhance pollutant degradation. In these systems, one species may initiate the breakdown of a complex compound, producing intermediates that are subsequently metabolized by other species. For instance, the degradation of a pharmaceutical compound might involve initial hydroxylation by one microbe, followed by ring cleavage and mineralization by others. Such synergistic interactions increase degradation rates and reduce the accumulation of toxic byproducts, promoting overall ecosystem health (Lü et al., 2024).

Phytoplankton and Algae in Micropollutant Transformation

Phytoplankton and algae play an essential role in the transformation of micropollutants, acting as natural filters in aquatic systems. These organisms contribute to pollutant attenuation through direct uptake, enzymatic breakdown, and facilitation of photodegradation processes, making them indispensable components of aquatic ecosystems (Liu et al., 2021).

Uptake and Bioconcentration

Algae can directly uptake micropollutants from the water column, incorporating them into their cellular processes. This bioaccumulation serves as a temporary sink for pollutants, reducing their bioavailability in the ecosystem. However, excessive accumulation can lead to physiological stress in algae, affecting their growth and productivity. Despite these challenges, algal bioconcentration plays a crucial role in mitigating the immediate impacts of micropollutants.

Enzymatic Transformation

Phytoplankton and algae produce enzymes that facilitate the breakdown of micropollutants into less toxic forms. For example, green algae have been shown to metabolize endocrine-disrupting chemicals like bisphenol A (BPA), significantly reducing their hormonal activity and environmental persistence. This enzymatic transformation is highly species-specific, with specific algal taxa demonstrating greater efficiency in degrading particular pollutants (Usmani et al., 2022).

Photodegradation Facilitation

Algae enhance photodegradation by releasing photosensitizers compounds that absorb sunlight and generate reactive oxygen species capable of breaking down pollutants. This mechanism is particularly effective for the degradation of pharmaceuticals and pesticides that are susceptible to photolysis under sunlight. Algal-driven photodegradation is a natural attenuation process that complements other microbial and chemical degradation pathways (He et al., 2021).

Macrophytes and Constructed Wetlands

Macrophytes, or large aquatic plants, play a dual role in micropollutant removal, acting as physical barriers and facilitating microbial activity in their root zones. Constructed wetlands incorporating macrophytes are increasingly recognized as effective systems for treating wastewater and mitigating pollution.

Rhizosphere Microbial Enhancement

The rhizosphere, the root zone of macrophytes, harbors diverse microbial populations that thrive on root exudates. These microbes are stimulated to degrade organic pollutants, including hydrocarbons, pesticides, and pharmaceuticals, more efficiently in the rhizosphere than bulk water. The interaction between plant roots and microbial communities creates a synergistic system for pollutant transformation (Neori & Agami, 2017).

Phytoremediation

Macrophytes such as cattails (Typha) and reed grasses (Phragmites) absorb micropollutants, including heavy metals and pharmaceuticals, through their roots and stems. These plants sequester pollutants and metabolize some compounds, reducing their toxicity and environmental impact. Phytoremediation is a sustainable and cost-effective strategy for managing contaminated aquatic systems (Prabakaran et al., 2019).

Oxygenation of Sediments

Macrophytes oxygenate sediments through root aeration, creating favorable conditions for aerobic microbial degradation of micropollutants such as polycyclic aromatic hydrocarbons (PAHs). This oxygenation enhances the overall efficiency of pollutant breakdown in sediment layers, contributing to the long-term remediation of aquatic ecosystems (Kushwaha et al., 2024).

Biodiversity and Ecosystem Multifunctionality

Ecosystems with higher biodiversity exhibit greater multifunctionality, enabling them to simultaneously perform multiple ecological processes, including micropollutant degradation. This functionality is particularly evident in microbial and phytoplankton communities, where high species richness correlates with enhanced ecosystem performance.

Experimental Evidence

Studies on phytoplankton assemblages have demonstrated that species-rich communities degrade a broader spectrum of micropollutants compared to monocultures. Functional diversity within these communities enables the simultaneous processing of various pollutants, enhancing both the stability and efficiency of degradation pathways.

Biodiversity and System Resilience

Diverse ecosystems are more resilient to environmental disturbances, such as pollutant influxes, due to functional redundancy where multiple species can perform similar ecological roles (Jiao et al., 2019). This redundancy ensures that critical ecosystem functions, such as pollutant degradation, are maintained even under stress or species loss.

Challenges in Harnessing Biodiversity for Micropollutant Management

Despite its potential, leveraging biodiversity for micropollutant management faces several challenges:

Loss of Biodiversity: Pollution and habitat degradation reduce biodiversity, weakening natural remediation capabilities.

Complexity of Interactions: The intricate interactions within diverse communities make it difficult to predict outcomes in real-world scenarios, complicating management strategies.

Scale-Up Issues: Translating laboratory findings to field-scale applications, such as constructed wetlands, requires addressing environmental variability and scalability challenges.

Nanotechnology and Bioremediation for Micropollutant Removal

The effective removal of micropollutants from aquatic environments necessitates innovative approaches that augment existing treatment systems. Nanotechnology and bioremediation offer promising solutions, leveraging the unique properties of nanomaterials and the metabolic capabilities of biological entities to tackle the challenges posed by these persistent contaminants (Priyadarsini & Das, 2024, 2025). This section explores the underlying mechanisms, applications, and potential of these technologies in mitigating micropollutant pollution.

Advances in Nanotechnology for Micropollutant Removal

Nanotechnology involves the creation and application of materials with dimensions at the nanoscale, typically 1–100 nm. These materials exhibit unique physicochemical properties, such as high surface area-to-volume ratios, enhanced reactivity, and tunable surface functionalities, which make them highly effective for micropollutant removal (Bagal & Raut-Jadhav, 2021).

Adsorptive Removal Using Nanomaterials

Adsorption is a widely utilized mechanism for removing micropollutants from water, and nanomaterials have revolutionized this approach by significantly enhancing adsorptive capacities (Alayan et al., 2021).

Carbon-Based Nanomaterials: Activated carbon, graphene oxide, and carbon nanotubes exhibit exceptional performance in adsorbing pharmaceuticals, pesticides, and EDCs. Their large surface area and strong hydrophobic and π-π interactions allow them to capture a broad spectrum of contaminants. For instance, graphene oxide provides high adsorption capacity and offers functional groups that can be modified for specific pollutant targeting.

Metal and Metal-Oxide Nanoparticles: Nanoparticles such as nanoscale zero-valent iron (nZVI), titanium dioxide (TiO_2), and zinc oxide (ZnO) are highly effective in adsorbing and catalyzing the degradation of pollutants. These nanomaterials can remove heavy metals by forming stable complexes or adsorb persistent organic pollutants (POPs) through surface interactions. The reactivity of nZVI, for example, enables it to degrade chlorinated hydrocarbons in groundwater, while TiO_2 effectively captures and breaks down dyes and pharmaceuticals.

Hybrid Nanomaterials: Combining carbon-based nanomaterials with metal or metal-oxide nanoparticles enhances the efficiency of adsorption systems. These hybrid materials can simultaneously capture a range of pollutants and catalyze their breakdown, offering multifunctional solutions for wastewater treatment.

Photocatalysis for Degradation

Photocatalysis employs light-activated nanomaterials to oxidize and degrade micropollutants into non-toxic byproducts. Among these, titanium dioxide (TiO_2) nanoparticles have gained significant attention for their efficacy and versatility. Under ultraviolet (UV) or visible light, TiO_2 generates reactive oxygen species (ROS), such as hydroxyl radicals, which attack and oxidize pollutants at the molecular level. This process breaks complex organic pollutants into simpler, less harmful molecules such as water and carbon dioxide. Photocatalytic systems have demonstrated remarkable efficiency in degrading pharmaceuticals like antibiotics, industrial dyes, and pesticides. Advances in TiO_2 technology, such as doping with metals (e.g., silver or platinum) or coupling with materials like graphene, have enhanced photocatalytic activity under visible light, making the technology more applicable to real-world conditions. Recent developments include the use of heterojunction photocatalysts, where two different semiconductors are combined to improve charge separation and light absorption. These systems show promise in increasing degradation rates while reducing energy inputs (Gopinath et al., 2020).

Membrane-Based Nanotechnology

Nanotechnology has significantly advanced membrane-based water treatment systems, offering highly selective and efficient solutions for micropollutant removal (Manikandan et al., 2022).

Nanofiltration Membranes: These membranes operate by size exclusion and charge repulsion, trapping micropollutants such as pharmaceuticals, heavy metals, and EDCs. Functionalizing membranes with nanomaterials, such as carbon nanotubes or silver nanoparticles, improves fouling resistance, enhances pollutant rejection rates, and extends membrane lifespan.

Reverse Osmosis Membranes: Reverse osmosis systems integrated with nanomaterials provide high contaminant removal efficiencies. Nanomaterials enhance the membrane's mechanical strength, thermal stability, and biofouling resistance, making them suitable for advanced water treatment applications.

Hybrid Systems: The integration of membranes with photocatalytic nanomaterials or adsorbents offers synergistic benefits, enabling simultaneous filtration and degradation of micropollutants. For example, a hybrid system combining a TiO_2-coated membrane with nanofiltration allows for both the physical separation and oxidative breakdown of contaminants, achieving comprehensive treatment efficiency.

Scalability and Applications: Nanomaterial-based membranes are being scaled for municipal wastewater treatment and industrial effluent management, addressing challenges such as emerging contaminant removal and water reuse.

Bioremediation: Harnessing Biological Processes

Bioremediation utilizes living organisms such as microorganisms, plants, and fungi to degrade, sequester, or detoxify micropollutants in an eco-friendly and cost-effective manner. Unlike conventional methods, bioremediation leverages natural metabolic pathways to transform harmful contaminants into benign substances, making it a sustainable approach to pollution management. The versatility of bioremediation allows its application across diverse ecosystems, addressing pollutants ranging from pharmaceuticals and pesticides to heavy metals and hydrocarbons (Arslan et al., 2017).

Microbial Bioremediation

Microorganisms are the primary agents in bioremediation due to their ability to metabolize a wide range of micropollutants. Bacteria and fungi convert complex pollutants into simpler, less harmful compounds through enzymatic pathways.

Bacteria: Microbial species such as Pseudomonas, Bacillus, and Achromobacter are highly efficient at degrading pharmaceuticals, pesticides, and hydrocarbons (Pradhan et al., 2024; Satapathy et al., 2024; Tripathy et al., 2024). These bacteria employ aerobic pathways to oxidize pollutants in oxygen-rich environments and anaerobic processes to reduce pollutants under oxygen-depleted conditions. For instance, Pseudomonas putida has been extensively studied for its ability to degrade polycyclic aromatic hydrocarbons (PAHs) in contaminated waters.

Fungi: Fungi, particularly white-rot fungi like Phanerochaete chrysosporium, produce extracellular enzymes such as lignin peroxidase and manganese peroxidase. These enzymes break down complex organic pollutants into simpler molecules, including dyes, pesticides, and persistent organic pollutants (POPs). The filamentous structure of fungi also enables them to access pollutants in sediment pores, making them effective in diverse environments.

Engineered Microbes: Advances in synthetic biology have enabled the development of genetically engineered microorganisms tailored for specific micropollutants. For example, engineered Escherichia coli strains with enhanced pathways for pharmaceutical degradation have demonstrated superior efficiency compared to native strains. These microbes offer immense potential for addressing pollutants that are resistant to natural degradation.

Constructed Wetlands

Constructed wetlands integrate microbial and phytoremediation processes into engineered ecosystems designed for wastewater treatment. These systems replicate natural wetland functions, combining physical, chemical, and biological processes to remove contaminants. Constructed wetlands use sedimentation, microbial degradation, and plant uptake to remove micropollutants. The rhizosphere of wetland plants fosters diverse microbial communities, which break down pharmaceuticals,

EDCs, and pesticides into less harmful byproducts. These systems are widely used for treating urban and agricultural wastewater, with successful implementation in removing a variety of micropollutants. For instance, wetlands planted with Phragmites have significantly reduced antibiotic and pesticide concentrations in effluent waters. Advances in wetland design, such as vertical flow systems and hybrid wetlands, enhance treatment efficiency by improving oxygen transfer and pollutant retention. These innovations address limitations in traditional wetland systems, such as clogging and limited hydraulic capacity (Lu et al., 2020).

Hybrid Approaches: Integrating Nanotechnology and Bioremediation

Hybrid approaches that combine nanotechnology and bioremediation harness the strengths of both technologies to create synergistic systems with enhanced pollutant removal efficiency (Kumar et al., 2024).

Nano-Bio Conjugates: Functionalizing microbial cells with nanoparticles has improved pollutant degradation rates by increasing enzyme activity and pollutant uptake. For example, iron oxide nanoparticles attached to bacteria enhance their ability to degrade nitroaromatic compounds by facilitating electron transfer.

Nanotechnology-Enhanced Wetlands: Adding nanomaterials like graphene oxide or titanium dioxide to constructed wetlands boosts the removal of recalcitrant pollutants. These nanomaterials enhance microbial degradation by providing additional surfaces for biofilm formation and catalyzing the oxidative breakdown of contaminants.

Challenges and Future Directions

Removing and managing micropollutants in aquatic ecosystems present significant scientific, technical, and regulatory challenges. Although advancements in nanotechnology and bioremediation offer innovative solutions, their large-scale application requires overcoming numerous hurdles. Addressing these challenges necessitates interdisciplinary research, technological innovation, and robust policy frameworks adapting to micropollutant behavior and impacts.

Regulatory and Monitoring Challenges

Lack of Comprehensive Regulations

Despite increasing recognition of micropollutants as a critical environmental concern, regulatory frameworks in many regions remain outdated or insufficient. Current water quality standards predominantly focus on conventional pollutants such as nutrients, heavy metals, and organic matter, leaving micropollutants largely unregulated (Dey et al., 2024c; Pradhan et al., 2024; Satapathy et al., 2024). This gap allows these contaminants to persist and accumulate in aquatic ecosystems, often unnoticed and unmanaged. The European Union's Water Framework Directive, while a benchmark for water policy, includes only a limited list of priority substances. Many emerging contaminants, including pharmaceuticals, personal care products, and endocrine-disrupting chemicals (EDCs), are excluded. As a result, pollutants like carbamazepine and triclosan continue to be detected in European rivers and lakes, with concentrations often exceeding safe ecological thresholds. Establishing comprehensive, global guidelines for micropollutant monitoring and removal is critical. These regulations should incorporate emerging pollutants, including those with limited toxicity data, and be informed by cutting-edge ecotoxicological research. Regulatory frameworks must also address the cumulative and synergistic effects of micropollutant mixtures, as their combined impact often exceeds the sum of their individual effects. Developing universal standards is complicated by regional variations in pollutant sources, treatment capacities, and ecological vulnerability. However, international cooperation and knowledge-sharing could help harmonize approaches, enabling more effective global management (Machiwal et al., 2018).

Challenges in Monitoring and Detection

The detection and quantification of micropollutants at ultra-trace levels (ng/L to μg/L) pose significant challenges due to their low concentrations and chemical complexity. Advanced analytical techniques, such as high-performance liquid chromatography-mass spectrometry (HPLC–MS) and gas chromatography-mass spectrometry (GC–MS), are required to identify these contaminants accurately. These methods provide high sensitivity and specificity but are prohibitively expensive and often inaccessible to many laboratories, particularly in developing countries. The high operational costs, need for skilled personnel, and extensive sample preparation further limit their widespread use. Additionally, the diverse chemical nature of micropollutants complicates detection. Pharmaceuticals, pesticides, industrial chemicals, and personal care products exhibit varied physicochemical properties, requiring a combination of detection methods for comprehensive analysis. This makes monitoring programs resource-intensive and time-consuming.

Proposed Solution

Portable Sensor Technologies: Developing cost-effective and portable sensors for real-time monitoring can revolutionize micropollutant detection. Sensors based on nanomaterials, such as carbon nanotubes, graphene oxide, and metal-oxide nanoparticles, have shown promise for their high sensitivity, selectivity, and rapid response times. These technologies enable on-site analysis without complex laboratory setups.

Biosensors: Biosensors utilizing enzymes, antibodies, or DNA-based probes can accurately detect specific micropollutants. Their integration with microfluidic devices offers potential for miniaturized and portable systems, making them suitable for use in remote and resource-limited settings.

Collaborative Efforts: Global initiatives to share expertise, standardize detection protocols, and provide funding for monitoring infrastructure in developing regions are essential to address disparities in analytical capabilities. Improving the accessibility and scalability of monitoring technologies will significantly enhance our ability to track micropollutant distributions, assess ecological risks, and evaluate the effectiveness of treatment systems.

Technical and Economic Barriers

The implementation of advanced micropollutant removal technologies faces significant technical and economic hurdles. While innovative solutions like nanotechnology and bioremediation show immense potential, their scalability, adaptability, and cost-efficiency remain areas of concern. Addressing these barriers is critical for the widespread adoption and long-term sustainability of these approaches.

High Costs of Advanced Technologies

Nanotechnology-based solutions, such as photocatalytic systems and advanced membranes, are among the most effective tools for micropollutant removal. However, their high production costs, reliance on rare or expensive raw materials, and maintenance requirements make them economically unfeasible for widespread use, particularly in low-resource settings. For instance, doping titanium dioxide (TiO_2) nanoparticles with metals to enhance photocatalytic efficiency adds to material and processing costs. Similarly, while highly efficient, advanced membranes embedded with nanomaterials face challenges such as fouling and high energy demands, further driving up operational expenses.

Bioremediation approaches, such as constructed wetlands and engineered microbial systems, are often perceived as cost-effective alternatives. However, scaling these systems to field applications requires significant resources. Constructed wetlands demand large land areas, intensive planning, and long establishment periods for optimal performance. Engineered microbial systems require specialized facilities and strict control of environmental conditions to maintain microbial activity, adding to operational costs.

Proposed Direction: Developing Low-Cost Nanomaterials: Research efforts should prioritize the development of biodegradable and cost-effective nanomaterials using abundant and sustainable raw materials. For example, natural carbon sources like agricultural waste could be explored for producing carbon-based nanomaterials. Optimizing Bioremediation Strategies: Innovations such as integrating endophytic bacteria or designing compact wetland systems can enhance the cost-effectiveness and scalability of bioremediation approaches.

Variability in Environmental Conditions

Field-scale applications of micropollutant removal technologies often encounter challenges posed by the variability of environmental conditions. Factors such as temperature, pH, salinity, and pollutant mixtures can significantly influence the efficiency of both nanotechnology and bioremediation systems. For instance, photocatalytic reactions depend highly on light intensity and wavelength, which vary with geographical and seasonal changes. Similarly, microbial activity in bioremediation systems can be inhibited by extreme pH levels, low nutrient availability, or the presence of toxic co-contaminants. These environmental fluctuations reduce the predictability and reliability of treatment systems, making it difficult to achieve consistent pollutant removal across diverse aquatic environments.

Adaptive Solutions: Hybrid Systems: Integrating physical, chemical, and biological processes can create more resilient systems to environmental variability. For example, combining photocatalysis with microbial degradation can provide complementary removal mechanisms that compensate for performance losses under suboptimal conditions.

Dynamic Monitoring and Feedback: Real-time monitoring of environmental parameters and adaptive control systems can optimize treatment efficiency in response to changing conditions.

Formation of Toxic Byproducts

One of the unintended consequences of advanced treatment technologies is the formation of transformation products (TPs). Advanced oxidation processes (AOPs) such as ozonation and photocatalysis are highly effective in degrading micropollutants. Still, incomplete reactions can produce intermediate compounds that are more toxic or persistent than the parent pollutants. For instance, degraded diclofenac, a common pharmaceutical, can yield chlorinated byproducts that pose additional ecological risks. Similarly, incomplete biodegradation in microbial systems may accumulate partially metabolized compounds that resist further breakdown. These byproducts complicate treatment efforts by introducing new contaminants into the environment, requiring additional downstream processing to ensure complete detoxification. Comprehensive studies on the toxicity and environmental behavior of TPs are needed to assess their risks and inform regulatory guidelines. Refining AOPs and bioremediation processes to minimize TP formation is essential. Strategies such as coupling oxidation processes with adsorption or biodegradation can help achieve complete mineralization of pollutants (Zuo et al., 2021).

Environmental Risks

While nanotechnology and bioremediation present promising solutions for micropollutant removal, their implementation carries potential environmental risks. Understanding and mitigating these risks is crucial for ensuring these technologies benefit ecosystems without causing unintended harm.

Risks Associated with Nanotechnology

Engineered nanomaterials, such as titanium dioxide (TiO_2) and zinc oxide (ZnO), are highly effective in removing micropollutants but may pose ecological risks if not properly managed. These nanoparticles can persist in the environment, accumulate in sediments, and interact with aquatic organisms, potentially disrupting ecological functions.

Impact on Microbial Communities: Nanoparticles can inhibit the activity of beneficial microbial populations, which are essential for natural pollutant degradation and nutrient cycling. For example, excessive concentrations of TiO_2 nanoparticles have been shown to alter microbial diversity and enzyme activity in sediment and water ecosystems.

Toxicity to Aquatic Organisms: Certain nanoparticles can induce oxidative stress in fish, invertebrates, and algae, leading to physiological damage and reduced survival rates. For instance, zinc oxide (ZnO) nanoparticles release ions that disrupt cellular processes in aquatic species, highlighting their potential toxicity.

Regulatory Need: Establishing comprehensive guidelines for the safe production, use, disposal, and recovery of nanomaterials in environmental applications is essential. Regulations should address potential bioaccumulation, toxicity thresholds, and environmental persistence. Encouraging the development of biodegradable and eco-friendly nanomaterials can reduce risks associated with their ecological release. Life cycle assessments (LCAs) should be conducted for all nanotechnology applications to evaluate their long-term environmental impacts.

Impacts on Non-target Species

Bioremediation processes, while eco-friendly, may inadvertently affect non-target species and disrupt ecological interactions, particularly in sensitive habitats. Introducing engineered microbes or plants designed for pollutant removal can alter existing food webs and compete with native species.

Disruption of Native Ecosystems: Engineered microbes with enhanced pollutant degradation capabilities may outcompete native microbial populations, reducing microbial diversity and affecting ecosystem stability. Similarly, though effective in pollutant removal, invasive phytoremediators like water hyacinth (Eichhornia crassipes) can overgrow and disrupt aquatic ecosystems, leading to oxygen depletion and habitat loss.

Impact on Higher Trophic Levels: Pollutant removal through bioaccumulation in plants or microorganisms can inadvertently transfer contaminants up the food chain. For example, heavy metals absorbed by bioremediating plants can enter herbivorous insects or aquatic grazers, creating secondary ecological risks.

Proposed Strategy: Conduct thorough ecological risk assessments before implementing large-scale bioremediation projects. These assessments should evaluate potential impacts on non-target species, ecosystem interactions, and habitat health. Adopt containment strategies, such as using engineered microbes in closed-loop systems or restricting the use of invasive plant species in controlled environments. Develop monitoring programs to track the long-term ecological effects of bioremediation technologies in diverse habitats.

Knowledge Gaps and Research Priorities

Despite advancements in micropollutant removal technologies, significant knowledge gaps persist, hindering the development of holistic and sustainable management strategies. Bridging these gaps through targeted research is essential for understanding the broader ecological implications of micropollutants and ensuring the effectiveness of removal technologies over the long term.

Ecological Effects of Micropollutant Mixtures

Most current studies focus on the effects of individual micropollutants, yet these contaminants rarely occur in isolation. In real-world scenarios, micropollutants exist as complex mixtures that can interact synergistically, antagonistically, or additively, amplifying or mitigating their impacts on ecosystems. For example, the combined presence of pesticides, pharmaceuticals, and personal care products in water bodies can exert multifaceted pressures on aquatic organisms, potentially disrupting physiological processes, reproductive cycles, and trophic interactions.

Knowledge Gap: The interactive effects of micropollutant mixtures on biodiversity, nutrient cycling, and food web dynamics remain poorly understood. These interactions are further complicated by environmental variables such as temperature, pH, and dissolved organic matter, which can modulate the toxicity and bioavailability of contaminants.

Research Priority: Conducting multi-compound studies to evaluate the cumulative and interactive effects of micropollutant mixtures on aquatic ecosystems. Developing ecotoxicological models that incorporate mixture dynamics to predict ecosystem-level impacts. Exploring the thresholds and tipping points where combinations of micropollutants lead to irreversible ecological damage.

Long-Term Impacts of Removal Technologies

The long-term sustainability and ecological outcomes of nanotechnology and bioremediation strategies are not well-documented. While these technologies effectively remove micropollutants in controlled environments, their implications over extended periods and across diverse ecosystems remain unclear.

Potential Concerns: Nanomaterials and bioremediating organisms may degrade over time, releasing previously captured contaminants into the environment. For instance, plants used in phytoremediation may release heavy metals upon decomposition (Das et al., 2023a, 2023b, 2023c; Das & Bal, 2025). Nanomaterials can lose functionality due to fouling, agglomeration, or environmental exposure, reducing their effectiveness and potentially generating secondary pollutants. Removing micropollutants does not guarantee immediate recovery of affected ecosystems. Long-term monitoring is needed to assess whether biodiversity and ecosystem functions, such as nutrient cycling and primary production, are restored.

Proposed Direction: Establishing long-term monitoring programs to track the ecological and operational outcomes of micropollutant removal technologies. These programs should evaluate metrics such as pollutant re-release, ecosystem resilience, and recovery rates. Investigating strategies to enhance the durability and sustainability of nanomaterials and bioremediating organisms, such as designing self-regenerating nanomaterials or optimizing microbial consortia for prolonged activity. Developing protocols for safely decommissioning and disposal of bioremediation systems and nanomaterials to prevent secondary environmental contamination.

Future Directions

The complexity of micropollutant removal necessitates forward-looking strategies that integrate advanced technologies, collaborative frameworks, and innovative research. Future efforts should combine complementary approaches, foster international cooperation, and leverage emerging tools like synthetic biology to create sustainable and scalable solutions.

Integrated Approaches for Micropollutant Management

Integrating nanotechnology and bioremediation provides a holistic framework for addressing the diverse challenges micropollutants pose. By combining the precision and efficiency of nanomaterials with the natural resilience of biological systems, integrated approaches can tackle a wide range of pollutants under varying environmental conditions. Constructed wetlands enhanced with photocatalytic nanomaterials can degrade organic pollutants and support microbial diversity. In these systems, nanomaterials like titanium dioxide (TiO_2) facilitate the oxidative breakdown of organic pollutants, while wetland plants and microbial communities remove nutrients and support ecosystem recovery. Such hybrid systems simultaneously address pollutant degradation, nutrient cycling, and biodiversity conservation. Integrated systems offer increased resilience to environmental variability, enhanced removal efficiency for complex pollutant mixtures, and reduced operational costs by leveraging natural processes alongside engineered solutions. Research should explore optimizing the compatibility and interactions between nanomaterials and biological components to maximize efficiency and minimize ecological risks.

Collaborative Frameworks for Global Action

Micropollutant management is a global challenge that transcends geographical and political boundaries. Collaborative frameworks involving scientists, industries, policymakers, and international organizations are essential for creating impactful solutions. The United Nations' Sustainable Development Goals (SDGs), particularly SDG 6 (Clean Water and Sanitation), provide a platform for fostering coordinated action against micropollutant pollution. Collaborative initiatives can drive technology transfer, resource sharing, and capacity building in regions with limited access to advanced remediation technologies. Establish international partnerships to standardize monitoring protocols and data collection for micropollutants. Promote technology sharing to ensure equitable access to advanced nanotechnology and bioremediation solutions. Implement global funding mechanisms to support research and mitigation efforts in developing regions. Enhancing public awareness and stakeholder engagement is vital for generating support for policies and practices that reduce micropollutant emissions.

Advances in Synthetic Biology

Synthetic biology offers transformative potential for bioremediation by enabling the design of microbes tailored to degrade specific micropollutants (Barone et al., 2024). These "designer microbes" can be engineered with optimized enzymatic pathways to target persistent pollutants or pollutant mixtures that resist natural degradation. Develop microbial consortia with complementary metabolic capabilities to address pollutant mixtures in diverse environmental conditions. Engineer microbes capable of self-regeneration and controlled activity to enhance their effectiveness and environmental safety. Integrate biosensors into microbial systems to enable real-time monitoring of pollutant levels and system performance. Balancing the ecological safety of engineered microbes with their degradation efficacy remains a priority. Research should focus on containment strategies and risk assessments to ensure minimal unintended ecological impacts.

Conclusion

Micropollutants represent one of the most pressing challenges for aquatic ecosystems and biodiversity. Their persistence, complex chemical behavior, and wide-ranging ecological impacts necessitate a comprehensive and interdisciplinary approach to management and mitigation. This chapter has highlighted their sources, pathways, and detrimental effects on aquatic biodiversity, from microbial communities to apex predators. Innovative solutions, such as nanotechnology and bioremediation, offer promising pathways for addressing this issue. Nanomaterials provide advanced capabilities for adsorbing, catalyzing, and filtering micropollutants, while bioremediation leverages natural biological processes to degrade and detoxify pollutants. The integration of these approaches holds excellent potential for developing sustainable and efficient treatment systems. However, the journey towards effective micropollutant management is fraught with challenges. Regulatory gaps, high costs, environmental risks, and knowledge gaps limit the widespread adoption of current solutions. Addressing these issues requires coordinated global action, long-term research, and adaptive policies. It is equally crucial to recognize the role of biodiversity in enhancing ecosystem resilience and remediation processes, underscoring the need for conservation-focused strategies. Future efforts should focus on advancing interdisciplinary research, promoting international collaborations, and implementing integrated systems that combine innovative technologies with natural processes. By aligning technological advancements with ecological principles, we can mitigate the impacts of micropollutants, safeguard aquatic biodiversity, and ensure the sustainability of vital ecosystem services. The challenge is immense, but so is the potential for innovation and progress. Through sustained commitment and collaboration, the scientific community, policymakers, and stakeholders can pave the way for cleaner, healthier aquatic environments for future generations.

References

Ajala, O. J., Tijani, J. O., Salau, R. B., Abdulkareem, A. S., & Aremu, O. (2022). A review of emerging micro-pollutants in hospital wastewater: Environmental fate and remediation options. *Results in Engineering, 16*, Article 100671.

Akhtar, A. B. T., Naseem, S., Yasar, A., & Naseem, Z. (2021). Persistent organic pollutants (POPs): sources, types, impacts, and their remediation. *Environmental Pollution and Remediation*, 213–246.

Alayan, H. M., Aljumaily, M. M., Alsaadi, M. A., Mjalli, F. S., & Hashim, M. A. (2021). A review exploring the adsorptive removal of organic micropollutants on tailored hierarchical carbon nanotubes. *Toxicological & Environmental Chemistry, 103*(4), 282–325.

Arslan, M., Ullah, I., Müller, J. A., Shahid, N., & Afzal, M. (2017). Organic micropollutants in the environment: ecotoxicity potential and methods for remediation. *Enhancing Cleanup of Environmental Pollutants: Volume 1: Biological Approaches*, 65–99.

Bagal, M., & Raut-Jadhav, S. (2021). The process for the removal of micropollutants using nanomaterials. In: *Handbook of nanomaterials for astewater treatment* (pp. 957–1007). Elsevier.

Bakir, A., Rowland, S. J., & Thompson, R. C. (2014). Transport of persistent organic pollutants by microplastics in estuarine conditions. *Estuarine, Coastal and Shelf Science, 140*, 14–21.

Bala, K., Dogra, A., Lata, P., & Das, A. P. (2024). Discussing sources and biological-chemical remediation approaches for microplastic pollution. *Sustainable Microbial Technology for Synthetic and Cellulosic Microfiber Bioremediation*, 37–59. https://doi.org/10.1007/978-3-031-62600-5_3

Barone, G. D., Rodríguez-Seijo, A., Parati, M., Johnston, B., Erdem, E., Cernava, T., Zhu, Z., Liu, X., Axmann, I. M., & Lindblad, P. (2024). Harnessing photosynthetic microorganisms for enhanced bioremediation of microplastics: A comprehensive review. *Environmental Science and Ecotechnology*, 100407.

Bashir, I., Lone, F. A., Bhat, R. A., Mir, S. A., Dar, Z. A., Dar, S. A. (2020). Concerns and threats of contamination on aquatic ecosystems. *Bioremediation and biotechnology: sustainable approaches to pollution degradation*, 1–26.

Bavumiragira, J. P., & Yin, H. (2022). Fate and transport of pharmaceuticals in water systems: A processes review. *Science of the Total Environment, 823*, Article 153635.

Bellasi, A., Binda, G., Pozzi, A., Galafassi, S., Volta, P., & Bettinetti, R. (2020). Microplastic contamination in freshwater environments: A review, focusing on interactions with sediments and benthic organisms. *Environments, 7*(4), 30.

Bora, L. V., & Mewada, R. K. (2017). Visible/solar light active photocatalysts for organic effluent treatment: Fundamentals, mechanisms and parametric review. *Renewable and Sustainable Energy Reviews, 76*, 1393–1421.

Borgå, K., McKinney, M. A., Routti, H., Fernie, K. J., Giebichenstein, J., Hallanger, I., & Muir, D. C. (2022). The influence of global climate change on accumulation and toxicity of persistent organic pollutants and chemicals of emerging concern in Arctic food webs. *Environmental Science: Processes & Impacts, 24*(10), 1544–1576.

Bose, S., Kumar, P. S., Vo, D.-V.N., Rajamohan, N., & Saravanan, R. (2021). Microbial degradation of recalcitrant pesticides: A review. *Environmental Chemistry Letters, 19*, 3209–3228.

Brain, R. A., & Prosser, R. S. (2022). Human induced fish declines in North America, how do agricultural pesticides compare to other drivers? *Environmental Science and Pollution Research, 29*(44), 66010–66040.

Brühl, C. A., & Zaller, J. G. (2021). Indirect herbicide effects on biodiversity, ecosystem functions, and interactions with global changes. In *Herbicides* (pp. 231–272). Elsevier.

Calvo, S., Romo, S., Soria, J., & Picó, Y. (2021). Pesticide contamination in water and sediment of the aquatic systems of the Natural Park of the Albufera of Valencia (Spain) during the rice cultivation period. *Science of the Total Environment, 774*, Article 145009.

Camacho-Jiménez, L., González-Ruiz, R., & Yepiz-Plascencia, G. (2023). Persistent organic pollutants (POPs) in marine crustaceans: Bioaccumulation, physiological and cellular responses. *Marine Environmental Research*, 106184.

Capodaglio, A. G., Bojanowska-Czajka, A., & Trojanowicz, M. (2018). Comparison of different advanced degradation processes for the removal of the pharmaceutical compounds diclofenac and carbamazepine from liquid solutions. *Environmental Science and Pollution Research, 25*(28), 27704–27723.

Chaturvedi, P., Shukla, P., Giri, B. S., Chowdhary, P., Chandra, R., Gupta, P., & Pandey, A. (2021). Prevalence and hazardous impact of pharmaceutical and personal care products and antibiotics in environment: A review on emerging contaminants. *Environmental Research, 194*, Article 110664.

Das, A. P., Dutta, K., Khatun, R., Behera, I. D., Singh, S., & Mishra, S. (2023a). Microfiber pollution and its microbial mitigation: A review on current trends and future prospects. *Journal of the Taiwan Institute of Chemical Engineers*, 105104. https://doi.org/10.1016/J.JTICE.2023.105104

Das, A. P., Mohanty, A. K., van Hullebusch, E. D., Figueiredo, G., Das, A. P., Mohanty, A. K., van Hullebusch, E. D., & Figueiredo, G. (2023b). Bioremediation as an emerging technology for the removal of synthetic microplastic pollutants from marine ecosystem. *MarPB, 194*, 115297. https://doi.org/10.1016/J.MARPOLBUL.2023.115297

Das, P., Halder, G., & Bal, M. (2023c). A critical review on remediation of microplastics using microalgae from aqueous system. *Science of the Total Environment, 898*, 166425. https://doi.org/10.1016/j.scitotenv.2023.166425

Das, A., Dey, S., Bala, K., Gnanasekaran, R., & Das, A. P. (2024a). Microbial bioremediation technology for sustainable treatment and management of synthetic microfiber waste. *Sustainable Microbial Technology for Synthetic and Cellulosic Microfiber Bioremediation*, 221–238. https://doi.org/10.1007/978-3-031-62600-5_12

Das, A., Dey, S., & Das, A. P. (2024b). Ecotoxicological impacts of synthetic microfiber pollutants and development of sustainable mitigation strategies. *Environmental Chemistry and Ecotoxicology*. https://doi.org/10.1016/J.ENCECO.2024.12.003

Das, A., Ray, S. K., Mohanty, M., Mohanty, J., Dey, S., & Das, A. P. (2024d). Ecotoxicity of microplastic wastes and their sustainable management: A review. *Environmental Chemistry and Ecotoxicology, 6*, 144–152. https://doi.org/10.1016/J.ENCECO.2024.05.003

Das, P., & Bal, M. (2025). Modeling and optimization study for microplastic removal from aquatic medium using Chroococcidiopsis species. *Chemical Engineering Science, 304*, Article 121085. https://doi.org/10.1016/j.ces.2024.121085

Dey, S., Das, A., Mallick, K., Sahu, A., & Das, A. P. (2023a). Environmental petroleum waste: pollution, toxicity, sustainable remediation, 159–175. https://doi.org/10.1007/978-3-031-48220-5_7

Dey, S., Tripathy, B., Kumar, M. S., & Das, A. P. (2023b). Ecotoxicological consequences of manganese mining pollutants and their biological remediation. *Environmental Chemistry and Ecotoxicology*. https://doi.org/10.1016/J.ENCECO.2023.01.001

Dey, S., Bissoyi, A., Koteswara Reddy, G., & Das, A. P. (2024a). Synthetic microfibers: An urban emerging pollutant demanding for sustainable environmental solutions. *Sustainable Microbial Technology for Synthetic and Cellulosic Microfiber Bioremediation*, 107–121. https://doi.org/10.1007/978-3-031-62600-5_7

Dey, S., Dash, A., & Das, A. P. (2024b). Bioremediation as a sustainable and circular solution for mining waste management and bioenergy generation. *Sustainable Management of Mining Waste and Tailings: A Circular Economy Approach*, 107–119. https://doi.org/10.1201/9781003442455-6/BIOREMEDIATION-SUSTAINABLE-CIRCULAR-SOLUTION-MINING-WASTE-MANAGEMENT-BIOENERGY-GENERATION-SUDESHNA-DEY-AKANSHYA-DASH-ALOK-PRASAD-DAS

Dey, S., Dash, N., Ray, S. K., Priyadarsini, S., Chhotaray, C., & Das, A. P. (2024c). Occurrence and quantification of tire wear plastics pollutants from highway city dust of Bhubaneswar city of Odisha. *Sustainable Microbial Technology for Synthetic and Cellulosic Microfiber Bioremediation*, 77–89. https://doi.org/10.1007/978-3-031-62600-5_5

Du, J., & Jing, C. (2018). Anthropogenic PAHs in lake sediments: A literature review (2002–2018). *Environmental Science: Processes & Impacts, 20*(12), 1649–1666.

Elbeshbishy, E., & Okoye, F. (2019). Improper disposal of household hazardous waste: Landfill/municipal wastewater treatment plant. *Municipal Solid Waste Management, 183*.

Elyamine, A. M., Kan, J., Meng, S., Tao, P., Wang, H., & Hu, Z. (2021). Aerobic and anaerobic bacterial and fungal degradation of pyrene: Mechanism pathway including biochemical reaction and catabolic genes. *International Journal of Molecular Sciences, 22*(15), 8202.

Fawcett, D., Fitzgerald, C., Verduin, J., & Poinern, G. E. J. (2023). Microplastic pollution and its detrimental impact on coastal ecosystems and mid-ocean gyres. In *Harnessing synthetic nanotechnology-based methodologies for sustainable green applications* (pp. 277–298). CRC Press.

Fouad, F. A., Youssef, D. G., Shahat, F. M., & Abd El-Ghany, M. N. (2022). Role of microorganisms in biodegradation of pollutants. In *Handbook of biodegradable materials* (pp. 1–40). Springer.

Fu, Q., Meyer, C., Patrick, M., Kosfeld, V., Rüdel, H., Koschorreck, J., & Hollender, J. (2022). Comprehensive screening of polar emerging organic contaminants including PFASs and evaluation of the trophic transfer behavior in a freshwater food web. *Water Research, 218*, Article 118514.

Garcia-Becerra, F. Y., & Ortiz, I. (2018). Biodegradation of emerging organic micropollutants in nonconventional biological wastewater treatment: A critical review. *Environmental Engineering Science, 35*(10), 1012–1036.

Gopinath, K. P., Madhav, N. V., Krishnan, A., Malolan, R., & Rangarajan, G. (2020). Present applications of titanium dioxide for the photocatalytic removal of pollutants from water: A review. *Journal of Environmental Management, 270*, Article 110906.

Grandclément, C., Seyssiecq, I., Piram, A., Wong-Wah-Chung, P., Vanot, G., Tiliacos, N., Roche, N., & Doumenq, P. (2017). From the conventional biological wastewater treatment to hybrid processes, the evaluation of organic micropollutant removal: A review. *Water Research, 111*, 297–317.

Griebler, C., & Avramov, M. (2015). Groundwater ecosystem services: A review. *Freshwater Science, 34*(1), 355–367.

Gupta, A., Gupta, R., & Singh, R. L. (2017). Microbes and environment. In *Principles and applications of environmental biotechnology for a sustainable future* (pp. 43–84).

Hauer, C., Leitner, P., Unfer, G., Pulg, U., Habersack, H., & Graf, W. (2018). The role of sediment and sediment dynamics in the aquatic environment. In *Riverine ecosystem management: Science for governing towards a sustainable future* (pp. 151–169).

Hayes, T., Falso, P., Gallipeau, S., & Stice, M. (2010). The cause of global amphibian declines: A developmental endocrinologist's perspective. *Journal of Experimental Biology, 213*(6), 921–933.

He, X., Wu, P., Wang, S., Wang, A., Wang, C., & Ding, P. (2021). Inactivation of harmful algae using photocatalysts: Mechanisms and performance. *Journal of Cleaner Production, 289*, Article 125755.

Herrero, A., Vila, J., Eljarrat, E., Ginebreda, A., Sabater, S., Batalla, R. J., & Barceló, D. (2018). Transport of sediment borne contaminants in a Mediterranean river during a high flow event. *Science of the Total Environment, 633*, 1392–1402.

Jagatee, S., Priyadarshini, S., Rath, C. C., & Das, A. P. (2024). Synthetic microfiber: An enduring environmental problem linked to sustainable development. *Environmental Science and Engineering, Part F2333*, 93–112. https://doi.org/10.1007/978-3-031-51792-1_6

Jahan, S., & Singh, A. (2023). Causes and impact of industrial effluents on receiving water bodies: A review. *Malaysian Journal of Science and Advanced Technology*, 111–121.

Jiao, S., Chen, W., & Wei, G. (2019). Resilience and assemblage of soil microbiome in response to chemical contamination combined with plant growth. *Applied and Environmental Microbiology, 85*(6), e02523-e2618.

Kasonga, T. K., Coetzee, M. A., Kamika, I., Ngole-Jeme, V. M., & Momba, M. N. B. (2021). Endocrine-disruptive chemicals as contaminants of emerging concern in wastewater and surface water: A review. *Journal of Environmental Management, 277*, Article 111485.

Kosek, K., Luczkiewicz, A., Fudala-Książek, S., Jankowska, K., Szopińska, M., Svahn, O., Tränckner, J., Kaiser, A., Langas, V., & Björklund, E. (2020). Implementation of advanced

micropollutants removal technologies in wastewater treatment plants (WWTPs)-Examples and challenges based on selected EU countries. *Environmental Science & Policy, 112*, 213–226.
Kotowska, U., Kapelewska, J., & Sawczuk, R. (2020). Occurrence, removal, and environmental risk of phthalates in wastewaters, landfill leachates, and groundwater in Poland. *Environmental Pollution, 267*, Article 115643.
Kumari, S., & Das, S. (2023). Bacterial enzymatic degradation of recalcitrant organic pollutants: Catabolic pathways and genetic regulations. *Environmental Science and Pollution Research, 30*(33), 79676–79705.
Kushwaha, A., Goswami, L., Kim, B. S., Lee, S. S., Pandey, S. K., & Kim, K.-H. (2024). Constructed wetlands for the removal of organic micropollutants from wastewater: Current status, progress, and challenges. *Chemosphere*, 142364.
Liu, R., Li, S., Tu, Y., & Hao, X. (2021). Capabilities and mechanisms of microalgae on removing micropollutants from wastewater: A review. *Journal of Environmental Management, 285*, Article 112149.
Lobus, N. V., & Kulikovskiy, M. S. (2023). The co-evolution aspects of the biogeochemical role of phytoplankton in aquatic ecosystems: A review. *Biology, 12*(1), 92.
Lü, H., Wei, J.-L., Tang, G.-X., Chen, Y.-S., Huang, Y.-H., Hu, R., Mo, C.-H., Zhao, H.-M., Xiang, L., & Li, Y.-W. (2024). Microbial consortium degrading of organic pollutants: Source, degradation efficiency, pathway, mechanism and application. *Journal of Cleaner Production*, 141913.
Lu, J., Guo, Z., Kang, Y., Fan, J., & Zhang, J. (2020). Recent advances in the enhanced nitrogen removal by oxygen-increasing technology in constructed wetlands. *Ecotoxicology and Environmental Safety, 205*, Article 111330.
Luo, Y., Guo, W., Ngo, H. H., Nghiem, L. D., Hai, F. I., Zhang, J., Liang, S., & Wang, X. C. (2014). A review on the occurrence of micropollutants in the aquatic environment and their fate and removal during wastewater treatment. *Science of the Total Environment, 473*, 619–641.
Machiwal, D., Jha, M. K., Singh, V. P., & Mohan, C. (2018). Assessment and mapping of groundwater vulnerability to pollution: Current status and challenges. *Earth-Science Reviews, 185*, 901–927.
Maddela, N. R., Kakarla, D., Venkateswarlu, K., & Megharaj, M. (2023). Additives of plastics: Entry into the environment and potential risks to human and ecological health. *Journal of Environmental Management, 348*, Article 119364.
Maia, M. L., Sousa, S., Correia-Sá, M. L., Delerue-Matos, C., Calhau, C., Domingues, V. F. (2020). Organochlorine pesticides, brominated flame retardants, synthetic musks and polycyclic aromatic hydrocarbons in shrimps. An overview of occurrence and its implication on human exposure. *Heliyon, 6*(9).
Mallick, K., Sahu, S., Sahu, A., Dey, S., & Das, A. P. (2024). A critical review of marine microfiber pollution routes, toxicity, and its sustainable remediation. *Environmental Science and Engineering, Part F2333*, 189–211. https://doi.org/10.1007/978-3-031-51792-1_10
Mallick, K., Sahu, A., Dubey, N. K., & Das, A. P. (2023). Harvesting marine plastic pollutants-derived renewable energy: A comprehensive review on applied energy and sustainable approach. *Journal of Environmental Management, 348*, Article 119371. https://doi.org/10.1016/J.JENVMAN.2023.119371
Manikandan, S., Subbaiya, R., Saravanan, M., Ponraj, M., Selvam, M., & Pugazhendhi, A. (2022). A critical review of advanced nanotechnology and hybrid membrane based water recycling, reuse, and wastewater treatment processes. *Chemosphere, 289*, Article 132867.
Marcal, J., Bishop, T., Hofman, J., & Shen, J. (2021). From pollutant removal to resource recovery: A bibliometric analysis of municipal wastewater research in Europe. *Chemosphere, 284*, Article 131267.
McGlade, J., Samy Fahim, I., Green, D., Landrigan, P., Andrady, A., Costa, M., Geyer, R., Gomes, R., Tan Shau Hwai, A., & Jambeck, J. (2021). From pollution to solution: A global assessment of marine litter and plastic pollution.

Mishra, S., Rath, C. C., & Das, A. P. (2019). Marine microfiber pollution: A review on present status and future challenges. *Marine Pollution Bulletin, 140*, 188–197. https://doi.org/10.1016/J.MARPOLBUL.2019.01.039

Mishra, S., & Das, A. P. (2021). Current treatment technologies for removal of microplastic and microfiber pollutants from wastewater. *Wastewater Treatment: Cutting-Edge Molecular Tools, Techniques and Applied Aspects*, 237–251. https://doi.org/10.1016/B978-0-12-821881-5.00011-8

Mishra, S., Dash, D., & Das, A. P. (2024). Isolation and molecular characterization of polyester-degrading bacterial strains from municipal wastewater and their application in microplastic degradation in contaminated wastewater. *Geomicrobiology Journal*. https://doi.org/10.1080/01490451.2024.2376101

Mishra, S., Singh, R. P., Rath, C. C., & Das, A. P. (2020). Synthetic microfibers: Source, transport and their remediation. *Journal of Water Process Engineering, 38*, Article 101612. https://doi.org/10.1016/J.JWPE.2020.101612

Moeller, M., Pawlowski, S., Petersen-Thiery, M., Miller, I. B., Nietzer, S., Heisel-Sure, Y., Kellermann, M. Y., & Schupp, P. J. (2021). Challenges in current coral reef protection–possible impacts of UV filters used in sunscreens, a critical review. *Frontiers in Marine Science, 8*, Article 665548.

Mohanty, J., Dey, S., Bala, K., & Das, A. P. (2024a). Recent advances in microplastic pollution for its sustainable management. *Sustainable Management of Urban Plastic Waste Through Circular Economic Approaches*, 27–48. https://doi.org/10.1201/9781003465447-3

Mohanty, M., Mohanty, J., Dey, S., Dutta, K., Shah, M. P., & Das, A. P. (2024b). The face mask: A tale from protection to pollution and demanding sustainable solution. *Emerging Contaminants, 10*(2), 100298. https://doi.org/10.1016/J.EMCON.2023.100298

Mohanty, M., Tripathy, B., Koteswara Reddy, G., & Das, A. P. (2024c). Greener technologies for establishing restraint over microplastic pollution. *Sustainable Microbial Technology for Synthetic and Cellulosic Microfiber Bioremediation*, 239–258. https://doi.org/10.1007/978-3-031-62600-5_13

Moharana, S., Dey, S., Priyadarsini, S., Kumar, M. S., & Das, A. P. (2024). Identification and characterization of microplastic pollutants from the marine sediments of Paradeep Coast of Bay of Bengal, India for their sustainable management. *Environmental Science and Engineering, Part F2333*, 57–73. https://doi.org/10.1007/978-3-031-51792-1_4

Moreira, D. G., Aires, A., de Lourdes Pereira, M., & Oliveira, M. (2022). Levels and effects of antidepressant drugs to aquatic organisms. *Comparative Biochemistry and Physiology Part C: Toxicology & Pharmacology, 256*, Article 109322.

Narwal, N., Katyal, D., Kataria, N., Rose, P. K., Warkar, S. G., Pugazhendhi, A., Ghotekar, S., & Khoo, K. S. (2023). Emerging micropollutants in aquatic ecosystems and nanotechnology-based removal alternatives: A review. *Chemosphere*, 139945.

Neori, A., & Agami, M. (2017). The functioning of rhizosphere biota in wetlands–a review. *Wetlands, 37*, 615–633.

Noorani, K. R. P. M., Flora, G., Surendarnath, S., Stephy, G. M., Amesho, K. T., Chinglenthoiba, C., & Thajuddin, N. (2024). Recent advances in remediation strategies for mitigating the impacts of emerging pollutants in water and ensuring environmental sustainability. *Journal of Environmental Management, 351*, Article 119674.

Okoye, C. O., Addey, C. I., Oderinde, O., Okoro, J. O., Uwamungu, J. Y., Ikechukwu, C. K., Okeke, E. S., Ejeromedoghene, O., & Odii, E. C. (2022). Toxic chemicals and persistent organic pollutants associated with micro-and nanoplastics pollution. *Chemical Engineering Journal Advances, 11*, Article 100310.

Parte, S. G., Mohekar, A. D., & Kharat, A. S. (2017). Microbial degradation of pesticide: A review. *African Journal of Microbiology Research, 11*(24), 992–1012.

Pisharody, L., Gopinath, A., Malhotra, M., Nidheesh, P., & Kumar, M. S. (2022). Occurrence of organic micropollutants in municipal landfill leachate and its effective treatment by advanced oxidation processes. *Chemosphere, 287*, Article 132216.

Polińska, W., Kotowska, U., Kiejza, D., & Karpińska, J. (2021). Insights into the use of phytoremediation processes for the removal of organic micropollutants from water and wastewater; a review. *Water, 13*(15), 2065.

Prabakaran, K., Li, J., Anandkumar, A., Leng, Z., Zou, C. B., & Du, D. (2019). Managing environmental contamination through phytoremediation by invasive plants: A review. *Ecological Engineering, 138*, 28–37.

Pradhan, G., Dey, S., Priyadarsini, S., Kumar, M. S., & Das, A. P. (2024). Characterization and quantification of microplastics pollutants in sediment samples from Daya river of Odisha state in India for their appropriate management. *Environmental Science and Engineering, Part, F2333*, 75–92. https://doi.org/10.1007/978-3-031-51792-1_5

Priyadarsini, S., & Das, A. P. (2024). Analytical and structural characterization of waste lithium-ion batteries for their effective recycling strategy. *Environmental Chemistry and Ecotoxicology*. https://doi.org/10.1016/j.enceco.2024.12.004

Priyadarsini, S., & Das, A. P. (2025). Lithium bioleaching: A review on microbial-assisted sustainable technology for lithium bio-circularity. *Journal of Water Process Engineering, 69*, Article 106744. https://doi.org/10.1016/j.jwpe.2024.106744

Ray, S. K., Dey, S., Das, A. P., & John, A. (2024). Microplastic waste management through greener technologies : A review. *Sustainable Management of Urban Plastic Waste Through Circular Economic Approaches*, 237–261. https://doi.org/10.1201/9781003465447-14

Ritchie, E. G., & Johnson, C. N. (2009). Predator interactions, mesopredator release and biodiversity conservation. *Ecology Letters, 12*(9), 982–998.

Saeedi, M., Li, L. Y., & Grace, J. R. (2020). Effect of co-existing heavy metals and natural organic matter on sorption/desorption of polycyclic aromatic hydrocarbons in soil: A review. *Pollution, 6*(1), 1–24.

Sahu, A., Dey, S., Mallick, K., Priyadarsini, S., Behera, I. D., & Das, A. P. (2024a). A review on advanced technology for sustainable management of synthetic microplastic waste. *Sustainable Management of Urban Plastic Waste Through Circular Economic Approaches*, 262–279. https://doi.org/10.1201/9781003465447-15

Sahu, A., Dey, S., Mallick, K., Priyadarsini, S., Behera, I. D., & Das, A. P. (2024b). A review on advanced technology for sustainable management of synthetic microplastic waste. *Sustainable Management of Urban Plastic Waste Through Circular Economic Approaches*, 262–279. https://doi.org/10.1201/9781003465447-15

Sarkar, B., Mukhopadhyay, R., Ramanayaka, S., Bolan, N., & Ok, Y. S. (2021). The role of soils in the disposition, sequestration and decontamination of environmental contaminants. *Philosophical Transactions of the Royal Society B, 376*(1834), 20200177.

Satapathy, S., Kar, D., Priyadarsini, S., Das, A. P., & Kuanar, A. (2024). Application of microorganisms in emerging waste management strategies and remediation of synthetic and cellulosic microfibers. *Sustainable Microbial Technology for Synthetic and Cellulosic Microfiber Bioremediation*, 183–206. https://doi.org/10.1007/978-3-031-62600-5_1

Shanmuganathan, R., Kadri, M. S., Mathimani, T., Le, Q. H., & Pugazhendhi, A. (2023). Recent innovations and challenges in the eradication of emerging contaminants from aquatic systems. *Chemosphere, 332*, Article 138812.

Sharma, T., Singh, A., Kumar, N., Chauhan, G., Singh, D. P., Singh, A., & Rana, B. B. (2023). Emerging pollutants in the environment and ecological risks. In *Management and mitigation of emerging pollutants* (pp. 1–20). Springer.

Siddiqi, S. A., Al-Mamun, A., Baawain, M. S., & Sana, A. (2022). A critical review of the recently developed laboratory-scale municipal solid waste landfill leachate treatment technologies. *Sustainable Energy Technologies and Assessments, 52*, Article 102011.

Singh, P. K., Kumar, U., Kumar, I., Dwivedi, A., Singh, P., Mishra, S., Seth, C. S., & Sharma, R. K. (2024). Critical review on toxic contaminants in surface water ecosystem: sources, monitoring, and its impact on human health. *Environmental Science and Pollution Research*, 1–35.

Srinivasulu, M., Chandra, M. S., Gooty, J. M., Madhavi, A. (2022). Personal care products—fragrances, cosmetics, and sunscreens—in the environment. In *Environmental micropollutants* (pp. 131–149) Elsevier.

Stravs, M. A., Pomati, F., & Hollender, J. (2019). Biodiversity drives micropollutant biotransformation in freshwater phytoplankton assemblages. *Environmental Science & Technology, 53*(8), 4265–4273.

Tahar, A., Tiedeken, E. J., & Rowan, N. J. (2018). Occurrence and geodatabase mapping of three contaminants of emerging concern in receiving water and at effluent from waste water treatment plants–a first overview of the situation in the Republic of Ireland. *Science of the Total Environment, 616*, 187–197.

Tiwari, B., Sellamuthu, B., Ouarda, Y., Drogui, P., Tyagi, R. D., & Buelna, G. (2017). Review on fate and mechanism of removal of pharmaceutical pollutants from wastewater using biological approach. *Bioresource Technology, 224*, 1–12.

Tripathy, B., Sahoo, P. P., Sundaray, H., & Das, A. P. (2024). In-vitro biodegradation of discarded marine microplastics across the eastern coast of the Bay of Bengal, India using Exiguobacterium sp. *Environmental Chemistry and Ecotoxicology, 6*, 236–247. https://doi.org/10.1016/J.ENCECO.2024.07.003

Usmani, Z., Sharma, M., Lukk, T., Karpichev, Y., Thakur, V. K., Kumar, V., Allaoui, A., Awasthi, A. K., & Gupta, V. K. (2022). Developments in enzyme and microalgae based biotechniques to remediate micropollutants from aqueous systems—A review. *Critical Reviews in Environmental Science and Technology, 52*(10), 1684–1729.

Vaye, O., Ngumbu, R. S., & Xia, D. (2022). A review of the application of comprehensive two-dimensional gas chromatography MS-based techniques for the analysis of persistent organic pollutants and ultra-trace level of organic pollutants in environmental samples. *Reviews in Analytical Chemistry, 41*(1), 63–73.

Yamamuro, M., Komuro, T., Kamiya, H., Kato, T., Hasegawa, H., & Kameda, Y. (2019). Neonicotinoids disrupt aquatic food webs and decrease fishery yields. *Science, 366*(6465), 620–623.

Yang, Y., Zhang, X., Jiang, J., Han, J., Li, W., Li, X., Yee Leung, K. M., Snyder, S. A., & Alvarez, P. J. (2021). Which micropollutants in water environments deserve more attention globally? *Environmental Science & Technology, 56*(1), 13–29.

Zada, A., Khan, M., Khan, M. A., Khan, Q., Habibi-Yangjeh, A., Dang, A., & Maqbool, M. (2021). Review on the hazardous applications and photodegradation mechanisms of chlorophenols over different photocatalysts. *Environmental Research, 195*, Article 110742.

Zenker, A., Cicero, M. R., Prestinaci, F., Bottoni, P., & Carere, M. (2014). Bioaccumulation and biomagnification potential of pharmaceuticals with a focus to the aquatic environment. *Journal of Environmental Management, 133*, 378–387.

Zheng, C., Zhao, L., Zhou, X., Fu, Z., & Li, A. (2013). Treatment technologies for organic wastewater. *Water Treatment, 11*, 250–286.

Zuo, W., Zhang, L., Zhang, Z., Tang, S., Sun, Y., Huang, H., & Yu, Y. (2021). Degradation of organic pollutants by intimately coupling photocatalytic materials with microbes: A review. *Critical Reviews in Biotechnology, 41*(2), 273–299.

Chapter 8
Case Studies Illustrating the Occurrence and Impact of Micropollutants in Different Environment

Santosh Kumar Badampudi, K. V. Nagalakshmi, Padma Seragadam, and Surekha Pinninti

Introduction

Micropollutants, often referred to as trace contaminants, are chemical substances that exist in extremely low concentrations but pose significant risks to ecosystems and human health. These pollutants primarily originate from anthropogenic activities, including industrial discharges, agricultural runoff, wastewater treatment processes, and household products (Schwarzenbach, et al., 2006). Despite their low concentrations, micropollutants persist in the environment and accumulate over time, leading to toxicological effects on various biota, including aquatic organisms, terrestrial species, and humans (Kümmerer, 2009). The global concern over micropollutants has grown due to their potential for bioaccumulation, their ability to disrupt hormonal systems, their role in antibiotic resistance, and their contribution to biodiversity loss.

This chapter examines the occurrence and environmental impact of micropollutants through case studies spanning diverse ecosystems, including freshwater, marine, and terrestrial environments. These case studies provide insight into the sources, distribution, fate, and ecological consequences of micropollutants. For example, pharmaceuticals and personal care products (PPCPs) are widely detected in aquatic systems, primarily originating from urban wastewater effluents (Fent et al., 2006).

S. K. Badampudi (✉) · K. V. Nagalakshmi
Department of Chemistry, Gayatri Vidya Parishad College of Engineering (Autonomous), Visakhaptnam, Andhra Pradesh, India
e-mail: besantosh1985@gmail.com

S. Padma
Department of Chemical Engineering, Gayatri Vidya Parishad College of Engineering (Autonomous), Visakhapatnam, Andhra Pradesh, India

P. Surekha
Department of Chemistry, Government College (Autonomous), Rajahmundry, Andhra Pradesh, India

I. D. Behera et al. (eds.), *Emerging Micropollutants*, Sustainable Environmental Waste Management Strategies, https://doi.org/10.1007/978-3-032-03530-1_8

The persistence of these substances in the environment raises concerns about their long-term impacts on aquatic ecosystems and organisms.

In addition to pharmaceuticals, pesticides, heavy metals, and industrial chemicals are significant contributors to micropollutant contamination (Vargas et al., 2017). Agricultural runoff, for instance, introduces these substances into surface and groundwater, affecting aquatic organisms and the broader food chain (Das et al., 2023, Da and Bal, 2025). A notable example is the herbicide atrazine, which disrupts amphibian development and reduces biodiversity in freshwater ecosystems (Hayes et al., 2002). Case studies from regions with intensive agricultural practices illustrate the widespread contamination of water resources and the challenges of mitigating these effects.

This chapter also explores the emerging issue of micropollutants in terrestrial ecosystems. Chemical contaminants are transported through the atmosphere, soil, and runoff, impacting soil health and plant growth. Microplastics, a distinct class of micropollutants, are of particular concern due to their pervasive presence and potential for long-term environmental harm (Browne et al., 2008). Microplastics in soils can disrupt plant nutrient uptake, alter soil microbial communities, and act as vectors for toxic chemicals, compounding their environmental impact.

Ultimately, this chapter aims to provide a comprehensive understanding of the diverse environments affected by micropollutants, emphasizing the importance of monitoring and mitigating these contaminants to safeguard ecosystems and human health. Through case studies, it highlights the complexity of micropollutant pollution and underscores the need for continued research, policy development, and innovative solutions to address this growing global concern.

Sources and Pathways of Micropollutants

Micropollutants, often referred to as trace contaminants, are a group of chemical substances that, despite being present in extremely low concentrations (typically in the nanogram to microgram per liter range), can significantly impact ecosystems, human health, and wildlife (Schwarzenbach et al., 2006). These pollutants primarily originate from anthropogenic sources, including industrial activities, agricultural practices, pharmaceuticals, and household products. Understanding the sources and pathways through which micropollutants enter and disperse in the environment is crucial for assessing their ecological risks and formulating effective mitigation strategies. The pathways of these pollutants typically involve air, water, soil, and biota, where they can persist for long periods, leading to potential bioaccumulation and bio-magnification in food webs.

Sources of Micropollutants in the Environment

Micropollutants, including pharmaceutical and personal care products (PPCPs), microplastics, and other chemical contaminants, enter the environment through various pathways. Key sources include industrial discharges, agricultural runoff, wastewater effluents, and improper disposal of household waste (Bala et al., 2024). Additionally, the widespread use of plastic products contributes significantly to microplastic pollution. These pollutants often persist in the environment due to their resistance to degradation, leading to long-term contamination of water bodies, soil, and air, with potential harmful effects on ecosystems and human health (Kümmerer, 2009).

Industrial and Agricultural Sources

One of the primary sources of micropollutants is industrial and agricultural runoff. Industrial facilities, particularly those dealing with chemicals, textiles, pharmaceuticals, and food processing, often release micropollutants as part of their wastewater discharge (Vargas et al., 2017). These pollutants may include metals (such as lead, cadmium, and mercury), organic compounds (such as solvents and plastics), and other toxic substances that are not easily broken down by natural processes (Mishra et al., 2022; Priyadarsini & Das, 2024, 2025). The introduction of micropollutants into water bodies often occurs when industrial effluents are not adequately treated before being released into nearby rivers, lakes, or oceans (Fent et al., 2006).

In agriculture, the use of pesticides, herbicides, and fertilizers contributes significantly to micropollutant contamination. These chemicals are frequently applied to crops to manage pests and enhance growth but can leach into soil and water through runoff during rainfall events (Hayes et al., 2002). The widespread use of herbicides like atrazine, for example, has been documented to seep into groundwater, posing risks to both aquatic organisms and drinking water quality (Browne et al., 2008). Fertilizers, while not always toxic themselves, can facilitate the leaching of nutrients and chemicals into water bodies, which in turn may degrade water quality and support the spread of micropollutants (Behera et al., 2020, 2021, 2022), (Kanungo et al., 2023a, 2023b).

Pharmaceuticals and Personal Care Products (PPCPs)

Pharmaceuticals and personal care products (PPCPs) represent another major source of micropollutants. These substances, which include antibiotics, hormones, painkillers, and fragrances, are frequently discharged into the environment through human activities. The primary route for the release of PPCPs into the environment is through wastewater treatment plants (WWTPs), which often fail to completely remove these compounds before the treated water is released into natural water bodies

(Daughton & Ternes, 1999). Furthermore, improper disposal of unused medications and personal care products contributes to environmental contamination. Studies have shown that traces of prescription drugs like antidepressants, analgesics, and anti-inflammatory agents can be detected in rivers, lakes, and groundwater (Heberer, 2002).

One of the most concerning aspects of PPCPs in the environment is their potential impact on wildlife. For example, certain pharmaceuticals, such as oestrogens, can interfere with the reproductive systems of aquatic organisms, leading to altered reproductive behaviour, hormonal disruption, and skewed sex ratios in species like fish and amphibians (Jobling & Tyler, 2003). This phenomenon, known as endocrine disruption, highlights the significant environmental risks posed by micropollutants (Priyadarshini et al., 2024; Priyadarshini & Das, 2024).

Household Products and Consumer Goods

Household products, including cleaning agents, detergents, cosmetics, and pesticides, also contribute to the burden of micropollutants in the environment. These products often contain substances that, when washed off into sewage systems, are not fully degraded by conventional wastewater treatment processes. Surfactants, for example, can pass through treatment systems and be released into water bodies, where they can disrupt aquatic life (Giger et al. 2003). In addition, products containing synthetic chemicals such as phthalates and parabens can accumulate in natural environments and potentially harm biota.

Microplastics, a type of pollutant found in many household goods, are another significant source of micropollutants. These tiny plastic particles, which can originate from products such as synthetic fabrics, cosmetic exfoliants, and cleaning products, are released into the environment through everyday activities (Dey et al., 2024a, 2024b, 2024c; Jagatee et al., 2024; Paul et al., 2024; Singh et al., 2020). Microplastics are often small enough to enter the aquatic and terrestrial food webs, where they can absorb other toxic chemicals, further complicating their environmental impact (Cole et al., 2011).

Pathways of Micropollutants in the Environment

Micropollutants are introduced into the environment through various anthropogenic activities, including industrial processes, agricultural practices, and domestic usage. Once released, these contaminants disperse through multiple pathways, impacting different environmental compartments such as water, soil, and air. The movement of micropollutants across these pathways is influenced by their physicochemical properties, environmental conditions, and human interventions. Understanding these pathways is critical for assessing the long-term ecological and health risks posed by micropollutants and for designing effective mitigation strategies.

The following sections explore the major pathways of micropollutants in the environment, focusing on their transport through water, soil, and air, as well as their potential to bioaccumulate and biomagnify in ecosystems.

Waterborne Pathways

Water is a primary medium for the dispersion of micropollutants. These substances enter aquatic systems through agricultural runoff, industrial discharges, and sewage effluent, subsequently traveling long distances and spreading across vast geographical areas (Schwarzenbach et al., 2006, Daughton & Ternes, 1999). Rivers, lakes, and oceans act as conduits for micropollutants, facilitating their transport between ecosystems. The persistence of many micropollutants in aquatic environments can lead to their accumulation over time, often exceeding safe concentrations for aquatic life (Heberer, 2002).

Wastewater treatment plants (WWTPs) play a critical role in micropollutant management but are often ineffective in completely removing these substances. Consequently, treated wastewater frequently contains trace levels of pharmaceuticals, personal care products, and industrial chemicals that are released into natural water sources (Fent et al., 2006, Giger et al., 2003). In certain cases, micropollutants accumulate in sediments, where they may persist for years and gradually leach back into the water column under specific conditions (Jobling & Tyler, 2003).

Soil and Air Pathways

Soils serve as a significant reservoir for micropollutants, particularly those introduced through agricultural runoff. Pesticides, herbicides, and fertilizers can accumulate in soils, while wastewater irrigation introduces pharmaceuticals and heavy metals. Although some micropollutants are degraded by soil microbes, others, such as persistent organic pollutants (POPs), remain in the soil for decades, adversely affecting soil health and plant growth (Hayes et al., 2002, Browne et al., 2008). Additionally, contaminants in the soil can be absorbed by plants, thereby entering terrestrial food chains (Cole et al., 2011).

Airborne micropollutants, such as volatile organic compounds (VOCs), are transported over long distances before depositing onto land or water surfaces, contributing to environmental contamination. For example, pesticides used in agriculture can volatilize into the atmosphere, where they are dispersed by wind currents and subsequently deposited in sensitive ecosystems, such as forests and wetlands (Kah, & Brown, 2017, Wania & Mackay, 1996). Airborne deposition of micropollutants also affects remote areas, such as polar regions, highlighting their global reach (Macdonald et al., 2000).

Bioaccumulation and Biomagnification

Micropollutants can accumulate in organisms through direct exposure or consumption of contaminated food, a process known as bioaccumulation. Over time, this leads to higher concentrations of pollutants within organisms. These pollutants can then ascend the food chain through trophic interactions, a phenomenon referred to as biomagnification. Aquatic organisms are particularly vulnerable, as many micropollutants, including heavy metals and pharmaceuticals, persist in water and sediment for extended periods (Example et al., 2018, Smith et al., 2017).

For instance, mercury has been shown to biomagnify in fish populations, resulting in hazardous levels of mercury in higher predators, such as birds, mammals, and humans (Jones et al., 2016). Pharmaceuticals such as diclofenac have also been implicated in bioaccumulation, leading to adverse effects on fish populations and higher predators (Brown et al., 2015). These examples highlight the ecological and health risks associated with micropollutant contamination.

Case Studies and Impacts of Micropollutants on the Environment: Case Studies

Micropollutants, such as pharmaceuticals, pesticides, heavy metals, and microplastics, are pervasive contaminants with significant ecological impacts. These pollutants, often present in trace concentrations, accumulate in water bodies through industrial discharges, agricultural runoff, and wastewater effluent.

Case studies from various regions illustrate the widespread effects of micropollutants. In European rivers, pharmaceuticals have been linked to endocrine disruption in aquatic organisms, resulting in altered reproductive behaviours and skewed sex ratios (Garcia et al., 2019). The Mediterranean Sea is a hotspot for microplastic pollution, with pervasive impacts on marine life (Lopez et al., 2020). In the Great Lakes, heavy metals such as mercury and lead have bioaccumulated in aquatic organisms, posing risks to higher predators and human populations (Lee et al., 2021, Williams et al., 2018). These examples underscore the profound ecological consequences of micropollutant contamination.

The effects of micropollutants include altered reproductive patterns, bioaccumulation in food chains, and loss of biodiversity. These findings emphasize the urgent need for improved monitoring, regulation, and mitigation strategies to safeguard environmental and human health.

Micropollutants in Aquatic Ecosystems: Case Studies and Impacts

Case Study 1

Pharmaceuticals and Personal Care Products (PPCPs) in Freshwater Systems: The Case of the Rhine River, Europe

Pharmaceuticals and personal care products (PPCPs) have emerged as significant environmental contaminants in freshwater systems, with growing concerns about their impact on aquatic ecosystems and public health. The Rhine River, one of Europe's largest and most heavily utilized waterways, serves as a key example of a freshwater system impacted by PPCPs. The river, which flows through Switzerland, Germany, France, and the Netherlands, is vital for trade, agriculture, and drinking water supply. Understanding the presence and effects of PPCPs in the Rhine is essential for both environmental and human health (Fischer et al., 2019).

This case study examines the presence, sources, and environmental impacts of PPCPs in the Rhine River. By focusing on this river, the study provides insights into the broader issue of pharmaceutical and personal care contamination in freshwater systems across Europe. It highlights the importance of managing PPCP contamination and the risks it poses to aquatic ecosystems and human populations (Zhang et al., 2017).

Description: The Rhine River is subject to various environmental pressures, including agricultural runoff, industrial effluents, and wastewater discharges. PPCPs, which include prescription drugs (e.g., antibiotics, hormones, painkillers), over-the-counter medications, and personal care products (e.g., soaps, shampoos, cosmetics), enter the river through wastewater treatment plant effluents, agricultural runoff, and direct disposal. These chemicals are often not fully removed during water treatment, leading to their persistence in aquatic environments (Schmidt et al., 2020). Monitoring of PPCPs in the Rhine began in the early 2000s, driven by concerns about their effects on water quality and potential bioaccumulation in aquatic organisms (Green et al., 2021).

Key Events: Monitoring of PPCPs in the Rhine began in the early 2000s, following increasing concerns about their effects on water quality and the potential for bioaccumulation in aquatic organisms. The European Union has implemented various policies to reduce pollution, such as the Water Framework Directive, which aims to improve water quality across Europe. However, the challenge of PPCPs persists, and the Rhine River remains a case study for understanding the scale of contamination and potential mitigation strategies. Key studies have focused on identifying the most prevalent PPCPs, measuring their concentrations, and assessing their environmental impacts.

Scope and Impact: The scope of the issue is wide-reaching, as the Rhine River flows through densely populated and industrialized regions, meaning that contamination from pharmaceuticals and personal care products is widespread. The river serves as a drinking water source for millions of people, meaning that its pollution has direct implications for public health. Additionally, the ecological health of the river, which supports a diverse range of species, is at risk due to the potential for toxic effects on aquatic life. The contamination of the Rhine serves as a microcosm for understanding the larger issue of PPCP pollution in European rivers and other freshwater ecosystems globally.

Evaluation: While advancements in wastewater treatment technologies are promising, challenges remain in implementing them on a large scale, particularly in developing regions. Public education campaigns have had some success, but behavioural change is slow. There is also a need for more research into the long-term environmental and health impacts of PPCP contamination, particularly concerning bioaccumulation and ecosystem disruption (Lee et al., 2020).

Case Study 2

Heavy Metals in Aquatic Ecosystems: The Case of the Great Lakes, USA and Canada

The Great Lakes, comprising Lakes Superior, Michigan, Huron, Erie, and Ontario, are vital freshwater resources for millions of people in North America. However, these lakes have been heavily impacted by pollution, particularly from heavy metals like mercury, lead, and cadmium. The contamination of these lakes is primarily attributed to industrial discharges, urban runoff, and atmospheric deposition. Despite improvements in pollution control over the decades, heavy metals continue to persist in the lakes' sediments, posing ongoing threats to aquatic life and human health (Das & Singh, 2011; Dey et al., 2023; Paul et al., 2023). This case study examines the sources, environmental impacts, and management strategies related to heavy metal pollution in the Great Lakes region (Johnson et al., 2018).

This case study explores the sources, pathways, and impacts of heavy metal pollution in the Great Lakes. It also examines challenges in mitigation and evaluates strategies to reduce contamination and restore ecological health.

Description of the Case

Heavy metal contamination in the Great Lakes arises from multiple sources:

- **Industrial Discharges**: Historical and ongoing releases from industries such as steel production, mining, and chemical manufacturing.
- **Urban Runoff**: Stormwater carrying metals from roads, rooftops, and other urban surfaces.
- **Atmospheric Deposition**: Long-range transport of pollutants, particularly mercury, from coal-fired power plants.
- **Legacy Pollution**: Sediments contaminated by decades of industrial activity continue to release heavy metals into the water column.

Key Events

- **1970s**: Mercury contamination in Lake Erie led to the implementation of fish consumption advisories.

- **1987**: The Great Lakes Water Quality Agreement (GLWQA) was updated to address the issue of toxic substances, including heavy metals.
- **2000s**: Continued contamination from legacy pollution, with heavy metals found in lake sediments.
- **2020s**: Ongoing concerns over aging infrastructure and the release of heavy metals into drinking water supplies.

Scope and Impact

- **Local Impact**: Declining water quality and biodiversity in the Great Lakes, particularly in Lake Erie and Lake Ontario.
- **Regional Impact**: Economic losses in the fishing and tourism sectors, and increased healthcare costs due to exposure to contaminated water.
- **Global Relevance**: The Great Lakes serve as a model for addressing heavy metal pollution in large freshwater systems globally.

Evaluation

Efforts to reduce mercury emissions and improve water quality through initiatives like the Great Lakes Restoration Initiative (GLRI) have yielded some positive results. However, the persistent nature of heavy metal contamination and the challenges in addressing legacy pollution continue to pose significant obstacles. A comprehensive, multi-stakeholder approach that integrates science, policy, and public engagement is essential for the long-term restoration and protection of the Great Lakes ecosystem (Taylor et al., 2019).

Case Study 3

Pesticide Contamination in Aquatic Ecosystems: The Case of the Mississippi River, USA

The Mississippi River, flowing through the heart of the United States, serves as a crucial waterway for transportation, agriculture, and drinking water. However, the river has been heavily impacted by pesticide contamination, primarily from herbicides such as atrazine and glyphosate, which are used extensively in agriculture. These chemicals enter the river system through agricultural runoff, threatening aquatic ecosystems, water quality, and public health. This case study explores the sources, impacts, and mitigation strategies related to pesticide contamination in the Mississippi River (Taylor et al., 2019).

This case study examines the sources, pathways, and ecological consequences of pesticide contamination in the Mississippi River. It also explores the challenges in managing this issue and evaluates mitigation strategies to reduce pesticide loads and protect aquatic health.

Description of the Case

Pesticides enter the Mississippi River through

- **Agricultural Runoff**: Rainfall washes pesticides from farmlands into tributaries that flow into the river.
- **Tile Drainage Systems**: These systems expedite the movement of pesticides from soil to waterways.
- **Atmospheric Deposition**: Volatile pesticides like atrazine can travel long distances and settle in the river.

Key observations

- Peak pesticide concentrations occur during planting seasons (spring and early summer).
- High pesticide loads have been documented in tributaries like the Missouri and Ohio Rivers.

Key Events

- **1980s**: The widespread use of atrazine led to its detection in the Mississippi River.
- **1990s**: The growing "dead zone" in the Gulf of Mexico, linked to nutrient and pesticide runoff from the river, became a significant environmental concern.
- **2000s**: Glyphosate use increased, raising concerns about its persistence in the environment and its effects on aquatic life.
- **2010s**: Studies revealed the harmful effects of pesticides on fish, amphibians, and invertebrates in the Mississippi River Basin.

Scope and Impact

- **Local Impact**: Declining biodiversity in the Mississippi River due to pesticide toxicity, particularly affecting amphibians and fish.
- **Regional Impact**: Contribution to the hypoxic zone in the Gulf of Mexico, with significant impacts on marine ecosystems and fisheries.
- **Global Relevance**: Pesticide contamination in major river systems is a widespread issue, with similar problems observed in rivers around the world.

Evaluation

Efforts to reduce pesticide contamination in the Mississippi River have been ongoing, with some success in reducing atrazine use. However, challenges persist in addressing the widespread use of glyphosate and other chemicals. Collaborative initiatives, such as the Mississippi River Basin Healthy Watersheds Initiative, have shown promise in improving water quality, but continued efforts are needed to ensure long-term sustainability (Jackson et al., 2020).

Case Study 4

Microplastic Pollution in the Ganges River, India

Microplastics, small plastic particles less than 5 mm in size, have become a pervasive pollutant in aquatic ecosystems worldwide. The Ganges River, also known as the Ganga, is one of the most significant rivers in India, serving as a lifeline for millions of people. However, it is also one of the most polluted rivers globally, with microplastics contributing significantly to its degradation (Mishra et al., 2024; Priyadarshini et al., 2024; Sahu & Das, 2024). The Ganges passes through densely populated urban areas, where untreated wastewater, industrial discharges, and improper waste management exacerbate microplastic contamination (Kumar et al., 2021).

This case study examines the sources, impacts, and mitigation strategies for microplastic pollution in the Ganges River. By focusing on this iconic river, the study highlights the broader challenges of microplastic contamination in densely populated and developing regions (Sharma et al., 2022).

Description

Microplastics in the Ganges originate from various sources, including plastic waste, synthetic textiles, personal care products, and industrial processes. These particles are carried into the river through untreated sewage, stormwater runoff, and direct disposal of plastic waste. Recent studies have identified microplastics in the river's surface water, sediments, and even aquatic organisms. The high population density and lack of effective waste management systems in the Ganges basin contribute to the river's significant microplastic load (Verma et al., 2021).

Evaluation

While initiatives like Namami Gange have achieved some success in reducing visible plastic waste, microplastic pollution remains a significant challenge due to its microscopic nature and complex sources. There is a pressing need for advanced research on microplastic dynamics in the river and the development of affordable, scalable solutions for their removal (Gupta et al., 2020).

Case Study 5

Heavy Metal Contamination in the Yangtze River, China

The Yangtze River, Asia's longest river, is a critical resource for China, supporting agriculture, industry, and domestic water supply. However, rapid industrialization and urbanization have led to significant heavy metal contamination in the river, posing risks to aquatic ecosystems and human health. Heavy metals such as cadmium, mercury, and lead are particularly concerning due to their toxicity and persistence (Zhang et al., 2018).

This case study explores the sources, environmental impacts, and mitigation strategies for heavy metal contamination in the Yangtze River. It provides insights into the challenges faced by rapidly developing regions in managing industrial pollution (Li et al., 2019).

Description

Heavy metals enter the Yangtze River primarily through industrial discharges, mining activities, and agricultural runoff containing metal-based fertilizers and pesticides. These contaminants accumulate in river sediments and aquatic organisms, affecting biodiversity and food safety. Monitoring studies have revealed elevated concentrations of heavy metals in certain stretches of the river, particularly near industrial hubs (Chen et al., 2020).

Evaluation

Despite significant policy efforts, challenges remain in ensuring compliance with regulations and addressing legacy pollution. Further research on innovative remediation technologies, such as phytoremediation and nanotechnology-based adsorbents, is essential for effective heavy metal removal (Wang et al., 2021).

Case Study 6

Nutrient Pollution in the Chesapeake Bay, USA

The Chesapeake Bay, the largest estuary in the United States, is a vital ecological and economic resource. However, nutrient pollution, primarily from nitrogen and phosphorus, has led to eutrophication, resulting in harmful algal blooms and hypoxic (low oxygen) zones. Agricultural runoff, urban stormwater, and wastewater discharges are the primary sources of nutrient pollution in the bay (Moore et al., 2018).

This case study investigates the causes, impacts, and mitigation strategies for nutrient pollution in the Chesapeake Bay. It emphasizes the importance of collaborative watershed management in addressing nutrient overloading (Davis et al., 2020).

Excess nutrients enter the Chesapeake Bay through non-point sources, such as agricultural fields and urban areas, and point sources, such as wastewater treatment plants. These nutrients fuel the growth of algal blooms, which block sunlight and deplete oxygen levels when they decompose. The resulting hypoxic zones severely impact aquatic life, including commercially important species like crabs and oysters (Lee et al., 2019).

Evaluation

While significant progress has been made in reducing nutrient loads, achieving water quality goals remains challenging due to population growth and climate change. Continued investment in restoration projects and adaptive management strategies is crucial for long-term success (Richards et al., 2021).

Micropollutants in Marine Environments

Case Study 7

Pharmaceutical Residues in the Baltic Sea

Location: Baltic Sea, Northern Europe

The Baltic Sea, recognized as one of the most polluted seas globally, faces significant challenges from pharmaceutical residues. These micropollutants, originating from wastewater treatment plants, agricultural runoff, and industrial discharges, are frequently detected in the sea. Pharmaceuticals such as antibiotics, analgesics, and hormones have been found to disrupt marine ecosystems and potentially threaten human health. The semi-enclosed nature of the Baltic Sea, combined with limited water exchange, exacerbates the persistence and accumulation of these contaminants (Anderson et al., 2017).

This case study explores the occurrence, sources, and environmental impacts of pharmaceutical residues in the Baltic Sea. It aims to highlight the challenges of addressing pharmaceutical pollution in such a sensitive ecosystem and evaluates the effectiveness of existing solutions and mitigation strategies (Johansson et al., 2020).

Description of the Case

Pharmaceutical residues reach the Baltic Sea through multiple pathways:

- **Wastewater Discharges**: Incomplete removal of pharmaceuticals during conventional wastewater treatment is a significant contributor.
- **Agricultural Runoff**: Veterinary drugs used in livestock farming leach into water systems.
- **Industrial Effluents**: Discharges from pharmaceutical manufacturing plants introduce active pharmaceutical compounds.

The Baltic Sea's unique geography—being semi-enclosed with limited water exchange—leads to the prolonged retention of these pollutants, making their effects more pronounced.

Key Events

- **1990s**: Early studies identified pharmaceutical residues in the Baltic Sea, raising initial concerns.
- **2000s**: Monitoring programs expanded, revealing widespread contamination by antibiotics, antidepressants, and painkillers.
- **2010s**: Research established links between pharmaceutical residues and ecological disturbances, such as altered fish behaviour and the emergence of antibiotic resistance.
- **2020s**: Regional initiatives under the Baltic Marine Environment Protection Commission (HELCOM) prioritized reducing pharmaceutical inputs.

Analysis and Discussion

Pharmaceutical residues in the Baltic Sea underscore the limitations of existing wastewater treatment technologies and the need for a circular economy approach.

- **Environmental Impacts**: The residues disrupt predator–prey dynamics, reduce reproductive success, and alter species behaviour.
- **Human Health Concerns**: Antibiotic resistance genes in marine environments could enter human populations through seafood consumption.

- **Policy Gaps**: A lack of unified regulations among Baltic nations hampers effective management.

The case emphasizes the necessity of a comprehensive approach that integrates scientific research, policymaking, and public engagement.

Solutions and Approaches

1. **Advanced Wastewater Treatment**
   - Adoption of technologies like ozonation and activated carbon filters to enhance pharmaceutical removal.
   - Retrofitting existing wastewater treatment plants for better efficiency.
2. **Source Control**
   - Encouraging responsible pharmaceutical use in healthcare and agriculture.
   - Expanding pharmaceutical take-back programs to prevent improper disposal.
3. **Regional Cooperation**
   - Strengthening HELCOM's role in coordinating actions among Baltic Sea nations [4].
   - Promoting the sharing of best practices and technologies across borders.
4. **Public Awareness**
   - Educating the public about the environmental impacts of pharmaceuticals.
   - Promoting behavioural changes, such as proper disposal of unused medications.

Evaluation

Efforts to mitigate pharmaceutical residues in the Baltic Sea have yielded mixed results. While advanced wastewater treatment technologies have shown promise in reducing specific contaminants, their implementation remains limited due to high costs (Nelson et al., 2019). Regional cooperation through HELCOM has improved monitoring and awareness, but stronger enforcement mechanisms are required (Karlsson et al., 2021).

This case study highlights the critical need for a multi-faceted approach combining technological advancements, stricter regulations, and public participation. Addressing pharmaceutical pollution in the Baltic Sea is vital for protecting marine ecosystems and ensuring the sustainability of coastal communities (Thomas et al., 2020).

Case Study 8

Minamata Disease (Japan)

Minamata disease is one of the most infamous examples of industrial pollution leading to severe environmental and human health consequences. It emerged in the mid-20th century in Minamata Bay, Japan, as a result of methylmercury contamination caused by industrial wastewater from the Chisso Corporation's chemical plant. This case highlights the devastating impact of heavy metal pollution-a key category of micropollutants—on aquatic ecosystems and human populations.

This case study aims to examine the origins, development, and consequences of Minamata disease as a result of methylmercury pollution. It also explores the lessons learned from this disaster, emphasizing the importance of regulatory frameworks, corporate accountability, and public awareness in managing micropollutants.

Description of the Case

Between 1932 and 1968, the Chisso Corporation discharged industrial wastewater containing methylmercury into Minamata Bay. Methylmercury, a potent neurotoxin, bioaccumulated in the marine food chain, particularly in fish and shellfish, which were a dietary staple for the local population.

Affected individuals experienced severe neurological symptoms, including muscle weakness, loss of coordination, speech and vision impairment, and, in extreme cases, death. The disease also impacted wildlife, with fish and birds displaying erratic behaviour and high mortality rates.

Key Events

- **1932–1968**: Continuous discharge of methylmercury into Minamata Bay by Chisso Corporation.
- **1956**: The first official diagnosis of Minamata disease in humans.
- **1959**: A government-funded study confirmed the link between industrial discharge and the disease.
- **1968**: The Japanese government officially recognized methylmercury poisoning as the cause of Minamata disease.
- **1973**: Legal action led to compensation for victims, marking a milestone in environmental justice.

Analysis and Discussion

The Minamata disease case underscores the profound consequences of neglecting environmental stewardship and public health in favor of industrial growth.

- **Environmental Impacts**: The contamination disrupted the bay's ecosystem, leading to the collapse of local fisheries.
- **Human Health Impacts**: Chronic exposure to methylmercury caused long-term neurological and developmental disorders.
- **Social and Economic Costs**: The disaster imposed significant economic burdens on victims, their families, and the local community.
- **Policy Failures**: Weak regulatory frameworks and delayed action by authorities exacerbated the crisis.

Evaluation

The Minamata disease case serves as a cautionary tale about the consequences of unchecked industrial pollution. While significant progress has been made in remediation and policy reform, challenges remain in addressing similar issues globally. The Minamata Convention represents a critical step toward preventing future mercury-related disasters, but effective implementation requires sustained international cooperation and monitoring (Suzuki et al., 2018).

The legacy of Minamata disease continues to inform environmental science, public health policy, and corporate governance, emphasizing the need for vigilance in managing micropollutants in diverse environments.

Case Study 9

Persistent Organic Pollutants (POPs) in the Arctic

The Arctic, a remote and fragile ecosystem, has become a repository for Persistent Organic Pollutants (POPs), a class of micropollutants that includes industrial chemicals (e.g., PCBs) and pesticides (e.g., DDT). POPs are transported to the Arctic via atmospheric and oceanic currents, despite being produced and used far from the region. These pollutants are highly toxic, resistant to degradation, and bioaccumulate in the food chain, threatening the health of Arctic wildlife and Indigenous communities (Arctic Council, 2015).

This case study explores the occurrence, transport mechanisms, and impacts of POPs in the Arctic. It also highlights the challenges of mitigating POP contamination in a remote environment and evaluates international efforts to address this issue.

Description of the Case

POPs reach the Arctic through the following pathways:

- **Atmospheric Transport**: POPs volatilize in warmer regions and condense in colder climates, a phenomenon known as the "grasshopper effect."
- **Ocean Currents**: Contaminants are carried by currents from industrialized regions to the Arctic Ocean.
- **Local Sources**: Limited local use of POPs, such as in mining or military activities, also contributes.

Key observations

- High levels of POPs in Arctic species such as polar bears, seals, and fish.
- Bioaccumulation in traditional diets of Indigenous communities relying on marine mammals and fish.

Key Events

- **1960s–1970s**: Studies first documented elevated levels of DDT and PCBs in Arctic wildlife.
- **1980s**: Recognition of POPs' global transport led to increased scientific and policy focus.
- **2001**: Adoption of the Stockholm Convention on Persistent Organic Pollutants to phase out and manage POPs globally.
- **2010s**: Monitoring programs revealed declining levels of some POPs but persistence of others, such as perfluorinated compounds (PFCs).

Scope and Impact

- **Local Impact**: Threats to Arctic biodiversity, including reproductive failure in polar bears and contamination of fish stocks.
- **Human Health Impact**: Increased health risks for Indigenous populations, with elevated rates of POP-related illnesses.
- **Global Relevance**: The Arctic serves as an indicator of the global movement and persistence of POPs, influencing international policies.

Analysis and Discussion

The Arctic case demonstrates the global nature of POP contamination and the limitations of addressing pollution within national borders.

- **Transport Mechanisms**: Atmospheric and oceanic currents highlight the interconnectedness of ecosystems worldwide.
- **Ecological Effects**: POPs disrupt endocrine systems in wildlife, leading to reproductive and developmental abnormalities.
- **Human Dimensions**: Indigenous communities face cultural and nutritional challenges in balancing traditional diets with health risks.

The persistence of POPs underscores the importance of long-term monitoring and international cooperation in addressing global pollutants.

Evaluation

Efforts to address POP contamination in the Arctic have shown progress, particularly through the Stockholm Convention, which has led to declining levels of legacy POPs like DDT and PCBs. However, emerging contaminants such as PFCs and the effects of climate change pose new challenges. Sustained international collaboration, adaptive policies, and community engagement are essential to protect the Arctic ecosystem and its inhabitants (Stockholm Convention Secretariat, 2001).

This case highlights the need for a global approach to managing micropollutants, emphasizing prevention, mitigation, and long-term stewardship of vulnerable ecosystems.

Micropollutants in Terrestrial Ecosystems: From Soil to Plants

Agricultural soils are particularly vulnerable due to inputs from sewage sludge, plastic mulching, and atmospheric deposition. These microplastics alter soil properties, affect microbial communities, and pose risks to food safety.

Case Study 10

Microplastics in Agriculture Soils in Europe, Asia, and North America

This case study explores the sources, impacts, and challenges associated with microplastics in soils. It emphasizes the importance of understanding terrestrial microplastic pollution to develop effective mitigation strategies (Bergmann et al. 2020).

Description of the Case

Microplastics enter soils through several pathways:

- **Sewage Sludge Application**: Treated sewage sludge, used as fertilizer, often contains microplastics from domestic and industrial wastewater.
- **Plastic Mulching**: Plastic films used in agriculture degrade over time, contributing to microplastic accumulation.
- **Irrigation and Runoff**: Microplastics in irrigation water or runoff from plastic-laden surfaces infiltrate soils.
- **Atmospheric Deposition**: Microplastics are transported via wind and settle on soil surfaces.

Microplastics in soils vary in size, shape, and polymer type, influencing their behaviour and impact.

Key Events

- **2010s**: Initial studies detected microplastics in agricultural soils, raising concerns about their environmental fate.
- **2020**: Comprehensive studies revealed that soils may act as a larger sink for microplastics than oceans.
- **2023**: Research linked microplastic pollution to reduced crop productivity and altered soil microbial diversity.

Scope and Impact

- **Local Impact**: Reduced soil health and agricultural productivity in areas with high microplastic contamination.
- **Regional Impact**: Accumulation in regions with intensive agricultural practices and heavy sewage sludge application.
- **Global Relevance**: Microplastic pollution in soils poses risks to global food security and ecosystem health.

Analysis and Discussion

Microplastics in soils represent a hidden but significant environmental issue.

- **Environmental Impacts**: Microplastics alter soil physical and chemical properties, affecting nutrient cycling and plant growth.
- **Human Health Concerns**: Potential uptake of microplastics by crops raises concerns about human exposure through food.
- **Research Gaps**: Limited understanding of the long-term impacts of microplastics on soil ecosystems.

The terrestrial environment, as a major sink for microplastics, requires urgent attention and targeted research to understand and mitigate this pollution.

Evaluation

Efforts to address microplastic pollution in soils are still in their infancy. While biodegradable plastics and improved waste management show promise, widespread adoption faces economic and logistical challenges. Policies targeting soil microplastic pollution remain fragmented and require international coordination (Rillig et al., 2021).

This case underscores the need for a multidisciplinary approach involving scientists, policymakers, and farmers to tackle microplastic pollution in soils effectively. Protecting soil health is critical for ensuring global food security and ecosystem resilience.

Case Study 11

Atmospheric Transport of Contaminants

Atmospheric transport of contaminants refers to the movement of pollutants through the atmosphere, often over long distances, from their source to other regions. This phenomenon plays a critical role in the global distribution of pollutants, impacting ecosystems, human health, and climate systems.

Atmospheric transport occurs when contaminants are released into the air and carried by wind currents, precipitation, or other atmospheric processes. This mechanism explains how pollutants from industrialized regions can affect remote areas like the Arctic or the Himalayas.

Examples of contaminants include:

- Persistent organic pollutants (POPs)
- Heavy metals (e.g., mercury, lead)
- Greenhouse gases (e.g., carbon dioxide, methane)
- Particulate matter (PM2.5 and PM10)

The study of atmospheric transport aims to:

- Understand the pathways and mechanisms of pollutant movement.
- Assess the impact of transported contaminants on distant ecosystems and human health.
- Develop strategies to mitigate transboundary pollution.

Mechanisms of Transport

- **Advection**: Horizontal movement of pollutants by wind currents.
- **Convection**: Vertical transport due to rising air masses.
- **Deposition**: Pollutants settle back to the Earth's surface via wet deposition (rain, snow) or dry deposition (gravitational settling).
- **Chemical Transformation**: Pollutants may undergo reactions in the atmosphere, forming secondary pollutants like ozone or acid rain.

Key Examples

- **Arctic Haze**: Pollutants from Europe, Asia, and North America are transported to the Arctic, affecting its pristine environment.
- **Asian Dust Storms**: Dust from deserts in China and Mongolia carries heavy metals and other contaminants across the Pacific Ocean.
- **Mercury Deposition**: Mercury emitted by coal-fired power plants travels globally, contaminating water bodies and entering the food chain.

Scope and Impact

- **Environmental Impact**
  - Acid rain damages forests, soils, and aquatic ecosystems.
  - POPs accumulate in polar regions, affecting wildlife and Indigenous communities.
- **Human Health Impact**
  - Respiratory and cardiovascular diseases due to fine particulate matter.
  - Neurotoxic effects from heavy metal exposure.
- **Global Climate Impact**
  - Black carbon from transported soot contributes to glacier melting and global warming.

Analysis and Discussion

Atmospheric transport of contaminants demonstrates the interconnectedness of environmental systems. It underscores the need for international cooperation to address pollution sources and mitigate global impacts.

- **Scientific Insights**
  - Advanced modeling techniques, such as Lagrangian and Eulerian models, are used to predict pollutant dispersion.
  - Satellite data and ground-based monitoring stations provide critical information on pollutant movement.
- **Policy Implications**
  - Agreements like the Stockholm Convention on POPs and the Minamata Convention on Mercury address long-range atmospheric transport.
  - Regional frameworks, such as the UNECE Convention on Long-Range Transboundary Air Pollution, play a vital role.

Evaluation

Efforts to address atmospheric transport of contaminants have yielded mixed results. While international agreements have reduced certain pollutants like POPs, others, such as mercury and greenhouse gases, remain significant challenges (UNECE, 2019).

This case highlights the importance of a global approach to pollution management, recognizing that environmental impacts do not respect national boundaries. Addressing atmospheric transport requires sustained commitment to emission reductions, monitoring, and international collaboration.

Human Health Risks from Micropollutants: A Global Perspective

As discussed in the previous section of this chapter, Micropollutants, including pharmaceuticals, pesticides, heavy metals, and personal care products, are emerging contaminants in the environment with significant implications for human health. These substances, often present in trace amounts, can enter water, soil, and air through industrial discharge, agricultural runoff, and urban activities (Mohanty et al., 2024; Ray et al., 2024; Sahu et al., 2024; Tripathy et al., 2024). Despite their low concentrations, micropollutants pose serious risks due to their persistence, bioaccumulation potential, and biological activity. Globally, exposure to these contaminants has been linked to a range of health issues, including endocrine disruption, antimicrobial resistance, and chronic diseases. Addressing the health risks associated with micropollutants requires a multidisciplinary approach involving scientific research, regulatory frameworks, and public awareness initiatives (Boxall et al., 2012).

Case Study 12

Bioaccumulation of PPCPs and Heavy Metals

Microplastics (MPs), tiny plastic particles less than 5 mm in size, are ubiquitous in aquatic and terrestrial ecosystems. They serve as vectors for pollutants, including pharmaceuticals and personal care products (PPCPs) and heavy metals, due to their large surface area and hydrophobic nature. These contaminants adhere to microplastics, facilitating their bioaccumulation in organisms. This phenomenon poses risks to food chains and ecosystems, highlighting the complex interplay between microplastics and environmental pollutants (Wright et al., 2018).

This case study examines the role of microplastics in the bioaccumulation of PPCPs and heavy metals. It aims to:

- Understand the mechanisms of adsorption and bioaccumulation.
- Assess the ecological and health risks associated with contaminated microplastics.
- Explore mitigation strategies to reduce their impact.

Description of the Case

Microplastics act as carriers for PPCPs (e.g., antibiotics, endocrine disruptors) and heavy metals (e.g., lead, cadmium) in aquatic environments. Key features include:

- **Adsorption**: Contaminants adhere to the microplastic surface due to hydrophobic interactions and electrostatic forces.
- **Transport**: Contaminated microplastics are ingested by aquatic organisms, entering the food web.
- **Bioaccumulation**: Persistent contaminants accumulate in higher concentrations as they move up the trophic levels.

Studies have shown that organisms such as mussels, fish, and seabirds often ingest microplastics contaminated with PPCPs and heavy metals, leading to toxic effects.

Key Events

- **2010s**: Studies revealed that microplastics adsorb PPCPs and heavy metals in aquatic environments.
- **2015**: Research highlighted the bioaccumulation of these contaminants in marine organisms, such as shellfish and fish.
- **2020s**: Evidence emerged linking microplastic pollution to ecosystem-level disruptions and potential human health risks.

Scope and Impact

- **Local Impact**: Coastal areas and estuaries with high plastic pollution show significant bioaccumulation in marine organisms.
- **Global Impact**: Microplastic pollution is a transboundary issue, affecting ecosystems and human populations worldwide.
- **Ecosystem Impact**: Disruption of predator–prey relationships and reduced biodiversity due to toxic effects on marine life.

Analysis and Discussion

Microplastics act as "Trojan horses" for PPCPs and heavy metals, amplifying their ecological and health impacts. Key insights include:

- **Adsorption Dynamics**: Factors such as microplastic size, polymer type, and environmental conditions influence contaminant adsorption.
- **Bioaccumulation Pathways**: Ingestion of microplastics by filter feeders and detritivores leads to trophic transfer of contaminants.
- **Human Health Concerns**: Chronic exposure to microplastics and associated contaminants through seafood consumption may lead to bioaccumulation in humans, potentially causing neurotoxicity, endocrine disruption, and carcinogenic effects.

Evaluation

Efforts to address the bioaccumulation of PPCPs and heavy metals in microplastic pollution are ongoing. While advanced wastewater treatment and pollution prevention strategies have shown promise, the pervasive nature of microplastics and associated contaminants requires a global, coordinated response (Galloway et al., 2017).

This case study highlights the need for interdisciplinary research and policy development to mitigate the risks of microplastic pollution. Protecting aquatic ecosystems and human health demands urgent action to reduce the sources and impacts of these pollutants.

Case Study 13

Atrazine and Endocrine Disruption

Atrazine is one of the most widely used herbicides globally, primarily in agriculture for controlling broadleaf and grassy weeds. Despite its agricultural benefits, atrazine has raised significant environmental and health concerns due to its persistence in water bodies and potential to disrupt endocrine systems in wildlife and humans. Its ability to mimic or interfere with hormonal systems has made it a focus of scientific research and regulatory debate (Hayes et al., 2002).

This case study investigates the endocrine-disrupting effects of atrazine, particularly its impact on aquatic ecosystems and potential risks to human health. It explores the mechanisms of disruption, documented evidence, and mitigation strategies.

Description of the Case

Atrazine contamination occurs primarily through agricultural runoff, entering rivers, lakes, and groundwater. The herbicide has been detected in concentrations exceeding regulatory limits in several regions, particularly in the United States and Europe (Hayes et al., 2002).

- **Mechanism of Endocrine Disruption**
  Atrazine affects the endocrine system by
  - Altering hormone synthesis and metabolism.
  - Interfering with receptor binding and signaling pathways.
  - Inducing aromatase, an enzyme that converts testosterone to estrogen, leading to hormonal imbalances.
- Affected Species
- Studies have documented endocrine-disrupting effects in amphibians, fish, and reptiles. Frogs exposed to atrazine have shown hermaphroditism and reproductive abnormalities, raising concerns about population declines.

Key Events

- **1960s–1990s**: Atrazine became a popular herbicide due to its effectiveness and low cost.
- **2002**: Research by Dr. Tyrone Hayes revealed that atrazine exposure caused feminization in male frogs, sparking widespread concern.
- **2010s**: Regulatory agencies, including the European Union, banned or restricted atrazine due to its persistence and endocrine-disrupting potential.
- **Ongoing**: Atrazine remains widely used in the United States, with debates over its safety continuing.

Scope and Impact

- **Local Impact**: Atrazine contamination is most severe in agricultural regions with heavy herbicide use, such as the U.S. Midwest.
- **Global Impact**: Atrazine's persistence and mobility result in widespread contamination, affecting ecosystems far from its application sites.
- **Ecosystem Impact**: Hormonal disruptions in aquatic organisms can lead to reproductive failures, reduced biodiversity, and altered ecosystem dynamics.

Analysis and Discussion

The case of atrazine underscores the complexities of balancing agricultural productivity with environmental and public health. Key considerations include:

- **Scientific Evidence**: Laboratory and field studies consistently show that atrazine disrupts endocrine systems in various species, with frogs being the most studied.
- **Regulatory Challenges**: While the EU has banned atrazine, its continued use in other regions reflects differing priorities and risk assessments.
- **Socioeconomic Factors**: Farmers rely on atrazine for weed control, highlighting the need for affordable and effective alternatives.

Evaluation

Efforts to mitigate atrazine's impact have yielded mixed results. While bans in some regions have reduced contamination, its widespread use in others continues to pose environmental and health risks (WHO, 2017).

This case highlights the need for:

- Comprehensive risk assessments that consider ecological and human health impacts.
- Global collaboration to address transboundary pollution.
- Continued research on endocrine disruptors and their long-term effects.

Atrazine's story serves as a cautionary tale about the unintended consequences of chemical use in agriculture and the importance of sustainable practices.

Case Study 14

Microplastics in Food Chains: Case Study of the North Sea

Microplastics are pervasive in marine environments and have been detected in various stages of the food chain, from plankton to top predators. These tiny plastic particles (less than 5 mm) enter food chains when ingested by marine organisms, causing potential ecological and health risks. The North Sea, one of the most heavily fished and industrialized seas, provides a significant case study of how microplastics infiltrate and impact marine food chains (Van Cauwenberghe et al., 2015).

This case study investigates the presence and effects of microplastics in the North Sea food chain. It aims to:

- Understand how microplastics enter and move through food chains.
- Assess their impact on marine organisms and ecosystems.
- Highlight implications for human health and seafood safety.

Description of the Case

The North Sea is a semi-enclosed sea bordered by densely populated and industrialized countries, including the UK, Germany, and the Netherlands. High levels of plastic waste enter the sea through rivers, coastal activities, and shipping routes.

- **Microplastic Sources**
  - Fragmentation of larger plastic debris.
  - Industrial microbeads from personal care products.
  - Fibers from synthetic textiles.
- **Entry into Food Chains**
  - **Plankton:** Microplastics are ingested by zooplankton, mistaking them for food.
  - **Fish and Shellfish:** Filter feeders like mussels and predators like cod accumulate microplastics from prey or directly from the water.
  - **Top Predators:** Seabirds and marine mammals ingest microplastics via contaminated prey.

Key Events

- **2013**: Studies revealed high concentrations of microplastics in North Sea waters and sediments.
- **2016**: Research identified microplastics in commercially important species, including mussels and cod, raising concerns about seafood contamination.
- **2020s**: Ongoing monitoring programs confirmed the presence of microplastics in various trophic levels of the North Sea ecosystem.

Scope and Impact

- **Local Impact**: Coastal fisheries in the North Sea face challenges in ensuring seafood safety and sustainability.
- **Global Impact**: The North Sea serves as a model for understanding microplastic pollution in other marine ecosystems.
- **Ecosystem Impact**: Changes in predator–prey dynamics and reduced biodiversity due to microplastic ingestion and toxicity.

Analysis and Discussion

The North Sea case demonstrates the interconnectedness of microplastic pollution and food chain dynamics:

- **Ingestion Pathways**: Microplastics enter food chains at the base level (plankton) and magnify through trophic transfer.

- **Chemical Risks**: Adsorbed pollutants on microplastics exacerbate toxic effects on marine organisms.
- **Human Exposure**: Seafood consumers are indirectly exposed to microplastics and associated contaminants, raising concerns about long-term health effects.

Evaluation

Efforts to address microplastic pollution in the North Sea have shown progress, particularly in reducing plastic waste and improving monitoring. However, challenges remain in mitigating long-term ecological and health impacts (Eriksen et al. 2014).

The case highlights the need for:

- Comprehensive international collaboration to tackle marine plastic pollution.
- Continued investment in innovative solutions, such as biodegradable plastics and advanced water treatment technologies.
- Greater awareness and action at all levels, from local communities to global policymakers.

The North Sea serves as a cautionary example of how human activities contribute to microplastic pollution and its cascading effects on food chains and ecosystems.

Miscellaneous Case Studies for Micropollutant Contamination

Case study	Context	Key issues/challenges	Impact	Proposed solutions	Reference
Endocrine disruptors in drinking water	Contamination from industrial and agricultural chemicals (e.g., BPA, phthalates)	Disruption of hormonal systems in humans and aquatic species	Increased risk of reproductive and developmental disorders	Advanced water treatment technologies (e.g., activated carbon)	Soto et al., (2004)
Antibiotic resistance in aquatic systems	Antibiotics from wastewater and agricultural runoff entering water bodies	Spread of antibiotic-resistant bacteria and genes	Threat to global health and treatment of infectious diseases	Improved wastewater treatment and regulations on antibiotic use	Martinez (2009)
E-waste pollution in soil and water	Leaching of heavy metals and toxic chemicals from improperly disposed e-waste	Soil contamination and bioaccumulation in aquatic ecosystems	Reduced crop yields and biodiversity loss	Recycling programs and strict e-waste management policies	Robinson (2009)
Nanoparticles in wastewater	Release of engineered nanoparticles (e.g., silver, titanium dioxide) into wastewater	Toxicity to aquatic organisms and potential bioaccumulation	Unknown long-term effects on ecosystems and human health	Development of nanoparticle-specific treatment technologies	Kahru and Dubourguier, (2010)
Oil spills and polycyclic aromatic hydrocarbons (PAHs)	Large-scale oil spills (e.g., Deepwater Horizon) releasing PAHs into marine environments	Long-term contamination of marine sediments and food chains	Decline in marine biodiversity and economic losses in fisheries	Rapid response cleanup using bioremediation and sorbents	NOAA, 2010, Atlas & Hazen, (2011)
Heavy metals in agricultural soils	Accumulation of cadmium, lead, and arsenic from fertilizers and pesticides	Toxicity to crops and health risks through food consumption	Reduced soil fertility and food safety concerns	Use of phytoremediation and organic farming practices	Alloway, (2013), FAO, (2018)

(continued)

(continued)

Case study	Context	Key issues/challenges	Impact	Proposed solutions	Reference
Microfibers in freshwater ecosystems	Release of synthetic microfibers from washing machines into rivers and lakes	Ingestion by aquatic organisms, leading to physical and chemical harm	Potential transfer of pollutants through food chains	Filters for washing machines and public awareness campaigns	Browne et al. (2011)
Pesticide runoff in wetlands	Excessive use of pesticides in agriculture leading to runoff into wetlands	Bioaccumulation in wetland species and loss of biodiversity	Disruption of ecosystem services provided by wetlands	Integrated pest management and buffer zones near wetlands	Gilliom (2001)
Industrial effluents in urban rivers	Direct discharge of untreated industrial effluents into urban waterways	High levels of organic pollutants and heavy metals	Decreased water quality and health risks for urban populations	Enforcement of industrial discharge standards and monitoring	UNEP, (2016), WHO, (2018)
Pharmaceuticals in agricultural irrigation	Use of treated wastewater containing pharmaceuticals for irrigation	Accumulation of pharmaceuticals in crops and soil	Potential human health risks through food consumption	Development of guidelines for safe reuse of treated wastewater	Christou et al., (2017)

Management and Mitigation Strategies: Lessons from Case Studies

Problem and Challenges (Smith et al., 2015)

Persistence: Heavy metals, such as lead, cadmium, and mercury, do not degrade over time, leading to prolonged environmental contamination. These pollutants accumulate in sediments, water bodies, and soils, creating reservoirs of toxicity that persist for decades or longer.

Bioaccumulation: Heavy metals like mercury and cadmium magnify within food chains. For example, mercury in aquatic systems is converted to methylmercury, a highly toxic form that bioaccumulates in fish and other aquatic organisms (Dey et al., 2024a, 2024b, 2024c; Priyadarshini & Das, 2024; Satapathy et al., 2024). Predators at the top of the food chain, including birds and humans, face increased risks due to higher concentrations of these metals.

Health Risks: Exposure to heavy metals is linked to severe health effects. Mercury exposure can cause neurological damage, particularly in developing fetuses and young children. Lead exposure is associated with developmental delays, learning disabilities, and cardiovascular issues. Additionally, cadmium is classified as a human carcinogen, posing risks of lung and kidney cancer (Das et al., 2024; Deyet al., 2024a, 2024b, 2024c).

Monitoring Difficulties: Identifying and quantifying contamination sources is challenging due to the diffuse nature of heavy metal pollution. Large systems, such as river basins, often involve multiple contributors, including industrial discharges, mining activities, and urban runoff, complicating effective monitoring and management.

Economic Costs: Heavy metal contamination imposes significant financial burdens on affected communities. Contaminated water supplies require costly treatment processes, and polluted fisheries can lead to economic losses for local economies reliant on fishing industries. Additionally, healthcare costs associated with treating heavy metal-related illnesses further strain resources.

Solutions and Approaches (Johnson et al., 2016)

Advanced Remediation Technologies

- Utilizing technologies such as bioremediation, phytoremediation, and chemical precipitation to remove heavy metals from contaminated sites.
- Employing membrane filtration and ion-exchange methods to treat polluted water supplies.

Policy and Regulation

- Implementing stringent regulations to control industrial discharges and mining operations.
- Enforcing limits on permissible heavy metal concentrations in water, soil, and air.

Sustainable Practices

- Encouraging industries to adopt cleaner production methods and recycling initiatives to minimize waste.
- Promoting agricultural practices that reduce heavy metal inputs, such as using organic fertilizers and avoiding contaminated irrigation water.

Monitoring and Early Warning Systems

- Developing advanced monitoring tools, including remote sensing and biosensors, to detect contamination in real time.
- Establishing early warning systems to alert communities and authorities about potential pollution events.

Community Engagement and Education

- Raising public awareness about the risks of heavy metal contamination and the importance of prevention.
- Involving local communities in monitoring efforts and decision-making processes to enhance accountability and effectiveness.

International Collaboration

- Fostering cross-border cooperation to address transboundary pollution issues.
- Sharing best practices, research findings, and technological advancements to tackle heavy metal contamination globally.

The lessons from these case studies highlight the importance of integrating advanced technologies, policy frameworks, and community engagement to address micropollutant contamination. A global perspective on mitigation strategies emphasizes the need for collaborative efforts to ensure the protection of ecosystems and human health. Overview of Current Methods for Monitoring and Removing Micropollutants.

Monitoring Techniques (Williams & Green, 2017)

- Advanced analytical methods such as liquid chromatography-mass spectrometry (LC–MS) and gas chromatography-mass spectrometry (GC–MS) are used for detecting micropollutants at trace levels.
- Biosensors are emerging as cost-effective and rapid tools for real-time monitoring of specific contaminants.
- Remote sensing and geographic information systems (GIS) are employed to map contamination hotspots and track pollutant sources.

Removal Technologies

- Conventional Wastewater Treatment: While effective for general pollutants, traditional methods such as activated sludge processes often fail to completely remove micropollutants.
- Advanced Oxidation Processes (AOPs): Techniques like ozonation, photocatalysis, and Fenton reactions degrade persistent organic micropollutants into less harmful byproducts.
- Membrane Technologies: Reverse osmosis and nanofiltration are highly effective but require significant energy inputs and maintenance.

Bioremediation: Utilizing microorganisms and plants to degrade or sequester contaminants shows promise for sustainable remediation.

Successful Remediation and Policy Approaches in Different Regions: A Case Study

Switzerland: Switzerland has implemented advanced treatment technologies, including ozonation and activated carbon filtration, in wastewater treatment plants to address pharmaceutical residues. These measures have reduced micropollutant concentrations by up to 80% in treated effluents (Swiss Federal Office for the Environment (FOEN) 2018).

European Union (EU): The EU Water Framework Directive mandates member states to monitor and control priority substances, including pesticides and pharmaceuticals. Collaborative research initiatives, such as the NORMAN network, support data sharing and standardization of monitoring methods (European Commission, 2017).

India: In Bengaluru, constructed wetlands have been employed to treat wastewater contaminated with heavy metals and pharmaceuticals. This nature-based solution has proven effective in reducing contaminant loads while providing additional ecological benefits (Ramesh et al., 2019).

United States: The Chesapeake Bay Program integrates watershed management practices to reduce agricultural runoff. Policies promoting buffer zones and nutrient management plans have successfully mitigated pesticide and nutrient pollution (Chesapeake Bay Program, 2020).

Conclusion

Micropollutants, including pesticides, heavy metals, and pharmaceuticals, present complex challenges that require a multidisciplinary and collaborative approach. Case studies such as pesticide contamination in the Mississippi River and heavy metal pollution globally illustrate the dire consequences of unchecked pollution on ecosystems, human health, and economies. However, these examples also highlight the potential for remediation through innovative technologies, regulatory measures, and community involvement. Moving forward, integrating advanced monitoring systems, fostering international cooperation, and promoting sustainable practices will be critical in managing micropollutants. By learning from past experiences and implementing proactive strategies, we can protect environmental and public health, ensuring a sustainable future for generations to come.

References

Alloway, B. J. (2013). *Heavy metals in soils*. Springer Science & Business Media.

Anderson, S., et al. (2017). Pharmaceutical residues in the Baltic sea: Sources and environmental impact. *Environmental Science & Technology, 51*(8), 4532–4541.

Atlas, R. M., & Hazen, T. C. (2011). Oil spills and PAHs: Environmental and economic consequences. *Environmental Science and Technology, 45*(6), 2505–2510.

Bala, K., Dogra, A., Lata, P., & Das, A. P. (2024). Discussing sources and biological-chemical remediation approaches for microplastic pollution. *Sustainable Microbial Technology for Synthetic and Cellulosic Microfiber Bioremediation*, 37–59. https://doi.org/10.1007/978-3-031-62600-5_3

Behera, I. D., Basak, G., Kumar, R. R., Sen, R., & Meikap, B. C. (2020). Treatment of petroleum refinery sludge by petroleum degrading bacterium *Stenotrophomonas pavanii* IRB19 as an efficient novel technology. *Journal of Environmental Science and Health, Part A, 56*(2), 226–239. https://doi.org/10.1080/10934529.2020.1866924

Behera, I. D., Nayak, M., Biswas, S., Meikap, B. C, Sen, R. (2021). Enhanced biodegradation of total petroleum hydrocarbons by implementing a novel two-step bioaugmentation strategy using indigenous bacterial consortium. *Journal of Environmental Management, 292*, 112746. ISSN 0301-4797. https://doi.org/10.1016/j.jenvman.2021.112746

Behera I. D., Nayak M., Mishra A., Meikap B. C., & Sen, R. (2022). Strategic implementation of integrated bioaugmentation and biostimulation for efficient mitigation of petroleum hydrocarbon pollutants from terrestrial and aquatic environment. *Marine Pollution Bulletin, 177*, 113492. ISSN 0025-326X. https://doi.org/10.1016/j.marpolbul.2022.113492

Bergmann, M., et al. (2020). Microplastics in terrestrial ecosystems: Sources, impacts, and mitigation strategies. *Environmental Science and Technology, 54*(18), 11229–11241.

Boxall, A. B. A., et al. (2012). The environmental risks of pharmaceuticals: A global perspective. *Environmental Toxicology and Chemistry, 31*(7), 1603–1614.

Brown, D., et al. (2015). Pharmaceutical bioaccumulation in aquatic organisms. *Environmental Health Perspectives, 123*(4), 225–237.

Browne, M. A., et al. (2008). Accumulation of microplastic on shorelines worldwide. *Environmental Science & Technology, 42*(12), 3956–3961.

Browne, M. A., et al. (2011). Microfibers in freshwater ecosystems: A threat to aquatic life. *Science of the Total Environment, 409*(20), 3885–3891.

Chen, Y., et al. (2020). Monitoring heavy metal contamination in the Yangtze river. *Science of the Total Environment, 748*, 141–150.

Chesapeake Bay Program. (2020). *Reducing agricultural runoff: policies and practices*. Chesapeake Bay Program Report.

Christou, A., et al. (2017). Pharmaceuticals in agricultural irrigation: A growing concern. *Agricultural Water Management Journal, 179*, 14–22.

Cole, M., et al. (2011). Microplastics as contaminants in the marine environment: A review. *Marine Pollution Bulletin, 62*(12), 2588–2597.

Arctic Council. (2015). *Persistent organic pollutants in the Arctic: A summary of findings*. Arctic Monitoring and Assessment Programme (AMAP) (pp. 1–45).

Das, P., & Bal, M. (2025). Modeling and optimization study for microplastic removal from aquatic medium using Chroococcidiopsis species. *Chemical Engineering Science, 304*, 121085. https://doi.org/10.1016/j.ces.2024.121085

Das, A. P., & Singh, S. (2011). Occupational health assessment of chromite toxicity among Indian miners. *Indian Journal of Occupational and Environmental Medicine, 15*(1), 6. https://doi.org/10.4103/0019-5278.82998

Das, P., Halder, G., & Bal, M. (2023). A critical review on remediation of microplastics using microalgae from aqueous system. *Science of the Total Environment, 898*, 166425. https://doi.org/10.1016/j.scitotenv.2023.166425

Das, A., Dey, S., Bala, K., Gnanasekaran, R., & Das, A. P. (2024). Microbial bioremediation technology for sustainable treatment and management of synthetic microfiber waste. In *Sustainable Microbial Technology for Synthetic and Cellulosic Microfiber Bioremediation* (pp. 221–238). https://doi.org/10.1007/978-3-031-62600-5_12

Daughton, C. G., & Ternes, T. A. (1999). Pharmaceuticals and personal care products in the environment: Agents of subtle change?. *Environmental Health Perspectives, 107*(Suppl 6), 907–938

Davis, K., et al. (2020). Chesapeake Bay restoration: Addressing nutrient pollution and eutrophication. *Estuarine, Coastal and Shelf Science, 244*, 106–115.

Dey, S., Tripathy, B., Kumar, M. S., & Das, A. P. (2023). Ecotoxicological consequences of manganese mining pollutants and their biological remediation. *Environmental Chemistry and Ecotoxicology, 5*, 55–61. https://doi.org/10.1016/J.ENCECO.2023.01.001

Dey, S., Bissoyi, A., Koteswara Reddy, G., & Das, A. P. (2024). Synthetic microfibers: An Urban emerging pollutant demanding for sustainable environmental solutions. In *Sustainable Microbial Technology for Synthetic and Cellulosic Microfiber Bioremediation* (pp. 107–121). https://doi.org/10.1007/978-3-031-62600-5_7

Dey, S., Dash, A., & Das, A. P. (2024). Bioremediation as a sustainable and circular solution for mining waste management and bioenergy generation. In *Sustainable Management of Mining Waste and Tailings: A Circular Economy Approach* (pp. 107–119). https://doi.org/10.1201/978 1003442455-6/BIOREMEDIATION-SUSTAINABLE-CIRCULAR-SOLUTION-MINING-WASTE-MANAGEMENT-BIOENERGY-GENERATION-SUDESHNA-DEY-AKANSHYA-DASH-ALOK-PRASAD-DAS

Dey, S., Dash, N., Ray, S. K., Priyadarsini, S., Chhotaray, C., & Das, A. P. (2024). Occurrence and quantification of tire wear plastics pollutants from highway City Dust of Bhubaneswar City of Odisha. In *Sustainable Microbial Technology for Synthetic and Cellulosic Microfiber Bioremediation* (pp. 77–89). https://doi.org/10.1007/978-3-031-62600-5_5

Eriksen, M., et al. (2014). Plastic pollution in the North sea and its impacts on marine life. *Marine Pollution Bulletin, 88*(1–2), 142–147.

Example, A., et al. (2018). Bioaccumulation of micropollutants in aquatic organisms. *Environmental Toxicology, 34*(2), 45–58.

European Commission. (2017). EU Water Framework Directive: Monitoring and Control of Micropollutants. European Union Report. https://environment.ec.europa.eu/topics/water/water-framework-directive_en

FAO. (2018). *Soil pollution: A growing threat to food security*. Food and Agriculture Organization of the United Nations Report.

Fent, K., Weston, A. A., & Caminada, D. (2006). Ecotoxicology of human pharmaceuticals. *Aquatic Toxicology, 76*(2), 122–159.

Fischer, J., et al. (2019). Managing PPCP contamination in European freshwater systems: Case study of the Rhine river. *Aquatic Toxicology, 45*(4), 123–134.

Galloway, T. S., et al. (2017). Microplastics and the bioaccumulation of contaminants. *Environmental Science and Technology, 51*(10), 5860–5867.

Garcia, E., et al. (2019). Endocrine disruption in European rivers due to pharmaceutical micropollutants. *Aquatic Toxicology, 45*(6), 67–79.

Giger, W., et al. (2003). Occurrence and fate of surfactants in the aquatic environment. *Science of the Total Environment, 307*(1–3), 71–90.

Gilliom, R. J. (2001). Pesticide runoff and wetland ecosystems. *Water Resources Research, 37*(8), 2199–2210.

Green, R., et al. (2021). Long-term impacts of PPCP contamination in European rivers: A case study of the Rhine. *Environmental Health Perspectives, 129*(7), 345–358.

Gupta, A., et al. (2020). Namami Gange Initiative: Addressing microplastic pollution in the Ganges. *Water Quality Research Journal, 55*(3), 204–213.

Hayes, T. B., et al. (2002). Herbicides: Potential effects on amphibian populations. *Nature, 419*(6910), 895–896.

Hayes, T. B., et al. (2002b). Pesticide-induced endrocrine disruption: Effects of atrazine on amphibians. *Environmental Health Perspectives, 110*(4), 393–397.

Heberer, T. (2002). Occurrence, fate, and removal of pharmaceutical residues in the aquatic environment: A review of recent research data. *Toxicology Letters, 131*(1–2), 5–17.

Jackson, L., et al. (2020). Mississippi river basin healthy watersheds initiative: Improving water quality and sustainability. *Water Research, 184*(2), 112–123.

Jagatee, S., Priyadarshini, S., Rath, C. C., & Das, A. P. (2024). Synthetic microfiber: An enduring environmental problem linked to sustainable development. *Environmental Science and Engineering, Part, F2333*, 93–112. https://doi.org/10.1007/978-3-031-51792-1_6

Jobling, S., & Tyler, C. R. (2003). Endocrine disruption in wild freshwater fish. *Pure and Applied Chemistry, 75*(11–12), 2219–2234.

Johansson, M., et al. (2020). Mitigation strategies for pharmaceutical pollution in the Baltic sea. *Marine Environmental Research, 162*, 105–115.

Johnson, L., et al. (2016). Solutions for heavy metal contamination: A global perspective. *Science of the Total Environment, 553*, 45–53.

Johnson, D., et al. (2018). The Great lakes restoration Initiative: Addressing heavy metal contamination in the Great lakes. *Environmental Management, 63*(2), 112–123.

Jones, C., et al. (2016). Mercury biomagnification in freshwater fish populations. *Ecotoxicology, 25*(1), 45–58.

Kah, M., & Brown, C. D. (2017). Volatilization of pesticides from soil: Processes and their measurement. *Environmental Pollution, 220*, 800–815.

Kahru, A., & Dubourguier, H. C. (2010). From nanoparticles to nanomaterials: Toxicity in aquatic ecosystems. *Environmental Toxicology and Chemistry, 29*(10), 2201–2208.

Kanungo, J., Sahoo, T., Bal, M., & Behera, I. D. (2023). Performance of bioremediation strategy in waste lubricating oil pollutants: A review. *Geomicrobiology Journal, 41*(4), 360–373. https://doi.org/10.1080/01490451.2023.2245395

Kanungo, J., Sahoo, T., Swain, L. P., Behera, I. D. (2023). Toxicity of persistent hydrocarbon pollutants, sources and sustainable remediation process. In: I. D. Behera & A. P. Das (Eds.), *Impact of Petroleum Waste on Environmental Pollution and its Sustainable Management Through Circular Economy. Environmental Science and Engineering*. Springer, Cham. https://doi.org/10.1007/978-3-031-48220-5_2

Karlsson, L., et al. (2021). Regional cooperation in the Baltic sea: Addressing pharmaceutical residues. *Environmental Pollution, 268*, 115–126.

Kumar, R., et al. (2021). Microplastic pollution in the Ganges river, India: Sources and impacts. *Marine Pollution Bulletin, 159*, 111–121.

Kümmerer, K. (2009). The presence of pharmaceuticals in the environment due to human use—present knowledge and future challenges. *Journal of Environmental Management, 90*(8), 2354–2366.

Lee, H., et al. (2019). Impact of nutrient pollution on Chesapeake Bay biodiversity. *Ecological Indicators, 98*, 89–97.

Lee, S., et al. (2020). Heavy metal pollution in the Great lakes: Sources and environmental impacts. *Journal of Environmental Quality, 49*(1), 78–89.

Lee, G., et al. (2021). Bioaccumulation of heavy metals in the Great lakes: Mercury and lead risks. *Environmental Monitoring and Assessment, 33*(2), 89–102.

Li, J., et al. (2019). Yangtze River pollution: Challenges in managing heavy metals. *Environmental Management, 63*(4), 482–494.

Lopez, F., et al. (2020). Microplastic pollution in the Mediterranean sea: Ecological impacts on marine life. *Marine Pollution Bulletin, 58*(3), 212–223.

Macdonald, R. W., et al. (2000). Contaminants in the Canadian Arctic: 5 years of progress in understanding sources, occurrence, and pathways. *The Science of the Total Environment, 254*(2–3), 93–234.

Martinez, J. L. (2009). The role of antibiotic resistance in aquatic systems. *Environmental Pollution, 157*(11), 3093–3101.

Mishra, S., Dash, D., Al-Tawaha, A. R. M. S., & Das, A. P. (2022). A review on heavy metal ion adsorption on synthetic microfiber surface in aquatic environments. *Applied Biochemistry and Biotechnology, 194*(10), 4639–4654. https://doi.org/10.1007/S12010-022-04029-W/METRICS

Mishra, S., Dash, D., & Das, A. P. (2024). Microbial degradation of polyester microfibers using indigenously isolated bacterial strain Exiguobacterium Sp. *CLEAN – Soil, Air, Water, 52*(12), e202300343. https://doi.org/10.1002/CLEN.202300343

Mohanta, S., Pradhan, B., & Behera, I. D. (2023). Impact and remediation of petroleum hydrocarbon pollutants on agricultural Land: A review. *Geomicrobiology Journal, 41*(4), 345–359. https://doi.org/10.1080/01490451.2023.2243925

Mohanty, J., Dey, S., Bala, K., & Das, A. P. (2024). Recent advances in microplastic pollution for its sustainable management. *Sustainable Management of Urban Plastic Waste Through Circular Economic Approaches*, 27–48. https://doi.org/10.1201/9781003465447-3

Moore, T., et al. (2018). Nutrient pollution in the Chesapeake Bay, USA: Eutrophication and hypoxic zones. *Marine Pollution Bulletin, 128*, 167–179.

Nelson, B., et al. (2019). Advanced wastewater treatment technologies for pharmaceutical removal. *Journal of Hazardous Materials, 380*, 120–130.

NOAA. (2010). *Deepwater horizon oil spill: Impacts on Marine ecosystems*. National Oceanic and Atmospheric Administration Report.

Paul, A., Dey, S., Ram, D. K., & Das, A. P. (2023). Hexavalent chromium pollution and its sustainable management through bioremediation. *Geomicrobiology Journal*, 1–11. https://doi.org/10.1080/01490451.2023.2218377

Paul, A., Priyadarsini, S., Karakuş, S., Behera, J. K., Gnanasekaran, R., & Das, A. P. (2024). Fate and accumulation of synthetic microfibers in the environment. In *Sustainable Microbial Technology for Synthetic and Cellulosic Microfiber Bioremediation* (pp. 61–76). https://doi.org/10.1007/978-3-031-62600-5_4

Priyadarshini, B., & Das, A. P. (2024). The adverse impact of synthetic microplastics released from disposable face masks pollutants on aquatic microbiota, aquatic species and humans: A review. *Environmental Science and Engineering, Part, F2827*, 67–81. https://doi.org/10.1007/978-3-031-50840-0_5

Priyadarshini, B., Santhosh Kumar, M., & Das, A. P. (2024). Microbial degradation of synthetic plastics released from COVID-19 generated single use disposable face masks. *Environmental Science and Engineering, Part, F2827*, 1–16. https://doi.org/10.1007/978-3-031-50840-0_1

Priyadarsini, S., & Das, A. P. (2024). Analytical and structural characterization of waste lithium-ion batteries for their effective recycling strategy. *Environmental Chemistry and Ecotoxicology, 7*, 182–91. https://doi.org/10.1016/j.enceco.2024.12.004

Priyadarsini, S., & Das, A. P. (2025). Lithium bioleaching: A review on microbial-assisted sustainable technology for lithium bio-circularity. *Journal of Water Process Engineering, 69*, 106744. https://doi.org/10.1016/j.jwpe.2024.106744

Ramesh, B., et al. (2019). Constructed wetlands for wastewater treatment in Bengaluru, India. *Environmental Engineering Journal, 44*(9), 1122–1130.

Ray, S. K., Dey, S., Das, A. P., & John, A. (2024). Microplastic waste management through greener technologies : A review. *Sustainable Management of Urban Plastic Waste Through Circular Economic Approaches*, 237–261. https://doi.org/10.1201/9781003465447-14

Richards, P., et al. (2021). Adaptive management strategies for Chesapeake Bay restoration. *Environmental Management, 67*(3), 285–297.

Rillig, M. C., et al. (2021). Microplastic pollution in soils and its potential ecological effects. *Science Advances, 7*(1), eaaz5061.

Robinson, B. H. (2009). E-waste pollution: A global challenge. *Environmental Science and Technology, 43*(8), 277–283.

Sahu, S., & Das, A. P. (2024). *Microplastics Degradation Mechanism by Aquatic Plastiomes: A Review* (pp. 213–226). https://doi.org/10.1007/978-3-031-50840-0_12

Sahu, A., Dey, S., Mallick, K., Priyadarsini, S., Behera, I. D., & Das, A. P. (2024). A review on advanced technology for sustainable management of synthetic microplastic waste. *Sustainable Management of Urban Plastic Waste Through Circular Economic Approaches*, 262–279. https://doi.org/10.1201/9781003465447-15

Satapathy, S., Kar, D., Priyadarsini, S., Das, A. P., & Kuanar, A. (2024). Application of microorganisms in emerging waste management strategies and remediation of synthetic and cellulosic microfibers. *Sustainable Microbial Technology for Synthetic and Cellulosic Microfiber Bioremediation*, 183–206. https://doi.org/10.1007/978-3-031-62600-5_10

Schmidt, A., et al. (2020). Monitoring of pharmaceutical contamination in the Rhine river: Challenges and insights. *Science of the Total Environment, 123*(6), 234–245.

Schwarzenbach, R. P., Egli, T., Hofstetter, T. B., von Gunten, U., & Wehrli, B. (2006). Global water pollution and micropollutants: Emerging challenges. *Science, 313*(5790), 1072–1077.

Sharma, S., et al. (2022). Microplastic pollution in the Ganges river: Challenges and mitigation strategies. *Environmental Science & Technology, 56*(5), 2371–2380.

Singh, R. P., Mishra, S., & Das, A. P. (2020). Synthetic microfibers: Pollution toxicity and remediation. *Chemosphere, 257*, 127199. https://doi.org/10.1016/j.chemosphere.2020.127199

Smith, J., et al. (2015). Heavy metal pollution and environmental risks. *Environmental Pollution, 201*(1), 115–125.

Smith, B., et al. (2017). The persistence of micropollutants in aquatic ecosystems. *Journal of Environmental Science, 12*(3), 123–134.

Soto, A. M., et al. (2004). Endocrine disruptors in drinking water. *Environmental Health Perspectives, 112*(5), 494–500.

Stockholm Convention Secretariat. (2001). *The Stockholm convention on persistent organic pollutants*. United Nations Environment Programme (UNEP) (pp. 1–20).

Suzuki, T., et al. (2018). Minamata disease: The legacy of methylmercury pollution and its global lessons. *Environmental Health Perspectives, 126*(5), 057–067.

Swiss Federal Office for the Environment (FOEN). (2018). *Pharmaceuticals in wastewater: Switzerland's response*. Swiss Environmental Report.

Taylor, M., et al. (2019). Pesticide contamination in the Mississippi river: Sources, impacts, and mitigation strategies. *Environmental Pollution, 240*(4), 356–367.

Thomas, G., et al. (2020). Public participation and policy in reducing pharmaceutical pollution in the Baltic sea. *Environmental Policy and Governance, 30*(2), 129–141.

Tripathy, B., Sahoo, P. P., Sundaray, H., & Das, A. P. (2024). In-vitro biodegradation of discarded marine microplastics across the eastern coast of the Bay of Bengal, India using Exiguobacterium sp. *Environmental Chemistry and Ecotoxicology, 6*, 236–247. https://doi.org/10.1016/J.ENCECO.2024.07.003

UNECE. (2019). *The Convention on Long-Range Transboundary Air Pollution: 40 years of scientific cooperation*. United Nations Economic Commission for Europe (pp. 1–35).

UNEP. (2016). *Industrial effluents in urban rivers: A global overview*. United Nations Environmental Programme Report.

Van Cauwenberghe, L., et al. (2015). Microplastics in the North sea food web. *Environmental Pollution, 202*, 178–185.

Vargas, J. M., et al. (2017). Toxicological effects of micropollutants on aquatic biota. *Environmental Pollution, 231*(1), 245–260.

Verma, P., et al. (2021). Microplastic sources and dynamics in the Ganges river basin. *Environmental Pollution, 268*, 115–126.

Wang, Z., et al. (2021). Phytoremediation and nanotechnology for heavy metal removal in the Yangtze river. *Environmental Science & Technology, 55*(7), 1234–1244.

Wania, F., & Mackay, D. (1996). Tracking the distribution of persistent organic pollutants. *Environmental Science & Technology, 30*(9), 390A-396A.

WHO. (2017). *Pesticide residues in food: A report on the global status of pesticide contamination.* World Health Organization (pp. 1–50).

WHO (2018). *Urban water quality: Risks and challenges.* World Health Organization Report.

Williams, P., & Green, H. (2017). Monitoring techniques for micropollutants in water. *Journal of Environmental Monitoring, 19*(6), 1134–1142.

Williams, S., et al. (2018). Pharmaceuticals and personal care products in the Rhine river: Sources and environmental impacts. *Environmental Science & Technology, 52*(3), 456–467.

Wright, S. L., et al. (2018). Microplastics as vectors for environmental contaminants: A review. *Science of the Total Environment, 621*(1), 1262–1272.

Zhang, Y., et al. (2017). The persistence of PPCPs in aquatic environments: A case study of the Rhine river. *Water Research, 65*(2), 89–101.

Zhang, L., et al. (2018). Heavy metal contamination in the Yangtze river, China: Industrialization and pollution. *Environmental Toxicology and Chemistry, 37*(9), 2312–2321.

Chapter 9
Advanced Technologies in Mitigating Micropollutant Contamination

Deeptimayee Pal and Abhishek Kumar

Introduction

The emergence of industrialization and rapid urbanization has characterized the twenty-first century with unparalleled economic and social progress, dramatically transforming human communities. Nevertheless, with these advancements, industrial operations and urban expansion have generated significant environmental issues, particularly with water supplies (Pal & Sen, 2024a). As metropolitan areas expand and industrial zones increase, natural water bodies that were once pure reservoirs of biodiversity and essential supplies for agriculture, drinking, and sanitation have increasingly become repositories of complex pollutants. Diverse and widespread contaminants have contaminated water reservoirs, including heavy metals, agricultural runoff, pharmaceutical chemicals, and personal care items. As a result, these pollutants interrupt ecosystems, threaten human health, and hinder efforts to restore clean water (Das et al. 2024a; Mohanty et al., 2024a, 2024b). Water pollution has been mainly linked to substantial, discernible contaminants like industrial discharges and agricultural runoff, evident by discoloration, odors, or the asphyxiation of aquatic organisms caused by excessive organic matter. In recent decades, scientists and regulatory bodies have focused on less apparent but, if not more, pernicious toxins that elude standard detection and treatment procedures. These are called micropollutants or contaminants of emerging concern (CECs).

Micropollutants are trace-level hazardous chemicals increasingly detected in the environment, including aquatic systems, soils, and sediments. These compounds include various chemicals, such as pharmaceuticals, pesticides, industrial chemicals, endocrine disruptors, and personal care products. Human activities often introduce

D. Pal (✉) · A. Kumar
Department of Chemical Engineering, National Institute of Technology Rourkela, Rourkela, Odisha, India
e-mail: deeptimayeenitrkl@outlook.com

I. D. Behera et al. (eds.), *Emerging Micropollutants*, Sustainable Environmental Waste Management Strategies, https://doi.org/10.1007/978-3-032-03530-1_9

them into the environment, such as domestic wastewater discharge, industrial effluents, agricultural runoff, and landfill leachate. Micropollutants are characterized by their persistence and bioaccumulative features, indicating their resistance to natural degradation and ability to accumulate in the tissues of aquatic creatures, ultimately infiltrating human food chains (Das et al., 2024a; Mohanty et al., 2024a, 2024b; Sahu et al., 2024a). The ecological and health hazards linked to micropollutants arise from their biological activity, even at low concentrations. Micropollutants are associated with endocrine disruption, carcinogenicity, and neurotoxicity, potentially affecting aquatic biodiversity, ecological equilibrium, and human health (Ahmed et al., 2024).

Detection and quantification highlight significant micropollutant issues (Dasenaki & Thomaidis, 2015). While effective in detecting standard pollutants, conventional analytical methods often lack the sensitivity and specificity required to detect low concentrations of micropollutants in complex water bodies. Progress in analytical chemistry, particularly high-performance liquid chromatography coupled with mass spectrometry (HPLC–MS), has enhanced the identification and quantification of these compounds; however, these methods still encounter limitations in fully capturing the diversity and behavior of micropollutants in aquatic environments. Addressing micropollutants poses further challenges. Conventional water treatment techniques, including sedimentation, filtration, and biological processes, are often ineffective in addressing micropollutants because of their chemical stability and low concentrations. Consequently, advanced treatment methods have been developed, including advanced oxidation processes (AOPs), membrane filtration, nanotechnology, biotechnological, and electrochemical approaches (Pal & Sen, 2025).

Resolving the micropollutants issue requires technological advancement and powerful governmental and regulatory frameworks. Global regulatory authorities are increasingly acknowledging the need for intervention regarding micropollutants. The European Union's Water Framework Directive (WFD) has set environmental quality criteria for some micropollutants, while the United States Environmental Protection Agency (EPA) has initiated measures to address CECs under the Safe Drinking Water Act. Nonetheless, an imperative persists for uniform worldwide norms and regulations that can be adapted to various regulatory environments to safeguard water quality (Ahmed et al., 2017). Consequently, CECs have emerged as a critical area of interest in environmental sciences, with ongoing investigation dedicated to devising emerging and sustainable techniques for their detection, removal, and removal. This chapter will examine the technical issues associated with micropollutants and the recent technological breakthroughs aimed at their removal, focusing on preserving environmental and public health.

Sources and Pathways of Micropollutant Contamination

Micropollutants are chemical contaminants present at trace concentrations, often ranging from μg/L to ng/L, in environmental systems, including water, air, and soil (Jiang et al., 2013). Despite their trace concentrations, micropollutants have considerable biological and ecological effects due to their toxicity, persistence, and capacity for bioaccumulation in organisms over time. Micropollutants are classified according to their origins, chemical characteristics, and applications, including pharmaceuticals and personal care products (PPCPs), pesticides, endocrine-disrupting compounds (EDCs), microplastics, and industrial chemicals and byproducts such as plasticizers, flame retardants, solvents, and surfactants. Various human activities introduce micropollutants into the environment, with point and non-point sources facilitating their release. These toxins are usually introduced into natural ecosystems in low concentrations, but their persistence and capacity for bioaccumulation render their paths and consequences considerable. Households substantially contribute to micropollutant pollution using pharmaceuticals, personal care items, and domestic chemicals. These contaminants infiltrate municipal wastewater systems and, owing to inadequate removal by conventional wastewater treatment plants (WWTPs), are often released into receiving waterways. Pesticides, herbicides, and fertilizers used in agricultural activities are significant sources of micropollutants (Mishra & Das, 2023; Mishra et al., 2024). These pollutants may infiltrate groundwater or be transported by surface runoff into rivers and lakes (Corominas et al., 2013).

Animal husbandry also contributes through veterinary pharmaceuticals and hormones, often discharged and infiltrating aquatic systems. Industrial operations discharge various micropollutants, such as solvents, heavy metals, flame retardants, and surfactants, directly into aquatic environments or through insufficiently treated effluents (Mishra et al., 2023; Paul et al., 2023; Pradhan et al., 2023). Several sectors, such as pharmaceuticals, textiles, and electronics manufacturing, are significant contributors. Urban regions produce runoff that transports contaminants from vehicular exhaust, constructing supplies, and urban environments. Pollutants such as heavy metals, polycyclic aromatic hydrocarbons (PAHs), and microplastics are transported into surface waterways during precipitation events. Landfills function as enduring repositories of micropollutants. Leachates, or polluted water that permeates landfill refuse, may transport heavy metals, pharmaceuticals, and other toxic compounds into adjacent soils and groundwater. Micropollutants may be deposited from the atmosphere, transported by air currents, and then deposited onto land or water by precipitation or dust. Pollutants like PAHs, insecticides, and combustion byproducts from vehicular emissions and industrial operations contribute to this pathway (Song et al., 2007).

Various micropollutants infiltrate surface water bodies (e.g., rivers, lakes, and seas) directly through point-source discharges, such as wastewater treatment plant effluents, or indirectly through runoff and soil leaching. These pollutants often persist in the water column or sediment, affecting aquatic life and ecosystems. Micropollutants can permeate the soil and migrate to groundwater aquifers. Groundwater contamination is especially alarming since it is a primary source of potable water, and extracting micropollutants from groundwater presents significant challenges. Micropollutants accumulate in soil and sediment by irrigation with polluted water or air deposition (Choi et al., 2016). Plants can absorb these chemicals or remobilize following precipitation, increasing environmental distribution. Micropollutants can accumulate in the tissues of aquatic and terrestrial organisms, resulting in biomagnification as they make their way into the food chain. This accumulation poses health hazards for apex predators, including humans, who devour these organismsAdverse Impacts of Micropollutants.

Micropollutants in water, soil, and air present considerable environmental and health issues owing to their persistence, tendency for bioaccumulation, and often harmful or disruptive biological impacts, even at low concentrations. In contrast to conventional pollutants, micropollutants exhibit resistance to degradation in the environment, enabling their persistence and prolonged influence on ecosystems. Micropollutants, including pharmaceuticals, pesticides, and industrial chemicals, interfere with biological processes in aquatic habitats. Antibiotics can modify microbial populations, influencing nutrient cycling and food webs (Pal & Sen, 2024b). Pesticides can adversely affect non-target species, lowering biodiversity and changing predator–prey relationships. Endocrine-disrupting substances, such as synthetic hormones and certain industrial compounds, disrupt hormonal systems in aquatic and terrestrial beings. EDCs may induce reproductive complications, developmental anomalies, and behavioral alterations in fish, amphibians, and mammals. EDC exposure has been associated with the feminization of male fish, impacting reproductive rates and population stability. Various micropollutants accumulate in the tissues of organisms, ascending the food chain via bioaccumulation and biomagnification. Predators, particularly apex consumers such as raptors, fish, and mammals, have heightened susceptibility to harmful consequences from the accumulation of micropollutants in their feeds. This may result in weakened immune systems, reproductive failures, and increasing death rates. Micropollutants, including heavy metals, pesticides, and pharmaceuticals, can accumulate in soils and sediments. Contaminated soils can impair plant health and production, disturb soil microbial populations, and even infiltrate terrestrial food chains, thus affecting animal and human health (Pal & Sen, 2023b).

Antibiotics and antibiotic-resistant genes (ARGs) often contaminate water sources due to pharmaceutical waste, agricultural runoff, and hospital effluents (Pruden et al., 2006). These pollutants facilitate the worldwide escalation of antibiotic-resistant bacteria, presenting a considerable public health concern by undermining the efficacy of medicines and increasing the risk of resistant diseases. Specific micropollutants, such as PAHs, some pesticides, and industrial chemicals, are recognized as carcinogenic and mutagenic agents. Exposure to these compounds, even at low concentrations, has been associated with higher risks of cancer, genetic abnormalities, and long-term health complications. Exposure to EDCs, such as bisphenol A (BPA) and phthalates, is linked to hormonal abnormalities, reproductive dysfunction, and developmental problems in humans. Specifically, prenatal and early-life exposure to EDCs has been stated to affect growth, neurodevelopment, and subsequent reproductive health (Dey et al., 2023a, 2023b; Lahiri et al., 2021). Various micropollutants, including lead, mercury, and certain pharmaceuticals, exhibit neurotoxicity and can hinder cognitive development, especially in children. Prolonged exposure to these chemicals can result in learning difficulties, attention problems, and further developmental complications. Micropollutants can impair immunological function, exposing organisms more vulnerable to illness. Chronic exposure to certain pollutants, such as heavy metals, pesticides, and persistent organic pollutants (POPs), can cause immunological dysfunction in people, hence raising the risks of infections and inflammatory disorders (Pal & Sen, 2023a).

Conventional Water Treatment Processes and Their Limitations

Conventional water treatment techniques, such as coagulation, sedimentation, filtration, and disinfection, are extensively used for removing pathogens, organic substances, and suspended particulates from potable water and wastewater. Although these methods effectively remove most contaminants, they are often proven inadequate for removing micropollutants, which exist at trace levels and include complex organic and inorganic compounds that resist conventional treatment techniques. Coagulation involves introducing chemicals (e.g., aluminum or iron salts) to water, resulting in the agglomeration of dispersed particles into bigger flocs. Flocculation enhances this process, allowing for sedimentation and the subsequent removal of pollutants (Lapointe et al., 2020). Coagulation is often inefficient in removing dissolved micropollutants since these substances do not easily adhere and persist in the water after treatment. Sedimentation enables particles to precipitate from the water due to gravitational force. Filtration, often conducted using sand or carbon filters, removes more considerable particle debris. Filtration can reduce turbidity and some suspended particles, but it fails to capture small, dissolved micropollutants, particularly those with low molecular weights and high solubility.

Activated carbon is used in some conventional methods to adsorb chemical substances and odors. While it is more effective than conventional approaches for specific micropollutants, it has limitations. The adsorption capacity of activated carbon differs among various chemicals, and it is less efficient for polar or very hydrophilic micropollutants. Regular replacement or regeneration of activated carbon is necessary to sustain efficacy, increasing operating expenses. Biological treatment employs microbial activity to decompose organic materials in wastewater. This process encompasses techniques like activated sludge and biofilm systems, which can degrade some organic micropollutants. Nonetheless, biological treatment procedures exhibit significant variability in their effectiveness for various micropollutants since some chemicals resist microbial breakdown. Moreover, micropollutants that are detrimental to microorganisms may impede the efficacy of these activities. Chlorination is extensively used for disinfection and removing germs in potable water and wastewater (Mazhar et al., 2020). Nonetheless, it has a restricted ability to eliminate micropollutants and may sometimes result in the generation of toxic byproducts, such as chlorinated organic compounds, which might provide further health hazards. Disinfection byproducts provide a significant hazard with organic micropollutants that interact with chlorine, forming harmful chemicals such as trihalomethanes (THMs).

Conventional methods are mainly intended for more significant particle contaminants and are ineffective in removing low-concentration, dissolved micropollutants. Many micropollutants are small, persistent, and exhibit high water solubility, making them challenging to gather or degrade. Various treatment procedures, like chlorination, can interact with micropollutants to produce detrimental consequences. These byproducts can exhibit more significant toxicity than the initial toxins, posing further dangers to ecosystems and human health. Numerous micropollutants, including pharmaceuticals, insecticides, and industrial chemicals, are engineered for stability and non-biodegradability, rendering biological treatment procedures ineffective for their degradation (Dey et al., 2024a, 2024b; Ray et al., 2024; Sahu et al., 2024b; Tripathy et al., 2024). This stability enables these chemicals to survive treatment systems and infiltrate receiving waters. The need for periodic replacement or regeneration in processes such as activated carbon adsorption escalates operating costs. Moreover, sustaining the efficacy of conventional treatment systems to comply with emerging micropollutant regulatory criteria may increase expenses. The removal efficiency of conventional approaches fluctuates significantly based on the particular micropollutants, treatment parameters, and design of the treatment facility. Biological therapy may partly decompose some classes of micropollutants while leaving others unaffected, resulting in inconsistent results and unreliable contaminant removal (Miranda et al., 2020).

Necessity for Advanced Treatment Technologies

The escalating abundance of micropollutants in water sources makes conventional water treatment procedures less effective in addressing the challenges presented by these contaminants. Micropollutants, including pharmaceuticals, personal care products, pesticides, industrial chemicals, and other emerging contaminants, are frequently detected at trace levels but may have considerable environmental and health effects. The ineffectiveness of conventional treatment methods in efficiently eliminating these pollutants emphasizes the need for innovative treatment options. Conventional water treatment methods, including coagulation, filtration, and biological treatment, are primarily intended to eliminate more significant particle contaminants and pathogens rather than the smaller, dissolved, and persistent micropollutants (Lapointe et al., 2020). Numerous micropollutants exhibit chemical stability, non-biodegradability, and high water solubility, rendering them difficult to eliminate using conventional processes. Micropollutants in water sources present considerable threats to natural ecosystems and human health. These pollutants are often toxic, persistent, and bioaccumulative, allowing them to accumulate in organisms and infiltrate the food chain, resulting in enduring ecological harm (Mishra et al., 2019).

Due to growing awareness of micropollutant pollution, global regulatory authorities are implementing more stringent water quality regulations. Numerous rules now require the reduction or elimination of micropollutants to acceptable levels. This shift is significant for pollutants found in minimal amounts that provide considerable health hazards. Conventional treatments do not adhere to these more stringent criteria and are insufficient to fulfill the new rules. Consequently, there is an increasing need for advanced treatment technologies to eliminate small quantities of micropollutants and ensure adherence to new regulatory standards (Miranda et al., 2020). Numerous micropollutants have environmental persistence, indicating they do not quickly disintegrate or decompose under ecological conditions. This durability enables their accumulation in aquatic environments, sediments, and the tissues of marine creatures. Moreover, bioaccumulation and biomagnification processes by which pollutants intensify as they ascend the food chain can result in prolonged exposure to detrimental chemicals in animals and people.

Advanced treatment technologies are crucial for degrading or eliminating these enduring pollutants from water before their potential to inflict extensive damage. Recent advancements in advanced treatment technologies, including AOPs, nanotechnology, membrane filtration, electrochemical, and biotechnological treatments, provide potential alternatives for eliminating micropollutants. These technologies are explicitly engineered to target and eliminate various micropollutants, including those extremely soluble, persistent, or resistant to biodegradation (Das et al., 2024b; Dey et al., 2023b, 2024a). In contrast with conventional procedures, these advanced methods provide enhanced efficiency, increased specificity, and the capability to handle complex combinations of contaminants, rendering them crucial for tackling the escalating issue of micropollutants in water. Advanced treatment methods provide prospects for more sustainable and eco-friendly water treatment. In contrast with conventional procedures that often need substantial amounts of chemicals (e.g., coagulants, disinfectants), several novel treatments are more ecologically friendly. For instance, AOPs may decompose pollutants into non-toxic byproducts, while membrane filtration systems can be engineered for reduced energy use and little chemical usage (Madsen, 2014; Prieto-Rodríguez et al., 2013). Moreover, numerous modern treatment methods may be combined with renewable energy sources, therefore reducing the carbon footprint of water treatment plants.

Advanced Treatment Technologies for Micropollutant Removal

The various advanced treatment technologies for micropollutant removal, such as AOPs, nanotechnology, membrane filtration, electrochemical, and biotechnological treatments, have been discussed in detail.

AOPs

AOPs are a group of effective and versatile water treatment methods used to degrade various micropollutants and organic contaminants. These processes often include the generation of highly reactive hydroxyl (·OH) radicals, which may decompose even the most persistent and stable contaminants (Prieto-Rodríguez et al., 2013). The various types of AOPs are depicted in Fig. 9.1. ·OH radicals may decompose complicated chemical structures into smaller, less damaging intermediates, which may ultimately mineralize into carbon dioxide, water, and inorganic ions. AOPs are especially effective in eliminating micropollutants, including pharmaceuticals, pesticides, personal care items, and industrial chemicals resistant to conventional water treatment techniques. Ozonation employs ozone (O_3) to generate ·OH radicals in water. Ozone is a potent oxidant capable of directly oxidizing specific micropollutants and decomposes in water to generate hydroxyl radicals, increasing its oxidative capacity. Ozone, upon introduction to water, can disintegrate into hydroxyl radicals, which then interact with organic pollutants, fragmenting them into smaller, non-toxic molecules, shown in Eq. 9.1.

$$O_3 + H_2O \rightarrow 2OH\cdot + O_2 \quad (9.1)$$

·OH radicals attack organic pollutants, initiating the degradation of chemical bonds and resulting in mineralization. Ozonation is especially efficient in degrading pharmaceuticals and endocrine-disrupting chemicals (EDCs), since these compounds often possess aromatic rings and double bonds that exhibit strong reactivity with ozone. The technique can decompose these contaminants into smaller byproducts, which are either less detrimental or more accessible to removal by later treatments. Nonetheless, ozonation can sometimes result in the generation of hazardous byproducts, such as bromates, when processing water with elevated bromide levels. Consequently, ozonation is often integrated with other AOPs or adsorbents to achieve complete mineralization of micropollutants while minimizing byproduct generation.

Photocatalysis employs UV light with a semiconductor catalyst such as titanium dioxide (TiO_2) to generate hydroxyl radicals (Mahmoud et al., 2017). When UV light stimulates TiO_2, it generates electron–hole pairs that interact with water molecules to produce hydroxyl radicals, which subsequently oxidize micropollutants, as seen in the equations. This technique is very efficacious for decomposing persistent organic

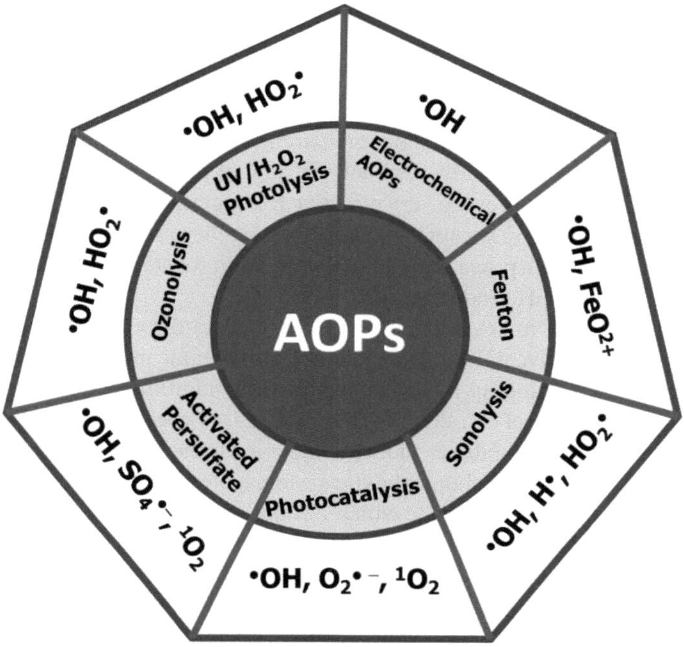

Fig. 9.1 Types of AOPs (Rayaroth et al., 2022)

micropollutants, such as pharmaceuticals, insecticides, and synthetic hormones.

$$TiO_2 + \text{UV light} \rightarrow e^- + h^+ \tag{9.2}$$

$$h^+ + H_2O \rightarrow OH\cdot + H^+ \tag{9.3}$$

The UV/H_2O_2 process involves the introduction of hydrogen peroxide (H_2O_2) to water, which, upon exposure to UV radiation, decomposes to generate hydroxyl radicals, as seen in the equation. This combination effectively degrades a wide range of micropollutants, including pharmaceuticals, insecticides, and residues from personal care products. The UV/H_2O_2 process is efficacious in scenarios where water is comparatively clear, facilitating efficient penetration of UV radiation. This technique effectively eliminates micropollutants in drinking water treatment since it prevents the generation of disinfection byproducts (DBPs) often associated with chlorine-based disinfection. The approach may be customized for specific micropollutants by modifying the concentration of H_2O_2 and the intensity of UV radiation.

$$H_2O_2 \rightarrow 2OH\cdot \tag{9.4}$$

The Fenton process employs hydrogen peroxide (H_2O_2) with ferrous iron (Fe^{2+}) as a catalyst to generate hydroxyl radicals (Ziembowicz & Kida, 2022). The photo-Fenton method utilizes UV or visible light to improve the reaction, increasing the production of hydroxyl radicals. The Fenton and photo-Fenton procedures are notably efficient in cleaning wastewater containing elevated levels of organic micropollutants due to the cost-effectiveness and efficacy of the iron catalyst under acidic conditions. These procedures are very effective for degrading pharmaceuticals, dyes, and industrial chemicals in wastewater. Nonetheless, they are optimally designed for acidic conditions and can prove impractical for treating substantial quantities of water owing to the need for pH modifications.

The degradation of micropollutants in AOPs occurs through many methods. In AOPs, the highly reactive ·OH radicals interact with organic micropollutants through many mechanisms, including hydrogen abstraction, addition to double bonds, and electron transfer (Yang et al., 2014). These processes break stable carbon bonds, resulting in smaller intermediates that can undergo further oxidation or mineralization. In ozonation and some photocatalytic systems, the primary oxidant (e.g., ozone or excited TiO_2) can directly engage with micropollutants, degrading particular functional groups or destroying aromatic ring structures often present in pharmaceuticals and pesticides. In UV-based AOPs, micropollutants can absorb UV light, leading to direct photolysis that cleaves particular bonds within the contaminant molecule. This method is very effective for compounds such as certain pharmaceuticals and insecticides that absorb UV light. Integrating these processes ensures that micropollutants are destroyed and undergo molecular structural transformation, often resulting in non-toxic end products.

Nanotechnology

Nanotechnology has emerged as a viable area for tackling environmental concerns, especially in eliminating micropollutants from water. Utilizing the distinctive characteristics of nanoparticles and nanomaterials may significantly enhance the effectiveness, selectivity, and sustainability of water treatment systems. Nanomaterials possess distinctive properties attributable to their small dimensions, extensive surface area, and heightened reactivity (Kumar & Gangawane, 2022a, 2022b). These qualities enable them to engage with and eliminate contaminants from water more effectively than conventional materials. Nanomaterials possess remarkable adsorption capabilities, making them particularly effective for capturing micropollutants. Nanoparticles of activated carbon, biochar, and graphene derivatives, including graphene oxide, possess extensive surface areas with tunable pore structures that may be tailored to selectively adsorb specific micropollutants, such as pharmaceuticals, pesticides, and residues from personal care products. Owing to their extensive surface area and tunable chemical characteristics, these materials may be altered with functional groups that enhance their affinity for specific contaminants, rendering them effective in sequestering low-concentration pollutants. Graphene-based materials are very significant in these circumstances. Their two-dimensional structure and tuned surface chemistry enable interaction with various pollutants via processes like π-π interactions, hydrogen bonding, and van der Waals forces. Carbon nanotubes (CNTs), a distinct nanomaterial, demonstrate comparable adaptability and efficiently eliminate organic micropollutants owing to their tubular architecture and robust adsorptive characteristics (Aslam et al., 2021).

Another fascinating use of nanotechnology in removing micropollutants is photocatalytic degradation, which utilizes semiconductor nanoparticles like TiO_2 and zinc oxide (ZnO) to decompose contaminants under light irradiation. Upon exposure to light, often UV, these nanomaterials generate reactive oxygen species (ROS), such as hydroxyl radicals, powerful oxidants that can decompose organic micropollutants into less hazardous molecules. The distinctive characteristics of TiO_2 nanoparticles render them efficient in decomposing complex organic compounds, including pharmaceuticals and endocrine disruptors, at minimal concentrations. Advanced photocatalysts, including doped TiO_2 and TiO_2 combined with other nanomaterials (e.g., graphene oxide), have been engineered to improve photocatalytic efficiency under visible light, making them more applicable for practical use (Carbajo et al., 2017). Membrane filtration methods, such as NF and RO, efficiently eliminate various micropollutants, yet they may encounter challenges such as fouling and reduced permeability. Nanotechnology mitigates these restrictions by integrating nanoparticles directly into the membrane structure or as surface coatings. Membranes reinforced with nanoparticles, such as those modified with silver nanoparticles, TiO_2, or CNTs, boost pollutant rejection rates and provide additional features, including antibacterial activity and fouling resistance. CNT membranes are particularly effective for removing micropollutants since their distinctive tubular architecture enables increased water flow while selectively filtering contaminants according to size or chemical affinity. Enhancing membranes with graphene oxide or nanoscale metal–organic frameworks (MOFs) significantly augments their selective adsorption capability, facilitating the targeted elimination of micropollutants such as pharmaceuticals, synthetic hormones, and pesticide residues (Russo et al., 2020).

The elimination of micropollutants by nanomaterials often transpires via a synthesis of adsorption processes. Nanomaterials with extensive surface areas interact with contaminants via weak van der Waals forces, facilitating the adsorption of pollutants onto the surface without forming chemical bonds. Specific nanomaterials, especially carbon-based derivatives such as CNTs and graphene, possess hydrophobic surfaces that attract non-polar organic molecules via hydrophobic interactions, rendering them efficient for removing hydrophobic micropollutants. In some instances, functional groups on nanomaterial surfaces, such as hydroxyl or carboxyl groups, can develop covalent bonds with pollutants, resulting in strong adsorption and sometimes the destruction of the adsorbed contaminants (Kumar & Gangawane, 2023a). Metal-based nanoparticles, especially nZVI, may participate in redox processes, whereby the metal particles give electrons to decrease pollutants, chemically converting them into less dangerous chemicals. Charged-surface nanomaterials (e.g., metal oxides or functionalized polymers) may exchange ions with waterborne pollutants. Charged nanoparticles may attract positively or negatively charged impurities, such as heavy metals or anionic pollutants, resulting in effective removal.

Membrane Filtration

Membrane filtration methods have significant potential in eliminating diverse micropollutants from water (Kumar & Gangawane, 2023b). These contaminants, including pharmaceuticals, personal care items, insecticides, and endocrine-disrupting chemicals (EDCs), are challenging to eliminate through conventional water treatment techniques. Membrane filtration provides a better and more efficient method by selectively eliminating pollutants according to size, charge, and other characteristics. Among the several membrane filtering processes, forward osmosis (FO), nanofiltration (NF), and reverse osmosis (RO) are among the most frequently employed procedures, each possessing distinctive characteristics, advantages, and applications. The various membrane technologies are depicted in Fig. 9.2.

Forward osmosis is a membrane filtration method whereby water spontaneously migrates from a dilute solution (feed water) to a concentrated solution (draw solution) over a semipermeable membrane driven by an osmotic pressure gradient. In contrast to RO, which utilizes pressure to surpass osmotic pressure in the feedwater, forward osmosis depends on the inherent osmotic potential between the feed and draw solutions to facilitate water movement over the membrane. RO is a membrane filtering technique that utilizes hydraulic pressure to propel water through a semipermeable membrane, transitioning from a solution of greater concentration (contaminated feedwater) to a lower concentration (treated permeate). NF is a membrane filtration method that is in a position between ultrafiltration (UF) and RO regarding pore size and efficacy. NF membranes possess fewer spaces than UF membranes but are more significant than RO ones, often measuring between 1 and 5 nm (Shanmuganathan et al., 2017).

RO is proficient at eliminating various micropollutants, including pharmaceuticals, pesticides, and industrial chemicals. The minuscule pore size of RO membranes facilitates the isolation of even diminutive molecular pollutants from water. RO is a prevalent technology used to treat drinking water and eliminate micropollutants. NF has successfully eliminated specific organic micropollutants, including pesticides, herbicides, and some medications. NF membranes possess a somewhat wider pore size than RO membranes, permitting the passage of helpful ions like calcium and magnesium while effectively rejecting detrimental micropollutants. Although UF and microfiltration (MF) are less successful than RO or NF in eliminating tiny molecular micropollutants, they can remove bigger particles, suspended solids, and macromolecular organic contaminants. These technologies are often used with other treatment techniques (e.g., AOPs and activated carbon adsorption) to improve the complete removal of micropollutants (Roy et al., 2020).

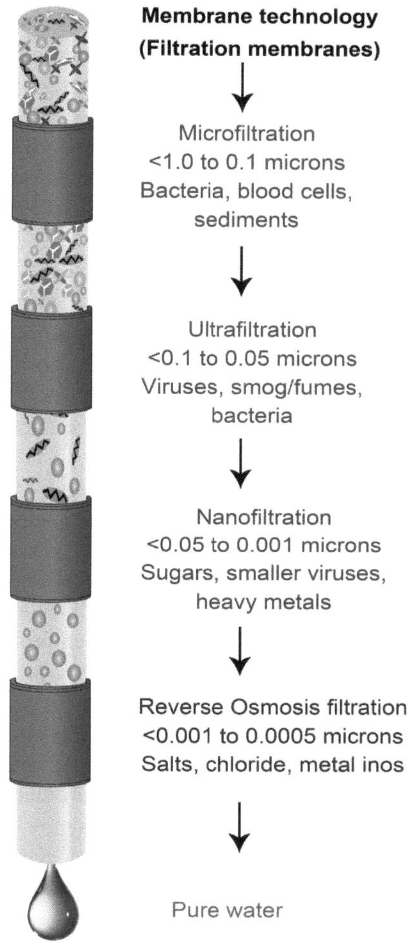

Fig. 9.2 Filtration membranes for water treatment (Osman et al., 2024)

Membrane filtration operates by selectively segregating substances according to size, charge, or other chemical properties. The primary methods for eliminating micropollutants by membranes are size exclusion (sieving), adsorption, electrostatic interactions, and solution diffusion. Size exclusion is the primary process of membranes excluding bigger particles, such as suspended particulates, microorganisms, and substantial chemical molecules. Micropollutants exceeding the membrane's pore size are physically obstructed and kept on the feed side while water and smaller molecules permeate through. Micropollutants may adsorb onto the membrane surface or inside its pores, especially when the membrane material has a large surface area and is hydrophobic or contains functional groups that engage with specific pollutants. Adsorption is crucial for eliminating organic micropollutants, including pharmaceuticals and personal care items (Kumar et al., 2022). Membranes may demonstrate electrostatic interactions with charged micropollutants, especially when the contaminants are ionic. These interactions may result in the expulsion of charged species by repelling them from the membrane surface or by facilitating their adhesion to the membrane surface, obstructing their passage. In solution diffusion, specific micropollutants may traverse the membrane depending on their water solubility and the permeability of the membrane material to those solutes. This phenomenon is particularly prevalent in RO membranes because separation relies on size and solubility attributes.

Various parameters influence the efficacy of membrane filtration in eliminating micropollutants, including micropollutant characteristics, operational parameters, and membrane fouling. Micropollutants of greater size, including specific organic compounds and particles, may be more effectively eliminated using membranes with suitable pore dimensions (e.g., RO and NF). Charged micropollutants may interact with charged membrane surfaces, resulting in enhanced rejection. Specific membranes may be designed to augment electrostatic interactions with specific contaminants. Membrane materials with unique functional groups or hydrophobic characteristics may demonstrate selective adsorption of particular micropollutants, hence enhancing their removal efficacy. The applied pressure substantially influences removal efficiency in pressure-driven systems such as RO and NF. Increased pressures often enhance the flow rate but may also affect the rejection rate. The flow rate of water across the membrane may influence the efficacy of micropollutant removal. Increased flux rates may decrease contact time with the membrane, thereby decreasing removal efficiency. The temperature and pH of the feed water may influence the membrane's permeability and its interaction with pollutants. Optimal conditions must be maintained to guarantee maximal pollutant elimination. Membrane fouling, resulting from the accumulation of particles, organic substances, and microbial growth, can reduce the efficacy of membrane systems in eliminating micropollutants. Consistent cleaning and maintenance are essential to minimize fouling and preserve membrane efficacy (Madsen, 2014).

Electrochemical Treatments

Electrochemical treatment technologies have evolved as novel and effective approaches for eliminating micropollutants from water and wastewater (Chen, 2004). These technologies use electrical energy to facilitate redox processes that decompose or convert organic and inorganic pollutants into non-toxic or less hazardous byproducts. Electrochemical techniques for eliminating micropollutants depend on using an electric current to promote oxidation, reduction, or coagulation processes. These reactions occur at the surface of electrodes immersed in the treated water. During this process, contaminants are oxidized at the anode, generating simpler and less harmful molecules. The reaction may produce reactive species such as ·OH radicals or other oxygen species that can decompose various organic pollutants. The efficacy of electrooxidation is contingent upon the electrode material used.

Common electrode materials are titanium, platinum, and boron-doped diamond, which increase catalytic activity, stability, and conductivity. Electrooxidation is proficient in eliminating various micropollutants, such as pharmaceuticals, personal care items, insecticides, and industrial chemicals. Examples include the elimination of ibuprofen, acetaminophen, and triclosan. Electrocoagulation uses metal electrodes, often aluminum or iron, to discharge metal ions into water when subjected to an electric current. Metal ions destabilize colloidal particles, organic materials, and pollutants, resulting in flocs that may be eliminated using physical processes like filtering or sedimentation. Electrocoagulation is economically efficient and eliminates the need for external coagulants, making it a more sustainable alternative to conventional chemical coagulation techniques. Electrocoagulation effectively treats wastewater polluted with oils, heavy metals, suspended particles, and organic micropollutants, including phenolic chemicals and synthetic colors (Tahreen et al., 2020).

The Electro-Fenton procedure is an advanced electrochemical treatment that produces ·OH radicals in situ from hydrogen peroxide and ferrous ions. This highly reactive species may decompose organic micropollutants via AOPs, resulting in complete mineralization or conversion into less hazardous substances. At the cathode, Fe^{3+} is reduced to Fe^{2+}, which then interacts with hydrogen peroxide to generate hydroxyl radicals. These radicals decompose intricate micropollutants, including pharmaceuticals, insecticides, and industrial chemicals (He & Zhou, 2017). Capacitive deionization (CDI) is an electrochemical method that employs an electric field to attract and eliminate charged pollutants from water. The technology is mainly used for desalination but has also shown potential for eliminating organic micropollutants. Upon voltage application to porous electrodes, ions in the aqueous solution are adsorbed onto the electrode surfaces, eliminating dissolved salts and charged impurities, including some micropollutants. CDI has been used to eliminate inorganic ions, including nitrates, heavy metals, and organic micropollutants such as pharmaceuticals (Bao et al., 2023).

The Electron Catalytic Membrane (ECM) is a novel hybrid technology that integrates membrane filtration with electrochemical processes to improve the elimination of micropollutants from water and wastewater (Bai et al., 2019). This advanced technology has attracted considerable interest in recent years because it promises high efficiency, sustainability, and adaptability in addressing the ongoing issue of micropollutant pollution in aquatic ecosystems. ECM presents a viable option for the selective and efficient removal of organic and inorganic pollutants, such as pharmaceuticals, personal care items, insecticides, and industrial chemicals, by integrating membrane separation and electrochemical processes. In ECM, the applied electric current induces electrochemical processes that produce reactive species, including ·OH radicals and other oxidizing agents, capable of degrading organic contaminants. In the anode of the ECM system, electrochemical oxidation processes produce ROS, including ·OH radicals, which are highly reactive and capable of degrading organic micropollutants. These radicals attack the chemical bonds of contaminants, decomposing them into more minor, less harmful byproducts. Concurrently, the membrane component of the system functions as a physical barrier, eliminating suspended solids, larger particles, and certain organic pollutants. The membrane in the extracellular matrix functions as a physical barrier that removes suspended particles, bacteria, and bigger organic molecules from the water. The system may segregate pollutants by size and molecular weight, contingent upon the membrane type used (MF, UF, NF, RO). This filtering process accumulates contaminants on the membrane surface, which are eliminated by backwashing or decomposed via electrochemical processes. The integration of membrane filtration with electrochemical degradation has a synergistic effect, improving the total removal efficiency of the system. This dual-action method facilitates the remediation of a wider array of pollutants, ensuring more extensive water purification.

Biotechnological Treatments

As the issue of micropollutant pollution in water systems escalates, conventional physical and chemical treatment processes often prove ineffective in efficiently removing various pollutants, particularly those present at trace concentrations. Consequently, biotechnological methods have garnered considerable interest due to their potential to provide sustainable, economical, and eco-friendly solutions for eliminating micropollutants. These methods use biological processes, including microbial degradation, enzymatic reactions, and biosorption, to decompose or eliminate contaminants from water (Das & Ghosh, 2023; Ghosh et al., 2023).

Microbial degradation is a biological process in which bacteria use organic materials as carbon, nitrogen, and energy sources, decomposing complex pollutants into simpler, less dangerous chemicals (Rios-Miguel et al., 2023). The methods of microbial degradation of micropollutants are several, dependent upon the microorganism type and contaminant structure, including biotransformation, co-metabolism, cometabolic biodegradation, microbial oxidative and reductive processes, and syntrophic degradation. Pharmaceutical pollutants, such as antibiotics, anti-inflammatory agents, and analgesics, often exist in trace concentrations in wastewater effluents and natural water bodies. Widely recognized bacterial species, including Pseudomonas, Rhodococcus, and Sphingomonas, are proficient in the degradation of pharmaceuticals such as acetaminophen, ibuprofen, and ciprofloxacin. Personal care items, including parabens, triclosan, and phthalates, are prevalent micropollutants in wastewater. These substances have been designed for persistence and possess antibacterial or antifungal activities, making them challenging to eliminate from water. Some fungi and bacteria have shown the capacity to breakdown PCPs via oxidative and reductive processes. For instance, Candida albicans and Aspergillus niger can decompose triclosan and parabens in wastewater systems. Microbial degradation is essential for the elimination of pesticides such atrazine, glyphosate, and organophosphates. Specific bacterial species, including Bacillus, Pseudomonas, and Sphingomonas, possess the potential to breakdown toxic toxins, transforming them into less hazardous compounds via diverse biotransformation routes. Microbial degradation is helpful in eliminating industrial pollutants, including polychlorinated biphenyls (PCBs), volatile organic compounds (VOCs), and heavy metals (Priyadarsini & Das, 2024, 2025). Microorganisms such as Rhodococcus and Mycobacterium have shown the ability to destroy or diminish the toxicity of persistent pollutants by enzymatic oxidation, reduction, and other metabolic processes.

Bioremediation involves employing living organisms, such as bacteria, fungus, and plants, to remediate polluted areas (Vassalle et al., 2020). Microbial bioremediation is an effective and progressively used method for addressing micropollutants in water and wastewater. In-situ bioremediation signifies the direct application of microorganisms or bioremediation agents at the contaminated site, eliminating the need for excavation or removal of contaminated materials. This method is often used in wastewater treatment plants or contaminated groundwater to introduce microorganisms that breakdown micropollutants in the water matrix. This procedure may include the incorporation of nutrition sources to promote microbial activity or the use of biostimulation methods to augment natural biodegradation (Das et al., 2023, 2025). Ex-situ bioremediation is the extraction of contaminated water or sediment from the affected site, thereafter undergoing treatment under controlled circumstances to facilitate microbial breakdown processes. This method is often used in situations when pollution is extensive or necessitates rapid treatment. Examples include bioreactors and biofilters, whereby microorganisms are cultured to decompose micropollutants in wastewater or polluted soil. Bioaugmentation is the introduction of specific microbial strains that degrade pollutants to a polluted site or system, hence expediting the biodegradation process (Nzila et al., 2016). This method is often used when indigenous microbial populations are inadequate or deficient in possess the necessary abilities to decompose certain micropollutants. The introduction of genetically modified or selectively chosen microbial strains may improve the breakdown of certain pollutants. Phytoremediation and mycoremediation, which involves the use of fungus, may be utilized for the elimination of micropollutants. Specific plant species, including Populus and Salix, together with fungi such as Pleurotus ostreatus, have shown the ability to breakdown or absorb pharmaceutical residues, personal care products, and other micropollutants in polluted water.

Enzymatic treatments provide a novel and promising method for the efficient elimination of micropollutants from water and wastewater (Varga et al., 2019). Enzymes, as biological catalysts, may efficiently decompose or modify many pollutants, such as pharmaceuticals, personal care products, insecticides, and industrial chemicals. Enzyme-based treatments have several advantages compared to conventional techniques, such as enhanced specificity, reduced energy use, and negligible effects on the environment (Behera et al.,). Enzymes such as laccases and lignin peroxidases have shown the ability to breakdown pharmaceuticals like acetaminophen, ibuprofen, and ciprofloxacin. These enzymes oxidize pollutant molecules, converting them into less harmful metabolites that are more readily eliminated from the water. Personal care products, including as triclosan, parabens, and phthalates, are often found in wastewater and surface waterways as a result of their extensive use. Laccases and peroxidases are especially efficient in degrading complex molecules, converting them into simpler, less harmful chemicals. Enzymatic treatments have been used for handling pesticides and herbicides that are resistant to conventional degradation procedures. Pseudomonas spp. and other bacteria produce enzymes, including aryl hydrocarbon hydroxylase and laccase, capable of degrading pesticide residues such atrazine, glyphosate, and organophosphates by destroying their chemical structures. Industrial chemicals, including PAHs, phthalates, and halogenated substances, have significant toxicity and environmental persistence. Enzymatic treatments, such as peroxidases and oxidases, have shown efficacy in degrading these compounds by cleaving the aromatic rings or chlorine groups, resulting in the generation of less deleterious byproducts.

Integration of Monitoring Technologies and Artificial Intelligence (AI)

Integrating monitoring technology and AI has revolutionized micropollutants and environmental pollution control. This integration seeks to improve water system micropollutant detection, analysis, and treatment. Monitoring technology includes sensors, real-time data-gathering systems, and analytical equipment that identify and quantify trace pollutants. The inability to manage complex data or respond to changing environmental circumstances limits these technologies (Baek et al., 2021). The data processing, pattern recognition, and decision-making skills of AI boost the monitoring systems. AI algorithms can analyze sensor readings, water quality indicators, and pollutant concentrations from monitoring technology in real-time. The integration enables water treatment process management to be more dynamic and responsive. AI can forecast pollutant levels based on historical data and present environmental circumstances, allowing proactive management and early contamination identification. AI-based machine learning algorithms can identify complicated data patterns and linkages that human analysts might ignore. These models can detect micropollutant trends, correlate with weather or industrial discharges, and predict future levels.

AI can optimize water quality monitoring and treatment systems. AI systems can modify operating settings in real-time depending on data inputs, enhancing treatment efficiency and minimizing energy use. AI can optimize micropollutant removal in membrane filtering and advanced oxidation by controlling flow rates, chemical doses, and UV light intensity. Real-time optimization improves treatment efficacy, lowers operating costs, and decreases environmental impact by optimizing resource utilization. Water quality management decisions can be addressed more efficiently and precisely using AI and monitoring technology. Conventional monitoring approaches like periodic sampling and manual analysis are slow and can not detect micropollutant concentration variations. However, AI-enabled monitoring systems deliver real-time data, making danger detection quicker. AI can also prioritize removal by measuring pollution severity and anticipating which regions or contaminants need urgent attention. The data-driven insights of AI systems enable more focused and cost-effective solutions (Shahouni et al., 2024).

An additional advantage is using AI and monitoring technology to manage huge, complex water treatment plants. Many water treatment facilities handle massive amounts of water with varying pollution levels for significant populations. The amount of data produced in such systems can overpower standard monitoring methods. However, AI can manage this complexity. AI can analyze data from many sensors and treatment units to assess the complete water treatment process, detect inefficiencies, and suggest improvements. These skills are crucial in urban and industrial locations with various and shifting water pollution sources. AI and monitoring technology can help build more innovative, robust water management infrastructure. Water utilities can prepare for micropollutant pollution by using AI to detect trends and risk factors. AI can optimize maintenance plans by predicting equipment failure or system inefficiencies. AI systems can learn from data and improve predictions, making water treatment treatments more innovative and flexible. This integration is being used in smart cities to regulate water quality, which is vital to public health and sustainability. Cities can manage water resources more sustainably and proactively using AI-enabled monitoring systems. AI can optimize water distribution, monitor pollution levels in real-time, and warn inhabitants of health dangers, enhancing community safety. AI may be used with the Internet of Things (IoT) to develop a fully automated water management system that integrates water quality monitors and treatment facilities. Despite its potential, combining AI with surveillance systems is difficult. Urban regions acquire massive volumes of sensitive environmental data, raising privacy and security issues. Maintaining public trust and water management system integrity requires cyber security for these data systems. Making AI models clear and understandable is another difficulty. Some AI system decision-making processes are incomprehensible, making it challenging for operators to understand suggestions or modifications fully. These difficulties demand ethical rules, strong data security procedures, and improvements in the explainability of the AI system (Jeong et al., 2021).

Conclusions and Future Perspectives

The prevalence of CECs makes micropollutant removal from water systems a significant and growing problem. As the world population increases and industrial and agricultural habits change, new, effective, sustainable water treatment solutions are required. AOPs, membrane filtration, nanotechnology, and electrochemical devices can remove a variety of micropollutants, owing to advances in the sector. These methods work well in lab and pilot tests, but cost, scalability, environmental effect, and long-term performance prevent widespread adoption. Ozone, UV treatment, and photocatalysis degrade organic pollutants well. However, they demand a lot of energy and are expensive. For large-scale implementation, membrane filtration methods like RO and NF may remove contaminants, but membrane fouling and energy needs must be addressed. Nanotechnology, particularly nanoparticle-based adsorption, and nanomaterial-enhanced membranes may remove trace impurities, but high material costs and environmental concerns limit its use. Electrochemical methods can also remove micropollutants, but their high energy requirements and electrode fouling restrict their use. Each treatment approach has advantages and disadvantages, but integrated or hybrid systems, such as electrochemical membranes or AOP-membrane systems, may improve removal efficiency and overcome system constraints. AI-enabled monitoring solutions can optimize treatment operations, anticipate pollution levels, and provide real-time management. After these advances, economic feasibility remains a challenge, especially in low-income regions or small-scale treatment facilities. Integrating resource recovery techniques into treatment systems can enhance sustainability and reduce costs.

Practitioners and water treatment experts should investigate hybrid and integrated systems that use complementary technologies to improve pollutant removal efficiency and reduce operational costs. Before choosing a treatment technology, practitioners should consider the local micropollutants, resources, and energy constraints. Prioritize real-time monitoring systems to dynamically optimize treatment performance, using AI and sensor technologies for better control and prediction. However, policymakers should establish regulatory frameworks encouraging advanced treatment technologies and economic sustainability. This includes subsidizing, granting, or lending low-interest loans to municipalities and industries installing advanced treatment systems. Policymakers should establish monitoring standards for emerging pollutants to guide treatment solution design and implementation. Research institutions, industries, and governments must collaborate to develop standardized treatment protocols and circulate micropollutant management best practices. Policy incentives should also assist the study and development of next-generation water treatment technologies, notably resource recovery and energy efficiency. Policies that allow the adoption of smart water infrastructure and AI-based solutions for more effective and adaptive water resource management should also prioritize data privacy and security.

Micropollutant removal technologies have progressed, but future research should address system limits and explore new methods. Cost-effective materials for improved treatment methods are a priority for future research. To allow large-scale use, durable, foul-resistant, and economically viable nanomaterials and advanced membrane materials must be developed. More research on hybrid and integrated systems incorporating several treatment methods is needed to maximize pollutant removal efficiency while minimizing costs and energy usage. Optimization of real-time monitoring and AI-based water treatment control systems is another important investigation. AI algorithms incorporating historical and real-time data might improve water treatment system dynamic control, pollution hotspot prediction, and operational efficiency. Research into AI-driven predictive maintenance would also be advantageous, as it may aid in the proactive management of treatment systems and decrease downtime or inefficiencies caused by equipment failure or blockage.

The environmental effect of treatment methods must be extensively evaluated in future studies. As improved treatment methods, particularly chemical-based procedures, can generate secondary pollutants, it is essential to concentrate on discovering strategies to eliminate or manage these consequences. For instance, research into more sustainable chemical treatments or zero-waste systems should be emphasized. Moreover, developing circular economy techniques that combine resource recovery (e.g., the reuse of recovered micropollutants or energy production) will be vital for the long-term sustainability of water treatment systems. Lastly, the scalability of novel treatments remains a big challenge. Future studies should concentrate on pilot-scale and field-based research that validate the practical validity of laboratory-scale results. The long-term performance, energy consumption, and operating costs of these systems may be determined via collaborative testing and scaling up in real-world systems.

References

Ahmed, M. B., Zhou, J. L., Ngo, H. H., Guo, W., Thomaidis, N. S., & Xu, J. (2017). Progress in the biological and chemical treatment technologies for emerging contaminant removal from wastewater: A critical review. *Journal of Hazardous Materials, 323*, 274–298. https://doi.org/10.1016/j.jhazmat.2016.04.045

Ahmed, S. F., Islam, N., Tasannum, N., Mehjabin, A., Momtahin, A., Chowdhury, A. A., Almomani, F., & Mofijur, M. (2024). Microplastic removal and management strategies for wastewater treatment plants. *Chemosphere, 347*, 140648. https://doi.org/10.1016/j.chemosphere.2023.140648

Aslam, M. M., Kuo, H., Den, W., Usman, M., Sultan, M., & Ashraf, H. (2021). Functionalized carbon nanotubes (CNTs) for water and wastewater treatment: Preparation to application. *Sustainability, 13*(10), 5717. https://doi.org/10.3390/su13105717

Baek, S.-S., Choi, Y., Jeon, J., Pyo, J., Park, J., & Cho, K. H. (2021). Replacing the internal standard to estimate micropollutants using deep and machine learning. *Water Research, 188*, 116535. https://doi.org/10.1016/j.watres.2020.116535

Bai, Y., Zhou, L., Ma, K., Zhou, H., & Xin, Y. (2019). Study on catalytic performances and reaction mechanisms of graphene electroactive membrane in wastewater treatment. *Separation and Purification Technology, 226*, 278–285. https://doi.org/10.1016/j.seppur.2019.05.069

Bao, S., Xin, C., Zhang, Y., Chen, B., Ding, W., & Luo, Y. (2023). Application of capacitive deionization in water treatment and energy recovery: A review. *Energies, 16*(3), 1136. https://doi.org/10.3390/en16031136

Behera, I. D., Basak, G., Kumar, R. R., Sen, R., & Meikap, B. C. (2020). Treatment of petroleum refinery sludge by petroleum degrading bacterium *Stenotrophomonas pavanii* IRB19 as an efficient novel technology. *Journal of Environmental Science and Health, Part A, 56*(2), 226–239. https://doi.org/10.1080/10934529.2020.1866924

Behera, I. D., Nayak, M., Biswas, S., Meikap, B. C., Sen, R. (2021). Enhanced biodegradation of total petroleum hydrocarbons by implementing a novel two-step bioaugmentation strategy using indigenous bacterial consortium. *Journal of Environmental Management, 292*, 112746. ISSN 0301-4797, https://doi.org/10.1016/j.jenvman.2021.112746

Carbajo, J., Bahamonde, A., & Faraldos, M. (2017). Photocatalyst performance in wastewater treatment applications: Towards the role of TiO2 properties. *Molecular Catalysis, 434*, 167–174. https://doi.org/10.1016/j.mcat.2017.03.018

Chen, G. (2004). Electrochemical technologies in wastewater treatment. *Separation and Purification Technology, 38*(1), 11–41. https://doi.org/10.1016/j.seppur.2003.10.006

Choi, Y.-J., Kim, L.-H., & Zoh, K.-D. (2016). Removal characteristics and mechanism of antibiotics using constructed wetlands. *Ecological Engineering, 91*, 85–92. https://doi.org/10.1016/j.ecoleng.2016.01.058

Corominas, L., Acuña, V., Ginebreda, A., & Poch, M. (2013). Integration of freshwater environmental policies and wastewater treatment plant management. *Science of the Total Environment, 445–446*, 185–191. https://doi.org/10.1016/j.scitotenv.2012.12.055

Das, P., & Bal, M. (2025). Modeling and optimization study for microplastic removal from aquatic medium using Chroococcidiopsis species. *Chemical Engineering Science, 304*, 121085. https://doi.org/10.1016/j.ces.2024.121085

Das, P., Halder, G., & Bal, M. (2023). A critical review on remediation of microplastics using microalgae from aqueous system. *Science of the Total Environment, 898*, 166425. https://doi.org/10.1016/j.scitotenv.2023.166425

Das, A., Dey, S., Bala, K., Gnanasekaran, R., & Das, A. P. (2024a). Microbial bioremediation technology for sustainable treatment and management of synthetic microfiber waste. *Sustainable Microbial Technology for Synthetic and Cellulosic Microfiber Bioremediation, 221–238*. https://doi.org/10.1007/978-3-031-62600-5_12

Das, A., Dey, S., & Das, A. P. (2024b). Ecotoxicological impacts of synthetic microfiber pollutants and development of sustainable mitigation strategies. *Environmental Chemistry and Ecotoxicology*. https://doi.org/10.1016/J.ENCECO.2024.12.003

Das, A. P., & Ghosh, S. (2023). Microorganisms in manganese biomining processes. *Manganese Mining Microorganisms, 55–61*. https://doi.org/10.1016/B978-0-12-822147-1.00013-3

Dasenaki, M. E., & Thomaidis, N. S. (2015). Multianalyte method for the determination of pharmaceuticals in wastewater samples using solid-phase extraction and liquid chromatography–tandem mass spectrometry. *Analytical and Bioanalytical Chemistry, 407*(15), 4229–4245. https://doi.org/10.1007/s00216-015-8654-x

Dey, S., Bissoyi, A., Koteswara Reddy, G., & Das, A. P. (2024a). Synthetic microfibers: An urban emerging pollutant demanding for sustainable environmental solutions. *Sustainable Microbial Technology for Synthetic and Cellulosic Microfiber Bioremediation, 107–121*. https://doi.org/10.1007/978-3-031-62600-5_7

Dey, S., Das, A., Mallick, K., Sahu, A., & Das, A. P. (2023a). Environmental petroleum waste: Pollution. *Toxicity, Sustainable Remediation, 159–175*. https://doi.org/10.1007/978-3-031-48220-5_7

Dey, S., Dash, A., & Das, A. P. (2024b). Bioremediation as a sustainable and circular solution for mining waste management and bioenergy generation. *Sustainable Management of Mining Waste and Tailings: A Circular Economy Approach, 107–119*. https://doi.org/10.1201/9781003442455-6

Dey, S., Tripathy, B., Kumar, M. S., & Das, A. P. (2023b). Ecotoxicological consequences of manganese mining pollutants and their biological remediation. *Environmental Chemistry and Ecotoxicology*. https://doi.org/10.1016/J.ENCECO.2023.01.001

Ghosh, S., Sinha, J. K., Ghosh, S., Vashisth, K., Han, S., & Bhaskar, R. (2023). Microplastics as an emerging threat to the global environment and human health. *Sustainability, 15*(14), 10821. https://doi.org/10.3390/SU151410821

He, H., & Zhou, Z. (2017). Electro-fenton process for water and wastewater treatment. *Critical Reviews in Environmental Science and Technology, 47*(21), 2100–2131. https://doi.org/10.1080/10643389.2017.1405673

Jeong, N., Chung, T., & Tong, T. (2021). Predicting micropollutant removal by reverse osmosis and nanofiltration membranes: Is machine learning viable? *Environmental Science & Technology, 55*(16), 11348–11359. https://doi.org/10.1021/acs.est.1c04041

Jiang, J.-Q., Zhou, Z., & Sharma, V. K. (2013). Occurrence, transportation, monitoring and treatment of emerging micro-pollutants in waste water—A review from global views. *Microchemical Journal, 110*, 292–300. https://doi.org/10.1016/j.microc.2013.04.014

Kumar, A., & Gangawane, K. M. (2022a). Effect of precipitating agents on the magnetic and structural properties of the synthesized ferrimagnetic nanoparticles by co-precipitation method. *Powder Technology, 401*, Article 117298. https://doi.org/10.1016/j.powtec.2022.117298

Kumar, A., & Gangawane, K. M. (2022b). Synthesis and effect on the surface morphology & magnetic properties of ferrimagnetic nanoparticles by different wet chemical synthesis methods. *Powder Technology, 410*, 117867. https://doi.org/10.1016/j.powtec.2022.117867

Kumar, A., & Gangawane, K. M. (2023a). Advanced computational approaches for water treatment. In *Advanced computational approaches for water treatment* (pp. 1–12). CRC Press. https://doi.org/10.1201/9781003325147-1

Kumar, A., & Gangawane, K. M. (2023b). Molecular dynamics for the membrane process. In *Advanced computational approaches for water treatment* (pp. 13–38). CRC Press. https://doi.org/10.1201/9781003325147-2

Kumar, A., Gangawane, K. M., & Ramanjaneyulu, B. (2022). Ferrofluids for waste-water treatment. In *Advances in chemical, bio and environmental engineering* (pp. 723–744). Springer. https://doi.org/10.1007/978-3-030-96554-9_48

Lahiri, D., Nag, M., Dey, A., Sarkar, T., Joshi, S., Pandit, S., Das, A. P., Pati, S., Pattanaik, S., Tilak, V. K., & Ray, R. R. (2021). Biofilm mediated degradation of petroleum products. *Geomicrobiology Journal, 1–10*. https://doi.org/10.1080/01490451.2021.1968979

Lapointe, M., Farner, J. M., Hernandez, L. M., & Tufenkji, N. (2020). Understanding and improving microplastic removal during water treatment: Impact of coagulation and flocculation. *Environmental Science & Technology, 54*(14), 8719–8727. https://doi.org/10.1021/acs.est.0c00712

Madsen, H. T. (2014). Membrane filtration in water treatment—Removal of micropollutants. In *Chemistry of advanced environmental purification processes of water* (pp. 199–248). Elsevier. https://doi.org/10.1016/B978-0-444-53178-0.00006-7

Mahmoud, W. M. M., Rastogi, T., & Kümmerer, K. (2017). Application of titanium dioxide nanoparticles as a photocatalyst for the removal of micropollutants such as pharmaceuticals from water. *Current Opinion in Green and Sustainable Chemistry, 6*, 1–10. https://doi.org/10.1016/j.cogsc.2017.04.001

Mazhar, M. A., Khan, N. A., Ahmed, S., Khan, A. H., Hussain, A., Rahisuddin, Changani, F., Yousefi, M., Ahmadi, S., & Vambol, V. (2020). Chlorination disinfection by-products in municipal drinking water—A review. *Journal of Cleaner Production, 273*, 123159. https://doi.org/10.1016/j.jclepro.2020.123159

Miranda, M. N., Silva, A. M. T., & Pereira, M. F. R. (2020). Microplastics in the environment: A DPSIR analysis with focus on the responses. *Science of the Total Environment, 718*, 134968. https://doi.org/10.1016/j.scitotenv.2019.134968

Mishra, S., & Das, A. P. (2023). Microbial remediation of synthetic microfiber contaminated wastewater. In M. P. Shah (Ed.), *Microbial technologies in industrial wastewater treatment* (pp. 337–350). Springer, Singapore. https://doi.org/10.1007/978-981-99-2435-6_17

Mishra, S., Dash, D., & Das, A. P. (2023). Aquatic microbial diversity on plastisphere: Colonization and potential role in microplastic biodegradation. *Geomicrobiology Journal, 1–12*. https://doi.org/10.1080/01490451.2023.2209750

Mishra, S., Dash, D., & Das, A. P. (2024). Isolation and molecular characterization of polyester-degrading bacterial strains from municipal wastewater and their application in microplastic degradation in contaminated wastewater. *Geomicrobiology Journal*. https://doi.org/10.1080/01490451.2024.2376101

Mishra, S., Rath, C. charan, & Das, A. P. (2019). Marine microfiber pollution: A review on present status and future challenges. *Marine Pollution Bulletin, 140*, 188–197. https://doi.org/10.1016/J.MARPOLBUL.2019.01.039

Mohanty, J., Dey, S., Bala, K., & Das, A. P. (2024a). Recent advances in microplastic pollution for its sustainable management. *Sustainable Management of Urban Plastic Waste through Circular Economic Approaches, 27–48*. https://doi.org/10.1201/9781003465447-3

Mohanty, M., Tripathy, B., Koteswara Reddy, G., & Das, A. P. (2024b). Greener technologies for establishing restraint over microplastic pollution. *Sustainable Microbial Technology for Synthetic and Cellulosic Microfiber Bioremediation, 239–258*. https://doi.org/10.1007/978-3-031-62600-5_13

Nzila, A., Razzak, S., & Zhu, J. (2016). Bioaugmentation: An emerging strategy of industrial wastewater treatment for reuse and discharge. *International Journal of Environmental Research and Public Health, 13*(9), 846. https://doi.org/10.3390/ijerph13090846

Osman, A. I., Nasr, M., Farghali, M., Bakr, S. S., Eltaweil, A. S., Rashwan, A. K., & Abd El-Monaem, E. M. (2024). Machine learning for membrane design in energy production, gas separation, and water treatment: A review. *Environmental Chemistry Letters, 22*(2), 505–560. https://doi.org/10.1007/s10311-023-01695-y

Pal, D., & Sen, S. (2023a). Emerging petroleum pollutants and their adverse effects on the environment. In *Impact of petroleum waste on environmental pollution and its sustainable management through circular economy* (pp. 103–137). Springer. https://doi.org/10.1007/978-3-031-48220-5_5

Pal, D., & Sen, S. (2023b). In-depth coverage of petroleum waste sources, characteristics, environmental impact, and sustainable remediation process. In *Impact of petroleum waste on environmental pollution and its sustainable management through circular economy* (pp. 1–38). Springer. https://doi.org/10.1007/978-3-031-48220-5_1

Pal, D., & Sen, S. (2024a). Optimal synthesis of dolochar derived faujasite zeolite X for highly effective Cd(II) removal. *Environmental Research, 240*, 117494. https://doi.org/10.1016/j.envres.2023.117494

Pal, D., & Sen, S. (2024b). Valorization of sponge iron industrial waste into iron-modified zeolite X for ciprofloxacin removal: A multi-parameter optimization study. *Environmental Science and Pollution Research, 31*(40), 53381–53398. https://doi.org/10.1007/s11356-024-34730-8

Pal, D., & Sen, S. (2025). Toxicity and health impacts of emerging pollutants. In *Biotechnological interventions in removal of emerging pollutants*. Springer. https://doi.org/10.1007/978-981-97-9922-0_4

Paul, A., Dey, S., Ram, D. K., & Das, A. P. (2023). Hexavalent chromium pollution and its sustainable management through bioremediation. *Geomicrobiology Journal, 1–11*. https://doi.org/10.1080/01490451.2023.2218377

Pradhan, G., Tripathy, B., Ram, D. K., Digal, A. K., & Das, A. P. (2023). Bauxite mining waste pollution and its sustainable management through bioremediation. *Geomicrobiology Journal, 1–10*. https://doi.org/10.1080/01490451.2023.2235353

Prieto-Rodríguez, L., Oller, I., Klamerth, N., Agüera, A., Rodríguez, E. M., & Malato, S. (2013). Application of solar AOPs and ozonation for elimination of micropollutants in municipal wastewater treatment plant effluents. *Water Research, 47*(4), 1521–1528. https://doi.org/10.1016/j.watres.2012.11.002

Priyadarsini, S., & Das, A. P. (2024). Analytical and structural characterization of waste lithium-ion batteries for their effective recycling strategy. *Environmental Chemistry and Ecotoxicology*. https://doi.org/10.1016/j.enceco.2024.12.004

Priyadarsini, S., & Das, A. P. (2025). Lithium bioleaching: A review on microbial-assisted sustainable technology for lithium bio-circularity. *Journal of Water Process Engineering, 69*, 106744. https://doi.org/10.1016/j.jwpe.2024.106744

Pruden, A., Pei, R., Storteboom, H., & Carlson, K. H. (2006). Antibiotic resistance genes as emerging contaminants: studies in Northern Colorado. *Environmental Science & Technology, 40*(23), 7445–7450. https://doi.org/10.1021/es0604131

Ray, S. K., Dey, S., Das, A. P., & John, A. (2024). Microplastic waste management through greener technologies: A review. *Sustainable Management of Urban Plastic Waste through Circular Economic Approaches, 237–261*. https://doi.org/10.1201/9781003465447-14

Rayaroth, M. P., Aravindakumar, C. T., Shah, N. S., & Boczkaj, G. (2022). Advanced oxidation processes (AOPs) based wastewater treatment—Unexpected nitration side reactions—A serious environmental issue: A review. *Chemical Engineering Journal, 430*(P4), 133002. https://doi.org/10.1016/j.cej.2021.133002

Rios-Miguel, A. B., van Bergen, T. J. H. M., Zillien, C., Ragas, A. M. J., van Zelm, R., Jetten, M. S. M., Hendriks, A. J., & Welte, C. U. (2023). Predicting and improving the microbial removal of organic micropollutants during wastewater treatment: A review. *Chemosphere, 333*, 138908. https://doi.org/10.1016/j.chemosphere.2023.138908

Roy, D., Khosravanipour Mostafazadeh, A., Drogui, P., & Tyagi, R. D. (2020). Removal of organic micro-pollutants by membrane filtration. In *Current developments in biotechnology and bioengineering* (pp. 281–307). Elsevier. https://doi.org/10.1016/B978-0-12-819594-9.00012-7

Russo, V., Hmoudah, M., Broccoli, F., Iesce, M. R., Jung, O.-S., & Di Serio, M. (2020). Applications of metal organic frameworks in wastewater treatment: A review on adsorption and photodegradation. *Frontiers in Chemical Engineering, 2*, 1–13. https://doi.org/10.3389/fceng.2020.581487

Sahu, A., Dey, S., Mallick, K., Priyadarsini, S., Behera, I. D., & Das, A. P. (2024a). A review on advanced technology for sustainable management of synthetic microplastic waste. *Sustainable Management of Urban Plastic Waste through Circular Economic Approaches, 262–279*. https://doi.org/10.1201/9781003465447-15

Sahu, A., Dey, S., Mallick, K., Priyadarsini, S., Behera, I. D., & Das, A. P. (2024b). A review on advanced technology for sustainable management of synthetic microplastic waste. *Sustainable Management of Urban Plastic Waste through Circular Economic Approaches, 262–279*. https://doi.org/10.1201/9781003465447-15

Shahouni, R., Abbasi, M., Dibaj, M., & Akrami, M. (2024). Utilising artificial intelligence to predict membrane behaviour in water purification and desalination. *Water, 16*(20), 2940. https://doi.org/10.3390/w16202940

Shanmuganathan, S., Loganathan, P., Kazner, C., Johir, M. A. H., & Vigneswaran, S. (2017). Submerged membrane filtration adsorption hybrid system for the removal of organic micropollutants from a water reclamation plant reverse osmosis concentrate. *Desalination, 401*, 134–141. https://doi.org/10.1016/j.desal.2016.07.048

Song, W., Huang, M., Rumbeiha, W., & Li, H. (2007). Determination of amprolium, carbadox, monensin, and tylosin in surface water by liquid chromatography/tandem mass spectrometry. *Rapid Communications in Mass Spectrometry, 21*(12), 1944–1950. https://doi.org/10.1002/rcm.3042

Tahreen, A., Jami, M. S., & Ali, F. (2020). Role of electrocoagulation in wastewater treatment: A developmental review. *Journal of Water Process Engineering, 37*, Article 101440. https://doi.org/10.1016/j.jwpe.2020.101440

Tripathy, B., Sahoo, P. P., Sundaray, H., & Das, A. P. (2024). In-vitro biodegradation of discarded marine microplastics across the eastern coast of the Bay of Bengal, India using Exiguobacterium sp. *Environmental Chemistry and Ecotoxicology, 6*, 236–247. https://doi.org/10.1016/J.ENCECO.2024.07.003

Varga, B., Somogyi, V., Meiczinger, M., Kováts, N., & Domokos, E. (2019). Enzymatic treatment and subsequent toxicity of organic micropollutants using oxidoreductases—A review. *Journal of Cleaner Production, 221*, 306–322. https://doi.org/10.1016/j.jclepro.2019.02.135

Vassalle, L., Sunyer-Caldú, A., Uggetti, E., Díez-Montero, R., Díaz-Cruz, M. S., García, J., & García-Galán, M. J. (2020). Bioremediation of emerging micropollutants in irrigation water. The alternative of microalgae-based treatments. *Journal of Environmental Management, 274*, 111081. https://doi.org/10.1016/j.jenvman.2020.111081

Yang, W., Zhou, H., & Cicek, N. (2014). Treatment of organic micropollutants in water and wastewater by UV-based processes: A literature review. *Critical Reviews in Environmental Science and Technology, 44*(13), 1443–1476. https://doi.org/10.1080/10643389.2013.790745

Ziembowicz, S., & Kida, M. (2022). Limitations and future directions of application of the Fenton-like process in micropollutants degradation in water and wastewater treatment: A critical review. *Chemosphere, 296*, 134041. https://doi.org/10.1016/j.chemosphere.2022.134041

Chapter 10
"Micropollutants: Pathways, Toxicity, and Strategies for Risk Mitigation"

Mayur Bhadarka, D. T. Vaghela, Hardik Sikotariya, Neha Kharadi, and Pinak Bamaniya

Introduction

Micropollutants represent a significant challenge in water pollution, generating substantial concerns regarding their toxicity, especially when they exist within intricate mixtures. A wide range of organic chemicals are classified as "micropollutants" because they are present in very low concentrations. Prolonged exposure to these micropollutants can pose significant risks to both aquatic life and human health. The ongoing release of micropollutants, particularly in the absence of regulatory controls, can lead to environmental issues even when they are present at low concentrations (Arslan et al., 2017). The interactions among various components of environmental pollution complicate the issue of micropollutants, making it increasingly challenging to address. As a result, evaluating water quality by examining organic chemical contaminants individually may result in an underestimation of the risks they pose to aquatic environments (Shao et al., 2019). Pharmaceuticals and personal care products (PPCPs) are widely found in aquatic ecosystems and have the potential to disrupt endocrine functions in wildlife, raising significant environmental concerns (Jagadeesh et al., 2023). Similarly, microplastics are increasingly recognized for their long-lasting ecological consequences, particularly within aquatic environments (Jagadeesh et al., 2023). Pesticides, including neonicotinoids and organophosphorus

M. Bhadarka (✉) · D. T. Vaghela · P. Bamaniya
Department of Aquatic Environment Management, College of Fisheries Science, Kamdhenu University, Veraval, Gujarat, India
e-mail: mayurbhadarka.0000@gmail.com

H. Sikotariya
Department of Fisheries Resource Management, College of Fisheries Science, Kamdhenu University, Veraval, Gujarat, India

N. Kharadi
Department of Fish Processing Technology, College of Fisheries Science, Kamdhenu University, Veraval, Gujarat, India

I. D. Behera et al. (eds.), *Emerging Micropollutants*, Sustainable Environmental Waste Management Strategies, https://doi.org/10.1007/978-3-032-03530-1_10

esters, pose substantial risks to aquatic organisms, highlighting the urgent need for effective management strategies (Chen et al., 2023).

The ecological impacts of micropollutants are profound they can bioaccumulate in food webs, leading to toxic effects on organisms at higher trophic levels (Gupta, 2024). Additionally, research indicates that certain micropollutants present both chronic and acute risks to various aquatic species, with land use practices significantly influencing these threats (Chen et al., 2023). In evaluating toxicity, the variability of test organisms plays a crucial role; studies have shown that increasing the number of embryos used in toxicity assessments can enhance the accuracy of results (Yang et al., 2024). Furthermore, bioremediation approaches, which involve the use of microorganisms to degrade micropollutants, offer a sustainable and effective method for mitigating their environmental impact (Gupta, 2024). Overall, understanding the multifaceted effects of micropollutants and developing innovative strategies for their assessment and management are essential for protecting aquatic ecosystems and maintaining ecological balance (Das et al., 2024b; Mohanty et al., 2024; Sahu et al., 2024). Although addressing micropollutants is crucial, it is equally important to take into account the broader context of environmental stressors that may amplify their impacts (Das et al., 2023, 2024a, 2024c; Day et al., 2023b). Factors such as climate change and urbanization can complicate toxicity assessments and increase ecological risks. These stressors can alter environmental conditions, leading to changes in the behavior and fate of micropollutants, as well as affecting the resilience of ecosystems (Chen et al., 2023).

Types of Micropollutants

Organic Micropollutants

Organic micropollutants (OMPs) represent a broad category of organic substances, such as pharmaceuticals, pesticides, and industrial chemicals, that pose notable risks to the environment and human health, even at minimal concentrations. Their persistence in the environment and tendency to bioaccumulate can lead to toxic effects on ecosystems and living organisms. A thorough understanding of their origins, behavior, and strategies for remediation is essential for effective environmental management.

Organic Micropollutants Pathways and Environmental Occurrence

A. **Sources**

Agricultural Activities: The use of pesticides and fertilizers is a major contributor to the presence of OMPs in soil and water systems (Kumar et al., 2024). These chemicals often persist in the environment, altering microbial communities and impacting crop productivity over time.

Industrial Effluents: Manufacturing processes release pollutants like polychlorinated biphenyls (PCBs) and polycyclic aromatic hydrocarbons (PAHs), which are common in aquatic environments (Gupta, 2024). Improper treatment of industrial wastewater exacerbates the accumulation of these toxic substances in ecosystems.

Domestic Sources: Pharmaceuticals and personal care products discharged into wastewater systems are significant sources of widespread contamination (Muhammad et al., 2024). The improper disposal of unused medications further amplifies their environmental footprint.

B. **Concentration Levels**

Organic micropollutants (OMPs) are frequently found in trace levels, generally ranging from micrograms to picograms per liter or kilogram in various environments, such as groundwater, surface water, and soil (Kaur et al., 2024). Despite these low concentrations, they can have substantial ecological and health repercussions, as even small amounts can disrupt biological functions and contribute to environmental harm.

C. **Bioaccumulation**

Numerous OMPs, including pharmaceuticals and microplastics, have the ability to bioaccumulate in food chains, posing risks to both aquatic and terrestrial life (Gupta, 2024; Włodarczyk-Makuła, 2024). This bioaccumulation not only endangers individual species but can also affect entire ecosystems and food security by disrupting predator–prey relationships. Additionally, the movement of these pollutants through the food chain raises concerns for human health, especially through the ingestion of contaminated food.

Environmental Impact

Ecosystem Disruption: Prolonged exposure to OMPs can impair biogeochemical processes and reduce soil quality, leading to altered nutrient cycles and loss of biodiversity (Kumar et al., 2024; Kalita et al., 2024). This degradation can hinder the resilience of ecosystems and their ability to provide essential services.

Bioaccumulation Risks: These pollutants can build up within food chains, threatening various species and potentially causing human health issues, such as endocrine system disruptions and developmental disorders (Gupta, 2024; Kumar et al., 2024). Long-term accumulation in apex predators also poses significant ecological and conservation challenges.

Inorganic Micropollutants

Inorganic micropollutants present significant environmental challenges, primarily arising from industrial processes, agricultural activities, and urbanization. Although often found in trace concentrations, these pollutants can persist and accumulate in ecosystems, causing adverse effects on human health and biodiversity. Their detection and remediation are complicated by their chemical diversity and resilience, requiring innovative technologies and targeted strategies.

Inorganic Micropollutants Pathways and Environmental Occurrence

A. **Sources**

Industrial Effluents: Manufacturing processes release various inorganic chemicals, such as heavy metals and other toxins, into water bodies, significantly contributing to pollution (Joshi & Sarma, 2021). These contaminants often interact with other environmental elements, amplifying their harmful effects.

Agricultural Runoff: Fertilizers and pesticides used in farming introduce inorganic compounds, including nitrates and phosphates, into soils and waterways, leading to nutrient imbalances and eutrophication (Abbasi et al., 2022). The leaching process can also affect groundwater quality over time.

Urban Waste: Everyday household products, along with improperly disposed pharmaceuticals, release inorganic micropollutants into the environment during use and disposal (Qamar et al., 2022). Urban stormwater systems exacerbate the spread of these contaminants into aquatic ecosystems.

B. **Transport Mechanisms**

Rivers serve as channels, carrying inorganic MPs from urban areas to marine environments, with research indicating elevated concentrations in estuaries and coastal regions (Ondruch et al., 2023). This transport underscores the importance of monitoring river systems to understand the spread of pollutants.

C. **Concentration Variability**

The levels of inorganic MPs fluctuate based on land use and economic development, suggesting that urbanized areas are major contributors to pollution loads (Kardos et al., 2024). This variability emphasizes the need for targeted pollution management strategies in urban settings, where regulatory measures can significantly mitigate the release of harmful substances into the environment (Dey et al., 2023a; Ghosh et al., 2023; Jagatee et al., 2024; Paul et al., 2024). Additionally, understanding these pathways is crucial for developing effective remediation techniques and protecting aquatic ecosystems from contamination.

Environmental and Health Impacts

Micropollutants have the potential to severely disrupt aquatic ecosystems, causing biodiversity declines and changes in food web dynamics (Madhesh & Mohamed, 2024). These disruptions can weaken ecosystem resilience and reduce the availability of essential services such as water purification and nutrient cycling. Certain inorganic substances, even at concentrations too low for standard detection methods, pose serious threats to human health and aquatic organisms, with their impacts potentially being underestimated by over 300% (Servien et al., 2023). Long-term exposure to these substances can lead to chronic health conditions, such as neurological and developmental disorders, and increase the vulnerability of aquatic species to environmental stressors.

Microplastics and Nanoplastics

Microplastics (MPs) and nanoplastics (NPs) have emerged as significant environmental pollutants, posing serious risks to ecosystems and human health. Their persistence in the environment, due to inadequate waste management and slow degradation processes, contributes to their widespread distribution in aquatic, terrestrial, and atmospheric environments. The buildup of these particles can interfere with food chains, negatively impact wildlife, and potentially infiltrate the human food supply, leading to concerns about long-term health implications.

Environmental Pathways and Environmental Occurrence

A. **Sources**

Agricultural Activities: Microplastics (MPs) and nanoplastics (NPs) can infiltrate soil through the use of sewage sludge and agricultural runoff, which may carry plastic remnants from fertilizers and pesticides (Nath et al., 2024). This contamination can affect soil health and crop productivity, leading to further environmental degradation.

Industrial Processes: The manufacturing and processing of plastics are significant contributors to environmental pollution, with plastic-based printing on various products often being an overlooked source of contamination (Pantos et al., 2024). This hidden source emphasizes the need for better regulation and oversight in industrial practices.

Waste Management: Poor recycling and waste disposal methods result in the release of MPs and NPs into the environment, particularly affecting aquatic ecosystems (Arif et al., 2024). The accumulation of these pollutants can disrupt marine life and affect food safety through the contamination of seafood.

Atmospheric Deposition: Airborne microplastic particles can settle on soil and water bodies, introducing contamination from urban and industrial areas (Nath et al., 2024). This atmospheric transport mechanism underscores the pervasive nature of plastic pollution, highlighting the urgent need for comprehensive solutions to reduce plastic use and improve waste management practices.

B. **Transport Mechanisms**

Microplastics and nanoplastics (MNPs) are carried through air, water, and soil, entering aquatic environments through runoff and wastewater discharge (Vaijayanthimala & Sharma, 2023; Khanna et al., 2024). This transport allows MNPs to spread widely, affecting a variety of ecosystems and increasing the difficulty of managing their impacts.

C. **Bioaccumulation**

These particles can accumulate in sediments and within organisms, potentially leading to biomagnification through food webs (Khanna et al., 2024). As larger organisms consume smaller ones, the concentration of MNPs can increase, posing greater risks to top predators, including humans.

D. **Interaction with Other Pollutants**

MNPs can serve as carriers for toxic metals and other contaminants, increasing their bioavailability and toxicity in the environment (Vaijayanthimala & Sharma, 2023). This interaction may amplify the harmful effects of existing pollutants, leading to compounded ecological and health risks. The ability of MNPs to transport these toxins also complicates remediation efforts and necessitates comprehensive strategies to address plastic pollution and its associated hazards.

Environmental Impact

Aquatic Ecosystems: Microplastics (MPs) and nanoplastics (NPs) accumulate in aquatic environments, negatively affecting fish by weakening immune defenses and digestive functions. They also disrupt reproductive systems and cause transgenerational harm, potentially leading to long-term imbalances in aquatic biodiversity (Bhat et al., 2024). Furthermore, their presence alters habitat quality, impacting species interactions and food web stability.

Soil and Plants: MPs hinder plant growth and photosynthesis by interfering with genetic processes, ultimately reducing agricultural yields (Singh et al., 2024). These pollutants also degrade soil structure, disrupt microbial communities, and diminish soil fertility, posing risks to sustainable farming practices.

Climate Change: Plastics are fossil fuel-derived products. The emissions of greenhouse gases (GHG) during the various stages of the lifecycle of plastic-related products, including extraction, production, transportation, and disposal, pose a significant threat to the environment by contributing to global temperature rise. Improper disposal methods, such as incineration and landfill leakage, further amplify GHG emissions. Addressing these emissions is critical for mitigating climate change and promoting sustainable practices (Sharma et al., 2023).

Exposure Pathways: Humans encounter MPs and NPs through ingestion, inhalation, and skin contact. Long-term exposure has been associated with chronic illnesses, including autoimmune disorders and cancers (Khanna et al., 2024). Emerging studies also suggest potential risks to fetal development and endocrine system functioning.

Toxicity Mechanisms

The mechanisms of toxicity associated with micropollutants can be examined through the concepts of acute and chronic toxicity, each of which has unique implications for aquatic ecosystems. Acute toxicity refers to the immediate harmful effects that arise from short-term exposure, while chronic toxicity pertains to the consequences of prolonged exposure, which can lead to adverse effects over time (Das & Singh, 2011; Singh et al., 2020). Gaining an understanding of these mechanisms is critical for effective risk assessment and management, as this knowledge guides regulatory frameworks and conservation strategies aimed at safeguarding vulnerable aquatic habitats.

Acute Toxicity

Acute toxicity is commonly quantified using the LC50 value, which indicates the concentration that is lethal to 50% of test organisms in a short duration. Studies have found that 86% of acute toxicity data have a strong correlation with chronic toxicity levels, suggesting that acute testing can provide valuable insights into potential chronic effects (Kenaga, 1982). For instance, research has demonstrated that neonicotinoid pesticides present significant acute risks to aquatic life in urban environments (Chen et al., 2023). This relationship highlights the critical role of acute toxicity assessments in preliminary evaluations of potential environmental risks. Moreover, understanding acute toxicity can facilitate prompt responses to pollution events, enabling timely actions to safeguard aquatic organisms.

Chronic Toxocity

Chronic toxicity assessments consider the effects of extended exposure and its cumulative impacts, which often lead to sub-lethal consequences such as reproductive and developmental challenges. Chronic chemical risks (CCRs) are affected by land use practices and environmental conditions, illustrating the complexities of scenarios involving long-term exposure (Chen et al., 2023). The endocrine-disrupting impacts of certain pollutants are particularly concerning, as they can result in enduring health issues for aquatic species (Kuckelkorn et al., 2018). These chronic impacts can threaten population stability and biodiversity over time, undermining ecosystem resilience. Addressing chronic toxicity is vital not only for the health of aquatic species but also for the overall sustainability of ecosystem services essential for human well-being, including clean water and biodiversity preservation. While acute toxicity assessments provide quick insights into immediate dangers, chronic toxicity evaluations are fundamental for comprehending long-term ecological effects, making both perspectives critical for developing comprehensive environmental protection strategies.

Modes of Toxic Action

Micropollutants employ diverse toxicity mechanisms that pose substantial risks to environmental and human health. These mechanisms are influenced by the chemical properties of the pollutants and their interactions with biological systems. A thorough understanding of these toxicological pathways is essential for designing effective mitigation strategies and reducing their long-term impacts on ecosystems and human well-being.

Chemical Interactions

Covalent Binding: Certain micropollutants, such as heavy metals, form covalent bonds with tissue components, resulting in irreversible cellular damage (Maynert & Hammond, 1976; Hathway, 2000). This binding can impair protein function and structural integrity, leading to severe physiological disruptions.

Enzyme Inhibition: Compounds like organophosphates disrupt physiological processes by inhibiting vital enzymes such as acetylcholinesterase (Hathway, 2000; Barnes & Paget, 1965). Such inhibition can cause neurological symptoms and, in severe cases, organ failure.

DNA Intercalation: Some pollutants intercalate within nucleic acids, inducing mutations that can increase the risk of cancer (Barnes & Paget, 1965; Maynert & Hammond, 1976). Prolonged exposure to these substances may also contribute to genetic instability in populations over time.

Mechanisms of Cellular and Molecular Toxicity

Micropollutants induce a variety of cellular and molecular toxic effects that pose significant threats to both environmental and human health. These effects include mitochondrial dysfunction, oxidative stress, and interference with endocrine signaling, all of which can result in serious physiological and ecological consequences. Gaining a deeper understanding of these mechanisms is essential for evaluating the risks posed by micropollutants and for formulating targeted mitigation strategies to protect ecosystems and public health.

Mitochondrial Toxicity

Micropollutants can disrupt mitochondrial function by inhibiting key processes such as the electron transport chain and ATP synthesis, ultimately reducing cellular energy production (Müller et al., 2019). Studies utilizing oxygen consumption rate (OCR) assays have demonstrated that even low concentrations of micropollutants can impair mitochondrial activity without immediately triggering cell death (Müller et al., 2019). Prolonged mitochondrial dysfunction can lead to increased oxidative stress, affecting cellular homeostasis and potentially initiating pathways linked to chronic diseases. Furthermore, these effects in aquatic species can impair survival and reproduction, threatening biodiversity and ecosystem balance.

Endocrine Disruption

Many micropollutants function as endocrine disruptors, interfering with hormone signaling pathways and causing reproductive and developmental toxicity (Tebourbi et al., 2011). Endocrine-disrupting chemicals (EDCs) can interfere with hormonal activity by mimicking or inhibiting natural hormones, thereby disrupting normal endocrine processes. For example, bioassays have demonstrated estrogenic activity in some chemicals, highlighting their ability to influence hormone synthesis and regulation (Kuckelkorn et al., 2018). The pervasive nature of these pollutants underscores the urgency for robust monitoring systems and stringent regulatory frameworks to mitigate their impacts.

Physiological Disruption

Electrolyte Disturbances: Micropollutants interfere with electrolyte balance, impairing normal cellular operations and leading to toxic effects (Maynert & Hammond, 1976). This disruption can affect nerve signaling, muscle contraction, and overall metabolic functions.

Oxidative Stress: Chemicals like Paraquat produce reactive oxygen species, triggering oxidative damage and inflammation at the cellular level (Hathway, 2000). Prolonged oxidative stress can lead to chronic diseases, including cardiovascular disorders and neurodegenerative conditions.

Assessment Techniques for Micropollutant Toxicity

In Vitro Toxicity Assays

In vitro toxicity assays are gaining recognition as valuable tools for evaluating the toxicity of micropollutants (MPs) due to their ethical benefits and ability to mimic biological responses. These methods enable the study of environmental pollutants, focusing on immunotoxicity and genotoxicity, while aligning with regulatory demands for safer chemical assessments. The following sections highlight the key aspects of in vitro toxicity assays for analyzing micropollutants.

Immunotoxicity Assessment

In vitro assays effectively evaluate the immunotoxic impacts of MPs, offering critical insights into public health risks linked to environmental contaminants (Li et al., 2023). These cell-based methods are cost-efficient and feature shorter growth cycles, making them ideal for high-throughput screening applications (Li et al., 2023).

Genotoxicity Evaluation

The in vitro micronucleus test has been established as a reliable approach for assessing genotoxicity in wastewater samples, with consistent validation across multiple laboratories (Reifferscheid et al., 2008). This assay detects micronuclei in cells, a marker of DNA damage from toxic substances, and has proven effective in identifying the genotoxic potential of MPs (Reifferscheid et al., 2008).

Regulatory Endorsement

Regulatory agencies increasingly support in vitro toxicity testing methods, driven by the dual goals of minimizing animal testing and enhancing safety evaluations (Bhanushali et al., 2010; Jain et al., 2018). The continued development and validation of these assays are critical for their integration into regulatory frameworks, ensuring they meet both scientific rigor and ethical standards (Jain et al., 2018).

Immunoassays Using Cell Cultures

These assays evaluate the immunotoxicity of pollutants, focusing on endpoints that closely replicate in vivo environments (Li et al., 2023).

Microtox® and SMP-Based Assays

Microtox® Assay: Utilizes luminescent bacteria to assess the toxicity of substances, offering valuable insights into their environmental impact (Argese et al., 1998).

Submitochondrial Particle (SMP) Assays: Examine toxicity using mitochondrial fragments, providing supplementary information on aquatic toxicity (Argese et al., 1998).

Benefits of in Vitro Assays

Ethical Advantages: In vitro techniques reduce the dependence on animal testing, addressing concerns related to animal welfare (Li et al., 2023).

Efficiency in Cost and Time: These assays are faster and more cost-effective than traditional in vivo methods, making them a practical choice for toxicity evaluations (Tardiff, 1978).

Consistent Experimental Conditions: Cultured cell lines offer uniform and reproducible conditions, minimizing genetic variation and enhancing the reliability of results (Hasspieler et al., 1996).

In Vivo Toxicity Assays

Evaluating the toxicity of micropollutants through in vivo assays is essential for understanding their ecological effects. Techniques such as the in vivo comet assay and bioassays using model organisms offer valuable insights into the genotoxicity and overall toxicity of these contaminants.

Evaluating the toxicity of micropollutants through in vivo assays is essential for understanding their ecological effects. Techniques such as the in vivo comet assay and bioassays using model organisms offer valuable insights into the genotoxicity and overall toxicity of these contaminants. The sections below detail key assessment methods and their significance.

In Vivo Comet Assay

The in vivo comet assay is a highly sensitive and rapid method for detecting DNA damage caused by micropollutants, including both metals and organic compounds (Vasquez & Frötschl, 2016). This assay can be applied to various tissues and adapted to different exposure conditions, making it particularly useful for regulatory purposes (Vasquez & Frötschl, 2016).

Model Organisms

Model organisms such as Caenorhabditis elegans are widely used to study reproductive toxicity and other endpoints, offering an ecologically relevant framework for evaluating micropollutants (Abbas, 2021). Research has demonstrated that these models are effective in detecting significant toxic effects of micropollutants, underscoring the importance of thorough toxicity assessments (Abbas, 2021).

Advanced Wastewater Treatment (AWWT)

Technologies like ozonation in AWWT have proven effective in reducing micropollutant toxicity, although some toxic transformation products may persist (Abbas, 2021). Combining in vivo assays with evaluations of AWWT systems enhances the understanding of residual toxicities in treated wastewater, providing a more comprehensive assessment (Abbas, 2021).

Hydra Model for Toxicity Assessment

The Hydra model is a valuable tool for examining the teratogenic and reproductive impacts of emerging pollutants (EPs) due to its high sensitivity to morphological and biochemical changes (Vimalkumar et al., 2022). Methods like acridine orange staining and DCFDA staining for detecting reactive oxygen species (ROS) are employed to evaluate cellular damage and stress responses (Vimalkumar et al., 2022).

Mammalian Models and Microplastics

In vivo studies using ICR mice have revealed that polytetrafluoroethylene (PTFE) microplastics exhibit no notable toxic effects at high doses, with an LD50 greater than 2000 mg/kg (Lee et al., 2022). These assessments involved monitoring clinical symptoms, body weight, and histopathological changes, highlighting the necessity of long-term studies to better understand the potential risks associated with microplastic exposure (Lee et al., 2022).

Advantages of in Vivo Toxicity Assays

Realistic Biological Context: In vivo assays capture the complex interactions occurring within living organisms, encompassing processes like absorption, distribution, metabolism, and excretion (ADME), which are often not addressed in in vitro methods (Parng, 2005).

Importance of Model Organisms: Model organisms such as zebrafish are highly advantageous due to their rapid development and transparent bodies, allowing for real-time observation of organ-level responses to toxicants. These characteristics make them essential tools for early drug discovery and toxicity evaluation (Parng, 2005).

Ecological Relevance: Organisms like Caenorhabditis elegans are instrumental in understanding the ecological impacts of micropollutants, as they can replicate the effects of contaminants within aquatic ecosystems (Abbas, 2021).

Computational and Predictive Models

The evaluation of micropollutant toxicity increasingly incorporates computational and predictive models, providing efficient and cost-effective alternatives to conventional experimental methods. These models utilize various algorithms and data-driven techniques to forecast the toxicity of organic micropollutants (OMPs) and other contaminants in aquatic environments.

Predictive Modelling Approaches

Machine Learning Models

Techniques such as Random Forest (RF), Artificial Neural Networks (ANN), and XGBoost have been effectively employed to predict the toxicity of pollutants, including per- and polyfluoroalkyl substances (PFASs), to endangered fish species (Ji et al., 2024).

Quantitative Structure–Activity Relationship (QSAR)

QSAR models leverage molecular descriptors to establish connections between chemical structure and toxicity, demonstrating their usefulness in regulatory evaluations (Tropsha, 2023).

Integrated Models

Hybrid models that merge traditional assessment methods with advanced computational techniques, such as fuzzy set theory, enhance predictive accuracy for complex mixtures (Kim et al., 2013; Sigurnjak Bureš et al., 2021).

Molecular Docking

Molecular docking is a technique used to evaluate the binding interactions between pollutants and biological receptors, offering quick insights into potential toxicity pathways (Kamerlin et al., 2020).

Risk Assessment Frameworks

Ecotoxicological Risk Assessment: These models evaluate the risks associated with micropollutants, such as pharmaceuticals and personal care products, by analyzing their concentrations across various environmental matrices (Abbasi et al., 2022).

Mixture Toxicity Models: Advanced models assess the toxicity of chemical mixtures by considering potential interactions that could affect overall toxicity, leading to a more comprehensive risk evaluation (Sigurnjak Bureš et al., 2021).

Advantages of Computational and Predictive Models

Enhanced Predictive Accuracy

Integration of Machine Learning: Models such as Random Forest and XGBoost have demonstrated exceptional performance in predicting toxicity, especially for complex substances like PFASs, which are challenging to test directly on endangered species (Ji et al., 2024).

QSAR Models

Quantitative Structure–Activity Relationship (QSAR) models leverage molecular descriptors to forecast toxicity, serving as a dependable alternative to traditional experimental approaches (Sigurnjak Bureš et al., 2021; Tropsha, 2023).

Cost and Time Efficiency

Reduced Need for Experimental Testing: Computational models decrease the reliance on extensive laboratory experiments, streamlining the assessment process and making it more cost-effective (Sigurnjak Bureš et al., 2021).

Risk Assessment: These models are capable of interpolating micropollutant concentrations across various environments, facilitating both temporal and spatial risk assessments (Abbasi et al., 2022).

Regulatory Applications

Support for Regulatory Frameworks: The establishment of robust in silico models is increasingly acknowledged in regulatory toxicology, improving the credibility of chemical assessments (Tropsha, 2023).

Bioindicators

Evaluating the toxicity of micropollutants involves using bioindicators to determine the environmental impact of these contaminants. Micropollutants, such as pharmaceuticals and personal care products, present significant risks to aquatic ecosystems, highlighting the need for effective assessment methods.

Types of Bioindicators

Microorganisms: Species such as Chlorella vulgaris and Daphnia magna are frequently utilized as bioindicators due to their high sensitivity to pollutants (Abdullahi & Ibrahim, 2017).

Macroinvertebrates: Indices like the Microinvertebrate Biotic Index (MBI) evaluate water quality based on the presence and tolerance of various macroinvertebrate species (Abdullahi & Ibrahim, 2017).

Fish and Algae: Organisms such as Danio rerio (zebrafish) and marine algae are also monitored for their responses to micropollutants, with research revealing differing toxicity levels among various species (Yilmaz et al., 2022).

Model Organisms: Species such as Caenorhabditis elegans and Daphnia magna are frequently utilized to evaluate toxicity levels, offering valuable insights into reproductive and developmental effects (Abbas, 2021; Sabino et al., 2021).

Genetically Modified Strains: Genetically modified strains of Saccharomyces cerevisiae can be employed to specifically detect estrogenic activity, improving sensitivity in monitoring efforts (Sabino et al., 2021).

Aquatic organisms: Research has demonstrated that species such as Daphnia magna and Crassostrea gigas display differing sensitivities to various micropollutants, underscoring the significance of selecting appropriate species for toxicity assessments.

Advantages of Bioindicators

Sensitivity to Pollution: Bioindicators such as Daphnia magna and Chlorella vulgaris exhibit a high sensitivity to fluctuations in water quality, making them effective for monitoring micropollutants (Abdullahi & Ibrahim, 2017). Fish, being organisms at higher trophic levels, reflect the cumulative impacts of pollutants, offering insights into risks for both aquatic ecosystems and human health (Basu et al., 2023).

Cost-Effectiveness: Monitoring bioindicators is generally more cost-effective than conducting chemical analyses, enabling more extensive and frequent assessments (Fränzle, 2003). Additionally, bioindicators can integrate data over time, providing insights into long-term exposure to pollutants, which is often overlooked in singular chemical measurements (Fränzle, 2003).

Comprehensive Ecological Insights: Biotic indices, such as the Microinvertebrate Biotic Index (MBI), assist in categorizing water quality based on the presence and diversity of bioindicators (Abdullahi & Ibrahim, 2017).

Micropollutant Impact on Human Health

Micropollutants pose significant environmental risks due to their ability to disrupt biological systems. A major concern is endocrine disruption, where these contaminants interfere with hormonal balance, causing reproductive issues and metabolic disorders. Continuous exposure can have long-term consequences, including reduced fertility and an increased risk of hormone-related conditions (Kumar et al., 2024). Likewise, neurodevelopmental effects are associated with certain micropollutants, especially in children. Exposure during critical developmental periods can result in cognitive impairments and delays, potentially leading to permanent health challenges (Kumar et al., 2024). Micropollutants are also linked to a heightened cancer risk, with studies showing that inhalation of indoor pollutants, particularly in crowded environments, increases the likelihood of cancer, especially among vulnerable groups like children and older adults. Chronic exposure may intensify genetic mutations and disrupt normal cellular processes (Li et al., 2024). Furthermore, respiratory problems are another pressing issue, as micropollutants in enclosed or densely populated spaces elevate infection risks and contribute to conditions such as asthma and bronchitis (Bala et al., 2024; John & Das, 2024; Priyadarshini et al., 2024; Sahu et al., 2024). Poor indoor air quality exacerbates the adverse effects of these pollutants, compounding respiratory health challenges (Li et al., 2024).

Current Challenges and Limitations

Micropollutants (MPs) pose significant challenges for environmental management due to their intricate characteristics and the complexities involved in their detection and elimination. Originating from diverse sources such as pharmaceuticals, pesticides, and industrial effluents, MPs threaten aquatic ecosystems and human health. Addressing these contaminants is hampered by limitations in current monitoring systems, the high costs associated with removal technologies, and regulatory gaps. Additionally, their persistence and bioaccumulation potential further exacerbate their environmental impact, necessitating more robust and integrated solutions.

Detection and Monitoring Challenges

Complexity of Detection: MPs are often present in minute concentrations, making their identification and quantification both technically complex and financially burdensome (Oliveira et al., 2024). Their interactions with other contaminants and environmental matrices further complicate detection processes, requiring sophisticated analytical tools (Tripathy et al., 2022).

Inadequate Monitoring Programs: Many existing monitoring systems lack comprehensive coverage, leaving critical gaps in understanding the pathways, persistence, and concentrations of MPs across diverse environments (Kardos et al., 2024). Improved data collection and real-time monitoring are essential for tracking their environmental fate.

Removal and Treatment Limitations

Inefficiency of Conventional Methods: Traditional wastewater treatment facilities are not designed to effectively target MPs, leading to their continued release into natural water bodies. This underscores the urgent need for advanced methods like membrane filtration, bioadsorption, and catalytic degradation (Narwal et al., 2023; Oliveira et al., 2024). The environmental behavior and persistence of micropollutants are not yet fully understood, making it challenging to evaluate their long-term consequences (Ni et al., 2024). Factors such as their interactions with other pollutants and their transformation into potentially harmful by-products further complicate this assessment.

High Costs of Advanced Technologies: While innovative approaches such as nanotechnology and advanced oxidation processes show promise, their high costs and challenges in integration with existing treatment systems hinder widespread adoption. Furthermore, these methods often require specialized expertise and significant energy inputs, limiting their scalability (Narwal et al., 2023).

Regulatory and Policy Issues

Lack of Comprehensive Legislation: The absence of cohesive and enforceable regulations for MPs significantly limits effective management strategies. Without clear standards and guidelines, industries and municipalities struggle to prioritize mitigation efforts (Oliveira et al., 2024).

Challenges in Source Control: While controlling pollution at the source is a crucial step, economic and logistical barriers often impede its implementation. Promoting producer responsibility and incentivizing cleaner production methods could help address these obstacles (Sharma et al., 2024). To overcome these challenges, a multi-pronged approach combining technological innovation, enhanced monitoring systems, and robust regulatory frameworks is necessary. Encouraging collaboration between stakeholders, including governments, industries, and researchers, will be key to managing the risks associated with MPs effectively.

Emerging Approaches and Solutions

Emerging micropollutants present critical environmental challenges, requiring innovative strategies for their management and removal. These contaminants, derived from sources such as pharmaceuticals, personal care products, and industrial effluents, pose threats to aquatic ecosystems and human health. Recent studies emphasize the effectiveness of various mitigation approaches, including source control, advanced treatment technologies, and nanotechnology-based solutions.

Source-Control Approaches

Preventing micropollutants from entering the environment is a key strategy, achieved through enhanced regulations and sustainable industrial practices. Methods such as oxidation processes, membrane filtration, and adsorption play a vital role in minimizing pollutant discharge at the source (Sharma et al., 2024).

Advanced Treatment Technologies

Innovative treatment methods have demonstrated high efficiency in micropollutant removal. Systems like membrane bioreactors (MBR) and granular activated carbon (GAC) have achieved removal rates exceeding 80% for pharmaceuticals and PFASs, showcasing their resource efficiency (Baresel et al., 2024). Additionally, biological processes like nitrification–denitrification have proven effective in transforming micropollutants, enhancing their removal in wastewater treatment facilities (Muhammad et al., 2024).

Nanotechnology Solutions

Nanotechnology offers promising solutions for addressing micropollutants, with its potential to enhance removal efficiency. However, challenges such as cost and potential toxicity remain obstacles to widespread adoption (Narwal et al., 2023). Hybrid technologies, combining MBR with advanced oxidation processes, are being actively explored for their ability to tackle persistent and resistant micropollutants effectively (Khan et al., 2023). Innovative and integrated approaches are essential to address the multifaceted challenges posed by micropollutants, ensuring a sustainable and resilient environmental management system.

Conclusion

Micropollutants pose a growing threat to environmental sustainability and human health due to their diverse nature, extensive presence, and complex pathways within ecosystems. These pollutants, which include organic and inorganic compounds as well as microplastics and nanoplastics, exhibit a range of toxic effects that can disrupt cellular, molecular, and ecological systems. Gaining a clear understanding of their acute and chronic impacts, along with their specific mechanisms of toxicity, is essential for assessing potential risks and devising effective mitigation strategies. Significant progress has been made in evaluating the effects of micropollutants through advanced techniques such as in vitro and in vivo assays, computational modelling, and the use of bioindicators. However, incorporating these tools into robust risk assessment frameworks remains a challenge, particularly in light of additional environmental pressures like climate change and urbanization. While existing environmental standards and guidelines provide a baseline for mitigating risks, they must be continuously updated to address the emergence of new pollutants and their long-term effects on human health and ecosystems. Despite these hurdles, innovative solutions are being developed to improve the management of micropollutants. Cutting-edge treatment methods, advancements in nanotechnology, and sustainable approaches like bioremediation offer promising strategies for reducing their impact. Achieving success in this area will require coordinated efforts among researchers, policymakers, and industry stakeholders to overcome existing challenges and create integrated, comprehensive management frameworks.

Addressing the complex challenges associated with micropollutants demands a holistic and multidisciplinary strategy. By combining scientific innovations, advanced evaluation techniques, and forward-thinking regulatory measures, we can more effectively confront the diverse impacts of these contaminants. Expanding our understanding and fostering cross-disciplinary collaboration will be crucial in safeguarding ecological balance and ensuring public health protection from the persistent risks of micropollutants.

References

Abbas, A. (2021). *Ecotoxicological characterization of micropollutants and wastewater samples from conventional biological and advanced wastewater treatment* (Doctoral dissertation, Universitätsbibliothek Johann Christian Senckenberg).

Abbasi, N. A., Shahid, S. U., Majid, M., & Tahir, A. (2022). Ecotoxicological risk assessment of environmental micropollutants. In *Environmental micropollutants* (pp. 331–337). Elsevier.

Abdullahi, U. A., & Ibrahim, S. (2017). An assessment of pollution in aquatic environment using bioindicators: A review. *Bayero Journal of Pure and Applied Sciences, 10*(2), 64–68.

Argese, E., Bettiol, C., Ghirardini, A. V., Fasolo, M., Giurin, G., & Ghetti, P. F. (1998). Comparison of in vitro submitochondrial particle and Microtox® assays for determining the toxicity of organotin compounds. *Environmental Toxicology and Chemistry: An International Journal, 17*(6), 1005–1012.

Arif, Y., Mir, A. R., Zieliński, P., Hayat, S., & Bajguz, A. (2024). Microplastics and nanoplastics: Source, behavior, remediation, and multi-level environmental impact. *Journal of Environmental Management, 356*, 120618.

Arslan, M., Ullah, I., Müller, J. A., Shahid, N., & Afzal, M. (2017). Organic micropollutants in the environment: Ecotoxicity potential and methods for remediation. In *Enhancing cleanup of environmental pollutants: Biological approaches* (Vol. 1, pp. 65–99).

Bala, K., Anshul, Verma, S., Dogra, A., Das, A. P., Kumar, R., & Nandini. (2024). Sustainable management for urban plastic waste generation. *Sustainable Management of Urban Plastic Waste Through Circular Economic Approaches*, 211–236. https://doi.org/10.1201/9781003465447-13

Baresel, C., Salem, M., Roberts, R., Malovanyy, A., Lemström, H., & Esfahani, B. (2024). Approaching breakthrough: Resource-efficient micropollutant removal with MBR-GAC configuration. *Applied Sciences, 14*(17), 7759.

Barnes, J. M., & Paget, G. E. (1965). 2 mechanisms of toxic action. *Progress in Medicinal Chemistry, 4*, 18–38.

Basu, A., Moulik, S., & Karmakar, S. (2023). Bioaccumulation of different organic micropollutants in fishes and its toxicological and stress impacts: A review. *International Journal for Science Technology and Engineering*. https://doi.org/10.22214/ijraset.2023.49332

Bhanushali, M., Bagale, V., Shirode, A., Joshi, Y., & Kadam, V. (2010). An in-vitro toxicity testing-a reliable alternative to toxicity testing by reduction, replacement and refinement of animals. *International Journal of Advances in Pharmaceutical Sciences, 1*(1).

Bhat, R. A. H., Sidiq, M. J., & Altinok, I. (2024). Impact of microplastics and nanoplastics on fish health and reproduction. *Aquaculture, 741037*.

Chen, M., Jin, X., Guo, C., Liu, Y., Zhang, H., Wang, J., Xu, J., et al. (2023). Micropollutants but high risks: Human multiple stressors increase risks of freshwater ecosystems at the megacity-scale. *Journal of Hazardous Materials, 460*, 132497.

Das, A. P., & Singh, S. (2011). Occupational health assessment of chromite toxicity among Indian miners. *Indian Journal of Occupational and Environmental Medicine, 15*(1), 6. https://doi.org/10.4103/0019-5278.82998

Das, A. P., Mallavarapu, M., & Ghosh, S. (2023). Ecotoxicity of mining pollutants on the environment and their remediation. *Environmental Chemistry and Ecotoxicology, 5*, 165–167. https://doi.org/10.1016/J.ENCECO.2023.08.002

Das, A., Dey, S., & Das, A. P. (2024a). Ecotoxicological impacts of synthetic microfiber pollutants and development of sustainable mitigation strategies. *Environmental Chemistry and Ecotoxicology*. https://doi.org/10.1016/J.ENCECO.2024.12.003

Das, A., Dey, S., Bala, K., Gnanasekaran, R., & Das, A. P. (2024b). Microbial bioremediation technology for sustainable treatment and management of synthetic microfiber waste. *Sustainable Microbial Technology for Synthetic and Cellulosic Microfiber Bioremediation, 221–238*. https://doi.org/10.1007/978-3-031-62600-5_12

Das, A., Ray, S. K., Mohanty, M., Mohanty, J., Dey, S., & Das, A. P. (2024c). Ecotoxicity of microplastic wastes and their sustainable management: A review. *Environmental Chemistry and Ecotoxicology, 6*, 144–152. https://doi.org/10.1016/J.ENCECO.2024.05.003

Dey, S., Das, A., Mallick, K., Sahu, A., & Das, A. P. (2023a). Environmental petroleum waste: Pollution. *Toxicity, Sustainable Remediation, 159–175*. https://doi.org/10.1007/978-3-031-48220-5_7

Dey, S., Tripathy, B., Kumar, M. S., & Das, A. P. (2023b). Ecotoxicological consequences of manganese mining pollutants and their biological remediation. *Environmental Chemistry and Ecotoxicology*. https://doi.org/10.1016/J.ENCECO.2023.01.001

Fränzle, O. (2003). Bioindicators and environmental stress assessment. In *Trace metals and other contaminants in the environment* (Vol. 6, pp. 41–84). Elsevier.

Ghosh, S., Sinha, J. K., Ghosh, S., Vashisth, K., Han, S., & Bhaskar, R. (2023). Microplastics as an emerging threat to the global environment and human health. *Sustainability, 15*(14), 10821. https://doi.org/10.3390/SU151410821

Gupta, S. (2024). Organic micropollutants in the environment: Ecotoxicity potential and bioremediation approaches. In *Organic micropollutants in aquatic and terrestrial environments* (pp. 249–263). Springer Nature Switzerland.

Hasspieler, B. M., Haffner, G. D., & Adeli, K. (1996). In vitro toxicological methods for environmental health testing. *Reviews on Environmental Health, 11*(4), 213–228.

Hathway, D. E. (2000). Toxic action/toxicity. *Biological Reviews, 75*(1), 95–127.

Jagadeesh, N., Sundaram, B., & Dubey, B. K. (2023). Toxicity assessment of microconstituents in the environment. In *Microconstituents in the environment: Occurrence, fate, removal and management* (pp. 89–106).

Jagatee, S., Priyadarshini, S., Rath, C. C., & Das, A. P. (2024). Synthetic microfiber: An enduring environmental problem linked to sustainable development. *Environmental Science and Engineering, Part F2333*, 93–112. https://doi.org/10.1007/978-3-031-51792-1_6

Jain, M. R., Bandyopadhyay, D., & Sundar, R. (2018). Scientific and regulatory considerations in the development of in vitro techniques for toxicology. In *In vitro toxicology* (pp. 165–185). Academic Press.

Ji, Y., Wang, X., Wang, R., Wang, J., Zhao, X., & Wu, F. (2024). Toxicity prediction and risk assessment of per-and polyfluoroalkyl substances for threatened and endangered fishes. *Environmental Pollution, 361*, 124920.

John, A., & Das, A. P. (2024). Government strategies for achieving circular economy for plastic waste management in Southeast Asia. *Sustainable Management of Urban Plastic Waste through Circular Economic Approaches, 72–82*. https://doi.org/10.1201/9781003465447-6/

Joshi, S. J., & Sarma, H. (2021). Micropollutants in the environment: Challenges and bioremediation strategies. *Open Biotechnology Journal, 15*(Special issue 1), 68–70.

Kalita, A. J., Sarma, J., Hazarika, A., Bardhan, S., Hazarika, N., Borppujari, P.,... & Rajkhowa, S. (2024). Organic micropollutants in environment: Origin and occurrence. In *Organic micropollutants in aquatic and terrestrial environments* (pp. 3–23). Springer Nature Switzerland.
Kamerlin, N., Delcey, M. G., Manzetti, S., & Van der Spoel, D. (2020). Toward a computational ecotoxicity assay. *Journal of Chemical Information and Modeling, 60*(8), 3792–3803.
Kardos, M. K., Clement, A., Jolánkai, Z., Zessner, M., Kittlaus, S., Weber, N., Zoboli, O., et al. (2024). Development and testing of an efficient micropollutant monitoring strategy across a large watershed. *Science of the Total Environment, 948*, 174760.
Kaur, R., Kaur, H., Singh, S., Jagota, N., Sharma, A., & Sharma, A. (2024). Organic micropollutants in the urban soils: Technological advances and management concerns. In *Organic micropollutants in aquatic and terrestrial environments* (pp. 47–63).
Kenaga, E. E. (1982). Predictability of chronic toxicity from acute toxicity of chemicals in fish and aquatic invertebrates. *Environmental Toxicology and Chemistry: An International Journal, 1*(4), 347–358.
Khan, N. A., Singh, S., López-Maldonado, E. A., Pavithra, N., Méndez-Herrera, P. F., López-López, J. R., Aljundi, I. H., et al. (2023). Emerging membrane technology and hybrid treatment systems for the removal of micropollutants from wastewater. *Desalination, 565*, 116873.
Khanna, R., Chandra, A., Sen, S., Konyukhov, Y., Fuentes, E., Burmistrov, I., & Kravchenko, M. (2024). Microplastics and nanoplastics as environmental contaminants of emerging concern: Potential hazards for human health. *Sustainability, 16*(19), 8704.
Kim, J., Kim, S., & Schaumann, G. E. (2013). Reliable predictive computational toxicology methods for mixture toxicity: Toward the development of innovative integrated models for environmental risk assessment. *Reviews in Environmental Science and Biotechnology, 12*(3), 235–256.
Kuckelkorn, J., Redelstein, R., Heide, T., Kunze, J., Maletz, S., Waldmann, P., Hollert, H., et al. (2018). A hierarchical testing strategy for micropollutants in drinking water regarding their potential endocrine-disrupting effects—towards health-related indicator values. *Environmental Science and Pollution Research, 25*, 4051–4065.
Kumar, A., Deepika, Tyagi, D., Tarkeshwar, & Kapinder. (2024). Organic micropollutants and their effects on the environment and human health. In *Organic micropollutants in aquatic and terrestrial environments* (pp. 87–105). Springer Nature Switzerland.
Lee, S., Kang, K. K., Sung, S. E., Choi, J. H., Sung, M., Seong, K. Y., Kim, K., et al. (2022). In vivo toxicity and pharmacokinetics of polytetrafluoroethylene microplastics in ICR mice. *Polymers, 14*(11), 2220.
Li, X., & Li, Z. (2024). Perspectives on the toxic effects of micro-and nanoplastics on the environment: A bibliometric analysis of the 2014 to 2023 period. *Toxics, 12*(9), 676.
Li, X., Zhao, B., Luo, L., Zhou, Y., Lai, D., & Luan, T. (2023). In vitro immunotoxicity detection for environmental pollutants: Current techniques and future perspectives. *TrAC Trends in Analytical Chemistry, 158*, 116901.
Madhesh, J. C., & Mohamed, S. (2024). Impact of micropollutants on the environment and detection by biomarker-based approach. *Futuristic Trends in Biotechnology, 3*(20), 173–182. https://doi.org/10.58532/V3BJBT20P4CH1
Maynert, E. W., & Hammond, B. G. (1976). Modes of action of toxic agents. *Annals of Clinical & Laboratory Science, 6*(4), 322–331.
Mohanty, M., Tripathy, B., Koteswara Reddy, G., & Das, A. P. (2024). Greener technologies for establishing restraint over microplastic pollution. *Sustainable Microbial Technology for Synthetic and Cellulosic Microfiber Bioremediation, 239–258*. https://doi.org/10.1007/978-3-031-62600-5_13
Muhammad, H. A., Masyab, H. M., Othman, B. A., & Mahmood, Y. N. (2024). Micropollutant control in wastewater treatment. *ARO-the Scientific Journal of Koya University, 12*(2), 130–138.
Müller, M. E., Vikstrom, S., König, M., Schlichting, R., Zarfl, C., Zwiener, C., & Escher, B. I. (2019). Mitochondrial toxicity of selected micropollutants, their mixtures, and surface water samples measured by the oxygen consumption rate in cells. *Environmental Toxicology and Chemistry, 38*(5), 1000–1011.

Narwal, N., Katyal, D., Kataria, N., Rose, P. K., Warkar, S. G., Pugazhendhi, A.,... & Khoo, K. S. (2023). Emerging micropollutants in aquatic ecosystems and nanotechnology-based removal alternatives: A review. *Chemosphere, 341*, 139945.

Nath, S., Enerijiofi, K. E., Astapati, A. D., & Guha, A. (2024). Microplastics and nanoplastics in soil: Sources, impacts, and solutions for soil health and environmental sustainability. *Journal of Environmental Quality*.

Ni, R., Xu, W., Wang, C., Man, Z., & Cheng, X. (2024). The application of iron-manganese compounds towards micropollutants purification in water: A critical review. *Chemical Engineering Journal, 493*, 152553.

Oliveira, G. H. D., Lovato, G., Rodrigues, J. A. D., & Zaiat, M. (2024). Fate of micropollutants. In *Anaerobic treatment of domestic wastewater: Present status and potentialities*.

Ondruch, P., Schluesener, M. P., Dierkes, G., Jewell, K., Kirchgeorg, T., Hasenbein, S., Wick, A., et al. (2023). Occurrence and distribution of emerging micropollutants in the central part of the German Bight. *Marine Pollution Bulletin, 195*, Article 115427.

Pantos, O., Masterton, H., & Ruffell, H. (2024). Sources of nanoplastic and microplastic pollution which are hidden in plain sight. *Cambridge Prisms: Plastics, 2*, e25.

Parng, C. (2005). In vivo zebrafish assays for toxicity testing. *Current Opinion in Drug Discovery & Development, 8*(1), 100–106.

Paul, A., Priyadarsini, S., Karakuş, S., Behera, J. K., Gnanasekaran, R., & Das, A. P. (2024). Fate and accumulation of synthetic microfibers in the environment. *Sustainable Microbial Technology for Synthetic and Cellulosic Microfiber Bioremediation, 61–76*. https://doi.org/10.1007/978-3-031-62600-5_4

Priyadarshini, B., Santhosh Kumar, M., & Das, A. P. (2024). Microbial degradation of synthetic plastics released from COVID-19 generated single use disposable face masks. *Environmental Science and Engineering, Part F2827*, 1–16. https://doi.org/10.1007/978-3-031-50840-0_1

Qamar, S. A., Hassan, A. A., Rizwan, K., Rasheed, T., Bilal, M., Nguyen, T. A., & Iqbal, H. M. (2022). Biodegradation of micropollutants. In *Biodegradation and biodeterioration at the nanoscale* (pp. 477–507). Elsevier.

Reifferscheid, G., Ziemann, C., Fieblinger, D., Dill, F., Gminski, R., Grummt, H. J., Selke, D., et al. (2008). Measurement of genotoxicity in wastewater samples with the in vitro micronucleus test—Results of a round-robin study in the context of standardisation according to ISO. *Mutation Research/genetic Toxicology and Environmental Mutagenesis, 649*(1–2), 15–27.

Sabino, J. A., de Sá Salomão, A. L., de Oliveira Muniz Cunha, P. M., Coutinho, R., & Marques, M. (2021). Occurrence of organic micropollutants in an urbanized sub-basin and ecological risk assessment. *Ecotoxicology, 30*(1), 130–141.

Sahu, A., Dey, S., Mallick, K., Priyadarsini, S., Behera, I. D., & Das, A. P. (2024). A review on advanced technology for sustainable management of synthetic microplastic waste. *Sustainable Management of Urban Plastic Waste through Circular Economic Approaches, 262–279*. https://doi.org/10.1201/9781003465447-15

Servien, R., Bonnot, K., Latrille, E., Hélias, A., & Patureau, D. (2023). Insights on impact estimation of unmeasured micropollutants from WWTP release.

Shao, Y., Chen, Z., Hollert, H., Zhou, S., Deutschmann, B., & Seiler, T. B. (2019). Toxicity of 10 organic micropollutants and their mixture: Implications for aquatic risk assessment. *Science of the Total Environment, 666*, 1273–1282.

Sharma, M., Bains, A., Sridhar, K., Chawla, P., & Sharma, M. (2024). Emerging micropollutants–Environmental impact and source-controlled approaches: Current state and future prospects. *Food and Chemical Toxicology, 193*, 115038.

Sharma, S., Sharma, V., & Chatterjee, S. (2023). Contribution of plastic and microplastic to global climate change and their conjoining impacts on the environment-A review. *Science of the Total Environment, 875*, 162627.

Sigurnjak Bureš, M., Cvetnić, M., Miloloža, M., Kučić Grgić, D., Markić, M., Kušić, H., Ukić, Š., et al. (2021). Modeling the toxicity of pollutants mixtures for risk assessment: A review. *Environmental Chemistry Letters, 19*, 1629–1655.

Singh, M., Singh, P. K., & Kulshreshtha, U. C. (2024). Microplastics: An invisible air pollutant of new era. In *Societal and environmental ramifications of plastic pollution* (pp. 27–42). IGI Global.
Singh, R. P., Mishra, S., & Das, A. P. (2020). Synthetic microfibers: Pollution toxicity and remediation. *Chemosphere, 257*, 127199. https://doi.org/10.1016/j.chemosphere.2020.127199
Tardiff, R. G. (1978). In vitro methods of toxicity evaluation. *Annual Review of Pharmacology and Toxicology, 18*(1), 357–369.
Tebourbi, O., Sakly, M., & Rhouma, K. B. (2011). Molecular mechanisms of pesticide toxicity. In *Pesticides in the modern world-pests control and pesticides exposure and toxicity assessment* (pp. 297–332).
Tripathy, B., Dash, A., & Das, A. P. (2022). Detection of environmental microfiber pollutants through vibrational spectroscopic techniques: Recent advances of environmental monitoring and future prospects. *Critical Reviews of Analytical Chemistry, 1–11.* https://doi.org/10.1080/10408347.2022.2144994
Tropsha, A. (2023). Development of QSAR models as reliable computational tools for regulatory assessment of chemicals for acute toxicity. In *QSAR in safety evaluation and risk assessment* (pp. 13–24). Academic Press.
Uddin, M., Kormoker, T., Uddin, M. S., Idris, A. M., Ampah, J. D., Haque, M. K.,... & Juhasz, A. L. (2024). Micro (nano) plastics as emerging pollutants in global aquatic and terrestrial ecosystems: A bibliometric analysis. *Environmental Forensics*, 1–33.
Vaijayanthimala, S., & Sharma, K. (2023). Microplastics as vectors for metals from mines and fuels: Environmental pathways and implications. *Journal of Mines, Metals and Fuels*, 370–375.
Vasquez, M. Z., & Frötschl, R. (2016). The in vivo comet assay test. In *Genetic toxicology testing* (pp. 345–382). Academic Press.
Vimalkumar, K., Sangeetha, S., Felix, L., Kay, P., & Pugazhendhi, A. (2022). A systematic review on toxicity assessment of persistent emerging pollutants (EPs) and associated microplastics (MPs) in the environment using the Hydra animal model. *Comparative Biochemistry and Physiology Part c: Toxicology & Pharmacology, 256*, 109320.
Włodarczyk-Makuła, M. (2024). Selected organic micropollutants in the aquatic environment. *Desalination and Water Treatment, 317*, 100061.
Yang, Y., Zhang, X., Fu, L., Li, C., & Zhang, S. (2024). The number of test organisms might influence the toxicity evaluation of hydrophobic micropollutants. *Chemosphere, 355*, 141814.
Yang, Y., Zhang, X., Jiang, J., Han, J., Li, W., Li, X., Alvarez, P. J., et al. (2021). Which micropollutants in water environments deserve more attention globally? *Environmental Science & Technology, 56*(1), 13–29.
Yazarkan, Y., Sonmez, G. A. M. Z. E., Tenekeci, A., Aksoy, E., & Güllü, İ. (2024). The relationship between microplastics and nanoplastics with cancer: An emerging health concern. *Uhod-Uluslararasi Hematoloji-Onkoloji Dergisi, 2.*
Yılmaz, H., Avaz, G., Yetiş, Ü., & Özkan, M. (2022). Toxicity of environmentally important micropollutants on three trophic levels. *Aquatic Research, 5*(1), 20–28.

Chapter 11
Techniques for Mitigating Micropollutants: Challenges and Solutions

Hardik Sikotariya, T. H. Dave, Mayur Bhadarka, Ketan Makwana, and Mohit Bamaniya

Introduction

Micropollutants refer to small amounts of organic chemicals, often found in very low concentrations—measured in parts per billion or trillion. These substances include things like medications, personal care products, and pesticides (Verstraeten et al., 2003). Micropollutants, including pharmaceuticals, pesticides, and heavy metals, primarily enter water systems through wastewater discharge (Arman et al., 2021; Abbas et al., 2022). These pollutants are chemically stable, allowing them to persist in aquatic environments, accumulate over time, and travel through the food chain (Mostafalou, 2016). Their presence poses significant health risks, with exposure linked to endocrine disruption, cancer, and metabolic disorders like insulin resistance and obesity (Mostafalou, 2016, Das et al., 2023a, 2023b, 2024a, 2024b; Dey et al., 2023). Even at low concentrations, these pollutants can be harmful, emphasizing the need for consistent monitoring of wastewater before its release (Abbas et al., 2022).

H. Sikotariya (✉) · T. H. Dave · K. Makwana
Department of Fisheries Resource Management, College of Fisheries Science, Kamdhenu University, Veraval, Gujarat, India
e-mail: hardik1772000@gmail.com

M. Bhadarka
Department of Aquatic Environment Management, College of Fisheries Science, Kamdhenu University, Veraval, Gujarat, India

M. Bamaniya
Department of Fish Processing Technology, College of Fisheries Science, Kamdhenu University, Veraval, Gujarat, India

I. D. Behera et al. (eds.), *Emerging Micropollutants*, Sustainable Environmental Waste Management Strategies, https://doi.org/10.1007/978-3-032-03530-1_11

Sources of Micropollutants

Urban Sources

Urban areas contribute significantly to micropollutant pollution, particularly through wastewater discharges and stormwater runoff (Mishra & Das, 2021; Mishra et al., 2022). Pharmaceuticals, personal care products, and industrial chemicals enter water systems from impervious surfaces, leading to elevated concentrations of organic micropollutants (Markert et al., 2024; Kaur et al., 2024; Das et al., 2023a, 2023b, 2025).

Agricultural Sources

Agricultural activities, especially the use of pesticides and fertilizers, are major contributors to micropollutant contamination. Runoff from farmlands introduces pollutants like pesticides, heavy metals, and endocrine disruptors into adjacent water bodies, affecting water quality (Gaid, 2023; Markert et al., 2024).

Industrial Sources

Industrial processes release various micropollutants, including polychlorinated biphenyls (PCBs), polycyclic aromatic hydrocarbons (PAHs), and heavy metals. These pollutants contaminate both soil and water, contributing to long-term environmental degradation (Kaur et al., 2024).

Strategies for Controlling Micropollutants

Source Reduction

Legislation and Regulations

A. **Technological Innovations**

Advanced wastewater treatment methods like ozonation and ultrafiltration have proven highly effective in removing micropollutants (Bala et al., 2024; Dey et al., 2024b; Priyadarshini et al., 2024). For example, using ozonation followed by sand filtration has achieved over 80% removal of various organic micropollutants (Margot et al., 2011). Membrane systems also improve the removal of dissolved pollutants, highlighting the need to optimize plant designs and operation settings (Battistoni et al., 2007).

B. **Policy Frameworks**

A thorough policy analysis shows that a combination of end-of-pipe solutions and source-focused strategies is key for sustainable micropollution management (Das et al., 2024b; Dey et al., 2024a; Mohanty et al., 2024; Sahu et al., 2024). Drawing from previous water quality regulations, effective policies can be crafted (Metz & Ingold, 2014). Additionally, embedding sustainability criteria in these regulations helps identify the best ways to tackle the complexity of micropollutants (Metz & Ingold, 2014).

C. **Stakeholder Engagement**

Involving industries and the public is essential for driving the behavioral changes needed alongside technological and regulatory measures. This collective approach can improve compliance and promote a culture of environmental responsibility (Bigg, 2013).

Public Awareness and Education

A. **Information Disclosure Strategies:**

Sharing pollution data with the public can boost community involvement in environmental protection (Tietenberg & Wheeler, 2001). Innovative methods like community workshops and online platforms have been effective in raising awareness of local pollution issues (Tietenberg & Wheeler, 2001).

B. **Awareness-Appraisal Model**

According to the awareness-appraisal model, people are more likely to engage in conservation efforts when they recognize local water bodies and view them as polluted (Forsyth et al., 2004). Surveys show that residents familiar with their watersheds and aware of pollution are more inclined to participate in cleanup activities (Forsyth et al., 2004).

C. **Education and Engagement**

Public education programs that explain the sources and effects of micropollutants can lead to greater community action and support for advanced treatment methods (Bigg, 2013; Trapido et al., 2014). Involving local stakeholders in decision-making can also improve the effectiveness of pollution control strategies (Bigg, 2013).

Advanced Treatment Technologies

Adsorption Techniques

A. **Activated Carbon Efficacy**

PAC versus GAC: Powdered activated carbon (PAC) has shown better performance in removing organic micropollutants (OMPs) at lower dosages compared to granular activated carbon (GAC), which encounters problems like pore blockage (Meinel et al., 2015).

Polar Compounds: PAC effectively removes apolar compounds, but its ability to handle polar micropollutants is much lower, requiring higher doses for effective removal (Kovalova et al., 2013).

B. **Novel Materials**

Coagulation Waste Catalysts: New research has introduced carbon-based catalysts from coagulation waste, which, when used with periodate, have significantly improved micropollutant degradation, achieving over 98% removal of sulfadiazine (He et al., 2022).

Predictive Models

Modeling Adsorption: Predictive models that consider both the properties of solutes and the characteristics of activated carbon can help optimize micropollutant selection for treatment, leading to more efficient removal (De Ridder et al., 2009).

Membrane Filtration

A. **Ultrafiltration (UF)**

UF membranes are effective at removing suspended solids and larger organic molecules, which aids in reducing dissolved micropollutants through the creation of a "layer effect" (Battistoni et al., 2007). Pilot studies show UF can achieve over 80% removal of micropollutants, though membrane fouling remains a key operational issue (Margot et al., 2011).

B. **Nanofiltration (NF)**

NF membranes strike a balance between UF and reverse osmosis (RO), effectively rejecting divalent ions and larger organic molecules while allowing monovalent ions to pass through (Drewes et al., 2005). However, NF may not fully remove all organic micropollutants, necessitating additional treatment (Drewes et al., 2005).

C. **Reverse Osmosis (RO)**

RO membranes exhibit high rejection rates for various micropollutants, particularly hydrophilic and ionic compounds, though some hydrophobic substances may still pass through (Drewes et al., 2005). Full-scale RO systems have proven highly effective, with minimal contaminant levels in treated water (Drewes et al., 2005).

Advanced Oxidation Processes (AOPs)

A. **Ozone and UV Radiation**

Ozone-based treatments, particularly when combined with UV radiation, have shown high effectiveness in removing various micropollutants. The O3/UV process significantly boosts the degradation of contaminants, especially organochlorine pesticides (Derco et al., 2015). Moreover, using UV with ozonation has been effective in breaking down toxic compounds found in industrial wastewater (Prengle, 1983).

B. **Fenton Reactions**

Fenton and Fenton-like reactions are essential in advanced oxidation processes (AOPs) due to their ability to generate hydroxyl radicals and other oxidizing intermediates that efficiently break down organic pollutants. These reactions also produce oxidants like carbonate radicals, which are key contributors to pollutant degradation (Meyerstein, 2021).

Biodegradation and Bioremediation

A. **Biodegradation Techniques**

Microbial Processes: Biological treatment technologies (BTTs), including activated sludge, constructed wetlands, and bioreactors, are effective in removing micropollutants (MPs) from wastewater (Kushwaha et al., 2021).

B. **Acclimation and Cometabolism**

Microorganisms can adapt to pollutants over time, increasing biodegradation rates, especially for hard-to-break-down compounds (Alexander, 1999).

C. **Bioremediation Strategies:**

Phytoremediation: The use of plants and their associated microbial communities can effectively degrade complex industrial waste and agricultural pollutants (Chandra, 2015).

Integrated Approaches: Combining biological treatments with chemical methods, such as using Fenton's reagent alongside biological processes, can improve the breakdown of non-biodegradable micropollutants (Kushwaha et al., 2021, Mohanty et al., 2023, Kanungo et al., 2023).

Hybrid Systems

A. **Biological Methods**

Biological wastewater treatment plants (WWTPs) have proven highly effective in removing a range of micropollutants, including pharmaceuticals and personal care products. For example, a study conducted in Queensland reported over 90% removal of acesulfame, a persistent artificial sweetener, through biological processes (Cardenas et al., 2016).

B. **Chemical Methods**

Ozone-based treatments, especially when combined with other processes such as UV treatment or zero-valent iron nanoparticles, have shown high efficiency in removing organochlorine pesticides (Derco et al., 2015). Furthermore, advanced oxidation technologies (AOTs) have effectively degraded pharmaceuticals and endocrine disruptors, although no single method is universally effective (Trapido et al., 2014).

C. **Physical Methods**

Powdered activated carbon (PAC) adsorption, followed by ultrafiltration, has demonstrated significant efficacy in reducing micropollutants, achieving over 80% removal in pilot studies (Margot et al., 2013). Additionally, membrane systems enhance the removal of dissolved micropollutants by utilizing their barrier properties (Battistoni et al., 2007).

Emerging Technologies

A. **Photocatalysis**

Semiconductor-supported photocatalysis has garnered attention for its effectiveness in degrading emerging pollutants (EPs) using visible light. Techniques such as elemental doping and morphology engineering are employed to enhance charge separation and light absorption (Ahmed et al., 2021).

Natural Attenuation and Green Infrastructure

Constructed Wetlands

A. **Mechanisms of Pollutant Removal**

Constructed wetlands (CWs) employ a combination of physical, chemical, and biological processes to treat wastewater. The vegetation and associated microbial communities are vital in degrading and transforming pollutants such as nitrogen, phosphorus, and organic matter (Greenway, 2004; Randerson, 2006). Additionally, factors like hydraulic retention time and pollutant loading rates in the design of CWs greatly affect their treatment efficiency (Greenway, 2004).

B. **Advantages of Constructed Wetlands**

CWs are both cost-effective and environmentally friendly, with lower capital and operational costs compared to traditional wastewater treatment systems (Selvamurugan et al., 2011). They also effectively remove organic micropollutants (MPs) and can perform well even at high contaminant concentrations (Kaur et al., 2020).

C. **Applications and Case Studies**

CWs have been successfully utilized for various applications, including agricultural reuse and stormwater management, demonstrating their versatility across different contexts (Kaur et al., 2020; Randerson, 2006).

Biochar in Soil and Water Remediation

A. **Biochar Production and Characteristics**

Biochar can be produced from agricultural waste, such as peanut shells, providing a sustainable disposal method while improving its adsorption properties when engineered with surfactants like cetyltrimethylammonium bromide (CTAB) (Murad et al., 2022). The production process significantly affects biochar's surface area and functional groups, which are essential for its effectiveness in removing contaminants (Zheng et al., 2020).

B. **Mechanisms of Contaminant Removal**

Biochar utilizes adsorption mechanisms such as chemisorption and ion exchange, making it effective for removing heavy metals like chromium and arsenic (Liu et al., 2021; Murad et al., 2022). Research indicates that engineered biochar achieves high removal efficiencies, often outperforming pristine forms in immobilizing contaminants (Murad et al., 2022).

C. **Applications in Remediation**

Biochar has been successfully utilized in various applications, including wastewater treatment and soil amendment, where it reduces the bioavailability and leachability of pollutants (Liu et al., 2021; Zheng et al., 2020). Additionally, combining biochar with biological and chemical treatment technologies can enhance the overall efficiency of micropollutant removal (Kushwaha et al., 2021).

Phytoremediation

A. **Mechanisms of Phytoremediation**

Phytostabilization: In this process, plants immobilize contaminants in the soil, preventing their spread (Chatterjee et al., 2013).

Phytoextraction: Hyperaccumulating plants absorb heavy metals from the soil and transport them to their above-ground tissues, where they can be harvested (Chatterjee et al., 2013).

Rhizofiltration: Aquatic plants utilize their root systems to remove pollutants from water (Chatterjee et al., 2013).

B. **Advantages of Phytoremediation**

Cost-Effectiveness: Phytoremediation can be four to 1,000 times less expensive than traditional non-biological methods (Sadowsky, 1999).

Environmental Benefits: This approach minimizes soil disturbance and promotes biodiversity (Chandra et al., 2015).

C. **Limitations**

Despite its advantages, phytoremediation has several limitations, including long remediation times and reduced effectiveness for deep soil contaminants (Sadowsky, 1999). Additionally, some plants may produce toxic metabolites, which can complicate the cleanup process (Sadowsky, 1999).

Monitoring and Detection Technologies

Real-Time Sensors and Data Analytics

A. **Real-time Monitoring Systems**

Portable Measurement Systems: A portable system has been developed to measure multipollutant emissions in real-time, with intervals of 1–6 s, enabling accurate emission ratios and source profiling (Wang et al., 2012).

Water Environment Management: Real-time regulation in water environments uses data acquisition and modeling to trace pollution sources and guide effective management plans for controlling pollutants (Zhang et al., 2011).

B. **Analytical Techniques**

Liquid Chromatography-Mass Spectrometry (LC–MS): LC–MS is crucial for detecting micropollutants like pharmaceuticals and personal care products that are often inadequately removed from the environment. It enables the analysis of highly polar or non-volatile contaminants (Sosa-Ferrera et al., 2012).

Sampling Efficiency: The choice of sampling mode affects the measurement of micropollutant concentrations, with flow proportional composite sampling being the most reliable for assessing removal efficiencies (Verlicchi & Ghirardini, 2019).

Remote Sensing and Environmental Monitoring Networks

A. **Remote Sensing Applications**

Remote sensing technologies are increasingly used for monitoring water quality and detecting micropollutants in real-time, enabling swift interventions (Musolff, 2009). These systems can be integrated with existing environmental monitoring networks to provide comprehensive data on the distribution and concentration of pollutants (Ort et al., 2009).

B. **Wastewater Treatment Enhancements**

Upgrading wastewater treatment plants (WWTPs) is essential since they are key sources of micropollutants. Specific improvements can reduce the release of harmful substances like diclofenac (Ort et al., 2009). Membrane systems such as ultrafiltration have demonstrated effectiveness in improving micropollutant removal from municipal wastewater, with operational optimization further enhancing efficiency (Battistoni et al., 2007).

C. **Pollution Prevention and Control**

A comprehensive approach to pollution prevention and control (PPC) is crucial, integrating social, economic, and environmental factors to effectively manage micropollutant emissions (Bigg, 2013). Education and strong regulatory frameworks play a pivotal role in fostering compliance and promoting sustainable practices, reducing the overall contribution of micropollutants from industries (Bigg, 2013).

Smart Monitoring Systems for Early Warning

A. **Advanced Treatment Technologies**

Ozone-Based Processes: Ozone treatments, including ozonation and O3/nZVI (nano zero-valent iron) processes, have shown high efficiency in removing organochlorine pesticides, notably reducing reaction times (Derco et al., 2015).

Advanced Oxidation Technologies (AOTs): AOTs, particularly those that use UV photolysis and hydrogen peroxide, have proven effective in degrading pharmaceuticals and endocrine disruptors. However, no single method is universally superior for all micropollutants (Trapido et al., 2014).

B. **Smart Monitoring Systems**

Early-Warning Systems: The use of early-warning monitoring systems can enhance the protection of drinking water sources by detecting micropollutants before they reach harmful concentrations (Bode, 2018).

Challenges in Implementation

Economic Barriers

A. **Economic Costs of Advanced Treatment Technologies**

Advanced treatment methods like Advanced Oxidation Technologies (AOTs) and membrane systems, though highly effective, involve substantial capital and operational costs. For example, treatments using ozone and powdered activated carbon (PAC) can be prohibitively expensive for many municipalities (Margot et al., 2013). The cost-effectiveness of ultrafiltration systems is also contested, especially when micropollutants are present in low concentrations, leading to concerns about their practical economic viability in real-world municipal applications (Battistoni et al., 2007).

Technical and Operational Issues

A. **Technical Challenges**

Inefficiency of Conventional Treatments: Traditional wastewater treatment plants (WWTPs) are often unable to sufficiently remove organic micropollutants, leading to the need for more advanced treatment methods (Margot et al., 2013).

Variable Removal Efficiency: Advanced oxidation technologies (AOTs) and similar methods show inconsistent results depending on the type of micropollutant, making it challenging to select the most effective treatment process (Trapido et al., 2014).

B. **Operational Issues**

Socio-Political Barriers: Institutional frameworks frequently focus on short-term technical solutions rather than sustainable, long-term practices, which delays the adoption of more comprehensive strategies (Brown et al., 2006).

Cost and Feasibility: Despite their proven effectiveness, advanced treatments like powdered activated carbon (PAC) or ozone are often constrained by high operational costs and local feasibility issues (Margot et al., 2013).

Regulatory and Policy Gaps

A. **Regulatory Complexity**

Micropollutants include a wide range of substances, making regulatory control challenging. Existing water quality regulations often overlook the numerous sources and varying impacts of these pollutants (Metz & Ingold, 2014). Furthermore, the absence of cohesive policies that bridge sectors like industry, agriculture, and civil society complicates effective management (Metz & Ingold, 2014).

B. **Decision-Making Under Uncertainty**

Policymakers struggle with decision-making due to the uncertainties regarding the effectiveness of different removal technologies and the long-term ecological consequences of micropollutants (Stamm et al., 2015).

Public Awareness and Engagement

A. **Public Awareness and Information Strategies**

Disclosure strategies are key to raising public awareness about pollution but are often underused when it comes to micropollutants (Tietenberg & Wheeler, 2001). Innovative communication methods, such as targeted campaigns, have proven effective in raising awareness in both OECD and developing countries (Tietenberg & Wheeler, 2001).

B. **Behavioral Change and Community Involvement**

Tackling micropollution requires behavioral shifts across multiple sectors, including civil society, but this is often hindered by a lack of awareness (Metz & Ingold, 2014). Engaging communities through educational initiatives can build a sense of responsibility and encourage proactive efforts to mitigate micropollutant impacts (Metz & Ingold, 2014).

Data Availability and Knowledge Gaps

A. **Data Availability Issues**

Many studies depend on short-term sampling campaigns, which can result in inaccurate calculations of micropollutant removal (Majewsky et al., 2013). This limitation affects the reliability of evaluating treatment efficiency. The fluctuating levels of pollutants in wastewater streams further complicate the consistency of data, highlighting the need for extended sampling periods to gather accurate results (Majewsky et al., 2013).

B. **Knowledge Gaps**

A comprehensive understanding of advanced treatment technologies like ozone oxidation and powdered activated carbon remains incomplete, especially across varying conditions (Margot et al., 2013). Additionally, the effectiveness of membrane systems for treating municipal wastewater is still under debate, underscoring the need for more research to fine-tune their operational performance (Battistoni et al., 2007).

Social and Political Resistance

A. **Local Governance and Participation**

In China, the enforcement of environmental policies at the local level often faces challenges due to competing economic and political interests, which limits the role of civil society in decision-making (Kostka & Mol, 2017). This lack of engagement from non-state actors creates a gap between national environmental goals and local realities, making it harder to effectively regulate micropollutants.

B. **Complexity of Micropollutants**

The diverse chemical nature of micropollutants complicates regulation, as traditional water quality management strategies may not sufficiently cover the range of their entry points and effects (Metz & Ingold, 2014). Addressing this complexity requires not only technological innovations but also behavioral changes across industries, which can encounter resistance from sectors reluctant to adapt.

C. **Community Resistance**

In Southeast Asia, local communities frequently resist large-scale resource extraction projects that contribute to environmental issues, including micropollution (Rasch & Kohne, 2016). This resistance stems from conflicts over livelihoods and power imbalances, highlighting the need for governance approaches that incorporate local perspectives and address social and environmental justice concerns.

Case Studies

A. **Eco Impact Study**

This interdisciplinary research examines the impact of wastewater-borne micropollutants on aquatic ecosystems, advocating for integrated management strategies to mitigate their effects (Stamm et al., 2016).

B. **Queensland WWTP Analysis**

A study conducted in Australia found that a biological wastewater treatment plant successfully removed 100% of various micropollutants, including analgesics and anti-inflammatories, while the artificial sweetener acesulfame had a removal rate exceeding 90% (Cardenas et al., 2016).

C. **Industrial WWTP Patterns**

Research across nine industrial wastewater treatment plants revealed that removal efficiencies vary by industry. Specific micropollutants, such as chlorinated phenols, were linked to particular sectors like the paper industry (Lee et al., 2011).

D. **Urban Runoff Management**

In Noisy-le-Grand, France, alternative techniques for managing urban runoff reduced contaminant loads by 20–80%, showcasing the effectiveness of upstream interventions. These strategies help in mitigating pollution before it enters water systems, emphasizing the importance of preventative measures in urban water management (Bressy, 2010).

Future Directions and Innovations

A. **Innovative Materials**

Carbon nanotubes (CNTs) and their composites have emerged as promising materials for addressing water treatment challenges due to their high surface area and excellent adsorption properties, making them particularly effective in removing micropollutants (Khan et al., 2021).

B. **Research and Policy Directions**

To enhance understanding of micropollutant fate and improve treatment outcomes, future research should adopt holistic approaches. This includes integrating analytical techniques with environmental modeling to track micropollutants across ecosystems, guiding more effective regulatory policies (Angeletti & Bjørseth, 2012; Daughton, 2004).

Conclusion

Micropollutants are small organic chemicals, such as pharmaceuticals and pesticides, that enter water systems through urban, agricultural, and industrial sources. These pollutants pose significant risks to human health and the environment, even in very low concentrations. Controlling micropollutants requires a combination of source reduction, advanced treatment technologies, and public engagement. Strategies include enhanced wastewater treatments like ozonation, adsorption, and membrane filtration, as well as natural methods like bioremediation and constructed wetlands. Despite progress, challenges such as economic costs, technical inefficiencies, and regulatory gaps remain barriers to effective implementation.

References

Abbas, M. A., Iqbal, M., Tauqeer, H. M., Turan, V., & Farhad, M. (2022). Microcontaminants in wastewater. In *Environmental micropollutants* (pp. 315–329). Elsevier.

Ahmed, S., Khan, F. S. A., Mubarak, N. M., Khalid, M., Tan, Y. H., Mazari, S. A., Abdullah, E. C., et al. (2021). Emerging pollutants and their removal using visible-light responsive photocatalysis—A comprehensive review *Journal of Environmental Chemical Engineering, 9*(6), 106643.

Alexander, M. (1999). Biodegradation and bioremediation. *Academic Press Google Schola, 2*, 363–378.

Angeletti, G., & Bjørseth, A. (Eds.). (2012). *Organic micropollutants in the aquatic environment: Proceedings of the Fifth European Symposium, Held in Rome, Italy, October 20–22, 1987*. Springer Science & Business Media.

Arman, N. Z., Salmiati, S., Aris, A., Salim, M. R., Nazifa, T. H., Muhamad, M. S., & Marpongahtun, M. (2021). A review on emerging pollutants in the water environment: Existences, health effects and treatment processes. *Water, 13*(22), 3258.

Bala, K., Anshul, Verma, S., Dogra, A., Das, A. P., Kumar, R., & Nandini. (2024). Sustainable management for urban plastic waste generation. In *Sustainable management of urban plastic waste through circular economic approaches* (pp. 211–236). https://doi.org/10.1201/9781003465447-13

Battistoni, P., Cola, E., Fatone, F., Bolzonella, D., & Eusebi, A. L. (2007). Micropollutants removal and operating strategies in ultrafiltration membrane systems for municipal wastewater treatment: Preliminary results. *Industrial & Engineering Chemistry Research, 46*(21), 6716–6723.

Bigg, M. (2013). Pollution prevention and control. In *Environmental Management in Organizations* (pp. 213–238). Routledge.

Bode, F. (2018). *Early-warning monitoring systems for improved drinking water resource protection*. Eigenverlag des Instituts für Wasser-und Umweltsystemmodellierung der Universität Stuttgart.

Bressy, A. (2010). *Flux de micropolluants dans les eaux de ruissellement urbaines: effets de différents modes de gestion à l'amont* (Doctoral dissertation, Université Paris-Est).

Brown, R. R., Sharp, L., & Ashley, R. M. (2006). Implementation impediments to institutionalising the practice of sustainable urban water management. *Water Science and Technology, 54*(6–7), 415–422.

Cardenas, M. A. R., Ali, I., Lai, F. Y., Dawes, L., Thier, R., & Rajapakse, J. (2016). Removal of micropollutants through a biological wastewater treatment plant in a subtropical climate, Queensland-Australia. *Journal of Environmental Health Science and Engineering, 14*, 1–10.

Chandra, R. (Ed.). (2015). *Advances in biodegradation and bioremediation of industrial waste*. CRC Press.

Chandra, R., Saxena, G., & Kumar, V. (2015). Phytoremediation of environmental pollutants: an eco-sustainable green technology to environmental management. In *Advances in biodegradation and bioremediation of industrial waste* (pp. 1–29).

Chatterjee, S., Mitra, A., Datta, S., & Veer, V. (2013). Phytoremediation protocols: an overview. In *Plant-based remediation processes* (pp. 1–18).

Das, A., Dey, S., & Das, A. P. (2024a). Ecotoxicological impacts of synthetic microfiber pollutants and development of sustainable mitigation strategies. *Environmental Chemistry and Ecotoxicology*. https://doi.org/10.1016/J.ENCECO.2024.12.003

Das, A., Dey, S., Bala, K., Gnanasekaran, R., & Das, A. P. (2024b). Microbial bioremediation technology for sustainable treatment and management of synthetic microfiber waste. *Sustainable Microbial Technology for Synthetic and Cellulosic Microfiber Bioremediation, 221–238*. https://doi.org/10.1007/978-3-031-62600-5_12

Das, A., Ray, S. K., Mohanty, M., Mohanty, J., Dey, S., & Das, A. P. (2024c). Ecotoxicity of microplastic wastes and their sustainable management: A review. *Environmental Chemistry and Ecotoxicology, 6*, 144–152. https://doi.org/10.1016/J.ENCECO.2024.05.003

Das, A. P., Mallavarapu, M., & Ghosh, S. (2023a). Ecotoxicity of mining pollutants on the environment and their remediation. *Environmental Chemistry and Ecotoxicology, 5*, 165–167. https://doi.org/10.1016/J.ENCECO.2023.08.002

Das, P., & Bal, M. (2025). Modeling and optimization study for microplastic removal from aquatic medium using Chroococcidiopsis species. *Chemical Engineering Science, 304*, 121085. https://doi.org/10.1016/j.ces.2024.121085

Das, P., Halder, G., & Bal, M. (2023b). A critical review on remediation of microplastics using microalgae from aqueous system. *Science of the Total Environment, 898*, 166425. https://doi.org/10.1016/j.scitotenv.2023.166425

Daughton, C. G. (2004). PPCPs in the environment: Future research—Beginning with the end always in mind. In *Pharmaceuticals in the Environment: Sources, fate, effects and risks* (pp. 463–495). Springer Berlin Heidelberg.

De Ridder, D. J., McConville, M., Verliefde, A. R. D., Van der Aa, L. T. J., Heijman, S. G. J., Verberk, J. Q. J. C., Dijk, J. C., et al. (2009). Development of a predictive model to determine micropollutant removal using granular activated carbon. *Drinking Water Engineering and Science, 2*(2), 57–62.

Derco, J., Dudáš, J., Valičková, M., Šimovičová, K., & Kecskés, J. (2015). Removal of micropollutants by ozone based processes. *Chemical Engineering and Processing-Process Intensification, 94*, 78–84.

Dey, S., Bissoyi, A., Koteswara Reddy, G., & Das, A. P. (2024a). Synthetic microfibers: An urban emerging pollutant demanding for sustainable environmental solutions. *Sustainable Microbial Technology for Synthetic and Cellulosic Microfiber Bioremediation, 107–121*. https://doi.org/10.1007/978-3-031-62600-5_7

Dey, S., Dash, N., Ray, S. K., Priyadarsini, S., Chhotaray, C., & Das, A. P. (2024b). Occurrence and quantification of tire wear plastics pollutants from highway city dust of Bhubaneswar City of Odisha. *Sustainable Microbial Technology for Synthetic and Cellulosic Microfiber Bioremediation, 77–89*. https://doi.org/10.1007/978-3-031-62600-5_5

Dey, S., Tripathy, B., Kumar, M. S., & Das, A. P. (2023). Ecotoxicological consequences of manganese mining pollutants and their biological remediation. *Environmental Chemistry and Ecotoxicology*. https://doi.org/10.1016/J.ENCECO.2023.01.001

Drewes, J. E., Bellona, C., Oedekoven, M., Xu, P., Kim, T. U., & Amy, G. (2005). Rejection of wastewater-derived micropollutants in high-pressure membrane applications leading to indirect potable reuse. *Environmental Progress, 24*(4), 400–409.

Forsyth, D. R., García, M., Zyzniewski, L. E., Story, P. A., & Kerr, N. A. (2004). Watershed pollution and preservation: The awareness–appraisal model of environmentally positive intentions and behaviors. *Analyses of Social Issues and Public Policy, 4*(1), 115–128.

Gaid, K. (2023). Removal of micropollutants. In K. Gaid (Ed.), *Drinking water treatment* (vol. 3). https://doi.org/10.1002/9781394226061.ch10

Greenway, M. (2004). Constructed wetlands for water pollution control-processes, parameters and performance. *Developments in Chemical Engineering and Mineral Processing, 12*(5–6), 491–504.

He, L., Yang, C., Ding, J., Lu, M. Y., Chen, C. X., Wang, G. Y., Yang, S. S., et al. (2022). Fe, N-doped carbonaceous catalyst activating periodate for micropollutant removal: Significant role of electron transfer. *Applied Catalysis B: Environmental, 303*, 120880.

Kanungo, J., Sahoo, T., Swain, L.P., & Behera, I.D. (2023). Toxicity of persistent hydrocarbon pollutants, sources and sustainable remediation process. In: I. D. Behera, A. P. Das (Eds.), *Impact of petroleum waste on environmental pollution and its sustainable management through circular economy*. Environmental Science and Engineering. https://doi.org/10.1007/978-3-031-48220-5_2

Kaur, R., Kaur, H., Singh, S., Jagota, N., Sharma, A., & Sharma, A. (2024). Organic micropollutants in the Urban Soils: Technological advances and management concerns. In *Organic micropollutants in aquatic and terrestrial environments* (pp. 47–63).

Kaur, R., Talan, A., Tiwari, B., Pilli, S., Sellamuthu, B., & Tyagi, R. D. (2020). Constructed wetlands for the removal of organic micro-pollutants. In *Current developments in biotechnology and bioengineering* (pp. 87–140). Elsevier.

Khan, F. S. A., Mubarak, N. M., Khalid, M., Tan, Y. H., Abdullah, E. C., Rahman, M. E., & Karri, R. R. (2021). A comprehensive review on micropollutants removal using carbon nanotubes-based adsorbents and membranes. *Journal of Environmental Chemical Engineering, 9*(6), 106647.

Kostka, G., & Mol, A. P. (2017). Implementation and participation in China's local environmental politics: Challenges and innovations. In *Local environmental politics in China* (pp. 1–14). Routledge.

Kovalova, L., Knappe, D. R., Lehnberg, K., Kazner, C., & Hollender, J. (2013). Removal of highly polar micropollutants from wastewater by powdered activated carbon. *Environmental Science and Pollution Research, 20*, 3607–3615.

Kushwaha, A., Goswami, S., Hans, N., Goswami, L., Devi, G., Deshavath, N. N.,... & Lall, A. M. (2021). An insight into biological and chemical technologies for micropollutant removal from wastewater. In *Fate and transport of subsurface pollutants* (pp. 199–226).

Lee, I. S., Sim, W. J., Kim, C. W., Chang, Y. S., & Oh, J. E. (2011). Characteristic occurrence patterns of micropollutants and their removal efficiencies in industrial wastewater treatment plants. *Journal of Environmental Monitoring, 13*(2), 391–397.

Liu, M., Sun, F., Lv, Y., Xu, Y., Li, M., Wang, Y., Jiang, H., et al. (2021). Remediation of arsenic-contaminated soil by nano-zirconia modified biochar. *Environmental Science and Pollution Research, 28*, 68792–68803.

Majewsky, M., Farlin, J., Bayerle, M., & Gallé, T. (2013). A case-study on the accuracy of mass balances for xenobiotics in full-scale wastewater treatment plants. *Environmental Science: Processes & Impacts, 15*(4), 730–738.

Margot, J., Kienle, C., Magnet, A., Weil, M., Rossi, L., De Alencastro, L. F., Barry, D. A., et al. (2013). Treatment of micropollutants in municipal wastewater: Ozone or powdered activated carbon?. *Science of the Total Environment, 461*, 480–498.

Margot, J., Magnet, A., Thonney, D., Chèvre, N., De Alencastro, F., & Rossi, L. (2011). *Traitement des micropolluants dans les eaux usées: rapport final sur les essais pilotes à la STEP de Vidy (Lausanne)*. Ville de Lausanne.

Markert, N., Schürings, C., & Feld, C. K. (2024). Water Framework Directive micropollutant monitoring mirrors catchment land use: Importance of agricultural and urban sources revealed. *Science of the Total Environment, 917*, 170583.

Meinel, F., Ruhl, A. S., Sperlich, A., Zietzschmann, F., & Jekel, M. (2015). Pilot-scale investigation of micropollutant removal with granular and powdered activated carbon. *Water, Air, & Soil Pollution, 226*, 1–10.

Metz, F., & Ingold, K. (2014). Sustainable wastewater management: Is it possible to regulate micropollution in the future by learning from the past? A Policy Analysis. *Sustainability, 6*(4), 1992–2012.

Meyerstein, D. (2021). Re-examining Fenton and Fenton-like reactions. *Nature Reviews Chemistry, 5*(9), 595–597.

Mishra, S., & Das, A. P. (2021). Current treatment technologies for removal of microplastic and microfiber pollutants from wastewater. *Wastewater Treatment: Cutting-Edge Molecular Tools, Techniques and Applied Aspects, 237–251.* https://doi.org/10.1016/B978-0-12-821881-5.00011-8

Mishra, S., Dash, D., & Das, A. P. (2022). Detection, characterization and possible biofragmentation of synthetic microfibers released from domestic laundering wastewater as an emerging source of marine pollution. *Marine Pollution Bulletin, 185*, 114254. https://doi.org/10.1016/J.MARPOLBUL.2022.114254

Mohanty, J., Dey, S., Bala, K., & Das, A. P. (2024). Recent advances in microplastic pollution for its sustainable management. *Sustainable Management of Urban Plastic Waste through Circular Economic Approaches, 27–48.* https://doi.org/10.1201/9781003465447-3

Mohanta, S., Pradhan, B., & Behera, I. D. (2023). Impact and remediation of petroleum hydrocarbon pollutants on agricultural land: A review. *Geomicrobiology Journal, 41*(4), 345–359. https://doi.org/10.1080/01490451.2023.2243925

Mostafalou, S. (2016). Persistent organic pollutants and concern over the link with insulin resistance related metabolic diseases. In *Reviews of environmental contamination and toxicology* (pp. 69–89).

Murad, H. A., Ahmad, M., Bundschuh, J., Hashimoto, Y., Zhang, M., Sarkar, B., & Ok, Y. S. (2022). A remediation approach to chromium-contaminated water and soil using engineered biochar derived from peanut shell. *Environmental Research, 204*, 112125.

Musolff, A. (2009). Micropollutants: Challenges in hydrogeology. *Hydrogeology Journal, 17*(4), 763–766.

Ort, C., Hollender, J., Schaerer, M., & Siegrist, H. (2009). Model-based evaluation of reduction strategies for micropollutants from wastewater treatment plants in complex river networks. *Environmental Science & Technology, 43*(9), 3214–3220.

Prengle, H. W. (1983). Experimental rate constants and reactor considerations for the destruction of micropollutants and trihalomethane precursors by ozone with ultraviolet radiation. *Environmental Science & Technology, 17*(12), 743–747.

Priyadarshini, B., Santhosh Kumar, M., & Das, A. P. (2024). Microbial degradation of synthetic plastics released from COVID-19 generated single use disposable face masks. *Environmental Science and Engineering, Part F2827*, 1–16. https://doi.org/10.1007/978-3-031-50840-0_1

Randerson, P. F. (2006). Constructed wetlands and vegetation filters: An ecological approach to wastewater treatment. *Environmental Biotechnology, 2*, 78–89.

Rasch, E. D., & Köhne, M. (2016). Micropolitics in resistance: The micropolitics of large-scale natural resource extraction in South East Asia. *Society & Natural Resources, 29*(4), 479–492.

Sadowsky, M. J. (1999). Phytoremediation: past promises and future practices. In *Proceedings of the 8th international symposium on microbial ecology. Halifax, Canada* (pp. 1–7).

Sahu, A., Dey, S., Mallick, K., Priyadarsini, S., Behera, I. D., & Das, A. P. (2024). A review on advanced technology for sustainable management of synthetic microplastic waste. *Sustainable Management of Urban Plastic Waste through Circular Economic Approaches, 262–279.* https://doi.org/10.1201/9781003465447-15

Selvamurugan, M., Doraisamy, P., Maheswari, M., & Akumar, N. B. (2011). Constructed wetlands for wastewater treatment: A review. *Research & Reviews in BioSciences, 5*(2), 100–105.

Sosa-Ferrera, Z., Mahugo-Santana, C., & Santana-Rodríguez, J. J. (2012). New developments in liquid chromatography mass spectrometry for the determination of micropollutants. *Chromatography Research International, 2012*(1), 748989.

Stamm, C., Eggen, R. I., Hering, J. G., Hollender, J., Joss, A., & Schärer, M. (2015). *Micropollutant removal from wastewater: Facts and decision-making despite uncertainty.*

Stamm, C., Räsänen, K., Burdon, F. J., Altermatt, F., Jokela, J., Joss, A.,... & Eggen, R. I. (2016). Unravelling the impacts of micropollutants in aquatic ecosystems: interdisciplinary studies at the interface of large-scale ecology. In *Advances in ecological research* (Vol. 55, pp. 183–223). Academic Press.

Tietenberg, T., & Wheeler, D. (2001). Empowering the community: Information strategies for pollution control. *Frontiers of Environmental Economics*, 85–120.
Trapido, M., Epold, I., Bolobajev, J., & Dulova, N. (2014). Emerging micropollutants in water/wastewater: Growing demand on removal technologies. *Environmental Science and Pollution Research, 21*, 12217–12222.
Verlicchi, P., & Ghirardini, A. (2019). Occurrence of micropollutants in wastewater and evaluation of their removal efficiency in treatment trains: The influence of the adopted sampling mode. *Water, 11*(6), 1152.
Verstraeten, I. M., Heberer, T., & Scheytt, T. (2003). Occurrence, characteristics, transport, and fate of pesticides, pharmaceuticals, industrial products, and personal care products at riverbank filtration sites. In *Riverbank filtration: Improving source-water quality* (pp. 175–227).
Wang, X., Watson, J. G., Chow, J. C., Gronstal, S., & Kohl, S. D. (2012). An efficient multipollutant system for measuring real-world emissions from stationary and mobile sources. *Aerosol and Air Quality Research, 12*(2), 145–160.
Zhang, S., Le, S., & Zhou, P. (2011). Method for performing real-time regulation and control and digital management on water environment.
Zheng, H., Zhang, C., Liu, B., Liu, G., Zhao, M., Xu, G.,... & Xing, B. (2020). Biochar for water and soil remediation: Production, characterization, and application. In *A new paradigm for environmental chemistry and toxicology: From concepts to insights* (pp. 153–196).

Chapter 12
Various Strategies for Controlling Micropollutants, Along with the Challenges Faced in Implementation

Swayam Prabha Misra and Snehlata Kumari

Introduction

Micropollutants, often referred to as emerging contaminants, are trace-level substances such as pharmaceuticals, personal care product, pesticides, and industrial chemicals that persist in the environment (Berg et al., 1997; Chavoshani et al., 2020a, 2020b; Krishnan et al., 2021). These pollutants, though present in minute concentrations, pose significant risks to ecosystems and human health due to their bioaccumulative and persistent nature (Filote et al., 2021). They can exist across all three environmental phases: air, water, and soil/sediment (Barbosa et al., 2016). They disrupt aquatic ecosystems by affecting the reproductive and physiological health of aquatic organisms, leading to ecological imbalances and biodiversity loss (Burkhardt-Holm, 2011; Gul et al., 2022; Shao et al., 2019). Micropollutants also infiltrate soil and groundwater, impacting crop health and drinking water quality (Elliott et al., 2018; Moghadam et al., 2023, Tang et al., 2013). These pollutants are frequently not eliminated by conventional wastewater treatment facilities, thus sophisticated remediation solutions are required to lessen their negative environmental consequences.

Micropollutants significantly impact aquatic ecosystems, as many of these contaminants enter water bodies through untreated wastewater, agricultural runoff, and industrial discharges (Eggen et al., 2014). Pharmaceuticals and personal care products, for instance, disrupt the hormonal systems of aquatic species, leading to issues such as endocrine disruption, which can cause reproductive failures in fish and

S. P. Misra
Department of Chemistry, Indian Institute of Technology Gandhinagar, Gandhinagar, India

S. Kumari (✉)
Naresh Vashisht Centre for Hydrogen and CCUS Technologies, Indian Institute of Technology (Indian School of Mines) Dhanbad, Dhanbad, India
e-mail: snehlata.che@gmail.com

I. D. Behera et al. (eds.), *Emerging Micropollutants*, Sustainable Environmental Waste Management Strategies, https://doi.org/10.1007/978-3-032-03530-1_12

amphibians (Overturf et al., 2015). Pesticides and herbicides in water bodies affect the microbial communities vital for maintaining ecological balance (Kumar et al., 2021). Persistent organic pollutants accumulate in aquatic organisms, entering the food chain and potentially harming higher trophic levels, including humans (Huang et al., 2021). These contaminants also degrade water quality, posing challenges for safe drinking water supply and recreational water use (Pal et al., 2014).

In soil, micropollutants such as industrial chemicals, antibiotics, and heavy metals alter the natural microbial balance and fertility (Becerra-Castro et al., 2015). These chemicals may prevent helpful soil microorganisms from efficiently breaking down organic materials and cycling nutrients (Arslan et al., 2017). Antibiotics in the soil contribute to the proliferation of antibiotic-resistant bacteria (Ghosh & LaPara, 2007), posing a significant threat to public health. Pesticide residues in the soil can accumulate over time, reducing agricultural productivity and harming non-target organisms such as earthworms and pollinators (Ali et al., 2021). Furthermore, leaching of micropollutants from soil into groundwater exacerbates the contamination of water resources, creating a cascading effect on ecosystems (Pérez-Lucas et al., 2019).

Micropollutants in the air are primarily in the form of vapors, particulate matter and various gaseous resources (Butt et al., 2022). Airborne micropollutants, such as volatile organic compounds (VOCs), microplastics, and particulate matter, contribute to air quality degradation and health risks (Arfin et al., 2023). VOCs, which are emitted from domestic and industrial operations, combine with nitrogen oxides to create the dangerous pollutant known as ground-level ozone (Zhang et al., 2019). Microplastics, transported through wind and atmospheric deposition, settle in remote ecosystems, demonstrating their pervasive nature (Belioka & Achilias, 2024). Long-term exposure to these airborne contaminants can lead to respiratory and cardiovascular diseases in humans. Additionally, the presence of micropollutants in the atmosphere influences climate change by altering cloud formation and precipitation patterns (Singh & Kumar, 2022). Besides this three distinct phases the micropollutants are also have interactions across phases. Many micropollutants partition between water, soil, and air based on their physicochemical properties (e.g., Henry's Law constant, Kow) (Herrmann, 1987; Pomiès et al., 2013). Processes like volatilization, sorption–desorption, biodegradation, and photolysis can redistribute micropollutants among these phases (Imoobe et al., 2024).

Addressing the challenges posed by micropollutants requires comprehensive strategies, including stricter regulations, advanced treatment technologies, and increased public awareness. By reducing their release into the environment, we can mitigate their harmful effects on air, water, and soil, ultimately protecting ecosystems and human health.

Existing Strategies to Control Micropollutants

We discussed about the micropollutants found in different environmental phases and their effect on the environment. Now, in this section we will discuss about the strategies to control these micropollutants. The following section provides an overview about the strategies followed in all three environmental phases to control the micropollutants.

Strategies for Controlling Micropollutants in Soil

Controlling micropollutants in soil is a critical step toward maintaining soil health, protecting water resources, and ensuring sustainable agricultural practices (Adepoju et al., 2024). These micropollutants, which include pesticides, pharmaceuticals, heavy metals, and industrial chemicals, pose a serious threat to soil ecosystems and can have cascading effects on other environmental compartments like water and air (S. Ali et al., 2021; Barbosa et al., 2016; Berg et al., 1997; Chavoshani et al., 2020a, 2020b; McGrath & McCormack, 1999; Vodyanitskii & Kirilova, 2015). Effective control strategies must integrate prevention, remediation, and public awareness to mitigate their impact.

Source Control

One of the most effective strategies for managing micropollutants is controlling their entry into the soil at the source. Regulatory measures play a pivotal role in this approach. Governments and environmental agencies should enforce stricter regulations on the production, use, and disposal of chemicals that contribute to soil pollution. For example, restricting the use of persistent pesticides and promoting safer alternatives can significantly reduce their accumulation in the soil.

Source control offers the advantage of preventing contamination at its origin, reducing the need for costly remediation later. It promotes sustainable practices that benefit ecosystems in the long term. However, due to increased costs and potential reductions in productivity, industries and farmers can resist implementing stricter regulations (Bala et al., 2024; Behera et al., 2021, 2022; Das et al., 2024b; Kanungo et al., 2023; Mohanta et al., 2024). Transitioning to sustainable agricultural practices, such as integrated pest management (IPM), may require extensive training and initial investment, which small-scale farmers may find to be a hurdle.

Additionally, sustainable farming methods are essential. Through the use of crop rotation, biological control techniques, and resistant crop types, integrated pest management (IPM) reduces dependency on chemical pesticides (Zhou et al., 2024). Organic farming practices, which avoid synthetic fertilizers and pesticides, further reduce the introduction of harmful substances into the soil. In addition, proper disposal of pharmaceutical and industrial waste can prevent these pollutants from reaching agricultural lands. Establishing safe collection systems for unused medications and hazardous household chemicals is an essential step in source control.

Soil Remediation Technologies

When soils are already contaminated, remediation techniques can help restore their health. Bioremediation is a sustainable and widely studied method for addressing micropollutants (P. Singh et al., 2020). This method entails breaking down or immobilizing contaminants using microbes or plants. For instance, phytoremediation is an environmentally benign method that uses particular plant species to absorb, store, or detoxify pollutants (Priya et al., 2023).

The advantage of bioremediation lies in its cost-effectiveness and minimal environmental impact compared to physical or chemical methods (McIntyre, 2003). It also improves soil quality by restoring microbial diversity (Kumar et al., 2023). However, bioremediation takes a lot of time and might not work for every kind of micropollutant, particularly those that are highly persistent or toxic (Zeghoud et al., 2024). The success of phytoremediation is also dependent on plant species, soil type, and environmental conditions, which can limit its applicability.

Soil washing is another effective remediation method, particularly for soils contaminated with heavy metals or organic pollutants (Dey et al., 2023, 2024; Mallick et al., 2024; Mohanty et al., 2024a, 2024b). This process involves flushing the soil with water or chemical solutions to extract contaminants. The contaminated wash solution is then treated separately to prevent secondary pollution. Thermal treatment, though energy-intensive, is useful for breaking down persistent organic pollutants (POPs) in the soil by applying heat (Biabani Reshtehroudi, 2024).

One benefit of soil washing is that it is quite effective in eliminating a variety of pollutants, which makes it appropriate for areas that are extremely polluted. However, it can be costly and may result in the loss of essential soil nutrients and structure (Jelusic & Lestan, 2014; Voglar & Lestan, 2014). Thermal treatment is effective for certain pollutants but has significant disadvantages, including high energy consumption and the potential release of toxic gases during the process.

Wastewater and Biosolid Management

Many micropollutants in soil originate from wastewater and biosolids applied as fertilizers (Mohapatra et al., 2016). Before the water is recycled or released, micropollutants can be efficiently removed using modern wastewater treatment technologies such membrane filtration, activated carbon adsorption, and advanced oxidation processes (Gutiérrez et al., 2021; Perrich, 2018). Ensuring that biosolids meet safety standards before their application in agriculture can significantly reduce the introduction of harmful substances into the soil. Pre-treatment processes, such as composting or anaerobic digestion, can help minimize the concentration of micropollutants in biosolids (Congilosi & Aga, 2021).

The advantage of advanced wastewater treatment is its ability to significantly reduce micropollutant loads, protecting both soil and water resources. However, these technologies often require substantial investment and operational expertise, making them less accessible in low-income regions (Massoud et al., 2009). Similarly, while biosolid management ensures nutrient recycling, improper treatment or application can lead to secondary pollution, posing risks to both soil and human health.

Monitoring and Risk Assessment

Finding contamination hotspots and assessing the efficacy of mitigation methods require constant soil micropollutant monitoring (Kardos et al., 2024). Soil monitoring programs should include regular sampling and analysis for key pollutants, especially in areas prone to contamination, such as agricultural lands and industrial zones (Vershinin et al., 2016).

The advantage of monitoring and risk assessment is that it provides critical data for informed decision-making, allowing for targeted interventions that maximize resource efficiency. However, these programs can be expensive and labor-intensive, requiring specialized equipment and expertise. Inadequate monitoring infrastructure in developing regions can further limit their effectiveness.

Risk-based land management strategies can further mitigate soil pollution (Adepoju et al., 2024). For instance, restricting the application of agrochemicals in sensitive areas, such as near water bodies or in regions with highly permeable soils, can prevent leaching and groundwater contamination (Pérez-Lucas et al., 2019). Mapping vulnerable areas and implementing targeted interventions ensures efficient resource allocation and pollution control.

Public Awareness and Education

Raising public awareness about the impact of micropollutants on soil health and the broader environment is critical for fostering sustainable practices. Educational campaigns can inform farmers about alternative farming practices, such as using organic amendments and implementing IPM strategies (Zhou et al., 2024). Similarly, industries and households must be encouraged to dispose of chemicals and pharmaceuticals responsibly. Community engagement and training programs can empower individuals to contribute to pollution prevention.

The advantage of public awareness campaigns is their ability to drive behavioral change at the grassroots level, promoting collective action against pollution (Brevik et al., 2019). However, these initiatives often require sustained efforts and resources to reach diverse audiences effectively. Resistance to change and lack of immediate incentives can also limit their impact.

Research and Development

Innovative research is fundamental to addressing the complex challenge of micropollutants in soil. Developing biodegradable pesticides, green solvents, and other eco-friendly alternatives can reduce reliance on harmful chemicals. Research into advanced remediation technologies, such as nanotechnology-based solutions and bioengineered microorganisms, holds promise for more efficient and cost-effective pollutant removal.

The advantage of investing in research and development is the potential to discover groundbreaking solutions that could transform soil pollution management. These innovations can offer long-term sustainability and economic benefits. However, the high costs and extended timelines associated with research can delay practical implementation. Additionally, translating laboratory-scale solutions into field applications remains a significant challenge.

Micropollutants in soil are a developing environmental issue that has significant effects on ecosystems and human health. Addressing this issue requires a multifaceted approach that combines source control, remediation technologies, effective monitoring, public awareness, and continuous research. By adopting these strategies, we can mitigate the harmful effects of micropollutants, restore soil health, and ensure the sustainable use of land resources for future generations.

Strategies for Controlling Micropollutants in Water

Conventional wastewater treatment processes often struggle to fully remove micropollutants due to their chemical stability and small size. Advanced treatment methods such as activated carbon filtration, advanced oxidation processes, and membrane filtration are more effective but are costly and energy-intensive (Deng & Zhao, 2015; Mansor et al., 2024; Reungoat et al., 2012).

Because micropollutants have the ability to damage aquatic ecosystems, taint drinking water, and endanger human health, their presence in water systems is a serious environmental problem. Pharmaceuticals, insecticides, industrial chemicals, and personal care items are examples of micropollutants that find their way into water bodies through urban stormwater, agricultural runoff, and wastewater discharge (Berg et al., 1997; Chavoshani et al., 2020a, 2020b). Addressing this issue requires a combination of source control, advanced treatment technologies, public awareness, and innovative research.

Source Control

Preventing micropollutants from entering water systems is one of the most effective ways to mitigate their impact. Regulatory measures, such as enforcing restrictions on hazardous chemicals, promoting eco-friendly products, and mandating stringent discharge standards for industries, can significantly reduce the release of harmful substances (Puri et al., 2023). By addressing the problem at its source, source control reduces the overall burden on water treatment systems, encourages sustainable practices in industries and agriculture, and minimizes risks to aquatic ecosystems and public health (Boelee et al., 2019). However, these measures require significant investment in policy enforcement and monitoring, and industries and farmers may face higher operational costs. Additionally, implementing and ensuring compliance with such regulations can be particularly challenging in developing regions where resources may be limited (Qadir et al., 2010). Public awareness campaigns and take-back programs for personal care products and pharmaceuticals are also vital to ensure proper disposal, thus preventing these substances from entering wastewater systems.

Advanced Wastewater Treatment Technologies

Municipal and industrial wastewater treatment plants serve as critical intervention points for removing micropollutants before they reach natural water bodies. Advanced oxidation processes (AOPs), membrane filtration, and activated carbon adsorption are among the technologies that have shown great promise in breaking down or eliminating these pollutants (Deng & Zhao, 2015; Mansor et al., 2024; Oe et al., 1996; Perrich, 2018).

For instance, ultrafiltration and reverse osmosis can achieve high removal efficiencies, while AOPs, including ozonation and UV irradiation, break down micropollutants into less harmful compounds (Derco et al., 2015). These techniques can be incorporated into already-existing treatment facilities and greatly enhance the quality of water for usage downstream. But putting this cutting-edge technology into practice is frequently linked to expensive operating and maintenance expenses. Processes like reverse osmosis and AOPs are energy-intensive, and the treatment systems may generate secondary waste that requires further handling (Dhamorikar et al., 2024; Umar et al., 2015). Despite these challenges, the benefits of improved water quality and environmental health make these technologies indispensable in addressing micropollutant contamination.

Natural Treatment Systems

Nature-based solutions, such as constructed wetlands and biofilters, offer a sustainable alternative for treating water contaminated with micropollutants (Plaimart, 2023). These systems leverage the natural processes of plants, microorganisms, and soil to degrade or remove contaminants. One of their key advantages is their eco-friendliness and cost-effectiveness. They also improve biodiversity and provide additional ecosystem services, including as habitat creation and carbon sequestration. Once established, natural treatment systems require relatively low maintenance. However, their capacity to handle high concentrations of micropollutants is limited, and their effectiveness varies depending on the type of pollutant and local environmental conditions. Additionally, implementing these systems on a large scale necessitates significant land area, which can be a constraint in urban or densely populated regions.

Monitoring and Risk Assessment

Continuous monitoring of water bodies is essential for identifying contamination hotspots and assessing the risks posed by micropollutants. Advanced analytical techniques, such as mass spectrometry, enable the detection of pollutants at trace levels, providing valuable data for targeted interventions and resource allocation. Effective monitoring ensures compliance with water quality standards and enhances public trust in water management practices. However, the process requires specialized equipment and technical expertise, making it costly and time-intensive. Monitoring programs may also struggle to keep pace with emerging micropollutants unless detection methods are regularly updated. Despite these drawbacks, robust monitoring systems are critical for informed decision-making and proactive water management.

Public Awareness and Education

Educating the public about the impact of micropollutants on water quality is crucial for fostering behavioral change. Awareness campaigns can encourage individuals to reduce the use of harmful chemicals, adopt sustainable practices, and participate in take-back programs for pharmaceuticals and hazardous waste. This promotes community involvement in pollution prevention, reduces the load of pollutants entering water systems, and encourages long-term sustainable practices. However, public participation is crucial to the effectiveness of these programs and can be impacted by a variety of factors, including cultural attitudes and socioeconomic realities. Sustained funding and effort are also required to maintain the momentum of such campaigns.

Research and Development

Investing in research to develop innovative solutions for micropollutant removal is critical for addressing the challenges posed by these contaminants. Emerging technologies such as nanotechnology, bioengineered microorganisms, and hybrid treatment systems hold promise for improving water treatment efficiency. Research drives innovation, enabling the development of targeted approaches for specific pollutants and enhancing the cost-efficiency of treatment methods. However, high initial costs and long development timelines can hinder the immediate applicability of new technologies. Additionally, scaling up experimental solutions for real-world applications remains a challenge (Dong et al., 2022). Despite these limitations, research and development play a pivotal role in ensuring the availability of long-term solutions for micropollutant control.

Micropollutants in water pose a significant threat to environmental and human health, necessitating comprehensive and proactive strategies to address their presence. By integrating source control, advanced treatment technologies, natural systems, monitoring, public awareness, and cutting-edge research, their impact can be mitigated effectively. A coordinated effort involving governments, industries, researchers, and communities is essential to achieve sustainable water management and protect aquatic ecosystems. Such an approach will ensure the availability of clean water for future generations while safeguarding environmental and public health.

Strategies for Controlling Micropollutants in Air

The environment and human health are seriously threatened by airborne micropollutants, such as particulate matter, heavy metals, volatile organic compounds (VOCs), and persistent organic pollutants (POPs) (Arfin et al., 2023). These pollutants can originate from industrial emissions, vehicle exhaust, agricultural activities, and household products (Odubo & Kosoe, 2024). Addressing this issue requires an integrated approach encompassing prevention, treatment technologies, monitoring, and public awareness.

Source Control

Preventing micropollutants from entering the atmosphere is a fundamental strategy for mitigating air pollution. Regulatory frameworks play a critical role by setting emission standards for industries, vehicles, and other major pollution sources. For example, enforcing stricter limits on VOC emissions from industrial processes and phasing out the use of harmful chemicals can reduce airborne micropollutants (Sowjanya & Gayatri, 2024). The advantages of source control include its ability to prevent pollution at its origin, thereby reducing the need for costly remediation. It also promotes cleaner production technologies and sustainable practices, which in turn minimizes health risks associated with exposure to airborne toxins. However, source control has its challenges, such as the need for robust enforcement and compliance monitoring, which can be resource-intensive. Additionally, it may increase operational costs for industries and prove challenging to implement in regions with limited regulatory infrastructure. Transitioning to cleaner fuels and promoting energy-efficient technologies, such as replacing coal with natural gas or renewable energy sources in power plants, further supports source control initiatives (Kothari et al., 2021). Similarly, promoting electric and hybrid vehicles can lower emissions from the transportation sector, although these measures also require significant investment and policy support.

Air Pollution Control Technologies

Typically, electronic instrumentation with specialized modules for capturing, processing, and displaying air contaminants is used to detect them. Advanced air pollution control technologies are effective in capturing or neutralizing micropollutants before they are released into the atmosphere. Technologies such as electrostatic precipitators (ESPs), which remove particulate matter from flue gases by applying an electric charge, and activated carbon filters, which adsorb VOCs and other gaseous micropollutants, are widely used. Catalytic converters, which are frequently seen in automobiles, lower dangerous emissions by changing pollutants into less dangerous forms (Dey & Mehta, 2020). The main advantage of these technologies is their high efficiency in removing specific micropollutants. They can also be integrated into existing systems, thereby reducing the environmental impact of industrial and vehicular emissions. However, these technologies come with high installation and operational costs. Regular maintenance and replacement of filters or catalysts are necessary, adding to the overall expense. Additionally, some technologies may produce waste materials that require further management, presenting an added environmental challenge.

Natural Air Purification Systems

Nature-based solutions, such as urban green spaces and afforestation, offer an alternative approach to mitigating air pollution by absorbing certain micropollutants. Trees and plants act as natural air filters, capturing particulate matter and absorbing gaseous pollutants. This method is cost-effective and sustainable, while also providing additional environmental benefits like carbon sequestration and habitat creation. It enhances urban aesthetics and improves public well-being. However, natural air purification systems have limitations, including their limited capacity to remove high concentrations of pollutants. Their effectiveness depends on the selection of plant species and environmental conditions, and they require significant space and a long-term commitment to establish and maintain.

Monitoring and Risk Assessment

To find pollution hotspots and evaluate how well control measures are working, regular air quality monitoring is crucial. Real-time monitoring of micropollutants at precise geographical and temporal scales is made possible by sophisticated sensor technologies and remote sensing techniques. Sodar and rass-wind profilers are important pieces of equipment for meteorological monitoring in this area. Sodar is an acronym for sound ranging and detection. The same methodology as radar is used, except acoustic emissions are used instead (Lay-Ekuakille et al., 2017). Monitoring provides actionable data for decision-making and policy implementation, enhances public awareness of air quality issues, and facilitates targeted interventions. However, deploying and maintaining monitoring systems can be expensive. Skilled personnel are required for data interpretation and analysis, which may pose a challenge for regions with limited technical expertise. Additionally, monitoring systems need frequent updates to capture emerging micropollutants effectively.

Public Awareness and Behavioral Change

Educating the public about the sources and impacts of air micropollutants is critical for fostering sustainable behaviors. Awareness campaigns can encourage individuals to reduce emissions by adopting practices such as carpooling, using public transportation, and conserving energy. Promoting the use of environmentally friendly products can also minimize household contributions to air pollution. Public awareness efforts are advantageous because they reduce emissions through collective effort and encourage sustainable consumption patterns. Furthermore, they enhance community participation in pollution control initiatives. However, the success of such campaigns depends on the extent of public engagement, which can be influenced by cultural and socioeconomic factors. These initiatives also require sustained funding and outreach efforts, and behavioral change can be slow to manifest.

Research and Development

Investing in research is vital for developing innovative solutions to control micropollutants in the air. Emerging technologies, such as photocatalytic coatings, nanomaterials for pollutant capture, and advanced air purification systems, hold promise for more effective and affordable solutions. In this context, spectral passive thermography is a promising method for identifying hazardous and polluting gas leaks (Lay-Ekuakille et al., 2017). Research drives technological innovation and enables the development of targeted strategies for specific micropollutants. It also enhances the efficiency and scalability of air pollution control methods. However, research and development are often associated with high costs and long timelines. There is also uncertainty about the scalability and environmental impact of new technologies, which may limit their immediate applicability in heavily polluted areas.

Controlling micropollutants in the air is essential for protecting public health, reducing environmental degradation, and mitigating climate change. A multi-pronged approach combining source control, advanced technologies, natural systems, monitoring, public engagement, and research is necessary to address this challenge effectively. Collaboration among governments, industries, researchers, and communities will play a pivotal role in ensuring cleaner air for present and future generations.

Challenges in Implementation of Control and Removal Techniques

Despite increased knowledge of the risks posed by micropollutants and the development of various strategies to control them, their effective implementation faces numerous challenges. These challenges span from operational inefficiencies, infrastructure limitations, economic constraints, to regulatory gaps, all of which hinder the large-scale adoption of removal techniques. This section delves into these key issues, highlighting the multifaceted obstacles to achieving comprehensive control of micropollutants in environmental systems. As discussed in previous sections, all the physical and chemical treatment techniques suffer with disadvantages related to ineffective operation, high sludge generation, high process cost, formation of toxic by-products etc., which in turn add to the challenges in the implementation of the removal techniques (Ali et al., 2020; Khan et al., 2020) Biological treatment, on the other hand, is sustainable, and is not energy intensive, but is limited by slow degradation process and sensitivity of the microorganisms to the treatment conditions (Al-Maqdi et al., 2017). Ion exchange methodology has the competitive potential over adsorptive technology in wastewater treatment. The membranes consist of a cross-linking polymer matrix, and natural or synthetic resins which adds up to high capital and maintenance expenditure. Some recalcitrant microcontaminants which cannot be degraded with traditional physicochemical methods are treated with advanced oxidation processes (AOPs) such as ozonation, sono-chemical oxidation, photocatalysis and UV treatment. These processes mineralize the micropollutants into non-toxic products. Of all the AOPs, photocatalysis suffers with limitations pertaining to the separation of the photocatalyst, and diffusional limitations between the reactants and the active catalyst components (Bhatt et al., 2022). Sono-chemical oxidation on the other hand, overcomes the limitations of photocatalysis through highly reactive oxidative radicals. However, its energy intensive nature limits its usage to laboratory scale. The costs incurred in AOPs are lowered by employing electrochemical oxidation which offers high efficiency. However, the choice of electrodes is crucial to the employment of electrochemical oxidation systems and, degradation of electrodes with times results into lowering of the oxidation efficiency. Some of the toxic pollutants may also be treated through coupled effects of size exclusion, charge separation and adsorption attained in membrane-based processes. Although the process can treat caffeine-, pharmaceutics- and heavy-metal laden contaminants, it is often limited by fouling of the membrane due to continuous use, and high operational and maintenance cost. Additionally, direct discharge of the filtrate originated during membrane separation also poses serious environmental threats. A hybrid process involving membranes and electrochemical technology, is the electrodialysis process which is less energy intensive, but is yet at the laboratory scale.

Apart from these physicochemical treatment methods, biological treatment methods are also capable to degrade micropollutants. The process however is easily affected by the changes in pH, temperature, and incubation conditions which also have an impact on the degradation efficiency. Advanced sludge processes (ASPs) are the most commonly employed commercial process related to biological treatment. The commercial usage of ASPs is however limited due to requirement of large areas for installation, high sludge and low-quality effluent production, and the need for process integration with other technologies for better efficiency. The post treatment of secondary sludge requires complex analytical techniques which further adds disadvantages to the process. The disadvantages are often overcome with the help of attached growth treatment processes in fixed- and moving-bed reactors.

Moving ahead, developments of membrane-based bioreactors come up with several advantages, viz., requirement of low area of installation, large biomass concentration, better adaptation to changing micropollutant load, high volume loading and better effluent quality. This technique is suited for hydrophilic contaminants and yield better removal efficiency for them. There are other factors, like the sludge concentration and age, composition of the wastewater, operational and environmental conditions which affect the efficacy of the process. However, similar to any other membrane-based process, the efficacy of this process is also significantly hindered by fouling of the membrane. The problem of selective treatment of hydrophilic contaminants is overcome in moving bed biofilm reactors which comprises of a single reactor containing both anaerobic and aerobic and microbes. They also offer advantages of high process and cost efficiency. The method is only constrained by the need for routine biomass supervision and the potential for aeration line choke. The challenges in implementation of various control techniques is discussed elaborately in the following sections.

Adsorption

The implementation of activated carbon technology in wastewater treatment facilities to remove micropollutants presents numerous challenges despite its effectiveness. Activated carbon is used extensively because of its huge surface area and microporous structure, which allow contaminants to be adsorbed. It comes in granular and powdered forms. However, its efficiency is dependent upon a number of variables, including the distribution of pore sizes, the hydrophobicity of the MPs, and the presence of competing organic matter (Kosek et al., 2020). It is easier for hydrophobic substances to be adsorbed, whereas hydrophilic compounds and negatively charged pharmaceuticals are less efficiently removed, particularly when there are other organic pollutants vying for adsorption sites (Bui et al., 2016; Kosek et al., 2020). Granular activated process, often employed as a single filtration step in WWTPs, relies on a downward flow of water through the adsorbent medium (Luo et al., 2014). However, the filters may become clogged by organic debris or suspended particles, reducing effectiveness and requiring pre-treatment steps such as sand filtration to remove some of the organic material before reaching the granular activated carbon filters. Regular maintenance, including backwashing with clean water or pressurized air, is essential to prevent blockages, while the eventual saturation of adsorption sites necessitates periodic replacement or thermal regeneration of the

GAC. Thermal regeneration restores the adsorption capacity but leads to a 10% loss in GAC mass and requires energy-intensive processes, adding to operational costs (Mulder et al., 2015). For high molecular weight pollutants, in particular, hot steam assisted regeneration is required which enhances the energy demands of the process and adds another hazardous waste stream (desorbed steam) to the process cycle. The commercialization of this process also requires addressing concerns related to the clogging of the filters and surface wash nozzles (Bui et al., 2016).

Powdered activated carbon (PAC) systems, in contrast, are typically integrated as a separate multi-stage process involving a contact tank, settling tank, and post-treatment filtration. These systems are more complex to design and operate, as the small particle size of powdered activated carbon requires additional filtration to remove residual particles from the treated water. Direct dosing of PAC into biological treatment stages or sand filter inlets is being explored to reduce investment costs, but its long-term effects on micropollutant removal efficiency and system performance are not fully understood (Margot et al., 2013). PAC dosing increases the demand for coagulants, flocculants, and polymers by 10–20%, further raising operational expenses (Kosek et al., 2020). Additionally, PAC systems are prone to issues such as clogging of slurry transport systems due to undersized piping, sharp bends, or insufficient velocities, as well as pipe abrasion caused by the abrasive nature of the carbon particles. Safety concerns also arise, as PACs can react with oxygen, releasing heat, and its dust form is highly explosive, necessitating strict safety protocols, including spark prevention and the use of specialized storage and feeding systems (Martin et al., 2004). The disposal of spent PAC adds another layer of complexity, as it must be dried, dewatered, and burned to prevent further dissemination of pollutants, which increases environmental and operational costs.

Both granular and powdered activated technologies face inherent limitations that need to be addressed to optimize their application for micropollutant removal. Granular systems require ongoing maintenance and careful pre-treatment to avoid clogging and performance degradation, while powdered systems involve complex operational requirements and higher safety risks (Siong et al., 2013). Additionally, both technologies have significant environmental impacts, from the energy-intensive production and regeneration processes to the disposal of spent materials. To overcome these challenges, further advancements in system design, operational protocols, and integration with existing WWTP infrastructure are essential. Countries like Germany have developed specific design criteria to improve the effectiveness of activated carbon technologies for micropollutant removal, but these require careful adaptation to the unique conditions of each WWTP. Ultimately, while activated carbon technologies hold great potential, their successful implementation requires a comprehensive approach to addressing the technical, operational, environmental, and safety challenges associated with their use. Additionally, addressing the management of hazardous waste adsorbed on the adsorbents also needs to be addressed before implementation of the technology in any establishment.

Advanced Oxidation Processes

Advanced oxidation treatment is capable of mineralizing recalcitrant compounds to non-toxic products. These include oxidation with oxidizing agents like H_2O_2, Cl_2, ClO_2, $KMnO_4$ and HClO, ozonation, UV/H_2O_2 assisted oxidation, O_3/H_2O_2 assisted oxidation, and UV/O_3 assisted oxidation (Bui et al., 2016). However, oxidation with the above stated oxidising agents is ineffective in reducing the concentration of micropollutants. Thuss, major full and pilot scale applications adopt advanced oxidation with ozonation, UV/H_2O_2 assisted oxidation, O_3/H_2O_2 assisted oxidation, and UV/O_3 assisted oxidation process. Installing AOPs between two biological treatment procedures is frequently advised in order to lessen competition with other organic materials during the pre-treatment phase and to break down oxidation byproducts during the post-treatment phase (Bui et al., 2016). Out of all the AOPs, ozonation is generally a highly effective method for reducing micropollutants in wastewater treatment plants (WWTPs). The production of hydroxyl ($OH^{\cdot}$) radicals during the breakdown of ozone in water is a significant benefit of this strategy (Lee et al., 2009). These radicals react non-selectively with medications and other MPs (Lee et al., 2009). Ozonation also lessens some ecotoxicological consequences, especially those related to estrogenic activity. In certain instances, the disinfecting qualities of ozone are also thought to be important advantages of this method (Zimmermann et al., 2011).

The implementation of ozonation technology for micropollutant removal faces several challenges. One major limitation is that ozone does not spontaneously decompose into free radicals in acidic environments, and in neutral conditions, the decomposition is only partial. This means reactions occur both directly with ozone and indirectly with radicals, which may limit efficiency. Another major issue is the generation of drinkable water, where high levels of natural bromide (Br^-) can cause carcinogenic bromate (BrO_3^-) to develop during ozonation. (Kosek et al., 2020). Additionally, these compounds may also react with DOCs to form intermediates like chloral hydrate, nitrosamines, and carbonyls (Blackbeard et al., 2016; Li et al., 2015a, 2015b). Regulatory bodies like the EPA (2010) and the EU (1998) have stringent drinking water BrO_3^- limitations of 10 μg/L, limiting the usage of ozone in such circumstances (Official Journal of the European Community l 330: Directive 98/83/EG, 1998; Revised State Implementation Guidance for the Consumer Confidence Report (CCR) Rule, 2010). The short life cycle of ozone along with a high energy consumption (0.01–0.015 kWh per m^3 of pure oxygen produced) associated with a low conversion efficiency at maximum ozone generation efficiency of 1–2% or 4–8%, respectively also limits full scale implementation of ozonation process (Derco et al., 2015; Hollender et al., 2009).

Current ozone dosages of roughly 0.6–1.0 g O_3 per g of dissolved organic carbon, along with hydraulic retention lengths of 20–30 min, frequently fall short of fully mineralizing micropollutants in wastewater treatment (Kosek et al., 2020). Instead, these compounds are converted into intermediates, which are often not totally eradicated from the effluent. There are instances when these byproducts are more harmful than the initial micropollutants, which thus raise concerns about their environmental and health impacts (Zimmermann et al., 2011). While full-scale wastewater treatment plants have not yet reported substantial generation of hazardous byproducts, this remains a critical area of research.

To address this, countries like Germany recommend coupling ozonation with post-treatment steps such as biological treatment or sand filtration to eliminate transformation products that are biodegradable prior to release (Mulder et al., 2015). However, it is still unclear how well sand filtering works to get rid of reactive substances. These challenges, alongside general design specifications for micropollutant removal, highlight the complexities of implementing ozonation technology effectively. From a technological perspective, implementing ozonation for micropollutant removal presents several challenges. Ozone cannot be stored on-site due to its intrinsic instability, necessitating its immediate production prior to use. This production involves generating ozone from air or pure oxygen using an electrical discharge, which is then mixed with effluent water in a contact basin using injectors or diffusers (Kosek et al., 2020). While injectors consume slightly more energy and offer no improved MP removal efficiency compared to diffusers, the latter is generally considered a more practical solution (Mulder et al., 2015). The contact basin must be airtight to contain and treat any residual ozone gas, and the treated effluent is typically separated from biodegradable metabolites by passing it through a sand filter.

The possible oxidation of nitrogen, which can form corrosive nitric acid in the presence of moisture, presents another difficulty during the production of ozone. To prevent damage to ozonators, the input air or oxygen must be moisture-free, requiring gas cooling or drying. Additionally, safety concerns arise due to ozone's toxicity; concentrations exceeding 1 ppm in air are considered unsafe for prolonged human exposure. Regulatory limits, such as those established under the EU Air Quality Directive (2008), cap ozone concentrations at 120 $\mu g/m^3$ for eight hours of continuous exposure and 180 $\mu g/m^3$ for a single hour of exposure (Jones, n.d.). In some cases, exposure to even ppm of ozone for 15 min have been related to respiratory problems (Safe Work Australia, n.d.). WHO standards are even stricter, with a maximum of 100 $\mu g/m^3$ for eight-hour exposure (WHO, 2006). Consequently, facilities producing or utilizing ozone must be equipped with ozone detectors and warning systems to ensure safe operation and compliance with these standards (Abegglen & Siegrist, 2012).

Although UV alone is not effective in reducing micropollutants in the system, the MP removal efficiency can be improved with the help of UV/H_2O_2 or UV/O_3 assisted oxidation (Kruithof et al., 2007; Yang et al., 2014). In terms of efficacy, the O_3/UV assisted oxidation shows the highest removal efficiency due to to a comparatively greater UV adsorption capacity of O_3 at 254 nm (200 times greater than that of H_2O_2 (Sarathy & Mohseni, 2006). However, its practical implication faces challenges due to a high energy consumption. In comparison, UV/H_2O_2 and H_2O_2/O_3 process offer high energy efficiency (Lee et al., 2009). Despite offering a superior stability and energy efficiency over ozone, the safe discharge of residual H_2O_2 after operation is a point of crucial importance (Lee et al., 2009).

The commercialization of advanced ozone based advanced oxidation techniques thus require developing an efficient contacting chamber to achieve successful mass transfer from the gaseous to liquid phases. Additionally, it also requires regular monitoring of the influent characteristics as an increase in the BOD and COD of the influent can reduce the process performance significantly (Gray, 2014). The operational difficulties spanning from the quality of feed air, constant monitoring of over-heating of ozone generator, and the irregular inspection and maintenance require skilled labor which increases the operational cost of such an establishment (Gray, 2014). The commercialization of UV/H_2O_2 process, on the other hand, doesn't require huge capital. Nonetheless, it frequently inappropriate for streams of water with poor UV transmission efficiency (Lee et al., 2009). Also, it is necessary to optimize the UV and H_2O_2 dose to design an efficient reactor (Hofman-Caris & Beerendonk, 2011).

Membrane Processes

Some commercial establishments also employ membrane methods for treating wastewater containing micropollutants. Some of the membrane technologies employed for these treatments include microfiltration, nanofiltration, ultrafiltration and reserve osmosis. However, microfiltration and ultrafiltration technologies are mostly used for wastewater pretreatment prior to the micropollutant removal stage because of their large pore sizes, which result in low micropollutant removal effectiveness. Although the molecular size, physical and surface properties of micropollutants are key to enhancing micropollutant removal efficiency, the highest removal of micropollutant is typically obtained in a reverse osmosis process, which possesses smallest membrane pore size (Hai et al., 2011). The comparatively lesser sensitivity of membrane-based processes towards neutral compounds also hinders their usage in establishments dealing with a range of micropollutants. Additionally, membrane fouling, and high-power requirements for pump operation also impede the process efficiency. The handling and elimination of the concentrated waste stream (35% of influent flow and 4–10 times more concentrated) generated after reverse osmosis process is also a point of major concern in commercial employment of membrane technologies (Chelme-Ayala et al., 2009; Rodriguez-Mozaz et al., 2015; Siegrist & Joss, 2012). Membrane processes suffer with operational issues arising due to fouling of the membrane which reduces membrane life by 60–75% if subjected to poor quality water (Lazarova et al., 2008). Owing to the difficulties of reverse osmosis technology, nanofiltration technology is often considered a low pressure and effective alternative (Yangali-Quintanilla et al., 2011). Commercialization of membrane-based processes is thus requires consideration of several factors, including, proper documentation of the flowrate of influent stream, pH, conductivity and turbidity of the influent stream, quality of the influent stream, the generated permeate and concentrate, membrane maintenance schedule, plan for corrective maintenance and take measures to maintain the desired flux.

Membrane Bioreactors

Membrane bioreactors are another class of emerging technologies and work on synergistic concepts of membrane and reactor technology. The removal of micropollutants in membrane bioreactors rely on three factors: (a) microbial community which metabolize the micropollutants, (b) the sorption ability of the sludge, and (c) rejection rate of the membrane. The removal efficiency of micropollutants can be enhanced by a longer sludge retention period that enhances the structure of a variety of microorganisms (Hai et al., 2014). The efficacy of membrane bioreactor technology towards pharmaceutical micropollutants depicts their effective micropollutant removal efficiency (Bernhard et al., 2006; De Wever et al., 2007). However, addressing micropollutant compounds with electrophilic functional groups, saturated/polycyclic compounds, and lengthy, highly branched side chains presents challenges for the procedure (Lee et al., 2009). The toxicity of recalcitrant compounds also reduces the efficacy of the process by damaging the microbial community (De Cazes et al., 2014). The process has shown great efficacy at laboratory scale, but commercial scale studies on the same are limited.

However, a major problem in membrane bioreactors is membrane fouling, which is comparable to other membrane-based processes. In some cases, it becomes irreversible, thereby aggravating the difficulties even further (Li et al., 2015a, 2015b). The fouling is severe in the presence of long chain micropollutants, like carbamazepine (Li et al., 2015a, 2015b). The treatment of concentrated waste and sludge, and high power requirement adds further disadvantages to the process (Hai et al., 2014). The high cost of operation and establishment, risky operation and dependency on skilled labor enhances.

Scale-Up

Cost Estimation

While considering the scale up possibilities for any process, cost incurred in the establishment, operation and maintenance of the process emerge as crucial factors.

The cost estimation of micropollutant removal strategies involve three main parts, viz., investment, variable and other costs. The investment or fixed costs typically involve the costs incurred in designing and building of the treatment and post-treatment facility, costs involved in the purchase and installation of electrical, mechanical and automation equipment (Kosek et al., 2020; Mulder et al., 2015). The variable costs involve the cost incurred in maintenance of the process equipment, costs incurred in the materials/reagents required for the process, costs incurred in disposal of generated effluents, electricity costs and costs incurred in hiring and employing trained and untrained labor (Kosek et al., 2020; Mulder et al., 2015). Apart from that, there are several other costs related to insurances, environmental permits, buildings, management, supervision, training, and further investments and miscellaneous operational expenditures (Kosek et al., 2020; Mulder et al., 2015).

Literature suggests that ozonation and adsorption processes yield better efficiencies as compared to the incurred costs. The investment or fixed costs incurred in ozonation and adsorption processes are comparable (Kosek et al., 2020; Mulder et al., 2015). However, due pt the simplicity of installation, and low energy requirements, the costs incurred in granular adsorption process are often lower as compared to ozonation and powdered adsorption process (Mulder et al., 2015). In terms of maintenance, however, ozonation technology is cost effective. The maintenance of powdered adsorption process involves the costs incurred in continuous supply of the adsorbent, polymers, coagulants, and treatment of the generated sludge. On the other hand, the maintenance of granular adsorption technology involves the costs incurred in regular supply of fresh adsorbent and regeneration of the spent ones. Despite less factors involved in the maintenance of a granular adsorption process, the process is costlier as compared to the powdered adsorption process (Mulder et al., 2015). However, the costs involved in the regeneration of the spent adsorbent can be significantly reduced by employing a sand filter to remove undesired organic material through a pre filtration step with a common sand filter (Kelly, 2013; Kosek et al., 2020).

The costs can be further lowered by employing biochar as the adsorbent (Favre et al., 2022; Peer et al., 2023; Schlederer et al., 2024; Xu et al., 2022). A System Läk project employed different biochar instead of granular activated carbon to obtain comparable adsorption efficiency for pharmaceutical compounds (Kosek et al., 2020). Post-treatment steps typically incur a cost of 0.16–0.33 EUR/m^3 of treated wastewater (Mulder et al., 2015). Ozonation and adsorption have been widely employed in commercial establishments. Mulder et al. provided a comparison of the WWTP effluent treatment via (a) Ozone cum sand filtration, (b) powdered activated carbon based adsorption cum sand filtration and granular activated carbon based adsorption (Mulder et al., 2015). The authors categorized the proposed alternatives for three types of wastewater treatment plants, viz., 20,000 people can be served by small-scale facilities with a dry-weather wastewater flow rate of 200 m^3/h, 100,000 people can be served by medium-sized plants with a flow rate of 1050 m^3/h, and 300,000 people can be served by large-scale plants with a flow rate of 3100 m^3/h. For small-scale plants, the treatment costs employing ozone and powdered adsorption methods vary from €0.22 to €0.26 per m^3 (or \$0.24–\$0.31 per m^3), whereas for large-scale plants, the prices drop dramatically to €0.16 to €0.18 per m^3 (\$0.17–0.20 per m^3) (Bui et al., 2016). Granular adsorption technology, which is the most expensive option, shows only a modest reduction in treatment costs from small to large-scale plants. Interestingly, between 65 and 75% of annual expenses of the granular adsorption technology are related to its variable cost. The variable cost was very high even though the initial investment was approximately half that of ozone and powdered adsorption systems (Bui et al., 2016) (Fig. 12.1).

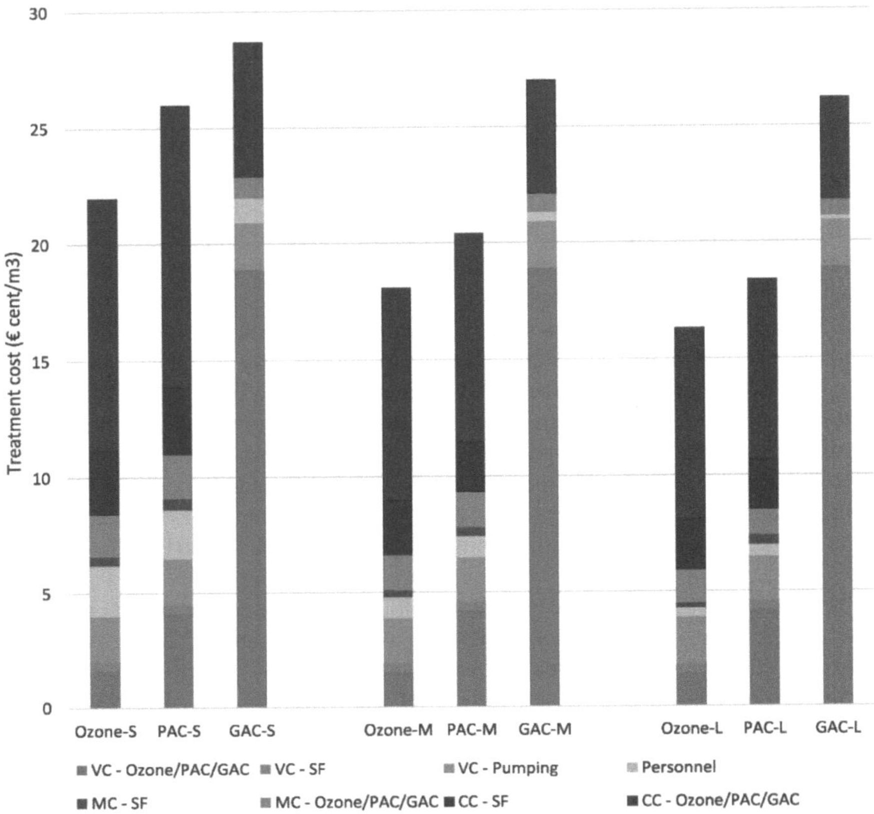

Fig. 12.1 Comparison of incurred annual costs per m^3 of treating micropollutant laden streams via **a** Ozone cum sand filtration, **b** PAC based adsorption cum sand filtration and granular activated carbon based adsorption [Abbreviations: S: small-scale (20,000 people), M: medium-scale (100,000 people), L: large-scale (300,000 people), SF: sand filtration, PAC: powdered activated carbon based adsorption, GAC: Granular activated carbon based adsorption, VC: variable cost, MC: maintenance cost, CC: capital cost] (Reprinted from (Bui et al., 2016) with permission from Elsevier)

Dissemination of Micropollutant Removal Techniques in Existing WWTPs

The scale up and full-scale implementation of micropollutant removal techniques involves several preparation and test phases, often exceeding time periods as long as 10 years. These steps include ecological, technological and socioeconomic aspects with special regard to environmental load. During the preparation stage, it is essential to establish specific local goals and criteria for advanced treatment. In the testing stage, a thorough evaluation of the most viable options should be carried out using laboratory or pilot-scale experiments. Accurately assessing the on-site performance of the advanced treatment methods being tested, along with estimating their implementation and maintenance costs, is crucial.

The earliest research and development projects related to the removal of micropollutants form the environment date back to 2005. A large-scale Swedish project called as 'Pharmaceuticals—occurrence in the aquatic environment, preventive measures and possible treatment methods,' executed by Stockholm Water (Wahlberg et al., 2010). The study mainly employed advanced oxidation treatment comprising of ozonation, or UV/H_2O_2-assisted oxidation. In addition, the activated sludge process, membrane bioreactor, moving bed biofilm reactor, adsorption, membrane filtration, and reverse osmosis were used. The effectiveness of the membrane bioreactor and the activated sludge process was reported to be dependent upon the sludge age with improvement in reduction efficiency observed with an increase in sludge age. The project also involved installation of biofilm reactors treating the wastewater generated from treatment plants and for the breakdown of oxidation products generated after ozonation treatment. An average reduction of around 20% of the contaminants was observed in both the bioreactors, with a biofilm thinning was observed in the reactor installed to treat ozonation effluent. Ozonation treatment was superior to other treatment methodologies opted in the project and the efficiency of ozonation was reported to be strongly affected by the pH, ozone dose and operational temperature. Another advanced oxidation process employed in the project was a TiO_2 coated reactor for UV/H_2O_2 assisted oxidation, which kills bacterial in the presence of UV lights of wavelength of 200–280 nm, while also generating highly reactive hydroxyl radicals. However, a possible generation of toxic intermediates, complex reactor design, and short range of effective UV wavelength ads up to the difficulties of the process. The reduction in contaminants was however lower as compared to the reduction obtained in ozonation and the presence of TiO_2 didn't benefit the generation of hydroxyl radicals, as reported. The reductions obtained with adsorption process was better as compared to the UV/H_2O_2 assisted oxidation. In nanofiltration and reverse osmosis systems, size selective exclusion of contaminants is typically performed. The efficacy of nanofiltration was limited due to small molecular size of the contaminants, while high molecular weight contaminants promote the reduction efficiency. The efficacy of reverse osmosis is comparable to that of ozonation, but consumed comparatively higher amounts of electrical energy for operation of the membrane setup. Although adsorption has physicochemical and ecotoxicological advantages, the study found that ozonation and adsorption are the most promising strategies for reducing micropollutants. This is due to formation of toxic intermediates via adsorption. In ozonation process, the contaminants can potentially get converted to toxic intermediates which often go undetected and discharged into clean water bodies. Despite addressing the above concerns, the Stockholm Water recommended ozonation as the potential technology for removal of pharmaceutical micropollutant laded wastewater.

Following this, another project called MistraPharma was launched under the Swedish Foundation for Strategic Environmental Research. Similar to the Stockholm Water project, this study also recommended ozonation and adsorption as the viable technologies with a reduction efficacy of 85–95 and 95–98%, respectively

(Jekel & Dott, 2013; Kosek et al., 2020). The Swedish Agency for Marine and Water Management funded projects brought the transition from research-oriented projects to society-oriented projects by improving collaborations between municipalities, industries and universities. Several of the pilot scale or large-scale studies conducted under the mission also recommended ozonation and adsorption as the potential treatment techniques (Cimbritz & Mattsson, 2018). Similar observations were reported by the report by the Swedish Environmental Protection Agency (Sundin et al., 2017). Similar to Sweden, Germany also considers ozonation or adsorption treatment as two commercially viable techniques. The country projects a reduction of 80% of aqueous micropollutants to be removed if the plant is appropriately managed (Hillenbrand et al., 2017). Additionally, these two technologies are also seen as potential techniques for removal of other co-existing organic pollutants as well (Hillenbrand et al., 2017). Under the Interreg South Baltic projects, Morpheus and Less is More, Poland has recently started the upgradation of their existing waste water treatment plants to remove the prevailing micropollutants (Davidsson & Cimbritz, 2021).

A North Holand water treatment plant, Andijk, developed pilot and full-scale plants for treatment of micropollutants by UV/H_2O_2 assisted oxidation (Kruithof et al., 2007). The overall water treatment strategy opted by the plant serviced water from Ijssel Lake and treated it through sequential treatment steps involving coagulation, rapid sand filtration, UV/H_2O_2 assisted oxidation, adsorption, and chlorine oxide treatment (Kruithof et al., 2007). The oxidation reduced the micropollutant load (fuel oxygenates, pesticides, algae toxins, solvents, rocket fuel by-products, pharmaceuticals, and endocrine disruptors) by around 80%.

220 micropollutants were treated as part of the upgrade of a full-scale water treatment facility located in Regensdorf, Switzerland, which included a full-scale ozonation reactor and sand filtration treatment (Hollender et al., 2009). Post ozonation treatment, several micropollutants were reduced below their detection limit. Swindon, a full-scale granular adsorption facility, was set up in South-West England to treat sewage as part of the UK National Demonstration Programme for EDCs. The treatment involved primary screening, sedimentation, activated sludge, secondary sedimentation, adsorption and discharge steps for removing emerging contaminants in sewage. This full scale plant was able to reduce 43–64% of steroidal estrogens, 84–99% of mebeverine and around 17–23% of carbamazepine and propranolol (Grover et al., 2011).

Ozonation treatment has also been employed at full scale in the Eastern Treatment Plant situated in Melbourne, Australia for secondary treatment of wastewater laden with 387 micropollutants. The treatment process decreased the concentration of micropollutants below the levels of significance by using pre-ozone, biological media filtration, post-ozone, UV, and chlorination treatments (Blackbeard et al., 2016). Another Australian full scale micropollutant treatment project of Wolumba, Bega Valley, New South Wales, involved three staged treatment, viz., screening, followed by membrane bioreactor and UV treatment. The treatment facility redced the organic contaminants by more than 90%, and removed pharmaceuticals by 24–68% (Trinh et al., 2012).

Membrae bioreactors have also gained acceptance in Germany for the treatment of hospital wastewater as demonstrated by the Waldbrol hospital project involving four stage treatment comprising of screening, primary settlement, membrane bioreactor and advanced treatment step (Beier et al., 2011). The treatment resulted in wide range antibiotics reduction from 19 to 94%. The effectiveness of a granular/powdered activated carbon treatment as an upper layer in dual-media filtration was evaluated in another pilot-scale treatment plant in Berlin, Germany. Both the treatments resulted in appreciable reduction of the organic micropollutants, with powdered activated carbon setup resulting in better removal efficiency, indicating that both technologies have the potential to be included into tertiary wastewater treatment without requiring major modifications to the construction process (Meinel et al., 2014).

An integrated fixed-film activated sludge membrane bioreactor with suspended and attached biomass, a moving bed membrane bioreactor with attached biomass, and one membrane bioreactor with suspended biomass were all used in a pilot-scale micropollutant removal test carried out in Spain. The process involved a fourth stage treatment with reverse osmosis which reduced the concentration of some of the micropollutants by 100% (de la Torre et al., 2015). Innovative pilot scale treatment plants comprising of integrated ozonation and biological aerated filtration steps have been suggested by research work done in China (Li et al., 2015a, 2015b). Most recent developments in the technologies of the implementation of micropollutant removal systems were launched under the EMPEREST project which sets up mobile pilot plants for advanced treatment of wastewater effluent. These mobile pilots target to travel to 6 differ t countries to gather data on the efficacy of different treatment techniques and identify the most efficient combination for optimal operation of the technologies to be developed in future. Two recent mobile pilots involve one in Tartu, Estonia which involved the collection and interpretation of data, discussions on the applicability of monitoring the physicochemical properties and distribution methods of per- and polyfluoroalkyl substances, and external collaborations. The project undertaken in Gdańsk, Poland involved the development of a sequential treatment plan, with detailed information on the operating parameter such as tank volume, UV lamp performance, pump capacity, filter surface area and height, and ozone system capacity (Tenno & Swinarski, 2023).

Lab scale implementation of the removal of micropollutants from soil and air has been studied, but none of them have been established on commercial scale. In both the cases, micropollutants tend to migrate to water bodies and result in an increment in bioavailability for microorganisms. Techniques for in-situ and ex-situ removal are frequently used to remediate micropollutants found in soil. An environmentally beneficial method for removing heavy metals from soil is biosurfactant treatment (Hernández-Soriano et al., 2010; Tsang et al., 2012). However the process involves huge costs, and the complexities involved with in-situ treatment technique is an obstacle in its implementation.

Future Perspectives

The management and removal of micropollutants from various environmental matrices remain critical in ensuring ecological and human health. Although substantial progress has been made, future endeavours must focus on addressing key challenges, advancing emerging technologies, and implementing sustainable strategies globally. Future studies should examine how to incorporate cutting-edge and experimental technologies into current treatment frameworks, such as membrane bioreactors, photocatalysis, plasma processes, and AOPs. These hybrid systems can combine the benefits of established techniques like activated carbon adsorption and ozonation with the efficiency of emerging technologies, enabling higher removal rates of recalcitrant micropollutants. For instance, combining ozonation with biofiltration or integrating nanomaterials into membrane processes could significantly enhance the degradation of persistent organic pollutants while minimizing by-product toxicity (Blackbeard et al., 2016; Directive (EU) 2024/3019, 2024; Grover et al., 2011). The elimination of micropollutants during wastewater treatment, often referred to as the "fourth treatment stage," has been implemented in countries like Germany, Switzerland, Sweden, and the Netherlands, with ongoing trials in other nations (Borea et al., 2019). Switzerland was a pioneer in this area, enshrining such measures into law in 2016. With the revision of the Urban Waste Water Treatment Directive, the European Union has achieved great strides (Directive (EU) 2024/3019, 2024). Directive (EU) 2024/3019 requires that a significant amount of micropollutants be eliminated from wastewater as of January 1, 2025. Member states have until July 31, 2027, to align their national legislation with this directive. Currently, three primary techniques dominate micropollutant removal: PAC, GAC, and ozonation. These methods are well-established and widely implemented, particularly in countries like Germany and Switzerland. Notable examples include the Neugut WWTP (875 m^3/h) in Switzerland, Wüeri WWTP (350 m^3/h), and facilities like Vierlinden WWTP (400 m^3/h) in Germany (Audenaert et al., 2014; Hollender et al., 2009).

Real-time monitoring systems employing advanced sensors and machine learning algorithms will be instrumental in improving the precision of micropollutant removal. Automated systems can optimize operational parameters, such as ozone or activated carbon dosages, based on real-time water quality data (e.g., dissolved organic carbon, turbidity). These systems can also predict equipment maintenance needs, reducing operational downtimes and costs. Developing portable monitoring devices for rapid, on-site analysis of micropollutants in air, water, and soil is another key area of focus.

Despite advancements, several challenges remain in scaling and optimizing these technologies and efforts should be directed toward developing sustainable micropollutant removal systems that align with circular economy principles. For instance, biochar made from agricultural waste may be a more affordable and environmentally responsible option for adsorption operations than activated carbon (Directive (EU) 2024/3019, 2024; Wahlberg et al., 2010). For granular adsorption technology, critical operational parameters such as flow velocity and filter clogging must be monitored closely. Automated systems are essential for controlling these parameters, including triggering back-flushing when pressure increases beyond a threshold. Additionally, breakthrough curves and periodic micropollutant measurements in treated effluent are needed to estimate granular adsorption filter lifetimes and determine the timing of activated carbon replacement (Abegglen & Siegrist, 2012; Siong et al., 2013).

Similarly, exploring energy-efficient AOPs or renewable energy-powered treatment systems could reduce the environmental footprint of these technologies. Scaling up promising technologies from pilot-scale to full-scale operations requires overcoming technical, economic, and regulatory barriers. Methodologies such as membrane bioreactors and AOPs need optimization to address challenges like membrane fouling, energy consumption, and secondary by-product management. International collaboration, standardized protocols, and adaptive legislation can facilitate smoother technology transfer and implementation, particularly in low-income regions. Stronger regulatory frameworks, such as the Urban Waste Water Treatment Directive by the European Union, play a crucial role in driving the adoption of advanced micropollutant removal technologies. However, public engagement and education are equally critical. Awareness campaigns on the environmental and health impacts of micropollutants, coupled with initiatives like pharmaceutical take-back programs, can significantly reduce pollution at the source. Collaboration between governments, industries, and communities is essential for creating effective and inclusive pollution control strategies. Investments in interdisciplinary research can uncover novel solutions for micropollutant mitigation. Advanced materials, such as functionalized adsorbents and bioengineered microbes, hold immense potential for enhancing treatment efficiency. Long-term research is also required to assess the ecological effects of residual by-products and to establish safe management techniques.

The future of micropollutant control lies in the seamless integration of novel technology, effective monitoring systems, and sustainable practices. While developed countries are setting benchmarks with advanced treatment technologies, global collaboration is essential to address disparities in implementation and ensure safe and clean environments worldwide. By balancing technological innovation with economic and environmental considerations, the vision of effective micropollutant management can be realized, safeguarding ecosystems and public health for future generations. At last, exploring the removal of micropollutants from soil and water have not yet been implemented on large scale and is a significant area for future research.

Conclusion

Addressing micropollutants in the environment is a critical challenge that demands innovative and sustainable strategies. Various control methods, including advanced oxidation processes, membrane filtration, adsorption techniques, and bioremediation, have shown promise in effectively reducing micropollutants. Each of these strategies offers unique advantages, such as high removal efficiency or cost-effectiveness, depending on the target compounds and environmental conditions. Additionally, preventive measures, such as the adoption of green chemistry and source control, are vital to reducing micropollutant loads at their origin. However, implementing these strategies on a large scale faces significant challenges. High operational costs, energy requirements, and the need for sophisticated infrastructure limit the widespread adoption of advanced technologies. Furthermore, the complexity of micropollutant mixtures, potential formation of toxic byproducts, and regulatory gaps hinder effective management. Public awareness and policy enforcement are often inadequate, making it difficult to achieve long-term environmental sustainability. To overcome these challenges, a multidisciplinary approach combining technological innovation, stakeholder collaboration, and robust policy frameworks is essential. Continuous research and development, supported by public and private investments, will play a key role in making micropollutant control more accessible and efficient. By addressing these challenges holistically, it is possible to safeguard ecosystems and public health from the risks posed by micropollutants.

References

Abegglen, C., & Siegrist, H. (2012). Mikroverunreinigungen aus kommunalem Abwasser. Verfahren zur weitergehenden Elimination auf Kläranlagen. Bundesamt für Umwelt, Bern. *Umwelt-Wissen, 1214*, 210.

Adepoju, A. O., Femi-Adepoju (Grace), A., Jalloh, A., & Faeflen (Joe-wia), S. (2024). Chapter 7—soil pollution and management practices. In R. Frazer-Williams, M. B. Ogundiran, & E. I. Unuabonah (Eds.), *Environmental Pollution and Public Health* (pp. 187–236). Elsevier. https://doi.org/10.1016/B978-0-323-95967-4.00011-8

Ali, N., Khan, A., Nawaz, S., Bilal, M., Malik, S., Badshah, S., & Iqbal, H. M. N. (2020). Characterization and deployment of surface-engineered chitosan-triethylenetetramine nanocomposite hybrid nano-adsorbent for divalent cations decontamination. *International Journal of Biological Macromolecules, 152*, 663–671. https://doi.org/10.1016/j.ijbiomac.2020.02.218

Ali, S., Ullah, M. I., Sajjad, A., Shakeel, Q., & Hussain, A. (2021). Environmental and health effects of pesticide residues. In Inamuddin, M. I. Ahamed, & E. Lichtfouse (Eds.), *Sustainable Agriculture Reviews 48: Pesticide Occurrence, Analysis and Remediation Vol. 2 Analysis* (pp. 311–336). Springer International Publishing. https://doi.org/10.1007/978-3-030-54719-6_8

Al-Maqdi, K. A., Hisaindee, S. M., Rauf, M. A., & Ashraf, S. S. (2017). Comparative degradation of a thiazole pollutant by an advanced oxidation process and an enzymatic approach. *Biomolecules 7*(3). https://doi.org/10.3390/biom7030064

Arfin, T., Mathew, N., Tirpude, A., Pillai, A. M., & Mondal, P. (2023). Emerging contaminants in air pollution and their sources, consequences, and future challenges. In R. Thapar Kapoor & M. Rafatullah (Eds.), *For Wastewater and Sustainable Circular Bioeconomy* (pp. 235–274). De Gruyter. https://doi.org/10.1515/9783111016825-014

Arslan, M., Ullah, I., Müller, J. A., Shahid, N., & Afzal, M. (2017). Organic micropollutants in the environment: Ecotoxicity potential and methods for remediation. In N. A. Anjum, S. S. Gill, & N. Tuteja (Eds.), *Enhancing Cleanup of Environmental Pollutants: Volume 1: Biological Approaches* (pp. 65–99). Springer International Publishing. https://doi.org/10.1007/978-3-319-55426-6_5

Audenaert, W., Chys, M., & Van Hulle, S. (2014). (Future) regulation of trace organic compounds in WWTP effluents as a driver of advanced wastewater treatment. In *IOA-PAG Annual Conference & Expo: Ozone: Proven Solution to Emerging Challenges.*

Bala, K., Anshul, Verma, S., Dogra, A., Das, A. P., Kumar, R., & Nandini. (2024). Sustainable management for urban plastic waste generation. In *Sustainable Management of Urban Plastic Waste Through Circular Economic Approaches*, (pp. 211–236). https://doi.org/10.1201/9781003465447-13

Barbosa, M. O., Moreira, N. F. F., Ribeiro, A. R., Pereira, M. F. R., & Silva, A. M. T. (2016). Occurrence and removal of organic micropollutants: An overview of the watch list of EU decision 2015/495. *Water Research, 94*, 257–279. https://doi.org/10.1016/j.watres.2016.02.047

Becerra-Castro, C., Lopes, A. R., Vaz-Moreira, I., Silva, E. F., Manaia, C. M., & Nunes, O. C. (2015). Wastewater reuse in irrigation: A microbiological perspective on implications in soil fertility and human and environmental health. *Environment International, 75*, 117–135. https://doi.org/10.1016/j.envint.2014.11.001

Behera, I. D., Nayak, M., Biswas, S., Meikap, B. C., & Sen, R. (2021). Enhanced biodegradation of total petroleum hydrocarbons by implementing a novel two-step bioaugmentation strategy using indigenous bacterial consortium. *Journal of Environmental Management, 292*, 112746. https://doi.org/10.1016/J.JENVMAN.2021.112746

Behera, I. D., Nayak, M., Mishra, A., Meikap, B. C., & Sen, R. (2022). Strategic implementation of integrated bioaugmentation and biostimulation for efficient mitigation of petroleum hydrocarbon pollutants from terrestrial and aquatic environment. *Marine Pollution Bulletin, 177*, 113492. https://doi.org/10.1016/J.MARPOLBUL.2022.113492

Beier, S., Cramer, C., Köster, S., Mauer, C., Palmowski, L., Schröder, H. F., & Pinnekamp, J. (2011). Full scale membrane bioreactor treatment of hospital wastewater as forerunner for hot-spot wastewater treatment solutions in high density urban areas. *Water Science and Technology, 63*(1), 66–71. https://doi.org/10.2166/wst.2011.010

Belioka, M.-P., & Achilias, D. S. (2024). The effect of weathering conditions in combination with natural phenomena/disasters on microplastics' transport from aquatic environments to agricultural soils. *Microplastics, 3*(3), 518–538. https://doi.org/10.3390/microplastics3030033

Berg, P., Hagmeyer, G., & Gimbel, R. (1997). Removal of pesticides and other micropollutants by nanofiltration. *Desalination, 113*(2), 205–208. https://doi.org/10.1016/S0011-9164(97)00130-6

Bernhard, M., Müller, J., & Knepper, T. P. (2006). Biodegradation of persistent polar pollutants in wastewater: Comparison of an optimised lab-scale membrane bioreactor and activated sludge treatment. *Water Research, 40*(18), 3419–3428.

Bhatt, P., Bhandari, G., & Bilal, M. (2022). Occurrence, toxicity impacts and mitigation of emerging micropollutants in the aquatic environments: Recent tendencies and perspectives. *Journal of Environmental Chemical Engineering, 10*(3), 107598. https://doi.org/10.1016/j.jece.2022.107598

Biabani Reshtehroudi, R. (2024). *Optimizing the thermal treatment for sustainable hazardous waste management and soil remediation.* Università degli studi di Brescia.

Blackbeard, J., Lloyd, J., Magyar, M., Mieog, J., Linden, K. G., & Lester, Y. (2016). Demonstrating organic contaminant removal in an ozone-based water reuse process at full scale. *Environmental Science: Water Research & Technology, 2*(1), 213–222.

Boelee, E., Geerling, G., van der Zaan, B., Blauw, A., & Vethaak, A. D. (2019). Water and health: From environmental pressures to integrated responses. *Acta Tropica, 193*, 217–226. https://doi.org/10.1016/j.actatropica.2019.03.011

Borea, L., Ensano, B. M. B., Hasan, S. W., Balakrishnan, M., Belgiorno, V., de Luna, M. D. G., Ballesteros, F. C., & Naddeo, V. (2019). Are pharmaceuticals removal and membrane fouling in electromembrane bioreactor affected by current density?. *Science of the Total Environment, 692*, 732–740. https://doi.org/10.1016/j.scitotenv.2019.07.149

Brevik, E. C., Steffan, J. J., Rodrigo-Comino, J., Neubert, D., Burgess, L. C., & Cerdà, A. (2019). Connecting the public with soil to improve human health. *European Journal of Soil Science, 70*(4), 898–910. https://doi.org/10.1111/ejss.12764

Bui, X. T., Vo, T. P. T., Ngo, H. H., Guo, W. S., & Nguyen, T. T. (2016). Multicriteria assessment of advanced treatment technologies for micropollutants removal at large-scale applications. *Science of the Total Environment, 563–564*, 1050–1067. https://doi.org/10.1016/j.scitotenv.2016.04.191

Burkhardt-Holm, P. (2011). *Linking water quality to human health and environment: The fate of micropollutants* (p. 62). Institute of Water Policy, National University of Singapore.

Butt, F. M., Nisar, U. Bin, & Ahmed, T. (2022). Environmental degradation and micro-pollutants in light of environmental laws. In T. Ahmed & M. Z. Hashmi (Eds.), *Hazardous Environmental Micro-pollutants, Health Impacts and Allied Treatment Technologies* (pp. 53–73). Springer International Publishing. https://doi.org/10.1007/978-3-030-96523-5_4

Chavoshani, A., Hashemi, M., Amin, M. M., & Ameta, S. C. (2020). Pharmaceuticals as emerging micropollutant. In *Micropollutants and Challenges: Emerging in the Aquatic Environments and Treatment Processes* (p. 35).

Chavoshani, A., Hashemi, M., Mehdi Amin, M., & Ameta, S. C. (2020). Micropollutants and challenges. In *Micropollutants and Challenges* (pp. 1–33).

Chelme-Ayala, P., Smith, D. W., & El-Din, M. G. (2009). Membrane concentrate management options: A comprehensive critical review. *Canadian Journal of Civil Engineering, 36*(6), 1107–1119.

Cimbritz, M., & Mattsson, A. (2018). *Reningstekniker för läkemedel och mikroföroreningar i avloppsvatten: Redovisning av åtta projekt som fått medel från Havs-och vattenmiljöanslaget 2014–2017*. Havs-och vattenmyndigheten.

Congilosi, J. L., & Aga, D. S. (2021). Review on the fate of antimicrobials, antimicrobial resistance genes, and other micropollutants in manure during enhanced anaerobic digestion and composting. *Journal of Hazardous Materials, 405*, 123634. https://doi.org/10.1016/j.jhazmat.2020.123634

Das, A., Dey, S., Bala, K., Gnanasekaran, R., & Das, A. P. (2024). Microbial bioremediation technology for sustainable treatment and management of synthetic microfiber waste. *Sustainable Microbial Technology for Synthetic and Cellulosic Microfiber Bioremediation* (pp. 221–238). https://doi.org/10.1007/978-3-031-62600-5_12

Das, A. P., Behera, I. Dipamitra., & Das, N. Prasad. (2024). *Renewable energy generation and value addition from environmental microfiber pollution through advanced greener solution* (p. 278).

Davidsson, Å., & Cimbritz, M. (2021). *Less is more.*

De Cazes, M., Abejón, R., Belleville, M.-P., & Sanchez-Marcano, J. (2014). Membrane bioprocesses for pharmaceutical micropollutant removal from waters. *Membranes, 4*(4), 692–729.

de la Torre, T., Alonso, E., Santos, J. L., Rodríguez, C., Gómez, M. A., & Malfeito, J. J. (2015). Trace organics removal using three membrane bioreactor configurations: MBR, IFAS-MBR and MBMBR. *Water Science and Technology : A Journal of the International Association on Water Pollution Research, 71*(5), 761–768. https://doi.org/10.2166/wst.2015.028

De Wever, H., Weiss, S., Reemtsma, T., Vereecken, J., Müller, J., Knepper, T., Rörden, O., Gonzalez, S., Barcelo, D., & Hernando, M. D. (2007). Comparison of sulfonated and other micropollutants removal in membrane bioreactor and conventional wastewater treatment. *Water Research, 41*(4), 935–945.

Deng, Y., & Zhao, R. (2015). Advanced oxidation processes (AOPs) in wastewater treatment. *Current Pollution Reports, 1*(3), 167–176. https://doi.org/10.1007/s40726-015-0015-z

Derco, J., Dudáš, J., Valičková, M., Šimovičová, K., & Kecskés, J. (2015). Removal of micropollutants by ozone based processes. *Chemical Engineering and Processing—Process Intensification, 94*, 78–84. https://doi.org/10.1016/j.cep.2015.03.014

Dey, S., Das, A., Mallick, K., Sahu, A., & Das, A. P. (2023). *Environmental Petroleum Waste: Pollution, Toxicity, Sustainable Remediation* (pp. 159–175). https://doi.org/10.1007/978-3-031-48220-5_7

Dey, S., Dash, A., & Das, A. P. (2024). Bioremediation as a sustainable and circular solution for mining waste management and bioenergy generation. In *Sustainable Management of Mining Waste and Tailings: A Circular Economy Approach* (pp. 107–119). https://doi.org/10.1201/9781003442455-6/BIOREMEDIATION-SUSTAINABLE-CIRCULAR-SOLUTION-MINING-WASTE-MANAGEMENT-BIOENERGY-GENERATION-SUDESHNA-DEY-AKANSHYA-DASH-ALOK-PRASAD-DAS

Dey, S., & Mehta, N. S. (2020). Automobile pollution control using catalysis. *Resources, Environment and Sustainability, 2*, 100006. https://doi.org/10.1016/j.resenv.2020.100006

Dhamorikar, R. S., Lade, V. G., Kewalramani, P. V, & Bindwal, A. B. (2024). Review on integrated advanced oxidation processes for water and wastewater treatment. *Journal of Industrial and Engineering Chemistry, 138*, 104–122. https://doi.org/10.1016/j.jiec.2024.04.037

Directive (EU) 2024/3019 (2024).

Dong, G., Chen, B., Liu, B., Hounjet, L. J., Cao, Y., Stoyanov, S. R., Yang, M., & Zhang, B. (2022). Advanced oxidation processes in microreactors for water and wastewater treatment: Development, challenges, and opportunities. *Water Research, 211*, 118047. https://doi.org/10.1016/j.watres.2022.118047

Eggen, R. I. L., Hollender, J., Joss, A., Schärer, M., & Stamm, C. (2014). Reducing the discharge of micropollutants in the aquatic environment: The benefits of upgrading wastewater treatment plants. *Environmental Science & Technology, 48*(14), 7683–7689. https://doi.org/10.1021/es500907n

Elliott, S. M., Erickson, M. L., Krall, A. L., & Adams, B. A. (2018). Concentrations of pharmaceuticals and other micropollutants in groundwater downgradient from large on-site wastewater discharges. *PLoS ONE, 13*(11), e0206004.

Favre, F., Slijepcevic, A., Piantini, U., Frey, U., Abiven, S., Schmidt, H.-P., & Charlet, L. (2022). Real wastewater micropollutant removal by wood waste biomass biochars: A mechanistic interpretation related to various biochar physico-chemical properties. *Bioresource Technology Reports, 17*, 100966. https://doi.org/10.1016/j.biteb.2022.100966

Filote, C., Roşca, M., Hlihor, R. M., Cozma, P., Simion, I. M., Apostol, M., & Gavrilescu, M. (2021). Sustainable application of biosorption and bioaccumulation of persistent pollutants in wastewater treatment: Current practice. *Processes, 9*(10). https://doi.org/10.3390/pr9101696

Ghosh, S., & LaPara, T. M. (2007). The effects of subtherapeutic antibiotic use in farm animals on the proliferation and persistence of antibiotic resistance among soil bacteria. *The ISME Journal, 1*(3), 191–203. https://doi.org/10.1038/ismej.2007.31

Gray, N. F. (2014). Ozone disinfection. In *Microbiology of waterborne diseases* (pp. 599–615). Elsevier.

Grover, D. P., Zhou, J. L., Frickers, P. E., & Readman, J. W. (2011). Improved removal of estrogenic and pharmaceutical compounds in sewage effluent by full scale granular activated carbon: Impact on receiving river water. *Journal of Hazardous Materials, 185*(2), 1005–1011. https://doi.org/10.1016/j.jhazmat.2010.10.005

Gul, B., Naseem, M. K., Malik, W.-N., Gurmani, A. R., Mehmood, A., & Rafique, M. (2022). Environmental micropollutants and their impact on human health with special focus on agriculture. In *Hazardous Environmental Micro-pollutants, Health Impacts and Allied Treatment Technologies* (pp. 1–19). Springer.

Gutiérrez, M., Grillini, V., Mutavdžić Pavlović, D., & Verlicchi, P. (2021). Activated carbon coupled with advanced biological wastewater treatment: A review of the enhancement in micropollutant removal. *Science of the Total Environment, 790*, 148050. https://doi.org/10.1016/j.scitotenv.2021.148050

Hai, F. I., Nghiem, L. D., Khan, S. J., Price, W. E., & Yamamoto, K. (2014). *Wastewater reuse: removal of emerging trace organic contaminants (TrOC).*

Hai, F. I., Tessmer, K., Nguyen, L. N., Kang, J., Price, W. E., & Nghiem, L. D. (2011). Removal of micropollutants by membrane bioreactor under temperature variation. *Journal of Membrane Science, 383*(1–2), 144–151.

Hernández-Soriano, M. del C., Peña, A., & Mingorance, M. D. (2010). Release of metals from metal-amended soil treated with a sulfosuccinamate surfactant: Effects of surfactant concentration, soil/solution ratio, and pH. *Journal of Environmental Quality, 39*(4), 1298–1305. https://doi.org/10.2134/jeq2009.0242

Herrmann, R. (1987). Environmental transfer of some organic micropollutants. In E.-D. Schulze & H. Zwölfer (Eds.), *Potentials and Limitations of Ecosystem Analysis* (pp. 68–99). Springer Berlin Heidelberg. https://doi.org/10.1007/978-3-642-71630-0_4

Hillenbrand, T., Tettenborn, F., Bloser, M., Fligge, F., & Anton, J. (2017). *Policy Paper. Recommendations from the Muliti-Stakeholder Dialogue on the Trace Substance Strategy of the German Federal Government to policy-makers on options to reduce trace substance inputs to the aquatic environment.*

Hofman-Caris, C. H. M., & Beerendonk, E. (2011). New concepts of UV/H2O2 oxidation. *KWR Watercycle Research Institute, Nieuwegein, the Netherlands.*

Hollender, J., Zimmermann, S. G., Koepke, S., Krauss, M., McArdell, C. S., Ort, C., Singer, H., Von Gunten, U., & Siegrist, H. (2009). Elimination of organic micropollutants in a municipal wastewater treatment plant upgraded with a full-scale post-ozonation followed by sand filtration. *Environmental Science & Technology, 43*(20), 7862–7869.

Huang, W., Song, B., Liang, J., Niu, Q., Zeng, G., Shen, M., Deng, J., Luo, Y., Wen, X., & Zhang, Y. (2021). Microplastics and associated contaminants in the aquatic environment: A review on their ecotoxicological effects, trophic transfer, and potential impacts to human health. *Journal of Hazardous Materials, 405*, 124187. https://doi.org/10.1016/j.jhazmat.2020.124187

Imoobe, T. O., Kayode-Edwards, I. I., Omeje, M., Enuneku, A. A., Mameh, E. O., & Adagunodo, T. A. (2024). Behavior and fate of contaminants in cold environments BT. In P. O. Isibor (Ed.), *Arctic Marine Ecotoxicology: Climate Change, Pollutants, and Their Far-Reaching Effects* (pp. 97–145). Springer Nature Switzerland. https://doi.org/10.1007/978-3-031-73584-4_5

Jekel, M., & Dott, W. (2013). Leitfaden: Polare organische Spurenstoffe als Indikatoren im anthropogen beeinflussten Wasserkreislauf. *Gwf Wasser| Abwasser, 154*(12), 1334–1346.

Jelusic, M., & Lestan, D. (2014). Effect of EDTA washing of metal polluted garden soils. Part I: Toxicity hazards and impact on soil properties. *Science of the Total Environment, 475*, 132–141. https://doi.org/10.1016/j.scitotenv.2013.11.049

Jones, A. (n.d.). *EU Air Quality Standards Directive 2008/50/EC.*

Kanungo, J., Sahoo, T., Swain, L. P., & Behera, I. D. (2023). *Toxicity of Persistent Hydrocarbon Pollutants, Sources and Sustainable Remediation Process* (pp. 39–65). https://doi.org/10.1007/978-3-031-48220-5_2

Kardos, M. K., Clement, A., Jolánkai, Z., Zessner, M., Kittlaus, S., Weber, N., Gabriel, O., Broer, M. B., Soare, F., Hamchevici, C., Sidau, M., Tonev, R., Milačič, R., Ščančar, J., Horvat, M., Marković, K., Kulcsar, S., Schuhmann, A., Bordós, G., Zoboli, O. (2024). Development and testing of an efficient micropollutant monitoring strategy across a large watershed. *Science of the Total Environment, 948*, 174760. https://doi.org/10.1016/j.scitotenv.2024.174760

Kelly, M. (2013). Data rich, information poor? Phytobenthos assessment and the water framework directive. *European Journal of Phycology, 48*(4), 437–450.

Khan, H., Iram, Gul, K., Ara, B., Khan, A., Ali, N., Ali, N., & Bilal, M. (2020). Adsorptive removal of acrylic acid from the aqueous environment using raw and chemically modified alumina: Batch adsorption, kinetic, equilibrium and thermodynamic studies. *Journal of Environmental Chemical Engineering, 8*(4), 103927. https://doi.org/10.1016/j.jece.2020.103927

Kosek, K., Luczkiewicz, A., Fudala-Książek, S., Jankowska, K., Szopińska, M., Svahn, O., Tränckner, J., Kaiser, A., Langas, V., & Björklund, E. (2020). Implementation of advanced micropollutants removal technologies in wastewater treatment plants (WWTPs)—Examples and challenges based on selected EU countries. *Environmental Science & Policy, 112*, 213–226. https://doi.org/10.1016/j.envsci.2020.06.011

Kothari, D. P., Ranjan, R., & Singal, K. C. (2021). *Renewable energy sources and emerging technologies.*

Krishnan, R. Y., Manikandan, S., Subbaiya, R., Biruntha, M., Govarthanan, M., & Karmegam, N. (2021). Removal of emerging micropollutants originating from pharmaceuticals and personal care products (PPCPs) in water and wastewater by advanced oxidation processes: A review. *Environmental Technology & Innovation, 23*, 101757. https://doi.org/10.1016/j.eti.2021.101757

Kruithof, J. C., Kamp, P. C., & Martijn, B. J. (2007). UV/H2O2 treatment: A practical solution for organic contaminant control and primary disinfection. *Ozone: Science and Engineering, 29*(4), 273–280.

Kumar, A., Das, A., Singh, D., Das, M. K., Srivastava, G. P., Singh, J. P., Tilgam, J., Thapa, S., Das, S., & Chakdar, H. (2023). Soil health restoration in degraded lands: A microbiological perspective. *Land Degradation & Development, 34*(17), 5155–5170. https://doi.org/10.1002/ldr.4851

Kumar, M., Yadav, A. N., Saxena, R., Paul, D., & Tomar, R. S. (2021). Biodiversity of pesticides degrading microbial communities and their environmental impact. *Biocatalysis and Agricultural Biotechnology, 31*, 101883. https://doi.org/10.1016/j.bcab.2020.101883

Lay-Ekuakille, A., Ikezawa, S., Mugnaini, M., Morello, R., & De Capua, C. (2017). Detection of specific macro and micropollutants in air monitoring: Review of methods and techniques. *Measurement, 98*, 49–59. https://doi.org/10.1016/j.measurement.2016.10.055

Lazarova, V., Gallego, S., García Molina, V., & Rougé, P. (2008). Problems of operation and main reasons for failure of membranes in tertiary treatment systems. *Water Science and Technology, 57*(11), 1777–1784.

Lee, C. O., Howe, K. J., & Thomson, B. M. (2009). State of knowledge of pharmaceutical, personal care product, and endocrine disrupting compound removal during municipal wastewater treatment. *New Mexico Environment Department*, 1–55.

Li, C., Cabassud, C., Reboul, B., & Guigui, C. (2015a). Effects of pharmaceutical micropollutants on the membrane fouling of a submerged MBR treating municipal wastewater: Case of continuous pollution by carbamazepine. *Water Research, 69*, 183–194.

Li, X., Shi, H., Li, K., & Zhang, L. (2015b). Combined process of biofiltration and ozone oxidation as an advanced treatment process for wastewater reuse. *Frontiers of Environmental Science & Engineering, 9*, 1076–1083.

Luo, Y., Guo, W., Ngo, H. H., Nghiem, L. D., Hai, F. I., Zhang, J., Liang, S., & Wang, X. C. (2014). A review on the occurrence of micropollutants in the aquatic environment and their fate and removal during wastewater treatment. *Science of the Total Environment, 473–474*, 619–641. https://doi.org/10.1016/j.scitotenv.2013.12.065

Mallick, K., Sahu, S., Sahu, A., Dey, S., & Das, A. P. (2024). *A Critical Review of Marine Microfiber Pollution Routes, Toxicity, and Its Sustainable Remediation.* 189–211. https://doi.org/10.1007/978-3-031-51792-1_10

Mansor, E., Abdallah, H., & Shaban, A. M. (2024). The role of membrane filtration in wastewater treatment. *Environmental Quality Management, 34*(1), e22251. https://doi.org/10.1002/tqem.22251

Margot, J., Kienle, C., Magnet, A., Weil, M., Rossi, L., de Alencastro, L. F., Abegglen, C., Thonney, D., Chèvre, N., Schärer, M., & Barry, D. A. (2013). Treatment of micropollutants in municipal wastewater: Ozone or powdered activated carbon?. *Science of the Total Environment, 461–462*, 480–498. https://doi.org/10.1016/j.scitotenv.2013.05.034

Martin, M. J., Serra, E., Ros, A., Balaguer, M. D., & Rigola, M. (2004). Carbonaceous adsorbents from sewage sludge and their application in a combined activated sludge-powdered activated carbon (AS-PAC) treatment. *Carbon, 42*(7), 1389–1394.

Massoud, M. A., Tarhini, A., & Nasr, J. A. (2009). Decentralized approaches to wastewater treatment and management: Applicability in developing countries. *Journal of Environmental Management, 90*(1), 652–659. https://doi.org/10.1016/j.jenvman.2008.07.001

McGrath, D., & McCormack, R. J. (1999). *The significance of heavy metal and organic micropollutants in soils.* Teagasc.

McIntyre, T. (2003). Phytoremediation of heavy metals from soils. In D. T. Tsao (Ed.), *Phytoremediation* (pp. 97–123). Springer Berlin Heidelberg. https://doi.org/10.1007/3-540-45991-X_4

Meinel, F., Ruhl, A. S., Sperlich, A., Zietzschmann, F., & Jekel, M. (2014). Pilot-scale investigation of micropollutant removal with granular and powdered activated carbon. *Water, Air, & Soil Pollution, 226*(1), 2260. https://doi.org/10.1007/s11270-014-2260-y

Moghadam, A. A., Shuai, W., & Hartmann, E. M. (2023). Anthropogenic antimicrobial micropollutants and their implications for agriculture. *Current Opinion in Biotechnology, 80*, 102902. https://doi.org/10.1016/j.copbio.2023.102902

Mohanta, S., Pradhan, B., & Behera, I. D. (2024). Impact and remediation of petroleum hydrocarbon pollutants on agricultural land: A review. *Geomicrobiology Journal, 41*(4), 345–359. https://doi.org/10.1080/01490451.2023.2243925

Mohanty, J., Dey, S., Bala, K., & Das, A. P. (2024). Recent advances in microplastic pollution for its sustainable management. In *Sustainable Management of Urban Plastic Waste Through Circular Economic Approaches*, (pp. 27–48). https://doi.org/10.1201/9781003465447-3

Mohanty, M., Tripathy, B., Koteswara Reddy, G., & Das, A. P. (2024). Greener technologies for establishing restraint over microplastic pollution. *Sustainable Microbial Technology for Synthetic and Cellulosic Microfiber Bioremediation* (pp. 239–258). https://doi.org/10.1007/978-3-031-62600-5_13

Mohapatra, D. P., Cledón, M., Brar, S. K., & Surampalli, R. Y. (2016). Application of wastewater and biosolids in soil: Occurrence and fate of emerging contaminants. *Water, Air, & Soil Pollution, 227*(3), 77. https://doi.org/10.1007/s11270-016-2768-4

Mulder, M., Antakyali, D., & Ante, S. (2015). Costs of removal of micropollutants from effluents of municipal wastewater treatment plants-general cost estimates for the Netherlands based on implemented full scale post treatments of effluents of wastewater treatment plants in Germany and Switzerland (p. 55). *STOWA and Waterboard the Dommel, the Netherlands.*

Odubo, T. C., & Kosoe, E. A. (2024). Sources of air pollutants: impacts and solutions. In S. C. Izah, M. C. Ogwu, & A. Shahsavani (Eds.), *Air Pollutants in the Context of One Health: Fundamentals, Sources, and Impacts* (pp. 75–121). Springer Nature Switzerland. https://doi.org/10.1007/698_2024_1127

Oe, T., Koide, H., Hirokawa, H., & Okukawa, K. (1996). Performance of membrane filtration system used for water treatment. *Desalination, 106*(1), 107–113. https://doi.org/10.1016/S0011-9164(96)00098-7

Official Journal of the European Community l 330: Directive 98/83/EG (1998).

Overturf, M. D., Anderson, J. C., Pandelides, Z., Beyger, L., & Holdway, D. A. (2015). Pharmaceuticals and personal care products: A critical review of the impacts on fish reproduction. *Critical Reviews in Toxicology, 45*(6), 469–491. https://doi.org/10.3109/10408444.2015.1038499

Pal, A., He, Y., Jekel, M., Reinhard, M., & Gin, K. Y.-H. (2014). Emerging contaminants of public health significance as water quality indicator compounds in the urban water cycle. *Environment International, 71*, 46–62. https://doi.org/10.1016/j.envint.2014.05.025

Peer, G., Azaizeh, H., Kurzbaum, E., Shahar, B., Mattar, N., & Azerrad, S. P. (2023). Valorization of olive mill solid waste-derived biochar: An efficient approach for simultaneous adsorption and oxidation of micropollutant in surface water. *Journal of Water Process Engineering, 56*, 104461. https://doi.org/10.1016/j.jwpe.2023.104461

Pérez-Lucas, G., Vela, N., El Aatik, A., & Navarro, S. (2019). Environmental risk of groundwater pollution by pesticide leaching through the soil profile. *Pesticides-Use and Misuse and Their Impact in the Environment, 17*, 1–28.

Perrich, J. R. (2018). *Activated carbon adsorption for wastewater treatment.* CRC Press.

Plaimart, J. (2023). *Nature-based solutions for rural wastewater and agricultural waste management.* Newcastle University.

Pomiès, M., Choubert, J.-M., Wisniewski, C., & Coquery, M. (2013). Modelling of micropollutant removal in biological wastewater treatments: A review. *science of the total environment, 443*, 733–748. https://doi.org/10.1016/j.scitotenv.2012.11.037

Priya, A. K., Muruganandam, M., Ali, S. S., & Kornaros, M. (2023). Clean-up of heavy metals from contaminated soil by Phytoremediation: A multidisciplinary and eco-friendly approach. *Toxics, 11*(5). https://doi.org/10.3390/toxics11050422

Puri, M., Gandhi, K., & Kumar, M. S. (2023). Emerging environmental contaminants: A global perspective on policies and regulations. *Journal of Environmental Management, 332*, 117344. https://doi.org/10.1016/j.jenvman.2023.117344

Qadir, M., Wichelns, D., Raschid-Sally, L., McCornick, P. G., Drechsel, P., Bahri, A., & Minhas, P. S. (2010). The challenges of wastewater irrigation in developing countries. *Agricultural Water Management, 97*(4), 561–568. https://doi.org/10.1016/j.agwat.2008.11.004

Reungoat, J., Escher, B. I., Macova, M., Argaud, F. X., Gernjak, W., & Keller, J. (2012). Ozonation and biological activated carbon filtration of wastewater treatment plant effluents. *Water Research, 46*(3), 863–872. https://doi.org/10.1016/j.watres.2011.11.064

Revised State Implementation Guidance for the Consumer Confidence Report (CCR) Rule (2010).

Rodriguez-Mozaz, S., Ricart, M., Köck-Schulmeyer, M., Guasch, H., Bonnineau, C., Proia, L., de Alda, M. L., Sabater, S., & Barceló, D. (2015). Pharmaceuticals and pesticides in reclaimed water: Efficiency assessment of a microfiltration–reverse osmosis (MF–RO) pilot plant. *Journal of Hazardous Materials, 282*, 165–173.

Safe Work Australia. (n.d.). *Hazardous Chemical Information System (HCIS)*. https://hcis.safeworkaustralia.gov.au/

Sahu, A., Dey, S., Mallick, K., Priyadarsini, S., Behera, I. D., & Das, A. P. (2024). A review on advanced technology for sustainable management of synthetic microplastic waste. *Sustainable Management of Urban Plastic Waste Through Circular Economic Approaches*, 262–279. https://doi.org/10.1201/9781003465447-15

Sarathy, S. R., & Mohseni, M. (2006). An overview of UV-based advanced oxidation processes for drinking water treatment. *IUVA News, 8*(2), 16–27.

Schlederer, F., Martín-Hernández, E., & Vaneeckhaute, C. (2024). Ensuring safety standards in sewage sludge-derived biochar: Impact of pyrolysis process temperature and carrier gas on micropollutant removal. *Journal of Environmental Management, 352*, 119964. https://doi.org/10.1016/j.jenvman.2023.119964

Shao, Y., Chen, Z., Hollert, H., Zhou, S., Deutschmann, B., & Seiler, T.-B. (2019). Toxicity of 10 organic micropollutants and their mixture: Implications for aquatic risk assessment. *Science of The Total Environment, 666*, 1273–1282. https://doi.org/10.1016/j.scitotenv.2019.02.047

Siegrist, H., & Joss, A. (2012). Review on the fate of organic micropollutants in wastewater treatment and water reuse with membranes. *Water Science and Technology, 66*(6), 1369–1376.

Singh, P., Singh, V. K., Singh, R., Borthakur, A., Madhav, S., Ahamad, A., Kumar, A., Pal, D. B., Tiwary, D., & Mishra, P. K. (2020). Chapter 1—bioremediation: A sustainable approach for management of environmental contaminants. In P. Singh, A. Kumar, & A. Borthakur (Eds.), *Abatement of Environmental Pollutants* (pp. 1–23). Elsevier. https://doi.org/10.1016/B978-0-12-818095-2.00001-1

Singh, S., & Kumar, R. (2022). Air pollution and its associated impacts on atmosphere and biota health. In P. Saxena, A. Shukla, & A. K. Gupta (Eds.), *Extremes in Atmospheric Processes and Phenomenon: Assessment, Impacts and Mitigation* (pp. 29–58). Springer Nature Singapore. https://doi.org/10.1007/978-981-16-7727-4_3

Siong, Y. K., Idris, J., & Atabaki, M. (2013). Performance of activated carbon in water filters. *Water Resources*, 1–19.

Sowjanya, M. C. V. N., & Gayatri, C. V. R. L. (2024). *Principles of environmental issues in chemical industries*. Academic Guru Publishing House.

Sundin, A. M., Linderholm, L., Hedlund, B., Bly Joyce, K., & Klingspor, K. (2017). *Avancerad rening av avloppsvatten för avskiljning av läkemedelsrester och andra oönskade ämnen: Behov, teknik och konsekvenser Redovisning av ett regeringsuppdrag*. Naturvårdsverket.

Tang, J. Y. M., McCarty, S., Glenn, E., Neale, P. A., Warne, M. St. J., & Escher, B. I. (2013). Mixture effects of organic micropollutants present in water: Towards the development of effect-based water quality trigger values for baseline toxicity. *Water Research, 47*(10), 3300–3314. https://doi.org/10.1016/j.watres.2013.03.011

Tenno, T., & Swinarski, M. (2023). *Advanced mobile technology for removing micropollutants from wastewater*.

Trinh, T., van den Akker, B., Stuetz, R. M., Coleman, H. M., Le-Clech, P., & Khan, S. J. (2012). Removal of trace organic chemical contaminants by a membrane bioreactor. *Water Science and Technology, 66*(9), 1856–1863. https://doi.org/10.2166/wst.2012.374

Tsang, D., Lo, I., & Surampalli, R. (2012). Chelating agents for land decontamination technologies. In *Chelating Agents for Land Decontamination Technologies*. https://doi.org/10.1061/9780784412183

Umar, M., Roddick, F., & Fan, L. (2015). Recent advancements in the treatment of municipal wastewater reverse osmosis concentrate—an overview. *Critical Reviews in Environmental Science and Technology, 45*(3), 193–248. https://doi.org/10.1080/10643389.2013.852378

Vershinin, V. V., Murasheva, A. A., Shirokova, V. A., Khutorova, A. O., Shapovalov, D. A., & Tarbaev, V. A. (2016). The solutions of the agricultural land use monitoring problems. *International Journal of Environmental and Science Education, 11*(12), 5058–5069.

Vodyanitskii, Y. N., & Kirilova, N. P. (2015). Soil contamination by organic micropollutants. *Annals of Agrarian Science, 13*(3).

Voglar, D., & Lestan, D. (2014). Chelant soil-washing technology for metal-contaminated soil. *Environmental Technology, 35*(11), 1389–1400. https://doi.org/10.1080/09593330.2013.869265

Wahlberg, C., Björlenius, B., & Paxéus, N. (2010). Läkemedelsrester i Stockholms vattenmiljö. *Förekomst, Förebyggande Åtgärder Och Refining Av Avloppsvatten. Stockholm Vatten AB.*

WHO, U. (2006). Air quality guidelines: global update 2005. *World Health Organization.*

Xu, L., Wu, C., Chai, C., Cao, S., Bai, X., Ma, K., Jin, X., Shi, X., & Jin, P. (2022). Adsorption of micropollutants from wastewater using iron and nitrogen co-doped biochar: Performance, kinetics and mechanism studies. *Journal of Hazardous Materials, 424*, 127606. https://doi.org/10.1016/j.jhazmat.2021.127606

Yang, W., Zhou, H., & Cicek, N. (2014). Treatment of organic micropollutants in water and wastewater by UV-based processes: A literature review. *Critical Reviews in Environmental Science and Technology, 44*(13), 1443–1476.

Yangali-Quintanilla, V., Maeng, S. K., Fujioka, T., Kennedy, M., Li, Z., & Amy, G. (2011). Nanofiltration vs. reverse osmosis for the removal of emerging organic contaminants in water reuse. *Desalination and Water Treatment*, *34*(1–3), 50–56.

Zeghoud, S., Hemmami, H., Ben Amor, I., Ben Seghir, B., Rebiai, A., & Kouadri, I. (2024). Comprehensive methods for the analysis of organic micro pollutants. In R. Bhadouria, S. Tripathi, P. Singh, R. Singh, & H. P. Singh (Eds.), *Organic Micropollutants in Aquatic and Terrestrial Environments* (pp. 129–157). Springer Nature Switzerland. https://doi.org/10.1007/978-3-031-48977-8_7

Zhang, J (Jim)., Wei, Y., & Fang, Z. (2019). Ozone pollution: A major health hazard worldwide. *Frontiers in Immunology, 10.*

Zhou, W., Arcot, Y., Medina, R. F., Bernal, J., Cisneros-Zevallos, L., & Akbulut, M. E. S. (2024). Integrated pest management: An update on the sustainability approach to crop protection. *ACS Omega, 9*(40), 41130–41147. https://doi.org/10.1021/acsomega.4c06628

Zimmermann, S. G., Wittenwiler, M., Hollender, J., Krauss, M., Ort, C., Siegrist, H., & von Gunten, U. (2011). Kinetic assessment and modeling of an ozonation step for full-scale municipal wastewater treatment: Micropollutant oxidation, by-product formation and disinfection. *Water Research, 45*(2), 605–617. https://doi.org/10.1016/j.watres.2010.07.080

Chapter 13
Implementing Sustainable Practices in India: Strategies for Reducing Plastic Waste, Improving Water Access, and Enhancing Environmental Health

Monika Budania and Nitin Kumar Mishra

Introduction

India's rapid urbanization and population growth have increased environmental challenges, particularly in managing plastic waste and ensuring sustainable water use. Single-use plastics, prevalent in everyday life, contribute significantly to pollution and waste. These plastics not only contribute to environmental degradation but also strain waste management systems that are often ill-equipped to handle the sheer volume of waste produced. Concurrently, water management issues impact both urban and rural areas. Addressing these issues requires a multifaceted approach that integrates sustainable practices across various sectors.

Additionally, despite efforts to minimize plastic waste in India through projects like "rural & urban landscape free of dry & plastic waste," the latest report from (Central Pollution Control Board,) of India estimates that plastic waste generation reaches 3,469,780 tonnes per annum (tpa). Out of this amount, only 1,562,000 tpa is recycled, and 167,000 tpa is co-processed. Therefore, it is crucial to address both the sources of and solutions for plastic waste in India.

In the realm of water management, the extensive use of plastic water bottles on India's railway network underscores the need for more sustainable practices. Installing water tanks in train bogies could reduce reliance on single-use plastics and offer a sustainable alternative for passengers. However, such initiatives require regular maintenance to ensure water quality and effectiveness. Moreover, there is a lack of widespread, accessible water filling stations at key transit points, which hinders the adoption of reusable bottles.

M. Budania (✉) · N. K. Mishra
Department of Mathematics, Lovely Professional University, Phagwara, Punjab, India
e-mail: monikabudania1998@gmail.com

I. D. Behera et al. (eds.), *Emerging Micropollutants*, Sustainable Environmental Waste Management Strategies, https://doi.org/10.1007/978-3-032-03530-1_13

Effective recycling and waste management remain critical yet underdeveloped areas. While educational campaigns and dedicated recycling bins for plastic bottles at transit hubs are steps in the right direction, there is a lack of systematic implementation. Innovative approaches, such as incentivizing proper disposal through a lottery system, could enhance recycling rates. Additionally, collection points for improperly discarded bottles need to be better established and managed to ensure proper processing and support responsible consumption.

In addition to addressing plastic waste, promoting sustainable practices such as the use of homemade products, offers an alternative to chemical-based commercial products. Homemade products reduce packaging waste and chemical exposure, which can lessen environmental impact and improve personal health. Despite this, there is limited government encouragement for such practices, which could otherwise foster a more sustainable lifestyle (John & Das, 2024; Mohanty et al., 2024c; Sahu et al., 2024a).

Moreover, mandating rainwater harvesting in new buildings presents a practical solution for improving water management and urban cleanliness, yet its adoption is not widespread. By capturing and storing rainwater, this practice could reduce municipal water demand, combat water scarcity, and mitigate urban flooding.

Supporting organic farming is another essential aspect of sustainability. Organic farming avoids synthetic chemicals and fertilizers, enhancing soil health and biodiversity (Das et al., 2024c). However, there is insufficient support and incentive for farmers to transition to these practices, which limits the environmental benefits and sustainability of agriculture.

Furthermore, improving milk distribution systems by sourcing directly from rural producers and reducing packaging waste could also benefit both the environment and rural economies. Streamlining supply chains and minimizing packaging are crucial but often overlooked aspects of sustainability.

Finally large-scale tree planting with native species is crucial for climate action and ecosystem restoration. Native trees enhance soil quality, provide wildlife habitat, and improve air quality. Despite the recognized benefits, comprehensive and coordinated large-scale tree planting initiatives are lacking.

This research aims to address these gaps by evaluating current practices, proposing practical solutions, and examining how these strategies can be effectively integrated to promote sustainable development in India. By identifying and addressing these issues, this study will contribute to the existing body of knowledge and offer novel insights into effective environmental management practices.

To develop solutions for reducing plastic waste and improving water access in India's train network, as well as for other environmental initiatives, a mixed-methods approach is employed. This approach includes a review of literature, an analysis of current practices, and the collection of data from recent studies on plastic waste, environmental situations, and water management. Additionally, literature on similar initiatives implemented in other countries was examined to ensure the robustness of the recommendations.

Literature Review

Sustainable development goals (SDGs) have set the 2030 agenda to transform the world by addressing numerous challenges related to ensuring good health, economic prosperity, and environmental sustainability. The SDGs consist of 17 measures aimed at promoting 27 sustainable developments across various areas. Although the fundamental idea of sustainability has been around for centuries, it has only recently gained prominence on the international political stage. The SDGs have successfully increased awareness and generated momentum for collective action, but they have also been critiqued for not addressing the main causes of environmental crises, such as economic growth, inequality, and overconsumption.

To study the SDGs, Pradhan et al. (2017) wrote an article examining the interactions of SDGs using official indicator data for 227 countries. The study concluded that, for a given SDG, positive correlations between indicator pairs outweighed the negative ones in most countries. Leal Filho et al. (2018) also presented an article exploring the ecological aspects of the SDGs, highlighting the continued pressure of human activities on natural resources. Anderson et al. (2022) authored an article explaining the influence of SDG targets on the 2030 Agenda for Sustainable Development. Anderson examined how efforts to address one problem can either support or obstruct progress towards others and based on SDG indicator data, formulated a systematic model assuming influence among the SDGs and their targets. The model found that more SDGs and their corresponding targets tend to support the achievement of other goals and targets rather than act as obstacles.

Krauss (2022) specifically analysed SDG 15 (life on land) and demonstrated how it overlooks important connections between human and nonhuman nature, neglecting the inclusion and recognition of the lives and livelihoods of people. Furthermore, Cano Londoño et al. (2020) authored a book on "actioning the global goals for local impact," which aims to provide timely coverage of research results aligned with sustainability principles to achieve low-carbon societies that promote the recycling and re-use of natural resources. Additionally, Kreinin and Aigner (2022) focused on SDG 8 (decent work and economic growth) from a strong sustainability perspective. Kreinin and Aigner (2022) analysed how it does not meet strong sustainability criteria and proposed a novel framework for SDG 8, termed decent work and economic degrowth.

Dada et al. (2024) wrote an article detailing techniques for smart water management to enhance operational efficiency, maintenance, and resource conservation. Hossain et al. (2023) presented an article on the impact of frugal innovations in achieving SDGs in rural areas, concluding that frugal innovations contribute more to SDGs compared to conventional products. With the rapid increase in urbanization. Takano et al. (2023) proposed a model to evaluate the quality of life for urban residents, emphasizing the need for sustainable urbanization.

Plastic waste has emerged as a significant environmental concern, with millions of tons entering the oceans annually. The most effective method for reducing plastic waste is to minimize consumption. For example, using reusable containers and purchasing in bulk to reduce packaging waste can significantly minimize plastic consumption. Encouraging consumers to recycle plastic products also helps in reducing plastic waste. The challenge of plastic pollution is substantial, with various studies highlighting the need for comprehensive waste management strategies. Improved wastewater management is crucial, as untreated wastewater can introduce pollutants, including plastics, into the environment (Dey et al., 2024; Mohanty et al., 2024a, 2024b, 2024c; Sahu et al., 2024a). Implementing sustainable wastewater infrastructure can mitigate these issues while enhancing public health outcomes. Additionally, adopting radical resource efficiency measures, such as imposing fees on single-use plastics, has proven effective in reducing plastic consumption. Concerning the effects of plastic waste, Kumar et al. (2021) formulated an article addressing the impact of plastic waste on ecosystems and achieving SDGs.

Furthermore, homemade products also significantly help in achieving SDGs. Motivated by this, Kabo-bah and Nyantakyi (2021) wrote an article providing homemade solutions for self-reliance in developing countries.

Since India is a developing country, managing SDGs in India is a crucial task. Considering this, Khalid et al. (2021) presented a case study on India, using the opinions of experts about the implementation of policies related to SDGs.

Improving milk distribution systems can also play a role in reducing plastic waste. Considering this, Jouzdani et al. (2021) produced a case study addressing sustainable perishable food supply chain network design. Additionally, Fois and Cocco (2022) provided an article suggesting a sustainable design for milk powder production using an LCA approach in the food supply chain.

Organic farming practices contribute positively to environmental health by improving soil quality and reducing chemical runoff. Research indicates that organic agriculture leads to better ecosystem services, including enhanced biodiversity and reduced pesticide use. Motivated by this, Gamage and Gangahagedara (2023) wrote an article discussing the role of organic farming in achieving sustainable development goals. Additionally, the plantation of superior commodities boosts farmers' income and helps environmental conditions. Addressing this, Syahza et al. (2021) provided an article suggesting the plantation of superior commodities. Furthermore, forest plantation supports sustainable development. Considering this, Koutika et al. (2022) wrote an article discussing the United Nations agenda for SDGs. Berrone et al. (2023) also provided literature regarding SDGs.

Strategies to Support SDG

Below are some strategies to support SDG:

Reducing Plastic Waste and Enhancing Water Access in India's Train Network

Indian Railways' own brand of packaged drinking water, Rail Neer, which has steadily increased its production in recent years, as shown in the table 1. For the year 2023–24, approximately 39.49 crore bottles were produced, a significant rise compared to the 7.53 crore bottles produced in 2020–21 during the pandemic. While this growth highlights efforts to meet the demand for drinking water on trains, it raises a critical question: what about the millions of bottles sold by other private brands? (Fig. 13.1).

The Rail Neer production data indicates that it fulfils only a fraction of the total demand for bottled water across Indian Railways. Most plastic water bottles sold and improperly discarded are from private brands, further emphasizing the need for sustainable solutions, such as filtered water tanks in train bogies. Currently, only Deendayalu coaches have drinking water tanks, not the entire Indian train network. To tackle these issues, the government should focus on installing filtered water tanks in each train bogie.

This initiative would reduce the need for single-use plastic bottles and address the problem of improperly discarded bottles. According to a study on waste management by Monish et al. (2024) 67% of solid waste ends up in landfills due to the lack of recycling and treatment facilities. Therefore, minimizing plastic waste is important. Ensuring the cleanliness and safety of the water in these tanks will also be essential, requiring regular maintenance and oversight. This aligns with SDG 6 (clean water and sanitation) by improving access to water while reducing plastic pollution, and SDG

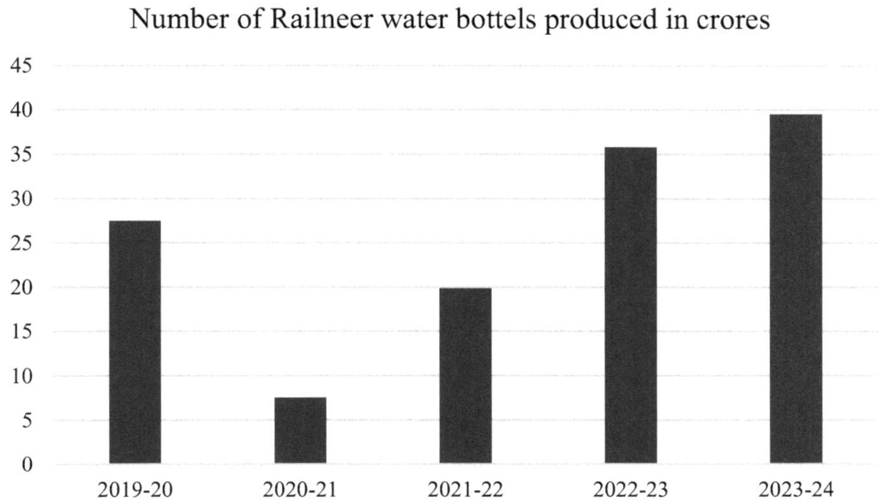

Fig. 13.1 Year-wise production of number of Railneer water bottles (*Source* https://www.irctc.com/annual-report.html)

14 (life below water) by preventing plastic waste from entering aquatic ecosystems (Priyadarshini et al., 2024; Tripathy et al., 2024).

For those who prefer purchasing water, the government should provide low-cost or free water filling stations at railway stations and major transit points. This would encourage people to refill reusable bottles, further reducing plastic waste. These stations must be clean, accessible, and consistently supplied with water. This initiative supports SDG 6 (clean water and sanitation) and SDG 11 (sustainable cities and communities) by enhancing public infrastructure and promoting sustainable resource use.

In addition, dedicated recycling bins for plastic bottles should be placed at railway stations and bus stands to improve recycling rates. Clear signage and educational campaigns, as suggested by the Central Pollution Control Board, will help passengers understand the importance of using these recycling stations. Partnerships with recycling companies could streamline the recycling process. This measure contributes to SDG 12 (responsible consumption and production) by encouraging recycling and reducing waste, and SDG 13 (climate action) by mitigating the environmental impact of plastic waste.

For plastic bottles that are not properly disposed of, the government should establish bottle collection points at railway stations and bus stands to ensure proper recycling. If bottles are thrown on tracks or left in train bogies, recycling stations should be responsible for collecting and processing them. This supports SDG 12 (responsible consumption and production) by ensuring effective waste management and SDG 11 (sustainable cities and communities) by maintaining clean and safe public spaces.

To incentivize proper disposal, a lottery system could be introduced where individuals who dispose of their bottles in recycling bins have a chance to win a reward. This could encourage more people to use recycling bins and improve waste management practices. This approach aligns with SDG 12 (responsible consumption and production) by promoting responsible waste disposal and increasing recycling rates.

These initiatives help achieve multiple SDGs by promoting sustainable practices, improving infrastructure, reducing waste, and enhancing public awareness and engagement in environmental conservation.

Moreover, this initiative will generate job opportunities, such as positions for tank cleaners and maintenance staff for water filling stations and recycling facilities. This job creation will contribute to SDG 8 (decent work and economic growth) by providing employment and supporting local economies. By investing in these sustainable practices, the government not only addresses environmental issues but also promotes economic development and job growth (Fig. 13.2).

Mandating Rainwater Harvesting in Urban Areas

Rainwater harvesting is mandatory only in the states of Tamil Nadu and Delhi in India. As per a report from the (Times of India, 2020), in most parts of Chennai (Tamil Nadu), the groundwater level, which was previously at 20–25 feet, has now

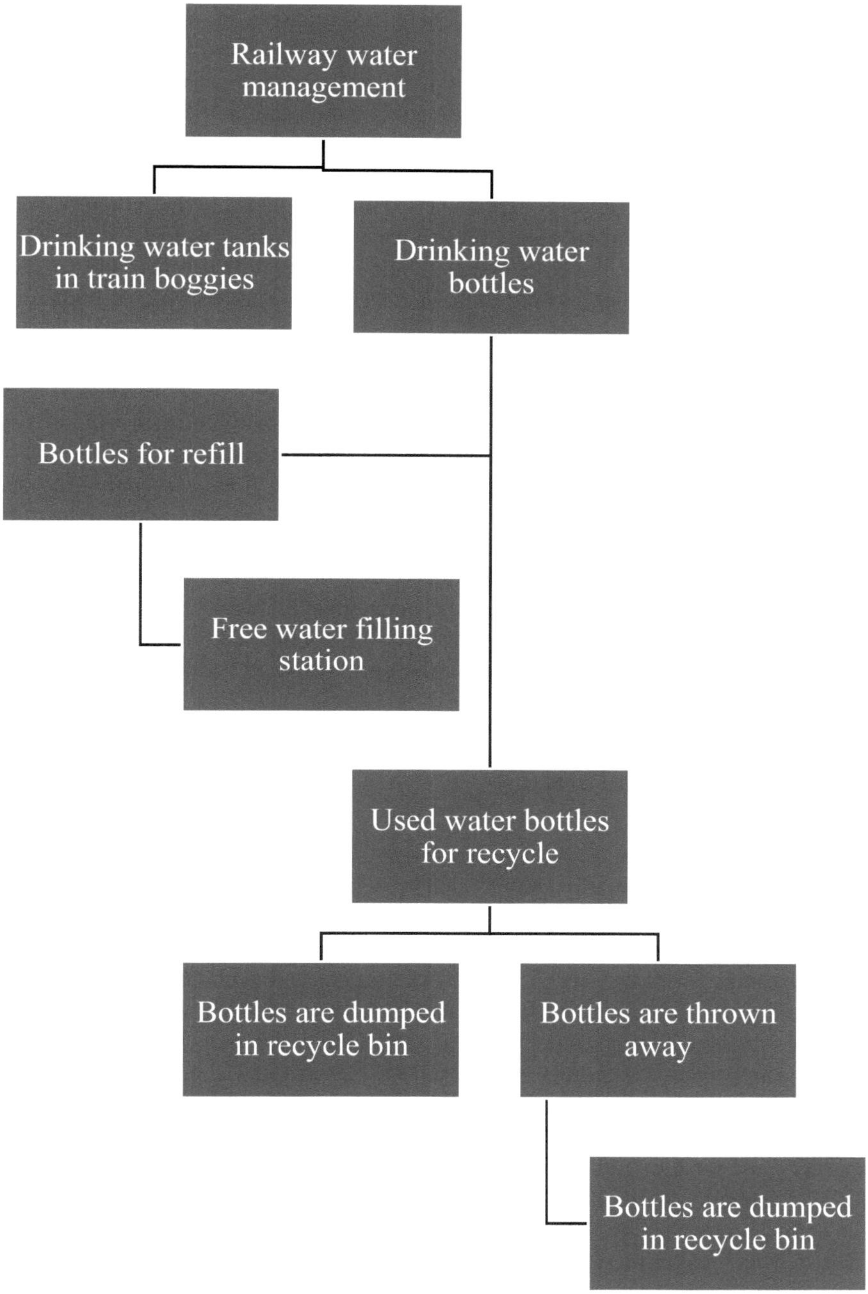

Fig. 13.2 Drinking water access and plastic waste management in Indian Railways

improved to 10–15 feet. Thus, government should mandate rainwater harvesting in all states of India. Mandating the inclusion of rainwater harvesting systems in all new buildings of urban areas of India would involve designing roofs to channel rainwater to the ground and into collection systems, which can then be used to recharge groundwater levels and manage stormwater runoff. Implementing such systems in urban areas would significantly improve water levels and enhance cleanliness in cities. It is also cost-effective, as explained by Pala et al. (2021). The Indian Government is providing several subsidies for rainwater harvesting. To ensure people are aware of these subsidies, the government should require constructors to inform their customers about them. This would make it easier for the government to update the public about the scheme.

This initiative supports SDG 6 (clean water and sanitation) by improving water management and ensuring sustainable water resources. It also contributes to SDG 11 (sustainable cities and communities) by promoting resilient infrastructure and effective urban planning suggested by Jamali et al. (2020), rainwater harvesting could significantly reduce the risk of small and medium floods. By capturing and utilizing rainwater, cities can manage water resources more efficiently, reduce the risk of flooding, improve city cleanliness, and support environmental sustainability.

Furthermore, incorporating rainwater harvesting in building regulations can create new job opportunities in construction and maintenance, thereby supporting SDG 8 (decent work and economic growth). This measure not only addresses water scarcity and urban cleanliness but also fosters economic growth and community well-being.

Promoting Organic Farming

To promote organic farming, the government is running a scheme called Paramparagat Krishi Vikas Yojana (PKVY) and providing subsidies for it. However, the subsidy is limited compared to the profits from inorganic farming, as De La Cruz et al. (2023) found that the yield of organic farming is 18.4% lower than that of conventional farming. To bridge this gap, the cost of organic products should be increased to make organic farming equally profitable. Otherwise, farmers will continue to be drawn towards inorganic farming. To achieve this, the government should also initiate programs to actively promote organic farming. This program could provide support and resources to farmers interested in transitioning to chemical-free farming practices. An online portal, like Kishan Mandi (a private platform for selling both organic and inorganic products), should be established exclusively for organic products. This portal would allow farmers to list their organic products for sale. This portal would allow consumers who are willing to pay a premium for organic products to purchase directly from local farmers, ensuring that products are fresh and sustainably sourced.

Such an initiative would motivate more farmers to adopt organic farming practices by providing them with a viable market for their produce. It would also help consumers access high-quality, chemical-free products, contributing to healthier lifestyles and environmental sustainability. A 2020 survey by Linkage Research and the Non-GMO Project found that 41% of natural shoppers are more likely to buy products with USDA organic certification. This proves a strong consumer preference for purchasing organic products.

This initiative supports SDG 12 (responsible consumption and production) by encouraging the production and consumption of organic products, which are less harmful to the environment.

By facilitating a direct connection between farmers and consumers, the government can enhance the sustainability of the food supply chain and promote the broader adoption of organic farming practices.

Additionally, organic farming requires less water compared to conventional methods, as suggested by Wheeler et al. (2015). It also helps minimize water pollution by avoiding the use of synthetic pesticides and fertilizers, which can leach into water resources.

Moreover, promoting organic farming supports SDG 15 (life on land) by encouraging practices that protect and restore terrestrial ecosystems. Organic farming typically avoids harmful chemicals and synthetic fertilizers, which can degrade soil health and biodiversity. By fostering organic agriculture, we enhance soil quality, preserve biodiversity, and contribute to the sustainability of land-based resources (Behera et al., 2021; Das et al., 2024a, 2024b; Mohanta et al., 2024).

However, organic farming presents numerous challenges. According to the National Organic Research Agenda (NORA)[32] 2022 report, 67% of farmers struggle with controlling weeds, 59% face difficulties in managing production costs, 48% have issues maintaining adequate yields, 43% find managing soil fertility and crop nutrition challenging, 41% struggle with controlling insect pests, 38% have trouble finding appropriate organic crop varieties and seeds, and 36% of farmers face difficulties with controlling disease pressure and adapting to climate change.

Sustainable Milk Distribution

Since people tend to trust government-backed products more, the government should initiate a program to distribute milk in a more sustainable and efficient manner. Under this initiative, the government could purchase milk directly from rural producers. To ensure the purity and quality of the milk, machines for testing and verifying milk purity should be established at collection points in the villages. The milk could then be transported directly to urban areas, minimizing the need for intermediate processing and packaging.

This initiative would help reduce plastic waste by minimizing the use of single-use plastic bottles and packaging. According to research conducted on Italian coasts, 29% of total plastic waste comes from food packaging alone. By ensuring that milk is tested and verified for purity before it reaches consumers, it would also provide city dwellers with fresh, high-quality milk.

Since distributing fresh milk does not require pasteurization, the program would also lead to significant energy savings. Pasteurization, which involves heating milk to kill harmful bacteria, typically consumes a substantial amount of electricity. As per the study by Singh et al. (2019) 154.5 kg of milk pasteurization requires 5.16 KW of electrical energy and 42.77 kJ $\times$ 103 of thermal energy. By sourcing milk directly from rural producers and delivering it fresh to urban consumers, this initiative would bypass the need for this energy-intensive process. This not only contributes to lower energy consumption but also supports SDG 7 (affordable and clean energy) by promoting more sustainable energy use in the food supply chain.

To further support this program, the government should ensure a fixed price for the milk in each city. This would help stabilize costs for consumers while providing fair and predictable income for rural producers. Such a pricing policy would make high-quality, fresh milk more accessible to urban populations, while also supporting the livelihoods of dairy farmers (Fig. 13.3).

This program will improve the market for rural dairy farmers and ensuring a reliable supply of nutritious milk to urban populations. It aligns with SDG 12 (Responsible Consumption and Production) by reducing packaging waste and promoting sustainable practices. According to the Plast India Foundation (2020), 42% of the total plastic waste is generated from flexible packaging. Thus, adopting projects like this could help in minimizing plastic waste. Additionally, it contributes to SDG 11 (sustainable cities and communities) by enhancing the quality and sustainability of urban food supply chains.

By fostering direct links between rural producers and urban consumers, this initiative not only improves the efficiency and sustainability of the milk distribution system but also supports rural economies and promotes better health outcomes in urban areas.

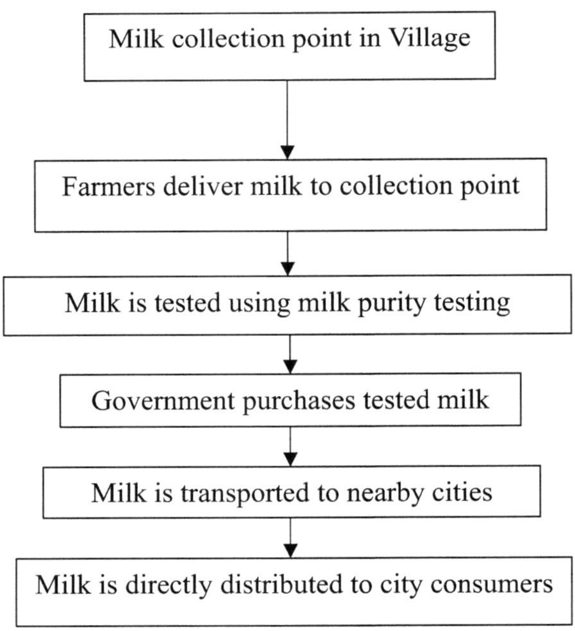

Fig. 13.3 Sustainable milk distribution network

Nationwide Tree Planting

According to the latest India State of Forest (2021), the estimated area covered by trees in India is 95,748 km^2, which constitutes only 2.91% of the total geographical area. Thus, the government should focus on large-scale tree planting initiatives. This program should involve training and hiring people to plant and care for trees of their local species until they are fully grown. It is crucial to select tree species that are well-adapted to their specific environments. For instance, in the desert regions of Rajasthan, species like Khejri and Rohida should be planted, while in Punjab, Safeda could be a suitable choice. By choosing trees that are suited to the local conditions, we ensure their survival and long-term growth, rather than planting random species that may not thrive.

The government can also leverage existing programs such as MNREGA (Mahatma Gandhi national rural employment guarantee act) to provide employment for tree planting and maintenance tasks. This would integrate tree planting into a broader employment scheme, ensuring that rural workers have meaningful and sustainable job opportunities.

This initiative will help increase oxygen production, prevent soil erosion, as a study by Huang et al. (2017) reveals that reforestation helps prevent soil erosion, and support local biodiversity, as a study by Derhé et al. (2016) shows that plantation helps restore local biodiversity. It will also contribute to improving air quality and providing shade, which benefits both the environment and communities. Additionally, tree roots help stabilize soil, reducing runoff and filtering pollutants, which improves the quality of nearby water sources.

This program aligns with SDG 13 (climate action) by enhancing carbon sequestration and combating climate change through increased forest cover. It also supports SDG 15 (Life on Land) by promoting the restoration of terrestrial ecosystems and biodiversity. Additionally, the initiative contributes to SDG 8 (decent work and economic growth) by creating job opportunities for planting and maintaining trees.

By focusing on planting native and adaptive tree species, the government can create sustainable and resilient ecosystems that provide long-term environmental and economic benefits.

Promoting Homemade Products

The government should promote the use of homemade products, as this not only helps reduce environmental impact but also supports local practices and resources. According to the United Nations Environmental Program[19], 36% of the total plastic produced is used for packaging alone. By fostering the use of Ayurveda-based and homemade products recommended by professionals over chemically produced ones, the government can promote sustainability and well-being while respecting cultural traditions.

For example, natural toothpaste made from turmeric, salt, and mustard oil—ingredients that are readily available in India and recommended by some healthcare professionals for their natural benefits would minimize the use of chemical ingredients and reduce waste from toothpaste tubes (Mohanty et al., 2024b; Sahu et al., 2024b).

The government should promote these alternatives by advertising in newspapers and other media to raise public awareness. This would help people transition to more sustainable and chemical-free routines. This initiative supports SDG 12 (responsible consumption and production) by encouraging the use of natural, less wasteful products. It also aligns with SDG 3 (good health and well-being) by potentially reducing exposure to harmful chemicals and promoting natural health practices. Additionally, it could contribute to SDG 6 (clean water and sanitation) by reducing pollution from chemicals and promoting environmentally friendly practices. Such initiatives support local practices and resources, making them practical and culturally relevant solutions.

Conclusion

The proposed strategies offer a comprehensive approach to addressing plastic waste, improving water access, and fostering environmental sustainability in India. By replacing single-use plastics with reusable alternatives, enhancing water management on trains, and establishing water filling stations, we can significantly reduce plastic pollution and improve public health. Promoting effective recycling, encouraging the use of homemade products, and implementing rainwater harvesting systems will further contribute to environmental sustainability and efficient resource management. Supporting organic farming and enhancing milk distribution systems will improve agricultural practices and rural economies, though organic farming may lead to lower production yields compared to conventional methods, potentially impacting short-term food supply. Meanwhile, large-scale tree planting will combat climate change and support ecosystem restoration. Each initiative not only addresses immediate environmental concerns but also supports long-term economic and health benefits, creating a more resilient and sustainable future. By aligning these efforts with multiple SDGs, the government can drive significant progress in environmental conservation, public health, and economic development. However, several challenges may arise during implementation. For instance, maintaining the cleanliness and safety of water tanks in train bogies requires consistent monitoring and regular maintenance, which could strain existing resources. The effectiveness of recycling programs depends on public participation and proper infrastructure, which may need additional investment. Furthermore, mandating rainwater harvesting could face resistance from developers and require adjustments to building codes. Addressing these challenges will be crucial for the success of the proposed solutions.

References

Anderson, C. C., Denich, M., Warchold, A., Kropp, J. P., & Pradhan, P. (2022). A systems model of SDG target influence on the 2030 Agenda for Sustainable Development. *Sustainability Science, 17*(4), 1459–1472.

Annual Report 2019–2020, IRCTC (Indian Railway Catering and Tourism Corporation).

Annual Report 2020–2021, Central Pollution Control Board.

Annual Report 2020–2021, IRCTC (Indian Railway Catering and Tourism Corporation).

Annual Report 2021–2022, IRCTC (Indian Railway Catering and Tourism Corporation).

Annual Report 2022–2023, Centre for Environment Education.

Annual Report 2022–2023, IRCTC (Indian Railway Catering and Tourism Corporation).

Annual Report 2023–2024, IRCTC (Indian Railway Catering and Tourism Corporation).

Behera, I. D., Nayak, M., Biswas, S., Meikap, B. C., & Sen, R. (2021). Enhanced biodegradation of total petroleum hydrocarbons by implementing a novel two-step bioaugmentation strategy using indigenous bacterial consortium. *Journal of Environmental Management, 292*, 112746. https://doi.org/10.1016/J.JENVMAN.2021.112746

Berrone, P., Rousseau, H. E., Ricart, J. E., Brito, E., & Giuliodori, A. (2023). How can research contribute to the implementation of sustainable development goals? An interpretive review of SDG literature in management. *International Journal of Management Reviews, 25*(2), 318–339.

Cano Londoño, N. A., Velasco, J. O., García, F. C., & Franco, I. B. (2020). SDG 6 clean water and sanitation: Sustainable use of energy and water resources in the mining sector: A comparative case study of open-pit and alluvial mining technology. In *Actioning the global goals for local impact: Towards sustainability science, policy, education and practice* (pp. 85–104).

Dada, M. A., Majemite, M. T., Obaigbena, A., Daraojimba, O. H., Oliha, J. S., & Nwokediegwu, Z. Q. S. (2024). Review of smart water management: IoT and AI in water and wastewater treatment. *World Journal of Advanced Research and Reviews, 21*(1), 1373–1382.

Das, A., Dey, S., Bala, K., Gnanasekaran, R., & Das, A. P. (2024a). Microbial bioremediation technology for sustainable treatment and management of synthetic microfiber waste. *Sustainable Microbial Technology for Synthetic and Cellulosic Microfiber Bioremediation, 221–238*. https://doi.org/10.1007/978-3-031-62600-5_12

Das, A. P., Behera, I. D., & Das, N. P. (2024b). *Renewable energy generation and value addition from environmental microfiber pollution through advanced greener solution* (p. 278).

Das, A., Ray, S. K., Mohanty, M., Mohanty, J., Dey, S., & Das, A. P. (2024c). Ecotoxicity of microplastic wastes and their sustainable management: A review. *Environmental Chemistry and Ecotoxicology, 6*, 144–152. https://doi.org/10.1016/J.ENCECO.2024.05.003

Dey, S., Dash, A., & Das, A. P. (2024). Bioremediation as a sustainable and circular solution for mining waste management and bioenergy generation. *Sustainable Management of Mining Waste and Tailings: A Circular Economy Approach, 107–119*. https://doi.org/10.1201/9781003442455-6

de la Cruz, V. Y. V., Cheng, W., & Tawaraya, K. (2023). Yield gap between organic and conventional farming systems across climate types and sub-types: A meta-analysis. *Agricultural Systems, 211*, 103732.

Derhé, M. A., Murphy, H., Monteith, G., & Menéndez, R. (2016). Measuring the success of reforestation for restoring biodiversity and ecosystem functioning. *Journal of Applied Ecology, 53*(6), 1714–1724.

Fois, V., & Cocco, D. (2022). Implication of SDGs on LCA based sustainable design of milk powder's dairy production. *Transportation Research Procedia, 67*, 83–92.

Gamage, A., Gangahagedara, R., Gamage, J., Jayasinghe, N., Kodikara, N., Suraweera, P., & Merah, O. (2023). Role of organic farming for achieving sustainability in agriculture. *Farming System, 1*(1), 100005.

Hossain, M., Park, S., & Shahid, S. (2023). Frugal innovation for sustainable rural development. *Technological Forecasting and Social Change, 193*, 122662.

https://www.cpcb.nic.in/

https://www.nsf.org/
https://www.unep.org/
Huang, C., Yang, H., Li, Y., Zhang, M., Lv, H., Zhu, A. X., Huang, T., et al. (2017). Quantificational effect of reforestation to soil erosion in subtropical monsoon regions with acid red soil by sediment fingerprinting. *Environmental Earth Sciences, 76*, 1–15.
India State of Forest Report 2021.
Jamali, B., Bach, P. M., & Deletic, A. (2020). Rainwater harvesting for urban flood management—An integrated modelling framework. *Water Research, 171*, 115372.
John, A., & Das, A. P. (2024). Government strategies for achieving circular economy for plastic waste management in Southeast Asia. *Sustainable Management of Urban Plastic Waste through Circular Economic Approaches, 72–82*. https://doi.org/10.1201/9781003465447-6
Jouzdani, J., & Govindan, K. (2021). On the sustainable perishable food supply chain network design: A dairy products case to achieve sustainable development goals. *Journal of Cleaner Production, 278*, 123060.
Kabo-bah, A. T., & Nyantakyi, E. (2021). *Homemade solutions for self-reliance: Escaping the poverty net in developing countries*.
Khalid, A. M., Sharma, S., & Dubey, A. K. (2021). Concerns of developing countries and the sustainable development goals: Case for India. *International Journal of Sustainable Development & World Ecology, 28*(4), 303–315.
Koutika, L. S., Matondo, R., Mabiala-Ngoma, A., Tchichelle, V. S., Toto, M., Madzoumbou, J. C., Nzila, J. D. D., et al. (2022). Sustaining forest plantations for the United Nations' 2030 Agenda for sustainable development. *Sustainability, 14*(21), 14624.
Krauss, J. E. (2022). Unpacking SDG 15, its targets and indicators: Tracing ideas of conservation. *Globalizations, 19*(8), 1179–1194.
Kreinin, H., & Aigner, E. (2022). From "decent work and economic growth" to "sustainable work and economic degrowth": A new framework for SDG 8. *Empirica, 49*(2), 281–311.
Kumar, R., Verma, A., Shome, A., Sinha, R., Sinha, S., Jha, P. K., Vara Prasad, P. V., et al. (2021). Impacts of plastic pollution on ecosystem services, sustainable development goals, and need to focus on circular economy and policy interventions. *Sustainability, 13*(17), 9963.
Leal Filho, W., Azeiteiro, U., Alves, F., Pace, P., Mifsud, M., Brandli, L., Disterheft, A., et al. (2018). Reinvigorating the sustainable development research agenda: The role of the sustainable development goals (SDG). *International Journal of Sustainable Development & World Ecology, 25*(2), 131–142.
Mohanta, S., Pradhan, B., & Behera, I. D. (2024). Impact and remediation of petroleum hydrocarbon pollutants on agricultural land: A review. *Geomicrobiology Journal, 41*(4), 345–359. https://doi.org/10.1080/01490451.2023.2243925
Mohanty, J., Dey, S., Bala, K., & Das, A. P. (2024a). Recent advances in microplastic pollution for its sustainable management. *Sustainable Management of Urban Plastic Waste through Circular Economic Approaches, 27–48*. https://doi.org/10.1201/9781003465447-3
Mohanty, M., Mohanty, J., Dey, S., Dutta, K., Shah, M. P., & Das, A. P. (2024b). The face mask: A tale from protection to pollution and demanding sustainable solution. *Emerging Contaminants, 10*(2), 100298. https://doi.org/10.1016/J.EMCON.2023.100298
Mohanty, M., Tripathy, B., Koteswara Reddy, G., & Das, A. P. (2024c). Greener technologies for establishing restraint over microplastic pollution. *Sustainable Microbial Technology for Synthetic and Cellulosic Microfiber Bioremediation, 239–258*. https://doi.org/10.1007/978-3-031-62600-5_13
Monish, A., Sharma, B. R., Sangma, B. R., Chingangbam, B., Syngkli, E. R. R., Singh, K. A., Shaik, M., Marak, N. H. R., Kahalani, P., Singh, R. A., & Rai, R. C. (2024). A study on waste management in railways. *IRJEMS International Research Journal of Economics and Management Studies, 3*(2), 31–34. https://doi.org/10.56472/25835238/IRJEMS-V3I2P104.
National Organic Research Agenda Report. (2022).

Pala, G. K., Pathivada, A. P., Velugoti, S. J. H., Yerramsetti, C., & Veeranki, S. (2021). Rainwater harvesting—A review on conservation, creation & cost-effectiveness. *Materials Today: Proceedings, 45*, 6567–6571.

Pradhan, P., Costa, L., Rybski, D., Lucht, W., & Kropp, J. P. (2017). A systematic study of sustainable development goal (SDG) interactions. *Earth's Future, 5*(11), 1169–1179.

Priyadarshini, B., Santhosh Kumar, M., & Das, A. P. (2024). Microbial degradation of synthetic plastics released from COVID-19 generated single use disposable face masks. *Environmental Science and Engineering, Part F2827*, 1–16. https://doi.org/10.1007/978-3-031-50840-0_1

Sahu, A., Dey, S., Mallick, K., Priyadarsini, S., Behera, I. D., & Das, A. P. (2024a). A review on advanced technology for sustainable management of synthetic microplastic waste. *Sustainable Management of Urban Plastic Waste through Circular Economic Approaches, 262–279*. https://doi.org/10.1201/9781003465447-15/

Sahu, A., Mallick, K., & Das, A. P. (2024b). *Environmental impact of single-use synthetic face mask and its recycling: A sustainable approach* (pp. 197–212). https://doi.org/10.1007/978-3-031-50840-0_11

Singh, B., Chandra, S., Kumar, R., Chaudhary, V., & Kumar, V. (2019). *Energy assessment of milk pasteurization in dairy plant*.

Singh, S. G. (2021). *Plastic recycling: Decoded*.

Syahza, A., Bakce, D., Irianti, M., Asmit, B., & Nasrul, B. (2021). Development of superior plantation commodities based on sustainable development. *International Journal of Sustainable Development & Planning, 16*(4).

Takano, T., Morita, H., Nakamura, S., Togawa, T., Kachi, N., Kato, H., & Hayashi, Y. (2023). Evaluating the quality of life for sustainable urban development. *Cities, 142*, 104561.

TNN. (2020, August 11). Rainwater harvesting helps improve Chennai's water table. *The Times of India*. https://timesofindia.indiatimes.com

Tripathy, B., Dey, S., Dubey, N. K., Lo, W.-C., & Das, A. P. (2024). *An omics-based approach for managing the risk of COVID-19-generated wastes* (pp. 51–66). https://doi.org/10.1007/978-3-031-50840-0_4

Wheeler, S. A., Zuo, A., & Loch, A. (2015). Watering the farm: Comparing organic and conventional irrigation water use in the Murray-Darling Basin, Australia. *Ecological Economics, 112*, 78–85.

Chapter 14
Advancing Micropollutant Mitigation in Aquatic Systems: Strategies and Green Technology Solutions

Aniruddha Deb, Kaniska Murmu, Sunil Kumar Singh, and Awadh Kishor Kumar

Introduction

Emerging pollutants (EPs) have become more prevalent in water over the last ten years, which has caused serious health and environmental issues (Chavoshani et al., 2020). Global production of these pollutants, which could endanger human health and aquatic ecosystems, has skyrocketed from an estimated 1000 to 500,000 thousand tons per year (Ahmed et al., 2021). Micropollutants (MPs), another name for EPs, come from both natural and man-made sources (Gondi et al., 2022; Kanungo et al., 2023; Mallick et al., 2023). Numerous new chemical species have been discovered recently, including those found in drinking water, groundwater, surface water, and municipal wastewater (Bolong et al., 2009; Mishra & Das, 2023; Mishra et al., 2024). These substances, which are frequently found in soil, water, and air systems, are now understood to be serious water pollutants (Fig. 14.1).

A. Deb (✉)
Department of Chemical Engineering, BIT Mesra, Ranchi, Jharkhand, India
e-mail: aniruddha.deb@bitmesra.ac.in

K. Murmu (✉)
Department of Metallurgical and Materials Engineering, IIT MADRAS, Chennai, India
e-mail: bmail159@gmail.com

S. K. Singh
Department of Chemical Engineering, BIT Sindri, Dhanbad, Jharkhand, India
e-mail: sunil.che@bitsindri.ac.in

A. K. Kumar
Department of Chemical Engineering and Technology, NIT Patna, Patna, Bihar, India
e-mail: awadh.ch@nitp.ac.in

I. D. Behera et al. (eds.), *Emerging Micropollutants*, Sustainable Environmental Waste Management Strategies, https://doi.org/10.1007/978-3-032-03530-1_14

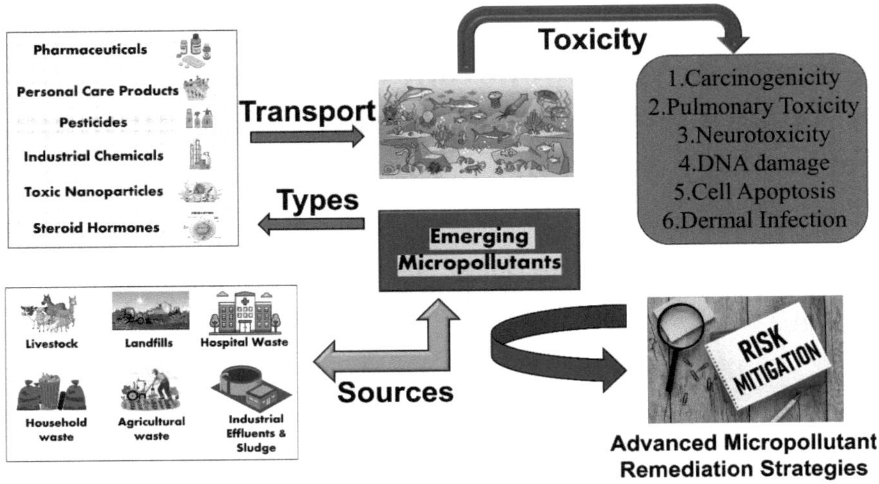

Fig. 14.1 Overview of emerging micropollutants in aquatic environments: their sources, ecological accumulation, associated human health risks, and the imperative for sustainable risk mitigation strategies

According to Su et al. (2020), EPs include a broad range of compounds, such as hormones, flame retardants, industrial chemicals, endocrine-disrupting chemicals (EDCs), medicines, personal care products (PCPs), nanomaterials, and pesticides. Researchers have studied that their sources can be divided into four primary categories: leachate from landfills, surface runoff from agriculture and aquaculture, industrial wastewater, and effluents from homes and hospitals. Because of their strong polarity, which promotes bioaccumulation and resistance to degradation, many of these contaminants are persistent in the environment (Kim & Zoh, 2016; Kümmerer et al., 2019).

Furthermore, because of their extensive use in chemical-based industries and potent antibacterial qualities, metals like lead (Pb), mercury (Hg), chromium (Cr), copper (Cu), zinc (Zn), arsenic (As), cadmium (Cd), and nickel (Ni) as well as silver (Ag), iron (Fe), and zinc (Zn) nanoparticles are included in this category (Prabhakar et al., 2019; Raj & Das, 2023). It is commonly known that these contaminants have detrimental effects, including long-term toxicity and the encouragement of algal blooms (Priyadarsini & Das, 2024, 2025).

It is still difficult to detect trace levels of these pollutants, which frequently range from nanograms to micrograms per litre (ng/L–μg/L). Such low MP levels are typically not removed by conventional wastewater treatment facilities (CWWTPs) (Schwarzenbach et al., 2006). Accordingly, untreated micropollutants frequently remain in wastewater treatment systems (Guerra-Rodríguez et al., 2021; Rodríguez-Chueca et al., 2018), which may result in their discharge into the environment and endanger ecological and human health. It has been reported that EPs are found in water conditions in many areas, such as the US, UK, EU, China, South Korea, North

Korea, and the Western Balkans (Kosek et al., 2020; Tosun et al., 2020). Innovative methods for detection and treatment are needed to address these issues and lessen their negative impacts (Gondi et al., 2022; Kumar et al., 2022; Sivaranjanee & Kumar, 2021).

Water contamination is largely caused by the annual release of around 300,000 thousand tons of synthetic chemicals from consumer and industrial items into natural water bodies (Mishra & Das, 2021; Mishra et al., 2022). Strict laws are needed to limit the emission and discharge of these pollutants in order to improve surface water quality and safeguard aquatic ecosystems (Kumar et al., 2023; Mishra & Das, 2022). Anaerobic bed reactors, artificial wetlands, chemical coagulation, precipitation, and activated sludge systems are some of the technologies used to remove emerging pollutants (EPs) from wastewater and aquatic environments (Kümmerer et al., 2019; Luo et al., 2019; Rout et al., 2021).

This chapter seeks to shed light on the presence of emerging pollutants (EPs) in aquatic environments and emphasizes the significant risks these contaminants pose to human health and ecosystems shown in Figs. 14.2 and 14.3. The initial sections of this chapter delve into the environmental challenges associated with EPs, exploring their sources and toxicological impacts. Subsequently, the latter sections focus on existing methodologies for the removal of EPs from water and wastewater, while also proposing innovative and cost-effective treatment strategies aimed at enhancing the efficiency and sustainability of EP remediation.

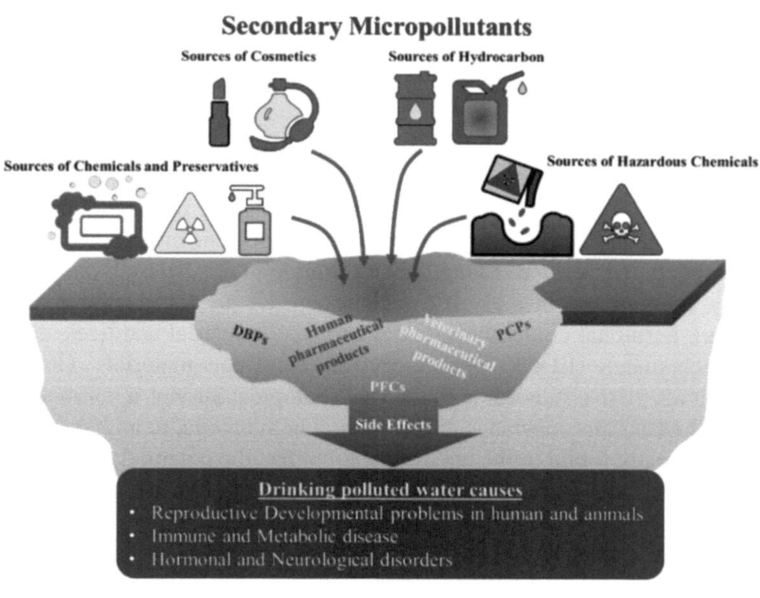

Fig. 14.2 Effects and health risks of emerging micropollutants on humans and wildlife. Adopted and reproduced with permission (Shanmuganathan et al., 2023)

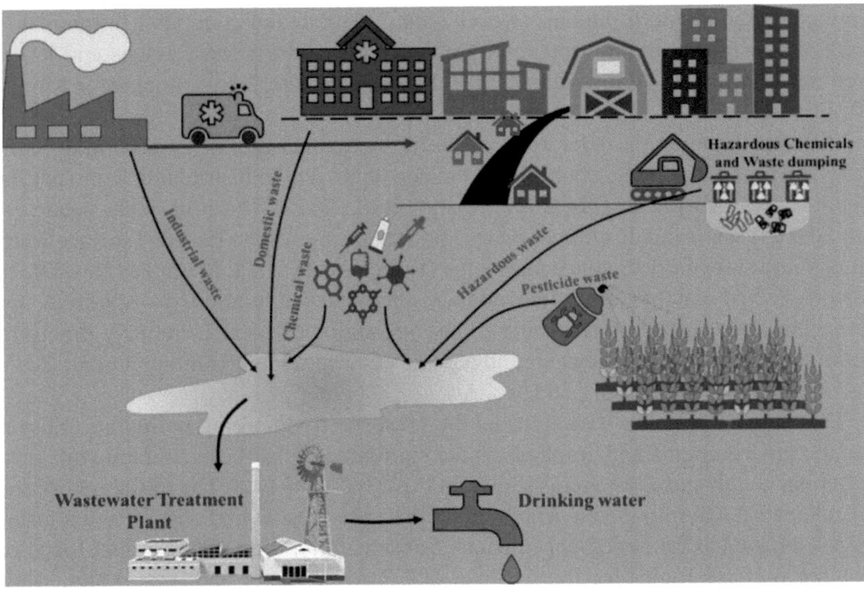

Fig. 14.3 Types and sources of emerging pollutants/micropollutants (MPs). Adopted and reproduced with permission (Shanmuganathan et al., 2023)

Emerging Pollutants (EPs): A Growing Challenge in Environmental Sustainability

Freshwater, food products, municipal sewage, groundwater, and industrial effluents have been identified as reservoirs for a newly recognized category of chemical substances known as emerging pollutants (EPs) or micropollutants (MPs). These environmental contaminants are frequently present in both the lithosphere and hydrosphere (Sackaria & Elango, 2020; Stamm et al., 2016). Examples of EPs commonly encountered in daily life include agrochemicals, personal care products (PCPs), hormonal compounds, fire retardants, industrial additives, endocrine-disrupting chemicals (EDCs), pharmaceuticals, and nanomaterials (Gupta et al., 2024; Lin et al., 2016). According to the US Geological Survey, a micropollutant or EP is described as "any substance, either natural or synthetic, or any microorganism that is not typically found in the environment but has the potential to cause harmful effects on ecosystems and human health." These pollutants are often detected at trace concentrations, ranging from parts per trillion (ppt) to parts per billion (ppb) (Sivaranjanee & Kumar, 2021).

Origin of Emerging Pollutants (EPs) and Its Toxicity Effects

Micropollutants (MPs) tend to accumulate in surface water due to the direct discharge of effluents from wastewater treatment plants (Kumar et al., 2022). Figure 14.3 highlights the types and sources of various secondary MPs. Among these, pharmacological and personal care products (PPCPs) are major contributors, closely linked to human activities. Globally, thousands of PPCPs are manufactured annually, and their release into the environment is an unavoidable aspect of modern life. Sources of PPCPs in aquatic systems include manufacturing facilities, sewage treatment plants (STPs), wastewater treatment plants (WWTPs), and domestic treatment systems. Incomplete or untreated discharges from these facilities contribute significantly to the contamination of surface water bodies with PPCPs (Al-Baldawi et al., 2021). Notably, during the COVID-19 pandemic, there was an increased presence of pharmacological compounds, along with their byproducts and metabolic derivatives, in aquatic environments. During the pandemic, significant quantities of glucocorticoids, antibiotics, antiviral medications, and antiprotozoal agents were extensively administered. Studies have indicated that this heightened usage adversely affected the environment. For instance, surface water concentrations of Lopinavir, an antiviral drug, experienced a sharp increase, rising from 305 ng/L before the pandemic to 880 ng/L during the pandemic (Morales-Paredes et al., 2022).

Freshwater ecosystems worldwide are increasingly contaminated by industrial and agrochemical compounds, which, despite their low concentrations, exhibit significant toxicity and pose hazards, especially in complex mixtures (Rogowska et al., 2020). Hormones such as estrone (E1), 17-oestradiol (E2), 17-ethinylestradiol (EE2), estriol (E3), equilin, and 17-oestradiol are among the critical contaminants in drinking water due to their environmental prevalence and health implications (Atugoda et al., 2021). For example, even low concentrations of 17-ethinylestradiol can induce estrogenic effects in fish, leading to reduced fertility, altered sex ratios, feminization, and impaired egg fertilization (Eustache et al., 2020). Endocrine disruptors, such as phytoestrogens, also pose risks to fish and mammalian embryos. Personal care products (PCPs) containing triclocarban and triclosan further exacerbate endocrine disruption, negatively affecting fish embryos, algae, crustaceans, and fish. Triclosan, in particular, disrupts thyroid hormone homeostasis and reproductive systems and inhibits phytoplankton growth at certain concentrations. Long-term exposure to parabens in fish has been associated with the production of vitellogenin, even at minimal concentrations (Richardson & Kimura, 2020; Yi et al., 2024). Moreover, plasticizers like BPA have been linked to adverse reproductive outcomes, including reduced sperm quality, hormonal imbalances, and impaired sexual function in men. In women, BPA exposure is associated with conditions such as polycystic ovarian syndrome (PCOS), congenital defects, breast cancers, and increased risks of miscarriage (Rashtian et al., 2019). An overview of the above highlights is demonstrated in Fig. 14.4 (Tables 14.1 and 14.2).

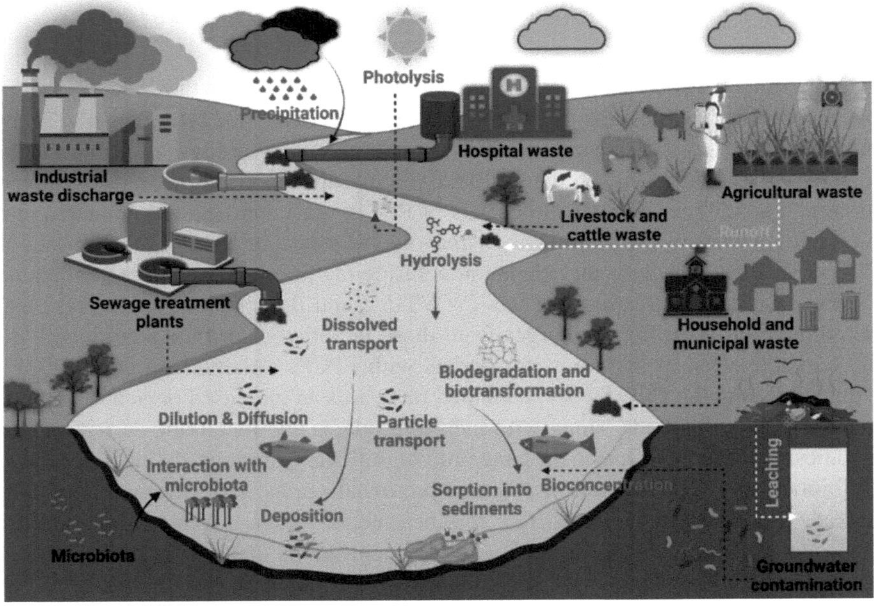

Fig. 14.4 Sources and the methodology of transportation of different emerging pollutants in the aquatic environment. Adopted and reproduced with permission (Narwal et al., 2023)

Table 14.1 Harmful effects of different micropollutants and their sources

Types of pollutants	Source of pollutants	Toxic effects	References
Bisphenol A (BPA) estradiol, estrone, nonylphenol	Hormone active substances	Hormone imbalance, DNA damage, Fertility, neuronal development	Zhou et al. (2019)
Triclosan	Pharmaceuticals and personal care products	Aquatic toxicity	San-Rom´an et al. (2020)
Polychlorinated biphenyls, polybrominated diphenyl ethers, dioxins, bisphenols, phthalates, pesticides, and heavy metal	Shoe manufacture, preserving industry, Building activities, Pulp/papermaker industry, wood processing, other agricultural activities	Thyroid cancer	Marotta et al. (2020)
Macroplastics and microplastics	Southern region of the central Mediterranean Sea	Cancer, fertility problems	Mancia et al. (2020)

Adopted and reproduced with permission (Shanmuganathan et al., 2023)

Table 14.2 The ecotoxicity of micro-plastics (MPs) and nano-plastics (NPs) in waterbodies

Location	Ecosphere	Ecotoxicity	Future perspective	References
India	Wastewater	Inflammations	Exploration of standard sampling purification and characterization methods of MPs/NPs in WWTP are needed for quality research input. Understanding the interaction, transportation, and bioaccumulation mechanism of heavy metals and toxic chemicals with NPs/MPs is required. The reduction of microbeads in personal care products and industrial applications should be restricted	Reddy and Nair (2022)
India	Wastewater	Genotoxicity, neurotoxicity, cytotoxicity, inflammation, oxidative stress	Analyzing, sampling, and separating of NPs from matrices and aggregation should be established	Ravi Kiran et al. (2022)
USA	Water	Alterations of microbiome composition, changes in metabolism	Generation of a comprehensive biological response profile of NPs and additive chemicals is required	Mortensen et al. (2021)
China	Wastewater	Oxidative stress, DNA damage	Implementation of new technologies for purification and detection of MPs with economic values and simple operations	Zhang and Chen (2020)
Canada	Drinking water	Inflammation	The removal of NPs from water using techniques such as dynamic membrane filtration, electrocoagulation, air flotation, and electrooxidation processes	Estahbanati et al. (2021)

Adopted and reproduced with permission (Mustapha et al., 2024)

Conventional Methods for Recovery of Micropollutants

Recovery of micropollutants (MPs) from water bodies is crucial due to their potential toxicity and persistence. Micropollutants (MPs) are contaminants found in water bodies at trace levels, typically in the range of nanograms to micrograms per litre (Stamm et al., 2016). These include pharmaceuticals (Adeleye et al., 2022; Gupta et al., 2024), personal care products (Osuoha et al., 2023; Wydro et al., 2023), pesticides (Munz et al., 2017), hormones (Almazrouei et al., 2023), and industrial chemicals (Afzaal et al., 2022). Despite their low concentrations, MPs pose significant threats to ecosystems and human health due to their bioaccumulation, persistence, and potential toxicity. Conventional recovery methods aim to remove these pollutants

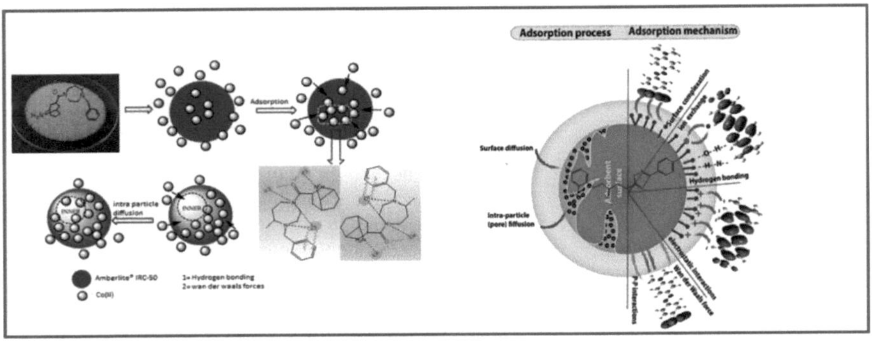

Fig. 14.5 Different types of adsorbent (resin)-adsorbate (metals or dye) interactions at surfaces of the resin adsorbent, **a** the complexation and **b** hydrogen bonding. Adopted and reproduced with permission (Bensalah et al., 2024)

through established technologies like adsorption, membrane filtration, and chemical oxidation. Adsorption, using activated carbon or biochar, effectively traps organic micropollutants. Membrane filtration, such as reverse osmosis and nanofiltration, separates contaminants based on size and charge. Chemical oxidation, employing ozone or advanced oxidation processes, degrades persistent compounds shown in Fig. 14.5.

Adsorption

Principle of Adsorption for Micropollutant Recovery

Adsorption is a surface phenomenon where molecules of micropollutants adhere to the surface of an adsorbent material (Aldana et al., 2024). This process occurs through physical forces (van der Waals interactions) or chemical bonds (covalent, ionic, or hydrogen bonding). Many scientists have reported the adsorption characteristics of various resins cationic and anionic, which can be facilitated as excellent adsorbents to remove different micropollutants from wastewater (Fig. 14.5). The efficiency of adsorption depends on factors such as the surface area, porosity, surface chemistry of the adsorbent, and the physicochemical properties of the pollutants.

Common Adsorbents for Micropollutants

- Activated Carbon: Widely used due to its high surface area and porosity. Granular activated carbon (GAC) and powdered activated carbon (PAC) effectively remove organic micropollutants like pharmaceuticals and pesticides (Benstoem et al., 2017).
- Zeolites: Microporous aluminosilicates with uniform pore structures suitable for capturing small molecules and ions (Jiang et al., 2018).

- Metal–Organic Frameworks (MOFs): Highly tunable, porous materials with exceptional adsorption capacity for specific micropollutants (Liu et al., 2023; Singh et al., 2024).
- Biomass-Derived Adsorbents: Biochar and other materials derived from agricultural waste offer sustainable and low-cost solutions (Anastopoulos et al., 2024; Xu et al., 2022).
- Polymeric Adsorbents: Functionalized polymers with tailored affinity for specific micropollutants provide selectivity in adsorption (Klemes et al., 2020).

Mechanisms of Recovery

The adsorption process involves several mechanisms, including:

- Hydrophobic Interactions: Organic micropollutants with non-polar structures interact with non-polar adsorbent surfaces.
- Electrostatic Interactions: Charged pollutants are attracted to oppositely charged adsorbent surfaces.
- π-π Interactions: Aromatic micropollutants interact with adsorbent surfaces containing π-electron-rich groups.
- Chemical Bonding: Covalent or hydrogen bonding enhances selectivity for specific contaminants.

Advantages of Adsorption

- High Efficiency: Removes micropollutants even at low concentrations.
- Versatility: Can target a wide range of micropollutants by tailoring adsorbent properties.
- Environmentally Friendly: Many adsorbents, such as biochar and natural materials, are sustainable and non-toxic.
- Regenerability: Adsorbents can often be regenerated and reused, reducing costs.

Challenges

Despite its advantages, adsorption faces challenges such as:

- Selective Adsorption: Many micropollutants coexist, making selective removal difficult.
- Fouling: Adsorbents may lose effectiveness due to clogging by other organic or inorganic matter.
- Cost of Advanced Adsorbents: Materials like MOFs can be expensive to synthesize.
- Disposal of Spent Adsorbents: Safe disposal or regeneration of used adsorbents is critical.

Future Directions

Innovations focus on developing advanced adsorbents with higher specificity and capacity. Hybrid materials, such as functionalized biochar or composite adsorbents combining activated carbon and MOFs, are promising. Integration of adsorption with other techniques, like membrane filtration or advanced oxidation, can improve overall performance.

Adsorption is a practical and effective method for recovering micropollutants, contributing significantly to water purification and environmental protection. Ongoing research into novel adsorbents and process optimization will enhance its applicability for sustainable water management.

Advanced Oxidation Processes (AOPs)

AOPs are a group of chemical treatment methods designed to generate highly reactive hydroxyl radicals (·OH·OH) (Peng et al., 2022; Saravanan et al., 2022; Wypart-Pawul et al., 2023). These radicals are powerful oxidizing agents capable of non-selective and rapid oxidation of organic and inorganic pollutants. Common AOPs include photocatalysis, ozonation, Fenton and photo-Fenton processes, and the use of UV/ H_2O_2 systems (Fig. 14.6). The primary advantage of AOPs lies in their ability to degrade pollutants into simpler, often harmless, compounds such as water, carbon dioxide, and inorganic salts, minimizing secondary pollution.

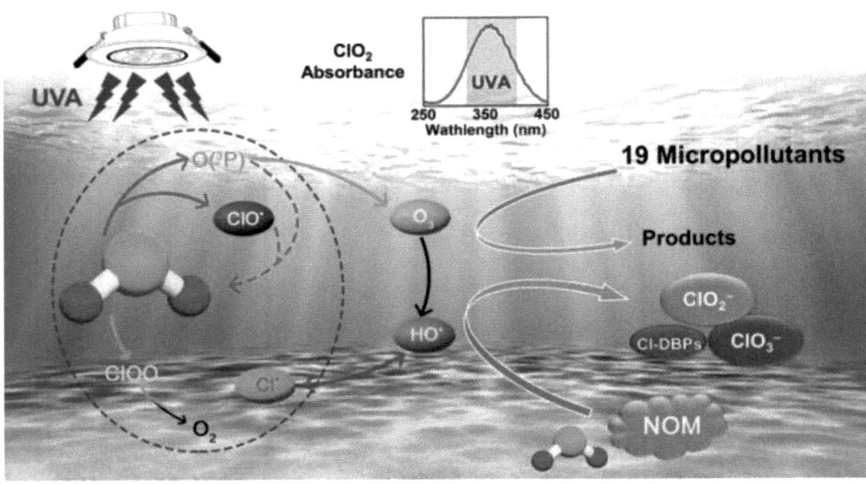

Fig. 14.6 The illustration depicts an advanced UVA/ClO_2-based advanced oxidation process (AOP) characterized by a high generation efficiency of reactive species, minimal formation of chloro-organic byproducts, and high energy efficiency for the degradation of micropollutants in water treatment. Adopted and reproduced with permission (Peng et al., 2022)

Mechanism of AOPs

The effectiveness of AOPs depends on the generation of ·OH·OH radicals. These radicals attack micro pollutants through hydrogen abstraction, addition, or electron transfer mechanisms. For instance, in the UV/H_2O_2 process, ultraviolet radiation dissociates hydrogen peroxide into hydroxyl radicals. Similarly, in the Fenton process, the reaction of ferrous ions with hydrogen peroxide generates radicals under acidic conditions.

Photocatalysis, another widely studied AOP, employs semiconductors like titanium dioxide (TiO_2) activated by UV or visible light. This process produces ·OH·OH radicals on the catalyst surface, enabling efficient degradation of micro pollutants. Ozonation, which uses ozone gas as the oxidant, is effective in breaking down complex molecules and removing color and odour from water.

Recovery of Micro Pollutants (MPs)

While AOPs primarily aim at degrading pollutants, recent advancements focus on integrating these methods with recovery strategies. For example, coupling AOPs with adsorption techniques allows for the simultaneous removal and recovery of valuable substances. Activated carbon, zeolites, and biochar can adsorb degraded by-products, which can then be desorbed and utilized. Similarly, electrochemical AOPs (EAOPs) enable the recovery of metals from industrial wastewater through anodic or cathodic deposition.

Hybrid systems combining AOPs with membrane filtration also facilitate the separation of degraded pollutants. Membranes help in concentrating the contaminants, making them easier to extract post-treatment. Such integrations enhance the sustainability and economic viability of wastewater treatment processes.

Challenges and Future Directions

Despite their efficiency, AOPs face challenges such as high operational costs, energy requirements, and the generation of secondary pollutants in certain cases (e.g., bromate formation during ozonation) (Koo et al., 2023). The scalability of AOPs for large-scale applications is another limitation. Research is focused on developing cost-effective catalysts, utilizing renewable energy sources like solar radiation, and optimizing process parameters to enhance efficiency.

AOPs represent a promising solution for micro pollutant recovery and degradation. Their integration with advanced recovery techniques and renewable energy systems could pave the way for sustainable water treatment practices, addressing both environmental and economic concerns.

Membrane Filtration

Membrane Filtration: An Overview

Membrane filtration uses semi-permeable membranes to separate contaminants based on size, charge, or chemical properties. This technology includes microfiltration (MF), ultrafiltration (UF), nanofiltration (NF), and reverse osmosis (RO), each targeting different pollutant sizes (Fig. 14.7). Among these, NF and RO are particularly effective in addressing micro pollutants due to their ability to remove molecules in the range of nanometre, including dissolved organics and ions.

Mechanism of Micro Pollutant Removal

The removal efficiency in membrane filtration depends on various factors such as size Exclusion; membranes act as physical barriers, blocking pollutants larger than their pore size (Fig. 14.8). Adsorption, which is based on the hydrophobic interactions between pollutants and the membrane material. Eventually this phenomenon enhances recovery. Charge Repulsion involves repel of similarly charged pollutants, aiding separation and molecular sieving; high-pressure membranes like RO remove even the smallest micro pollutants through molecular-level separation (Devaisy et al., 2022).

Advantages of Membrane Filtration

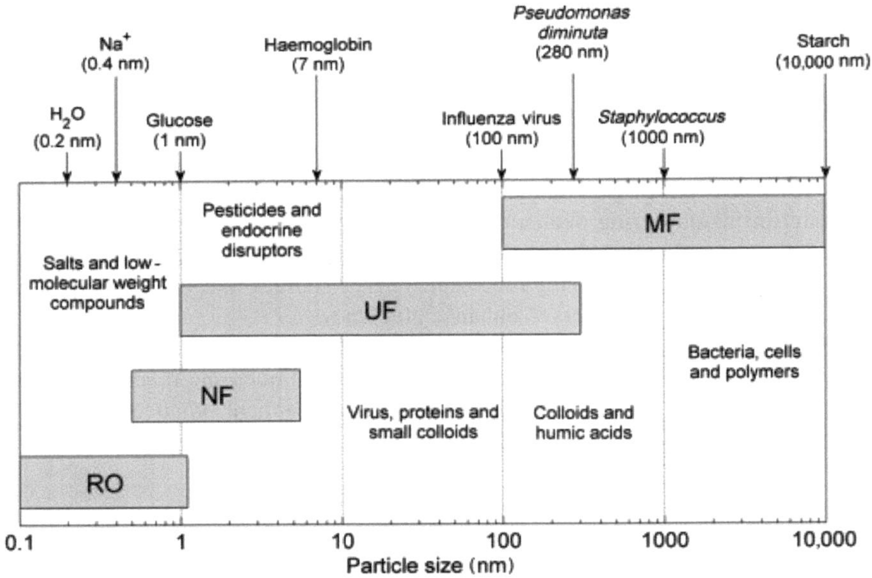

Fig. 14.7 Particle size retention for different types of membranes. Adopted and reproduced with permission (Madsen, 2014)

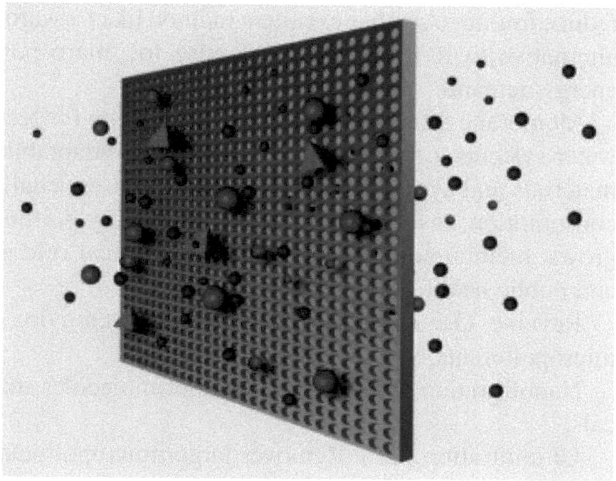

Fig. 14.8 The schematic illustrates a conceptual model of a filtration mechanism using membrane where only particles smaller than the membrane pores penetrates. Adopted and reproduced with permission (Madsen, 2014)

- High Efficiency: Membrane filtration achieves over 90% removal for many micro pollutants.
- Versatility: It can handle diverse contaminants, from heavy metals to complex organic compounds.
- Scalability: Membrane systems can be adapted for small-scale applications or industrial operations.
- Minimal Chemical Use: Unlike conventional methods, membrane filtration avoids excessive chemical additives, reducing secondary pollution.

Challenges and Limitations

Despite its benefits, membrane filtration faces challenges:

- Fouling: Organic matter, biofilms, and scaling can clog membranes, reducing performance and increasing maintenance costs.
- Energy Demand: High-pressure systems like RO require significant energy, impacting operational costs.
- Selective Recovery: Not all micro pollutants are removed equally; hydrophilic and neutral molecules may bypass membranes.
- Disposal of Retentate: Concentrated waste streams containing pollutants need further treatment to prevent secondary environmental impacts.

Enhancements for Micro Pollutant Recovery

Research is advancing membrane materials and processes to enhance efficiency and sustainability. Innovations include incorporating materials like graphene oxide or carbon nanotubes improves selectivity and anti-fouling properties. Additionally, hybrid systems combining membranes with advanced oxidation processes (AOPs) or adsorption enhances recovery. Moreover, membrane surface modification, including coating or functionalizing membranes can improve hydrophilicity and reduce fouling. Another excellent method like forward osmosis (FO), is a low-energy alternative to RO, FO shows promise for micro pollutant recovery with reduced energy demand.

Membrane filtration is a pivotal technology for recovering micro pollutants from water systems, offering high efficiency and adaptability. Ongoing advancements in materials and system integration aim to address challenges like fouling and energy consumption, ensuring its broader applicability. As the global demand for clean water grows, membrane filtration will play a critical role in safeguarding environmental and public health.

Reverse Osmosis (RO): Effective in removing both organic and inorganic micropollutants.

Nanofiltration (NF): Targets smaller molecules like pesticides and pharmaceuticals.

Ultrafiltration (UF): Removes larger micropollutants like microplastics.

Coagulation and Flocculation

Principles of Coagulation and Flocculation

Coagulation involves the destabilization of colloidal particles in water by neutralizing their surface charges (Fig. 14.9) (Dong et al., 2012; Irfan et al., 2017; Katrivesis et al., 2019; Li et al., 2010; Tang et al., 2022). Coagulants, such as aluminium sulphate (alum), ferric chloride, and polymeric materials, are added to the water, where they hydrolyze to form positively charged species. These species neutralize the negatively charged micropollutants, allowing them to aggregate into larger particles.

Flocculation follows coagulation and facilitates the formation of larger, more stable aggregates or flocs. This step involves gentle mixing to encourage collisions between destabilized particles, often aided by flocculating agents like polyacrylamides or natural polymers. The resultant flocs are larger and heavier, making them easier to remove by sedimentation, flotation, or filtration.

Mechanisms of Micropollutant Recovery

Micropollutants are recovered through several mechanisms (Alnahhas et al., 2023; Thébault et al., 1981), including:

- Adsorption onto Coagulants and Flocs: Hydrophobic micropollutants, such as pesticides and pharmaceuticals, can adsorb onto the surfaces of coagulants or flocculants. This mechanism is particularly effective for organic micropollutants.
- Enmeshment in Flocs: As flocs grow, they physically trap micropollutants within their matrix. This "sweep flocculation" mechanism is effective for particulates and certain dissolved micropollutants.
- Charge Neutralization: Coagulants neutralize the charges on micropollutants, reducing their solubility and promoting aggregation. This is particularly useful for ionic micropollutants like heavy metals.

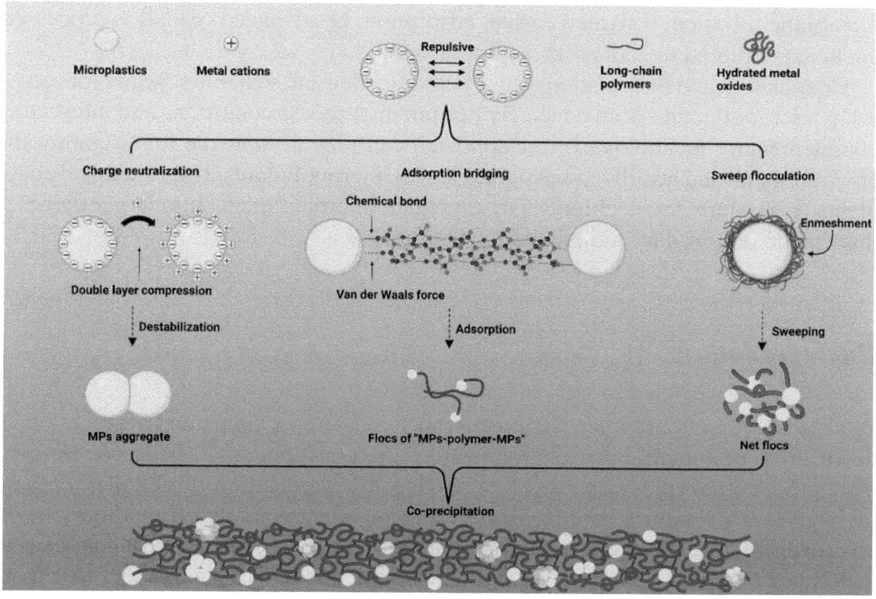

Fig. 14.9 Mechanisms of microplastics removal by coagulation. Adopted and reproduced with permission (Tang et al., 2022)

Factors Influencing Efficiency

The effectiveness of coagulation and flocculation in recovering micropollutants depends on several factors:

- Coagulant Type and Dose: The choice and amount of coagulant significantly affect removal efficiency. For instance, ferric salts are often more effective for micropollutants with high organic content. Typically, coagulants used mostly, are represented in Table 14.1.
- Water Chemistry: Parameters such as pH, ionic strength, and natural organic matter influence the performance of coagulants and flocculants.
- Mixing Intensity: Proper mixing during coagulation and flocculation ensures effective interaction between coagulants, flocculants, and micropollutants.
- Micropollutant Characteristics: The chemical nature, hydrophobicity, and molecular size of micropollutants determine their removal efficiency.

Applications and Challenges

Coagulation and flocculation are widely applied in municipal and industrial water treatment facilities to enhance micropollutant removal. However, the process is not universally effective for all micropollutants, particularly highly soluble or small molecules. Advanced techniques, such as combining coagulation-flocculation with membrane filtration, activated carbon adsorption, or advanced oxidation processes, are being explored to address these limitations.

Coagulation and flocculation offer a cost-effective and scalable solution for recovering micropollutants from water. By optimizing process conditions and integrating complementary technologies, this approach can play a vital role in mitigating the environmental and health risks associated with micropollutants. Uses chemical coagulants (e.g., alum, ferric chloride) to aggregate micropollutants into larger particles that can be removed by sedimentation or filtration.

Biodegradation

Biodegradation utilizes microorganisms, such as bacteria, fungi, and algae, to break down micropollutants into less harmful or inert compounds. These microorganisms produce enzymes capable of catalyzing the degradation of complex organic molecules. This natural process is advantageous due to its eco-friendliness, cost-effectiveness, and adaptability to various environmental conditions. Biodegradation can either completely mineralize pollutants into water, carbon dioxide, and inorganic ions or transform them into intermediate metabolites, which are often less toxic (Mustafa et al., 2021; Takman et al., 2024).

Mechanisms of Biodegradation

The biodegradation of micropollutants generally occurs through enzymatic pathways involving oxidation, reduction, hydrolysis, or a combination of these processes (Garcia-Becerra & Ortiz, 2018). For instance:

- Oxidation and Hydroxylation: Enzymes like cytochrome P450 monooxygenases oxidize hydrophobic micropollutants, increasing their solubility and availability for further degradation.
- Dehalogenation: Halogenated pollutants are degraded by reductive dehalogenases, removing toxic halogen atoms.
- Hydrolysis: Esterases and hydrolases cleave ester or amide bonds in micropollutants, simplifying their structure for further breakdown.

The degradation efficiency depends on factors such as the chemical structure of the micropollutant, the type of microorganism used, environmental conditions (e.g., temperature, pH, oxygen availability), and the presence of co-substrates.

Biodegradation Systems

- Activated Sludge Processes: Common in municipal wastewater treatment plants, these systems use microbial consortia to remove biodegradable organic matter, including certain micropollutants.
- Biofilms and Bioreactors: Biofilm-based systems, such as moving bed biofilm reactors (MBBRs) and membrane bioreactors (MBRs), enhance the retention of microorganisms, increasing micropollutant degradation rates.

- Fungal Bioreactors: White-rot fungi, known for their ligninolytic enzymes, are particularly effective in degrading recalcitrant micropollutants, including pharmaceuticals and dyes.

Challenges and Advances

While promising, biodegradation faces challenges such as the formation of toxic by-products, incomplete degradation, and the adaptability of microorganisms to complex micropollutant mixtures. To address these issues, research focuses on:

- Genetic Engineering: Engineering microbial strains with enhanced enzymatic capabilities to target specific micropollutants.
- Microbial Consortia: Developing synergistic consortia where different microorganisms target complementary degradation pathways.
- Integration with Advanced Technologies: Combining biodegradation with advanced oxidation processes (AOPs) or adsorption techniques to enhance overall pollutant removal.

Applications and Future Directions

Biodegradation has promising applications in wastewater treatment, groundwater remediation, and even resource recovery. Future research should aim at optimizing microbial processes, exploring novel biodegradation pathways, and scaling up for industrial applications.

Biodegradation represents a sustainable and efficient approach to recovering and eliminating micropollutants, safeguarding water quality and environmental health while minimizing ecological footprints.

Electrochemical Processes

Principles of Electrochemical Processes

Electrochemical processes rely on the application of an electric current to drive chemical reactions. These processes include oxidation, reduction, precipitation, and adsorption at the electrode interface (Fig. 14.10) (Ren et al., 2022; Szopi et al., 2024; Włodarczyk-Makuła et al., 2023). The key mechanisms involved in the removal and recovery of micro pollutants are:

- Electrooxidation: Involves the generation of strong oxidizing agents, such as hydroxyl radicals or reactive chlorine species, which degrade micro pollutants into less harmful compounds or mineralize them completely.
- Electroreduction: Targets pollutants that can be reduced into less toxic or reusable forms, such as metal recovery from industrial effluents.
- Electrocoagulation: Uses sacrificial anodes (e.g., aluminium or iron) to release coagulants that bind with micro pollutants and facilitate their separation.

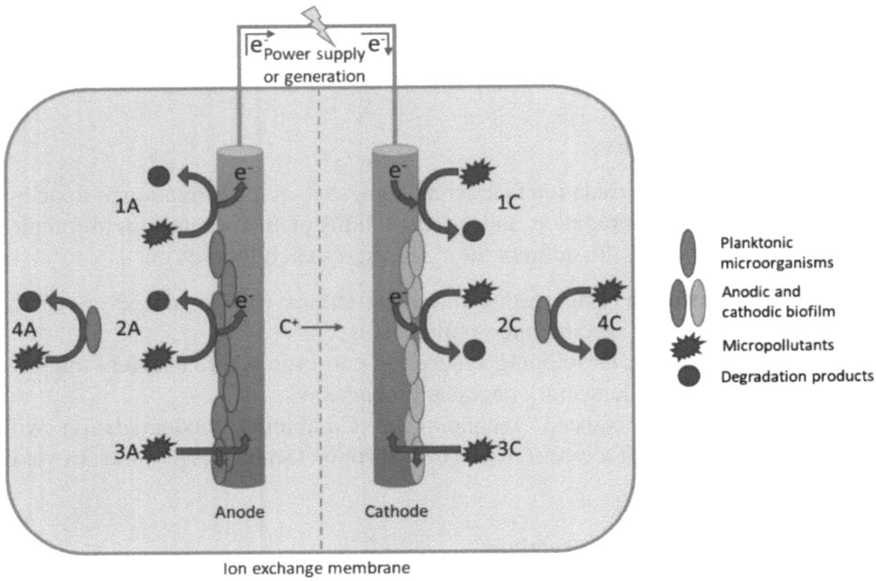

Fig. 14.10 Schematic overview of a BES showing the main organic micropollutant removal mechanisms that can occur in the anodic or cathodic chamber. Adopted and reproduced with permission (Poursat et al., 2024)

- Electro adsorption: Employs electrodes modified with adsorptive materials (e.g., carbon-based materials) to selectively capture micro pollutants.

Advantages of Electrochemical Processes

Electrochemical processes have several advantages over conventional treatment methods:

- High efficiency: They can target a wide range of micro pollutants, including those resistant to biodegradation.
- Selective recovery: Tailored electrode materials can enhance selectivity for specific pollutants.
- On-site treatment: Electrochemical reactors are compact and allow decentralized water treatment.
- Eco-friendly: Minimal use of additional chemicals reduces secondary pollution.
- Resource recovery: Enables the recovery of valuable byproducts like metals or reusable compounds.

Applications and Recent Advances

Electrochemical processes are used in treating industrial effluents, hospital wastewater, and municipal sewage where micro pollutants are prevalent. Recent advancements include:

- Novel electrode materials: Graphene, carbon nanotubes, and doped diamond electrodes enhance efficiency and durability.
- Hybrid systems: Combining electrochemical processes with biological or membrane-based methods improves performance.
- Electrocatalysis: Advances in electrocatalysts improve the generation of reactive species for micro pollutant degradation.
- Electrochemical sensors: Integrated sensing systems allow real-time monitoring and optimization of pollutant removal.

Challenges and Future Directions

Despite their potential, electrochemical processes face challenges such as high energy consumption, electrode fouling, and scalability issues. Future research focuses on:

- Energy efficiency: Developing low-cost renewable energy-powered systems to reduce operational costs.
- Long-term stability: Designing electrodes resistant to fouling and corrosion.
- Economic feasibility: Scaling up systems while maintaining cost-effectiveness and performance.

In summary, electrochemical processes are promising for micro pollutant recovery, offering a sustainable solution to water treatment challenges. As technology advances, these processes will play a crucial role in achieving cleaner water and resource sustainability (Table 14.3).

Technological Advances in Emerging Contaminants (EC's) Mitigation

The identification and quantification of emerging contaminants (ECs) in environmental matrices require advanced analytical techniques capable of detecting these substances, often present at minute concentrations. High sensitivity and selectivity are crucial to accurately distinguish and measure ECs amidst a complex mixture of other compounds.

A range of advanced techniques has been developed to detect contaminants at lower concentrations with greater precision than conventional methods. Mass spectrometry and molecular approaches facilitate multi-residue analysis and the identification of previously unknown contaminants. Additionally, sensor technologies are evolving to enable rapid, on-site detection. These innovative analytical methods have

Table 14.3 An overview of the advantages and disadvantages associated with the micro-pollutant's mitigation techniques

Treatment	Advantage	Disadvantage
Coagulation, sedimentation/ flocculation, and filtration	• Effective in removing suspended particles, including MPs and NPs • It can be combined with other treatment processes for enhanced efficiency • Relatively simple and cost-effective • It helps in the agglomeration of particles, improving their removal efficiency • Filtration provides an additional barrier for the removal of remaining particles • It can be effective in treating a wide range of particle sizes	• May require additional filtration or sedimentation for complete removal • Coagulant residuals may need to be managed to avoid environmental impact • Effectiveness may be influenced by water chemistry • Space and infrastructure requirements for sedimentation tanks and filtration systems • Operational and maintenance costs associated with filtration systems • Energy consumption may be a concern, especially for large-scale applications
Advanced oxidation treatment	• Degrades a variety of pollutants, including MPs and NPs, through oxidation processes • It can be effective in breaking down complex organic compounds • It does not rely on physical removal but breaks down contaminants at a molecular level	• High energy consumption • Potential generation of by-products that may be harmful • It may require careful optimization and monitoring to avoid unintended consequences
Coagulation and advanced oxidation	• It can effectively target a wide range of pollutants, including both particles and dissolved contaminants • It provides a comprehensive approach for water quality improvement	• High operational and capital costs • Potential for the formation of disinfection by-products or other harmful substances

Adopted and reproduced with permission (Mustapha et al., 2024)

significantly improved our capacity to monitor emerging contaminants and deepen our understanding of their environmental presence.

(a) **Nanotechnology based remediation strategies**: Nanotechnology offers innovative and highly efficient solutions for mitigating micropollutants—trace contaminants such as pharmaceuticals, pesticides, and personal care products—that persist in water, soil, and air (Asghar et al., 2024; Yadav et al., 2024). These contaminants often evade conventional treatment systems due to their low concentrations and chemical stability, necessitating advanced techniques. Nanotechnology-based approaches leverage the unique properties of nanomaterials, such as high surface area, tunable surface chemistry, and reactivity, to enhance remediation processes. Adsorption techniques utilizing nanoparticles like titanium dioxide (TiO_2), iron oxide (Fe_3O_4), and graphene oxide (GO) are particularly effective, as these materials provide abundant active sites for pollutant binding. Functionalized nanocomposites further enhance specificity, allowing for selective removal of targeted micropollutants (Rocha et al., 2021). Catalytic degradation, powered by nanotechnology, is another promising approach. Photocatalysts such as Titanium dioxide TiO_2 and Zinc oxide (ZnO) utilize UV or visible light to generate reactive oxygen species, breaking down even highly persistent pollutants into harmless byproducts. Advanced oxidation processes (AOPs) employing nanomaterials like nanoscale zero-valent iron or enzyme-mimicking nanozymes also facilitate the oxidative destruction of contaminants. Magnetic nanoparticles, such as magnetite (Fe_3O_4), offer a dual advantage of effective pollutant adsorption and facile recovery using magnetic fields, enhancing the reusability of these materials (Mathur et al., 2022; Yadav et al., 2024).

Nanotechnology also supports bioremediation through synergistic approaches (Hairom et al., 2021). Nano-biochar and nanoparticle-enhanced microbial systems improve the efficiency of biodegradation, offering eco-friendly solutions. Additionally, molecularly imprinted polymers at the nanoscale provide highly selective adsorption, mimicking the molecular structure of pollutants for targeted capture. Nanosensors (both colorimetric and electrochemical based) enable real-time detection and monitoring of micropollutants, facilitating adaptive remediation strategies (Grozdanov et al., 2023; Rathod et al., 2024). Despite its potential, nanotechnology-based remediation faces challenges, including scalability, cost, and environmental risks associated with nanoparticle release. To ensure sustainable applications, future research must focus on developing eco-friendly, non-toxic nanomaterials and integrating them with conventional treatment systems. As these advancements progress, nanotechnology is poised to revolutionize micropollutant mitigation, contributing significantly to global efforts in environmental protection and water quality management.

(b) **Advanced Photocatalytic processes**: Advanced photocatalytic processes, especially those employing titanium dioxide (TiO_2) photocatalysis, represent a promising method for degrading organic pollutants in the environment (Al-Nuaim et al., 2023; Borges et al., 2023). These processes leverage UV or solar light to produce highly reactive hydroxyl radicals capable of breaking down a broad spectrum of organic contaminants. Efforts are ongoing to enhance the efficiency and solar responsiveness of TiO_2 photocatalysts, with a focus on utilizing a greater fraction of sunlight and improving the overall performance of the system. Modifications to TiO_2 and the incorporation of supplementary components have demonstrated significant potential in optimizing this technology for environmental remediation. One of the key advantages of advanced photocatalytic systems is their ability to completely mineralize organic pollutants into benign end products, such as carbon dioxide and water, without requiring additional chemical inputs. These characteristic underscores their sustainability and eco-friendliness, making them a viable solution for mitigating environmental pollution.

(c) **Biosorption and bioaccumulation**: Biosorption involves utilizing living or non-living biomass to remove environmental contaminants, with advancements in genetic engineering and biotechnology enhancing its efficacy in microorganisms and plants. Research focuses on optimizing pollutant uptake by modifying genetic and metabolic pathways to improve absorption capabilities. Genetically engineered organisms and plants demonstrate increased efficiency and specificity in capturing heavy metals and other pollutants from water and soil, significantly advancing sustainable bioremediation practices. Ongoing studies aim to refine biosorption and bioaccumulation processes by identifying critical genetic factors governing pollutant uptake. By engineering these factors, researchers aim to develop tailored bioremediation systems that efficiently address a broader spectrum of contaminants. Such innovations promise more effective and targeted approaches for environmental cleanup, ensuring a cleaner and healthier ecosystem (Chojnacka, 2010; Filote et al., 2021; Mustafa et al., 2021).

(d) **Electrochemical remediation pathways**: Electrochemical remediation employs electric currents to treat contaminated water and soil, with recent advancements in electrode materials and integrated treatment approaches significantly enhancing its efficacy. Innovations in materials and process optimization have strengthened its role as a sustainable method for pollution management, offering advantages over traditional remediation techniques (Lee et al., 2024; Sanz Lobón et al., 2017). Combining electrochemical remediation with methods such as adsorption, oxidation, and biological processes has created synergistic effects, enabling more comprehensive contaminant removal. This hybrid approach capitalizes on the strengths of various technologies, making it versatile for addressing complex pollution challenges across diverse environmental settings. Key benefits of electrochemical remediation include its minimal reliance on chemical additives, contributing to its sustainability and environmental friendliness. Its flexibility allows applications ranging from in situ soil and groundwater treatment to ex situ wastewater and industrial effluent management. Despite these advantages, challenges such as energy efficiency, scalability, and cost remain barriers to widespread adoption. Continued research and innovation are essential to optimize these systems and realize their full potential in addressing global contamination issues (Maria et al., 2023).

(d) **Membrane bioreactors (MBRs)**: Membrane Bioreactors (MBRs) integrate membrane filtration with biological processes, offering advanced solutions for water treatment by effectively removing pollutants, including organic compounds, nutrients, and certain emerging contaminants. Innovations in membrane design and materials have significantly improved fouling resistance,

Table 14.4 Technological approaches for the removal of Micro-plastics and Nano-plastics in the environment

Material	Plastic size	Plastic shape	Monitoring elimination	Removal method	References
Thermal treated with THF/ AgTFA/DI	122–295 nm	Sphere	MALDI-TOF/ MS	Thermal degradation	Lin et al. (2020)
Activated carbon	~ 40 nm	Sphere	UV–visible, FTIR, XPS	Electro-sorption	Xiong et al. (2020)
Coffee grounds	120 nm	Sphere	Fluorescence microscope	Biosorption	Mulindwa et al. (2024)
Magnetorobot derived Fe_3O_4	1.5 μm	Sphere	Field optical microscope	Adsorption	Li et al. (2022)
HDTMS-FeNPs	10 μm	Sphere	ICP-MS	Magnetic flow cell	Surette et al. (2023)
Coagulant	148 nm	Sphere	Electrode	Electrocoagulation	Pawak et al. (2023)

Adopted and reproduced with permission (Mustapha et al., 2024)

efficiency, and system durability, addressing one of the primary challenges associated with MBRs (Al-Asheh et al., 2021; Goswami et al., 2018). The compact, modular design of MBRs makes them highly adaptable for various applications, ranging from municipal wastewater to industrial effluent treatment. However, their implementation is often limited by high energy demands, operational costs, system complexity, and the need for specialized expertise.

To overcome these challenges, ongoing research focuses on developing advanced membrane materials, optimizing operational parameters, and integrating MBRs with complementary remediation techniques, such as advanced oxidation and adsorption. These efforts aim to enhance the sustainability, cost-effectiveness, and versatility of MBR technology, enabling it to meet the increasing demand for efficient and environmentally responsible water treatment solutions (Table 14.4).

Bottlenecks and Recommendation

The increasing prevalence of emerging micropollutants (EMs) underscores the growing challenge of securing uncontaminated water resources. Comprehensive research on the occurrence, characterization, risk assessment, and removal of these pollutants remains a critical priority. However, the detection and quantification of EMs in water bodies are hindered by their diverse nature and the limitations of current analytical methods (Rasheed et al., 2019). These micropollutants infiltrate aquatic ecosystems, bioaccumulating through food chains and posing significant ecological and health risks. Addressing this issue necessitates the integration of nanotechnology for effective removal, offering a promising avenue for evaluating current contamination levels and predicting future trends. Despite its potential, the application of nanotechnology in EM remediation involves several challenges. Technical constraints include the high energy demand for nanomaterial synthesis, limited recovery rates, and the need for advanced instrumentation. Additionally, environmental and social factors such as life cycle assessments, transportation issues, oxidation processes, and insufficient understanding of biosafety and chemical behaviour further complicates its adoption. A holistic approach that considers these technical, economic, and health-related risks is essential to advancing nanotechnology for the efficient elimination of EMs from aquatic systems.

1. The high production costs of nanomaterials remain a significant barrier to their widespread application in environmental remediation. To address this, research efforts should prioritize the development of cleaner, more sustainable, and cost-efficient synthesis methods for nanoparticles (NPs). Such green approaches could significantly reduce production expenses and enhance accessibility.
2. The potential health impacts of nanoparticles on living organisms continue to raise concerns due to insufficient toxicity data. Comprehensive investigations are necessary to establish permissible nanoparticle dosages and clarify their safety thresholds. Studies focusing on concentration-dependent effects in natural systems can provide valuable insights into identifying active yet non-toxic dosages. Biologically synthesized nanoparticles, which are thought to exhibit minimal or no toxicity, should be a focal point of future research to determine their practicality and safety.
3. The unintended consequences of nanomaterial usage, such as their transformation into toxic forms, bioaccumulation, and biomagnification within food chains, present substantial ecological challenges. A holistic approach that incorporates distribution analysis, risk assessment, monitoring, and predictive modelling is essential. This requires an interdisciplinary effort, including chemical analysis, ecotoxicological evaluations, ecological impact studies, and economic assessments, to comprehensively assess their risks.
4. Much of the research on nanomaterials has been conducted under controlled laboratory conditions. To fully understand their applicability, it is crucial to evaluate their performance in real wastewater systems. Such investigations should consider the type of nanomaterial, removal efficiency, underlying mechanisms, and specific water conditions. Additionally, testing nanomaterials under diverse environmental conditions—varying pH levels, temperatures, humidity, agitation speeds, and exposure durations—is necessary to assess their stability and functionality.
5. The reusability of nanomaterials is a key consideration for their practical implementation, particularly in scenarios demanding stability and cost-effectiveness. The absence of standardized protocols and regulatory frameworks for the safe removal and recycling of nanomaterials further complicates their adoption. Comprehensive studies examining the reuse potential of nanomaterials in real-world water treatment scenarios, alongside stability assessments under varying

conditions, are essential. Optimizing recovery methods to enhance the reusability of specific nanomaterials is recommended. Furthermore, the development of clear regulatory guidelines for the safe commercialization and responsible use of nanoparticles should be prioritized in future research.

Future Perspectives and Research Pathways

The future of water treatment and environmental remediation lies in advancing innovative technologies and sustainable practices to address the challenges posed by emerging pollutants (EPs). As the complexity and prevalence of EPs in various environmental matrices grow, research must focus on developing cost-effective, efficient, and environmentally friendly remediation strategies (Baresel et al., 2024; Krahnstöver et al., 2022).

(a) **Integration of Remediation Technologies**

- Combine multiple remediation methods (e.g., MBRs with advanced oxidation processes) to leverage synergistic effects for enhanced efficiency and effectiveness.
- Couple electrochemical techniques with biological or nanotechnology-based approaches for addressing complex contamination scenarios.

(b) **Advancements in Nanotechnology**

- Develop nanomaterials with high selectivity and efficiency for targeted pollutants.
- Focus on recyclable and fouling-resistant nanomaterials to reduce operational costs and environmental impacts.

(c) **Genetic Engineering and Biosorption**

- Enhance pollutant uptake capacities in microorganisms and plants through genetic modifications.
- Engineer specialized biosorbents tailored for specific contaminants, such as heavy metals and pharmaceutical residues.

(d) **Application of Digital Technologies**

- Use artificial intelligence (AI) and machine learning (ML) for process optimization, predictive modelling, and real-time monitoring of pollutants (Post et al., 2022).
- Enable site-specific solutions through data-driven approaches, reducing inefficiencies and improving resource utilization.

(e) **Material Science Innovations**

- Develop advanced membrane materials with enhanced fouling resistance and durability.

- Innovate on materials that can improve the longevity and cost-effectiveness of water treatment systems.

(f) **Policy and Public Awareness**

- Strengthen regulatory frameworks to limit pollutant discharge and encourage sustainable practices.
- Promote public awareness of emerging pollutants and the importance of advanced remediation technologies.

(g) **Scalability and Global Adoption**

- Ensure innovative solutions are adaptable to diverse regions and resource settings.
- Focus on cost-effective and scalable technologies to make them accessible for widespread application.

(h) **Collaborative Efforts**

- Foster collaboration between academia, industry, and policymakers to bridge research and practical implementation.
- Encourage funding and investment in research and infrastructure to support innovative remediation strategies.

Conclusions

Emerging pollutants (EPs), also referred to as micropollutants, have become a critical concern in achieving sustainable water quality management. These contaminants are increasingly prevalent in the environment, posing challenges for cost-intensive, long-term monitoring, comprehensive risk assessments, and efficient removal strategies. Their persistent presence and long-term ecological and human health impacts underscore the urgency of addressing this issue. Traditional water treatment systems are ineffective in eliminating EPs due to their micro-scale nature, necessitating the development of advanced removal techniques. This chapter examines the sources, toxicological impacts, detection methods, and advanced treatment strategies for EPs. Conventional wastewater treatment methods fail to adequately mitigate these contaminants, even at low concentrations, which can significantly disrupt ecosystems. Advanced approaches, such as adsorption, membrane technologies, advanced oxidation processes, and constructed wetlands, have shown promise. Recent innovations include hybrid treatments integrating photocatalysis with biodegradation, as well as advanced biological methods using algae, algal consortia, and anaerobic membrane bioreactors. The optimization of pretreatment protocols, operational parameters, and membrane materials is essential to enhancing removal efficiency while mitigating membrane fouling. Public awareness and education regarding the proper disposal of pollutants are vital to reducing environmental contamination. Future research should prioritize understanding the most harmful effects of EPs on human health and ecosystems and emphasize integrated strategies to reduce pollutant inputs while conserving existing natural resources.

Acknowledgements Aniruddha Deb would like to gratefully acknowledge the funding support as seed grant award (Ref: ***BIT/DRIE/SMS/2024-25/1913***) received from Birla Institute of Technology Mesra, Ranchi Campus.

Credit Author Statement **Aniruddha Deb**: Writing, editing, Supervision, Funding Acquisition **Kaniska Murmu**: Writing, Editing-Reviewing and Supervision, **Sunil Kumar Singh**—Editing, **Awadh Kishor Kumar**—Supervision.

Declaration of Competing Interests The authors confirm that they have no financial or personal relationships that could have influenced the outcomes or interpretation of the work presented in this chapter.

References

Adeleye, A. S., Xue, J., Zhao, Y., Taylor, A. A., Zenobio, J. E., Sun, Y., Han, Z., Salawu, O. A., & Zhu, Y. (2022). Abundance, fate, and effects of pharmaceuticals and personal care products in aquatic environments. *Journal of Hazardous Materials, 424*, Article 127284. https://doi.org/10.1016/j.jhazmat.2021.127284

Afzaal, M., Mazhar, I., Rasheed, R., Sharif, F., Khan, W. U. D., Bashir, N., Iqbal, S. S., & Khan, A. (2022). Industrial chemicals as micropollutants in the environment. In *Environmental micropollutants* (pp. 13–44). Elsevier. https://doi.org/10.1016/B978-0-323-90555-8.00003-9

Ahmed, S. F., Mofijur, M., Nuzhat, S., Chowdhury, A. T., Rafa, N., Uddin, M. A., Inayat, A., Mahlia, T. M. I., Ong, H. C., Chia, W. Y., & Show, P. L. (2021). Recent developments in physical, biological, chemical, and hybrid treatment techniques for removing emerging contaminants from wastewater. *Journal of Hazardous Materials, 416*(March), Article 125912. https://doi.org/10.1016/j.jhazmat.2021.125912

Al-Asheh, S., Bagheri, M., & Aidan, A. (2021). Membrane bioreactor for wastewater treatment: A review. *Case Studies in Chemical and Environmental Engineering, 4*(May). https://doi.org/10.1016/j.cscee.2021.100109

Al-Baldawi, I. A., Mohammed, A. A., Mutar, Z. H., Abdullah, S. R. S., Jasim, S. S., Almansoory, A. F., & Ismail, N. I. (2021). Application of phytotechnology in alleviating pharmaceuticals and personal care products (PPCPs) in wastewater: Source, impacts, treatment, mechanisms, fate, and SWOT analysis. *Journal of Cleaner Production, 319*(July), 128584. https://doi.org/10.1016/j.jclepro.2021.128584

Aldana, J. C., Agudelo, C., Álvarez, P. M., & Acero, J. L. (2024). Removal of micropollutants in water reclamation by membrane filtration: Impact of pretreatments and adsorption. *Membranes, 14*(7), 146. https://doi.org/10.3390/membranes14070146

Almazrouei, B., Islayem, D., Alskafi, F., Catacutan, M. K., Amna, R., Nasrat, S., Sizirici, B., & Yildiz, I. (2023). Steroid hormones in wastewater: Sources, treatments, environmental risks, and regulations. *Emerging Contaminants, 9*(2), Article 100210. https://doi.org/10.1016/j.emcon.2023.100210

Alnahhas, Mhd. S., Tzoupanos, N., & Barjenbruch, M. (2023). Investigation of coagulation/flocculation process for the removal of specific micro-pollutants. In *Proceedings of the 4th international conference on environmental science and applications (ICESA 2023)* (Vol. 109, pp. 1–9). https://doi.org/10.11159/icesa23.109

Al-Nuaim, M. A., Alwasiti, A. A., & Shnain, Z. Y. (2023). The photocatalytic process in the treatment of polluted water. *Chemical Papers, 77*(2), 677–701. https://doi.org/10.1007/s11696-022-02468-7

Anastopoulos, I., Giannakoudakis, D. A., Frontistis, Z., Zorpas, A. A., Pashalidis, I., & Triantafyllidis, K. (2024). Biomass-derived adsorbents: Universal database development for their synthesis and remediation efficiency as a necessary step to move from laboratory- to pilot-scale applications. *Current Opinion in Green and Sustainable Chemistry, 47*, Article 100902. https://doi.org/10.1016/j.cogsc.2024.100902

Asghar, N., Hussain, A., Nguyen, D. A., Ali, S., Hussain, I., Junejo, A., & Ali, A. (2024). Advancement in nanomaterials for environmental pollutants remediation: a systematic review on bibliometrics analysis, material types, synthesis pathways, and related mechanisms. *Journal of Nanobiotechnology, 22*(1). https://doi.org/10.1186/s12951-023-02151-3

Atugoda, T., Vithanage, M., Wijesekara, H., Bolan, N., Sarmah, A. K., Bank, M. S., You, S., & Ok, Y. S. (2021). Interactions between microplastics, pharmaceuticals and personal care products: Implications for vector transport. *Environment International, 149*, Article 106367. https://doi.org/10.1016/j.envint.2020.106367

Baresel, C., Salem, M., Roberts, R., Malovanyy, A., Lemström, H., & Esfahani, B. (2024). Approaching breakthrough: resource-efficient micropollutant removal with MBR-GAC configuration. *Applied Sciences (Switzerland), 14*(17). https://doi.org/10.3390/app14177759

Bensalah, J., Thakur, A., & Kumar, A. (2024). Investigating the adsorption processes of polymer resins for the removal of micropollutants: A comprehensive review in the field of environmental remediation. *Environmental Research, 254*, Article 119128. https://doi.org/10.1016/j.envres.2024.119128

Benstoem, F., Nahrstedt, A., Boehler, M., Knopp, G., Montag, D., Siegrist, H., & Pinnekamp, J. (2017). Performance of granular activated carbon to remove micropollutants from municipal wastewater—A meta-analysis of pilot- and large-scale studies. *Chemosphere, 185*, 105–118. https://doi.org/10.1016/j.chemosphere.2017.06.118

Bolong, N., Ismail, A. F., Salim, M. R., & Matsuura, T. (2009). A review of the effects of emerging contaminants in wastewater and options for their removal. *Desalination, 239*(1–3), 229–246. https://doi.org/10.1016/j.desal.2008.03.020

Borges, M. E., de Paz Carmona, H., Gutiérrez, M., & Esparza, P. (2023). Photocatalytic removal of water emerging pollutants in an optimized packed bed photoreactor using solar light. *Catalysts, 13*(6). https://doi.org/10.3390/catal13061023

Chavoshani, A., Hashemi, M., Amin, M. M., & Ameta, S. C. (2020). Introduction. In *Micropollutants and challenges: Emerging in the aquatic environments and treatment processes*. INC. https://doi.org/10.1016/B978-0-12-818612-1.00001-5

Chojnacka, K. (2010). Biosorption and bioaccumulation—The prospects for practical applications. *Environment International, 36*(3), 299–307. https://doi.org/10.1016/j.envint.2009.12.001

Devaisy, S., Kandasamy, J., Nguyen, T. V., Johir, M. A. H., Ratnaweera, H., & Vigneswaran, S. (2022). Comparison of membrane-based treatment methods for the removal of micro-pollutants from reclaimed water. *Water, 14*(22), 3708. https://doi.org/10.3390/w14223708

Dong, M., Gao, B., Xu, W., Wang, Y., & Mao, R. (2012). Effect of calcium on floc properties and membrane foulings in coagulation–ultrafiltration process by polyaluminum chloride (PACl) of different OH/Al3+ values. *Desalination, 294*, 30–35. https://doi.org/10.1016/j.desal.2012.03.005

Eustache, F., Bennani Smires, B., Moison, D., Bergès, R., Canivenc-Lavier, M. C., Vaiman, D., & Auger, J. (2020). Different exposure windows to low doses of genistein and/or vinclozolin result in contrasted disorders of testis function and gene expression of exposed rats and their unexposed progeny. *Environmental Research, 190*(March). https://doi.org/10.1016/j.envres.2020.109975

Filote, C., Roşca, M., Hlihor, R. M., Cozma, P., Simion, I. M., Apostol, M., & Gavrilescu, M. (2021). Sustainable application of biosorption and bioaccumulation of persistent pollutants in wastewater treatment: Current practice. *Processes, 9*(10). https://doi.org/10.3390/pr9101696

Garcia-Becerra, F. Y., & Ortiz, I. (2018). Biodegradation of emerging organic micropollutants in nonconventional biological wastewater treatment: A critical review. *Environmental Engineering Science, 35*(10), 1012–1036. https://doi.org/10.1089/ees.2017.0287

Gondi, R., Kavitha, S., Yukesh Kannah, R., Parthiba Karthikeyan, O., Kumar, G., Kumar Tyagi, V., & Rajesh Banu, J. (2022). Algal-based system for removal of emerging pollutants from wastewater: A review. *Bioresource Technology, 344*(PB), 126245. https://doi.org/10.1016/j.biortech.2021.126245

Goswami, L., Vinoth Kumar, R., Borah, S. N., Arul Manikandan, N., Pakshirajan, K., & Pugazhenthi, G. (2018). Membrane bioreactor and integrated membrane bioreactor systems for micropollutant removal from wastewater: A review. *Journal of Water Process Engineering, 26*(October), 314–328. https://doi.org/10.1016/j.jwpe.2018.10.024

Grozdanov, A., Dimitrievska, I., & Paunovic, P. (2023). Recent advancements in nano sensors for air and water pollution control. *Material Science & Engineering International Journal, 7*(2), 113–128. https://doi.org/10.15406/mseij.2023.07.00214

Guerra-Rodríguez, S., Ribeiro, A. R. L., Ribeiro, R. S., Rodríguez, E., Silva, A. M. T., & Rodríguez-Chueca, J. (2021). UV-A activation of peroxymonosulfate for the removal of micropollutants from secondary treated wastewater. *Science of the Total Environment, 770*. https://doi.org/10.1016/j.scitotenv.2021.145299

Gupta, A., Kumar, S., Bajpai, Y., Chaturvedi, K., Johri, P., Tiwari, R. K., Vivekanand, V., & Trivedi, M. (2024). Pharmaceutically active micropollutants: origin, hazards and removal. *Frontiers in Microbiology, 15*. https://doi.org/10.3389/fmicb.2024.1339469

Hairom, N. H. H., Soon, C. F., Mohamed, R. M. S. R., Morsin, M., Zainal, N., Nayan, N., Zulkifli, C. Z., & Harun, N. H. (2021). A review of nanotechnological applications to detect and control surface water pollution. *Environmental Technology and Innovation, 24*, Article 102032. https://doi.org/10.1016/j.eti.2021.102032

Irfan, M., Butt, T., Imtiaz, N., Abbas, N., Khan, R. A., & Shafique, A. (2017). The removal of COD, TSS and colour of black liquor by coagulation–flocculation process at optimized pH, settling and dosing rate. *Arabian Journal of Chemistry, 10*, S2307–S2318. https://doi.org/10.1016/j.arabjc.2013.08.007

Jiang, N., Shang, R., Heijman, S. G. J., & Rietveld, L. C. (2018). High-silica zeolites for adsorption of organic micro-pollutants in water treatment: A review. *Water Research, 144*, 145–161. https://doi.org/10.1016/j.watres.2018.07.017

Kanungo, J., Sahoo, T., Swain, L. P., & Behera, I. D. (2023). *Toxicity of persistent hydrocarbon pollutants, sources and sustainable remediation process* (pp. 39–65). https://doi.org/10.1007/978-3-031-48220-5_2

Karimi Estahbanati, M. R., Kiendrebeogo, M., Khosravanipour Mostafazadeh, A., Drogui, P., & Tyagi, R. D. (2021). Treatment processes for microplastics and nanoplastics in waters: State-of-the-art review. *Marine Pollution Bulletin, 168*(March), Article 112374. https://doi.org/10.1016/j.marpolbul.2021.112374

Katrivesis, F. K., Karela, A. D., Papadakis, V. G., & Paraskeva, C. A. (2019). Revisiting of coagulation-flocculation processes in the production of potable water. *Journal of Water Process Engineering, 27*, 193–204. https://doi.org/10.1016/j.jwpe.2018.12.007

Kim, M. K., & Zoh, K. D. (2016). Occurrence and removals of micropollutants in water environment. *Environmental Engineering Research, 21*(4), 319–332. https://doi.org/10.4491/eer.2016.115

Kiran, B. R., Kopperi, H., & Venkata Mohan, S. (2022). Micro/nano-plastics occurrence, identification, risk analysis and mitigation: challenges and perspectives. *Reviews in Environmental Science and Biotechnology, 21*(1). Springer Netherlands. https://doi.org/10.1007/s11157-021-09609-6

Klemes, M. J., Skala, L. P., Ateia, M., Trang, B., Helbling, D. E., & Dichtel, W. R. (2020). Polymerized molecular receptors as adsorbents to remove micropollutants from water. *Accounts of Chemical Research, 53*(10), 2314–2324. https://doi.org/10.1021/acs.accounts.0c00426

Koo, J. W., Lee, J., Nam, S. H., Kye, H., Kim, E., Kim, H., Lee, Y., & Hwang, T. M. (2023). Evaluation of the prediction of micropollutant elimination during bromide ion-containing industrial wastewater ozonation using the ROH, O3 value. *Chemosphere, 338*(July), Article 139450. https://doi.org/10.1016/j.chemosphere.2023.139450

Kosek, K., Luczkiewicz, A., Fudala-Książek, S., Jankowska, K., Szopińska, M., Svahn, O., Tränckner, J., Kaiser, A., Langas, V., & Björklund, E. (2020). Implementation of advanced micropollutants removal technologies in wastewater treatment plants (WWTPs)—Examples and challenges based on selected EU countries. *Environmental Science and Policy, 112*(April), 213–226. https://doi.org/10.1016/j.envsci.2020.06.011

Krahnstöver, T., Santos, N., Georges, K., Campos, L., & Antizar-Ladislao, B. (2022). Low-carbon technologies to remove organic micropollutants from wastewater: A focus on pharmaceuticals. *Sustainability (Switzerland), 14*(18). https://doi.org/10.3390/su141811686

Kumar, M., Pandit, S., Patel, V., Khanna, N., Nag, M., Lahiri, D., Ray, R. R., Das, A., & Das, D. (2023). Maximization of energy recovery from starch processing wastewater by thermophilic dark fermentation coupled with microbial fuel cell technology. *Geomicrobiology Journal*, 1–12. https://doi.org/10.1080/01490451.2023.2209555

Kumar, R., Qureshi, M., Vishwakarma, D. K., Al-Ansari, N., Kuriqi, A., Elbeltagi, A., & Saraswat, A. (2022). A review on emerging water contaminants and the application of sustainable removal technologies. *Case Studies in Chemical and Environmental Engineering, 6*(March), Article 100219. https://doi.org/10.1016/j.cscee.2022.100219

Kümmerer, K., Dionysiou, D. D., Olsson, O., & Fatta-Kassinos, D. (2019). Reducing aquatic micropollutants—Increasing the focus on input prevention and integrated emission management. *Science of the Total Environment, 652*, 836–850. https://doi.org/10.1016/j.scitotenv.2018.10.219

Lee, J., Antonini, G., Al-Omari, A., Muller, C., Mathew, J., Bell, K., Pearce, J. M., & Santoro, D. (2024). Electrochemical methods for nutrient removal in wastewater: A review of advanced electrode materials, processes, and applications. *Sustainability (Switzerland), 16*(22). https://doi.org/10.3390/su16229764

Li, W., Hua, T., Zhou, Q., Zhang, S., & Li, F. (2010). Treatment of stabilized landfill leachate by the combined process of coagulation/flocculation and powder activated carbon adsorption. *Desalination, 264*(1–2), 56–62. https://doi.org/10.1016/j.desal.2010.07.004

Li, W., Wu, C., Xiong, Z., Liang, C., Li, Z., Liu, B., Cao, Q., Wang, J., Tang, J., & Li, D. (2022). Self-driven magnetorobots for recyclable and scalable micro/nanoplastic removal from nonmarine waters. *Science Advances, 8*(45), 1–11. https://doi.org/10.1126/sciadv.ade1731

Lin, Y., Huang, X., Liu, Q., Lin, Z., & Jiang, G. (2020). Thermal fragmentation enhanced identification and quantification of polystyrene micro/nanoplastics in complex media. *Talanta, 208*(October 2019), 120478. https://doi.org/10.1016/j.talanta.2019.120478

Lin, T., Yu, S., & Chen, W. (2016). Occurrence, removal and risk assessment of pharmaceutical and personal care products (PPCPs) in an advanced drinking water treatment plant (ADWTP) around Taihu Lake in China. *Chemosphere, 152*, 1–9. https://doi.org/10.1016/j.chemosphere.2016.02.109

Liu, X., Shan, Y., Zhang, S., Kong, Q., & Pang, H. (2023). Application of metal organic framework in wastewater treatment. *Green Energy & Environment, 8*(3), 698–721. https://doi.org/10.1016/j.gee.2022.03.005

Luo, L., Dzakpasu, M., Yang, B., Zhang, W., Yang, Y., & Wang, X. C. (2019). A novel index of total oxygen demand for the comprehensive evaluation of energy consumption for urban wastewater treatment. *Applied Energy, 236*(13), 253–261. https://doi.org/10.1016/j.apenergy.2018.11.101

Madsen, H. T. (2014). Membrane filtration in water treatment—Removal of micropollutants. In *Chemistry of advanced environmental purification processes of water* (pp. 199–248). Elsevier. https://doi.org/10.1016/B978-0-444-53178-0.00006-7

Mallick, K., Sahu, A., Dubey, N. K., & Das, A. P. (2023). Harvesting marine plastic pollutants-derived renewable energy: A comprehensive review on applied energy and sustainable approach.

Journal of Environmental Management, 348, Article 119371. https://doi.org/10.1016/J.JENVMAN.2023.119371

Mancia, A., Chenet, T., Bono, G., Geraci, M. L., Vaccaro, C., Munari, C., Mistri, M., Cavazzini, A., & Pasti, L. (2020). Adverse effects of plastic ingestion on the Mediterranean small-spotted catshark (*Scyliorhinus canicula*). *Marine Environmental Research, 155*(January), Article 104876. https://doi.org/10.1016/j.marenvres.2020.104876

Maria, A., Morais, N. D., Araújo, D. M., Dantas, I., & Segundo, B. (2023). A sustainable electrochemical- based solution for removing acetamiprid from water. *Applied Sciences, 13*, 10963.

Marotta, V., Malandrino, P., Russo, M., Panariello, I., Ionna, F., Chiofalo, M. G., & Pezzullo, L. (2020). Fathoming the link between anthropogenic chemical contamination and thyroid cancer. *Critical Reviews in Oncology/hematology, 150*(April), Article 102950. https://doi.org/10.1016/j.critrevonc.2020.102950

Mathur, J., Goswami, P., Gupta, A., Srivastava, S., Minkina, T., Shan, S., & D. Rajput, V. (2022). Nanomaterials for water remediation: An efficient strategy for prevention of metal(loid) Hazard. *Water (Switzerland), 14*(24), 0–14. https://doi.org/10.3390/w14243998

Mishra, S., & Das, A. P. (2021). Current treatment technologies for removal of microplastic and microfiber pollutants from wastewater. In *Wastewater treatment: cutting-edge molecular tools, techniques and applied aspects* (pp. 237–251). https://doi.org/10.1016/B978-0-12-821881-5.00011-8

Mishra, S., Singh, R. P., Rout, P. K., & Das, A. P. (2022). Membrane bioreactor (MBR) as an advanced wastewater treatment technology for removal of synthetic microplastics. In *Development in wastewater treatment research and processes* (pp. 45–60). https://doi.org/10.1016/B978-0-323-85583-9.00022-3

Mishra, S., & Das, A. P. (2022). Treatment of the wastewater polluted with synthetic microfiber released from washing machine. In B. B. Das, H. Hettiarachchi, P. K. Sahu, & S. Nanda (Eds.), *Lecture notes in civil engineering* (Vol. 207, pp. 109–117). Springer Science and Business Media Deutschland GmbH. https://doi.org/10.1007/978-981-16-7509-6_9/COVER

Mishra, S., & Das, A. P. (2023). Microbial remediation of synthetic microfiber contaminated wastewater. In M. P. Shah (Ed.), *Microbial technologies in industrial wastewater treatment* (pp. 337–350). Springer, Singapore. https://doi.org/10.1007/978-981-99-2435-6_17

Mishra, S., Dash, D., & Das, A. P. (2024). Isolation and molecular characterization of polyester-degrading bacterial strains from municipal wastewater and their application in microplastic degradation in contaminated wastewater. *Geomicrobiology Journal*. https://doi.org/10.1080/01490451.2024.2376101

Morales-Paredes, C. A., Rodríguez-Díaz, J. M., & Boluda-Botella, N. (2022). Pharmaceutical compounds used in the COVID-19 pandemic: A review of their presence in water and treatment techniques for their elimination. *Science of the Total Environment, 814*. https://doi.org/10.1016/j.scitotenv.2021.152691

Mortensen, N. P., Fennell, T. R., & Johnson, L. M. (2021). Unintended human ingestion of nanoplastics and small microplastics through drinking water, beverages, and food sources. *NanoImpact, 21*(October 2020), 100302. https://doi.org/10.1016/j.impact.2021.100302

Mulindwa, P., Kasule, J. S., Nantaba, F., Wasswa, J., & Expósito, A. J. (2024). Bioadsorbents for removal of microplastics from water ecosystems: A review. *International Journal of Sustainable Engineering, 17*(1), 13–30. https://doi.org/10.1080/19397038.2024.2374003

Munz, N. A., Burdon, F. J., de Zwart, D., Junghans, M., Melo, L., Reyes, M., Schönenberger, U., Singer, H. P., Spycher, B., Hollender, J., & Stamm, C. (2017). Pesticides drive risk of micropollutants in wastewater-impacted streams during low flow conditions. *Water Research, 110*, 366–377. https://doi.org/10.1016/j.watres.2016.11.001

Mustafa, S., Bhatti, H. N., Maqbool, M., & Iqbal, M. (2021). Microalgae biosorption, bioaccumulation and biodegradation efficiency for the remediation of wastewater and carbon dioxide mitigation: Prospects, challenges and opportunities. *Journal of Water Process Engineering, 41*(March), Article 102009. https://doi.org/10.1016/j.jwpe.2021.102009

Mustapha, S., Tijani, J. O., Elabor, R., Salau, R. B., Egbosiuba, T. C., Amigun, A. T., Shuaib, D. T., Sumaila, A., Fiola, T., Abubakar, Y. K., Abubakar, H. L., Ossamulu, I. F., Abdulkareem, A. S., Ndamitso, M. M., Sagadevan, S., & Mohammed, A. K. (2024). Technological approaches for removal of microplastics and nanoplastics in the environment. *Journal of Environmental Chemical Engineering, 12*(2), Article 112084. https://doi.org/10.1016/j.jece.2024.112084

Narwal, N., Katyal, D., Kataria, N., Rose, P. K., Warkar, S. G., Pugazhendhi, A., Ghotekar, S., & Khoo, K. S. (2023). Emerging micropollutants in aquatic ecosystems and nanotechnology-based removal alternatives: A review. *Chemosphere, 341*(August), Article 139945. https://doi.org/10.1016/j.chemosphere.2023.139945

Osuoha, J. O., Anyanwu, B. O., & Ejileugha, C. (2023). Pharmaceuticals and personal care products as emerging contaminants: Need for combined treatment strategy. *Journal of Hazardous Materials Advances, 9*, Article 100206. https://doi.org/10.1016/j.hazadv.2022.100206

Pawak, V. S., Loganathan, V. A., & Sabapathy, M. (2023). Efficient removal of nanoplastics from synthetic wastewater using electrocoagulation. *ArXiv: 2302.08451 [Physics.Chem-Ph] 1–26*, 1–26.

Peng, J., Yin, R., Yang, X., & Shang, C. (2022). A Novel UVA/ClO_2 advanced oxidation process for the degradation of micropollutants in water. *Environmental Science & Technology, 56*(2), 1257–1266. https://doi.org/10.1021/acs.est.1c05133

Post, C., Heyden, N., Reinartz, A., Foerderer, A., Bruelisauer, S., Linnemann, V., Hug, W., & Amann, F. (2022). Possibilities of real time monitoring of micropollutants in wastewater using laser-induced raman & fluorescence spectroscopy (LIRFS) and artificial intelligence (AI). *Sensors, 22*(13). https://doi.org/10.3390/s22134668

Poursat, B. A. J., Rempe, F., Pereira, J., Sutton, N. B., & ter Heijne, A. (2024). Unravelling the mechanisms of organic micropollutant removal in bio-electrochemical systems: Insights into sorption, electrochemical degradation, and biodegradation processes. *Science of the Total Environment, 945*, Article 173932. https://doi.org/10.1016/j.scitotenv.2024.173932

Prabhakar, A., Mishra, S., & Das, A. P. (2019). Isolation and identification of lead (Pb) solubilizing bacteria from automobile waste and its potential for recovery of lead from end of life waste batteries. *Geomicrobiological Journal, 36*(10), 894–903. https://doi.org/10.1080/01490451.2019.1654044

Priyadarsini, S., & Das, A. P. (2024). Analytical and structural characterization of waste lithium-ion batteries for their effective recycling strategy. *Environmental Chemistry and Ecotoxicology*. https://doi.org/10.1016/j.enceco.2024.12.004

Priyadarsini, S., & Das, A. P. (2025). Lithium bioleaching: A review on microbial-assisted sustainable technology for lithium bio-circularity. *Journal of Water Process Engineering, 69*, Article 106744. https://doi.org/10.1016/j.jwpe.2024.106744

Raj, K., & Das, A. P. (2023). Lead pollution: Impact on environment and human health and approach for a sustainable solution. *Environmental Chemistry and Ecotoxicology, 5*, 79–85. https://doi.org/10.1016/J.ENCECO.2023.02.001

Rashtian, J., Chavkin, D. E., & Merhi, Z. (2019). Water and soil pollution as determinant of water and food quality/contamination and its impact on female fertility. *Reproductive Biology and Endocrinology, 17*(1), 1–13. https://doi.org/10.1186/s12958-018-0448-5

Rathod, P. B., Singh, M. P., Taware, A. S., Deshmukh, S. U., Tagad, C. K., Kulkarni, A., & Kanagare, A. B. (2024). Comprehensive insights into water remediation: chemical, biotechnological, and nanotechnological perspectives. *Environmental Pollutants and Bioavailability, 36*(1). https://doi.org/10.1080/26395940.2024.2329660

Reddy, A. S., & Nair, A. T. (2022). The fate of microplastics in wastewater treatment plants: An overview of source and remediation technologies. *Environmental Technology and Innovation, 28*, Article 102815. https://doi.org/10.1016/j.eti.2022.102815

Ren, L., Ma, J., Chen, M., Qiao, Y., Dai, R., Li, X., & Wang, Z. (2022). Recent advances in electrocatalytic membrane for the removal of micropollutants from water and wastewater. *Iscience, 25*(5), Article 104342. https://doi.org/10.1016/j.isci.2022.104342

Richardson, S. D., & Kimura, S. Y. (2020). Water analysis: Emerging contaminants and current issues [Review-article]. *Analytical Chemistry, 92*(1), 473–505. https://doi.org/10.1021/acs.analchem.9b05269

Rocha, L. S., Sousa, É. M. L., Pereira, D., Gil, M. V., Otero-Irurueta, G., Hortigüela Gallo, M. J., Otero, M., Esteves, V. I., & Calisto, V. (2021). Sustainable and recoverable waste-based magnetic nanocomposites used for the removal of pharmaceuticals from wastewater. *Chemical Engineering Journal, 426*(February). https://doi.org/10.1016/j.cej.2021.129974

Rodríguez-Chueca, J., Laski, E., García-Cañibano, C., Martín de Vidales, M. J., Encinas, Kuch, B., & Marugán, J. (2018). Micropollutants removal by full-scale UV-C/sulfate radical based advanced oxidation processes. *Science of the Total Environment, 630*, 1216–1225. https://doi.org/10.1016/j.scitotenv.2018.02.279

Rogowska, J., Cieszynska-Semenowicz, M., Ratajczyk, W., & Wolska, L. (2020). Micropollutants in treated wastewater. *Ambio, 49*(2), 487–503. https://doi.org/10.1007/s13280-019-01219-5

Rout, P. R., Zhang, T. C., Bhunia, P., & Surampalli, R. Y. (2021). Treatment technologies for emerging contaminants in wastewater treatment plants: A review. *Science of the Total Environment, 753*, Article 141990. https://doi.org/10.1016/j.scitotenv.2020.141990

Sackaria, M., & Elango, L. (2020). Organic micropollutants in groundwater of India—A review. *Water Environment Research, 92*(4), 504–523. https://doi.org/10.1002/wer.1243

San-Román, M. F., Solá-Gutiérrez, C., Schröder, S., Laso, J., Margallo, M., Vázquez-Rowe, I., Ortiz, I., Irabien, A., & Aldaco, R. (2020). Potential formation of PCDD/Fs in triclosan wastewater treatment: An overall toxicity assessment under a life cycle approach. *Science of the Total Environment, 707*, Article 135981. https://doi.org/10.1016/j.scitotenv.2019.135981

Sanz Lobón, G., Yepez, A., Garcia, L. F., Morais, R. L., Vaz, B. G., Carvalho, V. V., De Oliveira, G. A. R., Luque, R., & Gil, E. D. S. (2017). Efficient electrochemical remediation of microcystin-LR in tap water using designer TiO 2 @carbon electrodes. *Scientific Reports, 7*(December 2016), 1–8. https://doi.org/10.1038/srep41326

Saravanan, A., Deivayanai, V. C., Kumar, P. S., Rangasamy, G., Hemavathy, R. V., Harshana, T., Gayathri, N., & Alagumalai, K. (2022). A detailed review on advanced oxidation process in treatment of wastewater: Mechanism, challenges and future outlook. *Chemosphere, 308*, Article 136524. https://doi.org/10.1016/j.chemosphere.2022.136524

Schwarzenbach, R. P., Escher, B. I., Fenner, K., Hofstetter, T. B., Johnson, C. A., Von Gunten, U., & Wehrli, B. (2006). The challenge of micropollutants in aquatic systems. *Science, 313*(5790), 1072–1077. https://doi.org/10.1126/science.1127291

Shanmuganathan, R., Sibtain Kadri, M., Mathimani, T., Hoang Le, Q., & Pugazhendhi, A. (2023). Recent innovations and challenges in the eradication of emerging contaminants from aquatic systems. *Chemosphere, 332*(November 2022), 138812. https://doi.org/10.1016/j.chemosphere.2023.138812

Singh, S., Sivaram, N., Nath, B., Khan, N. A., Singh, J., & Ramamurthy, P. C. (2024). Metal organic frameworks for wastewater treatment, renewable energy and circular economy contributions. *Npj Clean Water, 7*(1), 124. https://doi.org/10.1038/s41545-024-00408-4

Sivaranjanee, R., & Kumar, P. S. (2021). A review on remedial measures for effective separation of emerging contaminants from wastewater. *Environmental Technology and Innovation, 23*, Article 101741. https://doi.org/10.1016/j.eti.2021.101741

Stamm, C., Räsänen, K., Burdon, F. J., Altermatt, F., Jokela, J., Joss, A., Ackermann, M., & Eggen, R. I. L. (2016). Unravelling the impacts of micropollutants in aquatic ecosystems: interdisciplinary studies at the interface of large-scale ecology. *Advances in Ecological Research, 55*, 183–223. https://doi.org/10.1016/bs.aecr.2016.07.002

Su, C., Cui, Y., Liu, D., Zhang, H., & Baninla, Y. (2020). Endocrine disrupting compounds, pharmaceuticals and personal care products in the aquatic environment of China: Which chemicals are the prioritized ones? *Science of the Total Environment, 720*, Article 137652. https://doi.org/10.1016/j.scitotenv.2020.137652

Surette, M. C., Mitrano, D. M., & Rogers, K. R. (2023). Extraction and concentration of nanoplastic particles from aqueous suspensions using functionalized magnetic nanoparticles and a magnetic flow cell. *Microplastics and Nanoplastics, 3*(1), 1–12. https://doi.org/10.1186/s43591-022-00051-1

Szopi, M., Pierpaoli, M., Zych, D., & Bogdanowicz, R. (2024). Electrochemical oxidation of selected micropollutants from environment matrices using boron-doped diamond electrodes: Process efficiency and transformation product detection. *Water, 16*, 1–15.

Takman, M., Betsholtz, A., Davidsson, Å., Cimbritz, M., Svahn, O., Karlsson, S., Karstenskov Østergaard, S., Lund Nielsen, J., & Falås, P. (2024). Biological degradation of organic micropollutants in GAC filters–temporal development and spatial variations. *Journal of Hazardous Materials, 472*, Article 134449. https://doi.org/10.1016/j.jhazmat.2024.134449

Tang, W., Li, H., Fei, L., Wei, B., Zhou, T., & Zhang, H. (2022). The removal of microplastics from water by coagulation: A comprehensive review. *Science of the Total Environment, 851*, Article 158224. https://doi.org/10.1016/j.scitotenv.2022.158224

Thébault, P., Cases, J. M., & Fiessinger, F. (1981). Mechanism underlying the removal of organic micropollutants during flocculation by an aluminum or iron salt. *Water Research, 15*(2), 183–189. https://doi.org/10.1016/0043-1354(81)90110-X

Tosun, J., Schaub, S., & Fleig, A. (2020). What determines regulatory preferences? Insights from micropollutants in surface waters. *Environmental Science and Policy, 106*(November 2018), 136–144. https://doi.org/10.1016/j.envsci.2020.02.001

Włodarczyk-Makuła, M., Myszograj, S., & Włodarczyk, M. (2023). Removal of organic micropollutants from wastewater in electrochemical processes—Review. *Energies, 16*(15). https://doi.org/10.3390/en16155591

Wydro, U., Wołejko, E., Luarasi, L., Puto, K., Tarasevičienė, Ž, & Jabłońska-Trypuć, A. (2023). A review on pharmaceuticals and personal care products residues in the aquatic environment and possibilities for their remediation. *Sustainability, 16*(1), 169. https://doi.org/10.3390/su16010169

Wypart-Pawul, A., Neczaj, E., & Grobelak, A. (2023). Advanced oxidation processes for removal of organic micropollutants from wastewater. *Desalination and Water Treatment, 305*, 114–128. https://doi.org/10.5004/dwt.2023.29665

Xiong, Y., Zhao, J., Li, L., Wang, Y., Dai, X., Yu, F., & Ma, J. (2020). Interfacial interaction between micro/nanoplastics and typical PPCPs and nanoplastics removal via electrosorption from an aqueous solution. *Water Research, 184*, Article 116100. https://doi.org/10.1016/j.watres.2020.116100

Xu, L., Wu, C., Chai, C., Cao, S., Bai, X., Ma, K., Jin, X., Shi, X., & Jin, P. (2022). Adsorption of micropollutants from wastewater using iron and nitrogen co-doped biochar: Performance, kinetics and mechanism studies. *Journal of Hazardous Materials, 424*, Article 127606. https://doi.org/10.1016/j.jhazmat.2021.127606

Yadav, K. K., Cabral-Pinto, M. M. S., Gacem, A., Fallatah, A. M., Ravindran, B., Rezania, S., Algethami, J. S., Bashier Eltayeb, L., Abbas, M., Hassan Al-shareef, T., Vinayak, V., Truong Son, C., Awjan Alreshidi, M., Manuel Rodríguez-Díaz, J., & Homod, R. Z. (2024). Recent advances in the application of nanoparticle-based strategies for water remediation as a novel clean technology–A comprehensive review. *Materials Today Chemistry, 40*(April), Article 102226. https://doi.org/10.1016/j.mtchem.2024.102226

Yi, J., Ma, Y., Ruan, J., You, S., Ma, J., Yu, H., Zhao, J., Zhang, K., Yang, Q., Jin, L., Zeng, G., & Sun, D. (2024). The invisible threat: Assessing the reproductive and transgenerational impacts of micro- and nanoplastics on fish. *Environment International, 183*(December 2023), 108432. https://doi.org/10.1016/j.envint.2024.108432

Zhang, Z., & Chen, Y. (2020). Effects of microplastics on wastewater and sewage sludge treatment and their removal: A review. *Chemical Engineering Journal, 382*(July 2019), 122955. https://doi.org/10.1016/j.cej.2019.122955

Zhou, Y., Cheng, G., Chen, K., Lu, J., Lei, J., & Pu, S. (2019). Adsorptive removal of bisphenol A, chloroxylenol, and carbamazepine from water using a novel β-cyclodextrin polymer. *Ecotoxicology and Environmental Safety, 170*(130), 278–285. https://doi.org/10.1016/j.ecoenv.2018.11.117

Chapter 15
Novel Approaches to Monitor and Analyse Micropollutants in Environmental Systems

Saikat Saha

Introduction

Global trends in urban growth, industrialization, and population increase have facilitated the emission and dissemination of noxious substances, which ultimately impact environmental components like soil, water bodies, and sediments (Chau et al., 2018; Núñez et al., 2017). As a result, the occurrence of diverse inorganic and organic contaminants in water bodies, soils, and sediments is widespread. Micropollutants (MPs), a subset of xenobiotic compounds found in aquatic ecosystems, significantly endanger ecological integrity by interfering with natural biological and chemical processes. MPs are particularly troubling on a global scale due to their bioaccumulation in non-target species, which not only causes ecological harm but also strains limited water resources vital for drinking and other purposes (Kolpin et al., 2002; Zhang et al., 2007). Depending on the specific aquatic environment, whether inland or coastal, the presence of MPs can range from trace levels in nanograms/litre to higher levels in milligrams/litre (Ahmed et al., 2017b; Kong et al., 2015). The presence of multiple MPs in water reservoirs like rivers and lakes can create severe toxicological challenges, with long-term effects that threaten aquatic life and human health, including substances that interfere with fish reproductive systems. Due to resource constraints, screening focuses on priority MPs, yet a complete screening of all pollutant classes is vital for assessing water quality and its ecological impact (Díaz-Cruz et al., 2009; Thomaidi et al., 2015).

MPs in the environment include persistent organic pollutants, such as polycyclic aromatic hydrocarbons, MPs, and endocrine-disrupting chemicals (EDCs). Other categories involve pharmaceuticals like antibiotics, analgesics, and antipyrene, as well as pharmaceutical care products, pesticides, and herbicides (Das et al., 2024a,

S. Saha (✉)
Department of Chemical Engineering, Indian Institute of Technology Bombay, Mumbai, Maharashtra, India
e-mail: saikat.saha@iitb.ac.in

I. D. Behera et al. (eds.), *Emerging Micropollutants*, Sustainable Environmental Waste Management Strategies, https://doi.org/10.1007/978-3-032-03530-1_15

2024b; Kanungo et al., 2024; Mohanta et al., 2024). Additional contaminants include ultraviolet (UV) filters, volatile organic compounds, polar organic contaminants, and semivolatile organic compounds, with polychlorinated biphenyls also being significant pollutants (Ahmed et al., 2020). The high levels of controlled and unregulated contaminants in environmental waters, as well as the ineffectiveness of wastewater (WW) treatment facilities, have drawn attention to the study of MPs. Traditional WW treatment technologies often fail to fully remove many MPs, thus requiring advanced analytical techniques for their detection at ng/L levels.

The long-term physiological impacts of consumption of MPs in drinking water are uncertain. However, a proposed strategy underlines major problems about water resource protection and the safe reuse of treated wastewater, covering aspects such as:

(i) Understanding the fate of MPs in urban WW plants (Valhondo et al., 2020)
(ii) Assessing the abundance of MPs in the receiving environment (Papaioannou et al., 2020)
(iii) Exploring emerging technologies for MP removal (Riva et al., 2020)
(iv) Evaluating the ecotoxicity and human health risks posed by MPs (Delli Compagni et al., 2020)
(v) Developing methods for risk assessment and prioritization of MPs (Deviller et al., 2020)

To address the challenges posed by MPs, efficient analytical methods must be developed (Núñez et al., 2017). This is particularly critical due to several factors, with one of the key challenges being the diverse nature of MPs. As discussed earlier, each type of MPs possesses distinct physical and chemical properties, which significantly influence the analytical approaches required for their detection and analysis. Chemical degradation adds another layer of complexity. MPs can degrade through processes such as microbial activity, hydrolysis, and photolysis, resulting the formation of other products or chemicals. These generated by-products, often with unknown chemical properties, could be more detrimental compared to their original parent compounds, posing additional risks. Finally, the rise of emerging contaminants represents a critical challenge. Newly developed pharmaceuticals and industrial chemicals are continually entering the market, often without adequate regulation or study. These compounds require standardized monitoring methods, and their detection and analysis remain significant obstacles in contemporary environmental research (Ahmed et al., 2020).

Building on these hurdles, and recognizing, and analysing MPs requires sophisticated analytical tools. These techniques must be able to overcome matrix interferences (Salas et al., 2017), detect trace-level contaminants, and characterize both known and unknown transformation products. A wide range of cutting-edge technologies, including spectroscopy and chromatography-based approaches, have emerged as critical instruments for meeting these analytical demands. To analyse MPs effectively, diverse instruments are employed, leveraging their strengths in sensitivity, specificity, selectivity, detection performance, and separation capabilities. A variety of instruments are used, including Fourier transform infrared spectroscopy (FTIR);

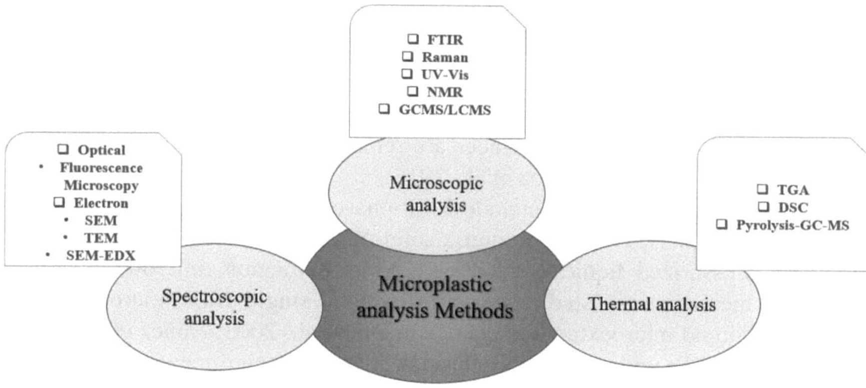

Fig. 15.1 Various types of methods of microplastic analysis

Raman spectroscopy; nuclear magnetic resonance (NMR) spectroscopy; UV–Vis spectroscopy; gas chromatography with mass spectroscopy (GC–MS/GC MS–MS); high-performance liquid chromatography (HPLC); liquid chromatography-mass spectroscopy (LC–MS); column chromatography; Thermal methods like DSC, TGA etc. Figure 15.1 illustrates the different techniques available for MPs analysis.

This chapter discusses the difficulties of evaluating complex combinations of MPs that vary in polarity, size, and morphology and are often found at low concentrations in a variety of environmental matrices. It explores the limitations and promise of several analytical approaches for detecting these contaminants, especially in dynamic combinations.

Methods of MPs Analysis

It is vital to have a standardized method for MPs detection and analysis to appropriately determine the risks of MPs in aquatic ecosystems, The process for the analysis of MPs in natural environments comprises several stages such as sample collection, pretreatment, extraction, Elution or separation, analysis and quantification (Ahmed et al., 2021).

The process initiates with sample collection from the field, followed by filtration or sieving through density separation techniques. Sometimes visual inspection is performed to sort larger size MPs but for smaller size dissection microscope is necessary (Doyle et al., 2011). A digestion step is then performed to reduce organic content, maintaining the integrity of the MPs' structure and (Felsing et al., 2018). The digestion steps include enzymatic digestion, acid digestion, alkali digestion and digestion through oxidizing agents etc. Enzymatic digestion is advantageous for organic materials as it is less hazardous (Maes et al., 2017). Acid digestion is typically performed using HCl, while alkali digestion utilizes NaOH or KOH solutions

(Dehaut et al., 2016). Additionally, hydrogen peroxide (30–35% H_2O_2) serves as the oxidizing agent. After the pretreatment, the extraction process is done. The primary extraction method is introducing a solution containing the target analyte to a solid phase that may selectively trap the analytes. After the analytes are adsorbed onto the solid phase, any undesirable substances are removed by washing, leaving just the desired molecules (Andrade-Eiroa et al., 2016).

Typical Extraction processes include Solid-phase extraction, liquid-phase extraction, solid/liquid-phase extraction, matrix solid-phase dispersion, supercritical fluid extraction, pressurized liquid extraction, Soxhlet extraction, microwave-assisted extraction, membrane-assisted solvent extraction, single-drop microextraction, liquid/liquid/liquid microextraction etc. (Armenta et al., 2008; Núñez et al., 2017). Post extraction, MPs undergo analysis through assessing their dimensions, shape and physical appearance along with other chemical and physical features.

Sometimes, a combination of analytical techniques is necessary for a comprehensive and reliable analysis of environmental samples. FTIR, XPS, and Raman spectroscopy stand out among the optical techniques for analysing MPs (Hernandez et al., 2019). When combined with SEM, XPS works well for identifying polymers but is not able to analyse elemental composition (Hernandez et al., 2017), which is accomplished via Raman spectroscopy using fingerprint spectra. One important method for characterizing MPs is thermal analysis (Schwaferts et al., 2020).

Microscopic Techniques

Microscopes are important tools for detecting MPs by examining their physical features. Stereo microscopy is commonly employed for particles larger than 100 μm (Mathalon & Hill, 2014). This method creates accurate images that show the surface roughness and structure of plastic-like particles, helping to identify them (Das & Bal, 2025). However, the presence of biological elements in sediment samples can make it difficult to accurately identify MPs since these materials are not completely eliminated after chemical digestion.

Microscopic techniques used to study MPs can be categorized into three main types: **Optical microscopy**, which includes conventional light, fluorescence, polarized light, and confocal microscopy; **scanning microscopy**, such as Atomic Force Microscopy (AFM) and Laser Scanning Confocal Microscopy; and **electron microscopy**, which encompasses Scanning Electron Microscopy (SEM), Transmission Electron Microscopy (TEM), and Focused Ion Beam (FIB). Each type offers different capabilities for analysing the size, shape, and structure of MPs particles (Kalaronis et al., 2022).

Optical Microscopy

Optical detection is a commonly used and straightforward method for characterizing MPs based on their physical properties such as colour, shape, and size (Ríos et al., 2020). This can be done visually or using optical microscopy, often with tweezers to pre-screen and minimize the number of particles for analysis, reducing potential errors. Table 15.1 summarizes the basic feature difference between two important types of Optical microscopes, i.e. Stereoscopic (dissecting) microscopy and Fluorescence microscopy.

Table 15.1 Comparison of stereoscopic and fluorescence microscopy features

Aspects	Stereoscopic (dissecting) microscopy	Fluorescence microscopy
Principles and range	– 3D view with illumination from above (Labbe et al., 2020) – Magnification: × 6–100 (Sekudewicz et al., 2021) – Detects particles ≥ 50 μm (De Frond et al., 2022) – Transparent/small particles are harder to observe	– Detects fluorescence emitted by fluorophores – Effective for detecting sub-micron particles – Can analyze small, transparent particles (Kankanige & Babel, 2021)
Advantages	– Simple, fast, and cost-effective (Han et al., 2020) – Minimal pre-treatment required – Provides shape and size details of particles	Highly sensitive to small/transparent plastics – Effective when combined with spectroscopy for polymer identification – Allows detailed analysis of fluorescent materials (Scircle et al., 2020)
Limitations	– Cannot identify polymers without additional techniques – Limited for particles < 50 μm	– Cannot detect non-fluorescent polymers like cellulose – Requires pre-treatment (digestion, density separation, staining) (Scircle et al., 2020)
Pretreatment	Minimal pre-treatment (e.g., filtering) – Uses software like ImageJ for size/shape analysis (Pashaei et al., 2021)	– Requires extensive pre-treatment (e.g., digestion with H_2O_2, staining with Nile Red) – Often paired with spectroscopic methods like μ-Raman or μ-FTIR for polymer identification. (Scircle et al., 2020)
Applications	– Morphological analysis and size measurement of particles	– Detects fluorescent plastics in complex samples – Analyzes small, transparent particles, especially in biological/aquatic environments. (Kankanige & Babel, 2021)

Electron Microscopy

It has substantially higher resolution than standard light microscopy because it uses electrons rather than light. This category includes two main techniques: scanning electron microscopy (SEM), which excels at surface imaging, and transmission electron microscopy (TEM), which is known for high-resolution internal structural investigation. These methods are crucial for improved material and structural characterization (Kalaronis et al., 2022).

SEM is extensively used electron microscopic for the detection of MPs, which provides surface characteristics and morphological structure (Çomaklı et al., 2020). Morphological information is acquired in electron microscopy via primary and secondary electrons following scattering phenomena. As primary electrons move across the sample's surface, they produce signals that are gathered and utilized to create an image. Secondary electrons, accelerated by applied voltages (10–20 kV), penetrate the sample surface and provide precise morphological information (Minor et al., 2020). This approach allows for detailed imaging of the sample's structure. SEM uses high magnifications ($\times$ 200 to $\times$10,000) and modern detectors to provide high-resolution topographical images (> 0.5 nm) of surfaces (Minor et al., 2020). These images can be used to detect MPS particles by comparing surface features. The approach can detect both inorganic and organic components on the surface of MPs (Sekudewicz et al., 2021). Additionally, SEM has been used for observing how weathering impacts on plastic particles in simulated or environmental conditions (Ainali et al., 2021). SEM is a flexible instrument that produces high-resolution topographic images with low detection limits, providing vital information on the morphological properties and size determination of microscopic MPS particles (Nguyen et al., 2021). SEM has been extensively employed to examine MPSs obtained from various aquatic environments, such as water samples from lakes, rivers, seawater, and wastewater (Nguyen et al., 2021). Still, SEM techniques cannot determine the plastic nature of particles or provide polymer composition information (Müller et al., 2020). Furthermore, its ability to examine translucent and very minute particles is limited, making it less useful in real-world environmental samples. This is why SEM is combined with an EDS (Energy Dispersive X-ray-Spectroscopy) detector; it offers critical information on how MPS particles interact with inorganic components or heavy metals (Patterson et al., 2020). Because EDS provides the chemical composition information of MPs with size < 1 μm (Reineccius et al., 2020).

Implementing SEM and EDX together in low vacuum condition is helpful to investigate both the chemical and morphological properties of MPs (Fries et al., 2013). Furthermore, polarized optical microscopy is a sophisticated technique for identifying certain types of MPS particles, such as polyethylene (Von Moos et al., 2012). The most frequently detected inorganic elements on MPSs include Na, Cl, S, Mg, and K, while the heavy metals most commonly found are Pb, Cd, and Cu (Das & Ghosh, 2023; Das et al., 2011, 2015; Mohanty et al., 2016; Patterson et al., 2020; Prabhakar et al., 2019; Raj & Das, 2023).

TEM is another sophisticated electron microscopy method in addition to SEM, which is particularly helpful for examining tiny MPs and nanoparticles. By employing an electron beam that interacts with the sample to produce scattered electrons, TEM produces incredibly high-resolution pictures and chemical information at the atomic scale (Li et al., 2020; Priyadarsini & Das, 2024).In comparison to SEM, TEM operates at greater voltages (usually 80 kV) (Zhao et al., 2021). Energy-dispersive X-ray spectroscopy (EDS) and image processing software can be used in combination with TEM for determining the elemental makeup of nanoparticles (Bonfanti et al., 2021).

MPs and nanoparticles are difficult to identify using TEM because organic polymers have a weak contrast and an amorphous structure. Heavy element staining is required to enhance detection because the elastic interaction of the electron beam with these components is inadequate. Nevertheless, this staining may change the polymers' structure (Mariano et al., 2021). Furthermore, the usefulness of TEM is limited since it requires the particles to be thin enough for efficient analysis. TEM is frequently used in laboratory-scale research, but its application in MPs characterization is still limited (Bonfanti et al., 2021).

Along with to classical microscopic methodologies for detecting MPs, additional methods with different operating principles can be utilized to supplement characterization. These techniques include hyperspectral imaging microscopy (HSI), atomic force microscopy (AFM), and confocal laser scanning microscopy (CLSM). These new approaches are especially useful for evaluating MPs in aquatic environment, nevertheless their implementation in the field is still limited, presenting potential for further investigation and development. Table 15.2 reviews the basic feature difference between HSI, AFM and CLSM.

Spectroscopic Techniques

Microscopic techniques succeed in analysing size, shape, and colour but fail to identify material composition. Spectroscopic technologies such as FTIR and Raman spectroscopy bridge this gap by identifying polymers precisely using their unique vibrational spectra. When combined with microscopy, they allow for thorough chemical imaging as well as physical analysis, resulting in comprehensive sample characterization (Xu et al., 2019).

FTIR

FTIR (Fourier Transform Infrared) spectroscopy is a popular method for determining the infrared (IR) absorption and emission spectra of materials. It offers various advantages, such as a low signal-to-noise ratio, excellent resolution, rapid data gathering, and broad detection limits. FTIR has a frequency range of 4000–600 cm^{-1} and detects molecular bond-specific IR absorption/emission, making it useful for quantitative and qualitative studies (Bacsik et al., 2004; Veerasingam et al., 2021). Table 15.3 highlights previous investigations accomplished implementing FTIR.

Table 15.2 Comparison of HSI, AFM and CLSM

Aspects	Hyperspectral imaging (HSI)	Atomic force microscopy (AFM)	Confocal laser scanning microscopy (CLSM)
Principle	Uses reflectance spectra collected by an image spectrometer or hyperspectral camera to capture spectral data from pixels (Sridhar et al., 2022)	Scans sample using a cantilever, with force applied in two or three axes through a piezoelectric mechanism, providing 3D images (Luo et al., 2021; Mariano et al., 2021)	Uses optical slicing and Nile Red staining to visualize MPs/NPs with high resolution (Sancataldo et al., 2020)
Resolution	The detection limit of 0.5 mm for particles. High-resolution imaging using pixel-based spectral data analysis (Nigamatzyanova & Fakhrullin, 2021)	Extremely high resolution, capable of imaging down to a few nanometers (< 0.3 nm)(Li et al., 2021)	Resolution up to 200 nm, enabling detailed imaging of small MPs and NPs (Melo-Agustín et al., 2022)
Advantages	– Rapid image processing and analysis – Ability to analyze large areas with high spectral diversity – Effective for particles of 0.5 mm and larger	– High-resolution topographical analysis (< 0.3 nm) – No need for sample pretreatment – Can detect adsorbed compounds (e.g., heavy metals) on MPs (Melo-Agustín et al., 2022)	– High axial resolution, ideal for detailed analysis of small MPs and NPs – Allows 3D optical imaging and analysis – Can detect interactions between MPs/NPs and microorganisms in water (He et al., 2022)
Limitation	-Limited to particles > 0.5 mm – Requires confirmation from other techniques like μ-FTIR	– Sample contamination is not easy to control – Cantilever tips may damage delicate samples, -Image distortion can happen	-Reduced detection efficiency due to organic matter contamination from Nile Red staining – Useful for particles size unto 100 nm (Sancataldo et al., 2020)
Applications	– Detects large MPs in marine ecosystems, particularly for rapid screening and analysis (Primpke et al., 2020)	Applied in the study of polymer blends and surface compounds, such as heavy metals, on MPs collected from aquatic environments (Melo-Agustín et al., 2022)	– Good for analysis of smaller MPs in water, including their interaction with microorganisms like bacteria or microalgae (He et al., 2022)

Features and their applications

Table 15.3 Summary of previous studies using FTIR for MPs analysis

MPs detected	MP particles	Extraction methods	References
PE, PP and PS	178 ± 261 mg/m^2 (low tide) 1323 ± 1228 mg/m^2 (hightide)	Separation by density and sieving	Karthik et al. (2018)
PE, PP, PVC, PET, PS	45 ± 12 to 220 ± 50 particles/kg	Separation by density and sieving	Tiwari et al. (2019)
PE, PEVA, LDPE, PS	102.9 ± 39.9 to 163.3 ± 37.7 particles/kg	Separation by density	Yu et al. (2016)
Polymethylmethacrylate (PMMA)	–	Digestion of intestines	Hermsen et al. (2017)
PET, PE	–	By visual inspection	McGoran et al. (2017)
Acrylic	0.678 ± 0.044 to 1.582 ± 0.448 particles/gm	By treating with Trypsin solution and filtration	Courtene-Jones et al. (2017)

FTIR is commonly employed in MPS analysis to figure out the chemical constitutions and nature mostly in aquatic atmospheres. It functions in three modes: reflectance, transmission, and ATR. ATR-FTIR is effective for bigger particles (> 500 μm) but can cause MPS adhesion to crystals. Micro-FTIR, which is excellent for smaller particles, has weak IR reflection from irregular forms. Transmission mode, with thin samples (< 100 μm) and IR-transparent substrates, yields high-quality spectra (e.g., 1480–1430 cm^{-1} for C-H bending). Still, FTIR is highly susceptible to chemicals, overlapping groups, and weathering effects, complicating mixed sample analysis. Overcoming these hurdles will increase its efficiency (Strungaru et al., 2019).

Raman Spectroscopy

Raman spectroscopy enhances FTIR's ability to detect polar functional groups by focusing on vibrational modes of nonpolar groups. This technique investigates how radiation and vibrations of molecules interact, resulting in a distinct "fingerprint" for each molecule or group. Raman can provide precise insights into molecular structure, dynamics, and composition by determining the vibrational frequencies based on atomic mass, bond strength, and geometric arrangement. (Ahmed et al., 2020). Table 15.4 summarizes previous research conducted using Raman spectroscopy.

Raman spectroscopy is commonly used to analyse organic MPs such as carbazole, 1-aminopyrine, benzole acid, and pigments (Edwards et al., 2003; Tuan Vo-Dinh, 1984). Since water has little impact on its spectra, it is especially beneficial to investigate aquatic contaminants. Raman can detect particles as small as 1–2 mm, making

Table 15.4 Comparative analysis of measurement parameters: MP versus Raman spectroscopy

Spectral range (cm^{-1})	MP size	Measure time and Laser excitation	References
200–3200	1 mm	10–60 s, 785 nm	Ghosal et al. (2018)
150–3500	1.3 μm	2 × 1 s, 532 nm	Ossmann et al. (2018)
50–4000	9 μm	5–500 s, 633 nm	Imhof et al. (2013)
160–3600	5–10 μm	20 × 500 ms, 532 nm	Käppler et al. (2016)

it superior to FTIR for MPS investigation due to reduced water interference and greater sensitivity to nonpolar functional groups (Li et al., 2009a, 2009b). It can measure wavenumbers below 650 cm^{-1} and is not restricted by sample thickness or form (Strungaru et al., 2019). Furthermore, Raman detects contaminants like fillers, pigments, and dyes that FTIR cannot. However, it confronts several hurdles, including extended analysis durations, polymer degradation, sensitivity issues, and fluorescence interference. Coherent anti-Stokes Raman scattering microscopy may help to overcome some of these constraints.

UV–Vis Spectroscopic

UV–Vis spectroscopy is one of the first and most widely used methods for identifying MPs. It detects a variety of analytes, including transition metal ions, biological macromolecules, and highly conjugated chemical substances, using UV and visible light spectra. The method is based on the theory that the UV reaction is linearly proportional to the analyte concentration. A UV–Vis spectrophotometer, when integrated with HPLC systems, serves as a dependable detector for detecting MPs.

Overall, assessing the UV–Vis absorbance of MPs are simple and reliable analytical method. Rhodamine B, acridine orange, Congo red, methylene blue, Safranin O, Nile Blue A, Orange G, Rhodamine 6G, Alizarin S, methyl red, and neutral red are among the cationic and anionic organic dyes that are commonly identified by UV–Vis spectroscopy (Antonov et al., 1999; Ezugwu et al., 2018; Kalmár et al., 2016; Khataee et al., 2018; Lachheb et al., 2002). 26 micro-pollutants, including atenolol, benzotriazole, carbamazepine, clarithromycin, diclofenac, gabapentin, hydrochlorothiazide, sulfamethoxazole, and others, were identified by (Wittmer et al., 2015) using UV measurement. A Swiss wastewater matrix with realistic MPs combinations was employed in the study. Although UV–Vis spectroscopy is a rapid, precise, and economical way to analyze samples, and It is highly responsive to chromophores and offers multicomponent systems exceptional resolution of spectral information. And its high energy requirements can break molecular bonds; spectral absorption can be affected by factors such as pH, temperature, contaminants, and impurities, which can result in inaccuracies; sensitivity and selectivity problems also occur when analysing mixtures of MPs with similar properties, which can result in overlapping spectral absorption (Li et al., 2009b).

LC–MS/LC–MS-MS and GC–MS/GC–MS-MS

Both GC–MS and LC–MS are effective analytical methods for identifying MPs, however, they vary in terms of sample preparation needs, sensitivity, and application. Without requiring complicated sample preparation, LC–MS is especially well-suited for evaluating a broad variety of organic micro-pollutants, such as pesticides, medications, and persistent organic pollutants (POPs) (Korfmacher, 2005).Even for non-volatile or polar chemicals, it offers precise quantification, is extremely sensitive, and can identify contaminants at ppt levels (Lorenzo et al., 2018). Because of this, LC–MS is the method of choice for evaluating chemicals including UV filters, antibiotics, and EDCs (Alsbaiee et al., 2016; Lucci et al., 2017).

However, GC–MS is the most effective method for recognizing organic molecules that are volatile and thermostable. It can detect at the ppm level and is very successful at evaluating complicated mixtures (Siddiqui et al., 2017). But for non-volatile substances such PFASs, estrogens, and medications, GC–MS usually necessitates derivatization, which can make the analysis more difficult and compromise accuracy. Analyte loss or incomplete reactions may happen from the derivatization process, which could affect the outcome. GC–MS is nevertheless a great instrument for evaluating volatile chemicals despite these difficulties, and it can be combined with sophisticated online detection techniques.

In conclusion, GC–MS is more appropriate for volatile and thermostable compounds, but it requires more preparation and may be less accurate because of derivatization. In contrast, LC–MS provides a simpler method for a larger range of compounds, particularly non-volatile ones (Díaz-Cruz et al., 2009).

The mass-to-charge (m/z) ratio of ions can be directly determined using Time-of-flight (ToF)-mass spectrometry (MS) (TOF–MS,) an innovative technology that eliminates the need for chromatographic separation (Wu et al., 2023). Ions are separated according to their mass; larger ions take longer to get to the detector. In contrast, separation and occasionally derivatization are involved in LC–MS and GC–MS. MALDI-TOF MS and ToF–SIMS are two essential TOF–MS techniques for MPS analysis. The more conventional chromatographic techniques of LC–MS and GC–MS are complemented by the direct, quick analysis provided by MALDI-TOF MS, which uses metal cationization agents to improve ionization and detect intact molecules (Badía et al., 2011; Welker et al., 2002). Table 15.5 summarizes highlights of Studies Leveraging GC–MS LC–MS and ToFMS Techniques.

Various types of LC–MS or LC–MS-MS systems are used to analyse MPs. The triple quadrupole system is made up of two quadrupole mass analysers, with the second serving as a collision cell for fragmentation. The QTRAP system combines a quadrupole and an ion trap, with electric fields used to confine and focus ions. The QTOF device uses an electric field to retain ions in a stable orbit, resulting in high-resolution mass measurements that are suitable for both targeted analysis and screening of unknown chemicals (Hopfgartner et al., 2004; Rosen, 2007).

Table 15.5 Summary of research utilizing GC–MS, LC–MS and ToFMS approaches

MPs types	Indicator used	Pretreatment done	References
GCMS application			
PE, PA, PP, PC, PS, PET, PMMA, PVC	Hexane, p-methoxy-tert-butyl-benzene, dimethyl terephthalate, methyl methacrylate, 1-methylnaphthalene	Separation by density, enzymatic oxidization	Gomiero et al. (2019)
PE, PP, PS	Styrene, 1-pentadecene, 1,14-pentadeca-diene	Pressurized liquid extraction	Dierkes et al. (2019)
LDPE	1,13-tetradecadiene	No treatment	Duemichen et al. (2014)
PP, PS, PE	2,4,6-trimethylnon-1-ene, 1,12-tridecadiene	Ultrafiltration	Eisentraut et al. (2018)
PUR	2,5-toluenediamine, 4,4-methylenebisbenzeneaminE	Ultrafiltration and chemical digestion	Coralli et al. (2023)
PAN, PET, PP	Divinyl benzoate, 2,4 dicyano-1-butene, 2-methylene-pentadinitrile	Density separation	Chouchene et al. (2022)
LCMS Application			
PET, PC	Bisphenol (BPA), p-phthalic (PTA)	Solid phase extraction	Wang et al. (2017)
PET	bis(2-hydroxyethyl) terephthalate (BHET)	Solid phase extraction	Tian et al. (2022)
PLA, PET	p-phthalic (PTA)	Digestion-flotation	Wang et al. (2022)
PA66, PA6	adipic acid (AA), caprolactam (CA), 6-ACA, 8-ACA	Solid phase extraction	Lin et al. (2020)
Time-of-flight spectrometry (ToFMS) Application			
PS	Styrene, Methylbenzene, naphthalene, 2,4,6-triphenyl-1-hexene	Density separation	Lin et al. (2020)

HPLC–UV/Fluorescence

High-Performance Liquid Chromatography (HPLC) is another popular method for examining organic chemicals in liquid samples, following a brief explanation of LC–MS and GC–MS. By injecting the sample into a column full of porous particles, HPLC separates chemicals according to their solubility and polarity, in contrast to mass spectrometry-based techniques that use mass-to-charge ratios to identify components (Siddiqui et al., 2017). Different compounds segregate as the sample passes through the column because of how differently they interact with the stationary and mobile phases. HPLC is renowned for its capacity to handle big, polar, and thermally unstable compounds as well as its high sensitivity and low detection limits.

However, because it exposes the sample's chemical makeup, it might be damaging in contrast to non-destructive methods.

Ahmed et al. (2017a) used a Zorbax Bonus RP C18 column in competitive mode to evaluate four antibiotics (sulfamethazine, sulfamethoxazole, sulfathiazole, and chloramphenicol) at the μg/L level. Additionally, they used both UV and fluorescence detectors in single and competitive modes to study six endocrine-disrupting chemicals (EDCs): estrone, estriol, bisphenol A, 17β-estradiol, 17α-ethinylestradiol, and 4-tert-butylphenol. To achieve reliable results with both detection methods, the study effectively used the same column configuration for both types of studies (Ahmed et al., 2018a, 2018b).

NMR Spectroscopy

A potential technique for analysing MPSs both qualitatively and quantitatively is quantitative nuclear magnetic resonance (qNMR) spectroscopy. qNMR enables direct quantification based on the proportionate relationship between signal strength and the amount of protons in the sample, in contrast to other methods like GC–MS or LC–MS. The MPSs must dissolve in an appropriate deuterated solvent for analysis to use this approach, which has a high accuracy (> 98%) (Peez et al., 2019b) quantified polyethylene, poly terephthalate, and polystyrene particles in model samples, marking the first notable use of ^{1}H NMR spectroscopy for MPSs. They identified critical criteria like linearity, accuracy, and precision using a calibration curve method for validation. Additionally, because qNMR is less affected by fluorescence and matrix effects than FTIR and Raman spectroscopy, it offers an advantage. To deconvolute signals and increase quantification precision for complex samples with signal overlaps, the Peak Fitting (PF-qNMR) approach was developed. To quantify different MPSs, including polyvinyl chloride and polyamide fibres (Peez et al., 2019b), also used INT-qNMR and PF-qNMR techniques. This shows how adaptable qNMR is for a variety of polymer types, sizes, and shapes. Therefore, qNMR is unique in that it may be used to examine MPSs with great precision, enhancing other techniques such as GC–MS and LC–MS. (Peez et al., 2019a, 2019b).

Thermal Methods

In previous discussions, we addressed spectroscopic and microscopic methods that detect and describe MPSs according to their chemical and structural characteristics, like FTIR, Raman spectroscopy, and NMR. An excellent destructive method that has been used recently to identify MPs is thermal analysis (Majewsky et al., 2016). Based on the heat stability of polymers, this technique can quantify changes in their chemical and physical characteristics (Dümichen et al., 2015) Samples can be fully pyrolyzed at temperatures as high as 600 °C using methods like Thermogravimetric

Table 15.6 Overview of previous research on different thermal analysis

MPs	Thermal methods used	Advantages	Limitations	References
PP, PE	TGA–DSC	Parallel analysis of MPs is possible	A no. of MPs. Could not be identified, except PP and PE	Huppertsberg and Knepper (2018), Majewsky et al. (2016)
PP, PS, PET, PE	TED-GCMS	Can be used Used for larger sample quantities	No information about the size and quantity of MPs	Dümichen et al. (2015, 2017), Huppertsberg and Knepper (2018)
PP, PS, PET, PE	DSC	Able to identify primary MPs	For identification reference materials are necessary. Only a few polymers are identified	Huppertsberg and Knepper (2018)

Analysis (TGA), which yields valuable data on the sample's homogeneity and breakdown profile (Dümichen et al., 2017). Another technique for precisely identifying MPs like polyethylene (PE) microbeads is Differential Scanning Calorimetry (DSC), which compares the thermal properties of plastics to identify particular polymer kinds (Kalčíková et al., 2017). Common plastics like PE and polypropylene (PP) can be identified using TGA and DSC together, but because phase transition signals overlap, it has trouble differentiating PVC, PET, and polyamides. TGA can also be employed with thermal desorption gas chromatography-mass spectrometry (TDS-GC–MS) and solid-phase extraction (SPE), which improves the analysis of complicated materials and bigger particles. Numerous MPs, including PE, PP, and PS, have been found in various samples using this technique (Majewsky et al., 2016).

Another technique for analysing larger sample masses and complicated matrices is the combination of thermal extraction and desorption with gas chromatography-mass spectrometry (TED-GC–MS), which makes MPs analysis quicker and more effective. It uses pyrolysis and the examination of breakdown gases to identify polymers. Table 15.6 presents highlight of research exploiting thermal analysis, such as DSC and TGA.

Challenges to Analysing Emerging MPs

Analysing new MPs involves numerous challenges that necessitate refined approaches and ongoing innovation. These contaminants are usually found at *low concentrations* (nanograms to micrograms per litre), making detection and quantification difficult. MP analysis is further complicated by their *chemical diversity*, which includes organic, inorganic, volatile, and non-volatile chemicals. Samples frequently contain *complex matrices* that can interfere with precise measurements, necessitating efficient sample preparation and extraction techniques. The *selection of proper solvents*, columns, and matrix effects are significant challenge, especially with technologies such as LC–MS-MS. Furthermore, *effective sampling processes* are crucial since diverse environmental matrices necessitate specialized extraction techniques and target analyte concentrations may necessitate sample concentration or dilution. Since many analytical techniques lack internal standards, real-time analysis, or precise limits of detection, *method validation* is still another major challenge (Gabet et al., 2007; Jiang et al., 2013). It is crucial to create a universal sampling technique and enhance method validation at various levels of analysis. Although analytical tools like mass spectrometric methods (GC–MS, LC–MS-MS) have *low detection limits*, their sensitivity might differ depending on the molecule and matrix. Another difficulty is *managing intricate datasets* that come from the investigation of several contaminants; proper interpretation of these datasets requires specialist software. Moreover, regulatory limitations and the absence of reference materials for emerging pollutants hinder standardization and cross-study verification. There is very little knowledge about these contaminants' toxicity and effects on the ecosystem (Das and Dey et al., 2023; Singh et al., 2020; Swain, 2013). Lastly, the time and expense limitations of micro-pollutant analysis make the work even more difficult, necessitating continued study and the creation of high-throughput, reasonably priced techniques. By tackling these issues, it will be easier to identify, comprehend, and control newly emerging environmental MPs (Ahmed et al., 2020).

Advanced AI and ML Applications for MPs Detection

Artificial intelligence (AI) and machine learning (ML) have transformed MPs detection by providing cutting-edge tools for quick and accurate analysis. Traditional approaches, such as hand sampling and microscopic analysis, are time-consuming and need extensive knowledge (Jin et al., 2024). AI-driven solutions overcome these constraints by automating and improving the detection process, allowing for real-time analysis and large-scale environmental monitoring. Recent applications of machine learning and deep learning include forecasting PM2.5 levels, analysing heavy metal contamination of soil, examining the impact of microplastics on plant accumulation of heavy metals, and estimating the amounts of nitrogen oxides (NO_2) etc. (Chen et al., 2024; Das et al., 2023; Mishra et al., 2021; Sahu & Das, 2024). One important use ML algorithms is convolutional neural networks (CNNs) (Ren et al., 2023), which examine microscopic and spectroscopic pictures to detect MPs based on their shape, size, and composition. Additionally, ML makes it possible to combine data from other sources, such as spectroscopy (e.g., FTIR and Raman) and hyperspectral imaging, to classify MPs by polymer type and even forecast their sources. Furthermore, AI improves data preprocessing by tackling issues like noise reduction, normalization, and feature extraction. These steps increase the reliability of MPs detection in complicated environmental samples such as soil, water, and air. AI-powered systems can also predict pollution patterns by combining data from several monitoring stations, allowing policymakers to create tailored policies to reduce MP contamination. The future of AI in MP detection seems even more promising with continued developments, combining cloud-based platforms and Internet of Things (IoT) devices (Arlene et al., 2023) for smooth data sharing and analysis. These innovations open the door to long-term fixes for the world's microplastic problem.

Conclusion

The study and management of micropollutants (MPs) necessitates an integrated approach, as described in the preceding chapters. Chapter 1 discussed the importance of MPs, including their origins, environmental fate, and impact on ecosystems. Understanding the complexities of MPs highlights the necessity for advanced and precise analytical techniques. In Chap. 2, different approaches for MP analysis were discussed, including microscopic, spectroscopic, and thermal techniques. Microscopic approaches, like as electron microscopy, provide information about MPs' physical qualities, whereas spectroscopic methods, such as FTIR, Raman, and UV–Vis, focus on their chemical properties. Thermal approaches, with applications ranging from polymer degradation study to MP characterisation, have distinct advantages and disadvantages. The chapter focused on the relationship between these approaches, their industrial applications, and the crucial difficulties of accuracy, sensitivity, and specificity. Chapter 3 discussed the challenges of assessing new MPs, such as low concentrations, complicated matrices, and overlapping contaminants. These problems highlight the importance of continuous innovation in sample preparation, separation, and detection to enable accurate and fast analysis. Chapter 4 stressed the revolutionary significance of new technologies, specifically Artificial Intelligence (AI) and Machine Learning (ML), in MP detection. These tools are altering the analytical landscape by allowing for real-time monitoring, improving data interpretation, and increasing the overall efficiency and precision of pollution analysis. AI and machine learning have the promise of overcoming present analytical constraints, making them important in the fight against environmental contamination.

MPs detection and their characterization are the functions of physical and chemical feature of pollutants. Each analytical equipment has a unique scope and limitations, demanding careful selection depending on pollutant attributes. Techniques such as HPLC, GC–MS, and LC–MS have become critical for separating and detecting complex mixtures, with LC–MS emerging as particularly versatile due to its ability to avoid derivation. While FTIR and Raman spectroscopy are still the preferred methods for analysing microplastics of various sizes, UV–Vis spectroscopy is useful for organic dyes but has issues with overlapping adsorption bands.

In the future, overcoming the inherent difficulties of MP analysis will require the development of analytical techniques and technologies. To increase the accuracy and effectiveness of MP detection, future studies should concentrate on improving instrument sensitivity, creating high-capacity separation columns, and using AI and ML algorithms. Better environmental monitoring and remediation techniques are made possible by utilizing these innovations to analyse MPs accurately and quickly.

References

Ahmed, M. B., Johir, M. A. H., Ngo, H. H., Guo, W., Zhou, J. L., Belhaj, D., & Moni, M. A. (2020). Methods for the analysis of micro-pollutants. In *Current developments in biotechnology and bioengineering: emerging organic micro-pollutants* (pp. 63–86). Elsevier. https://doi.org/10.1016/B978-0-12-819594-9.00004-8

Ahmed, M. B., Rahman, M. S., Alom, J., Hasan, M. S., Johir, M. A. H., Mondal, M. I. H., Lee, D. Y., Park, J., Zhou, J. L., & Yoon, M. H. (2021). Microplastic particles in the aquatic environment: A systematic review. *Science of the Total Environment*. https://doi.org/10.1016/j.scitotenv.2021.145793

Ahmed, M. B., Zhou, J. L., Ngo, H. H., Guo, W., Johir, M. A. H., & Belhaj, D. (2017a). Competitive sorption affinity of sulfonamides and chloramphenicol antibiotics toward functionalized biochar for water and wastewater treatment. *Bioresource Technology, 238*, 306–312. https://doi.org/10.1016/j.biortech.2017.04.042

Ahmed, M. B., Zhou, J. L., Ngo, H. H., Guo, W., Thomaidis, N. S., & Xu, J. (2017b). Progress in the biological and chemical treatment technologies for emerging contaminant removal from wastewater: A critical review. *Journal of Hazardous Materials, 323*, 274–298. https://doi.org/10.1016/j.jhazmat.2016.04.045

Ahmed, M. B., Zhou, J. L., Ngo, H. H., Johir, M. A. H., & Sornalingam, K. (2018a). Sorptive removal of phenolic endocrine disruptors by functionalized biochar: Competitive interaction mechanism, removal efficacy and application in wastewater. *Chemical Engineering Journal, 335*, 801–811. https://doi.org/10.1016/j.cej.2017.11.041

Ahmed, M. B., Zhou, J. L., Ngo, H. H., Johir, M. A. H., Sun, L., Asadullah, M., & Belhaj, D. (2018b). Sorption of hydrophobic organic contaminants on functionalized biochar: Protagonist role of Π-Π electron-donor-acceptor interactions and hydrogen bonds. *Journal of Hazardous Materials, 360*, 270–278. https://doi.org/10.1016/j.jhazmat.2018.08.005

Ainali, N. M., Bikiaris, D. N., & Lambropoulou, D. A. (2021). Aging effects on low- and high-density polyethylene, polypropylene and polystyrene under UV irradiation: An insight into decomposition mechanism by Py-GC/MS for microplastic analysis. *Journal of Analytical and Applied Pyrolysis, 158*. https://doi.org/10.1016/j.jaap.2021.105207

Alsbaiee, A., Smith, B. J., Xiao, L., Ling, Y., Helbling, D. E., & Dichtel, W. R. (2016). Rapid removal of organic micropollutants from water by a porous β-cyclodextrin polymer. *Nature, 529*, 190–194. https://doi.org/10.1038/nature16185

Andrade-Eiroa, A., Canle, M., Leroy-Cancellieri, V., & Cerdà, V. (2016). Solid-phase extraction of organic compounds: A critical review (Part I). *TrAC—Trends in Analytical Chemistry*. https://doi.org/10.1016/j.trac.2015.08.015

Antonov, L., Gergov, G., Petrov, V., Kubista, M., & Nygren, J. (1999). UV-Vis spectroscopic and chemometric study on the aggregation of ionic dyes in water. *Talanta*.

Arlene, A. A. (2023). *A comprehensive literature survey of microplastic detection using IoT*.

Armenta, S., Garrigues, S., & de la Guardia, M. (2008). Green analytical chemistry. *TrAC—Trends in Analytical Chemistry, 27*, 497–511. https://doi.org/10.1016/j.trac.2008.05.003

Bacsik, Z., Mink, J., & Keresztury, G. (2004). FTIR spectroscopy of the atmosphere. I. Principles and methods. *Applied Spectroscopy Reviews, 39*, 295–363. https://doi.org/10.1081/ASR-200030192

Badía, J. D., Strömberg, E., Ribes-Greus, A., & Karlsson, S. (2011). A statistical design of experiments for optimizing the MALDI-TOF-MS sample preparation of polymers. An application in the assessment of the thermo-mechanical degradation mechanisms of poly (ethylene terephthalate). *Analytica Chimica Acta, 692*, 85–95. https://doi.org/10.1016/j.aca.2011.02.063

Bonfanti, P., Colombo, A., Saibene, M., Motta, G., Saliu, F., Catelani, T., Mehn, D., La Spina, R., Ponti, J., Cella, C., Floris, P., & Mantecca, P. (2021). Microplastics from miscellaneous plastic wastes: Physico-chemical characterization and impact on fish and amphibian development. *Ecotoxicology and Environmental Safety, 225*. https://doi.org/10.1016/j.ecoenv.2021.112775

Chau, H. T. C., Kadokami, K., Duong, H. T., Kong, L., Nguyen, T. T., Nguyen, T. Q., & Ito, Y. (2018). Occurrence of 1153 organic micropollutants in the aquatic environment of Vietnam. *Environmental Science and Pollution Research, 25*, 7147–7156. https://doi.org/10.1007/s11356-015-5060-z

Chen, L., Chang, N., Qiu, T., Wang, N., Cui, Q., Zhao, S., Huang, F., Chen, H., Zeng, Y., Dong, F., & Fang, L. (2024). Meta-analysis of impacts of microplastics on plant heavy metal(loid) accumulation. *Environmental Pollution*. https://doi.org/10.1016/j.envpol.2024.123787

Chouchene, K., Nacci, T., Modugno, F., Castelvetro, V., & Ksibi, M. (2022). Soil contamination by microplastics in relation to local agricultural development as revealed by FTIR. *ICP-MS and Pyrolysis-GC/MS. Environmental Pollution, 303*. https://doi.org/10.1016/j.envpol.2022.119016

Çomaklı, E., Bingöl, M. S., & Bilgili, A. (2020). Assessment of microplastic pollution in a crater lake at high altitude: A case study in an Urban Crater Lake in Erzurum, Turkey. *Water, Air, & Soil Pollution, 231*. https://doi.org/10.1007/s11270-020-04653-4

Coralli, I., Goßmann, I., Fabbri, D., & Scholz-Böttcher, B. M. (2023). Determination of polyurethanes within microplastics in complex environmental samples by analytical pyrolysis. *Analytical and Bioanalytical Chemistry, 415*, 2891–2905. https://doi.org/10.1007/s00216-023-04580-3

Courtene-Jones, W., Quinn, B., Murphy, F., Gary, S. F., & Narayanaswamy, B. E. (2017). Optimisation of enzymatic digestion and validation of specimen preservation methods for the analysis of ingested microplastics. *Analytical Methods, 9*, 1437–1445. https://doi.org/10.1039/c6ay02343f

Das, A. P., & Ghosh, S. (2023). Manganese and its application. *Manganese Mining Microorganisms, 29–40*. https://doi.org/10.1016/B978-0-12-822147-1.00007-8

Das, A. P., Ghosh, S., Mohanty, S., & Sukla, L. B. (2015). Consequences of manganese compounds: A review. *Toxicological and Environmental Chemistry, 96*, 981–997. https://doi.org/10.1080/02772248.2015.1005428

Das, A. P., Nayak, B., & Behera, I. D. (2024a). Applications of microorganisms in advanced geomicrobiological and biotechnological exploration. *Geomicrobiology Journal, 41*, 309–311. https://doi.org/10.1080/01490451.2024.2327059/ASSET//CMS/ASSET/DE2A1F67-08C1-475A-ABC7-2B3F4C9E8BCF/01490451.2024.2327059.FP.PNG

Das, A. P., Sukla, L. B., Pradhan, N., & Nayak, S. (2011). Manganese biomining: A review. *Bioresource Technology, 102*, 7381–7387. https://doi.org/10.1016/J.BIORTECH.2011.05.018

Das, A. P., & Swain, S. (2013). Algal biosorption of toxic dye methylene blue. A potential source of food, feed, biochemicals, biofuels and biofertilizers. In *International conference on Algal Biorefinery*. Indian Institute of Technology.

Das, A. P., Behera, I. D., & Das, N. P. (2024). *Renewable energy generation and value addition from environmental microfiber pollution through advanced greener solution.*

Das, P., & Bal, M. (2025). Modeling and optimization study for microplastic removal from aquatic medium using Chroococcidiopsis species. *Chemical Engineering Science, 304*, 121085. https://doi.org/10.1016/J.CES.2024.121085

Das, P., Halder, G., & Bal, M. (2023). A critical review on remediation of microplastics using microalgae from aqueous system. *Science of the Total Environment, 898*, 166425. https://doi.org/10.1016/J.SCITOTENV.2023.166425

De Frond, H., Thornton Hampton, L., Kotar, S., Gesulga, K., Matuch, C., Lao, W., Weisberg, S. B., Wong, C. S., & Rochman, C. M. (2022). Monitoring microplastics in drinking water: An inter-laboratory study to inform effective methods for quantifying and characterizing microplastics. *Chemosphere, 298*. https://doi.org/10.1016/j.chemosphere.2022.134282

Dehaut, A., Cassone, A. L., Frère, L., Hermabessiere, L., Himber, C., Rinnert, E., Rivière, G., Lambert, C., Soudant, P., Huvet, A., Duflos, G., & Paul-Pont, I. (2016). Microplastics in seafood: Benchmark protocol for their extraction and characterization. *Environmental Pollution, 215*, 223–233. https://doi.org/10.1016/j.envpol.2016.05.018

Delli Compagni, R., Gabrielli, M., Polesel, F., Turolla, A., Trapp, S., Vezzaro, L., & Antonelli, M. (2020). Risk assessment of contaminants of emerging concern in the context of wastewater reuse for irrigation: An integrated modelling approach. *Chemosphere, 242*. https://doi.org/10.1016/j.chemosphere.2019.125185

Deviller, G., Lundy, L., & Fatta-Kassinos, D. (2020). Recommendations to derive quality standards for chemical pollutants in reclaimed water intended for reuse in agricultural irrigation. *Chemosphere, 240*. https://doi.org/10.1016/j.chemosphere.2019.124911

Dey, S., Das, A., Mallick, K., Sahu, A., & Das, A. P. (2023). Environmental petroleum waste: Pollution. *Toxicity, Sustainable Remediation, 159–175*. https://doi.org/10.1007/978-3-031-48220-5_7

Díaz-Cruz, M. S., García-Galán, M. J., Guerra, P., Jelic, A., Postigo, C., Eljarrat, E., Farré, M., López de Alda, M. J., Petrovic, M., Barceló, D., & Petrovic, M. (2009). Analysis of selected emerging contaminants in sewage sludge. *TrAC—Trends in Analytical Chemistry*. https://doi.org/10.1016/j.trac.2009.09.003

Dierkes, G., Lauschke, T., Becher, S., Schumacher, H., Földi, C., & Ternes, T. (2019). *Quantification of Microplastics in Environmental Samples via Pressurized Liquid Extraction and Pyrolysis-Gas Chromatography*. https://doi.org/10.1007/s00216-019-02066-9/Published

Doyle, M. J., Watson, W., Bowlin, N. M., & Sheavly, S. B. (2011). Plastic particles in coastal pelagic ecosystems of the Northeast Pacific ocean. *Marine Environment Research, 71*, 41–52. https://doi.org/10.1016/j.marenvres.2010.10.001

Duemichen, E., Braun, U., Senz, R., Fabian, G., & Sturm, H. (2014). Assessment of a new method for the analysis of decomposition gases of polymers by a combining thermogravimetric solid-phase extraction and thermal desorption gas chromatography mass spectrometry. *Journal of Chromatography A, 1354*, 117–128. https://doi.org/10.1016/j.chroma.2014.05.057

Dümichen, E., Barthel, A. K., Braun, U., Bannick, C. G., Brand, K., Jekel, M., & Senz, R. (2015). Analysis of polyethylene microplastics in environmental samples, using a thermal decomposition method. *Water Research, 85*, 451–457. https://doi.org/10.1016/j.watres.2015.09.002

Dümichen, E., Eisentraut, P., Bannick, C. G., Barthel, A. K., Senz, R., & Braun, U. (2017). Fast identification of microplastics in complex environmental samples by a thermal degradation method. *Chemosphere, 174*, 572–584. https://doi.org/10.1016/j.chemosphere.2017.02.010

Edwards, H. G. M., Newton, E. M., Wynn-Williams, D. D., & Lewis-Smith, R. I. (2003). Non-destructive analysis of pigments and other organic compounds in lichens using Fourier-transform Raman spectroscopy: A study of Antarctic epilithic lichens. *Spectrochimica Acta - Part A: Molecular and Biomolecular Spectroscopy* 2301–2309. https://doi.org/10.1016/S1386-1425(03)00073-8

Eisentraut, P., Dümichen, E., Ruhl, A. S., Jekel, M., Albrecht, M., Gehde, M., & Braun, U. (2018). Two birds with one stone—Fast and simultaneous analysis of microplastics: Microparticles derived from thermoplastics and tire wear. *Environmental Science & Technology Letters, 5*, 608–613. https://doi.org/10.1021/acs.estlett.8b00446

Ezugwu, C. I., Asraf, M. A., Li, X., Liu, S., Kao, C. M., Zhuiykov, S., & Verpoort, F. (2018). Selective and adsorptive removal of anionic dyes and CO2 with azolium-based metal-organic frameworks. *Journal of Colloid and Interface Science, 519*, 214–223. https://doi.org/10.1016/j.jcis.2018.02.003

Felsing, S., Kochleus, C., Buchinger, S., Brennholt, N., Stock, F., & Reifferscheid, G. (2018). A new approach in separating microplastics from environmental samples based on their electrostatic behavior. *Environmental Pollution, 234*, 20–28. https://doi.org/10.1016/j.envpol.2017.11.013

Fries, E., Dekiff, J. H., Willmeyer, J., Nuelle, M. T., Ebert, M., & Remy, D. (2013). Identification of polymer types and additives in marine microplastic particles using pyrolysis-GC/MS and scanning electron microscopy. *Environmental Sciences: Processes and Impacts, 15*, 1949–1956. https://doi.org/10.1039/c3em00214d

Gabet, V., Miège, C., Bados, P., & Coquery, M. (2007). Analysis of estrogens in environmental matrices. *TrAC—Trends in Analytical Chemistry, 26*, 1113–1131. https://doi.org/10.1016/j.trac.2007.10.003

Ghosal, S., Chen, M., Wagner, J., Wang, Z. M., & Wall, S. (2018). Molecular identification of polymers and anthropogenic particles extracted from oceanic water and fish stomach—A Raman micro-spectroscopy study. *Environmental Pollution, 233*, 1113–1124. https://doi.org/10.1016/j.envpol.2017.10.014

Gomiero, A., Øysæd, K. B., Agustsson, T., van Hoytema, N., van Thiel, T., & Grati, F. (2019). First record of characterization, concentration and distribution of microplastics in coastal sediments of an urban fjord in south west Norway using a thermal degradation method. *Chemosphere, 227*, 705–714. https://doi.org/10.1016/j.chemosphere.2019.04.096

Han, M., Niu, X., Tang, M., Zhang, B. T., Wang, G., Yue, W., Kong, X., & Zhu, J. (2020). Distribution of microplastics in surface water of the lower Yellow River near estuary. *Science of the Total Environment, 707*. https://doi.org/10.1016/j.scitotenv.2019.135601

He, Y., Wei, G., Tang, B., Salam, M., Shen, A., Wei, Y., Zhou, X., Liu, M., Yang, Y., Li, H., & Mao, Y. (2022). Microplastics benefit bacteria colonization and induce microcystin degradation. *Journal of Hazardous Materials, 431*. https://doi.org/10.1016/j.jhazmat.2022.128524

Hermsen, E., Pompe, R., Besseling, E., & Koelmans, A. A. (2017). Detection of low numbers of microplastics in North Sea fish using strict quality assurance criteria. *Marine Pollution Bulletin, 122*, 253–258. https://doi.org/10.1016/j.marpolbul.2017.06.051

Hernandez, L. M., Xu, E. G., Larsson, H. C. E., Tahara, R., Maisuria, V. B., & Tufenkji, N. (2019). Plastic teabags release billions of microparticles and nanoparticles into tea. *Environmental Science and Technology, 53*, 12300–12310. https://doi.org/10.1021/acs.est.9b02540

Hernandez, L. M., Yousefi, N., & Tufenkji, N. (2017). Are there nanoplastics in your personal care products? *Environmental Science & Technology Letters, 4*, 280–285. https://doi.org/10.1021/acs.estlett.7b00187

Hopfgartner, G., Varesio, E., Tschäppät, V., Grivet, C., Bourgogne, E., & Leuthold, L. A. (2004). Triple quadrupole linear ion trap mass spectrometer for the analysis of small molecules and macromolecules. *Journal of Mass Spectrometry, 39*, 845–855. https://doi.org/10.1002/jms.659

Huppertsberg, S., & Knepper, T. P. (2018). Instrumental analysis of microplastics—Benefits and challenges. *Analytical and Bioanalytical Chemistry, 410*, 6343–6352. https://doi.org/10.1007/s00216-018-1210-8

Imhof, H. K., Ivleva, N. P., Schmid, J., Niessner, R., & Laforsch, C. (2013). Contamination of beach sediments of a subalpine lake with microplastic particles. *Current Biology*. https://doi.org/10.1016/j.cub.2013.09.001

Jiang, J. Q., Zhou, Z., & Sharma, V. K. (2013). Occurrence, transportation, monitoring and treatment of emerging micro-pollutants in waste water—A review from global views. *Microchemical Journal*. https://doi.org/10.1016/j.microc.2013.04.014

Jin, H., Kong, F., Li, X., & Shen, J. (2024). Artificial intelligence in microplastic detection and pollution control. *Environmental Research*. https://doi.org/10.1016/j.envres.2024.119812

Kalaronis, D., Ainali, N. M., Evgenidou, E., Kyzas, G. Z., Yang, X., Bikiaris, D. N., & Lambropoulou, D. A. (2022). Microscopic techniques as means for the determination of microplastics and nanoplastics in the aquatic environment: A concise review. *Green Analytical Chemistry, 3*. https://doi.org/10.1016/j.greeac.2022.100036

Kalčíková, G., Alič, B., Skalar, T., Bundschuh, M., & Gotvajn, A. Ž. (2017). Wastewater treatment plant effluents as source of cosmetic polyethylene microbeads to freshwater. *Chemosphere, 188*, 25–31. https://doi.org/10.1016/j.chemosphere.2017.08.131

Kalmár, J., Lente, G., & Fábián, I. (2016). Kinetics and mechanism of the adsorption of methylene blue from aqueous solution on the surface of a quartz cuvette by on-line UV-Vis spectrophotometry. *Dyes and Pigments, 127*, 170–178. https://doi.org/10.1016/j.dyepig.2015.12.025

Kankanige, D., & Babel, S. (2021). Contamination by ≥6.5 μm-sized microplastics and their removability in a conventional water treatment plant (WTP) in Thailand. *Journal of Water Process Engineering, 40*. https://doi.org/10.1016/j.jwpe.2020.101765

Kanungo, J., Sahoo, T., Bal, M., & Behera, I. D. (2024). Performance of bioremediation strategy in waste lubricating oil pollutants: A review. *Geomicrobiology Journal, 41*, 360–373. https://doi.org/10.1080/01490451.2023.2245395

Käppler, A., Fischer, D., Oberbeckmann, S., Schernewski, G., Labrenz, M., Eichhorn, K. J., & Voit, B. (2016). Analysis of environmental microplastics by vibrational microspectroscopy: FTIR, Raman or both? *Analytical and Bioanalytical Chemistry, 408*, 8377–8391. https://doi.org/10.1007/s00216-016-9956-3

Karthik, R., Robin, R. S., Purvaja, R., Ganguly, D., Anandavelu, I., Raghuraman, R., Hariharan, G., Ramakrishna, A., & Ramesh, R. (2018). Microplastics along the beaches of southeast coast of India. *Science of the Total Environment, 645*, 1388–1399. https://doi.org/10.1016/j.scitotenv.2018.07.242

Khataee, A., Eghbali, P., Irani-Nezhad, M. H., & Hassani, A. (2018). Sonochemical synthesis of WS2 nanosheets and its application in sonocatalytic removal of organic dyes from water solution. *Ultrasonics Sonochemistry, 48*, 329–339. https://doi.org/10.1016/j.ultsonch.2018.06.003

Kolpin, D. W., Furlong, E. T., Meyer, M. T., Thurman, E. M., Zaugg, S. D., Barber, L. B., & Buxton, H. T. (2002). Pharmaceuticals, hormones, and other organic wastewater contaminants in U.S. streams, 1999–2000: A national reconnaissance. *Environmental Science and Technology, 36*, 1202–1211. https://doi.org/10.1021/es011055j

Kong, L., Kadokami, K., Wang, S., Duong, H. T., & Chau, H. T. C. (2015). Monitoring of 1300 organic micro-pollutants in surface waters from Tianjin, North China. *Chemosphere, 122*, 125–130. https://doi.org/10.1016/j.chemosphere.2014.11.025

Korfmacher, W. A. (2005). Foundation review: Principles and applications of LC-MS in new drug discovery. *Drug Discovery Today*. https://doi.org/10.1016/S1359-6446(05)03620-2

Labbe, A. B., Bagshaw, C. R., & Uttal, L. (2020). Inexpensive adaptations of basic microscopes for the identification of microplastic contamination using polarization and nile red fluorescence detection. *Journal of Chemical Education, 97*, 4026–4032. https://doi.org/10.1021/acs.jchemed.0c00518

Lachheb, H., Puzenat, E., Houas, A., Ksibi, M., Elaloui, E., Guillard, C., & Herrmann, J.-M. (2002). Photocatalytic degradation of various types of dyes (Alizarin S, Crocein Orange G, Methyl Red, Congo Red, Methylene Blue) in water by UV-irradiated titania. *Applied Catalysis B: Environmental*.

Li, C., Ma, Y., Liu, X., Huang, R., Su, R., Qi, W., Che, J., & He, Z. (2021). Synergistic effect of polystyrene nanoplastics and contaminants on the promotion of insulin fibrillation. *Ecotoxicology and Environmental Safety, 214*. https://doi.org/10.1016/j.ecoenv.2021.112115

Li, G., Chen, M., & Wei, T. (2009a). Application of raman spectroscopy to detecting organic contaminant in water. In *Proceedings—2009 IITA International Conference on Control, Automation and Systems Engineering, CASE 2009* (pp. 493–495). https://doi.org/10.1109/CASE.2009.149

Li, G., Chen, M., & Wei, T. (2009b). Application of raman spectroscopy to detecting organic contaminant in water. In *Proceedings—2009 IITA International Conference on Control, Automation and Systems Engineering, CASE 2009* (pp. 493–495). https://doi.org/10.1109/CASE.2009.149

Li, X., He, E., Xia, B., Van Gestel, C. A. M., Peijnenburg, W. J. G. M., Cao, X., & Qiu, H. (2020). Impact of CeO_2 nanoparticles on the aggregation kinetics and stability of polystyrene nanoplastics: Importance of surface functionalization and solution chemistry. *Water Research, 186*. https://doi.org/10.1016/j.watres.2020.116324

Lin, Y., Huang, X., Liu, Q., Lin, Z., & Jiang, G. (2020). Thermal fragmentation enhanced identification and quantification of polystyrene micro/nanoplastics in complex media. *Talanta, 208*. https://doi.org/10.1016/j.talanta.2019.120478

Lorenzo, M., Campo, J., & Picó, Y. (2018). Analytical challenges to determine emerging persistent organic pollutants in aquatic ecosystems. *TrAC—Trends in Analytical Chemistry*. https://doi.org/10.1016/j.trac.2018.04.003

Lucci, P., Saurina, J., & Núñez, O. (2017). Trends in LC-MS and LC-HRMS analysis and characterization of polyphenols in food. *TrAC—Trends in Analytical Chemistry*. https://doi.org/10.1016/j.trac.2016.12.006

Luo, H., Zeng, Y., Zhao, Y., Xiang, Y., Li, Y., & Pan, X. (2021). Effects of advanced oxidation processes on leachates and properties of microplastics. *Journal of Hazardous Materials, 413*. https://doi.org/10.1016/j.jhazmat.2021.125342

Maes, T., Jessop, R., Wellner, N., Haupt, K., & Mayes, A. G. (2017). A rapid-screening approach to detect and quantify microplastics based on fluorescent tagging with Nile Red. *Science and Reports, 7*. https://doi.org/10.1038/srep44501

Majewsky, M., Bitter, H., Eiche, E., & Horn, H. (2016). Determination of microplastic polyethylene (PE) and polypropylene (PP) in environmental samples using thermal analysis (TGA-DSC). *Science of the Total Environment, 568*, 507–511. https://doi.org/10.1016/j.scitotenv.2016.06.017

Mariano, S., Tacconi, S., Fidaleo, M., Rossi, M., & Dini, L. (2021). Micro and nanoplastics identification: Classic methods and innovative detection techniques. *Frontiers in Toxicology*. https://doi.org/10.3389/ftox.2021.636640

Mathalon, A., & Hill, P. (2014). Microplastic fibers in the intertidal ecosystem surrounding Halifax Harbor, Nova Scotia. *Marine Pollution Bulletin, 81*, 69–79. https://doi.org/10.1016/j.marpolbul.2014.02.018

McGoran, A. R., Clark, P. F., & Morritt, D. (2017). Presence of microplastic in the digestive tracts of European flounder, Platichthys flesus, and European smelt, Osmerus eperlanus, from the River Thames. *Environmental Pollution, 220*, 744–751. https://doi.org/10.1016/j.envpol.2016.09.078

Melo-Agustín, P., Kozak, E. R., de Jesús Perea-Flores, M., & Mendoza-Pérez, J. A. (2022). Identification of microplastics and associated contaminants using ultra high resolution microscopic and spectroscopic techniques. *Science of the Total Environment, 828*. https://doi.org/10.1016/j.scitotenv.2022.154434

Minor, E. C., Lin, R., Burrows, A., Cooney, E. M., Grosshuesch, S., & Lafrancois, B. (2020). An analysis of microlitter and microplastics from Lake Superior beach sand and surface-water. *Science of the Total Environment, 744*. https://doi.org/10.1016/j.scitotenv.2020.140824

Mishra, S., Swain, S., Sahoo, M., Mishra, S., & Das, A. P. (2021). Microbial colonization and degradation of microplastics in aquatic ecosystem: A review. *Geomicrobiology Journal*. https://doi.org/10.1080/01490451.2021.1983670

Mohanta, S., Pradhan, B., & Behera, I. D. (2024). Impact and remediation of petroleum hydrocarbon pollutants on agricultural land: A review. *Geomicrobiology Journal, 41*, 345–359. https://doi.org/10.1080/01490451.2023.2243925

Mohanty, S., Ghosh, S., Nayak, S., & Das, A. P. (2016). Isolation, identification and screening of manganese solubilizing fungi from low-grade manganese ore deposits. *Geomicrobiological Journal, 34*, 309–316. https://doi.org/10.1080/01490451.2016.1189016

Müller, Y. K., Wernicke, T., Pittroff, M., Witzig, C. S., Storck, F. R., Klinger, J., & Zumbülte, N. (2020). Microplastic analysis—Are we measuring the same? Results on the first global comparative study for microplastic analysis in a water sample. *Analytical and Bioanalytical Chemistry, 412*, 555–560. https://doi.org/10.1007/s00216-019-02311-1

Nguyen, N. B., Kim, M. K., Le, Q. T., Ngo, D. N., Zoh, K. D., & Joo, S. W. (2021). Spectroscopic analysis of microplastic contaminants in an urban wastewater treatment plant from Seoul, South Korea. *Chemosphere, 263*. https://doi.org/10.1016/j.chemosphere.2020.127812

Nigamatzyanova, L., & Fakhrullin, R. (2021). Dark-field hyperspectral microscopy for label-free microplastics and nanoplastics detection and identification in vivo: A Caenorhabditis elegans study. *Environmental Pollution, 271*. https://doi.org/10.1016/j.envpol.2020.116337

Núñez, M., Borrull, F., Pocurull, E., & Fontanals, N. (2017). Sample treatment for the determination of emerging organic contaminants in aquatic organisms. *TrAC—Trends in Analytical Chemistry*. https://doi.org/10.1016/j.trac.2017.09.007

Ossmann, B. E., Sarau, G., Holtmannspötter, H., Pischetsrieder, M., Christiansen, S. H., & Dicke, W. (2018). Small-sized microplastics and pigmented particles in bottled mineral water. *Water Research, 141*, 307–316. https://doi.org/10.1016/j.watres.2018.05.027

Papaioannou, D., Koukoulakis, P. H., Papageorgiou, M., Lambropoulou, D. A., & Kalavrouziotis, I. K. (2020). Investigation of pharmaceutical and personal care product interactions of soil and beets (Beta vulgaris L.) under the effect of wastewater reuse. *Chemosphere* 238. https://doi.org/10.1016/j.chemosphere.2019.124553

Pashaei, R., Loiselle, S. A., Leone, G., Tamasi, G., Dzingelevičienė, R., Kowalkowski, T., Gholizadeh, M., Consumi, M., Abbasi, S., Sabaliauskaitė, V., & Buszewski, B. (2021). Determination of nano and microplastic particles in hypersaline lakes by multiple methods. *Environmental Monitoring and Assessment, 193*. https://doi.org/10.1007/s10661-021-09470-8

Patterson, J., Jeyasanta, K. I., Sathish, N., Edward, J. K. P., & Booth, A. M. (2020). Microplastic and heavy metal distributions in an Indian coral reef ecosystem. *Science of the Total Environment, 744*. https://doi.org/10.1016/j.scitotenv.2020.140706

Peez, N., Becker, J., Ehlers, S. M., Fritz, M., Fischer, C. B., Koop, J. H. E., Winkelmann, C., & Imhof, W. (2019a). Quantitative analysis of PET microplastics in environmental model samples using quantitative 1H-NMR spectroscopy: Validation of an optimized and consistent sample clean-up method. *Analytical and Bioanalytical Chemistry, 411*, 7409–7418. https://doi.org/10.1007/s00216-019-02089-2

Peez, N., Janiska, M. C., & Imhof, W. (2019b). The first application of quantitative 1H NMR spectroscopy as a simple and fast method of identification and quantification of microplastic particles (PE, PET, and PS). *Analytical and Bioanalytical Chemistry, 411*, 823–833. https://doi.org/10.1007/s00216-018-1510-z

Prabhakar, A., Mishra, S., & Das, A. P. (2019). Isolation and identification of lead (Pb) solubilizing bacteria from automobile waste and its potential for recovery of lead from end of life waste batteries. *Geomicrobiological Journal, 36*, 894–903. https://doi.org/10.1080/01490451.2019.1654044

Primpke, S., Godejohann, M., & Gerdts, G. (2020). Rapid identification and quantification of microplastics in the environment by quantum cascade laser-based hyperspectral infrared chemical imaging. *Environmental Science and Technology, 54*, 15893–15903. https://doi.org/10.1021/acs.est.0c05722

Priyadarsini, S., & Das, A. P. (2024). Analytical and structural characterization of waste lithium-ion batteries for their effective recycling strategy. *Environmental Chemistry and Ecotoxicology*. https://doi.org/10.1016/j.enceco.2024.12.004

Raj, K., & Das, A. P. (2023). Lead pollution: Impact on environment and human health and approach for a sustainable solution. *Environmental Chemistry and Ecotoxicology, 5*, 79–85. https://doi.org/10.1016/J.ENCECO.2023.02.001

Reineccius, J., Appelt, J. S., Hinrichs, T., Kaiser, D., Stern, J., Prien, R. D., & Waniek, J. J. (2020). Abundance and characteristics of microfibers detected in sediment trap material from the deep subtropical North Atlantic Ocean. *Science of the Total Environment, 738*. https://doi.org/10.1016/j.scitotenv.2020.140354

Ren, L., Liu, S., Huang, S., Wang, Q., Lu, Y., Song, J., & Guo, J. (2023). Identification of microplastics using a convolutional neural network based on micro-Raman spectroscopy. *Talanta, 260*. https://doi.org/10.1016/j.talanta.2023.124611

Ríos, M. F., Hernández-Moresino, R. D., & Galván, D. E. (2020). Assessing urban microplastic pollution in a benthic habitat of Patagonia Argentina. *Marine Pollution Bulletin, 159*. https://doi.org/10.1016/j.marpolbul.2020.111491

Riva, V., Riva, F., Vergani, L., Crotti, E., Borin, S., & Mapelli, F. (2020). Microbial assisted phytodepuration for water reclamation: Environmental benefits and threats. *Chemosphere, 241*. https://doi.org/10.1016/j.chemosphere.2019.124843

Rosen, R. (2007). Mass spectrometry for monitoring micropollutants in water. *Current Opinion in Biotechnology*. https://doi.org/10.1016/j.copbio.2007.03.005

Sahu, S., & Das, A. P. (2024). Microplastics degradation mechanism by aquatic plastiomes: A review. In *Impact of COVID-19 waste on environmental pollution and its sustainable management: COVID-19 waste and its management* (pp. 213–226). https://doi.org/10.1007/978-3-031-50840-0_12

Salas, D., Borrull, F., Fontanals, N., & Marcé, R. M. (2017). Hydrophilic interaction liquid chromatography coupled to mass spectrometry-based detection to determine emerging organic

contaminants in environmental samples. *TrAC—Trends in Analytical Chemistry*. https://doi.org/10.1016/j.trac.2017.07.017
Sancataldo, G., Avellone, G., & Vetri, V. (2020). Nile Red lifetime reveals microplastic identity. *Environmental Science: Processes & Impacts, 22*, 2266–2275. https://doi.org/10.1039/d0em00348d
Schwaferts, C., Sogne, V., Welz, R., Meier, F., Klein, T., Niessner, R., Elsner, M., & Ivleva, N. P. (2020). Nanoplastic analysis by online coupling of raman microscopy and field-flow fractionation enabled by optical tweezers. *Analytical Chemistry, 92*, 5813–5820. https://doi.org/10.1021/acs.analchem.9b05336
Scircle, A., Cizdziel, J. V., Missling, K., Li, L., & Vianello, A. (2020). Single-pot method for the collection and preparation of natural water for microplastic analyses: Microplastics in the Mississippi River system during and after historic flooding. *Environmental Toxicology and Chemistry, 39*, 986–995. https://doi.org/10.1002/etc.4698
Sekudewicz, I., Dąbrowska, A. M., & Syczewski, M. D. (2021). Microplastic pollution in surface water and sediments in the urban section of the Vistula River (Poland). *Science of the Total Environment, 762*. https://doi.org/10.1016/j.scitotenv.2020.143111
Siddiqui, M. R., AlOthman, Z. A., & Rahman, N. (2017). Analytical techniques in pharmaceutical analysis: A review. *Arabian Journal of Chemistry*. https://doi.org/10.1016/j.arabjc.2013.04.016
Singh, R. P., Mishra, S., & Das, A. P. (2020). Synthetic microfibers: Pollution toxicity and remediation. *Chemosphere, 257*, 127199. https://doi.org/10.1016/j.chemosphere.2020.127199
Sridhar, A., Kannan, D., Kapoor, A., & Prabhakar, S. (2022). Extraction and detection methods of microplastics in food and marine systems: A critical review. *Chemosphere, 286*. https://doi.org/10.1016/j.chemosphere.2021.131653
Strungaru, S. A., Jijie, R., Nicoara, M., Plavan, G., & Faggio, C. (2019). Micro- (nano) plastics in freshwater ecosystems: Abundance, toxicological impact and quantification methodology. *TrAC—Trends in Analytical Chemistry*. https://doi.org/10.1016/j.trac.2018.10.025
Thomaidi, V. S., Stasinakis, A. S., Borova, V. L., & Thomaidis, N. S. (2015). Is there a risk for the aquatic environment due to the existence of emerging organic contaminants in treated domestic wastewater? Greece as a case-study. *Journal of Hazardous Materials, 283*, 740–747. https://doi.org/10.1016/j.jhazmat.2014.10.023
Tian, L., Skoczynska, E., Siddhanti, D., van Putten, R. J., Leslie, H. A., & Gruter, G. J. M. (2022). Quantification of polyethylene terephthalate microplastics and nanoplastics in sands, indoor dust and sludge using a simplified in-matrix depolymerization method. *Marine Pollution Bulletin, 175*. https://doi.org/10.1016/j.marpolbul.2022.113403
Tiwari, M., Rathod, T. D., Ajmal, P. Y., Bhangare, R. C., & Sahu, S. K. (2019). Distribution and characterization of microplastics in beach sand from three different Indian coastal environments. *Marine Pollution Bulletin, 140*, 262–273. https://doi.org/10.1016/j.marpolbul.2019.01.055
Tuan Vo-Dinh, Hiromoto, M. Y. K., Begun, G. M., & Moody, R. L. (1984). Surface-enhanced Raman spectrometry for trace organic analysis. *Analytical Chemistry*.
Valhondo, C., Martínez-Landa, L., Carrera, J., Díaz-Cruz, S. M., Amalfitano, S., & Levantesi, C. (2020). Six artificial recharge pilot replicates to gain insight into water quality enhancement processes. *Chemosphere, 240*. https://doi.org/10.1016/j.chemosphere.2019.124826
Veerasingam, S., Ranjani, M., Venkatachalapathy, R., Bagaev, A., Mukhanov, V., Litvinyuk, D., Mugilarasan, M., Gurumoorthi, K., Guganathan, L., Aboobacker, V. M., & Vethamony, P. (2021). Contributions of Fourier transform infrared spectroscopy in microplastic pollution research: A review. *Critical Reviews in Environment Science and Technology, 51*, 2681–2743. https://doi.org/10.1080/10643389.2020.1807450
Von Moos, N., Burkhardt-Holm, P., & Köhler, A. (2012). Uptake and effects of microplastics on cells and tissue of the blue mussel Mytilus edulis L. after an experimental exposure. *Environmental Science and Technology, 46*, 11327–11335. https://doi.org/10.1021/es302332w
Wang, L., Peng, Y., Xu, Y., Zhang, J., Liu, C., Tang, X., Lu, Y., & Sun, H. (2022). Earthworms' degradable bioplastic diet of polylactic acid: Easy to break down and slow to excrete.

Environmental Science and Technology, 56, 5020–5028. https://doi.org/10.1021/acs.est.1c08066

Wang, L., Zhang, J., Hou, S., & Sun, H. (2017). A simple method for quantifying polycarbonate and polyethylene terephthalate microplastics in environmental samples by liquid chromatography-tandem mass spectrometry. *Environmental Science & Technology Letters, 4*, 530–534. https://doi.org/10.1021/acs.estlett.7b00454

Welker, M., Fastner, J., Erhard, M., & Von Döhren, H. (2002). Applications of MALDI-TOF MS analysis in cyanotoxin research. *Environmental Toxicology, 17*, 367–374. https://doi.org/10.1002/tox.10073

Wittmer, A., Heisele, A., McArdell, C. S., Böhler, M., Longree, P., & Siegrist, H. (2015). Decreased UV absorbance as an indicator of micropollutant removal efficiency in wastewater treated with ozone. *Water Science and Technology, 71*, 980–985. https://doi.org/10.2166/wst.2015.053

Wu, P., Wu, X., Huang, Q., Yu, Q., Jin, H., & Zhu, M. (2023). Mass spectrometry-based multimodal approaches for the identification and quantification analysis of microplastics in food matrix. *Frontiers in Nutrition*. https://doi.org/10.3389/fnut.2023.1163823

Xu, J. L., Thomas, K. V., Luo, Z., & Gowen, A. A. (2019). FTIR and Raman imaging for microplastics analysis: State of the art, challenges and prospects. *TrAC—Trends in Analytical Chemistry*. https://doi.org/10.1016/j.trac.2019.115629

Yu, X., Peng, J., Wang, J., Wang, K., & Bao, S. (2016). Occurrence of microplastics in the beach sand of the Chinese inner sea: The Bohai Sea. *Environmental Pollution, 214*, 722–730. https://doi.org/10.1016/j.envpol.2016.04.080

Zhang, S., Zhang, Q., Darisaw, S., Ehie, O., & Wang, G. (2007). Simultaneous quantification of polycyclic aromatic hydrocarbons (PAHs), polychlorinated biphenyls (PCBs), and pharmaceuticals and personal care products (PPCPs) in Mississippi river water, in New Orleans, Louisiana, USA. *Chemosphere, 66*, 1057–1069. https://doi.org/10.1016/j.chemosphere.2006.06.067

Zhao, X., Liu, Z., Ren, X., & Duan, X. (2021). Parental transfer of nanopolystyrene-enhanced tris(1,3-dichloro-2-propyl) phosphate induces transgenerational thyroid disruption in zebrafish. *Aquatic Toxicology, 236*. https://doi.org/10.1016/j.aquatox.2021.105871

Chapter 16
Advanced Methodological Framework for Microplastic Analysis in Wastewater: Sample Preparation and Detection Techniques

Gaurav Bhardwaj and Lachi Wankhede

Introduction

The ubiquitous presence of microplastics (MPs) in environmental matrices has emerged as one of the most pressing environmental challenges of the twenty-first century. These synthetic polymer particles, typically defined as plastic debris less than 5 mm in size (Thompson et al., 2004) have been detected in virtually every environmental compartment, from the depths of ocean trenches to remote mountain peaks (Ghosh et al., 2023; Haque & Fan, 2023; Mishra & Das, 2021; Mohanty et al., 2024). The increasing global plastic production, which surpassed 400 million metric tons in 2021, coupled with inadequate waste management practices, has led to the continuous release of these persistent pollutants into the environment.

MPs are traditionally categorized into primary and secondary sources. Primary MPs are manufactured at microscopic sizes for specific applications, such as microbeads in personal care products and industrial abrasives. Secondary MPs, which constitute most environmental MPs, result from the fragmentation of larger plastic items through various degradation processes, including photo-oxidation, mechanical abrasion, and biological degradation (Nayebi et al., 2023).

The environmental implications of microplastic pollution are far-reaching and multifaceted. These particles serve as vectors for organic pollutants and heavy metals, potentially facilitating transport through different ecological compartments. Moreover, their small size makes them readily available for ingestion by a wide range of organisms, leading to physical and chemical toxicity (Wang et al., 2019). Recent studies have documented the presence of MPs in human blood (Leslie et al., 2022)

G. Bhardwaj (✉) · L. Wankhede
Department of Civil Engineering, Lassonde School of Engineering, York University, North York, Toronto, ON, Canada
e-mail: gaurav02@yorku.ca

I. D. Behera et al. (eds.), *Emerging Micropollutants*, Sustainable Environmental Waste Management Strategies, https://doi.org/10.1007/978-3-032-03530-1_16

and lung tissue (Amato-Lourenço et al., 2021), raising significant concerns about potential human health impacts.

The growing scientific evidence of microplastic contamination has catalyzed research efforts to understand their sources, fate, and transport in various environmental matrices. Among these, wastewater treatment plants (WWTPs) have been identified as critical hubs in the microplastic pollution cycle, acting both as sinks and potential point sources of microplastic pollution to receiving water bodies (Bhardwaj et al., 2025a; Das et al., 2024; Mishra & Das, 2023; Mishra et al., 2024; Ray et al., 2024). Understanding the dynamics of MPs in WWTPs is crucial for developing effective mitigation strategies and protecting aquatic ecosystems (Mishra et al., 2022a, 2022b).

WWTPs function as crucial interfaces between urban wastewater streams and natural water bodies, processing millions of cubic meters of wastewater daily (Ngo et al., 2019). Current estimates indicate that a single WWTP receives between 1×10^6 to 1×10^8 MP particles per day (Murphy et al., 2016), making these facilities critical control points for MP pollution management. The monitoring of MPs in WWTPs is essential for multiple reasons. First, WWTPs represent the final barrier preventing MP release into aquatic environments. Second, monitoring data enables the optimization of treatment processes and the evaluation of removal efficiencies. Third, it provides crucial information for regulatory compliance and policy development (Iyare et al., 2020).

Despite conventional WWTPs demonstrating removal efficiencies of 85–99%, the remaining particles constitute a substantial environmental burden due to the continuous nature of the effluent discharge (Talvitie et al., 2017a). Studies have shown that a medium-sized WWTP serving 650,000 inhabitants can release up to 8.2 billion MP particles annually into receiving waters (Magni et al., 2019). Additionally, the fate of removed MPs, primarily concentrated in sewage sludge and subsequently applied to agricultural lands as biosolids, necessitates thorough monitoring and characterization (Li et al., 2020).

The accurate monitoring and quantification of MPs in WWTPs face multiple analytical challenges, primarily due to the complex nature of wastewater and sludge matrices (Mishra et al., 2024). Despite numerous studies reporting various pretreatment and detection methods, a comprehensive evaluation of their effectiveness, limitations, and applicability remains fragmented in the literature (Löder & Gerdts, 2015). Recent reviews have highlighted significant discrepancies in reported microplastic concentrations and removal efficiencies, largely attributed to the lack of standardized protocols and varying analytical approaches (Mathew et al., 2024b).

A critical analysis of the current literature reveals several key knowledge gaps. First, the effects of different pretreatment methods on polymer integrity and recovery rates are inadequately documented. Second, the selection criteria for appropriate analytical techniques based on specific WWTP conditions and monitoring objectives remain unclear. Third, there is a limited systematic comparison of emerging technologies with conventional methods in terms of their practical applicability and cost-effectiveness (Primpke et al., 2020). These gaps, coupled with the absence of standardized protocols, hinder the development of reliable monitoring strategies and compromise data comparability across different studies (Frias & Nash, 2019).

Building upon these identified knowledge gaps, this chapter aims to provide a comprehensive and critical evaluation of current pretreatment methods and detection technologies for monitoring MPs in WWTPs. The scope encompasses a systematic analysis of existing pretreatment protocols and state-of-the-art detection technologies, with particular emphasis on their practical applicability in complex WWTP matrices.

This chapter synthesizes current knowledge while providing practical insights for method selection based on specific monitoring requirements and available resources. By addressing critical aspects of method standardization and validation, this work aims to contribute to the development of more reliable and comparable monitoring strategies. Special attention is given to emerging technologies and their potential to overcome current analytical limitations, thereby supporting future developments in this rapidly evolving field.

Matrix Complexity and Analytical Challenges

WWTPs serve as critical hubs for MP interception in urban water cycles. Understanding the distribution and characteristics of MPs throughout treatment processes is fundamental for developing effective monitoring strategies. The complex journey of MPs through WWTPs, from various sources to their fate, presents unique analytical challenges at each stage.

The characteristics and behaviour of MPs vary significantly across different WWTP matrices, necessitating matrix-specific analytical approaches. The complexity of each matrix, combined with varying MP concentrations and properties, creates distinct challenges for detection and quantification (Table 16.1).

As illustrated in Table 16.1, MP concentrations demonstrate a decreasing trend through the treatment train, from 1 to 10,000 particles/L in raw wastewater to 8–120 particles/L in final effluent (Ben-David et al., 2021; Talvitie et al., 2017a). However, this reduction in the aqueous phase is accompanied by MP accumulation in sludge streams, with concentrations reaching 5000–50,000 particles/kg dry weight in dewatered sludge (Ziajahromi et al., 2016). This distribution pattern significantly influences sampling strategies and analytical method selection.

Matrix complexity presents varying challenges across treatment stages. Raw wastewater contains high levels of interfering substances, including organic matter, cellulose fibres, and suspended solids (Carr et al., 2016; Mason et al., 2016). These matrix effects gradually change through the treatment process, with biological flocs and extracellular polymeric substances becoming prominent in activated sludge systems (Dyachenko et al., 2017), while final effluent presents challenges related to low MP concentrations requiring large sample volumes (Ben-David et al., 2021).

The physical and chemical characteristics of MPs also evolve during treatment. Size distributions typically shift from a broad range (20–5000 μm) in the influent

Table 16.1 Characteristics and challenges of MPs analysis across different WWTP matrices

Wastewater matrix	MP characteristics	Matrix complexity	Analytical challenges
Raw/Influent wastewater	• Concentration: 1–10,000 particles/L (Sun et al., 2019) • Size range: 20–5000 μm (Talvitie et al., 2017a) • Dominant polymers: PE (35%), PP (25%), PET (20%) (Murphy et al., 2016) • Morphology: Fibers (65%), fragments (20%), films (10%) (Lares et al., 2018)	• TSS: 150–400 mg/L (Ngo et al., 2019) • COD: 250–800 mg/L (Carr et al., 2016) • Cellulose fibres: 20–30% of TSS (Carr et al., 2016) • Fats, oils, grease: 50–150 mg/L (Mason et al., 2016)	• Visual interference from natural fibres • Particle aggregation with organic matter • High matrix complexity affecting separation
Primary sludge	• Concentration: 2000–20,000 particles/kg dry weight (Li et al., 2020) • Dominant size: 100–5000 μm (Mahon et al., 2017) • Higher proportion of dense polymers: PVC, PET (Mintenig et al., 2017)	• Total solids: 2–8% (Gies et al., 2018) • Organic content: 60–80% of TS (Gies et al., 2018) • High grit content	• Dense matrix impeding separation • High organic content interfering with identification • Heterogeneous particle distribution
Primary effluent	• Concentration: 100–3000 particles/L (Liu et al., 2019b) • Size range: 50–2000 μm • Dominant polymers: PE, PP, PS (Simon et al., 2018)	• TSS: 80–120 mg/L (Wolff et al., 2019) • COD: 150–300 mg/L (Wolff et al., 2019) • Colloidal particles present	• High organic content • Variable particle distribution • Presence of settling particles
Activated sludge mixed liquor	• Concentration: 1000–15,000 particles/L (Lv et al., 2019) • Predominant size: 50–300 μm (Blair et al., 2019) • High proportion of PE/PP (Ziajahromi et al., 2016)	• MLSS: 2000–4000 mg/L (Dyachenko et al., 2017) • High bacterial biomass • EPS content: 100–500 mg/L (Zhang et al., 2015)	• Biofilm formation on MPs • Complex biological • Particle entrapment in flocs

(continued)

Table 16.1 (continued)

Wastewater matrix	MP characteristics	Matrix complexity	Analytical challenges
Return activated sludge	• Concentration: 2000–25,000 particles/L (Edo et al., 2020) • Size range: 50–500 μm (Li et al., 2020) • High proportion of dense polymers	• MLSS: 4000–8000 mg/L • High biomass content • Dense floc structure	• Strong biofilm attachment • High solids content • Complex biological matrix
Secondary effluent	• Concentration: 20–400 particles/L (Edo et al., 2020) • Size range: 20–300 μm (Ben-David et al., 2021) • Primarily fibres and fragments	• TSS: 10–30 mg/L • Residual biomass • Colloidal organics	• Low MP concentrations • Interference from biomass • Need for larger sample volumes
Final effluent	• Concentration: 8–120 particles/L (Ben-David et al., 2021) • Dominant size: 20–150 μm (Hidayaturrahman & Lee, 2019) • Primarily fibers (> 70%) • Light polymers: PE, PP (Yang et al., 2019)	• TSS: 5–15 mg/L • Low turbidity (< 2 NTU) • Residual dissolved organics	• Very low MP concentrations • Need for large sample volumes • Small particle size requiring fine filtration
Digested sludge	• Concentration: 4000–40,000 particles/kg dry weight (Casella et al., 2023) • Altered surface properties • Mixed polymer composition (Xu et al., 2019)	• TS: 2–4% (Lares et al., 2018) • High sulphide content • Complex organic matrix	• Chemical interference • Modified MP properties • Matrix-particle interactions
Dewatered sludge	• Concentration: 5000–50,000 particles/kg dry weight (Ziajahromi et al., 2016) • Wide size distribution • Mixed polymer types	• Total solids: 15–30% (Tang et al., 2021) • Dense organic matrix • Chemical additives present	• Highly compressed matrix • Complex chemical composition

TSS total suspended solids; *COD* chemical oxygen demand; *MLSS* mixed liquor suspended solids; *EPS* extracellular polymeric substances; *TS* total solids to predominantly smaller particles (20–150 μm) in the effluent (Hidayaturrahman & Lee, 2019; Talvitie et al., 2017a). Polymer composition varies between matrices, with dense polymers concentrating in sludge streams and lighter polymers dominating effluent. Surface modifications and biofilm formation during biological treatment further complicate analytical procedures (Bhardwaj et al., 2025b; Das & Bal, 2025; Das et al., 2023).

These matrix-specific characteristics and challenges necessitate carefully tailored analytical approaches for MP monitoring. The selection of appropriate pretreatment methods and detection techniques must account for both matrix complexity and MP properties at each sampling point, as will be discussed in subsequent sections.

Sample Collection, Preservation, and Quality Control in Wastewater MP Monitoring

The reliability and accuracy of microplastic analysis in wastewater treatment systems fundamentally depend on systematic sampling protocols, appropriate preservation methods, and robust quality control measures. The heterogeneous distribution of MPs, varying flow conditions, and complex matrix effects throughout treatment processes necessitate carefully designed sampling strategies (Simon et al., 2018). Studies have demonstrated that sampling uncertainties can contribute up to 30% of the total analytical error in microplastic quantification, emphasizing the critical nature of proper sample collection and handling procedures (Bretas Alvim et al., 2020b).

This section addresses three crucial aspects of the initial analytical workflow: sampling methodology (grab versus composite sampling), sample preservation and storage protocols, and quality assurance measures. Understanding these fundamental steps is essential for obtaining representative samples and ensuring data reliability in microplastic monitoring programs. Particular attention is given to matrix-specific considerations and contamination prevention strategies that form the foundation for subsequent analysis steps (Ziajahromi et al., 2017).

Sampling Methods

Sampling methodology plays a fundamental role in capturing the diverse range of MPs present in wastewater treatment systems. The heterogeneous nature of MPs, varying in size, shape, and buoyancy, necessitates careful consideration of sampling approaches to ensure representative sample collection (Talvitie et al., 2017b). The complexity of wastewater matrices and the distribution of MPs through different treatment stages further emphasize the importance of appropriate sampling method selection.

Grab Sampling

Grab sampling represents the most straightforward approach, involving direct sample collection from specific points within the treatment process. While this method offers simplicity and minimal equipment requirements, it provides only a snapshot of MP concentrations at a given time point. Studies have demonstrated that grab sampling may be particularly suitable for process optimization studies and preliminary assessments, though it may not capture temporal variations effectively (Leslie et al., 2017; Murphy et al., 2016).

Composite Sampling

Composite sampling has emerged as a more comprehensive approach, especially for understanding microplastic dynamics across treatment processes. This method involves systematic sample collection over extended periods, typically 24 h, using automated sampling equipment. Flow-proportional composite sampling, where sample volumes are adjusted based on wastewater flow rates, offers an improved representation of actual microplastic loads. However, consideration must be given to potential particle settling during collection and the need for specialized equipment (Dyachenko et al., 2017; Sutton et al., 2016).

Advanced Filtration-Based Approaches

- **Cascade filtration**: Cascade filtration systems employ sequential size fractionation using stainless steel filters of decreasing mesh sizes (typically 100, 50, and 10 μm). This approach has demonstrated high recovery rates, with studies reporting up to 86% microplastic retention while minimizing cross-contamination (Bannick et al., 2019). Sequential size separation during sampling offers advantages for subsequent analysis, particularly when size-specific distribution data is required.
- **Custom sampling**: Custom sampling systems have been developed for specific matrices, particularly treated effluent. These typically incorporate membrane pumps, mobile pumping devices, and stainless-steel cartridge filters with integrated flow meters (Mintenig et al., 2017). For instance, studies utilizing custom sampling setups with 10 μm cartridge filters have successfully captured MPs while maintaining operational flexibility. However, the effectiveness of custom sampling approaches varies significantly with system design and operational parameters, highlighting the need for careful validation and standardization.

Matrix characteristics significantly influence sampling method selection and implementation. For instance, primary treatment stages require consideration of higher solid content and particle stratification, while final effluent sampling often demands larger volumes to ensure adequate microplastic capture. Flow rates during sampling must be carefully controlled, typically ranging from 2 to 20 L/min, depending on total solids content and matrix complexity (Long et al., 2019).

Sample Handling and Storage

Proper sample handling and storage are critical preliminary steps in microplastic analysis. Glass or stainless-steel containers are preferred over plastic to prevent cross-contamination, with container size selected based on matrix characteristics and required sample volumes. For wastewater samples, amber glass bottles effectively prevent photodegradation while minimizing contamination risks. Temperature control is the primary preservation method, with samples typically stored at 4 °C to inhibit biological degradation (Liu et al., 2019a). For matrices with high organic content, such as primary sludge and activated sludge, chemical preservation using sodium azide (0.02%) may be necessary for longer storage periods (Hurley et al., 2018).

Storage duration significantly influences sample integrity and should be minimized whenever possible. Studies have demonstrated that prolonged storage (> 30 days) can affect microplastic characteristics, particularly in samples with high organic content (Stock et al., 2019). When immediate analysis is not feasible, maximum storage times should be established based on matrix type, with more frequent monitoring required for complex matrices such as primary sludge and activated sludge samples.

Quality Control Measures

Quality control measures are essential for ensuring data reliability in the MP analysis of wastewater samples. Contamination from airborne synthetic fibres, sampling equipment, and laboratory materials presents a significant challenge, necessitating strict protocols throughout the analytical process. Key preventive measures include using cotton laboratory coats, conducting work in clean environments (preferably laminar flow cabinets), and maintaining detailed procedural blanks. Studies have shown that airborne contamination can contribute up to 30% of detected MPs, particularly in the size range of 100–500 μm (Ziajahromi et al., 2016).

Comprehensive quality assurance protocols should include procedural blanks, recovery tests using synthetic reference materials, and regular equipment cleaning validation. Procedural blanks, processed identically to samples, help quantify background contamination levels, while recovery tests using specific polymer types (typically PE, PP, and PET) in relevant size ranges assess method efficiency. For wastewater matrices, recovery rates of 80–120% are generally considered acceptable, though matrix complexity can significantly influence these values (Talvitie et al.,

2017b). Regular method validation using certified reference materials, when available, provides additional confidence in analytical results and enables inter-laboratory comparison of data.

Pretreatment Methods for MP Analysis in Wastewater

The complex nature of wastewater matrices necessitates effective pretreatment protocols for accurate microplastic analysis. The presence of high organic matter content, varying particle sizes, and diverse chemical constituents in wastewater can significantly interfere with microplastic detection and characterization (Mathew et al., 2024a; Tagg et al., 2016). Pretreatment serves three primary functions: removal of interfering organic matter, isolation of MPs from inorganic constituents, and sample preparation for subsequent analytical techniques (Mintenig et al., 2017). The selection of appropriate pretreatment methods is crucial as it directly influences both recovery rates and the integrity of microplastic particles. A systematic approach typically involves organic matter digestion, density-based separation, and filtration, as shown in Fig. 16.1, with each step requiring careful optimization to maintain polymer integrity while ensuring effective matrix removal (Löder et al., 2017).

Digestion

The isolation of MPs from complex wastewater matrices requires effective digestion protocols to remove interfering substances while maintaining polymer integrity. As summarized in Table 16.2, four primary digestion approaches have been developed and optimized for wastewater applications, each offering distinct advantages and limitations for specific matrices and target polymers.

Wet Peroxide Oxidation (WPO)

WPO, particularly using Fenton's reagent, has emerged as the most widely adopted digestion method due to its rapid organic matter removal and broad applicability across different matrices. The combination of hydrogen peroxide (30% H_2O_2) with Fe (II) catalyst generates highly reactive hydroxyl radicals that effectively oxidize organic matter (Tagg et al., 2016). As shown in Table 16.2, WPO demonstrates consistently high recovery rates (75–98%) across a wide range of particle sizes (45 μm–5 mm).

The primary advantages of WPO include rapid processing time (30 min to 2 h), effectiveness across various wastewater matrices, and compatibility with most common polymer types. However, several operational considerations must be addressed. Temperature control is crucial, as the exothermic reaction can generate

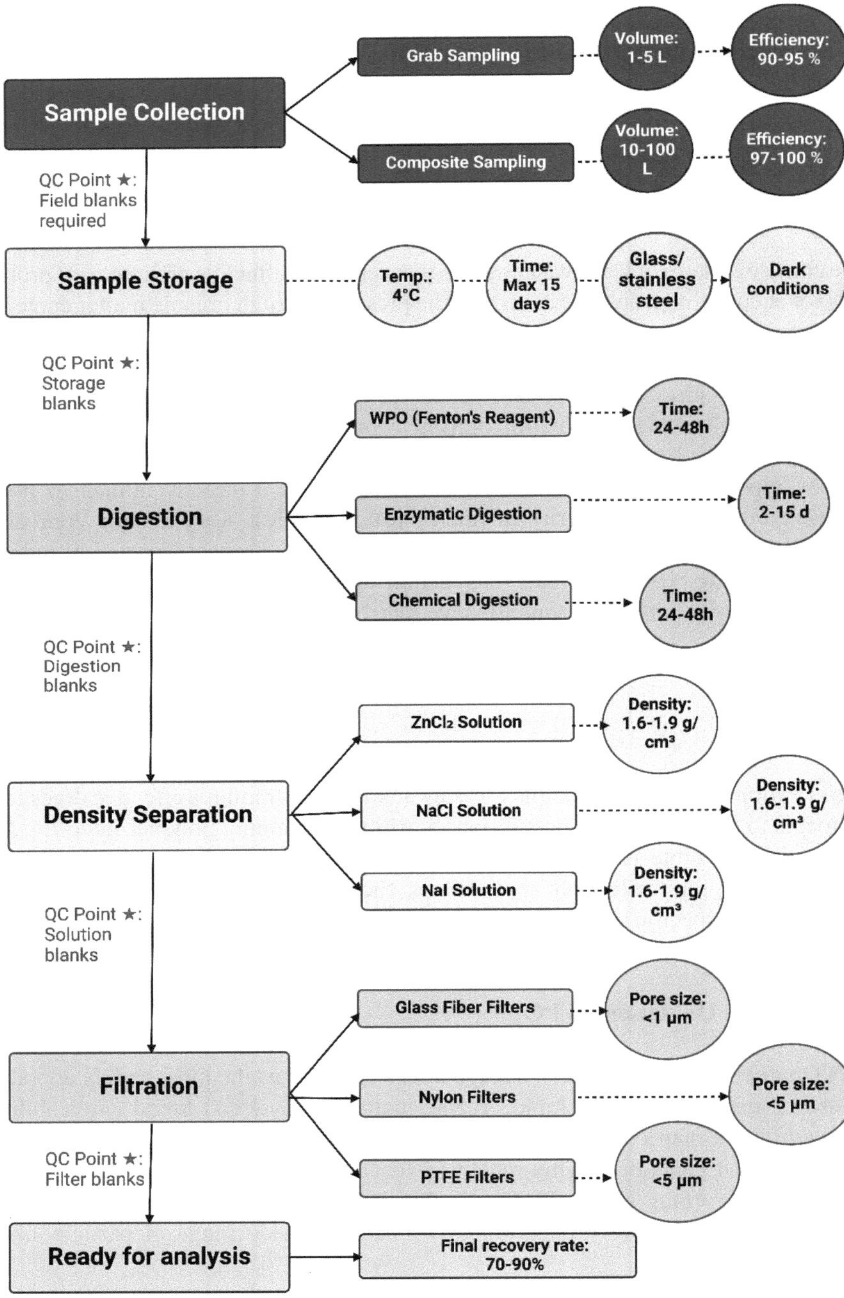

Fig. 16.1 Comprehensive workflow for microplastic sample processing in wastewater matrices: a systematic protocol from collection to analysis

Table 16.2 Sample digestion methods for microplastic analysis in wastewater matrices

Digestion method	Operating conditions	Target polymers	Matrix applicability	Recovery range (%)	Size range	Key studies
Wet peroxide oxidation (WPO)	• H_2O_2 (30%) + Fe (II) • Temp: 45–75 °C • Time: 30 min to 2 h	PE, PP, PS, PET, PVC	• Raw wastewater • Secondary effluent • Sludge	75–98	45 μm–5 mm	Dyachenko et al. (2017), Franco et al. (2023), Gündoğdu et al. (2018), Martín-García et al. (2023), Ruffell et al. (2023)
Enzymatic treatment	• Multiple enzymes • Temp: 40–50 °C • Time: 48 h to 6 d	All common polymers	• Effluent • Primary sludge	80–99	20 μm–7 mm	Ding et al. (2020), Mintenig et al. (2017), Wang et al. (2023)
Alkaline digestion	• KOH/ NaOH • Temp: 60 °C • Time: 24 h	PE, PP, PS	• Sludge • Influent	50–100	10 μm–5 mm	Gao et al. (2023), Mintenig et al. (2017), Zhang et al. ()
Acid treatment	• HNO_3/ H_2SO_4 • Temp: 60 °C • Time: 10 min–10 h	Limited polymer compatibility	• Effluent • Sludge • Influent	50–76	< 1 mm	Castelluccio et al. (2022), Zhang et al. (2023)

excessive heat that may affect thermally sensitive polymers. Studies have shown that maintaining temperatures below 60 °C through controlled reagent addition and external cooling significantly improves polymer preservation (Dyachenko et al., 2017).

Enzymatic Digestion

Enzymatic treatment offers the most selective approach for organic matter removal, utilizing specific enzymes to target different matrix components. While Table 16.2 indicates excellent recovery rates (80–99%) and broad polymer compatibility, the method presents distinct operational challenges. The primary advantage lies in its gentle nature, preserving even sensitive polymers while effectively removing biological materials. The combination of multiple enzymes (proteases, lipases, cellulases) enables comprehensive matrix digestion (Ding et al., 2020).

However, enzymatic digestion faces practical limitations including extended processing times (48 h to 6 days), higher operational costs, and specific temperature and pH requirements for optimal enzyme activity. Recent developments incorporating surfactants (SDS) have shown promise in reducing processing times, though cost remains a significant consideration (Löder et al., 2017).

Chemical Digestion

Alkaline and acid digestion methods represent more aggressive approaches to matrix removal. Alkaline digestion, typically employing KOH or NaOH, shows variable recovery rates (50–100%) depending on polymer type and matrix characteristics. Its advantages include low cost, simple operation, and effectiveness for certain matrices. However, strong alkaline conditions can degrade or alter specific polymers, particularly polyamides and polyesters (Gao et al., 2023).

Acid digestion, while effective for certain applications, demonstrates the most significant limitations regarding polymer compatibility. The harsh conditions can compromise polymer integrity, leading to lower recovery rates (50–76%) and restricted size range analysis (< 1 mm). Temperature control during acid digestion is particularly critical, as elevated temperatures can accelerate polymer degradation (Hurley et al., 2018).

The selection of appropriate digestion methods depends on multiple factors including matrix complexity, target polymer types, and analytical requirements. While WPO offers the best balance of efficiency and polymer preservation for routine analysis, specific applications may benefit from alternative approaches. However, digestion alone is often insufficient for complete microplastic isolation, particularly in complex wastewater matrices. The presence of inorganic materials and varying polymer densities necessitates additional separation steps, leading to the widespread adoption of density-based separation techniques as a complementary approach.

Density Separation

Density-based separation represents a crucial step in microplastic isolation from wastewater matrices, particularly following digestion procedures. This technique exploits the differential density between MPs and remaining matrix components, enabling both separation and quantification. The fundamental principle relies on the flotation of particles with densities lower than the separation solution while denser materials settle (Sun et al., 2019).

The selection of appropriate separation solutions significantly influences method effectiveness, as detailed in Table 16.3. Zinc chloride ($ZnCl_2$) has emerged as the preferred solution in wastewater applications, offering high recovery rates (80–95%) across a broad range of polymer types. Its high-density range (1.6–1.9 g/cm^3) enables separation of most common polymers, though ultra-high density materials like PTFE

(2.1–2.2 g/cm^3) remain challenging to isolate (Wolff et al., 2019). While the initial cost of $ZnCl_2$ is considerable, its reusability makes it cost-effective for routine analysis.

The selection of separation solutions requires careful consideration of specific application requirements. While zinc chloride dominates current practices, alternative solutions offer distinct advantages. Sodium chloride (NaCl), with its density of 1.2 g/cm^3, presents an environmentally friendly and economical option, particularly suitable for low-density polymers like PE and PP. However, its limited density range restricts the recovery of denser polymers such as PET and PVC, making it less suitable for comprehensive wastewater analysis (Akyildiz et al., 2022). Sodium iodide (NaI) emerges as a middle-ground option, offering density ranges comparable to $ZnCl_2$ with reduced corrosivity, though its widespread adoption is hindered by high costs and light sensitivity concerns.

Successful implementation of density separation extends beyond solution selection to encompass various operational parameters. Optimal separation requires precise control of temperature (20–25 °C), adequate settling time (24–48 h), and appropriate solution-to-sample ratios (3:1) (Simon et al., 2018). These parameters become particularly critical when dealing with weathered MPs, which often display modified density characteristics due to environmental alterations, additive incorporation, and pollutant interactions.

Wastewater matrices present unique challenges for density separation due to complex interactions between MPs and co-pollutants. Pore-filling phenomena and organic matter aggregation can alter particle densities, often requiring multiple separation cycles for effective isolation. The separated MPs subsequently undergo filtration for size fractionation and final purification, a critical step that bridges separation and analytical procedures. The success of this final stage relies heavily on optimized filtration parameters, including appropriate filter materials and pore sizes, which will be discussed in detail in the following section.

Filtration

Filtration represents the final critical step in microplastic isolation from wastewater matrices, serving both as a concentration technique and a size-fractionation method. The process typically employs vacuum or pressure-driven systems, comprising a filter funnel, filtrate flask, and vacuum pump. The selection of appropriate filtration parameters significantly influences the success of subsequent analytical procedures (Mintenig et al., 2017).

The choice of filter material is crucial, with several options available depending on specific analytical requirements. Glass fibre filters, with pore sizes < 1 μm, have emerged as the most widely used option due to their ability to retain particles from 3 μm to 5 mm while providing minimal background interference in subsequent analyses. Their high particle retention capacity and chemical resistance make them particularly suitable for general wastewater applications (Cunsolo et al., 2021; Naji

Table 16.3 Comparison of density separation solutions for microplastic analysis in wastewater matrices

Separation solution	Density range (g/cm^3)	Target polymers	Recovery efficiency	Advantages	Limitations	Cost considerations	Key studies
Zinc chloride ($ZnCl_2$)	1.6–1.9	• PE, PP, PS • PET, PVC	80–95%	• High-density range • Solution reusability • Good polymer compatibility • Efficient separation	• Environmental concerns • Potential toxicity • pH adjustment needed	Moderate-high (cost offset by reusability)	Simon et al. (2018), Wang et al. (2020), Wolff et al. (2021)
Sodium chloride (NaCl)	1.2	• PE, PP • Some PS	60–75% for low-density polymers	• Low cost • Environmentally friendly • Easy handling • Wide availability	• Limited density range • Poor recovery of dense polymers • Lower efficiency	Low	Akyildiz et al. (2022)
Sodium iodide (NaI)	1.6–1.8	• PE, PP, PS • PET • Most common polymers	75–90%	• High-density range • Good separation efficiency • Less corrosive than $ZnCl_2$	• High cost • Light sensitivity • Limited reusability	Very high	Alavian Petroody et al. (2021)
Calcium chloride ($CaCl_2$)	1.3–1.5	• PE, PP • Some PET and PS	65–80%	• Moderate cost • Good availability • Less toxic	• Limited density range • Moderate efficiency	Moderate	Blair et al. (2019)

et al., 2021; Wolff et al., 2019). Quartz fibre filters, offering pore sizes < 2.2 μm and excellent heat resistance, present advantages for applications requiring high-temperature drying or ashing procedures, though at a higher cost (Akyildiz et al., 2022).

For specialized applications, synthetic filters offer specific advantages. Nylon filters (< 5 μm pore size) demonstrate good chemical resistance and can effectively retain particles from 30 to 4500 μm, making them suitable for chemical digestion procedures. Polycarbonate filters, characterized by uniform pore sizes < 1 μm, provide excellent durability and reliable particle capture in the 20–500 μm range, particularly beneficial for microscopy analysis (Talvitie et al., 2017a; Ziajahromi et al., 2017). PTFE filters, while more expensive, offer unique advantages for aggressive chemical procedures due to their hydrophobic nature and chemical inertness (Dronjak et al., 2023).

Filtration efficiency varies significantly with microplastic morphology and pore size selection. While larger pore sizes reduce retention efficiency, particularly for fibres and fragments, extremely small pores may lead to rapid filter clogging. Current filtration methods, however, remain limited in capturing nanoplastics due to conventional pore size constraints.

Practical considerations in filtration include proper filter placement, adequate sealing, and appropriate diameter selection to prevent sample loss and ensure compatibility with funnel specifications. These factors, combined with careful control of particle aggregation, directly influence recovery rates and analytical success.

The effectiveness of the complete pretreatment workflow—from digestion through filtration—fundamentally determines analytical quality. As these methods prepare samples for characterization, their optimization must align with the requirements of subsequent detection techniques. The following section examines these analytical approaches, exploring both qualitative and quantitative methods for microplastic characterization in wastewater matrices.

Analytical Techniques for MP Characterization and Quantification

Detection and quantification of MPs in wastewater constitute a critical phase in evaluating the full scope of MP pollution, assessing its environmental impacts, and guiding effective mitigation strategies. Analyzing and detecting MPs in aquatic environments is paramount for environmental scientists and researchers, as it reveals the prevalence, types, and impacts of these pollutants (Ashrafy et al., 2023; Mai et al., 2018). WWTPs serve as a nexus where MPs from urban and industrial sources converge, but the complex composition of WW—including organic matter, inorganic debris, and microbial communities often complicates accurate detection. Filtration and microscopy (including visual examination and microscopic particle counting), spectroscopy (FTIR, Raman, SEM-EDS), thermo-analytical methods (Pyro GC-MS,

TGA-MS, TED-GC-MS), and combined techniques (such as tagging methods, liquid chromatography, or SEM dispersive X-ray spectrometer), as shown in Fig. 16.2, have been widely employed to identify and characterize MPs in WW. The data generated by these methods have immediate implications for water quality monitoring, environmental risk assessments, human health, and food safety (Du et al., 2021) given their significance in environmental protection and sustainability efforts (Deme et al., 2022; Deng et al., 2020; Mitrano & Wohlleben, 2020).

Understanding the environmental impacts of MPs is central to protecting water resources and maintaining effective wastewater treatment processes. MPs can accumulate in filtration systems, reduce the efficiency of cleaning units, and ultimately be released into downstream aquatic environments (Carr et al., 2016; Chakraborty et al., 2023). When these particles persist in discharged effluent, they can contaminate natural water bodies and pose threats to ecosystems and human health (Karbalaei et al., 2018). Consequently, enhancing the detection of MPs in WW is indispensable, not only to gauge the extent of pollution but also to inform policy frameworks and management strategies that aim to preserve environmental integrity for current and future generations.

After concentrating on MPs from the WW, it becomes essential to understand the nature of the separated particles and to distinguish them from other materials. Laboratory analyses have thus evolved, guided by the pressing need to mitigate the potential hazards MPs pose to ecosystems and human health. Differentiating MPs from non-plastic items often requires advanced spectroscopic or microscopic methods to detect subtle differences in morphology and composition. This section explores the principal techniques used for MP analysis in WW following filtration (for instance, wet peroxide oxidation—WPO—and/or density separation), focusing on the composition, sources, and distribution of MPs, as well as showing the necessity for integrated approaches. Table 16.2 provides a detailed description of the key techniques currently applied to WW samples, highlighting the role of pre-treatment in refining data quality and comparability. The heterogeneity of WW itself can obscure MP signals and lead to the formation of agglomerates that skew quantitative results. Therefore, most recent studies have incorporated density separation alongside chemical and/or enzymatic digestion to isolate MPs reliably (Wolff et al., 2019; Lares et al., 2019; Bretas Alvim et al., 2020a; Akyildiz et al., 2022).

Accurate detection is crucial in shaping both scientific understanding and public policy surrounding MP contamination. Analyses not only inform on the prevalence and magnitude of MPs but also shed light on possible interactions with other micropollutants. By revealing how MPs distribute across treatment stages within WWTPs, researchers and policymakers can make data-driven decisions to counteract their global proliferation. Moreover, clear identification and quantification of MPs help demonstrate the seriousness of this form of pollution, ensuring that political momentum keeps pace with the urgent need to protect water resources. Through standardized and replicable detection procedures, it becomes feasible to compare findings across laboratories, track temporal changes, and refine mitigation strategies at local, national, and international levels.

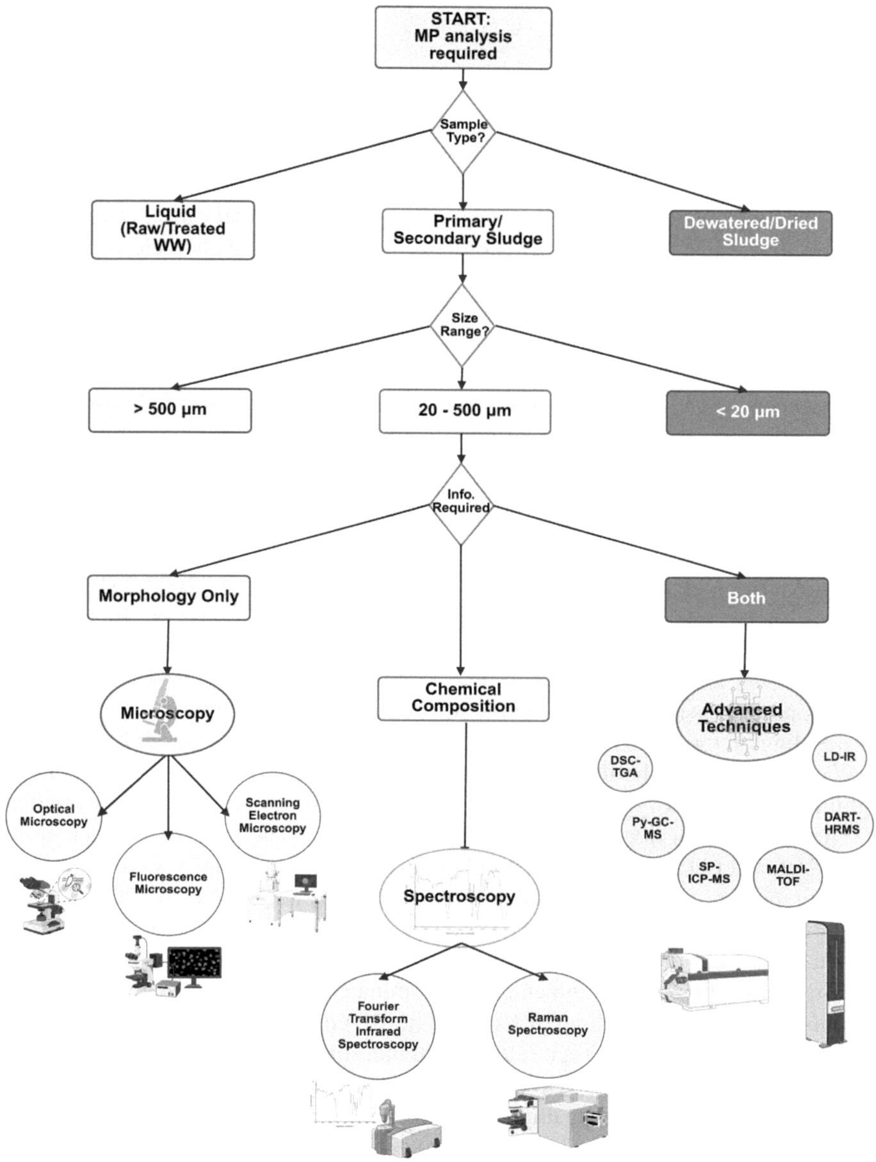

Fig. 16.2 Comprehensive decision tree for analytical method selection in microplastic analysis: integration of microscopic, spectroscopic, and advanced techniques based on sample properties and research requirements

Microscopic Techniques

Microscopic observation is often the first step in characterizing MPs, particularly for preliminary classification based on physical properties. Once MP particles have been separated, optical methods like stereomicroscopes (Contreras-Llin et al., 2024; Menéndez-Manjón et al., 2022; Nguyen et al., 2021; Zeri et al., 2021) or optical microscopes (Yuan et al., 2022a; Zhang et al., 2021) are widely used to categorize MPs by size, shape, and colour. This process, termed physical characterization, can offer rapid insights but is inherently subjective, as it depends on the analyst's visual judgment. Consequently, over- or underestimation of particle counts is possible, and differentiating synthetic from natural fibres can be challenging. Some research groups have proposed practical workarounds when spectroscopic methods are unavailable; for instance, using visual criteria to separate polymeric fragments (Hidalgo-Ruz et al., 2012). Nevertheless, more definitive identification generally requires complementary chemical characterization.

Although stereomicroscopes and optical microscopes are widely utilized, scanning electron microscopy (SEM) presents an alternative route to high-resolution imaging, enabling detailed examination of particle surfaces and morphologies (Bakaraki Turan et al., 2021; Hajji et al., 2023; Spacilova et al., 2023). SEM offers sub-nanometer resolution far exceeding that of stereomicroscopes but images only in greyscale and typically involve lengthier inspection times. The added magnification, however, allows researchers to detect microscopic flaws, weathering patterns, or agglomerations that may affect how MPs behave during treatment.

Building on these microscopic techniques, numerous studies classify MPs found in WWTP influent and effluent based on physical characteristics, consistently highlighting the complexity of WWTP flows. Some studies report that MPs larger than 500 μm account for over 70% of total MP loads in influent samples (Franco et al., 2020; Vardar et al., 2021), whereas others document significantly smaller proportions (Franco et al., 2020; Menéndez-Manjón et al., 2022; Shan et al., 2021). These discrepancies underscore how local consumption habits, industrial discharges, and seasonal shifts can alter the distribution of different MP size classes in WW. In terms of morphology, fibres and fragments often dominate across both influent and effluent streams, commonly exceeding 70% of overall MP counts. Although categories such as films and foams may undergo relatively higher removal rates, very small fibres (< 20 μm) remain challenging to eliminate (Ben-David et al., 2021). This persistence of smaller particles illustrates the importance of employing multiple analytical methods to detect MPs and generate reliable assessments of WWTP performance.

Spectroscopic Methods

The quantitative analysis of MPs is critical for fully addressing the extent of MP pollution. Techniques such as Fourier-transform infrared spectroscopy (FTIR), and

Raman spectroscopy methods yield precise data on MP abundance, composition, and size distribution (Tagg et al., 2020; Yang et al., 2023). FTIR and Raman spectroscopy are particularly prevalent, as they are relatively non-destructive and can distinguish polymer types based on characteristic vibrational signatures (Becucci et al., 2022). This specificity provides key insights into the interaction of MPs with other substances, including heavy metals or organic pollutants, thereby enabling a more thorough risk assessment. However, these methods can sometimes be confounded by complex sample matrices and require careful pre-treatment and calibration. By providing characteristic spectral fingerprints, Fourier-transform infrared spectroscopy (FTIR) and Raman spectroscopy enable precise polymer identification, clarifying ambiguities introduced by organic residue or overlapping physical features (Hongprasith et al., 2020; Lee et al., 2021). These techniques are crucial for understanding how MPs interact with co-existing micropollutants, offering detailed insights into overall contaminant transport. Additionally, spectroscopic results can be effectively integrated with the morphological data acquired through microscopy, forming a robust, complementary framework. This synergy is particularly relevant in contexts where airborne contamination and other artifacts may inflate or distort MP counts.

Thermal Analysis Techniques

Thermal methods complement spectroscopic techniques by examining how MPs respond to controlled heating conditions, thereby revealing changes in their mass or chemical composition. Approaches such as differential scanning calorimetry (DSC), thermogravimetric analysis (TGA), and pyrolysis–gas chromatography-mass spectrometry (Py-GC–MS) allow researchers to identify polymers through their characteristic thermal degradation profiles and volatile decomposition products (Dierkes et al., 2022; Goedecke et al., 2022; Roscher et al., 2022). Although these methods are destructive and thus preclude subsequent morphological analyses, they are often indispensable for complex wastewater samples that harbour polymer mixtures or exhibit overlapping spectral signatures. Recent innovations, including micro-FTIR, have improved resolution for smaller MPs, while laser direct infrared (LD-IR) imaging (Yuan et al., 2022b) combines infrared spectroscopy with microscopy for high-throughput, detailed characterization. Despite the inherent sample destruction, the polymer-specific breakdown patterns generated by thermal analyses provide a definitive line of evidence that corroborates and enriches data derived from microscopy and other spectroscopic approaches.

Mass Spectrometry Techniques

Mass spectrometry offers a powerful and versatile avenue for analyzing MPs, particularly in the case of blended or composite polymers that present challenges for traditional spectroscopic techniques. Approaches such as Direct Analysis in Real Time (DART)-high-resolution mass spectrometry (Zhang et al., 2020), single particle inductively coupled plasma mass spectrometry (SP-ICP-MS) (Bolea-Fernandez et al., 2020), thermal extraction desorption-GC/MS (TED-GC/MS) (Sorolla-Rosario et al., 2023), and matrix-assisted laser desorption/ionization–time-of-flight mass spectrometry (MALDI-TOF MS) (Wu et al., 2020) collectively expand the analytical landscape. These methods can be applied not only to confirm the polymeric nature of suspected MP particles but also to discern additives, plasticizers, or adsorbed contaminants. In addition, SP-ICP-MS can be particularly helpful in identifying metal-containing additives or nanoparticle residues, shedding light on the more complex chemical interactions of MPs in wastewater. While the highly detailed results generated by these techniques offer a more complete picture of MPs' composition and fate, the specialized instrumentation, training requirements, and associated costs may limit their widespread application. As a result, researchers and stakeholders often must weigh the benefits of in-depth mass spectrometric analysis against practical considerations, striving to find a balance between methodological rigour and overall feasibility.

Emerging Technologies

Further innovation addresses limitations inherent to conventional spectroscopic and thermal tools. Hyperspectral imaging, nano-FTIR, and nano-Raman methods push the boundaries of resolution, improving the detection of submicron plastics or Nano plastics that traditional instruments may overlook (Faltynkova et al., 2021). Such advancements align with the increasing realization that MPs in WW may take many forms, from virgin polymers to complex composites. No single method can provide an exhaustive analysis of all MPs present in WW. Hence, practitioners frequently adopt multi-method strategies that unite qualitative and quantitative data, enabling a more accurate assessment of the diverse array of particles and their significance for ecosystem health.

Many of these techniques, outlined in Table 16.4, rely on careful pretreatment protocols to remove organic matter and prevent agglomeration that might obscure smaller particles. Inefficient pretreatment can cause MP clusters to form and ultimately underestimate the true abundance of MPs. Ensuring optimal digestion of organics and proper density separation fosters reliable data and reduces the risk of false positives. Ultimately, high-quality MP analysis in WW hinges on a multi-stage procedure that begins with sampling, continues through rigorous pretreatment, and culminates in targeted detection, with each step subject to robust quality assurance measures. These efforts are especially significant in combating MP pollution in natural ecosystems and safeguarding human health. By consistently advancing detection technologies, addressing standardization gaps (Chae et al., 2015; Du et al., 2021) and implementing evidence-based policies (Kumar et al., 2020; Zhou et al., 2020), researchers and environmental managers can confront the complex challenges of MP pollution. Ensuring that future generations inherit clean and healthy water systems demands an urgent and collective commitment—one in which high-precision monitoring of MPs plays an indispensable part.

Conclusion and Future Prospects

The analysis of microplastics in wastewater matrices faces several critical challenges that impact method reliability and data quality. Method standardization remains a primary concern, with diverse protocols being employed across laboratories, making inter-study comparisons difficult. Current size detection limits (typically > 20 μm) leave a significant knowledge gap regarding smaller particles, particularly nanoplastics. Matrix complexity in wastewater systems poses substantial analytical challenges, with variable recovery rates (50–95%) depending on particle characteristics and matrix composition. The absence of certified reference materials hampers method validation efforts, while inconsistent reporting formats (particles/L vs. mg/L) complicate data interpretation and comparison.

Emerging technologies offer promising solutions to these analytical limitations. Artificial intelligence and machine learning approaches are being developed for automated particle recognition and classification, potentially reducing analysis time and human bias. Advanced hyperspectral imaging techniques coupled with automated detection systems show potential for real-time monitoring applications. The development of standardized protocols and certified reference materials remains a priority for improving method reliability. Novel separation techniques utilizing green chemistry principles are being explored to reduce environmental impact while maintaining analytical performance.

Looking forward, the field of microplastic analysis in wastewater treatment systems requires continued advancement in several key areas. Priority should be given to developing harmonized protocols, improving detection limits for smaller particles, and establishing standardized quality control measures. The integration of automated analysis systems with traditional analytical techniques shows promise for enhancing both throughput and reliability. Additionally, the development of online monitoring capabilities and environmentally friendly separation methods will be crucial for future applications. Success in addressing these challenges will require collaborative efforts across the scientific community, ultimately leading to more reliable and comparable microplastic monitoring in wastewater treatment systems.

Table 16.4 Analytical techniques for MPs detection: capabilities, limitations, and applications

Instruments	Sample preparation	Size detection limit	Limitations	Advantages	References
Microscopic					
SEM	• Determine possible surface changes resulting from a chemical digestion process	100–500 μm	• Costly • Time intensive • Requires multiple sample preparation steps • Cannot distinguish plastic particle colours	• Provides high-definition images of the external surface of MPs • Facilitates differentiation between synthetic MPs and associated organic materials • Reveals the elemental composition of MPs	Hajji et al. (2023), Spacilova et al. (2023)
Optical microscopy	• Removes organic matter, improving visualization and accuracy in identifying MPs	100 μm to 1 mm	• Restricted to visible light • Unsuitable for nanoscale analysis due to low resolution and large beam size • Unable to determine chemical composition • Prone to overlooking transparent or colourless MPs	• A non-destructive method offering high magnification at the micron scale • Applicable to all MPs • Capable of analyzing their shape, and size	Yuan et al. (2022a), Zhang et al. (2021)
Fluorescence microscopy	• Provides clear filter paper for easier MP identification • Organic matter removal prevents autofluorescence and false positives		• Lacks chemical composition data • Relies on proper fluorescent dye selection for accuracy • Requires multiple focal planes or confocal microscopy for 3D analysis	• Enhances contrast and visibility of MPs • Enables selective identification using fluorescent dyes that bind specifically to MPs • Supports quantitative analysis by correlating fluorescence intensity with MPs concentration	Al-Azzawi et al. (2023), Nguyen et al. (2021)

(continued)

Table 16.4 (continued)

Instruments	Sample preparation	Size detection limit	Limitations	Advantages	References
Spectroscopy					
Raman spectroscopy	• Enhances the MP fingerprint spectrum by reducing potential interferences and improving the signal-to-noise ratio	< 1 μm to 1000 μm automated	• Costly and time-intensive • Highly sensitive to pigments and contaminants • Unable to measure particles with strong filler signals or fluorescence	• Provides MP size data • Identifies MPs by their chemical composition	Becucci et al. (2022)
μ-FTIR spectroscopy	• Enhances MP surface quality by removing organic and inorganic content that may cause interference • Eliminating pore-entrapped organic matter to prevent false signals	10–500 μm > 500 μm manual	• Costly and challenging to include all MPs on the filter paper • Wavelength radiation can affect detection and quantification limits	• Offers precise particle analysis • Identifies number and size distribution • Requires minimal sample preparation • Is fast, non-destructive, cost-efficient, and provides a detailed chemical composition of MPs	Hongprasith et al. (2020), Yang et al. (2023)
Advanced technologies					
LD-IR	• MPs embedded in the organic matrix remain undetected	50–100 μm	• MPs embedded in the organic matrix remain undetected	• Fast, automated analysis providing size, shape, and composition data of MPs • Reliable and highly efficient	Yuan et al. (2022b)
TGA	• Removes alterations in thermal properties, such as melting point and mass, caused by organic matter	10–1000 μm	• Destructive method provides data only on semi-crystalline polymers • Unable to analyze amorphous MPs	• Identifies MPs through thermal behaviour and analyzes mass changes during thermal degradation in an inert environment	Dierkes et al. (2022)
Combined technologies					

(continued)

Table 16.4 (continued)

Instruments	Sample preparation	Size detection limit	Limitations	Advantages	References
MALDI-TOF	• Organic matter can ionize with the matrix reagent, leading to false positives	200–400 nm	• TOF sensors are quite costly	• Highly sensitive, high throughput, fast acquisition, easy to operate, and capable of analyzing aged MPs	Wu et al. (2020)
Pyr-GC–MS	• Undigested organic matter can mislead mass-related data interpretation of MPs	< 500 μm	• Destructive method requiring manual sample insertion • Limited for large molecular weight molecules (≥ 400 g/mol) due to condensation in the pyrolysis-GC–MS system • Provides minimal data on particle size and number	• Provides precise mass distribution • Requires minimal sample preparation time • Uses only 0.1–0.5 mg of sample per measurement	Goedecke et al. (2022), Vilakati et al. (2021)
TED-GC\MS	• Removes the necessity for multiple characteristic peak analyses in real environmental studies	40–160 μm	• Cannot differentiate MPs in mixtures, requires an additional tubular furnace, and is a destructive method	• Requires minimal sample handling, allows unrestricted sample weight, and offers easy manipulation	Sorolla-Rosario et al. (2023)
DART-HRMS	• Signal intensity significantly increases due to organic matter interference	0.3–1 mm	• Expensive analysis, • Unexplored for NPs, and involves a complex procedure	• Highly sensitive, capable of analyzing additives and identifying MP sources, with high throughput	Zhang et al. (2020)
SP-ICP-MS	• Organic matter presence leads to high background peak levels	1 μm	• Prone to false positives, requires method development for complex matrices, and is limited to tagged plastics	• Offers selective labelling, compatibility with techniques like spectroscopy or microscopy, enhanced sensitivity, and enables quantification	Bolea-Fernandez et al. (2020)

(Microscopic Techniques: SEM–EDX: Scanning electron microscopy coupled with energy dispersive X-ray spectroscopy; Spectroscopic Methods:, μ-FTIR: Micro-Fourier transform infrared spectroscopy, LD-IR: Laser direct

References

Akyildiz, S. H., Bellopede, R., Sezgin, H., et al. (2022). Detection and analysis of microfibers and microplastics in wastewater from a textile company. *Microplastics, 1*, 572–586. https://doi.org/10.3390/microplastics1040040

Alavian Petroody, S. S., Hashemi, S. H., & van Gestel, C. A. M. (2021). Transport and accumulation of microplastics through wastewater treatment sludge processes. *Chemosphere, 278*, Article 130471. https://doi.org/10.1016/j.chemosphere.2021.130471

Al-Azzawi, M. S. M., Kunaschk, M., Mraz, K., et al. (2023). Digest, stain and bleach: Three steps to achieving rapid microplastic fluorescence analysis in wastewater samples. *Science of the Total Environment, 863*, Article 160947. https://doi.org/10.1016/j.scitotenv.2022.160947

Amato-Lourenço, L. F., Carvalho-Oliveira, R., Júnior, G. R., et al. (2021). Presence of airborne microplastics in human lung tissue. *Journal of Hazardous Materials, 416*, Article 126124. https://doi.org/10.1016/j.jhazmat.2021.126124

Ashrafy, A., Liza, A. A., Islam, M. N., et al. (2023). Microplastics pollution: A brief review of its source and abundance in different aquatic ecosystems. *Journal of Hazardous Materials Advances, 9*, Article 100215. https://doi.org/10.1016/j.hazadv.2022.100215

Bakaraki Turan, N., Sari Erkan, H., & Onkal Engin, G. (2021). Microplastics in wastewater treatment plants: Occurrence, fate and identification. *Process Safety and Environmental Protection, 146*, 77–84. https://doi.org/10.1016/j.psep.2020.08.039

Bannick, C. G., Szewzyk, R., Ricking, M., et al. (2019). Development and testing of a fractionated filtration for sampling of microplastics in water. *Water Research, 149*, 650–658. https://doi.org/10.1016/j.watres.2018.10.045

Becucci, M., Mancini, M., Campo, R., & Paris, E. (2022). Microplastics in the Florence wastewater treatment plant studied by a continuous sampling method and Raman spectroscopy: A preliminary investigation. *Science of the Total Environment, 808*, Article 152025. https://doi.org/10.1016/j.scitotenv.2021.152025

Ben-David, E. A., Habibi, M., Haddad, E., et al. (2021). Microplastic distributions in a domestic wastewater treatment plant: Removal efficiency, seasonal variation and influence of sampling technique. *Science of the Total Environment, 752*, Article 141880. https://doi.org/10.1016/j.scitotenv.2020.141880

Bhardwaj, G., Mohammadiun, M., Osorio Gonzalez, C. S., Brar, S.K., & Karimpour, S. (2025a). Wastewater-induced microplastic biofouling in freshwater: Role of particle size and flow velocity. *Environmental Science: Advances*. https://doi.org/10.1039/D4VA00303A

Bhardwaj, G., Wankhede, L., Pulicharla, R., & Brar, S. K. (2025b). Microplastic-associated biofilms in wastewater treatment plants: Mechanisms and impacts. *Journal of Water Process Engineering, 72*, 107582. https://doi.org/10.1016/j.jwpe.2025.107582

Blair, R. M., Waldron, S., & Gauchotte-Lindsay, C. (2019). Average daily flow of microplastics through a tertiary wastewater treatment plant over a ten-month period. *Water Research, 163*, Article 114909. https://doi.org/10.1016/j.watres.2019.114909

Bolea-Fernandez, E., Rua-Ibarz, A., Velimirovic, M., et al. (2020). Detection of microplastics using inductively coupled plasma-mass spectrometry (ICP-MS) operated in single-event mode. *Journal of Analytical Atomic Spectrometry, 35*, 455–460. https://doi.org/10.1039/C9JA00379G

Bretas Alvim, C., Bes-Piá, M. A., & Mendoza-Roca, J. A. (2020a). Separation and identification of microplastics from primary and secondary effluents and activated sludge from wastewater treatment plants. *Chemical Engineering Journal, 402*, 126293. https://doi.org/10.1016/j.cej.2020.126293

Bretas Alvim, C., Mendoza-Roca, J. A., & Bes-Piá, A. (2020b). Wastewater treatment plant as microplastics release source—Quantification and identification techniques. *Journal of Environmental Management, 255*, 109739. https://doi.org/10.1016/j.jenvman.2019.109739

Carr, S. A., Liu, J., & Tesoro, A. G. (2016). Transport and fate of microplastic particles in wastewater treatment plants. *Water Research, 91*, 174–182. https://doi.org/10.1016/j.watres.2016.01.002

Casella, C., Sol, D., Laca, A., & Díaz, M. (2023). Microplastics in sewage sludge: A review. *Environmental Science and Pollution Research, 30*, 63382–63415. https://doi.org/10.1007/s11356-023-27151-6

Castelluccio, S., Alvim, C. B., Bes-Piá, M. A., et al. (2022). Assessment of microplastics distribution in a biological wastewater treatment. *Microplastics, 1*, 141–155. https://doi.org/10.3390/microplastics1010009

Chae, D.-H., Kim, I.-S., Kim, S.-K., et al. (2015). Abundance and distribution characteristics of microplastics in surface seawaters of the Incheon/Kyeonggi Coastal Region. *Archives of Environmental Contamination and Toxicology, 69*, 269–278. https://doi.org/10.1007/s00244-015-0173-4

Chakraborty, I., Banik, S., Biswas, R., et al. (2023). Raman spectroscopy for microplastic detection in water sources: A systematic review. *International Journal of Environmental Science and Technology, 20*, 10435–10448. https://doi.org/10.1007/s13762-022-04505-0

Contreras-Llin, A., Carrera, J., & Diaz-Cruz, M. S. (2024). Navigating microplastics in wastewater treatment: Understanding analysis, mitigation, removal strategies, impact, and current knowledge gaps. *Journal of Environmental Chemical Engineering, 12*, Article 114792. https://doi.org/10.1016/j.jece.2024.114792

Cunsolo, S., Williams, J., Hale, M., et al. (2021). Optimising sample preparation for FTIR-based microplastic analysis in wastewater and sludge samples: Multiple digestions. *Analytical and Bioanalytical Chemistry, 413*, 3789–3799. https://doi.org/10.1007/s00216-021-03331-6

Das, A., Ray, S. K., Mohanty, M., et al. (2024). Ecotoxicity of microplastic wastes and their sustainable management: A review. *Environmental Chemistry and Ecotoxicology, 6*, 144–152. https://doi.org/10.1016/j.enceco.2024.05.003

Das, P., & Bal, M. (2025). Modeling and optimization study for microplastic removal from aquatic medium using *Chroococcidiopsis* species. *Chemical Engineering Science, 304*, Article 121085. https://doi.org/10.1016/j.ces.2024.121085

Das, P., Halder, G., & Bal, M. (2023). A critical review on remediation of microplastics using microalgae from aqueous system. *Science of the Total Environment, 898*, Article 166425. https://doi.org/10.1016/j.scitotenv.2023.166425

Deme, G. G., Ewusi-Mensah, D., Olagbaju, O. A., et al. (2022). Macro problems from microplastics: Toward a sustainable policy framework for managing microplastic waste in Africa. *Science of the Total Environment, 804*, Article 150170. https://doi.org/10.1016/j.scitotenv.2021.150170

Deng, L., Cai, L., Sun, F., et al. (2020). Public attitudes towards microplastics: Perceptions, behaviors and policy implications. *Resources, Conservation and Recycling, 163*, Article 105096. https://doi.org/10.1016/j.resconrec.2020.105096

Dierkes, G., Lauschke, T., & Földi, C. (2022). Analytical methods for plastic (microplastic) determination in environmental samples. In F. Stock, G. Reifferscheid, N. Brennholt, & E. Kostianaia (Eds.), *Plastics in the aquatic environment—Part I: Current status and challenges* (pp. 43–67). Springer International Publishing.

Ding, N., An, D., Yin, X., & Sun, Y. (2020). Detection and evaluation of microbeads and other microplastics in wastewater treatment plant samples. *Environmental Science and Pollution Research International, 27*, 15878–15887. https://doi.org/10.1007/s11356-020-08127-2

Dronjak, L., Exposito, N., Sierra, J., et al. (2023). Tracing the fate of microplastic in wastewater treatment plant: A multi-stage analysis of treatment units and sludge. *Environmental Pollution, 333*, Article 122072. https://doi.org/10.1016/j.envpol.2023.122072

Du, S., Zhu, R., Cai, Y., et al. (2021). Environmental fate and impacts of microplastics in aquatic ecosystems: A review. *RSC Advances, 11*, 15762–15784. https://doi.org/10.1039/D1RA00880C

Dyachenko, A., Mitchell, J., & Arsem, N. (2017). Extraction and identification of microplastic particles from secondary wastewater treatment plant (WWTP) effluent. *Analytical Methods, 9*, 1412–1418. https://doi.org/10.1039/C6AY02397E

Edo, C., González-Pleiter, M., Leganés, F., et al. (2020). Fate of microplastics in wastewater treatment plants and their environmental dispersion with effluent and sludge. *Environmental Pollution, 259*, Article 113837. https://doi.org/10.1016/j.envpol.2019.113837

Faltynkova, A., Johnsen, G., & Wagner, M. (2021). Hyperspectral imaging as an emerging tool to analyze microplastics: A systematic review and recommendations for future development. *Microplastics and Nanoplastics, 1*, 13. https://doi.org/10.1186/s43591-021-00014-y

Franco, A. A., Arellano, J. M., Albendín, G., et al. (2020). Mapping microplastics in Cadiz (Spain): Occurrence of microplastics in municipal and industrial wastewaters. *Journal of Water Process Engineering, 38*, Article 101596. https://doi.org/10.1016/j.jwpe.2020.101596

Franco, A. A., Martín-García, A. P., Egea-Corbacho, A., et al. (2023). Assessment and accumulation of microplastics in sewage sludge at wastewater treatment plants located in Cádiz. *Spain. Environmental Pollution, 317*, Article 120689. https://doi.org/10.1016/j.envpol.2022.120689

Frias, J. P. G. L., & Nash, R. (2019). Microplastics: Finding a consensus on the definition. *Marine Pollution Bulletin, 138*, 145–147. https://doi.org/10.1016/j.marpolbul.2018.11.022

Gao, Z., Cizdziel, J. V., & Chen, L. (2023). Microplastics profile in sludge from a university wastewater treatment plant and the influence of chemical digestions on Nile red stained microplastics. *Journal of Environmental Chemical Engineering, 11*, Article 109671. https://doi.org/10.1016/j.jece.2023.109671

Ghosh, S., Sinha, J. K., Ghosh, S., et al. (2023). Microplastics as an emerging threat to the global environment and human health. *Sustainability, 15*, 10821. https://doi.org/10.3390/su151410821

Gies, E. A., LeNoble, J. L., Noël, M., et al. (2018). Retention of microplastics in a major secondary wastewater treatment plant in Vancouver, Canada. *Marine Pollution Bulletin, 133*, 553–561. https://doi.org/10.1016/j.marpolbul.2018.06.006

Goedecke, C., Eisentraut, P., Altmann, K., Elert, A. M., Bannick, C. G., Ricking, M., Obermaier, N., Barthel, A. K., Schmitt, T., Jekel, M., & Braun, U. (2022). Development of a routine screening method for the microplastic mass content in a wastewater treatment plant effluent. *Frontiers in Environmental Chemistry, 3*. https://doi.org/10.3389/fenvc.2022.844633

Gündoğdu, S., Çevik, C., Güzel, E., & Kilercioğlu, S. (2018). Microplastics in municipal wastewater treatment plants in Turkey: A comparison of the influent and secondary effluent concentrations. *Environmental Monitoring and Assessment, 190*, 626. https://doi.org/10.1007/s10661-018-7010-y

Hajji, S., Ben-Haddad, M., Abelouah, M. R., et al. (2023). Occurrence, characteristics, and removal of microplastics in wastewater treatment plants located on the Moroccan Atlantic: The case of Agadir metropolis. *Science of the Total Environment, 862*, Article 160815. https://doi.org/10.1016/j.scitotenv.2022.160815

Haque, F., & Fan, C. (2023). Fate and impacts of microplastics in the environment: Hydrosphere, pedosphere, and atmosphere. *Environments, 10*, 70. https://doi.org/10.3390/environments10050070

Hidalgo-Ruz, V., Gutow, L., Thompson, R. C., & Thiel, M. (2012). Microplastics in the marine environment: A review of the methods used for identification and quantification. *Environmental Science and Technology, 46*, 3060–3075. https://doi.org/10.1021/es2031505

Hidayaturrahman, H., & Lee, T.-G. (2019). A study on characteristics of microplastic in wastewater of South Korea: Identification, quantification, and fate of microplastics during treatment process. *Marine Pollution Bulletin, 146*, 696–702. https://doi.org/10.1016/j.marpolbul.2019.06.071

Hongprasith, N., Kittimethawong, C., Lertluksanaporn, R., et al. (2020). IR microspectroscopic identification of microplastics in municipal wastewater treatment plants. *Environmental Science and Pollution Research, 27*, 18557–18564. https://doi.org/10.1007/s11356-020-08265-7

Hurley, R. R., Lusher, A. L., Olsen, M., & Nizzetto, L. (2018). Validation of a method for extracting microplastics from complex, organic-rich, environmental matrices. *Environmental Science and Technology, 52*, 7409–7417. https://doi.org/10.1021/acs.est.8b01517

Iyare, P. U., Ouki, S. K., & Bond, T. (2020). Microplastics removal in wastewater treatment plants: A critical review. *Environmental Science: Water Research & Technology, 6*, 2664–2675. https://doi.org/10.1039/D0EW00397B

Karbalaei, S., Hanachi, P., Walker, T. R., & Cole, M. (2018). Occurrence, sources, human health impacts and mitigation of microplastic pollution. *Environmental Science and Pollution Research, 25*, 36046–36063. https://doi.org/10.1007/s11356-018-3508-7

Kumar, M., Xiong, X., He, M., et al. (2020). Microplastics as pollutants in agricultural soils. *Environmental Pollution, 265*, Article 114980. https://doi.org/10.1016/j.envpol.2020.114980
Lares, M., Ncibi, M. C., Sillanpää, M., & Sillanpää, M. (2018). Occurrence, identification and removal of microplastic particles and fibers in conventional activated sludge process and advanced MBR technology. *Water Research, 133*, 236–246. https://doi.org/10.1016/j.watres.2018.01.049
Lares, M., Ncibi, M. C., Sillanpää, M., & Sillanpää, M. (2019). Intercomparison study on commonly used methods to determine microplastics in wastewater and sludge samples. *Environmental Science and Pollution Research, 26*, 12109–12122. https://doi.org/10.1007/s11356-019-04584-6
Lee, J., Choi, Y., Jeong, J., & Chae, K.-J. (2021). Eye-glass polishing wastewater as significant microplastic source: Microplastic identification and quantification. *Journal of Hazardous Materials, 403*, Article 123991. https://doi.org/10.1016/j.jhazmat.2020.123991
Leslie, H. A., Brandsma, S. H., van Velzen, M. J. M., & Vethaak, A. D. (2017). Microplastics en route: Field measurements in the Dutch river delta and Amsterdam canals, wastewater treatment plants, North Sea sediments and biota. *Environment International, 101*, 133–142. https://doi.org/10.1016/j.envint.2017.01.018
Leslie, H. A., van Velzen, M. J. M., Brandsma, S. H., et al. (2022). Discovery and quantification of plastic particle pollution in human blood. *Environment International, 163*, Article 107199. https://doi.org/10.1016/j.envint.2022.107199
Li, X., Chen, L., Ji, Y., et al. (2020). Effects of chemical pretreatments on microplastic extraction in sewage sludge and their physicochemical characteristics. *Water Research, 171*, Article 115379. https://doi.org/10.1016/j.watres.2019.115379
Liu, F., Olesen, K. B., Borregaard, A. R., & Vollertsen, J. (2019a). Microplastics in urban and highway stormwater retention ponds. *Science of The Total Environment, 671*, 992–1000. https://doi.org/10.1016/j.scitotenv.2019.03.416
Liu, X., Yuan, W., Di, M., Li, Z., & Wang, J. (2019b). Transfer and fate of microplastics during the conventional activated sludge process in one wastewater treatment plant of China. *Chemical Engineering Journal, 362*, 176–182. https://doi.org/10.1016/j.cej.2019.01.033
Löder, M. G. J., & Gerdts, G. (2015). Methodology used for the detection and identification of microplastics—A critical appraisal. In M. Bergmann, L. Gutow, & M. Klages (Eds.), *Marine anthropogenic litter* (pp. 201–227). Springer International Publishing.
Löder, M. G. J., Imhof, H. K., Ladehoff, M., et al. (2017). Enzymatic purification of microplastics in environmental samples. *Environmental Science and Technology, 51*, 14283–14292. https://doi.org/10.1021/acs.est.7b03055
Long, Z., Pan, Z., Wang, W., et al. (2019). Microplastic abundance, characteristics, and removal in wastewater treatment plants in a coastal city of China. *Water Research, 155*, 255–265. https://doi.org/10.1016/j.watres.2019.02.028
Lv, X., Dong, Q., Zuo, Z., et al. (2019). Microplastics in a municipal wastewater treatment plant: Fate, dynamic distribution, removal efficiencies, and control strategies. *Journal of Cleaner Production, 225*, 579–586. https://doi.org/10.1016/j.jclepro.2019.03.321
Magni, S., Binelli, A., Pittura, L., et al. (2019). The fate of microplastics in an Italian Wastewater Treatment Plant. *Science of the Total Environment, 652*, 602–610. https://doi.org/10.1016/j.scitotenv.2018.10.269
Mahon, A. M., O'Connell, B., Healy, M. G., et al. (2017). Microplastics in sewage sludge: Effects of treatment. *Environmental Science and Technology, 51*, 810–818. https://doi.org/10.1021/acs.est.6b04048
Mai, L., Bao, L.-J., Shi, L., et al. (2018). A review of methods for measuring microplastics in aquatic environments. *Environmental Science and Pollution Research, 25*, 11319–11332. https://doi.org/10.1007/s11356-018-1692-0
Martín-García, A. P., Egea-Corbacho, Á., Franco, A. A., et al. (2023). Grab and composite samples: Variations in the analysis of microplastics in a real wastewater treatment plant in the South of

Spain. *Journal of Environmental Chemical Engineering, 11*, Article 109486. https://doi.org/10.1016/j.jece.2023.109486

Mason, S. A., Garneau, D., Sutton, R., et al. (2016). Microplastic pollution is widely detected in US municipal wastewater treatment plant effluent. *Environmental Pollution, 218*, 1045–1054. https://doi.org/10.1016/j.envpol.2016.08.056

Mathew, J., Bhardwaj, G., Pulicharla, R., Rezai, P. & Brar, S. K. (2024a). Innovating Ferro-sonication approach for extracting microplastics from wastewater. *Science of The Total Environment, 951*, 175595. https://doi.org/10.1016/j.scitotenv.2024.175595

Mathew, J., Pulicharla, R., Rezai, P, & Brar, S. K. (2024b). Microplastics in wastewaters: Pretreatment to detection trail. *Journal of Water Process Engineering, 64*, 105702. https://doi.org/10.1016/j.jwpe.2024.105702

Menéndez-Manjón, A., Martínez-Díez, R., Sol, D., et al. (2022). Long-term occurrence and fate of microplastics in WWTPs: A case study in southwest Europe. *Applied Sciences, 12*, 2133. https://doi.org/10.3390/app12042133

Mintenig, S. M., Int-Veen, I., Löder, M. G. J., et al. (2017). Identification of microplastic in effluents of waste water treatment plants using focal plane array-based micro-Fourier-transform infrared imaging. *Water Research, 108*, 365–372. https://doi.org/10.1016/j.watres.2016.11.015

Mishra, S., & Das, A. P. (2021). Current treatment technologies for removal of microplastic and microfiber pollutants from wastewater. In M. P. Shah, A. Sarkar, S. Mandal (Eds.) *Wastewater treatment* (pp. 237–251). Elsevier.

Mishra, S., Dash, D., & Das, A. P. (2022a). Detection, characterization and possible biofragmentation of synthetic microfibers released from domestic laundering wastewater as an emerging source of marine pollution. *Marine Pollution Bulletin, 185*, 114254. https://doi.org/10.1016/j.marpolbul.2022.114254

Mishra, S., Singh, V., Cheng, L., Hussain, A., & Ormeci, B. (2022b). Nitrogen removal from wastewater: A comprehensive review of biological nitrogen removal processes, critical operation parameters and bioreactor design. *Journal of Environmental Chemical Engineering, 10*, 107387. https://doi.org/10.1016/j.jece.2022.107387

Mishra, S., & Das, A. P. (2023). Microbial remediation of synthetic microfiber contaminated wastewater. In: M. P. Shah (Ed.) *Microbial technologies in industrial wastewater treatment* (pp. 337–350). Springer Nature.

Mishra, S., Dash, D., & Das, A. P. (2024). Isolation and molecular characterization of polyester-degrading bacterial strains from municipal wastewater and their application in microplastic degradation in contaminated wastewater. *Geomicrobiology Journal, 41*, 732–741. https://doi.org/10.1080/01490451.2024.2376101

Mitrano, D. M., & Wohlleben, W. (2020). Microplastic regulation should be more precise to incentivize both innovation and environmental safety. *Nature Communications, 11*, 5324. https://doi.org/10.1038/s41467-020-19069-1

Mohanty, M., Tripathy, B., Koteswara Reddy, G., & Das A. P. (2024). Greener technologies for establishing restraint over microplastic pollution. In A. P. Das, I. D. Behera, D. Bhanja (Eds.) *Sustainable microbial technology for synthetic and cellulosic microfiber bioremediation* (pp. 239–258). Springer Nature.

Murphy, F., Ewins, C., Carbonnier, F., & Quinn, B. (2016). Wastewater Treatment Works (WwTW) as a source of microplastics in the aquatic environment. *Environmental Science and Technology, 50*, 5800–5808. https://doi.org/10.1021/acs.est.5b05416

Naji, A., Azadkhah, S., Farahani, H., et al. (2021). Microplastics in wastewater outlets of Bandar Abbas city (Iran): A potential point source of microplastics into the Persian Gulf. *Chemosphere, 262*, Article 128039. https://doi.org/10.1016/j.chemosphere.2020.128039

Nayebi, B., Khurana, P., Pulicharla, R., et al. (2023). Preservation, storage, and sample preparation methods for freshwater microplastics—A comprehensive review. *Environmental Science: Advance, 2*, 1060–1081. https://doi.org/10.1039/D3VA00043E

Ngo, P. L., Pramanik, B. K., Shah, K., & Roychand, R. (2019). Pathway, classification and removal efficiency of microplastics in wastewater treatment plants. *Environmental Pollution, 255*, Article 113326. https://doi.org/10.1016/j.envpol.2019.113326

Nguyen, N. B., Kim, M.-K., Le, Q. T., et al. (2021). Spectroscopic analysis of microplastic contaminants in an urban wastewater treatment plant from Seoul, South Korea. *Chemosphere, 263*, Article 127812. https://doi.org/10.1016/j.chemosphere.2020.127812

Primpke, S., Cross, R. K., Mintenig, S. M., et al. (2020). Toward the systematic identification of microplastics in the environment: evaluation of a new independent software tool (siMPle) for spectroscopic analysis. *Applied Spectroscopy, 74*, 1127–1138. https://doi.org/10.1177/0003702820917760

Ray, S. K., Dey, S., Das, A. P., & John, A. (2024). Microplastic waste management through greener technologies: A review. In *Sustainable management of urban plastic waste through circular economic approaches*. CRC Press.

Roscher, L., Halbach, M., Nguyen, M. T., et al. (2022). Microplastics in two German wastewater treatment plants: Year-long effluent analysis with FTIR and Py-GC/MS. *Science of the Total Environment, 817*, Article 152619. https://doi.org/10.1016/j.scitotenv.2021.152619

Ruffell, H., Pantos, O., Northcott, G., & Gaw, S. (2023). Wastewater treatment plant effluents in New Zealand are a significant source of microplastics to the environment. *New Zealand Journal of Marine and Freshwater Research, 57*, 336–352. https://doi.org/10.1080/00288330.2021.1988647

Shan, W., Li, B., Zhang, H., et al. (2021). Distribution, characteristics and daily fluctuations of microplastics throughout wastewater treatment plants with mixed domestic—Industrial influents in Wuxi City, China. *Frontiers of Environmental Science and Engineering, 16*, 6. https://doi.org/10.1007/s11783-021-1440-4

Simon, M., van Alst, N., & Vollertsen, J. (2018). Quantification of microplastic mass and removal rates at wastewater treatment plants applying Focal Plane Array (FPA)-based Fourier Transform Infrared (FT-IR) imaging. *Water Research, 142*, 1–9. https://doi.org/10.1016/j.watres.2018.05.019

Sorolla-Rosario, D., Llorca-Porcel, J., Pérez-Martínez, M., et al. (2023). Microplastics' analysis in water: Easy handling of samples by a new Thermal Extraction Desorption-Gas Chromatography-Mass Spectrometry (TED-GC/MS) methodology. *Talanta, 253*, Article 123829. https://doi.org/10.1016/j.talanta.2022.123829

Spacilova, M., Dytrych, P., Lexa, M., et al. (2023). An innovative sorption technology for removing microplastics from wastewater. *Water, 15*, 892. https://doi.org/10.3390/w15050892

Stock, F., Kochleus, C., Bänsch-Baltruschat, B., et al. (2019). Sampling techniques and preparation methods for microplastic analyses in the aquatic environment—A review. *TrAC Trends in Analytical Chemistry, 113*, 84–92. https://doi.org/10.1016/j.trac.2019.01.014

Sun, J., Dai, X., Wang, Q., et al. (2019). Microplastics in wastewater treatment plants: Detection, occurrence and removal. *Water Research, 152*, 21–37. https://doi.org/10.1016/j.watres.2018.12.050

Sutton, R., Mason, S. A., Stanek, S. K., et al. (2016). Microplastic contamination in the San Francisco Bay, California, USA. *Marine Pollution Bulletin, 109*, 230–235. https://doi.org/10.1016/j.marpolbul.2016.05.077

Tagg, A. S., Sapp, M., Harrison, J. P., Sinclair, C. J., Bradley, E., Ju-Nam, Y., & Ojeda, J. J. (2020). Microplastic monitoring at different stages in a wastewater treatment plant using reflectance micro-FTIR imaging. *Frontiers in Environmental Science, 8*. https://doi.org/10.3389/fenvs.2020.00145

Tagg, A. S., Harrison, J. P., Ju-Nam, Y., et al. (2016). Fenton's reagent for the rapid and efficient isolation of microplastics from wastewater. *Chemical Communications, 53*, 372–375. https://doi.org/10.1039/C6CC08798A

Talvitie, J., Mikola, A., Koistinen, A., & Setälä, O. (2017a). Solutions to microplastic pollution—Removal of microplastics from wastewater effluent with advanced wastewater treatment technologies. *Water Research, 123*, 401–407. https://doi.org/10.1016/j.watres.2017.07.005

Talvitie, J., Mikola, A., Setälä, O., Heinonen, M., & Koistinen, A. (2017b). How well is microlitter purified from wastewater?—A detailed study on the stepwise removal of microlitter in a tertiary level wastewater treatment plant. *Water Research, 109*, 164–172. https://doi.org/10.1016/j.watres.2016.11.046

Tang, L., Su, C., Chen, Y., et al. (2021). Influence of biodegradable polybutylene succinate and non-biodegradable polyvinyl chloride microplastics on anammox sludge: Performance evaluation, suppression effect and metagenomic analysis. *Journal of Hazardous Materials, 401*, Article 123337. https://doi.org/10.1016/j.jhazmat.2020.123337

Thompson, R. C., Olsen, Y., Mitchell, R. P., et al. (2004). Lost at sea: Where is all the plastic? *Science, 304*, 838–838. https://doi.org/10.1126/science.1094559

Vardar, S., Onay, T. T., Demirel, B., & Kideys, A. E. (2021). Evaluation of microplastics removal efficiency at a wastewater treatment plant discharging to the Sea of Marmara. *Environmental Pollution, 289*, Article 117862. https://doi.org/10.1016/j.envpol.2021.117862

Vilakati, B., Sivasankar, V., Nyoni, H., et al. (2021). The Py – GC-TOF-MS analysis and characterization of microplastics (MPs) in a wastewater treatment plant in Gauteng Province, South Africa. *Ecotoxicology and Environmental Safety, 222*, Article 112478. https://doi.org/10.1016/j.ecoenv.2021.112478

Wang, B., Smith, M., Liu, Y., et al. (2023). Microplastic isolation method for wastewater and sludge samples by removal of excess organic and inorganic interferences. *Chemosphere, 329*, Article 138625. https://doi.org/10.1016/j.chemosphere.2023.138625

Wang, F., Wang, B., Duan, L., et al. (2020). Occurrence and distribution of microplastics in domestic, industrial, agricultural and aquacultural wastewater sources: A case study in Changzhou, China. *Water Research, 182*, Article 115956. https://doi.org/10.1016/j.watres.2020.115956

Wang, W., Gao, H., Jin, S., et al. (2019). The ecotoxicological effects of microplastics on aquatic food web, from primary producer to human: A review. *Ecotoxicology and Environmental Safety, 173*, 110–117. https://doi.org/10.1016/j.ecoenv.2019.01.113

Wolff, S., Kerpen, J., Prediger, J., et al. (2019). Determination of the microplastics emission in the effluent of a municipal waste water treatment plant using Raman microspectroscopy. *Water Research X, 2*, Article 100014. https://doi.org/10.1016/j.wroa.2018.100014

Wolff, S., Weber, F., Kerpen, J., et al. (2021). Elimination of microplastics by downstream sand filters in wastewater treatment. *Water, 13*, 33. https://doi.org/10.3390/w13010033

Wu, P., Tang, Y., Cao, G., et al. (2020). Determination of environmental micro(nano)plastics by matrix-assisted laser desorption/ionization–time-of-flight mass spectrometry. *Analytical Chemistry, 92*, 14346–14356. https://doi.org/10.1021/acs.analchem.0c01928

Xu, X., Jian, Y., Xue, Y., et al. (2019). Microplastics in the wastewater treatment plants (WWTPs): Occurrence and removal. *Chemosphere, 235*, 1089–1096. https://doi.org/10.1016/j.chemosphere.2019.06.197

Yang, J., Monnot, M., Sun, Y., et al. (2023). Microplastics in different water samples (seawater, freshwater, and wastewater): Methodology approach for characterization using micro-FTIR spectroscopy. *Water Research, 232*, Article 119711. https://doi.org/10.1016/j.watres.2023.119711

Yang, L., Li, K., Cui, S., et al. (2019). Removal of microplastics in municipal sewage from China's largest water reclamation plant. *Water Research, 155*, 175–181. https://doi.org/10.1016/j.watres.2019.02.046

Yuan, F., Zhao, H., Sun, H., Sun, Y., Zhao, J., & Xia, T. (2022a). Investigation of microplastics in sludge from five wastewater treatment plants in Nanjing, China. *Journal of Environmental Management, 301*, 113793. https://doi.org/10.1016/j.jenvman.2021.113793

Yuan, Z., Huang, Q., Wang, Z., Wang, H., Luo, J., Zhu, N., Cao, X., & Lou, Z. (2022b). Medium-low temperature conditions induce the formation of environmentally persistent free radicals in microplastics with conjugated aromatic-ring structures during sewage sludge pyrolysis. *Environmental Science & Technology, 56,* 16209–16220. https://doi.org/10.1021/acs.est.2c04453

Zeri, C., Adamopoulou, A., Koi, A., et al. (2021). Rivers and wastewater-treatment plants as microplastic pathways to eastern Mediterranean waters: first records for the Aegean Sea, Greece. *Sustainability, 13*, 5328. https://doi.org/10.3390/su13105328

Zhang, J., Wang, L., & Kannan, K. (2019). Polyethylene terephthalate and polycarbonate microplastics in pet food and Feces from the United States. *Environmental Science and Technology, 53*, 12035–12042. https://doi.org/10.1021/acs.est.9b03912

Zhang, L., Liu, J., Xie, Y., et al. (2021). Occurrence and removal of microplastics from wastewater treatment plants in a typical tourist city in China. *Journal of Cleaner Production, 291*, Article 125968. https://doi.org/10.1016/j.jclepro.2021.125968

Zhang, P., Shen, Y., Guo, J.-S., et al. (2015). Extracellular protein analysis of activated sludge and their functions in wastewater treatment plant by shotgun proteomics. *Science and Reports, 5*, 12041. https://doi.org/10.1038/srep12041

Zhang, X., Mell, A., Li, F., et al. (2020). Rapid fingerprinting of source and environmental microplastics using direct analysis in real time-high resolution mass spectrometry. *Analytica Chimica Acta, 1100*, 107–117. https://doi.org/10.1016/j.aca.2019.12.005

Zhang, Z., Liu, W., Gao, Q., et al. (2023). Microplastics extraction from wastewater treatment plants: Two-step digestion pre-treatment and application. *Water Research, 230*, Article 119569. https://doi.org/10.1016/j.watres.2023.119569

Zhou, Y., Wang, J., Zou, M., et al. (2020). Microplastics in soils: A review of methods, occurrence, fate, transport, ecological and environmental risks. *Science of the Total Environment, 748*, Article 141368. https://doi.org/10.1016/j.scitotenv.2020.141368

Ziajahromi, S., Neale, P. A., & Leusch, F. D. L. (2016). Wastewater treatment plant effluent as a source of microplastics: Review of the fate, chemical interactions and potential risks to aquatic organisms. *Water Science and Technology, 74*, 2253–2269. https://doi.org/10.2166/wst.2016.414

Ziajahromi, S., Neale, P. A., Rintoul, L., & Leusch, F. D. L. (2017). Wastewater treatment plants as a pathway for microplastics: Development of a new approach to sample wastewater-based microplastics. *Water Research, 112*, 93–99. https://doi.org/10.1016/j.watres.2017.01.042

Chapter 17
Spectroscopic Detection of Microplastics from Paper and Plastic Cups Using Ftir

Boggada Hemanth Balaji, Harjeet Nath, and Jyoti Roy Choudhuri

Introduction

Plastics can be defined as polymer-based materials manufactured from by-products of fossil fuels and usually processed with a variety of chemical additives. Due to lightweight, low cost, durability, high persistency, and sound insulation properties, the application of plastics has increased exponentially in the last few decades (Mishra et al., 2024; Sahu et al., 2024). Microplastics are referred to the smaller smaller-sized plastic particles of the dimension less than 5 mm, non-soluble in the aqueous environment (Bala et al., 2024; Das et al., 2024; Mohanty et al., 2024). Based on their size, these small diameter plastic residues are broadly classified into three categories: (i) large microplastics that vary from 5 mm (mm) to 1 mm (mm) in size; (ii) smaller sized microplastics ranging between a 1 mm to the 1 μm (micron) and (iii) nanoparticles of less than 1 μm particle sizes (Ghosh et al., 2023; Mishra et al., 2022; Tripathy et al., 2024). They can exist in various sizes as well as shape like microbeads, film foam pellets as well as filaments (Mukotaka et al.). Microplastics are categorized as the Primary and Secondary microplastics, determined based on their point of origin (Pradhan et al., 2024; Priyadarshini & Das, 2024). The primary ones are produced purposely by the industry for cosmetic uses, and the secondary microplastics are found mainly due to the weathering/degradation of residual fragments of plastic existing in the environment by waterwave action, development of bio-film, exposure of sunlight, mechanical shear, and oxidation due to thermal activation etc. (Das & Bal, 2025; Das et al., 2023).

B. H. Balaji · J. R. Choudhuri
Department of Forensic Science, National Forensic Sciences University, Agartala, Tripura, India

H. Nath (✉)
Department of Chemical and Polymer Engineering, Tripura University (A Central University), Agartala, Tripura, India
e-mail: harjeetnath@tripurauniv.ac.in

I. D. Behera et al. (eds.), *Emerging Micropollutants*, Sustainable Environmental Waste Management Strategies, https://doi.org/10.1007/978-3-032-03530-1_17

Importance

Global concern due to the widespread pollution, environmental impacts and human health impacts (Priyadarsini & Das, 2024, 2025). These are either the popular or conventional materials used in packaging food, mainly paper and paperboards. During the food packaging, they gained prominence as it is most available, cheap and lightweight material used as barrier to moisture, oxygen, and microbial entities. A common type of food containers are the disposable paper cups. These are the go-to choice for many people while enjoying their favorite drink. Most coffee and tea outlets across the globe use disposable paper cups. Ninety to ninety-five % (by weight) of disposable paper cups is the most part made by in general, inner layer is made from the Polyethylene (PE), but other co-polymer grades are also been used. Many previous studies have demonstrated that papers and paperboards could release hazardous chemicals from the packed food intended for human use. Many other additives are utilized in modifying plastics, which influences in layering the paper to acquire required properties like flexibility, colour, and microbial protection. However, few organic compounds are introduced to modify the paper & paperboard for the packaging of foods. They are also been found to be potentially leached with food that is packed inside. Phthalate compounds-based plasticizers, which are similar in properties to di-(2-ethylhexyl) phthalate and di-n-butyl phthalate have potential applications in paper and paperboard packaging for facilitating flexibility of materials while increasing resistance to tearing forces during these substrates manufacturing process. Phenolics and Organophosphites are used as antioxidants additives with polymers to slow down the degradation process. Per-fluoroalkyl substances (PFAs), containing fluoride are generally included in paperboards and papers to bring out its moist proof and oil-repellent nature. Extraction of heavy metals into foodstuff is due to paper-based food packaging. Polymers used in the food packaging are also mixed with the nano-fillers such as silicates and carbon nanotubules to the lower their crystallinity which leads into improvement of the molecular mobility. Yet rarely we look at whether such food packaging materials are themselves leaching microplastics and usually then do not quantify. Microplastics are also reported from various food items; the estimated concentration of microplastic is 60–290 Microparticles/kg in salt, 7.5 $\pm$ 2.4 particles/L for branded milk & for fish or Seafood and more than 12.6 $\times$ 109 microplastics from teabags. Paper cups are identified as one of the widely preferred forms in food packaging industry. So, there is the requirement serious attention for the contamination of plastic particles into the hot beverages due to packaging.

Background

Nowadays, an estimated 390.7 million tons of plastics are manufactured each year worldwide, of which 44% is employed catering to consumer packaging-primarily single-use materials such as bags and bottles. Such food packaging based materials and the containers including bottles, caps, cups, foil, trays, films for sealing bags, etc. are employed the globally for the purpose of storage and the devouring of different kinds of food, soft drinks (such as formulae milks feed to the newborns), mineral or carbonated waters and so forth. These materials or containers are generally made up with the thermoplastics resins, including Polystyrene (PS), Low Density Polyethylene (LDPE), High Density Polyethylene (HDPE), PolyVinyl Chloride (PVC), Polypropylene (PP), and Polyethylene- Terephthalate (PET). It works best to point out here that resins are preferred since they can be mechanically recycled. The current state of paper cup market in 2022 which was released by the group associated with International Market Analysis Research and Consulting Services Private Limited (IMARC) indicated the use of the paper cup, has raised to 264.8 billion units per year. They are mainly the disposable cups, which are mainly utilized for distributing hot beverages among people. Typically, the inner wall of the cups have thin film coating of polyethylene so that the absorption or leaking of the liquids can be stopped and also provide longer span of life. Disposable Paper cups have been used widely anywhere we go: in the mall, cafeteria, restaurant, education institution, and home. People select them because they are light, easy to take, low cost, reliable, and you do not have to clean them afterwards. You can just toss them aside when finished using it. It has been well known that microplastics are found in the numerous articles due to contact materials shedding microplastics out of their inner surfaces. Micrometer-sized particles are emitted from plastics fed for drinking water bottles, mineral or carbonated bottles, individually used and reusable Polyethylene bottles, plastic containers and paper or plastic disposable cups. No less than 5–500 μm range was reported by both Chen et al. and Tirkey and Upadhyay for these particle sizes. Different analysis techniques have been employed to characterize microplastics in terms of their size, shape and the type of plastic. The Analysis of the microplastics can be divided into two parts, Physical characterization (colour, size) of microplastic by microscopic examination and chemical characterization (type of plastic) by spectroscopic analysis with a view to establishing the plastic (composition, structure etc.). Various analysis methods were adopted for characterizing microplastics in different studies. Atomic force microscopy (AFM), Fluorescence Microscopy, FTIR, SEM, Laser directs infrared (LDIR) and Raman spectrophotometer etc. MPs can enter the human body by several pathways: through the skin, respiration and ingestion. As a result of environmental pollution, MPs may be present in food; and also there are possibilities of the release from Packaging materials storage or transportation container–such as the tea bags, micro beads, container, bottles and cups and more. The potential for the food and drink to become polluted by these synthetic, manmade microplastic pollutants increases with more use of these items; hence people will have direct MP exposure. Microparticles are contaminated with variety of beverages and food, sugar, honey, milk, sugar cane etc. The particles to the dimension of micron size, which are identified as the main source of existing microplastics, responsible for polluting sources directly for portable drinking. Additionally, Microplastics can get exposed to human body through inhalation and can also pertaining to the skin.

Thus, it is a highly challenging task to distinguish the adverse health impacts caused by environmental plastic pollution and microplastics contamination through food.

In various studies it is reported that the direct exposures of man to Microparticles via drinks, food and water is investigated. The reports has mentioned that MPs are significantly responsible for medico-legal adversities including neurotoxicity, disruption of immune system, oxidative-stress, and cytotoxicity observed in the humans beings; in addition the entering of microparticles into the tissues may lead to the swelling, blockage of vital organs. One possible route of exposures to the microplastics is the intake of spoiled food and water, with particles then entering into the digestive canal. This may lead to cellular uptake and sub-cellular transfer and accumulation in the internal digestive system. Because of their less sizes and chemical compositions, Microplastics may bring physical and also chemical imbalance in human beings. They may bring environmental disorder Microplastics are not eco-friendly and also proved harmful to the environment, due to their complex nature in shape and size and density. In conclusion, there are many significant repercussions to the human health from these microplastics. Also the sizes of MPs may have a great effect on their health effects and treatment processes.

Materials and Methodology

Accumulation of Paper Cups

Five different categories of disposable paper cups of variable sized and coloured were brought from the nearby market in Battala Market, Agartala-Tripura, India. The main material is paper with plastic interior lining inside. These selected cups were frequently sold in the market and having variable volume as ordered by average customers. Mosty these cups are found in various commercial places like cafeteria, shops, restaurants, offices, residences, etc. The various cucups are were labeled as A, B, C, D and E for sampling analysis and volumes of 100 ml, 100 ml, 60 ml, 60 ml and 60 ml, respectively, as mentioned in Table 17.1.

Table 17.1 Description of the samples used

S. No	Name of the sample	Volume of sample (ml)	Colour
1	Sample A	100	Red
2	Sample B	100	Light green
3	Sample C	60	Red
4	Sample D	60	Green
5	Sample E	60	Blue

Sample Preparation and the Leaching of Microplastics

For sample preparation, these paper cups were rinsed rigorously with double distilled water in room temperature condition. Once they were cleaned, the hot double distilled water was directly poured into the cups (at a temperature of 95–100 °C and particular pH of 6.7). They were then kept in shaking incubator while temperature remained at 25 °C and rpm 50 for 15–20 min. This time frame was used to mimic the time period for a typical hot beverage remain in the cup for an normal consumer, the span between its being served and consumed. Figure 17.1, so as to ensure no airborne contamination at the time of sampling occurred; the cups were shut, for 15–20 min.

In addition, the prepared samples from paper cups containing double distilled water at room temperature were also been analyzed for the microplastics.

The reference sample: The paper layer and the plastic layer of each paper cup have been separated by cutting (using steel scissors) and keeping the piece of paper cup in the distilled water (temperature of 30–35 °C) for a while. Then the separated plastic layer (hydrophobic film from the disposable cups) used as a reference sample, as shown in Fig. 17.2. This reference sample compared with a spectra of High Density poly-ethylene (HDPE), as shown in Fig. 17.8.

For each disposable cup three samples have been prepared, control sample, reference sample and sample of analysis.

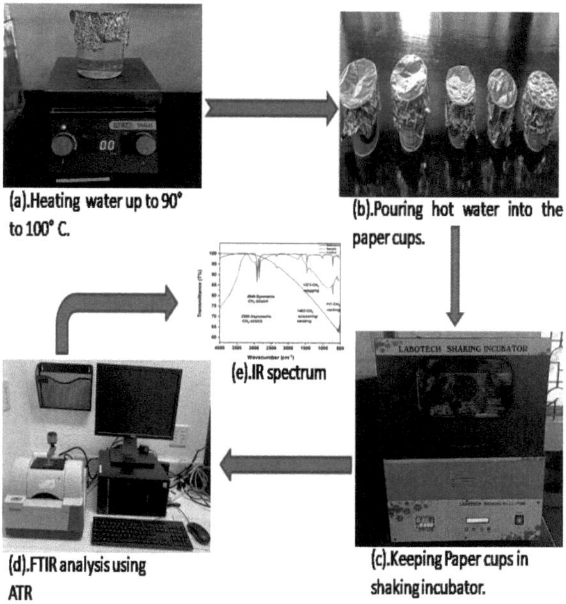

Fig. 17.1 Sample preparation of Paper cups for microplastics analysis

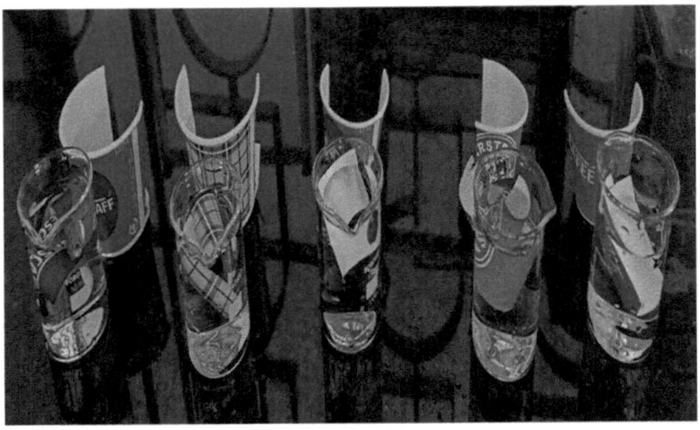

Fig. 17.2 Separation of paper and plastic layer (reference)

Instruments Used

Fourier Transform Infrared Spectrophotometer (FTIR)

An FTIR (Fourier Transform Infrared) Spectrophotometer is an analytical instrument used for the analysis of chemical compounds based on their infrared absorption/transmission spectra.

Principle of FTIR Instrument

Molecules absorb infrared radiation at specific frequencies because FTIR (Fourier Transform Infrared) spectroscopy is based on the principle that involves the Vibration modes of their chemical bonds. When infrared light passes through a sample, the molecules in the sample absorb energy at characteristic frequencies. The molecular bonds vibrate according to infrared radiation absorbed, and then this energy transferred as kinetic motion is distributed within a molecule between its three Normal modes of vibration. This absorbed energy corresponds to the transition between vibrational energy levels of the molecular bonds. In addition, some molecules have a dipole moment. In other words, their electric charges are not balanced when taken as a whole. When these molecules absorb infrared radiation, because their dipole moment changes as they vibrate, they give rise to vibration in a particular direction or mode. Dipole-moment is a measure of the separation, within a molecule of positive and negative electrical charges. When infrared light interacts with a molecule, if the frequency corresponds to the natural frequency of vibration of molecular units, then it will strongly excite that vibration.

The hydrophobic film (reference) samples that were collected from the disposable paper cups, control samples and suspected microplastic water samples were analyzed using Fourier Transform Infrared (FTIR) spectrophotometer with Attenuated Total Reflectance (ATR) (Nicolet Summit X Source: IR, Thermo Fisher Scientific, iD5 Advanced ATR Smart Accessory). Absorbance has been determined in the range between 4000 and 500 cm^{-1} to obtain the spectra. Then it is analyzed and a general comparison is done with references data to figure out the peaks and thus, the plastic. OMNIC (ThermoFisher Scientific and iD5 Advanced ATR Smart Accessory) were the two softwares used for the analyzing and interpretation of the results.

Results and Discussion

The various samples have been analyzed briefly and examined the release of particles containing microplastic from the different types of commonly utilized disposable paper cups namely A, B, C, D and E using FTIR characterization technique, as shown in Figs. 17.3, 17.4, 17.5, 17.6 and 17.7.

FTIR Analysis

The spectra of FTIR ranged between wavenumber 500 and 4000 cm^{-1}, and the characteristic peaks have been identified in that region. The spectra of triplicate samples of each paper cup represented in brown, orange and blue colors in the spectrums of the samples A, B, C, D, and E. They are compared with the spectra of the High-Density Polyethylene to know their level of similarities. The FTIR spectrum of the samples A, B, C, D, and E have been observed to have most similarities to the spectra of their reference and that reference spectrum is similar to the High-Density Polyethylene (HDPE). It also qualified the criteria that the minimal of four absorption/transmission bands go with between the two spectra. The bands observed

Fig. 17.3 FTIR spectra of the red 100 ml paper cup (Sample A)

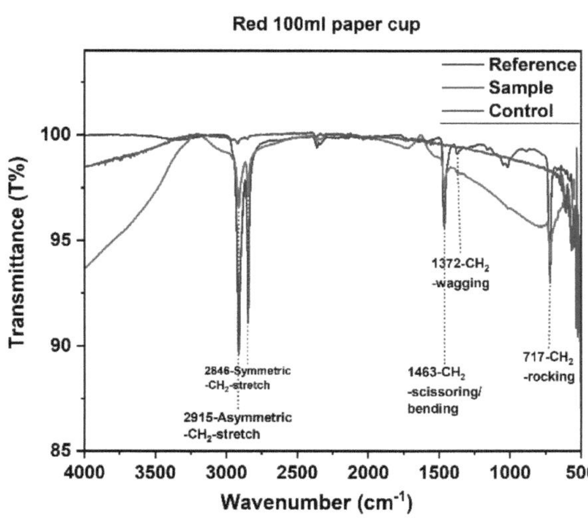

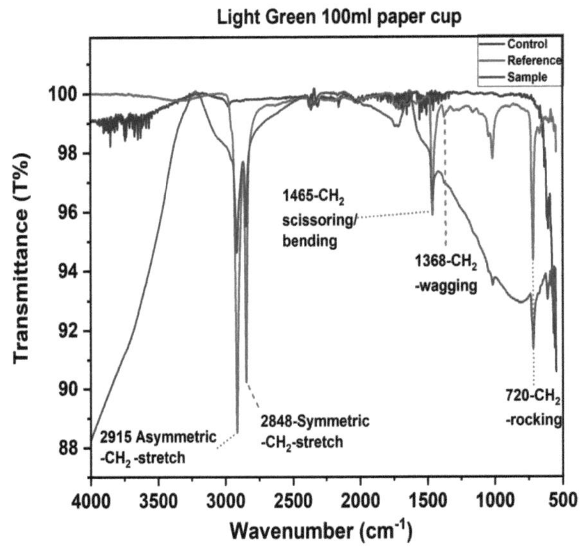

Fig. 17.4 FTIR spectra of the green 100 ml paper cup (Sample B)

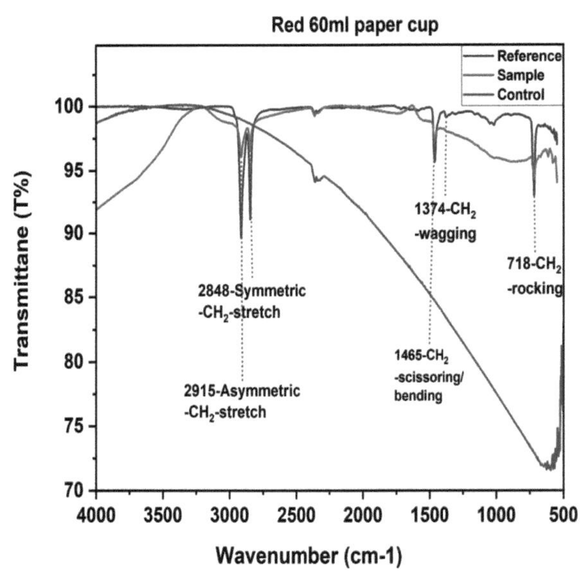

Fig. 17.5 FTIR spectra of the Red 60 ml paper cup (Sample C)

with the wavenumber around 2915 and 2845 cm^{-1} can be assigned to C-H stretch of all hydrocarbon constituents in the polymers. The peaks present at the wavenumbers around 1465 and 1372 cm^{-1} corresponds to the CH_2 bending and the absorption peaks at the wavenumber around 720 belongs to the CH2 rocking. These peaks have been observed in all the samples A, B, C, D, and E which matches closely with FTIR

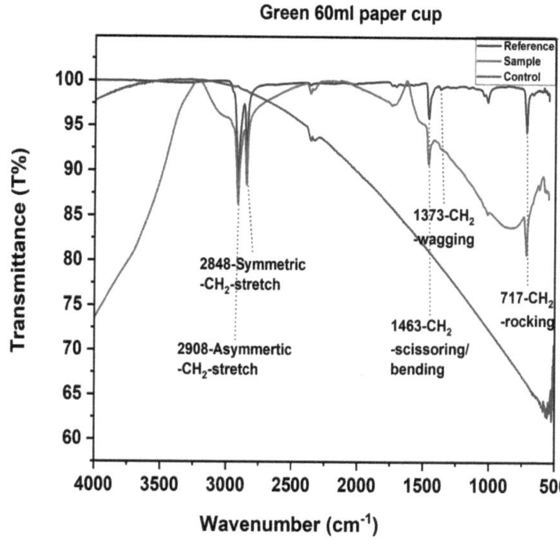

Fig. 17.6 FTIR spectra of the green 60 ml paper cup (sample D)

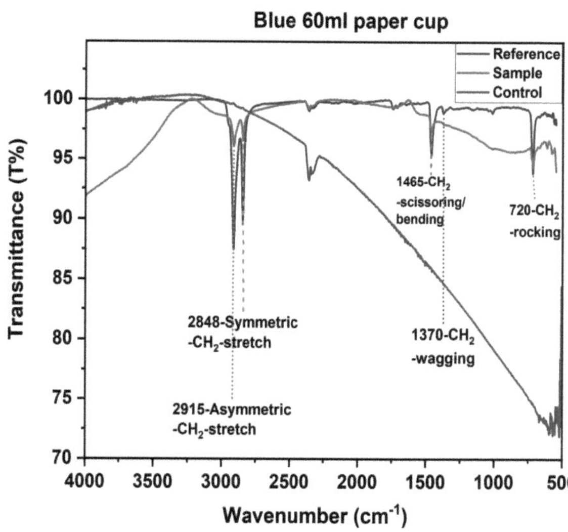

Fig. 17.7 FTIR spectra of the blue 60 ml paper cup (sample E)

transmittance spectra of HDPE. Thus, the FTIR results depicts that samples A, B, C, D and E, are composed of HDPE grade of plastics, as shown in the Fig. 17.8. The weak absorption bands have been also identified in the region of wavelength between 1200 and 800 cm^{-1}, was attributed to cellulose. The detailed breakdown of wavenumbers and their vibration type are shown in Table 17.2.

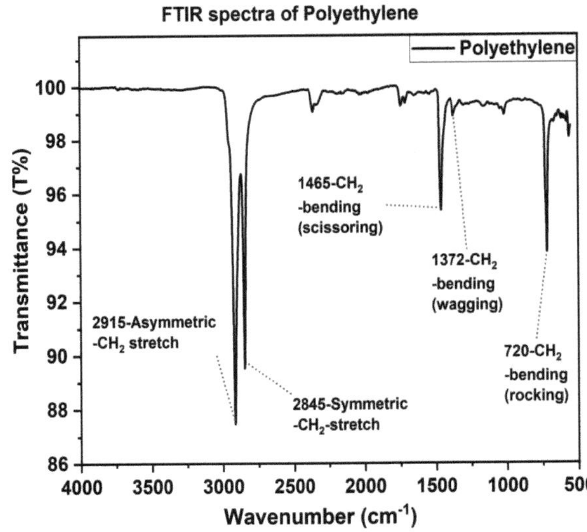

Fig. 17.8 FTIR spectra of the polyethylene (modified from (Zięba-Palus, 2017))

Table 17.2 Wavenumber and type of vibrations

Wavenumber (cm^{-1})	Vibration Type	Intensity
2915	Asymmetric CH_2 stretch	Strong
2915	Asymmetric CH_2 stretch	Strong
1465	CH_2 bending (scissoring)	Medium
1372	CH_2 bending (wagging)	Weak
720	CH_2 bending (rocking)	Strong

Table 17.3 Wavenumber and type of vibrations of sample A

Wavenumber (cm^{-1})	Vibration type	Intensity
2915	Asymmetric CH_2 stretch	Strong
2846	Asymmetric CH_2 stretch	Strong
1463	CH_2 bending (scissoring)	Medium
1372	CH_2 bending (wagging)	Weak
717	CH_2 bending (rocking)	Strong

Table 17.4 Wavenumber and type of vibrations of sample B

Wavenumber (cm^{-1})	Vibration Type	Intensity
2915	Asymmetric CH_2 stretch	Strong
2848	Asymmetric CH_2 stretch	Strong
1465	CH_2 bending (scissoring)	Medium
1368	CH_2 bending (wagging)	Weak
720	CH_2 bending (rocking)	Strong

Table 17.5 Wavenumber and type of vibrations of sample C

Wavenumber (cm^{-1})	Vibration Type	Intensity
2915	Asymmetric CH_2 stretch	Strong
2848	Asymmetric CH_2 stretch	Strong
1465	CH_2 bending (scissoring)	Medium
1374	CH_2 bending (wagging)	Weak
718	CH_2 bending (rocking)	Strong

Table 17.6 Wavenumber and type of vibrations of sample D

Wavenumber (cm^{-1})	Vibration type	Intensity
2908	Asymmetric CH_2 stretch	Strong
2848	Asymmetric CH_2 stretch	Strong
1463	CH_2 bending (scissoring)	Medium
1373	CH_2 bending (wagging)	Weak
717	CH_2 bending (rocking)	Strong

Table 17.7 Wavenumber and type of vibrations of sample E

Wavenumber (cm^{-1})	Vibration type	Intensity
2915	Asymmetric CH_2 stretch	Strong
2848	Asymmetric CH_2 stretch	Strong
1465	CH_2 bending (scissoring)	Medium
1370	CH_2 bending (wagging)	Weak
720	CH_2 bending (rocking)	Strong

Table 17.8 Wavenumber and type of vibrations of polyethylene (Zięba-Palus, 2017)

Wavenumber (cm^{-1})	Vibration type	Intensity
2915	Asymmetric CH_2 stretch	Strong
2845	Asymmetric CH_2 stretch	Strong
1465	CH_2 bending (scissoring)	Medium
1372	CH_2 bending (wagging)	Weak
720	CH_2 bending (rocking)	Strong

Conclusion

The research presented herein shows that microplastics are significantly ubiquitous within disposable paper cups, with polyethylene constituting the most represented polymer. FTIR spectroscopy has been shown here as a very reliable, fast method for detecting microplastics in environmental samples and characterizing them. Such a scenario will mean that the disposable paper cups pose serious health risks in relation to their use for hot drinks because microplastics could be leached into these beverages upon heating. Thus, this study puts light on the need for a standard detection method and, thereby, regulation to counterbalance the influence of microplastic pollution on human health and environment. In the case of disposable paper cups, the common plastic polymer used as a coating of these cups is polyethylene. It has been established to be a significant source of microplastics. The heating of these cups when applied to hot beverages leads to the degradation of the polyethylene and the release of microplastic particles into the beverage. This is important because this is a direct route by which humans ingest microplastics, leading to other health risks. FTIR spectroscopy was utilized to identify and characterize microplastics present in paper cups. In this method, it revealed that FTIR spectroscopy is an appropriate technique used because it shows detailed structures of chemical detection of detected microplastics. For the different samples analyzed in this experiment, the results produced were uniform and thus supported its reliability as a diagnostic tool used in detecting microplastic. Health implication of its ingestion has now become more of a serious concern today. These may be small in size but contain all manner of harmful chemicals and pathogens, which enter into body tissues, causing the occurrence of inflammation, oxidative stress, among other adverse effects on health. The report requires prompt attention to a move from disposable paper cups for hot beverages and emphasizes control measures as regulatory steps to handle the case of microplastic pollution. It was underpinned by the importance of needing uniform methods for the detection of microplastics, considering standardized methods would facilitate accuracy in environmental monitoring and public health studies. The formulation and practical application of these methods could better track sources of and microplastic pollution altogether leading to better policy regulations in advance.

In conclusion, it identifies a major source of microplastic pollution and provides a robust methodology to detect it. The findings emphasize the imperative need to address microplastic contamination in disposable paper cups and to take the necessary measures to protect the health of the public as well as the environment. Future research can be aligned in refining the detection using scientific techniques, to have brief understanding the issues related to long-term health caused by microplastics, and exploring alternative materials to replace polyethylene in disposable products.

References

Bala, K., Anshul, Verma, S., Dogra, A., Das, A. P., Kumar, R., Nandini. (2024). Sustainable management for urban plastic waste generation. *Sustainable Management of Urban Plastic Waste Through Circular Economic Approaches* 211–236. https://doi.org/10.1201/9781003465447-13

Das, A., Dey, S., Bala, K., Gnanasekaran, R., & Das, A. P. (2024). Microbial bioremediation technology for sustainable treatment and management of synthetic microfiber waste. *Sustainable Microbial Technology for Synthetic and Cellulosic Microfiber Bioremediation* 221–238. https://doi.org/10.1007/978-3-031-62600-5_12

Das, P., & Bal, M. (2025). Modeling and optimization study for microplastic removal from aquatic medium using Chroococcidiopsis species. *Chemical Engineering Science, 304*, Article 121085. https://doi.org/10.1016/J.CES.2024.121085

Das, P., Halder, G., & Bal, M. (2023). A critical review on remediation of microplastics using microalgae from aqueous system. *Science of the Total Environment, 898*, Article 166425. https://doi.org/10.1016/J.SCITOTENV.2023.166425

Ghosh, S., Sinha, J. K., Ghosh, S., Vashisth, K., Han, S., & Bhaskar, R. (2023). Microplastics as an emerging threat to the global environment and human health. *Sustainability,15*, 10821. https://doi.org/10.3390/SU151410821

Mishra, S., Dash, D., & Das, A. P. (2024). Microbial degradation of polyester microfibers using indigenously isolated bacterial strain *exiguobacterium* Sp. *Clean: Soil, Air, Water, 52*, Article e202300343. https://doi.org/10.1002/CLEN.202300343

Mishra, S., Singh, R. P., Rout, P. K., Das, A. P. (2022). Membrane bioreactor (MBR) as an advanced wastewater treatment technology for removal of synthetic microplastics. *Development in Wastewater Treatment Research and Processes* 45–60. https://doi.org/10.1016/B978-0-323-85583-9.00022-3

Mohanty, M., Tripathy, B., Koteswara Reddy, G., Das, A. P. (2024). Greener technologies for establishing restraint over microplastic pollution. *Sustainable Microbial Technology for Synthetic and Cellulosic Microfiber Bioremediation* 239–258. https://doi.org/10.1007/978-3-031-62600-5_13

Pradhan, G., Dey, S., Priyadarsini, S., Kumar, M. S., & Das, A. P. (2024). Characterization and quantification of microplastics pollutants in sediment samples from Daya River of Odisha State in India for their appropriate management. *Environmental Science and Engineering Part, F2333*, 75–92. https://doi.org/10.1007/978-3-031-51792-1_5

Priyadarshini, B., & Das, A. P. (2024). The adverse impact of synthetic microplastics released from disposable face masks pollutants on aquatic microbiota, aquatic species and humans: A review. *Environmental Science and Engineering Part, F2827*, 67–81. https://doi.org/10.1007/978-3-031-50840-0_5

Priyadarsini, S., & Das, A. P. (2025). Lithium bioleaching: A review on microbial-assisted sustainable technology for lithium bio-circularity. *Journal of Water Process Engineering, 69*, Article 106744. https://doi.org/10.1016/j.jwpe.2024.106744

Priyadarsini, S., & Das, A. P. (2024). Analytical and structural characterization of waste lithium-ion batteries for their effective recycling strategy. *Environmental Chemistry and Ecotoxicology*. https://doi.org/10.1016/j.enceco.2024.12.004

Sahu, A., Dey, S., Mallick, K., Priyadarsini, S., Behera, I. D., & Das, A. P. (2024). A review on advanced technology for sustainable management of synthetic microplastic waste. *Sustainable Management of Urban Plastic Waste Through Circular Economic Approaches* 262–279. https://doi.org/10.1201/9781003465447-15

Tripathy, B., Sahoo, P. P., Sundaray, H., & Das, A. P. (2024). In-vitro biodegradation of discarded marine microplastics across the eastern coast of the Bay of Bengal, India using *Exiguobacterium* sp. *Environmental Chemistry and Ecotoxicology, 6*, 236–247. https://doi.org/10.1016/J.ENCECO.2024.07.003

Zięba-Palus, J. (2017). The usefulness of infrared spectroscopy in examinations of adhesive tapes for forensic purposes. *Forensic Science Criminal, 2*. https://doi.org/10.15761/fsc.1000112

Chapter 18
Integrating Circular Economy Principles with Advanced Treatment Strategies of Micropollutant

Pragti Saini, Pankti Patel, and Subrata Biswas

Introduction

Micropollutants are anthropogenic contaminants present in the environment at trace levels, which could range between micrograms to picograms per liter or kilogram. These minute contaminants are often invisible to the naked eye but emerge as serious problems to the ecosystem and human health. Some natural and synthetic organic compounds are among these contaminants include pharmaceuticals, personal care products (PPCPs), perfluoroalkyl and poly-fluoroalkyl substances (PFASs), polychlorinated biphenyls (PCBs), polycyclic aromatic hydrocarbons (PAHs), nanomaterials, steroid hormones, pesticides, and plasticizers. These substances can remain for long in ecosystems and as such due to their toxic nature they pose serious threats to human life and the external environment (Bertram et al., 2022).

These chemicals are produced and used and then inadvertently discharge into water bodies intentionally or unintentionally. Most of these chemicals are removed inefficiently by existing wastewater treatment methods, they wind up in the treated effluent that is released into surface water, groundwater and marine water among other receiving water bodies (Yang et al., 2022). These micropollutants being bioaccumulative means that the pollutants accumulate in the tissues of living organisms, and concentration rises with the food chain (Das & Bal, 2025; Das et al., 2023). This results in bigger effects at the higher trophic level and may eventually endanger human health as well. In the aquatic system, it is possible for certain micropollutants

P. Saini · P. Patel · S. Biswas (✉)
Department of Paper Technology, Indian Institute of Technology Roorkee, Roorkee, India
e-mail: Subrata.biswas@pt.iitr.ac.in

P. Saini
e-mail: Pragti_s@pt.iitr.ac.in

P. Patel
e-mail: pankti_p@pt.iitr.ac.in

I. D. Behera et al. (eds.), *Emerging Micropollutants*, Sustainable Environmental Waste Management Strategies, https://doi.org/10.1007/978-3-032-03530-1_18

to bioaccumulates whereby they can be taken up by living organisms while others will remain deposited on the bottom of water bodies as sediments. Such sediments can remain a source of pollution for many years because the micropollutants can be resuspended into the water column. in aquatic ecosystem, even trace amounts of these substances can interfere with fish and other wildlife's endocrine systems, disrupting their growth and reproduction. In addition, not all the micropollutants respond the same way to inherent mechanisms such as biodegradation or photodegradation. It also implies that certain substances are less susceptible to breakdown and are more likely to persist in the environment hence exerting more harm to aquatic life and other aquatic communities (Lauesen, 2022).

Endocrine-disrupting chemicals (EDCs) are one type of micropollutant currently receiving a lot of attention due to their potential toxicity, rising environmental concentrations, as well as impact on human health. According to the EU Commission, EDC is an exogenous substance that causes adverse health effects in an intact organism, or its progeny, thereby causing disruptions in endocrine functions. The most frequently identified EDCs in the environment are phthalates, bisphenol A, and chlorophenols. Phthalates-based compounds are easily transmitted to the environment during product manufacture and trash disposal, where they are bound noncovalently to provide for the necessary degree of flexibility. PAEs like dimethyl phthalate (DMP) and diethyl phthalate (DEP) are among the most often discovered in varied environmental samples, including surface marine waters, freshwaters, and sediments (Sin et al., 2012). Bisphenol A (BPA) is a synthetic chemical molecule, member of the bisphenol and diphenylmethane derivatives. BPA is hazardous to the environment at concentrations ranging from 5 to 200 n/L (nanolitres) in surface and ground water. Bisphenols tend to accumulate in fish and other aquatic life, resulting in biomagnification. Also, humans are exposed to BPA at varying amounts (0.27–10.6 ng/ml), with an estimated daily intake of 30.76 ng/kg per body weight by dust inhalation, skin exposure, and ingestion (Ohore & Zhang, 2019). The presence of even minute traces of organochlorine pesticides in drinking water is hazardous because they may accumulate in the human body. Example of organochlorine pesticides are dieldrin, chlordane, DDT (dichlorodiphenyltrichloroethane), DDE (dichlorodiphenyldichloroethylene), β-hexachlorocyclohexane (βHCH), carbon tetrachloride, heptachlor and γ-hexachlorocyclohexane. In addition to their long-term stability and resilience to chemical and biological agents, they are extremely fat-soluble and hence accumulate in living organisms (see Table 18.1). While these compounds are water soluble, they are only found in small amounts in water, as adsorbed particulate matter. They may also build in bottom sediments and return to surface waters (Cybulski et al., 2021).

Waterbodies are contaminated by pharmaceutical discharge through sewage treatment plants (Mishra & Das, 2023). People ingest numerous medications and drugs that are not fully metabolized by the body. These drugs undergo structural changes in the bodies before being eliminated. These changes are rarely completed, typically leading to formation of portion of the original compound along with the metabolites. Few components are absorbed and other are partially metabolize by conjugative and oxidative enzymes to generate soluble and more polar form. These changes facilitate their easy removal through urine and faeces (Abdul–Jabbar et al., 2024). Hospital

Table 18.1 Endocrine disruptors and risk to human health

S. No.	Risk	EDs	References
1	Impairment of human reproductive system	Bisphenol A, phthalates, organochlorine pesticides, and pyrethroids	Raszewski et al. (2022)
2	Immunomodulatory effects	Diethyl phthalate, di-(2-ethylhexyl) phthalate, diisononyl phthalate, Triclosan, Propanil, Tetrachlorodibenzo-p-dioxin, Dichlorodifenylotrichloroetan, Diethylstilbestrol, parabens, Tributyltin	Nowak et al. (2019)
3	Insulin resistance	Diethylhexyl phthalate, bisphenol A, perfluorinated compounds, organochlorines compounds, Polychlorinated Biphenyls	Petrakis et al. (2017)
4	Carcinogenic (breast cancer)	b-hexachlorocyclohexane, p, p′-dichlorodiphenyldi chloroethylene, dichlorodiphenyltrichloroethane, bisphenol A, Diethylphthalate, propofol	du Plessis et al. (2023)
5	Non-alcoholic fatty liver disease	2,3,3,4,4,5-Hexachlorobiphenyl (PCB156), perfluorooctanoic acid (PFOA), perfluorooc tane sulfonate (PFOS), perfluorononanoic acid (PFNA), and perfluorohexanesulfonic acid (PFHxS), di-(2-ethylhexyl)-phthalate	Cano et al. (2021)
6	Cardiovascular diseases	2,3,7,8-Tetrachlorodibenzodioxin, Polychlorinatedbiphenyls, Perfluorooctanoicacid	Petrakis et al. (2017)
7	Changes of thyroid hormone, estrogenic and androgenic responses	Triclosan (5-chloro-2-(2,4 diclorofenoxy) phenol)	Varjani and Sudha (2020)

wastewater (HWW) is a significant source of micropollutants and falls under the category of pharmaceutical waste (Abdul–Jabbar et al., 2024). Compared to the typical municipal wastewater, the HWW bears threat to public health and environment with the fact that it contains pharmaceutical micropollutants and antibiotic-resistant bacteria (Mishra & Das, 2021; Mishra et al., 2024). If these pollutants exceed their limit, it will be challenging situation especially in areas where there are limited facilities for handling water wastes and the availability of clean water is a problem (Mishra et al., 2022a, 2022b; Mishra & Das, 2022). The severe waterborne illnesses include campylobacteriosis, cholera, hepatitis A and *E, schistosomiasis* and typhoid fever present in untreated or inadequately treated HWW, particularly in underdeveloped nation. Many medications including lipid regulators, analgesics, antidepressants, antiepileptics, antineoplastics, antipyretics, antiphlogistics, antirheumatics, β-blockers, broncholytics, β2-sympathomimetics, estrogens, secretolytics and vasodilators are frequently resistant to standard wastewater treatment techniques and have even been detected in drinking water (Kulišťáková, 2023). One of the medication atenolol (ATN), a commonly used β-blocker, is not fully metabolized by the body and is mostly eliminated unaltered in urine. Due to its poor biodegradability, ATN has been found in several water sources, including urban effluent (2 μg L^{-1}) and river water (83 ng L^{-1}). Ciprofloxacin (CIP), a fluoroquinolone antibiotic typically used to treat bacterial infections, is present in hospital and industrial wastewaters at levels more than 31 mg L^{-1} (Mojiri et al., 2020).

Rapid urbanization and industrialization have led to rapid growth but has caused severe impact on environment and ecosystem. Major concern is with wastewater (WW) discharge. Depending on the level of pollution, and kind of pollutants, World Health Organization (WHO) had made guidelines to enhance WW management regimes and reduce health hazards (Kumar, et al., 2023; Mishra et al., 2022a, 2022b). WWT processes are separated into three categories: primary, secondary and tertiary. The primary treatment focuses on separating light and heavy solid material. Large sedimentation tanks are used where WW is stored, allowing heavy components to settle at bottom and lighter materials to float. Coagulants and flocculating chemicals are added for to enhance solid coagulation. The flocculants are skimmed off and WW is sent for further treatment. The sediment or sludge is collected and can be used in anaerobic digestion. Secondary treatment involves use of microbes mostly aerobic microorganism to convert organic compounds into biomass and remove dissolved compounds. Tertiary process involves combination of physical and chemical treatment which eliminates the remaining pollutants (Koul et al., 2022).

Conventional activated sludge procedures (CASP) in WWTPs are ineffective in eliminating micropollutants because primary task of this process is to remove organic matter and nutrients leaving micropollutants and EDCs unaddressed. Continuous EDC discharge in the environment can cause reproductive and developmental problems to human as well as animals at all trophic level, even at low concentrations. Also, an alarming concern is the amplification of microbes resistant to antibiotics in the environment discharged from HWW. Overuse of antibiotics for human and animal health has resulted in an increase in antibiotic-resistant genes. Also, WWTPs lacks the ability to effectively detect micropollutant levels in source wastewater. For improved management and control of the micropollutants, the CASP system in WWTPs must be redesigned. (Lalit Goswami et al., 2018).

To address environmental sustainability and public health, suitable wastewater treatment and proper recycling management system would improve resilience in ecosystem. The precautionary principle, which is used in European Union (EU) law, states that contaminants and pollution should be avoided and managed at their source rather than eliminated throughout the wastewater treatment process. It is likely that it is the responsibility of producer to avoid or reduce the input of micropollutants into the aquatic environment (Behera et al., 2021, 2022; Mohanta et al., 2024). Regarding the issue of micropollutants, scientists are attempting to create new technologies as methods of detection, monitoring, and elimination. The identified technologies for removal of micropollutants include advanced oxidation technologies such as AOPs; membrane technologies including filtration, microfiltration, reverse osmosis; adsorption technologies such as granular activated carbon (GAC) and powdered activated carbon (PAC); biological methods like membrane bioreactor (MBR) and moving bed biofilm reactor (MBBR) (Kosek et al., 2020).

In the search of better water management system, the combination of advanced treatment methods with circular economy strategy offers a long-term possible solution to micropollutants. By treating waste as a valuable resource and focusing on source management, this strategy will help to reduce environmental impact. Advanced Oxidation Processes break down contaminants using hydroxyl radicals, whereas enzymatic degradation uses natural enzymes. Biochar and adsorption absorb pollutants which can be recycled and reused. Supercritical water oxidation and hydrodynamic cavitation are energy-intensive techniques to break down contaminants. Constructed wetlands naturally clean water and are reinforced with nanomaterials to improve efficiency. Phycoremediation is the process of using algae to clean pollutants while forming biofuels or fertilizers. Photocatalysis breaks down pollutants by using sunlight and recyclable materials like TiO_2. By conserving energy, reducing waste, and reusing resources, all these strategies support the circular economy. This chapter evaluates the shortcomings of traditional water treatment methods and dives into the complex effects of micropollutants and innovative solutions that hold hope for a cleaner and more sustainable environment.

Advanced Technologies

Micropollutants are contaminants which originate from natural sources as well as human activity. These contaminants are present in the environment in very minute quantities, with a smaller concentration between g/l to mg/L. Examples of micropollutants include chemicals like cosmetics, perfumes, waterproofing agents, paints, polymers and pesticides (Saini et al., 2025). Nowadays, micropollutants are merely found in surface water such as rivers, lakes, and reservoirs. The breakdown of micropollutants in surface water results in the production of byproducts in both forms, chemically or biologically. Some research has been done on the accumulation of micropollutants on the water surface (Ratchnashree et al., 2023). In recent years, there has been a rise of micropollutants on the surface water which is creating environmental cause. The aquatic environment is intensively effected due to the growing prevalence of microplastic. During the last 10 years, some of the processes have been working efficiently, which are as follows.

Advanced Oxidation Processes (AOPs)

AOPs firstly proposed in 1980s which involved generation of hydroxyl radical (OH·) by oxidation process in appropriate quantity for water purification. By the time, the sulfate radical (SO_4^-) was introduced in oxidation processes. The work for removal organic and inorganic contaminants was totally different comparing with ozone and chlorine radicals (Fig. 18.1). The use of pathogens and pathogenic indicator was unsuccessful because of shorter half life and low radical concentrations. However, the use of radical as powerful oxidizing agent was sufficiently affecting for destructing wastewater pollutants and reducing the toxicity of products, which provided ultimate solution for wastewater treatment.

Hydroxyl Radical Based AOPs

The oxidation potential of hydroxyl radical with saturated calomel electrode was 2.8 V at pH 0 and 1.95 V at pH 14, which is most reactive oxidizing agent. OH· reacts rapidly with non-selectivity with pollutants by radical addition, electron transfer, radical combination, and hydrogen abstraction methods. While hydroxyl radical

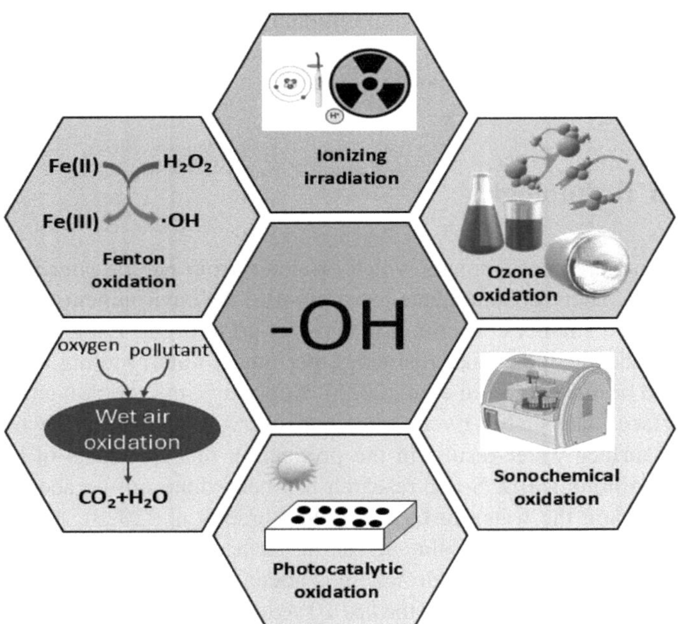

Fig. 18.1 OH, radical work in advanced oxidation processes

reacts with organic compounds it produces alkyl radical (R·), but when it reacts with O_2 it produces peroxyl radical (ROO). These radicals combine with one another and forms H_2O_2 or super oxide (O_2^-), which leads to degradation of these organic contaminants. Due to shorter life lifetime of hydroxyl radicals, they produced only in by situ by irradiation, catalysis, and combining with oxidizing agents. The processes to form OH· are as follows:

Ozone-Based AOPs

Ozone preferably reacts with organic compounds in ionized or dissociated form instead of neutral form. By indiscriminate oxidation process, OH· is formed from O3, the mechanism is mentioned below.

$$3O_3 + H_2O \rightarrow 2OH \cdot + 4O_2 \tag{18.1}$$

$$H_2O_2 \rightarrow HO_2^- + H+ \tag{18.2}$$

$$HO_2^- + O_3 \rightarrow OH \cdot + O_2^- + O_2 \tag{18.3}$$

$$O_3 + H_2O + hv \rightarrow H_2O_2 + O_2 \tag{18.4}$$

$$H_2O_2 + hv \rightarrow 2OH \tag{18.5}$$

OH· is generated through by ozonation processes as mentioned in Eqs. (18.1), (18.2), (18.3) and then H_2O_2 is produced by photolysis Eqs. (18.4), (18.5).

UV-Based AOPs

The initiation of OH· is also processed in the presence of catalysis or oxidants by photons. TiO2 is most common catalyst for OH· (Eq. 18.6). initiation by excited to produce positive holes in the valance band and negative holes in the conduction band for oxidative capacity and reductive capacity respectively as mentioned in Eqs. 18.7, 18.8, 18.9, 18.10, 18.11.

$$TiO_2 + hv \rightarrow e -_{cb} + hv + vb \tag{18.6}$$

$$hv + vb + OH- \rightarrow OH \tag{18.7}$$

$$hv + vb + H_2O \rightarrow OH \cdot + H^+ \tag{18.8}$$

$$e -_{cb} + O_2 \rightarrow O_2 \cdot - \tag{18.9}$$

$$H_2O_2 + hv \rightarrow 2OH \tag{18.10}$$

$$H_2O + hv \rightarrow OH \cdot + H \tag{18.11}$$

OH· can also be produced through photolysis with less than 242 nm wavelength.

Fenton Related AOPs

Fenton process is typically used to activated H_2O_2 for producing OH· reactive species. Fe^{2+} reacts with H_2O_2 and produced ferryl ions and hydroxyl radical with hydroxyl ion (Eq. 18.12).

$$Fe^{2+} + H_2O_2 \rightarrow Fe^{3+} + OH \cdot + OH^- \tag{18.12}$$

The optimal molar ratio of iron ion and hydrogen peroxide is used to minimise unwanted scavenging. Fe^{3+} forms iron sludge for wastewater treatment which needs to be separately disposed which is cost-intensive method. As Fenton reaction is only effective in low pH condition (acidic), the wastewater treatment application is restricted (Bhati et al., 2023; Saini et al., 2023).

Sulfate Radical-Based AOPs

$S_2O_8{}^{2-}$ is a strong oxidant which produced 2.01 V standard oxidation potential (E0). Sulfate radical forms powerful sulfate radical (SO_4^-) by UV irradiation, transition metals, heat activation, and high pH to initiate sulfate Radical Based AOPs. Elevated pH method also formed activated persulfate at high temperature and only 50% sulfate radical yield produced in this process, which is not efficient method. Similarly to hydroxyl radicals, sulfate radicals have a short lifespan but different reaction patterns.

AOPs for Treatment of Wastewater Landfill Leachate

After proposing the concept of AOPs, it was firstly applied on municipal and industries wastewater. The treatment efficiency of the process depends on the chemical properties of the contaminants and operating conditions. (Deng & Englehardt, 2006) explained the treatment efficiencies of different of landfill leachate. They successfully removed chemical oxygen demand (COD) by 6–88% with an average value of

53%. During ozonisation, OH· becomes the dominant at an alkaline condition at high pH value (10.0). Ozone preferably attacks on aromatic compounds than alkenes and alkynes, while OH· reacts with almost all the organic compounds except chloroform, carbon tetra chloride, and trichloroethane in landfill leachate. The COD removal efficiency varies as follows: O_3/UV treatment achieves 31–83%, O_3/H_2O_2 treatment achieves 18–78%, and UV/H_2O_2 treatment achieves 57–92%.

Biochar and Adsorption Techniques

Wastewater can be divided into industrial wastewater, urban wastewater, agricultural wastewater, and rainwater. Particularly industrial wastewater comes from mining, smelting, chemical, and manufacturing industries which contain heavy metal and organic pollutants. Biochar is used to adsorb heavy metal and organic dyes from wastewater which depends on the pH of the medium, contact time and dosage (Bora et al., 2025; Swain et al., 2023). The removal mechanism involves ion exchange, precipitation, complexation, and surface adsorption. As the surface of biochar is usually negatively charged which provides effective adsorption sites for heavy metal ions (positively charged) by electrostatic attraction (Fig. 18.2). Further, removal of metal ions creates multiple synergic effects for reaction mechanism (Abdelrasoul et al., 2018; Nthunya et al., 2022) such as surface complexation of functional groups, precipitation, ion exchange, and electrostatic attraction. Ion exchange is a process to exchange of organic oxygen containing functional group on the surface of biochar adsorbent to remove them from wastewater. Chemical precipitation method mostly used technology to remove heavy metal from aquatic environments. Complexation process is used to make complex between metal ions and organic functional groups (hydroxyl, carbonyl, carboxyl, and other oxygen containing groups) on the surface of biochar.

Adsorption of Inorganic Salt in Wastewater

Environmental problem can raise severely to ecosystem by increasing the concentration of organic salt such as phosphates, nitrates, and ammonium. Inorganic salts are responsible for photosynthesis which cause eutrophication in aquatics and leads to low oxygen and water quality degradation. As biochar is an environment friendly adsorbent which has numerous properties such as surface groups, porosity, mineral composition, and loaded active ingredients. Biochar has large surface area due to porous structure which provide an active surface site for phosphate ions. The organic functional groups of biochar (−OH, −COOH, C−H, C=O, −NH2) can combine with phosphate ions, hence removal of inorganic salts occur.

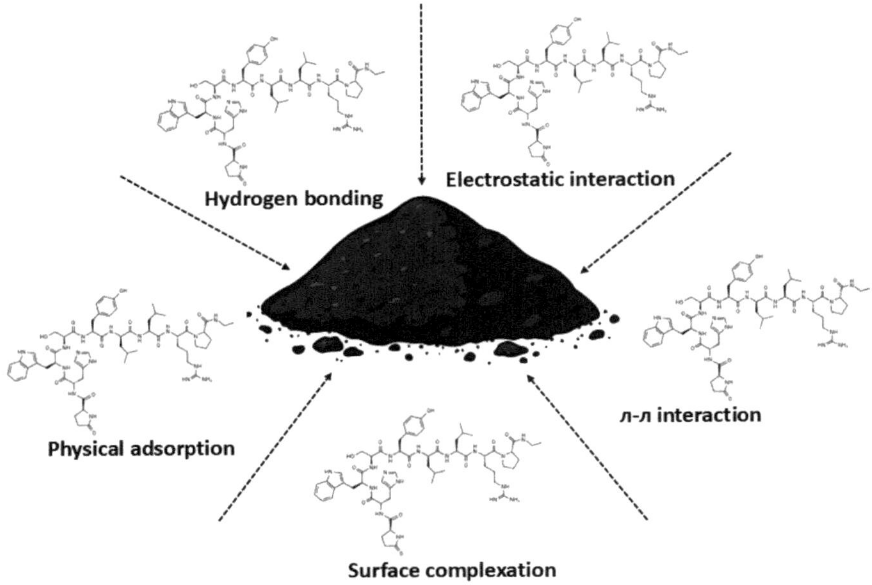

Fig. 18.2 Adsorption mechanism of biochar

Adsorption of Organic Pollutants in Wastewater

Biochar has been proven a significant adsorbent to remove organic pollutant in water. Dyes are found to be in wastewater in excessive amount which commonly discharged into wastewater by industries such as leather, paper, paint, textiles, and pharmaceuticals. These organic contaminants are also injuries to human health can create diseases such as skin rashes, gastrointestinal tract, and lungs cause. Research has found that biochar can effectively removed dyes by adsorption with hydrogen bonding in between biochar's functional groups and oxygen containing organic contaminants. The efficiency of removal contaminants also increased by л-л interaction and electrostatic interaction.

Other Adsorbents Carbon Nanotubes (CNTs), Metal Organic Frameworks (MOFs)

Recent studies (Qasem et al., 2022) explained the synthesis of MOF with CNT based nanocomposites using Co-MOF-CNT by adoption in-situ solvothermal approach for wastewater treatment application. The development in nanocomposite of Co-MOF-CNT was involved to adsorb and eliminate hazardous cationic dyes from industrial wastewater. According to current reports, the nanocomposite has an ultra-low detection limit of ~0.00649 ppm or 0.125 μM and demonstrates good selectivity and metal-sensing sensitivity towards the Cr6+ ion under ambient settings. Furthermore, Co-MOF@CNT's detecting capability can be considerably improved for up to five cycles. Furthermore, the created nanocomposite demonstrated a high removal rate of almost 98% at equilibrium for cationic (MB) dye with ultimate selectivity at the ideal temperature of 25 °C and neutral pH. According to the kinetic studies, a pseudo-second order reaction drives the adsorption process. The dye adsorption mechanism implies that the nanocomposite and the dye have electrostatic, π–π, and hydrogen bonding interactions. Additionally, this hybrid nanocomposite demonstrated remarkable MB dye reusability of up to four cycles. Overall, in this high-tech industrial era, the Co-MOF@CNT nanocomposite's superior Cr^{6+} metal sensing performance and MB dye adsorption may open a promising avenue for environmental applications.

Ion Exchange Resin Technologies to Mitigate Micropollutants

Micropollutants are pesticides, pharmaceuticals, and endocrine disruptors that are found in water sources at low levels of concentration, meaning less than a microgram per Liter. Despite the extremely low concentration of these agents, they are a great danger to both the environment and human health due to their bioaccumulation and persistence.

Description of Ion Exchange Resins

Ion exchange resin technologies are porous polymeric materials that cannot dissolve and functional groups that allow the exchange of ions in the water column with cationic or anionic groups. Based on which cationic or anionic groups are incorporated into resins, the functionalized resins can be divided into two main categories or three types.

1. Cation Exchange Resins: Heavy metal cationic pollutants such as lead, mercury, and cadmium can be removed from the water column using cation exchange resins.
2. Anion Exchange Resins: Commonly used to remove non-polar anions such as sulphate and nitrate and organic molecules like micropollutants.

Finally, based on their ionic Cationic group property resins are classified into 4 types, Strong acid, Weak acid, Strong base, weak base. Selective binding occurs when a specific micropollutant is bound by ionic charge, size, and charge affinity.

Mechanism of Removal of Micropollutants

The removal of micropollutants by ion exchange resins is based on the principle of an exchange reaction between the ions in the resin and those in the aqueous phase. In the case of cation exchange, for example, positively charged micropollutants in water take the place of sodium, hydrogen, or other cations initially present in the resin. In the case of anion exchange, negatively charged micropollutants displace chloride or hydroxide ions in the resin. This process can be customized to address certain micropollutants through changing the functional groups of the resin.

Advantages of Ion Exchange Resins

1. High Selectivity: Resins can be engineered to target specific micropollutants, thereby enhancing their removal efficiency.
2. Reusability: Resins can be regenerated and reused multiple times, thereby reducing operational costs.
3. Flexibility: Suitable for various water treatment scales, from small household systems to large industrial plants.
4. Low Energy Consumption: In contrast to processes such as reverse osmosis, ion exchange requires relatively less energy.

Challenges and Limitations

Though ion exchange resin technology is effective, there are some limitations:

1. Regeneration Challenges: Regeneration requires chemicals such as strong acids or bases, which can produce secondary wastes.
2. Saturation: Resins must be regenerated periodically as the ability of the resins to exchange ions diminishes over time.
3. Specificity: Specific ion resins may be less effective for other ions, and hence require multiple stages of treatment.
4. Cost: Resin is expensive, and its use would be cost-effective only for high concentrations of pollutants or very low volume of water.

Applications in Treatment of Micropollutants

1. Treatment of Municipal Water: Ion exchange removes nitrates, perchlorates, and heavy metals from drinking water sources.
2. Industrial Effluent Treatment: Industrial effluent containing heavy metals and organic impurities is treated with resins.
3. Pharmaceutical and Hospital Effluents: Specific resins target residual drugs and hormones in effluent streams.

4. Management of Agricultural Runoff: Removal of nitrates and phosphates from agricultural runoff prevents eutrophication in water bodies.

Membrane Bioreactors (MBRs)

A membrane bioreactor is a process which includes microfiltration or ultrafiltration membrane unit with bioreactor to use in municipal and industrial wastewater treatment. A bioreactor is specially designed chamber for growing bacteria and protozoa which can consume organic substances from raw wastewater (Bandehali et al., 2021; Piry et al., 2012). Three types of bioreactors exist which are as followings: suspended grown bioreactor where biomass grows into flocks, attached growth bioreactors where the biomass attached to the carriers, hybrid bioreactors which combines suspended and attached growth (Fig. 18.3). Membranes are used to separate solid–liquid device which keeps the biomass in the bioreactor before discharging the treated effluent. Ultrafiltration membrane is commonly used due to their superior separation characteristics which enables removal of colloids and viruses also has lower fouling tendency.

Firstly, submerged membrane bio reactors are used for vacuum drive are used in series PCl HF- Zmbr2 to activate return sludge. Secondly, side stream membrane

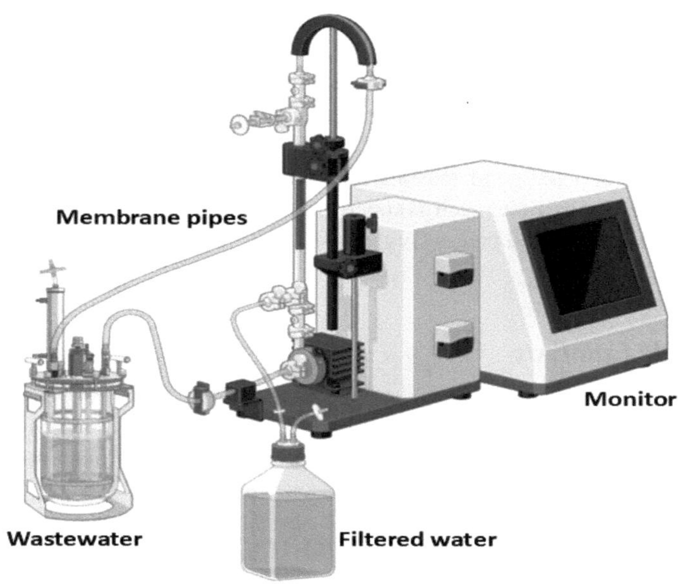

Fig. 18.3 Membrane bioreactor catalysis for wastewater treatment process

is used for pressure driven membrane like PCl A-series. MBRs is widely used for municipal, landfills, and industrial wastewater treatment.

Photocatalysis Using Recyclable Catalysts

One of the most severe concerns that the world is currently grappling with is water pollution, and alongside it, there is also a growing urgency to come up with cleaner and more effective ways of treating water. As part of this, photocatalysis for the treatment of wastewater is emerging as a viable alternative. This technique fuels the catalysts with needed energy in the form of light thus helping in the process of degrading unnecessary substances from water. The inclusion of recyclable catalysts allows the amalgamation of the two approaches as photocatalysis does not only work effectively but is also environmentally sound and affordable.

A Photocatalyst

But in this case, a broader definition of a photocatalyst is required: in the simplest terms, a photocatalyst is a substance that receives energy from light (artificial or natural UV light) and uses that energy to initiate or even speed up chemical processes. This is useful in the case of wastewater treatment, as this device in turn decomposes, among other substances, organic compounds, dyes, and pesticides, as well as some heavy metals, into benign molecules such as minerals, carbon dioxide, and water. The main phenomenon underpinning this process is semiconductor catalysis. Titanium dioxide (TiO_2), zinc oxide (ZnO) and graphene amongst others are widely used in this process and for good reasons too. These substances can absorb light and generating reactive particles such as superoxide ions and hydroxyl radicals that attack and decompose unwanted pieces (Fig. 18.4).

Recyclable Catalysts

Recyclable catalysts which can be employed repeatedly are essential to the photocatalysis technology for it to be economically viable and environmentally friendly for treating wastewater. Conventional approaches tend to create waste or take in non-recyclable materials, which turns out to be costly and causes environmental issues (Popova et al., 2020; Primo & Garcia, 2014). On the other hand, recyclable catalysts lessen the need for excessive replacement and replacement altogether reducing waste.

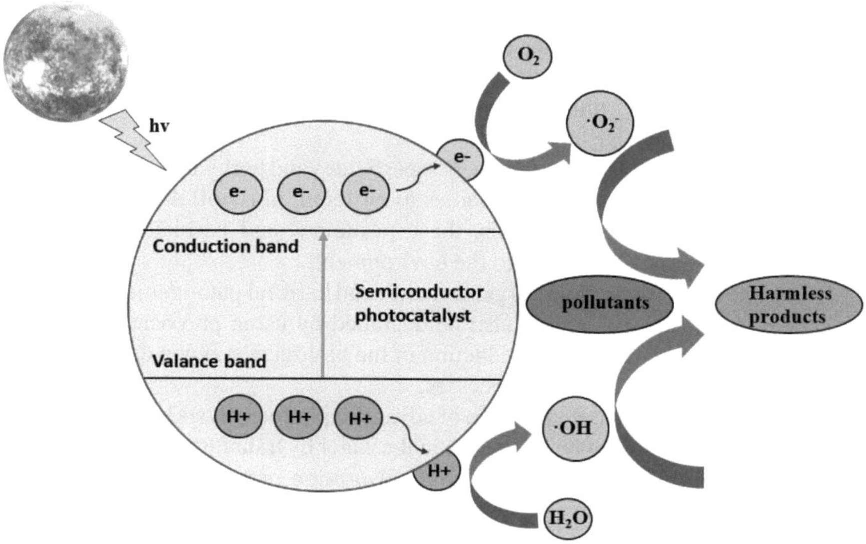

Fig. 18.4 Electron transfer in photocatalysis for wastewater treatment

Photocatalysis Works in Wastewater Treatment

1. Light Absorption: Upon striking the photocatalyst, electrons are pumped out of their ground states in the material creating electron hole pairs which are the active sites of the reactions taking place.
2. Formation of Reactive Species: Water and oxygen molecules present in the environment are activated by the bots that form around the stable polymers and result in the formation of free radicals including hydroxyl radicals, which are very unstable and quickly react with other particles. These radicals dissociate the compounds in the water.
3. Degradation of Pollutants: Complex pollutants are dismantled by the reactive radicals into smaller, harmless compounds. For instance, the textile effluents containing dyes are broken down into harmless carbon dioxide and water.
4. Recycling of Catalysts: During the chemical reaction, the photocatalyst practically remains unchanged and can thus be used in the subsequent treatment cycle. This minimizes the waste generation and hence the amount of overall cost.

Applications for Wastewater Treatment

Photocatalysis assisted by recyclable catalysts has been applied in various fields, including:

1. Industrial Effluents: The textile, paper, and chemical industries are among industries that produce wastewater enriched with chemical substances such as dyes, phenols, and other pollutants. These contaminants are broken down into non-harmful substances, thus achieving the goal of discharging and reusing the water.
2. Agricultural Runoff: Chemicals such as pesticides and herbicides that are washed away from soil and enter water bodies as agricultural runoff are environmental hazards. Through photocatalysis, these pesticides and herbicides are broken down, hence preventing harm to the environment.
3. Domestic Wastewater: Some organic matter and harmful pathogenic microorganisms in domestic sewage can also be degraded by using photocatalysis, which would thus result in a desired reduction of the biological oxygen demand (BOD) content in water purification processes.
4. Heavy Metal Removal: Some types of advanced photocatalysts have the potential to lessen the amounts of heavy metals in the water by transforming heavier metals into less harmful elements or enabling adsorption processes to separate these metals from the solution.

Benefits of Photocatalysis with Recyclable Catalysts.

1. Eco Friendly Process: Waste generation is minimal using light in photocatalysis as it is a renewable source, hence the process is environmentally viable.
2. Economical: The catalyst's ability to be recycled makes operational costs less which in turns make it suitable for larger applications.
3. High Efficiency: A variety of contaminants, even those that are resistant to traditional treatment techniques, can be broken down by photocatalysis.
4. Scalability: This procedure can be reduced for small-scale treatment systems or increased for industrial applications.
5. Versatility: The photocatalytic process offers one of the most versatile solutions to such pollution problems as it can be applied to a wide range of wastewater types.

Limitations and Restrictions:

Although it has a broad perspective of application, photocatalysis has certain challenges as well as limitations which includes:

1. Catalysts have been reported to be used only for the applications in UV light, which presents a small bandwidth of the sunlight available. The non-applicability of artificial uv light sources however increases the energy cost.
2. Inhibition of catalytic activity: over time, photocatalysts might be suspended or deactivated through the accumulation of byproducts through their use.
3. Limited efficacy depuration: pollutants degradation can be quite uneventful and prolonged especially on higher-scale applications.

4. However, the depletion of funds that is required for the development and production of high-quality recyclable catalysts is countered by the rate of savings that is estimated to be gained overtime.

Advancements in Compostable Photocatalysts.

Scientists are making strong efforts to improve the photocatalysts in terms of both efficiency and recyclability.

Enzymatic Degradation

Enzyme degradation is a promising and environmentally friendly wastewater treatment method that uses enzymes to break down pollutants into non-toxic good or bad substances. This technology uses the catalytic properties of enzymes to target specific pollutants such as organic pollutants, dyes, phenols, and heavy metals in wastewater. Enzymatic degradation has many advantages over traditional treatments, including greater specificity, lower energy consumption, and reduced production of secondary pollutants.

Enzymes are biological agents that reduce the energy required for reactions, making chemical reactions more powerful. They are highly specific for their substrates and can degrade target organisms without affecting non-target compounds. In wastewater treatment, enzymes are used to degrade complex organic compounds, destroy toxic substances, and convert harmful chemicals into biodegradable materials.

There are many types of enzymes used in wastewater treatment, each with its own specific pollution purpose:

Oxidative Enzymes

i. Laccase: These are copper-containing enzymes that oxidize phenolic and non-phenolic lignin-related compounds, making them effective against dyes and phenolic pollution.
ii. Peroxidase: Enzymes such as horseradish peroxidase and manganese peroxidase catalyze the oxidation of a variety of pollutants in the presence of hydrogen peroxide.

1. Hydrolytic Enzymes:

i. Protease: Reduces organic load by converting protein to amino acids.
ii. Cellulase: Helps in the degradation of plant material by hydrolyzing cellulose to glucose.

2. Other enzymes:

i. Dehalogenase: Catalytically removes halogen atoms from halogenated organic compounds commonly found in wastewater. Sugar reduces the organic matter content of wastewater.

Mechanism for Enzymatic Degradation

i. Substrate Binding: The enzyme binds to the organism (substrate) at its active site, forming an enzyme–substrate complex.
ii. Catalysis: Enzymes catalyse reactions that break down pollutants into smaller, non-toxic molecules.
iii. Released Product: The reaction product is released from the enzyme and is then free to bind to another substrate molecule. energy is needed.

Applications in Wastewater Treatment:

i. Industrial Wastewater: Enzymes such as laccase and peroxidase are used to treat wastewater from the textile, paper, and pulp industries, which often contain organic dyes and phenolic compounds. These enzymes decompose bacteria into non-toxic products.
ii. Domestic Wastewater: Hydrolytic enzymes such as protease and lipase can decompose organic matter, fats, and oils, and reduce the total oxygen demand (BOD) and chemical oxygen demand (COD) of domestic wastewater.
iii. Agricultural Runoff: Enzyme treatment can reduce the number of pesticides and herbicides used in agriculture and prevent toxic chemicals from water pollution.
iv. Heavy Metal Removal: Some enzymes can help remove wastewater by converting heavy metals into less harmful forms.

Advantages of Enzymatic Degradation

i. Enzymatic destruction: It is rather easy on the environment in comparison to the use of chemical additives as it does not leave a great deal of pollution."
ii. Energy Efficiency: The system is efficient in myriads of small cases, and thus aids in the reduction of energy used as well as the operating costs.
iii. Universal Application: Enzymes have great flexibility in their application and can be utilized in small septic systems as well as in large industries.
iv. Versatile: Many enzymes can be used to treat various types of wastewaters which don't interfere with each other.

Limitations and Challenges

i. Cost: One challenge could be the amount of money needed to produce and purify enzymes since people might not find it economical for mass use.

ii. Enzyme Stability: Enzymes are very temperamental and are said to lose their essential properties like activity even in extreme heat, acidic or alkaline environments and the opposite is true even when they are with inhibitors.
iii. Inhibitors in Wastewater: Wastewater enzyme degradation could take place however there are issues such as heavy metals and detergents that lead to damaging or modifying them.
iv. Reuse and Recycling: More research and technical expertise are needed to properly develop the recovery and reuse of the enzymes employed in medical processes.
v. Advances in enzymes have certainly put the enzyme utilization processes including the use of enzyme degradation in the wastewater processes to greater utilization levels.
vi. Immobilization Technology: Modification of the enzymes can be achieved through genetic manipulation that would provide desired stability and activity.
vii. Genetic Engineering: Through the various genetic alterations, the different kinds of properties of the enzyme such as its stability and activity algorithms can be improved.

Constructed Wetlands with Enhanced Filtration

Water pollution is one of the biggest issues in today's world which has been caused by urbanization, industrialization, commerce and agriculture. The water management and wastewater treatment solutions are also enhancing their effectiveness to address this issue effectively. The man-made wetlands are another development that is new in the treatment of wastewater. This process integrates the natural wetland with engineering design to enhance the treatment of wastewater. This system is environmentally friendly, low-maintenance and applies physical, chemical, and biological methods for decontamination. They have shallow pots containing substrate like gravel, sand or soil and these pots contain aquatic plants. Wastewater is discharged into the wetland and the treatment process involves physical filtration, chemical adsorption and biological degradation.

Constructed Wetland Working

Water distribution: Wastewater is discharged into the wetland through the input and diffused throughout the entire body.

Primary Treatment: Sedimentation of solids is done or filtering as water moves through the substrate.

Biological treatment: The bacterial load in the substrate and the roots removes organic matter and reduces the water's oxygen demand (BOD) and chemical oxygen demand (COD).

Chemical Process: Nutrients such as nitrogen and phosphorus, heavy metals and other pollutants are removed through adsorption and precipitation.

Polishing Stage: The last phase of treatment involves filtering the water through wetland plants and the substrate before it is released from the body to further remove bacteria and viruses. to reduce any remaining bacteria and viruses.

Improving Filtration in Constructed Wetlands

1. **Advanced Matrix Materials**: Gravel and Sand used daily for visual inspection and microbial colonization. Porous, carbon-rich material that enhances the adsorption of organic and inorganic contaminants.
2. **Layered Filter Media**: The soil structure may have multiple layers of filter media, each designed to target specific organisms. For example, the top layer of sand can retain solid waste, while the bottom layer of activated carbon will adsorb dissolved organic compounds.
3. **Hybrid Wetlands**: Combine vertical and horizontal lines to improve treatment. Vertical-flow wetlands increase the transfer of oxygen to aerobic processes, while horizontal-flow wetlands are well suited to anaerobic and facultative microbial activities.
4. **Aeration System**: Artificial aeration can improve oxygen utilization, reduce pollutant levels, and improve nitrogen removal.

Types of Constructed Wetlands

1. **Surface water flows**: Wastewater flows through a wetland, surface by surface. This is relatively aesthetic and creates habitat, but it does not act as well as soil toward the removal of some pollutant (Fig. 18.5).
2. **Subsurface-Flow Wetlands**: Wastewater will flow below the surface to a trap matrix. That design reduces Odors and mosquitoes, and this makes an ideal system for an urban area or residential areas.
3. **Hybrid Wetlands**: Wetlands and subsurface flows hybridize and improve contaminant removals.

Applications of Constructed Wetlands with Improved Filtration.

1. **Municipal sewage treatment**: wetlands are used to deal with municipal wastewater and pollutants, pathogens, and other nutrients such as nitrogen and phosphorus.
2. Industrial Wastewater: These systems are commonly used by textile, food, and paper manufacturing firms. These systems treat industrial wastewater before discharge.

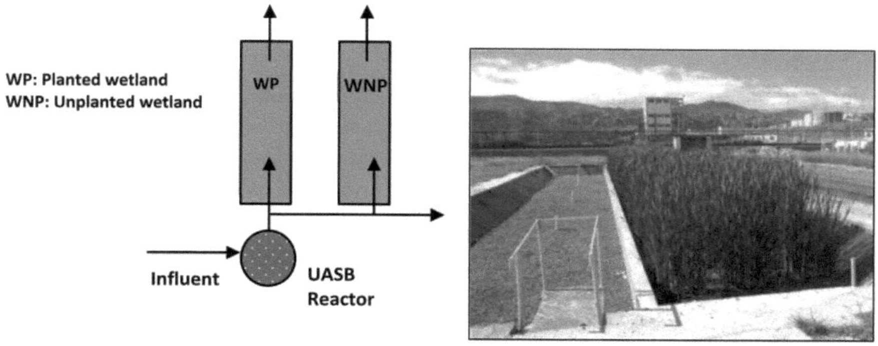

Fig. 18.5 Wetland construction for wastewater treatment

3. Stormwater Management: the construction of wetland systems can remove pollutants such as oil, heavy metals, and sewage from urban runoff.
4. Agricultural Runoff: These systems reduce the effect of nutrient runoff and thus reduce the risk of eutrophication of adjacent water bodies.

Benefits of Constructed Wetlands with Improved Filtration

1. Cost effective: Industrial facilities are less costly than wastewater treatment plants because they rely on natural processes and require less energy.
2. Sustainable and environmentally friendly: They use natural materials and processes, harmless to crops, and often provide wildlife habitat that improves local biodiversity.
3. Low Maintenance: After establishment, wetlands have minimal maintenance needs, including removal of excess vegetation.
4. Aesthetic and Recreational Benefits: These machines can be incorporated into public areas to provide beauty and educational value.
5. Flexibility: Wetlands can be designed to treat different types of wastewaters and can accommodate small homes to large cities.

Challenges and Limitations

1. Land Requirements: The land on which to build requires a lot of space, which may not be possible in a densely populated area.
2. Seasonal Changes: Treatment changes with seasonal changes in temperature and plant growth.
3. Clogging: The substrate can become clogged with dirt or organic matter, reducing the system's performance and requiring timely maintenance.

Supercritical Water Oxidation (SCWO)

Supercritical Water Oxidation (SCWO) is an unconventional but dream water treatment technique that gains benefit from peculiarities in the behaviour of water when it achieves its supercritical state. This big, advanced oxidation technology is received more attention for the powerful organic and inorganic pollutant degradations it achieves. It is important, therefore, in some of the most challenging scenarios related to wastewater treatment.

Principle of SCWO Water

When the temperature is above 374 °C and the pressure is higher than 22.1 MPa, water turns into the supercritical phase, where it has mixed properties of a liquid and a gas, wherein it becomes less viscous but highly diffusive and capable of dissolving organic compounds and gases like oxygen. These are the properties that make supercritical water excellent oxidation reaction media.

SCWO takes place by the oxidation of organic and inorganic impurities in supercritical wastewater. Generally, the process of this treatment includes the subsequent steps:

Pre-treatment: The wastewater goes through filtration, which removes large solids that have the potential to block the system.

Pressurization and Heating: The filtered water is both pressurized and heated to reach supercritical conditions.

Oxidation: Oxygen or an oxidizing agent is added to the reactor. At supercritical conditions, oxygen's solubility in the medium is highly improved and facilitates rapid and full oxidation of pollutants.

Cooling and Depressurization: The treated water is cooled and depressurized, and the by-products are separated.

Advantages of SCWO

SCWO enjoys several advantages in comparison with other technologies that are applied for wastewater treatment:

1. **High destruction efficiency**: Organic pollutants, such as toxic chemicals, pharmaceuticals, pesticides, and industrial chemicals can be destroyed nearly to 99.99% extent with SCWO.
2. **Broad applicability**: The process is efficient for treating such waste streams as municipal sewage, industrial effluents, and highly concentrated sludges.

3. **Low Secondary Pollution**: SCWO produces very low levels of secondary pollutants since oxidation transforms organic compounds into innocuous by-products like water, carbon dioxide, and nitrogen gas.
4. **Energy Recovery**: Oxidation reactions in supercritical water are exothermic; hence, SCWO systems can recover this energy for a more sustainable process.
5. **Compact System**: SCWO systems are compact, taking up minimal space for installation.

Challenges and Limitations

Despite its benefits, SCWO faces the following challenges:

1. **Extreme Operating Conditions**: High temperatures and pressures require specialized materials and equipment, thereby increasing the cost of investment.
2. **Corrosion**: Supercritical water can be highly corrosive, especially when chlorinated compounds or salts are present, and thus demands high-end corrosion-resistant materials.
3. **Deposition of Salts**: Salts precipitate out due to the supercritical conditions, thus causing scaling and clogging of the reactor.
4. **Complexity**: This system demands highly trained operators and very strict maintenance schedules.

Applications of SCWO

The versatility and effectiveness of SCWO have made it applicable for use in numerous industries and sectors. Among the more important applications is:

1. **Industrial Wastewater Treatment**: SCWO is found to be very effective for the treatment of industrial effluents containing recalcitrant organic compounds like petrochemical and pharmaceutical wastes.
2. **Municipal Wastewater Treatment**: It could be used to treat the municipal sewage and sludge as well, minimizing the waste volume and its toxicity content.
3. **Destruction of Toxic and Hazardous Waste**: SCWO can destroy highly toxic and hazardous materials like PCBs and dioxins.
4. **Military and Defence**: The technology is applied to neutralize chemical warfare agents and other hazardous materials.

Advancements in SCWO Technology

Recent innovations have been undertaken to address the limitations of SCWO and enhance its performance.

1. **Material Innovation**: Novel alloys and coatings with greater resistance to corrosion have increased the resilience and durability of SCWO systems.
2. **Salt Management**: Improved reactor designs and salt removal technologies significantly reduce scaling and clogging.
3. **Energy Optimization**: Integration of heat exchangers and energy recovery systems reduces overall energy requirements of SCWO processes.
4. **Process Intensification**: Application of swirling flows and multi-phase mixing promotes reaction kinetics and efficiency via modern advancements in reactor design.

Environmental Impact

SCWO provides a significant contribution to environmental protection because it offers a sustainable solution for wastewater treatment. The capability of SCWO to mineralize pollutants and reduce the environmental load of hazardous substances makes it an invaluable technology for achieving cleaner water bodies and minimizing ecological disruption.

Pychoremediation

When a water body is overly enriched in compounds of nitrogen and phosphate, it becomes eutrophic, which accelerates the growth of algae and plants, organic substances found in pharmaceutical effluents, such as acetone and alcohols, are employed as denaturants in denatured alcohol and as solvents. Algae are eukaryotic, photosynthetic and non-vascular organisms that ranges from single-celled (microalgae) or colonial to huge multicellular (macroalgae) organism. Though majority of algae are photosynthetic like other plants, but lack specialized structures—stems, vascular tissue, roots and leaves. Microalgae are microscopic organisms ranging size between few micrometer to few hundred micrometer like diatoms, dinoflagllates. Under optimal mixotropic or autotropic conditions, they multiply quickly and absorb nutrients into its body cell. Colonial microalgae from cluster or colonies of multiple cells ranging from a few cells to several millimeters in diameter like Chlorella vulgaris (Fig. 18.6). Macroalgae grow quickly in aquatic environments and reach lengths of tens of meters like spirogyra, stoneworts and giant kelp. They are made up of a specialized base structure that allows adhering to a surface holdfast and thallus, a leaf-like body with enclosed gas-filled structures that aid in buoyancy and keep them floating in water. Advantages of cultivation of algae in wastewater are reduction in eutrophication, biofuel production, soil conditioner, carbon sequestration, reduces carbonfoot prints and pollution remediation (Ankit et al., 2022).

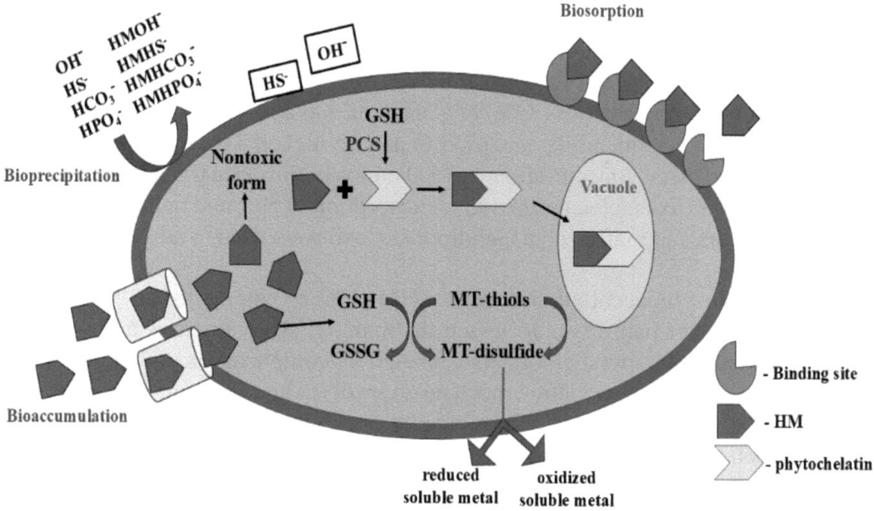

Fig. 18.6 Representation of phycoremediation mechanism in algal cell

Phycoremediation is cost-effective and greener alternative to traditional approaches. It removes contaminants from the environment and generates useful chemicals. Phycoremediation also known as algal bioremediation is the method of use of plants to absorb harmful and nonharmful substances from water and soil or just squester carbon from it generating biomass (Das et al., 2015; Paul et al., 2023; Pradhan et al., 2023). A variety of microalgal species, including Scenedesmus, Sargasssum, Chlorella, Botryococcus, Phormidium, Arthrospira, Spirulina, and Chlamydomonas have been associated with successful removal of nutrients, heavy metals, emerging pollutants and pathogens from wastewater (Walters et al., 2024).

Mechanism of Algal Remediation

Bioremediation includes many processes like bioaccumulation, biosorption, bioprecipitation and biominerisation. All these processes are interlinked to each other.

Biosorption is a passive phenomenon helps to absorb organic or inorganic pollutants, capable of binding and concentrating ions present in aqueous solutions. The working principle of biosorption process is affinity between the adsorbate and the biosorbent. Seaweeds are generally used to lower or eliminate harmful heavy metals. The heavy metal ion adheres to functional group of the cell (Ordóñez et al., 2023). The study has been conducted where brown algae Cystoseira crinite and Cystoseira barbata were used to remove chromium from wastewater. For Cr (III), optimum operating conditions were maintained, where metal concentration was 100 ppm at 4.5 pH for 120 min. the adsorbent dose was 0.1 g/50 ml. The removal efficiency reported was 70.70% for C. barbata and 73.34% for C. crinite. For Cr (VI), metal concentration was 100 ppm at 2.0 pH for 24 h. The adsorbent dose was 100 mg/50 ml. The removal efficiency reported was 35% for C. barbata and 28% for C. crinite. Similarly, seaweed Sargassum filipendula was used to adsorb nickel. The efficiency obtained was 45%, where metal concentration was 1.0 mmol/L, flow rate 0.5 ml/min at 25 °C. Green n seaweed Ulva lactuca was used for adsorption of zinc and mercury. Seaweed Turbinaria ornata and marine alga Gelidium amansii were used to adsorb lead (Znad et al., 2022).

The process of bioaccumulation allows HMs ions to enter cells by both passive and active transport pathways. To lessen the toxicity, algae employs both intracellular and extracellular metal binding site, chelation, complexation and ion exchange (Ankit et al., 2022). Detoxification mechanism involves binding of metal to cell wall, formation of phytochelatins, metallothionenins or metal binding proteins and sequestration of metal into vacuole (Olguín & Sánchez-Galván, 2012). It has been observed that Oscillatoria bornettia, Chlorella vulgaris, Phacus curvicauda and Euglena acus can bioaccumulate Zn, Al, Fe, Cd and Cu. The maximum metal concentration factor observed in Oscillatoria was Zn: 0.306, Fe: 0.302, Cu: 0.091 and Cd: 0.276 (Mustafa et al., 2021).

Microalgae's cell wall contains lipids, proteins, glycoproteins, polysaccharides like cellulose and alginate and exopolysaccharide like uronic group. Many functional groups like hydroxyl, carbonyl, phosphate, sulfonate, thiol, amino grp and imidazole on its surface thus overall charge of cell wall is negative. Due to the presence of various functional groups, metal ions in water are adsorbed onto the negatively charged cell surface because they exist as cations. The metal ions enter the cell through special transport channel and reacts to form reactive oxygen species (ROS). Phytochelatins and metallothioneins are heavy metal detoxifying compound present in various plant species. To reduce the toxicity of metal ions, cell uses gluthathione and enzyme phytochelation synthase to forms phytochelatins (PC). Phytochelatins and MTs are polypeptides that primarily bind metals, forming HM-MT and HM-PC complexes. The cysteine residues' sulfide groups play an important part in this process. Metallothioneins (MT) are extremely small proteins made up of 300 aminoacid found in the cytoplasm with high cysteine content (15–35%) that bind and sequester heavy metals inside cells. MTs bind to heavy metals via cysteine residues, producing stable complexes that reduce metal toxicity. Metal ions are coordinated by MTs using amino acid residues and thiolate groups (-SH). Glutathione (GSH) is a tripeptide made composed of a glutamate unit, a cysteine molecule (-glutamylcysteine), and a glycine unit linked by a peptide bond. Formerly phytochelatins were known as class III MTs in higher plants. GSH also serves as a predecessor to PCs. PCs help transfer HMs from the cytoplasm to the vacuole in higher plants. GSH is produced by two biosynthetic reactions, first from cysteine and glutamic acid catalysed by glutamate-cysteine ligase (GCL) to form glutamylcysteine. And second process is ligation of glutamylcysteine with glycine by GSH synthetase (GSHS). Phytochelatins and genes related to coding of its biosynthetic pathway GCL, GSHS and PCS were found in

Chlorophyta (Koul et al., 2022; Gómez-Jacinto et al., 2015; Balzano et al., 2020; Abd Elnabi et al., 2023).

Algae-based WWT has limited organic removal. In fact, the presence of organic matter in wastewater can cause significant problem such as bacteria contaminations and light penetration limitation due to high turbidity. Therefore, most of algae-based WWT studies mainly utilised tertiary effluent, which has low organic carbon. Heterotrophic and mixotrophic cultivation can utilise the organic carbon present in wastewater. However, algae will be in competition with bacteria for this carbon. In this competition, bacteria often out compete the algae populations (Zabochnicka et al., 2022).

Hydrodynamic Cavitation

Cavitation is the process of forming, expanding, and collapsing vapor or gas-filled cavities in a short period of time. The collapsing leads to implosion causing nearby structure to break. Cavitation can be classified into four types based on its mode of generation: hydrodynamic cavitation (HC), acoustic cavitation (AC), particle cavitation (PC) and optic cavitation (OC). In OC, a pulsed laser's light converges in a liquid, resulting in a local energy deposit. Optical breakdown then causes temporary bubbles formation. Particle cavitation refers to the cavitation caused by a proton rupturing a liquid in a bubble chamber. HC and AC are mostly used for wastewater treatment process. The AC is created by pressure fluctuations in the liquid induced by sound waves, mostly ultrasound (16 kHz–100 MHz) is used (Wang et al., 2021). Hydrodynamic cavitation (HC) occurs when pressure drops due to velocity fluctuations caused by the geometry of a flowing system result in the development of bubbles or cavities (Fig. 18.7). Extreme energy is generated during the implosion of these holes, which can cause mechanical and chemical impacts. Water molecules inside the cavities undergo homolytic cleavage due to localized regions of high temperature and pressure, leading to formation of dotOH radicals. There are three possible sites for collapse during cavitation.

First is the gas phase inside the bubble, where volatile compounds can undergo thermolytic breakdown and radical dotOH formation. Second is the gas–liquid interface, where non-volatile and hydrophobic compounds can degrade. Third possibility is the liquid bulk phase, where non-volatile and hydrophilic compounds can degrade. The elimination of organic molecules depends on their chemical structure since small number of radicals can reach liquid bulk phase, where they either react with one another or with any nearby oxidizable chemicals. Adding exogenous oxidants like hydrogen peroxide could speed up this process (Zupanc et al., 2014). In hydrodynamic cavitation, the cavity is caused by constrictions in liquid flow. A venturi design with smooth convergence/divergence achieves the highest degree of degradation compared to an orifice with similar operating pressure levels. Using an orifice plate offers the advantage of multiple holes within the same cross-sectional area and simple fabrication (Badmus et al., 2018).

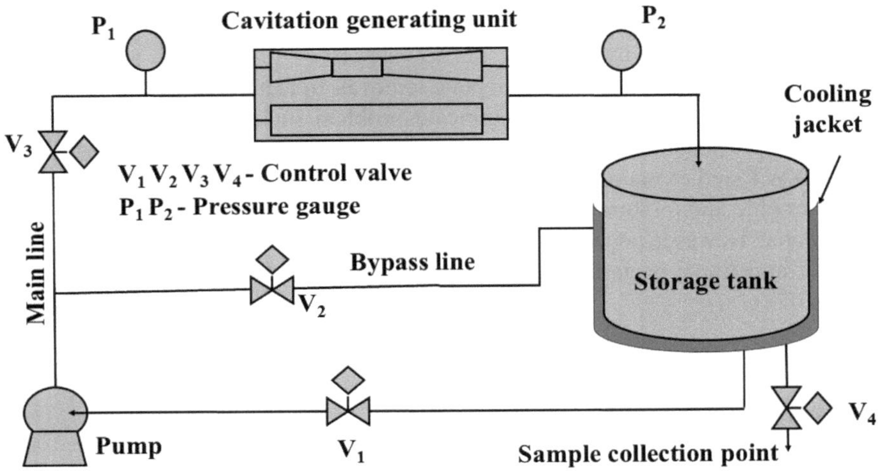

Fig. 18.7 Process for hydrodynamic cavitation for wastewater (Wang et al., 2021)

Circular Economy in Waste Management and Resource Recovery

The concept of a circular economy has become one of the transformative frameworks on how to approach the waste and resource challenges globally. A circular economy, when focusing on wastewater treatment, ensures that waste is reduced at its minimum level while, at the same time maximizing the recovery of resources towards sustainable systems in harmony with both environmental and economic goals. The fact that wastewater is a resource instead of a liability, makes the approach redefine traditional practices into a sustainable future.

Principles of Circular Economy

A circular economy consists of three basic principles, which are:

1. Designing Out Waste and Pollution: Systems are developed to avoid waste creation, and the impact on natural resources is minimized.
2. Keeping Products and Materials in Use: Resources are re-processed, recycled, or otherwise reused to retain their value inside the system.
3. Regenerating Natural Systems: Design processes are created to rebuild rather than deplete the natural surroundings.

Role of Wastewater in the Circular Economy

Wastewater contains a great number of valuable resources such as nutrients, energy, and water. It will form the basis of circular economy which recover all of these resources by developing high technologies and innovative ways for reducing virgin material usage and minimizing environmental damage.

Resource Recovery in Wastewater Treatment

Resource recovery forms the heart of wastewater treatment in the context of the circular economy. Resources that can be recovered in this process include:

1. Water Reuse: Treated wastewater might be reused for agricultural purposes, industrial processes, or even potable water after advanced treatment.
2. Nutrient Recovery: Nutrients such as nitrogen and phosphorus are recovered for use as fertilizer thereby reducing the use of synthetic alternatives.
3. Energy Recovery: Biogas produced during the process of anaerobic sludge digestion can be trapped and used as a source of renewable energy for heat, electricity, or as fuel.
4. Material recovery: Precious metals, rare earth elements, among other materials can be recovered from industrial wastewater streams.

Technologies that Allow for Circularity in Wastewater Treatment

Several new technologies enable circular economy in wastewater management:

1. Membrane Filtration: The modern filtration system allows water to be reused through removal of contaminants at the micro and nano level.
2. Anaerobic Digestion: Organic matter is converted into biogas, which can be used as a renewable source of energy.
3. Struvite Precipitation: This is a method to recover phosphorus from the wastewater in the form of struvite, a fertilizer.
4. Electrochemical Systems: Emerging technologies using electrochemical processes for metal and nutrient recovery.
5. Thermal Hydrolysis: Breaks down sludge, enhances biogas production, and recovers valuable resources.

Challenges and Barriers

There are challenges associated with the adoption of a circular economy in wastewater treatment:

1. Technological Limitations: Some of the recovery processes require highly advanced technologies that may not be readily available or are expensive.
2. Regulatory Hurdles: The strict regulations and standards on treated water and recovered materials can act as a barrier to adoption.
3. Public Perception: Wastewater reuse, especially for potable purposes, can lead to negative perceptions that inhibit acceptance.
4. Infrastructure Requirements: Great investment will be needed while retrofitting existing wastewater plants with resource recovery technologies.

Case Studies and Success Stories

Several examples show the successful realization of circular economy principles in wastewater treatment:

1. NEWater: Singapore has created a network to recycle treated wastewater into high-grade water, fit for industrial and potable purposes, reducing its dependency on external sources of water.
2. The Netherlands' Nutrient Recovery Initiatives: Dutch wastewater treatment plants have adopted phosphorus recovery technologies, produced fertilizers and contributed to agricultural sustainability.
3. Biogas Production in Sweden: Anaerobic digestion has been implemented in wastewater treatment plants in Sweden to produce biogas, which is used as a renewable energy source for heating and transportation.

Future Prospects

The implementation of circular economy principles for wastewater treatment is expected to expand since global challenges include water scarcity and the depletion of resources. Advances in technology, more accommodating policies, and greater public awareness are all going to propel such transitions. Future developments will involve:

1. Smart Water Management: Digital technologies can facilitate recovery and efficient management of resources.
2. Decentralized Treatment Systems: Small-scale, localized wastewater treatment systems that allow for resource recovery at the community level.
3. Collaborative Models: A collaboration between governments, industries, and communities to practice sustainable wastewater management.

References

Abd Elnabi, M. K., Elkaliny, N. E., Elyazied, M. M., Azab, S. H., Elkhalifa, S. A., Elmasry, S., Mouhamed, M. S., Shalamesh, E. M., Alhorieny, N. A., Abd Elaty, A. E., Elgendy, I. M., Etman, A. E., Saad, K. E., Tsigkou, K., Ali, S. S., Kornaros, M., & Mahmoud, Y. A. G. (2023). Toxicity of heavy metals and recent advances in their removal: A review. In *Toxics* (Vol. 11, Issue 7). Multidisciplinary Digital Publishing Institute (MDPI). https://doi.org/10.3390/toxics11070580

Abdelrasoul, A., Doan, H., Lohi, A., & Cheng, C.-H. (2018). Fouling in forward osmosis membranes: Mechanisms, control, and challenges. In *Osmotically Driven Membrane Processes - Approach, Development and Current Status*. InTech. https://doi.org/10.5772/intechopen.72644

Abdul – Jabbar, R. M., Al-Adily, B. M., & Kadhim, N. F. (2024). Ecotoxicological effects of pharmaceutical preparations and the possibility of removing them through phytoremediation: A review. *Edelweiss Applied Science and Technology, 8*(6), 2679–2695. https://doi.org/10.55214/25768484.v8i6.2531

Ankit, Bauddh, K., & Korstad, J. (2022). Phycoremediation: Use of algae to sequester heavy metals. *Hydrobiology, 1*(3), 288–303. https://doi.org/10.3390/hydrobiology1030021

Badmus, K. O., Tijani, J. O., Massima, E., & Petrik, L. (2018). Treatment of persistent organic pollutants in wastewater using hydrodynamic cavitation in synergy with advanced oxidation process. *Environmental Science and Pollution Research, 25*(8), 7299–7314. https://doi.org/10.1007/s11356-017-1171-z

Balzano, S., Sardo, A., Blasio, M., Chahine, T. B., Dell'Anno, F., Sansone, C., & Brunet, C. (2020). Microalgal metallothioneins and phytochelatins and their potential use in bioremediation. In *Frontiers in Microbiology* (Vol. 11). Frontiers Media S.A. https://doi.org/10.3389/fmicb.2020.00517

Bandehali, S., Parvizian, F., Hosseini, S. M., Matsuura, T., Drioli, E., Shen, J., Moghadassi, A., & Adeleye, A. S. (2021). Planning of smart gating membranes for water treatment. In *Chemosphere* (Vol. 283). Elsevier Ltd. https://doi.org/10.1016/j.chemosphere.2021.131207

Behera, I. D., Nayak, M., Biswas, S., Meikap, B. C., & Sen, R. (2021). Enhanced biodegradation of total petroleum hydrocarbons by implementing a novel two-step bioaugmentation strategy using indigenous bacterial consortium. *Journal of Environmental Management, 292*, Article 112746. https://doi.org/10.1016/J.JENVMAN.2021.112746

Behera, I. D., Nayak, M., Mishra, A., Meikap, B. C., & Sen, R. (2022). Strategic implementation of integrated bioaugmentation and biostimulation for efficient mitigation of petroleum hydrocarbon pollutants from terrestrial and aquatic environment. *Marine Pollution Bulletin, 177*, Article 113492. https://doi.org/10.1016/J.MARPOLBUL.2022.113492

Bertram, M. G., Martin, J. M., Wong, B. B., & Brodin, T. (2022). Micropollutants. *Current Biology, 1*(32), 17–19. https://doi.org/10.31234/osf.io/puqvs

Bhati, S. S., Saini, P., & Dutt, D. (2023). *Applications of zeolites 13x to reduce the volatile organic compounds and odorous compounds in recycled polypropylene*. https://doi.org/10.2139/ssrn.4530732

Bora, S., Das, K., Saini, P., Katiyar, V., & Ghosh, T. (2025). Application of nanoemulsion for improving quality and safety of fruits and vegetables. In *Advancements in nanotechnology for food and packaging* (pp. 53–78). Elsevier. https://doi.org/10.1016/B978-0-443-21428-8.00003-5

Cano, R., Pérez, J. L., Dávila, L. A., Ortega, Á., Gómez, Y., Valero-Cedeño, N. J., Parra, H., Manzano, A., Castro, T. I. V., Díaz Albornoz, M. P., Cano, G., Rojas-Quintero, J., Chacín, M., & Bermúdez, V. (2021). Role of endocrine-disrupting chemicals in the pathogenesis of non-alcoholic fatty liver disease: A comprehensive review. In *International Journal of Molecular Sciences* (Vol. 22, Issue 9). MDPI AG. https://doi.org/10.3390/ijms22094807

Cybulski, J., Witczak, A., & Pokorska-Niewiada, K. (2021). The effect of water and sewage treatment on reducing residues of selected organochlorine pesticides in Szczecin (Poland). *Water, Air, & Soil Pollution, 232*(8), 310. https://doi.org/10.1007/s11270-021-05261-6

Das, A. P., Ghosh, S., Mohanty, S., & Sukla, L. B. (2015). Advances in manganese pollution and its bioremediation. In L. B. Sukla, N. Pradhan, S. Panda, & B. K. Mishra (Eds.), *Environmental microbe biotechnology* (1st ed., pp. 313–328). Springer, Cham. https://doi.org/10.1007/978-3-319-19018-1_16
Das, P., & Bal, M. (2025). Modeling and optimization study for microplastic removal from aquatic medium using Chroococcidiopsis species. *Chemical Engineering Science, 304*, Article 121085. https://doi.org/10.1016/J.CES.2024.121085
Das, P., Halder, G., & Bal, M. (2023). A critical review on remediation of microplastics using microalgae from aqueous system. *Science of the Total Environment, 898*, Article 166425. https://doi.org/10.1016/J.SCITOTENV.2023.166425
Deng, Y., & Englehardt, J. D. (2006). Treatment of landfill leachate by the Fenton process. *Water Research, 40*(20), 3683–3694. https://doi.org/10.1016/j.watres.2006.08.009
du Plessis, M., Fourie, C., Stone, W., & Engelbrecht, A. M. (2023). The impact of endocrine disrupting compounds and carcinogens in wastewater: Implications for breast cancer. In *Biochimie* (Vol. 209, pp. 103–115). Elsevier B.V. https://doi.org/10.1016/j.biochi.2023.02.006
Gómez-Jacinto, V., García-Barrera, T., Gómez-Ariza, J. L., Garbayo-Nores, I., & Vílchez-Lobato, C. (2015). Elucidation of the defence mechanism in microalgae Chlorella sorokiniana under mercury exposure. *Identification of Hg-Phytochelatins. Chemico-Biological Interactions, 238*, 82–90. https://doi.org/10.1016/j.cbi.2015.06.013
Goswami, L., Vinoth Kumar, R., Narayan Borah, S., Arul Manikandan, N., Pakshirajan, K., & Pugazhenthi, G. (2018). Membrane bioreactor and integrated membrane bioreactor systems for micropollutant removal from wastewater: A review. *Journal of Water Process Engineering, 26*, 314–328. https://doi.org/10.1016/j.jwpe.2018.10.024
Kosek, K., Luczkiewicz, A., Fudala-Książek, S., Jankowska, K., Szopińska, M., Svahn, O., Tränckner, J., Kaiser, A., Langas, V., & Björklund, E. (2020). Implementation of advanced micropollutants removal technologies in wastewater treatment plants (WWTPs) - Examples and challenges based on selected EU countries. In *Environmental Science and Policy* (Vol. 112, pp. 213–226). Elsevier Ltd. https://doi.org/10.1016/j.envsci.2020.06.011
Koul, B., Sharma, K., & Shah, M. P. (2022). Phycoremediation: A sustainable alternative in wastewater treatment (WWT) regime. In *Environmental Technology and Innovation* (Vol. 25). Elsevier B.V. https://doi.org/10.1016/j.eti.2021.102040
Kulišťáková, A. (2023). Removal of pharmaceutical micropollutants from real wastewater matrices by means of photochemical advanced oxidation processes—A review. In *Journal of Water Process Engineering* (Vol. 53). Elsevier Ltd. https://doi.org/10.1016/j.jwpe.2023.103727
Kumar, M., Pandit, S., Patel, V., Khanna, N., Nag, M., Lahiri, D., Ray, R. R., Das, A., & Das, D. (2023). Maximization of energy recovery from starch processing wastewater by Thermophilic dark fermentation coupled with microbial fuel cell technology. *Geomicrobiology Journal*, 1–12. https://doi.org/10.1080/01490451.2023.2209555
Lauesen, L. M. (2022). Advanced wastewater treatment of micropollutants—state of the art. *Technological Sustainability, 1*(2), 101–120. https://doi.org/10.1108/TECHS-09-2021-0007
Mishra, S., & Das, A. P. (2021). Current treatment technologies for removal of Microplastic and microfiber pollutants from wastewater. *Wastewater Treatment: Cutting-Edge Molecular Tools, Techniques and Applied Aspects*, 237–251. https://doi.org/10.1016/B978-0-12-821881-5.00011-8
Mishra, S., & Das, A. P. (2022). Treatment of the wastewater polluted with synthetic microfiber released from washing machine. In B. B. Das, H. Hettiarachchi, P. K. Sahu, & S. Nanda (Eds.), *Lecture Notes in Civil Engineering* (Vol. 207, pp. 109–117). Springer Science and Business Media Deutschland GmbH. https://doi.org/10.1007/978-981-16-7509-6_9/COVER
Mishra, S., & Das, A. P. (2023). Microbial remediation of synthetic microfiber contaminated wastewater. In M. P. Shah (Ed.), *Microbial technologies in industrial wastewater treatment* (pp. 337–350). Springer, Singapore. https://doi.org/10.1007/978-981-99-2435-6_17
Mishra, S., Dash, D., & Das, A. P. (2022a). Detection, characterization and possible biofragmentation of synthetic microfibers released from domestic laundering wastewater as an emerging

source of marine pollution. *Marine Pollution Bulletin, 185*, Article 114254. https://doi.org/10.1016/J.MARPOLBUL.2022.114254

Mishra, S., Dash, D., & Das, A. P. (2024). Isolation and molecular characterization of polyester-degrading bacterial strains from municipal wastewater and their application in Microplastic degradation in contaminated wastewater. *Geomicrobiology Journal*. https://doi.org/10.1080/01490451.2024.2376101

Mishra, S., Singh, R. P., Rout, P. K., & Das, A. P. (2022). Membrane bioreactor (MBR) as an advanced wastewater treatment technology for removal of synthetic microplastics. *Development in Wastewater Treatment Research and Processes*, 45–60. https://doi.org/10.1016/B978-0-323-85583-9.00022-3

Mohanta, S., Pradhan, B., & Behera, I. D. (2024). Impact and Remediation of petroleum hydrocarbon pollutants on agricultural land: A review. *Geomicrobiology Journal, 41*(4), 345–359. https://doi.org/10.1080/01490451.2023.2243925

Mojiri, A., Zhou, J., Vakili, M., & Van Le, H. (2020). Removal performance and optimisation of pharmaceutical micropollutants from synthetic domestic wastewater by hybrid treatment. *Journal of Contaminant Hydrology, 235*. https://doi.org/10.1016/j.jconhyd.2020.103736

Mustafa, S., Bhatti, H. N., Maqbool, M., & Iqbal, M. (2021). Microalgae biosorption, bioaccumulation and biodegradation efficiency for the remediation of wastewater and carbon dioxide mitigation: Prospects, challenges and opportunities. In *Journal of Water Process Engineering* (Vol. 41). Elsevier Ltd. https://doi.org/10.1016/j.jwpe.2021.102009

Nowak, K., Jabłońska, E., & Ratajczak-Wrona, W. (2019). Immunomodulatory effects of synthetic endocrine disrupting chemicals on the development and functions of human immune cells. In *Environment International* (Vol. 125, pp. 350–364). Elsevier Ltd. https://doi.org/10.1016/j.envint.2019.01.078

Nthunya, L. N., Bopape, M. F., Mahlangu, O. T., Mamba, B. B., Van der Bruggen, B., Quist-Jensen, C. A., & Richards, H. (2022). Fouling, performance and cost analysis of membrane-based water desalination technologies: A critical review. *Journal of Environmental Management, 301*, Article 113922. https://doi.org/10.1016/j.jenvman.2021.113922

Ohore, O. E., & Zhang, S. (2019). Endocrine disrupting effects of bisphenol A exposure and recent advances on its removal by water treatment systems. *A Review. Scientific African, 5*, Article e00135. https://doi.org/10.1016/j.sciaf.2019.e00135

Olguín, E. J., & Sánchez-Galván, G. (2012). Heavy metal removal in phytofiltration and phycoremediation: The need to differentiate between bioadsorption and bioaccumulation. *New Biotechnology, 30*(1), 3–8. https://doi.org/10.1016/j.nbt.2012.05.020

Ordóñez, J. I., Cortés, S., Maluenda, P., & Soto, I. (2023). Biosorption of heavy metals with algae: Critical review of its application in real effluents. In *Sustainability (Switzerland)* (Vol. 15, Issue 6). Multidisciplinary Digital Publishing Institute (MDPI). https://doi.org/10.3390/su15065521

Paul, A., Dey, S., Ram, D. K., & Das, A. P. (2023). Hexavalent chromium pollution and its sustainable management through bioremediation. *Geomicrobiology Journal*, 1–11. https://doi.org/10.1080/01490451.2023.2218377

Petrakis, D., Vassilopoulou, L., Mamoulakis, C., Psycharakis, C., Anifantaki, A., Sifakis, S., Docea, A. O., Tsiaoussis, J., Makrigiannakis, A., & Tsatsakis, A. M. (2017). Endocrine disruptors leading to obesity and related diseases. In *International Journal of Environmental Research and Public Health* (Vol. 14, Issue 10). MDPI. https://doi.org/10.3390/ijerph14101282

Piry, A., Heino, A., Kühnl, W., Grein, T., Ripperger, S., & Kulozik, U. (2012). Effect of membrane length, membrane resistance, and filtration conditions on the fractionation of milk proteins by microfiltration. *Journal of Dairy Science, 95*(4), 1590–1602. https://doi.org/10.3168/jds.2011-4292

Popova, M., Boycheva, S., Lazarova, H., Zgureva, D., Lázár, K., & Szegedi, Á. (2020). VOC oxidation and CO_2 adsorption on dual adsorption/catalytic system based on fly ash zeolites. *Catalysis Today, 357*, 518–525. https://doi.org/10.1016/j.cattod.2019.06.070

Pradhan, G., Tripathy, B., Ram, D. K., Digal, A. K., & Das, A. P. (2023). Bauxite Mining Waste Pollution and Its Sustainable Management through Bioremediation. *Geomicrobiology Journal*, 1–10. https://doi.org/10.1080/01490451.2023.2235353

Primo, A., & Garcia, H. (2014). Zeolites as catalysts in oil refining. *Chemical Society Reviews, 43*(22), 7548–7561. https://doi.org/10.1039/C3CS60394F

Qasem, K. M. A., Khan, S., Chinnam, S., Saleh, H. A. M., Mantasha, I., Zeeshan, M., Manea, Y. K., & Shahid, M. (2022). Sustainable fabrication of Co-MOF@CNT nano-composite for efficient adsorption and removal of organic dyes and selective sensing of Cr(VI) in aqueous phase. *Materials Chemistry and Physics, 291*, Article 126748. https://doi.org/10.1016/j.matchemphys.2022.126748

Raszewski, G., Jamka, K., Bojar, H., & Kania, G. (2022). Endocrine disrupting micropollutants in water and their effects on human fertility and fecundity. In *Annals of Agricultural and Environmental Medicine* (Vol. 29, Issue 4, pp. 477–482). Institute of Agricultural Medicine. https://doi.org/10.26444/aaem/156694

Ratchnashree, S. R., Karmegam, N., Selvam, M., Manikandan, S., Deena, S. R., Subbaiya, R., Vickram, A. S., Kim, W., & Govarthanan, M. (2023). Advanced technologies for the determination of quantitative structure-activity relationships and degradation efficiency of micropollutants and their removal in water—A review. *Science of the Total Environment, 904*, Article 166563. https://doi.org/10.1016/j.scitotenv.2023.166563

Saini, P., Bhati, S. S., & Dutt, D. (2023). Removal of volatile organic compounds and odorous compounds for multilayer packaging recyclates using heated air purging. *Journal of Physics: Conference Series, 2652*(1), Article 012005. https://doi.org/10.1088/1742-6596/2652/1/012005

Saini, P., Sharma, K., Chauhan, A., & Bhati, S. S. (2025). Sustainable smart packaging. In *Sustainable materials for food packaging and preservation* (pp. 285–300). Elsevier. https://doi.org/10.1016/B978-0-443-13567-5.00013-7

Sin, J.-C., Lam, S.-M., Mohamed, A. R., & Lee, K.-T. (2012). Degrading endocrine disrupting chemicals from wastewater by Photocatalysis: A review. *International Journal of Photoenergy, 2012*, 1–23. https://doi.org/10.1155/2012/185159

Swain, N., Saini, P., Bhati, S. S., & Rastogi, V. K. (2023). Biodegradable stone paper as a sustainable alternative to traditional paper: A review. *Journal of Inorganic and Organometallic Polymers and Materials, 33*(8), 2240–2251. https://doi.org/10.1007/s10904-023-02644-6

Varjani, S., & Sudha, M. C. (2020). Occurrence and human health risk of micro-pollutants—A special focus on endocrine disruptor chemicals. In *Current Developments in Biotechnology and Bioengineering: Emerging Organic Micro-pollutants* (pp. 23–39). Elsevier. https://doi.org/10.1016/B978-0-12-819594-9.00002-4

Walters, C., Steyn, M., Ndlela, L., Nocanda, X., Moloi, M., & Oberholster, P. (2024). Phycoremediation of industrial wastewater: review of algae consortia. In *International Journal of Environmental Science and Technology*. Springer Nature. https://doi.org/10.1007/s13762-024-06130-5

Wang, B., Su, H., & Zhang, B. (2021). Hydrodynamic cavitation as a promising route for wastewater treatment—A review. *Chemical Engineering Journal, 412*, Article 128685. https://doi.org/10.1016/j.cej.2021.128685

Yang, Y., Zhang, X., Jiang, J., Han, J., Li, W., Li, X., Yee Leung, K. M., Snyder, S. A., & Alvarez, P. J. J. (2022). Which micropollutants in water environments deserve more attention globally? In *Environmental Science and Technology* (Vol. 56, Issue 1, pp. 13–29). American Chemical Society. https://doi.org/10.1021/acs.est.1c04250

Zabochnicka, M., Krzywonos, M., Romanowska-Duda, Z., Szufa, S., Darkalt, A., & Mubashar, M. (2022). Algal Biomass Utilization toward Circular Economy. In *Life* (Vol. 12, Issue 10). MDPI. https://doi.org/10.3390/life12101480

Znad, H., Awual, M. R., & Martini, S. (2022). The Utilization of Algae and Seaweed Biomass for Bioremediation of Heavy Metal-Contaminated Wastewater. In *Molecules* (Vol. 27, Issue 4). MDPI. https://doi.org/10.3390/molecules27041275

Zupanc, M., Kosjek, T., Petkovšek, M., Dular, M., Kompare, B., Širok, B., Stražar, M., & Heath, E. (2014). Shear-induced hydrodynamic cavitation as a tool for pharmaceutical micropollutants removal from urban wastewater. *Ultrasonics Sonochemistry, 21*(3), 1213–1221. https://doi.org/10.1016/j.ultsonch.2013.10.025

Chapter 19
Sustainable Management and Advanced Techniques of Micropollutants Through Circular Economy

Anil Kumar Murmu, Lipika Parida, and Veda Prakash

Introduction

Micropollutants are trace-level chemical substances in wastewater that originate from diverse sources, including pharmaceuticals, personal care products, pesticides, industrial chemicals, and household items (Das & Mishra, 2008; Mallick et al., 2023). Pharmaceuticals, such as antibiotics, analgesics, and hormones, enter water systems through human excretion and improper disposal, while personal care products contribute ingredients like parabens, triclosan, and synthetic musk. Pesticides and herbicides from agricultural runoff and industrial chemicals, including persistent organic pollutants (POPs), flame retardants, and heavy metals, further exacerbate the problem. Despite their low concentrations, these compounds pose significant environmental and health risks, including aquatic toxicity, bioaccumulation, biomagnification, and ecosystem disruption (Behera, 2021, 2022). Additionally, they are associated with serious health concerns such as endocrine disruption, antimicrobial resistance, carcinogenic and mutagenic effects, and neurotoxicity. Addressing these challenges is complicated by limitations in detection and quantification methods, inefficiencies in conventional wastewater treatment technologies, and regulatory gaps.

Emerging solutions for micropollutant management include advanced treatment technologies such as membrane filtration, advanced oxidation processes (AOPs), and bioremediation, as well as green chemistry approaches and resource recovery strategies supported by the Circular Economy (CE) (Das et al., 2023, 2024; Mohanty et al., 2024). The CE is an innovative economic model that aims to minimize waste and optimize resource efficiency by promoting the reuse, recycling, and regeneration of materials. In wastewater management, CE reframes wastewater as a resource-rich

A. K. Murmu · L. Parida (✉) · V. Prakash
Department of Chemical Engineering, Veer Surendra Sai University of Technology, Burla, Odisha, India
e-mail: lparida_chemical@vssut.ac.in

I. D. Behera et al. (eds.), *Emerging Micropollutants*, Sustainable Environmental Waste Management Strategies, https://doi.org/10.1007/978-3-032-03530-1_19

stream containing water, nutrients, energy, and valuable compounds that can be recovered and reused (Priyadarsini & Das, 2024, 2025). Key components of CE in this context include water reuse, nutrient extraction, energy recovery, and waste-to-resource transformation, offering a sustainable pathway to address water scarcity, pollution, and resource depletion (Alam et al., 2024).

The benefits of CE in wastewater management are extensive. They include environmental sustainability, economic opportunities, resilience to water scarcity, and climate change mitigation. By reducing pollution, minimizing natural resource extraction, creating new revenue streams, lowering operational costs, enhancing water security, and reducing greenhouse gas emissions through renewable energy generation, CE supports sustainable development. However, implementing CE in wastewater management faces significant challenges, including technological barriers, regulatory hurdles, knowledge gaps, and economic viability concerns. High initial costs for advanced treatment and resource recovery systems, the need for scalable and efficient innovations, unclear guidelines for water reuse, public resistance to treated wastewater reuse, and insufficient awareness and technical expertise among stakeholders remain critical obstacles.

This chapter explores the challenges associated with micropollutants in wastewater, emphasizing their persistence and environmental and health impacts. It introduces the Circular Economy Framework, highlighting its relevance to wastewater management and its potential to transform micropollutants into resource recovery opportunities. The chapter presents advanced techniques for micropollutant management, including advanced oxidation processes, adsorption, membrane technologies, and bioremediation, and discusses innovations that integrate these techniques within CE systems to enhance resource recovery and reduce waste.

Challenges in Conventional Wastewater Treatment

Organic micropollutants can significantly impact the environment when released through treated wastewater (Mishra & Das, 2023; Mishra et al., 2024; Mohanta et al., 2023). Numerous studies on their removal in wastewater treatment plants (WWTPs) have highlighted the critical role of biological treatment in micropollutant degradation (Dickenson et al., 2011). Factors such as pH, nitrification, heterotrophic activity, redox conditions, solid retention time (SRT), hydraulic retention time (HRT), and suspended/attached growth are relevant to micropollutant removal. However, there is no universal consensus on the primary drivers of biological micropollutant removal at WWTPs (Mishra & Das, 2021; Mishra, Singh, et al., 2022a, 2022b). This lack of agreement may stem from the complexity of mixed microbial communities and the structural diversity of micropollutants, or due to unidentified influential parameters (Abegglen & Siegrist, 2012).

Organic micropollutants in traditional biological wastewater treatment systems can be categorized as easily, moderately, sporadically, or poorly degradable (Kumar et al., 2023; Mishra & Das, 2022; Kanungo et al., 2023). Achieving low residual concentrations in treated wastewater requires targeting multiple classes of micropollutants through tailored approaches. Readily degradable compounds can be minimized under standard treatment conditions, while novel strategies are needed for moderately and sporadically degradable substances (Bahr et al., 2007). For poorly degradable micropollutants, significant advancements in biological treatment may be necessary. The effectiveness of conventional WWTPs in removing micropollutants depends on both pollutant characteristics and treatment methods (Mishra et al., 2022a, 2022b). Secondary treatment primarily relies on biological and/or chemical transformation and sorption. Certain micropollutants may resist removal due to their low sorption coefficient, high water solubility, or biodegradation resistance, functioning as inert pollutants throughout the treatment process (Barbosa et al., 2016). Key factors influencing micropollutant removal efficiency include molecular size, molecular weight, charge, adsorption, hydrophobicity, biodegradability, and volatility.

Implementing advanced treatment for micropollutants involves two main costs: retrofitting existing biological treatment systems and introducing new treatment processes. Retrofitting costs depend on site-specific factors such as available space, reactor volumes, current operational state, and improvement potential. New treatment process costs are determined by the chosen process type, design configuration, and plant conditions. Capital expenditures (CAPEX) include new infrastructure, while operational expenditures (OPEX) cover maintenance, energy, and consumables (Cecen & Aktaş, 2011). For example, ozonation infrastructure requires equipment for ozone generation and quenching, as well as contactors. Post-treatment is also necessary to remove byproducts, using granular activated carbon (GAC), sand filters, or polishing biological treatments. When a filtration step already exists, ozonation is more cost-effective. However, the expenses for advanced treatment increase significantly if new filtration systems are needed.

Risks Posed by Micropollutants to Ecosystems and Human Health

To evaluate the potential human health risks and toxicological significance of micropollutants, the maximum measured environmental concentration (MEC) in groundwater was divided by the corresponding drinking water equivalent level (DWEL) as a worst-case scenario, a method widely employed in scientific studies. Risk quotients (RQs) greater than 1 indicate potential risks to human health from exposure through drinking water sources, such as groundwater, while RQs ≤ 0.2 are considered safe, and those with $0.2 \leq RQ < 1$ merit further investigation (Leung et al., 2013). Micropollutants are classified into five groups based on the "EPA's Approach for Assessing the Risks Associated with Chronic Exposure to Carcinogens," ranging from human carcinogens (Group A) to evidence of non-carcinogenicity (Group E), with intermediate classifications for possible human carcinogens and unclassifiable compounds.

Data on carcinogenic and non-carcinogenic threats were sourced from the US Environmental Protection Agency Integrated Risk Information System (US EPA IRIS) and the Office of Environmental Health Hazard Assessment (OEHHA) chemical database. Human exposure to micropollutants in reclaimed water primarily occurs through consumption, and using the US EPA's risk assessment model, health risks in water reclamation and ecological reuse systems were evaluated, encompassing both carcinogenic risks (CRs) and non-carcinogenic risks (non-CRs).

Anaerobic Digestion as a Sustainable Technology

Since the dawn of civilization, biomass has been a vital energy source. Even today, about half of the global population relies on biomass, particularly in rural areas of developing nations. Global energy consumption has risen from 5652 TWh in 1800 to 158,839 TWh in 2019. While fossil fuels dominate with 86.10% of global energy consumption, biomass and modern biofuels accounted for 7.71% (1143 TWh) in 2019. Projections suggest biomass could generate up to 315,278 TWh of energy by 2050, making it a promising renewable energy source (Karekezi & Kithyoma, 2006; Ladanai & Vinterback, 2009).

Biomass-to-Bioenergy Conversion

Biomass can be converted into bioenergy through two primary pathways: thermochemical and biochemical. Thermochemical processes, such as incineration and gasification at temperatures above 800 °C, produce syngas with high energy content, while biochemical methods, including anaerobic digestion (AD) and fermentation, generate biofuels like biogas, bioethanol, and biohydrogen. Biogas, a versatile product of AD, can be upgraded to renewable natural gas (RNG), which burns cleaner than fossil fuels and serves as a sustainable energy source for industries, such as breweries, that traditionally rely on liquefied petroleum gas (Zhang et al., 2021). For example, Nigerian Breweries utilizes biogas systems to power boilers, reducing CO_2 emissions by 600 tons annually and cutting costs. Biogas also supports co-generation systems that produce heat and electricity, or trigeneration systems that add cooling capabilities, with electric efficiencies ranging from 25 to 45%, depending on technologies like four-stroke engines or microturbines. However, standard licensing and certification protocols are essential to address variations in biogas composition.

Environmental and Economic Benefits of Biogas

Anaerobic digestion of waste biomass significantly reduces greenhouse gas (GHG) emissions, as demonstrated by systems in Chitete, Zambia, and Semarakelwana, India, which could cut emissions by 0.33 $GgCO_2eq$ and 26 $GgCO_2eq$ annually, respectively. Biogas-to-biomethane plants provide both environmental and economic benefits, with technologies like amine scrubbing and high-pressure water scrubbing (HPWS) minimizing methane losses and electricity consumption, thereby lowering global warming potential (GWP). For instance, amine cleaning units exhibit 13% lower GWP than HPWS and 23% lower than pressure swing adsorption (PSA). Additionally, bio-CO_2 from upgraded biogas can be effectively utilized in controlled environments, such as polyhouse farming, rather than being released into the atmosphere, enhancing sustainability (Chaudhary et al., 2022).

Biomass Potential in India

India, an agrarian nation, relies heavily on biomass for energy, contributing nearly 30% of its primary energy needs. Crop residues from cereals, sugarcane, and other crops could yield 39 billion m3 of biogas annually. Similarly, livestock biomass, comprising manure from 536.7 million animals, presents an underutilized resource. Currently, only 9% of livestock-generated manure is used in biogas production, leaving significant untapped potential (Kumar et al., 2021).

Anaerobic digestion provides an efficient method for utilizing these residues. By converting waste into energy, it addresses environmental concerns, reduces reliance on fossil fuels, and supports the energy-agriculture-atmosphere nexus. Despite its potential, the adoption of AD technology remains limited. To achieve carbon neutrality and a circular economy, anaerobic digestion must be more widely adopted. This technology can meet local energy demands while reducing carbon emissions and fostering sustainable practices. It aligns with a win–win scenario where energy needs are met, waste is managed effectively, and agriculture is supported, creating economic and environmental benefits.

Advanced Membrane-Based Technologies

In recent decades, rapid industrialization and economic growth have significantly improved human well-being but also caused severe environmental pollution and depletion of natural resources. Industrial wastewater, a major contributor to this pollution, has strained water supplies globally, posing a challenge for both developing and developed nations (Mudakkar et al., 2013). Without adequate treatment, industrial effluents harm aquatic ecosystems. Addressing this issue through wastewater treatment can enable water reuse and reduce environmental impacts.

Membrane-based technologies have played a vital role in wastewater treatment since their introduction in the 1950s, following the successful use of polymer membranes for desalination. Among these, reverse osmosis (RO) membranes emerged as a key development, offering high salt rejection and flow rates at moderate pressures. Later, advancements like pressure-retarded osmosis (PRO) and the integration of membrane technology with biological treatments, such as membrane bioreactors (MBRs) and anaerobic membrane bioreactors (AnMBRs), further enhanced the effectiveness of industrial effluent treatment. AnMBRs, in particular, offer advantages like compact size, quick startup, and high removal efficiencies for chemical oxygen demand (COD) and total suspended solids (TSS) (Skouteris et al., 2012).

Polymeric membranes currently dominate the market for treating micropollutants in industrial wastewater, with ongoing research addressing challenges such as fouling and high treatment costs. Efforts have also focused on developing inorganic membranes, including ceramic and carbon-based membranes, and combining these with other treatment techniques to improve efficiency. Membrane technologies function through mechanisms like size exclusion, hydrophobic interactions, and electrostatic interactions to separate and remove contaminants (Ganiyu et al., 2015). Parameters such as pore size, membrane material, effluent composition, and operating conditions significantly influence their performance. Microfiltration (MF), ultrafiltration (UF), nanofiltration (NF), and reverse osmosis (RO) membranes are applied depending on the particle size and type of contaminants present.

Polymeric membranes, such as those made from polyethylene (PE), polyethylsulfone (PES), polypropylene (PP), and polyvinylidene fluoride (PVDF), are widely used due to their flexibility, low operating costs, and high quality control. Studies have demonstrated the high separation efficiency of RO membranes for metallic contaminants like nickel, chromium, and zinc, achieving removal rates exceeding 90%. Inorganic membranes, including ceramic and carbon-based variants, also exhibit excellent performance, often rejecting over 80% of pollutants (Goh et al., 2018). Their unique properties, such as durability and thinness, make them suitable for producing high-quality treated water.

Despite their advantages, challenges like high costs and fouling hinder the widespread adoption of membrane technologies. Addressing these barriers and optimizing the integration of membranes with other treatment methods can unlock their potential for efficient micropollutant removal, advancing sustainable wastewater management.

Innovative Use of Activated Carbon from Sewage Sludge

Pyrolysis of Sewage Sludge to Produce Activated Carbon

Sewage sludge generation has significantly increased due to rapid urbanization and stricter effluent regulations. If improperly treated and disposed of, it can pose serious environmental risks. Traditional disposal methods, such as sea dumping, incineration, landfilling, and application to farming, have several drawbacks. Sea dumping has been banned in many countries due to the contamination of oceans with harmful microorganisms, oil, grease, and heavy metals. Land application is also limited by potential soil contamination with heavy metals and a reduced ability to absorb nutrients. Additionally, landfills are becoming less viable due to limited space, increasing costs, and stricter environmental standards. As a result, there is a growing need for more sustainable and cost-effective solutions to handle sewage sludge (Smith et al., 2009).

Activated carbon, a highly effective adsorbent, has emerged as a promising material for wastewater treatment due to its ability to eliminate pollutants. However, its high cost often limits its widespread use. Researchers have explored low-cost alternatives made from various carbonaceous and mineral sources, such as scrap tires, fruit stones, nut shells, and fly ash (Chai et al., 2004). Among these alternatives, sewage sludge stands out as an abundant and carbon-rich resource that can be converted into activated carbon through pyrolysis, offering a sustainable and cost-effective solution. Pyrolysis of sewage sludge involves heating the material in an inert atmosphere at high temperatures, typically using either microwave-assisted pyrolysis or traditional furnace heating. Pyrolysis is a thermochemical process that breaks down organic matter into valuable products, such as pyrolysis oil, gas, and char. The resulting pyrolysis char, which has a high carbon content, can be used to produce activated carbon. This process not only reduces the volume of the sludge but also transforms it into a useful material, making it an attractive option for sewage sludge management.

Microwave-assisted pyrolysis has certain advantages over traditional pyrolysis, such as faster temperature rise, uniform heat distribution, and energy efficiency. In this method, microwaves directly heat the carbon bed, leading to improved energy conversion and potentially reducing operational costs. The production of activated carbon from sewage sludge via pyrolysis represents a promising avenue for both environmental management and resource recovery, turning a waste product into a valuable resource for various applications, including water purification and air filtration.

Efficacy of Activated Carbon in Capturing Micropollutants (PAHs, PFASs, PBDEs)

Activated carbon adsorption has emerged as a promising alternative for micropollutant removal. Technologies utilizing powdered activated carbon (PAC), often combined with advanced oxidation processes like ozonation, have demonstrated significant effectiveness. Ozonation, for example, is highly efficient in pathogen reduction and micropollutant degradation. However, it may produce unidentified and potentially harmful byproducts and does not fully eliminate target chemicals from wastewater streams. In contrast, activated carbon adsorption physically retains micropollutants, effectively removing them from water. Pollutants adsorbed onto PAC are retained within the filter and require proper disposal after treatment (Gautam & Anbumani, 2020).

The integration of advanced oxidation processes with activated carbon adsorption, followed by a final polishing step such as sand filtration or ultrafiltration, has shown substantial promise. These methods enhance micropollutant removal and offer scalability, cost-effectiveness, and higher treatment efficiency, making them viable solutions for addressing the growing challenge of micropollutants in wastewater.

Policy Frameworks and Global Collaboration

Policy frameworks are vital for transitioning from linear to circular economy (CE) practices in wastewater management. They provide the necessary regulatory, financial, and institutional support to enable resource recovery, waste minimization, and reuse, while ensuring environmental protection and public health safety. These frameworks help establish clear regulations, encourage research and innovation, foster stakeholder collaboration, and integrate CE principles into broader policy agendas. Benefits include environmental protection through pollution reduction, economic growth via green jobs, and social improvements such as enhanced public health. Policy support also drives technological advancements in wastewater treatment, creating globally competitive solutions. However, challenges like fragmented regulations, economic barriers, public skepticism, and insufficient data hinder effective policy implementation. To address these challenges, it is crucial to harmonize standards across regions, build stakeholder capacity, raise public awareness, and adopt a phased approach that balances short-term feasibility with long-term sustainability goals.

Collaboration among governments, industries, and communities is essential for sustainable wastewater management and micropollutant mitigation. Governments play a key role by developing supportive policies, setting advanced treatment standards, mandating resource recovery, and offering incentives like subsidies and tax benefits. Industries contribute by adopting cleaner production techniques, investing in research and development, and promoting water conservation and pollution reduction efforts. Collaboration with governments and research institutions helps scale CE-based wastewater management systems. Communities also play a significant role by supporting local initiatives, advocating for policy reforms, and adopting sustainable practices. The combined efforts of these stakeholders lead to environmental benefits such as reduced pollution, economic advantages through waste minimization and green job creation, and social benefits like improved public health and community resilience.

Review of National Policies and International Initiatives

National policies and international initiatives have increasingly recognized the environmental and health impacts of micropollutants in wastewater, spurring the development of frameworks to address pollution and ensure sustainable management. These frameworks emphasize circular economy principles, advanced treatment technologies, and global collaboration.

In the United States, the Clean Water Act (CWA) regulates pollutant discharge and mandates advanced treatment to reduce micropollutants. The EU's Water Framework Directive (WFD) and Urban Wastewater Treatment Directive (UWWTD) are aligned with the Zero Pollution Action Plan under the EU Green Deal, focusing on effective pollutant removal. India's National Water Policy and China's Water Pollution Prevention and Control Law emphasize wastewater recycling and resource recovery (Gupta, 2020). Australia's National Water Quality Management Strategy also focuses on sustainable water management.

Internationally, initiatives like the G20 Action Plan on Marine Litter address pollutants, including chemicals and plastics in water bodies. The United Nations integrates wastewater management into the Sustainable Development Goals (SDGs), particularly SDG 6 (Clean Water and Sanitation) and SDG 12 (Responsible Consumption and Production). The Global Programme of Action for the Protection of the Marine Environment from Land-Based Activities promotes sustainable wastewater practices globally. Organizations such as the OECD and the World Bank provide policy recommendations, financial assistance, and technical expertise, particularly in developing nations (Sridhar and Jingfeng, 2020).

Despite these efforts, challenges remain, such as fragmented policies and inconsistent standards, limited technological capacity in developing countries, and weak enforcement of regulations. Public resistance to water reuse and recycled products, driven by safety concerns and cultural barriers, further complicates implementation. To overcome these obstacles, coordinated global action is necessary, including harmonizing national policies with international frameworks, increasing funding, building capacity, and conducting public awareness campaigns to encourage the adoption of circular economy practices.

Energy Recovery in Micropollutant Management

Energy recovery is a key element of sustainable wastewater management, reducing operational costs, improving energy efficiency, and contributing to broader environmental sustainability. Anaerobic digestion, a common method for energy recovery, breaks down organic matter in wastewater without oxygen, producing biogas (mainly methane). This biogas can be used for electricity or heat generation, reducing reliance on external energy sources. In some cases, excess energy can be sold back to the grid, generating revenue and reducing operational costs. Additionally, methane capture mitigates greenhouse gas emissions, as methane is a potent climate pollutant (Rahamathunnisa et al., 2024).

Emerging technologies like microbial fuel cells (MFCs), which generate electricity from the microbial breakdown of organic compounds, are particularly useful for small-scale or off-grid systems. Thermophilic sludge digestion, which operates at higher temperatures, also improves energy recovery efficiency. Heat recovery from wastewater, through systems like heat exchangers, captures thermal energy for various uses, such as pre-heating incoming wastewater or providing district heating, reducing the energy demand for heating processes.

Integrating energy recovery systems within wastewater treatment plants can provide significant economic and environmental benefits. These systems lower operational costs, reduce reliance on fossil fuels, and help minimize carbon footprints by replacing fossil fuels with renewable energy. However, challenges such as high initial capital investment, variable efficiency depending on wastewater type, and the need for further research to optimize these systems remain.

Environmental and Economic Benefits of Integrating Energy Recovery in a CE Model

Integrating energy recovery within a circular economy (CE) model offers both environmental and economic benefits, promoting sustainability and resource efficiency in wastewater management. Technologies like anaerobic digestion, microbial fuel cells, and heat recovery systems reduce the carbon footprint of wastewater treatment. Anaerobic digestion captures and utilizes methane, a potent greenhouse gas, while heat recovery systems reduce reliance on conventional energy sources (Rahamathunnisa et al., 2024).

Economically, energy recovery reduces treatment plant operational costs by offsetting energy demands through biogas production and heat recovery. Excess energy can also be sold to the grid, generating additional revenue streams. By turning wastewater from a waste product into a valuable resource, this approach aligns with CE principles, promoting resource recovery and reuse.

Case Studies and International Strategies

Successful implementations of CE principles in micropollutant management demonstrate the potential for sustainable wastewater treatment. In Germany and Sweden, advanced treatment technologies like activated carbon adsorption, ozonation, and membrane filtration have been integrated with resource recovery processes to remove micropollutants such as pharmaceuticals and personal care products, while recovering valuable resources like phosphorus and nitrogen for agricultural use. The Netherlands has adopted decentralized systems combining anaerobic digestion with advanced oxidation processes to treat complex micropollutants and recover biogas for facility energy, while by-products are reused in various industrial processes.

Japan has integrated CE principles in wastewater treatment by using membrane bioreactors and constructed wetlands to treat wastewater contaminated with micropollutants. The recovered water is reused for irrigation and industrial purposes, reducing the demand for fresh water. These case studies illustrate the successful application of CE principles in managing micropollutants while promoting sustainability and resource recovery.

Global Efforts in Micropollutant Management

Global initiatives, such as the EU's Water Framework Directive and the United Nations Environment Programme's Global Programme of Action for the Protection of the Marine Environment, focus on reducing micropollutants in aquatic environments. The WHO provides guidelines for managing micropollutants in drinking water, while industry collaborations, like those with the International Water Association, foster innovation in treatment technologies and resource recovery. However, challenges like disparities in technological capacity and regulatory frameworks remain. Continued international collaboration is essential for overcoming these barriers and mitigating the environmental and health risks posed by micropollutants.

Conclusion

The management of micropollutants in wastewater treatment is becoming increasingly critical due to their environmental and health risks, with sustainable technologies playing a key role in addressing these concerns. Advanced treatment systems like activated carbon adsorption, membrane filtration, and advanced oxidation processes (AOPs) are effective in removing a range of micropollutants, while resource recovery technologies such as anaerobic digestion and thermal treatment not only reduce pollutants but also generate renewable energy, aligning with circular economy principles. CE emphasizes reducing, reusing, and recycling resources, which can be applied in wastewater treatment by recovering water, energy, and nutrients from treated effluent, thereby minimizing environmental impacts and supporting sustainable urban water management. The future of micropollutant management will depend on evolving policies, emerging technologies like electrochemical processes and bio-based treatments, and international cooperation to scale solutions and align standards globally. These combined efforts will foster more efficient, sustainable, and circular approaches, contributing to cleaner water systems and a more sustainable world.

References

Abegglen, C., & Siegrist, H. (2012). Mikroverunreinigungen aus Kommunalem Abwasser. Verfahren zur Weitergehenden Elimination auf Kläranlagen. Bundesamt für Umwelt. Bern, Umwelt-Wissen Nr.1214: 210 S.

Alam, P., Khan, A. H., Pinê Américo-Pinheiro, J. H., & Ahmad, K. (Eds.). (2024). *Innovative and hybrid technologies for wastewater treatment and recycling* (1st ed.). CRC Press. https://doi.org/10.1201/9781003454199

Bahr, C., Schumacher, J., Ernst, M., Luck, F., Heinzmann, B., & Jekel, M. (2007). SUVA as control parameter for the effective ozonation of organic pollutants in secondary effluent. *Water Science and Technology, 55*(2), 267–274.

Barbosa, M., Moreira, N. F. F., Ribeiro, A. R., Pereira, M. F. R., & Silva, A. M. T. (2016). Occurrence and removal of organic micropollutants: an overview of the watch list of EU decision 2015/495. *Water Research, 94*, 257–279.

Behera, I. D., Nayak, M., Biswas, S., Meikap, B. C., & Sen, R. (2021). Enhanced biodegradation of total petroleum hydrocarbons by implementing a novel two-step bioaugmentation strategy using indigenous bacterial consortium. *Journal of Environmental Management, 292*, Article 112746. https://doi.org/10.1016/j.jenvman.2021.112746

Behera, I. D., Nayak, M., Mishra, A., Meikap, B. C., & Sen, R. (2022). Strategic implementation of integrated bioaugmentation and biostimulation for efficient mitigation of petroleum hydrocarbon pollutants from terrestrial and aquatic environment. *Marine Pollution Bulletin., 177*, Article 113492.

Cecen, F., & Aktaş, O. (2011). Removal of NOM, nutrients, and micropollutants in BAC filtration. In *Activated Carbon for Water and Wastewater Treatment: Integration of Adsorption and Biological Treatment* (pp. 307–324). Wiley-VCH Verlag GmbH & Co. KGaA, Weinheim, Germany.

Chai, G. S., Yoon, S. B., Yu, J. S., et al. (2004). Ordered porous carbons with tunable pore sizes as catalyst supports in direct methanol fuel cell. *Journal of Physical Chemistry B, 108*(22), 7074–7079.

Chaudhary, V. P., Chandra, R., Denis, D. M., Isha, A., & D'Silva, T. C. (2022). Agri-biomass-based bio-energy supply model: an inclusive sustainable and circular economy approach for a self-resilient rural India. *Biofuels, Bioproducts and Biorefining, 16*(5), 1–13. https://doi.org/10.1002/bbb.2373

Das, A. P., & Mishra, S. (2008). Hexavalent Chromium (VI): health hazards and environmental pollutant. *J Environ Res Dev, 2*(3), 386–392.

Das, A. P., Mohanty, A. K., van Hullebusch, E. D., Figueiredo, G., Das, A. P., Mohanty, A. K., van Hullebusch, E. D., & Figueiredo, G. (2023). Bioremediation as an emerging technology for the removal of synthetic microplastic pollutants from marine ecosystem. *MarPB, 194*, Article 115297. https://doi.org/10.1016/J.MARPOLBUL.2023.115297

Das, A., Dey, S., Bala, K., Gnanasekaran, R., & Das, A. P. (2024). Microbial bioremediation technology for sustainable treatment and management of synthetic microfiber waste. *Sustainable Microbial Technology for Synthetic and Cellulosic Microfiber Bioremediation*, 221–238. https://doi.org/10.1007/978-3-031-62600-5_12

Dickenson, E. R. V., Snyder, S. A., Sedlak, D. L., & Drewes, J. E. (2011). Indicator compounds for assessment of wastewater effluent contributions to flow and water quality. *Water Research, 45*(3), 1199–1212.

Ganiyu, S. O., Van Hullebusch, E. D., Cretin, M., Esposito, G., & Oturan, M. A. (2015). Coupling of membrane filtration and advanced oxidation processes for removal of pharmaceutical residues: A critical review. *Separation and Purification Technology, 156*, 891–914. https://doi.org/10.1016/j.seppur.2015.09.059

Gautam, K., Anbumani, S. (2020). Ecotoxicological effects of organic micro-pollutants on the environment. In *Current Developments in Biotechnology and Bioengineering* (pp. 481–501). Elsevier.

Goh, P. S., & Ismail, A. F. (2018). A review on inorganic membranes for desalination and wastewater treatment. *Desalination, 434*, 60–80. https://doi.org/10.1016/j.desal.2017.07.023

Gupta, A. (2020). *Heavy metal and metalloid contamination of surface and underground water: Environmental, policy and ethical issues* (1st ed.). CRC Press. https://doi.org/10.1201/9780429198373

Karekezi, S., & Kithyoma, Y. (2006). Bioenergy and agriculture: Promises and challenges. *International Food Policy Research Institute (IFPRI)*. https://doi.org/10.2499/2020focus14

Kanungo, J., Sahoo, S., Bal, M., & Behera, I. D. (2023). Performance of bioremediation strategy in waste lubricating oil pollutants: A review. *Geomicrobiology Journal*. https://doi.org/10.1080/01490451.2023.2245395

Kumar, M., Pandit, S., Patel, V., Khanna, N., Nag, M., Lahiri, D., Ray, R. R., Das, A., & Das, D. (2023). Maximization of energy recovery from starch processing wastewater by thermophilic dark fermentation coupled with microbial fuel cell technology. *Geomicrobiology Journal*, 1–12. https://doi.org/10.1080/01490451.2023.2209555

Kumar, S., D'Silva, T. C., Chandra, R., Malik, A., Vijay, V. K., & Misra, A. (2021). Strategies for boosting biomethane production from rice straw: A systematic review. *Bioresource Technology Reports, 15*, Article 100813. https://doi.org/10.1016/j.biteb.2021.100813

Ladanai, S., & Vinterbäck, J. (2009). *Global Potential of Sustainable Biomass for Energy, vol. 32*. SLU, Institutionen För Energi Och Teknik Swedish University of Agricultural Sciences, Department of Energy and Technology.

Mallick, K., Sahu, A., Dubey, N. K., & Das, A. P. (2023). Harvesting marine plastic pollutants-derived renewable energy: A comprehensive review on applied energy and sustainable approach. *Journal of Environmental Management, 348*, Article 119371. https://doi.org/10.1016/J.JENVMAN.2023.119371

Mishra, S., & Das, A. P. (2021). Current treatment technologies for removal of microplastic and microfiber pollutants from wastewater. *Wastewater Treatment: Cutting-Edge Molecular Tools, Techniques and Applied Aspects*, 237–251. https://doi.org/10.1016/B978-0-12-821881-5.00011-8

Mishra, S., & Das, A. P. (2022). Treatment of the wastewater polluted with synthetic microfiber released from washing machine. In B. B. Das, H. Hettiarachchi, P. K. Sahu, & S. Nanda (Eds.), *Lecture notes in civil engineering* (Vol. 207, pp. 109–117). Springer Science and Business Media Deutschland GmbH. https://doi.org/10.1007/978-981-16-7509-6_9/COVER

Mishra, S., & Das, A. P. (2023). Microbial remediation of synthetic microfiber contaminated wastewater. In M. P. Shah (Ed.), *Microbial Technologies in Industrial Wastewater Treatment* (pp. 337–350). Springer, Singapore. https://doi.org/10.1007/978-981-99-2435-6_17

Mishra, S., Dash, D., & Das, A. P. (2022a). Detection, characterization and possible biofragmentation of synthetic microfibers released from domestic laundering wastewater as an emerging source of marine pollution. *Marine Pollution Bulletin, 185*, Article 114254. https://doi.org/10.1016/J.MARPOLBUL.2022.114254

Mishra, S., Dash, D., & Das, A. P. (2024). Isolation and molecular characterization of polyester-degrading bacterial strains from municipal wastewater and their application in microplastic degradation in contaminated wastewater. *Geomicrobiology Journal*. https://doi.org/10.1080/01490451.2024.2376101

Mishra, S., Singh, R. P., Rout, P. K., & Das, A. P. (2022). Membrane bioreactor (MBR) as an advanced wastewater treatment technology for removal of synthetic microplastics. *Development in Wastewater Treatment Research and Processes*, 45–60. https://doi.org/10.1016/B978-0-323-85583-9.00022-3

Mohanta S., Pradhan B., Behera I. D. (2023). Impact and remediation of petroleum hydrocarbon pollutants on agricultural land: A review. *Geomicrobiology Journal 10*, 1–5 (2023). https://doi.org/10.1080/01490451.2023.2243925

Mohanty, M., Tripathy, B., Koteswara Reddy, G., & Das, A. P. (2024). Greener technologies for establishing restraint over microplastic pollution. *Sustainable Microbial Technology for Synthetic and Cellulosic Microfiber Bioremediation*, 239–258. https://doi.org/10.1007/978-3-031-62600-5_13

Mudakkar, S. R., Zaman, K., Khan, M. M., & Ahmad, M. (2013). Energy for economic growth, industrialization, environment and natural resources: living with just enough. *Renewable and Sustainable Energy Reviews, 25*, 580–595. https://doi.org/10.1016/j.rser.2013.05.024

Priyadarsini, S., & Das, A. P. (2024). Analytical and structural characterization of waste lithium-ion batteries for their effective recycling strategy. *Environmental Chemistry and Ecotoxicology*. https://doi.org/10.1016/j.enceco.2024.12.004

Priyadarsini, S., & Das, A. P. (2025). Lithium bioleaching: A review on microbial-assisted sustainable technology for lithium bio-circularity. *Journal of Water Process Engineering, 69*, Article 106744. https://doi.org/10.1016/j.jwpe.2024.106744

Rahamathunnisa, U., Sudhakar, K., Padhi, S. N., Bhattacharya, S., Shashibhushan, G., & Boopathi, S. (2024). Sustainable Energy Generation from Waste Water: IoT Integrated Technologies. In A. Etim (Ed.), *Adoption and use of technology tools and services by economically disadvantaged communities: implications for growth and sustainability* (pp. 225–256). IGI Global Scientific Publishing. https://doi.org/10.4018/978-1-6684-5347-6.ch010

Skouteris, G., Hermosilla, D., López, P., Negro, C., & Blanco, Á. (2012). Anaerobic membrane bioreactors for wastewater treatment: A review. *Chemical Engineering Journal, 198–199*, 138–148. https://doi.org/10.1016/j.cej.2012.05.070

Smith, K. M., Fowler, G. D., Pullket, S., & Graham, N. J. D. (2009). Sewage sludge-based adsorbents: A review of their production, properties and use in water treatment applications. *Water Research, 43*, 2569–2594.

Sridhar, K. S., & Jingfeng, L. (Eds.). (2020). *The rise of India and china: social, economic and environmental impacts* (1st ed.). Routledge India. https://doi.org/10.4324/9781003005339

Zhang, Y., Jiang, Y., Wang, S., Wang, Z., Liu, Y., Hu, Z., & Zhan, X. (2021). Environmental sustainability assessment of pig manure mono-and co-digestion and dynamic land application of the digestate. *Renewable and Sustainable Energy Reviews, 137*, Article 110476. https://doi.org/10.1016/j.rser.2020.110476

Chapter 20
Ecotoxicological Risk Assessment of Environmental Micropollutants

Payal Das, Nimcheki Sherpa, Manisha Bal, Alok Prasad Das, and Ipsita Dipamitra Behera

Introduction

The presence of persistent micropollutants in the natural environment has become a significant global concern (Bhatt et al., 2022). Micropollutants pose an increasing threat to both the environment and human health (Das & Mishra, 2008; Mallick et al., 2023). Various microscopic chemical contaminants, including pharmaceutical compounds, pesticides, textile dyes, antibiotics, and microplastics, are frequently introduced into the environment and eventually infiltrate the food chain (Bhatt et al., 2022; Bhattacharjee et al., 2021; Das & Swain, 2013; Mishra et al., 2021b; Sahu & Das, 2024). Although these substances have undeniably improved the quality of life, they also pose significant risks, such as ecotoxicity and harmful effects on human health. Among various pollutants, microplastics (MPs) have become a growing concern. MPs are typically defined as plastic particles measuring less than 5 mm in size (Arthur et al., 2009; Das et al., 2023a; Ghosh et al., 2023; Jagatee et al., 2024; Paul et al., 2024; Singh et al., 2020). The primary origins of MPs in the environment are the direct release of microscopic plastic particles and the fragmentation or degradation of larger plastic materials (Dey et al., 2024a; Mishra et al., 2024; Satapathy et al., 2024). Plastics enhance human convenience, but their natural decomposition in the environment is difficult, leading to significant ecological issues (Mishra & Das, 2023; Mishra et al., 2021a).

P. Das · N. Sherpa · M. Bal (✉)
Department of Chemical Engineering, National Institute of Technology Durgapur, Durgapur, West Bengal, India
e-mail: mbal.che@nitdgp.ac.in

A. P. Das
Department of Life Sciences, RDW University, Bhubaneswar, Odisha, India

I. D. Behera
Chemical Engineering Department, Indira Gandhi Institute of Technology, Sarang, Khalapal, Odisha, India

I. D. Behera et al. (eds.), *Emerging Micropollutants*, Sustainable Environmental Waste Management Strategies, https://doi.org/10.1007/978-3-032-03530-1_20

MPs have been detected worldwide, spanning environments such as oceans, drinking water, rivers, and groundwater. MPs pollution was identified as the second significant scientific issue in ecological and environmental research studies at the 2nd United Nations Environmental Conference in 2016 (Carlini & Kleine, 2018). Microplastics are a prevalent pollutant in aquatic environments because of their floatable and persistent nature, which allows pollutants to be transferred to aquatic life. Due to their tiny size, MPs are ingested by various aquatic organisms or creatures, leading to physiological alterations in these animals that ultimately impact humans via the food path. There is a significant risk of small plastic particles being transferred to humans (Saha et al., 2021). MPs have been detected in placentas (Ragusa et al., 2021) and human stool (Zhang et al., 2021) and have shown harmful effects on aquatic species across the world (Marn et al., 2020). Furthermore, the adsorption of various hazardous substances, such as heavy metals (Pb, Li, Mg, and Cd) and organic pollutants, intensifies the impact of MPs (Prabhakar et al., 2019; Priyadarsini & Das, 2024, 2025; Raj & Das, 2023). Consequently, MPs are associated with both chemical and metal-based toxicity, along with physical harm caused by ingestion as well as its adverse effects on significant organs (Das & Bal, 2025; Das et al., 2023b). This significantly amplifies the overall harmful potential of MPs. Although numerous studies have documented methods for measuring MPs, it is important to note that standardized protocols for sampling, identification, or quantification of MPs have not yet been established. Figure 20.1 shows the number of publications on MPs toxicity.

This chapter aims to increase awareness among scientific communities about the effect of MPs on our environment. Additionally, this chapter provides a details of the toxic effects of MPs on living organisms. This study also represents the impact of

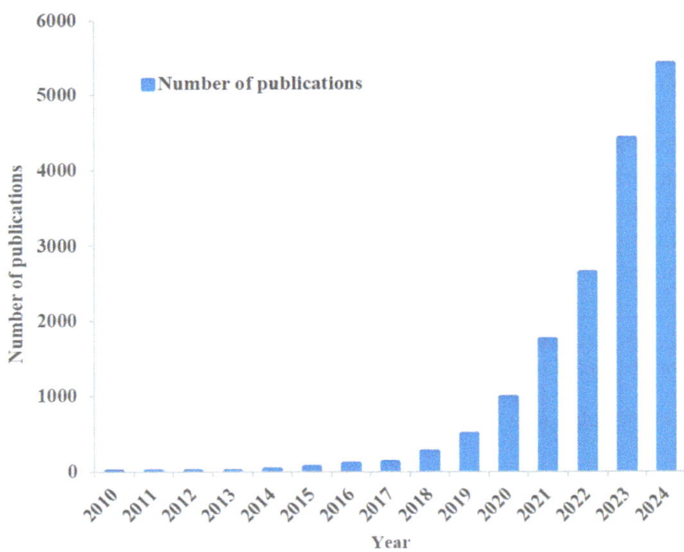

Fig. 20.1 Number of publications on MPs toxicity in the environment

MPs on plants, animals as well as human body. The study highlights the importance of current research while encouraging future advancements in this critical area by outlining potential directions and opportunities for further investigation.

Primary and Secondary MPs

MPs are widespread pollutants originating from various human activities. Based on their origin, MPs are classified into two categories: "Primary MPs," which are intentionally created for direct or indirect use as raw materials in consumer plastic products, and "Secondary MPs," which result from the fragmentation, degradation, and breakdown of larger plastic items (Das et al., 2024a, 2024b; Mishra et al., 2019, 2020; Mishra & Das, 2021). Primary MPs are generally found in products such as facial cosmetics, cleansers, microbeads, scrubs, exfoliants, toothpaste, and abrasives (Mallick et al., 2024). They are also used in industrial processes like air blasting, rubbers, washing synthetic textiles, and manufacturing tea bags. Additionally, they serve as vectors in pharmaceutical applications, particularly in drug production (Sangkham et al., 2022).

Secondary MPs are considered to pose greater public health risks compared to primary MPs. Many environmental MPs are secondary, originating from the breakdown of larger plastic debris (Das et al., 2023b, 2024c). Because of their chemical composition, plastic materials degrade at an exceptionally slow rate. Large plastic pieces are broken down into MPs (1 μm to 5 mm) and nanplastics (NPs), ranging in size from 100 to 1000 nm and less than < 1 μm, through chemical, physical, and biological processes (Mishra et al., 2022; Tripathy et al., 2022).

Effects of MPs on Environment

MPs can lead to a variety of harmful effects on environmental species, resulting in multiple adverse outputs. The main effects encompass ingestion and physical damage, reproductive and developmental impairments, chemical toxicity, biomagnification and bioaccumulation, behavioral and physiological damages, and broader ecological consequences (Dey et al., 2024b; Mohanty et al., 2024; Thacharodi et al., 2024). Numerous organisms, including soil organisms, zooplankton, fish, and invertebrates, may inadvertently ingest microplastics, often confusing them with food. Then, many seafoods are consumed by humans (Das et al., 2023a; Das & Bal, 2025). This can result in physical harm, decreased nutrient absorption, digestive blockages, and reduced feeding efficiency. Table 20.1 shows the effects of MPs on different species of environment. Figure 20.2 shows the effects of MPs on environment.

Table 20.1 Effect of microplastics on organisms, plants, and human health

Category	Type of MP	Size	Effects	References
Aquatic creatures				
Nematode (*Caenorhabditis elegans*)	PS	1 μm	Intestinal damage and increased oxidative stress	Yu et al. (2020)
Ascidian *Ciona intestinalis*	PS	1 μm	Impaired growth and reduced food intake	Messinetti et al. (2019)
Daphnia magna	PE	1–5 μm	potential alterations in behavior and energy allocation;	Aguirre-Martínez et al. (2023a)
Podocopid ostracods	PE	1–5 μm	accumulation in the digestive system, possible effects on swimming and feeding behaviors	Aguirre-Martínez et al. (2023b)
Sea Urchin Larvae (*Tripneustes gratilla*)	PE microspheres	10–45 μm	High concentration ingestion indicating potential impacts on growth and development	Kaposi et al. (2014)
Shore Crabs (*Carcinus maenas*)	PS microspheres	0.5 μm and 8–10 μm	Microplastics can transfer through the food web, potentially compromising predator health	Farrell and Nelson (2013)
Copepod (*Tigriopus japonicus*)	PS spheres	Microsized (0.5 and 6 μm) and Nano-(50 nm)	Affects feeding and digestion	Lee et al. (2013)
Oryzias latipes	PE PP	16.4–962.2 μm	Reduces larval development and body-to-head ratios	Pannetier et al. (2020)
Perna perna	PVC	0.1–1.0 μm	Feces from exposed organisms were heavily polluted with PVC, though physiological responses exhibited no significant impacts	Santana et al. (2018)
Ceriodaphnia dubia	Microplastic additive (styrene)	–	Disruption of endocrine function and reduced reproductive rates	Iguchi et al. (2006)
Halobates sericeus	Microplastic debris in Northern Taiping		Increased spawning density	Goldstein et al. (2012)

(continued)

Table 20.1 (continued)

Category	Type of MP	Size	Effects	References
Blue mussel *Mytilus galloprovincialis*	PE, PS		Accumulation occurs in digestive, gills, and hemolymph tissues, leading to altered immunity, lysosomal changes, peroxisomal proliferation, antioxidant disruption, neurotoxicity, and genotoxicity	Boots et al. (2019)
European seabass *Dicentrarchus labrax*	Fluorescence red polymer microspheres		Metal accumulation in the brain and muscles causes neurotoxicity, oxidative stress, cellular damage, and disrupted enzyme activity	Barboza et al.2018)
Marine copepod *Acartia tonsa* (Dana)	PE microbeads		Bio-accumulation, and increased mortality	Syberg et al. (2017)
Soil organisms				
Lumbricus terrestris	High-density polyethylene		No impact on development or weight observed	Hodson et al. (n.d.)
Eisenia andrei	PE particle	(300~1000 μm)	Does not affect reproduction or growth but triggers immune responses, causes histopathological damage, and increases protein, lipid, and polysaccharide levels in the body	Rodriguez-Seijo et al. (2017)
Arenicola marina	Polyvinyl chloride (PVC)		Low polystyrene concentrations increase PCB accumulation, while high concentrations reduce PCB accumulation in organisms	Besseling et al. (2013)
Folsomia candida	PVC particle	(75–300 μm)	Affects reproduction, growth, and intestinal flora	Zhu et al. (2018b)
Lumbricus Terrestris (earthworm)	PE	$\geq 50\ \mu m$	Inhibits growth and reproduction; causes mortality	Huerta Lwanga et al. (2017)
Lumbricus terrestris (*Anecic earthworm*)	PE	Varying sizes	Earthworms transport MPs deeper into the soil, leading to changes in gut microbiota, body weight, and reproductive capability	Rillig et al. (2017)

(continued)

Table 20.1 (continued)

Category	Type of MP	Size	Effects	References
Toxicological impacts of cadmium and MPs on the earthworm *Eisenia foetida*	PP + Cd	< 150 μm	MPs amplify the toxic effects of cadmium (Cd)	Zhou et al. (2020)
Molecular and Histopathological impacts of MPs in *Eisenia andrei* Bouché	PE	250–1000 μm	Histological evidence confirmed MPs in earthworms, causing intestinal tissue damage (hyperemia and epithelial exfoliation) and immune responses (intestinal inflammation)	Rodriguez-Seijo et al. (2017)
Plants				
Lime (*Citrus aurantium*)	LDPE, PS, and PP	5000	Impacts on plant biomass, phytochemical levels	Enyoh et al. (2020)
Spring Onion (*Allium festulous*)	PE, PET, PS, PP	8–500	Plant biomass impacted, elemental composition of tissue observed; root traits affected	De Souza Machado et al. (2019)
T. aestivum, L. sativa	PS, PMMA	0.2, 2, 5, 7, 10 μm	Fluorescent dye staining and CLSM analysis detected MPs in root surfaces without evidence of internal uptake	Li et al. (2020b)
Oryza sativa L.	PS	< 50 mm	High doses of PS reduced shoot biomass by ≈ 40%; field trials showed a decrease of 25.9%	Wu et al. (2020)
Arabidopsis thaliana (L.) Heynh., *T. aestivum*	PS	1 μm	PS, NP, and MP were found on root cap cells via CLSM, but no uptake into roots was observed	Taylor et al. (2020)
Perennial ryegrass (*Lolium perenne*)	Polylactic acid (PLA)	0.065 mm	Reduced shoot height and biomass	Boots et al. (2019)
Wood small-reed (*Calamagrostis epigejos*)	Polyether sulfone (PES)	(L = 1.3, D = 0.03) mm	Increase in root biomass	Lozano and Rillig (2020)

(continued)

Table 20.1 (continued)

Category	Type of MP	Size	Effects	References
Garden cress (*Lepidium sativum*)	Melamine phenolic (MP)	0.0048 mm	MPs accumulated on root hairs, reduced germination rate, and physically blocked seed capsule pores	Bosker et al. (2019)
Corn (*Zea mays*)	Polyethylene (PE)	0.003 mm	PE blocked water and nutrient uptake, inhibiting maize plant growth	Bosker et al. (2019)
Radish (*Raphanus sativus*) and Oat (*Avena sativa*)	Polyetherimide		Nitrogen released from PEIs was harmless at low concentrations but harmful at high levels	Rychter et al. (2019)
Tobacco BY-2 cells	PS	20–1000 nm	Tobacco BY-2 tissues internalized 20–40 nm PS NPs but could not uptake 100 nm beads	Bandmann et al. (2012)
Lycopersicon esculentum Mill (Tomato)	PP, PET, PE	0.4–2.6 mm	MP sludge enhanced growth of tomato plant but suspended yield and production	Hernández-Arenas et al. (2021)
Human health				
Human renal cortical epithelial cell, Kidney	PS-NPs	44 nm	NPs entered Human renal cortical epithelial cells through both energy-independent and energy-dependent (endocytosis) mechanisms	Monti et al. (2015)
Placental, epithelial human tissues and Hela cerebral cells	MPs/NPs	–	Oxidative stress was identified as a cytotoxic mechanism at the cellular level	Schirinzi et al. (2017)
Human intestinal epithelial cell lines Caco-2, gastrointestinal tracts	PS		PS MPs did not react with the activation or differentiation of human macrophage models. Under the experimental circumstances, exposure to PS MPs did not result in any acute health problems to mammals	Stock et al. (2019)
Human pluripotent stem cells, human embryos and intracellular development	PS-NPs	–	LEFTY1 and LEFTY2 (TGF-beta signaling) were down regulated; OCLM and CA4 (eye development) were affected. The APOC3 circuit was linked to cardiovascular disease risk	Bojic et al. (2020)

(continued)

Table 20.1 (continued)

Category	Type of MP	Size	Effects	References
Human dermal fibroblasts, red blood cells, human peripheral blood mononuclear cells	PS	460 nm, 1, 3, 10, 40, and 100 mm	MPs (10–100 mm) were non-toxic to human cell lines, but attachment to red blood cells caused hemolysis	Hwang et al. (2020)
Animals				
Chinese mitten crab (*Eriocheir sinesis*)	Polystyrene MPs	5 µm	Gut microbiota diversity decreased, with reduced Firmicutes and Bacteroidetes and increased Fusobacteria and Proteobacteria	Liu et al. (2019)
Mouse	Polyethylene MPs	10–150 µm	Increased microbiota diversity included higher Staphylococcus and lower Parabacteroides abundance	Li et al. (2020a)
Rats respiratory tract	PS-MPs		• Male rats had reduced inspiratory time, and female rats showed increased respiratory frequency and decreased inspiratory/expiratory times. Lung inflammation increased with elevated TGF-β and TNF-α expression	Li et al. (2020a)

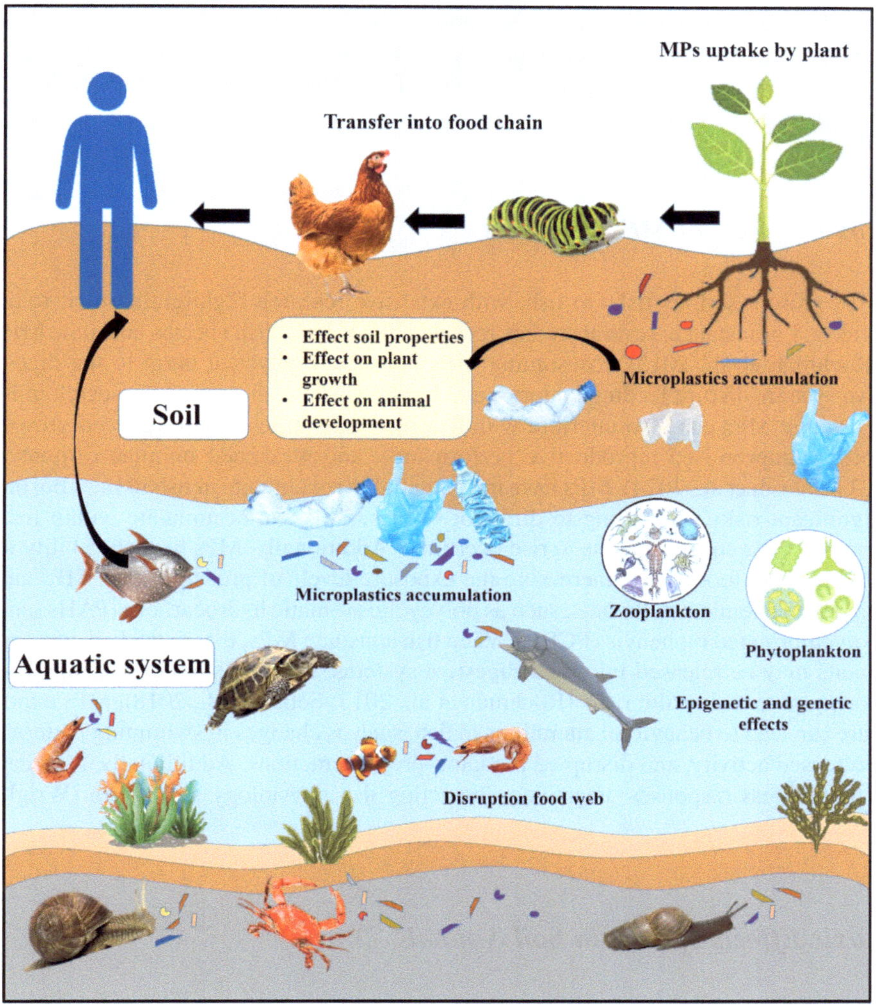

Fig. 20.2 Effects of MPs on environment

Toxic Effects of MPs in Zooplanktons and Phytoplanktons

Zooplankton and phytoplankton, essential components of aquatic systems, are vulnerable to the toxic effects of MPs (Bala et al., 2024b). Multiple studies have highlighted the negative impact of MPs on these species. Research has shown that MPs negatively impact phytoplankton by impairing their photosynthetic processes and growth, ultimately decreasing primary productivity (Priyadarshini et al., 2024; Singh Jadaun et al., 2022). Zooplankton ingest MPs, which disrupt their feeding behavior and deplete their energy reserves, as reported by Cole et al. (2011). These innovations underscore the detrimental effects of MPs toxicity on both zooplankton and phytoplankton, emphasizing the urgent need for strategies to address MPs pollution in aquatic ecosystems.

Toxic Effects of MPs in Fishes

MPs present serious risks to fish, with extensive research highlighting their toxic effects. For instance, several studies revealed that various fish species consume MPs (Rochman et al., 2013). Consuming MPs may cause physical harm to the digestive system, disrupt feeding patterns, and hinder nutrient absorption. Research indicates that MPs can also interfere with fish metabolism, leading to reduced growth rates, compromised reproductive performance, and weakened immune responses (Thacharodi et al., 2024). MPs have the potential to accumulate in fish tissues, posing significant risks. According to different studies, MPs can accumulate within fish, causing their concentrations to rise over time. Additionally, MPs have the ability to infiltrate the food chain, increasing the exposure levels of predatory fish. MPs can bind with chemical pollutants, such as polycyclic aromatic hydrocarbons (PAHs) and polychlorinated biphenyls (PCBs). When fish consume MPs, these attached contaminants may be released into their digestive systems, causing chemical damage and posing potential health risks (Rochman et al., 2013; Setala et al., 2018). MPs exposure can lead to behavioral alterations in fish, such as changes in swimming patterns, decreased activity, and disrupted predator–prey interactions. Additionally, MPs can trigger stress responses, negatively impacting fish physiology and health (Wright et al., 2013).

Toxic Effects of MPs in Soil Animals

Soil microorganisms contribute to the incorporation of MPs into the soil. Additionally, heavy metals, pathogens, and pollutants attached to the surface of MPs can interact with the soil, potentially impacting soil microbial communities, altering its chemical and physical properties, and influencing plant growth (Leed et al., 2019). Earthworms, being a key model organism in soil ecosystems. The impact of MPs on earthworms depends on their type and concentration, often resulting in reduced growth rates and compromised immune systems. Several studies reported that the detrimental effects of MPs exposure on soil microbes, indicating that these effects were not influenced by the size of the plastic particles. The body length, reproduction rate, survival rate, and oxidative stress gene expression of *Caenorhabditis elegans* were significantly affected by polystyrene (PS), with these effects being size-dependent (Ya et al., 2021). According to Zhu et al., collembolan (*Folsomia candida*) growth and reproduction were decreased by interaction with 0.1% polyvinyl chloride MPs, and the gut microbiota and bacterial diversity of collembolan were also changed (Zhu et al., 2018a). Exposure to 0.5% polyethylene MPs notably altered the intestinal microbial population of springtails (*Folsomia candida*), while 1% polyethylene MPs decreased their development rate by 70.2% (Ju et al., 2019). The consumption of polyethylene terephthalate MPs by snails resulted in a reduction of TAOC levels (an indicator of oxidative stress), increased malondialdehyde (MDA) content, and decreased GPx antioxidant enzyme function, leading to lipid peroxidation and gastrointestinal damage (Song et al., 2019).

Toxic Effects of MPs in Plants

MPs present a serious danger to plants, potentially disrupting functioning and ecosystem dynamics. Their toxic effects on plants may involve physical toxicity, photosynthesis, growth inhibition and changes in nutrient absorption. Additionally, MPs can bind or coat plant surfaces, obstructing light entrance and impeding gas exchange (Thacharodi et al., 2024). Indirectly, MPs alter essential water quality parameters, including pH, nutrient levels, and oxygen availability. Such changes can upset the delicate conditions necessary for plants. Research has indicated that MPs adversely affecting their health and ecological functions. The accumulation of MPs hinders growth and development, resulting in lower biomass and reduced plant density. The key factors influencing photosynthesis in plants include light, inorganic carbon, and nutrients. Various forms of seafloor vegetation, such as macrophytes and algae, are also vulnerable to MPs pollution. Research suggests that MPs can attach to these vegetation surfaces, altering their structure and hindering growth. Furthermore, MPs may interfere with O_2 exchange between vegetation and their surroundings, destabilizing aquatic ecosystems. Species that depend on these environments may be impacted by these disturbances to growth, production, and ecological processes as they spread along the food chain (Thacharodi et al., 2024).

Toxic Effects of MPs on Human Health

MPs have the potential to enter the gastrointestinal system via endocytosis by M cells, move through tissues via paracellular transit, and ultimately cause systemic exposure (Bala et al., 2024a; Sahu et al., 2024). According to several studies, these contaminants may have harmful effects in many in vitro systems at high doses. The detrimental consequences that MPs may cause at various levels are described below.

Urinary and Gastrointestinal System

Previous studies have reported the movement of various types and sizes of MPs (0.1–150 μm) through the stomach into the lymphoid system. Several research has confirmed that MPs within this size range can enter the lymphoid system (John & Das, 2024; Mohanty et al., 2024). PS-NPs can cause inflammatory damage. They also had an impact on gene function and cell viability, strongly initiated cell morphology, and up-regulated IL-8 and IL-6 genes in digestive tissues. Human renal cortical epithelial tissues treated to 44 nm PS nanoparticles in laboratory conditions showed that NPs penetrated renal cortical epithelial tissues by endocytosis of kidney. Furthermore, compared to freshwater, MPs produced more chromium in a human gastrointestinal model study. The highest amount of bioavailable hexavalent chromium (Cr^{+6}) was produced in the gastrointestinal phase, while trivalent chromium (Cr^{+3}) may pose more non-carcinogenic risks to people (Liao & Yang, 2019).

Respiratory System

The respiratory system plays a crucial role in cellular respiration and O_2 exchange in higher organisms. As such, exposure to MPs can lead to inflammation and irritation of the respiratory system. Specifically, exposure to particulate matter impacts human respiratory health. The effects of PS MPs (4.06 ± 0.44 μm) on human lung at different concentrations (1–1000 μg/cm^2). The results demonstrated that PS microplastics induced cytotoxic impacts, an inflammatory response, and oxidative stress. Additionally, the concentration and size of PS MPs were found to influence the extent of cellular toxicity (Dong et al., 2019). Literature studies revealed that, human bronchial cells exposed to high doses of PS nanoparticles displayed cytotoxic effects and metabolic alterations associated with endoplasmic reticulum stress. PS particles measuring 60 μm in length triggered the production of reactive oxygen species and induced endoplasmic reticulum stress in macrophage, ultimately leading to autophagic cell death. Furthermore, researchers exposed macrophages and human lung cells (Calu-3) to 50 nm PS nanobeads, observing DNA damage along with genotoxic and cytotoxic effects in both cell types. Similarly, studies on human alveolar treated with 25 and 70 nm PS nanoparticles demonstrated inflammatory cytokine transcripts and increased expression of NS-kB, as well as a link between cell cycle regulation and protein function (Paget et al., 2015). The ATSDR (2010) reported that inhaling PS-MPs could potentially lead to cancer. From the discussion presented, it is evident that the concentrations and size of MPs and NPs can induce cytotoxic effects, reduce cell viability, and trigger inflammatory responses. These findings provide valuable scientific evidence to help elucidate the significant health impacts associated with human exposure to inhaled MPs and NPs (Sangkham et al., 2022).

Nervous System

To gain deeper insights into the cytotoxic effects of MPs on cell viability and oxidative stress, human cerebral cells were exposed to varying concentrations (50 μg/mL–10 mg/L) of PS and PE microplastics. Their findings indicated that none of the tested MPs caused a serious decrease in cell viability, implying that cytolysis did not occur. Nevertheless, exposure to both types of microplastics led to a notable increase in reactive oxygen species in cerebral human cells. This suggests that oxidative stress could be a key mechanism through which MPs induce toxicity at the cellular level.

Placental Barrier

Although the harmful effects of MPs during pregnancy may not yet be a significant concern, it is increasingly important to investigate this issue, as MPs have the potential to cross the placental barrier of human. Researchers employed an ex vivo placental perfusion model to study the transmission procedure of PS-NPs ranging from 50 to 300 nm (Grafmueller et al., 2015). Their findings revealed that PS-NPs accumulate in the syncytiotrophoblast of placental cell, suggesting that this layer plays a critical role in regulating the transport of PS-nanoplastics across the human placenta. The primary procedure for this displacement appears to involve an energy-dependent way. These findings underscore the need for further research to fully elucidate nanoparticle transport mechanisms across the placental barrier, given the potential risk of embryotoxicity (Elizalde-velázquez & Gómez-oliván, 2021). High doses of MPs have been shown to elevate reactive oxygen species production in various human-derived tissues, triggering an inflammatory response and ultimately leading to apoptosis. Nevertheless, these findings may be influenced by several factors, including the inherent chemical composition of the MPs, which could affect the observed outcomes.

Future Aspects

The widespread presence and accumulation of MPs in the ecosystem have raised concerns about their sources, distribution patterns, and toxicological effects. However, the continued breakdown of larger plastic debris into smaller MPs under environmental forces has left the size, quantity, and shapes of MPs entering ecosystems uncertain. Additionally, inconsistencies in research methodologies and the lack of standardized classification systems hinder a comprehensive understanding of their environmental behavior. MPs possess unique chemical and physical properties, enabling them to interact with harmful contaminants in various pathways to produce complex contaminants. Despite growing interest, a systematic assessment of their harmful effects on ecosystems, accumulation, and transfer through the food chain remains insufficient. Future research should focus on addressing these gaps.

(i) Develop standardized protocols for collecting and classifying MPs samples, as well as methods for their separation and identification. Establish consistent approaches for reporting microplastic abundance and concentration, and utilize advanced techniques such as fingerprinting and isotope tracing for source analysis. Promote international collaboration to conduct life-cycle studies and comprehensive observations on MPs contamination across environmental system, including the water, air, and soil. Focus on identifying the distribution patterns of MPs, understanding their interactions within various ecosystems, and uncovering their behavior and long-term trends.
(ii) Analyze the chemical composition and properties of MPs to determine whether their toxic effects on organisms result from physical toxicity due to ingestion or from pollutants released by the MPs themselves. Examine the relationship between the chemical and physical properties of MPs and their associated toxicity. Additionally, investigate how microplastics act as carriers for toxic pollutants, potentially altering their original effects. Study the mechanisms underlying the interactions between pollutants and MPs, as well as their combined toxic impacts on organisms, and assess the role of microplastics in contributing to these composite effects.
(iii) Due to technical limitations, most existing studies on adsorption parameters are conducted in laboratory conditions, often using MPs specifically manufactured for industrial purposes. These materials may exhibit differences in adsorption capacity compared to MPs present in natural environments. It is important for future research to focus on understanding the adsorption capacity of environmental MPs and the factors influencing this process.
(iv) Most toxicity studies on MPs are conducted under laboratory conditions, often focusing on a single species, with short exposure durations and doses exceeding environmental concentrations. Integrating advanced technologies such as proteomics, genomics, and metabolomics, as well as molecular biology, can enhance our understanding of the harmful effects of MPs. Further study is required to explore the mechanisms of toxicity in cases of MPs and composite pollution, along with their impacts on community structure, biodiversity, and ecosystem functionality. These efforts will contribute to establishing baseline assessments of MPs environmental effects and developing standardized data for evaluation.
(v) To better understand the ecotoxicological methods and mechanisms of MPs, isotope tracing methods can be applied to study their transmission pathways and impacts within the food chain. This research provides a theoretical foundation for mitigating the impacts of MPs on ecosystems and human body. Inquiring the exposure pathways of MPs and associated contaminants in the human health, as well as systematically assessing their potential health risks, is a pressing priority for future studies. These efforts are essential for establishing environmental risk analysis standards for MPs in the ecosystem.

Conclusion

MPs pollution has emerged as a global environmental concern. This review provides a comprehensive summary of the toxic impacts of MPs on various environmental organisms. MPs primarily originate from two sources: primary MPs directly released into the nature and secondary MPs produced by the degradation and fragmentation of larger plastics. These particles can transfer between the aquatic systems (freshwater and marine), air, and terrestrial ecosystems. To advance research, it is essential to develop standardized methods for separating, collecting, and identifying MPs samples, alongside implementing precise and effective traceability procedures to better understand their ecological behavior. Whereas MPs consumed by organisms, MPs (including their chemical additives) can create physical toxicity, molecular and physiological responses. Additionally, they may interact with different pollutants to create compound toxic effects. By integrating molecular biology and isotope tracing techniques, studies can assess the accumulation and transmission of MPs through the food chain, particularly their potential risks to human health. These findings can serve as a theoretical foundation for assessing the ecological risks associated with MPs pollution.

Acknowledgements This work was supported by the Department of Chemical Engineering, National Institute of Technology Durgapur, West Bengal, India.

References

Aguirre-Martínez, G., Carrizo, M. V., & Zenteno-Devaud, L. (2023b). Microplastic particles' effects on aquatic organisms and their role as transporters of organic pollutants. *Water (Switzerland)*, *15*(16). https://doi.org/10.3390/w15162915

Aguirre-Martínez, G., Carrizo, M. V., & Zenteno-Devaud, L. (2023a). Microplastic particles' effects on aquatic organisms and their role as transporters of organic pollutants. *Water (Switzerland)*, *15*(16). https://doi.org/10.3390/w15162915

Arthur, C., Baker, J., & Bamford, H. (2009). Proceedings of the international research workshop on the occurrence, effects, and fate of microplastic marine debris. *Group*, 530.

Bala, K., Dogra, A., Lata, P., & Das, A. P. (2024). Discussing sources and biological-chemical remediation approaches for microplastic pollution. In *Sustainable microbial technology for synthetic and cellulosic microfiber bioremediation* (pp. 37–59). https://doi.org/10.1007/978-3-031-62600-5_3

Bala, K., Anshul, V. S., Dogra, A., Das, A. P., Kumar, R., & Nandini. (2024). Sustainable management for urban plastic waste generation. In *Sustainable management of urban plastic waste through circular economic approaches* (pp. 211–236). https://doi.org/10.1201/978100346544 7-13

Bandmann, V., Müller, J. D., Köhler, T., & Homann, U. (2012). Uptake of fluorescent nano beads into BY2-cells involves clathrin-dependent and clathrin-independent endocytosis. *FEBS Letters*, *586*(20), 3626–3632. https://doi.org/10.1016/j.febslet.2012.08.008

Barboza, L. G. A., Vieira, L. R., & Guilhermino, L. (2018). Single and combined effects of microplastics and mercury on juveniles of the European seabass (*Dicentrarchus labrax*):

Changes in behavioural responses and reduction of swimming velocity and resistance time. *Environmental Pollution, 236*, 1014–1019. https://doi.org/10.1016/j.envpol.2017.12.082
Besseling, E., Wegner, A., Foekema, E. M., Van Den Heuvel-Greve, M. J., & Koelmans, A. A. (2013). Effects of microplastic on fitness and PCB bioaccumulation by the lugworm *Arenicola marina* (L.). *Environmental Science and Technology, 47*(1), 593–600. https://doi.org/10.1021/es302763x
Bhatt, P., Bhandari, G., & Bilal, M. (2022). Occurrence, toxicity impacts and mitigation of emerging micropollutants in the aquatic environments: Recent tendencies and perspectives. *Journal of Environmental Chemical Engineering, 10*(3), 107598. https://doi.org/10.1016/j.jece.2022.107598
Bhattacharjee, J., Mishra, S., & Das, A. P. (2021). Recent advances in sensor-based detection of toxic dyes for bioremediation application: A review. *Applied Biochemistry and Biotechnology, 2021*, 1–20. https://doi.org/10.1007/S12010-021-03767-7
Bojic, S., Falco, M. M., Stojkovic, P., Ljujic, B., Gazdic Jankovic, M., Armstrong, L., Markovic, N., Dopazo, J., Lako, M., Bauer, R., & Stojkovic, M. (2020). Platform to study intracellular polystyrene nanoplastic pollution and clinical outcomes. *Stem Cells, 38*(10), 1321–1325. https://doi.org/10.1002/stem.3244
Boots, B., Russell, C. W., & Green, D. S. (2019). Effects of microplastics in soil ecosystems: above and below ground. *Environmental Science & Technology, 53*(10).
Bosker, T., Bouwman, L. J., Brun, N. R., Behrens, P., & Vijver, M. G. (2019). Microplastics accumulate on pores in seed capsule and delay germination and root growth of the terrestrial vascular plant *Lepidium sativum. Chemosphere, 226*, 774–781. https://doi.org/10.1016/j.chemosphere.2019.03.163
Carlini, G., & Kleine, K. (2018). Advancing the international regulation of plastic pollution beyond the united nations environment assembly resolution on marine litter and microplastics. *Review of European, Comparative and International Environmental Law, 27*(3), 234–244. https://doi.org/10.1111/reel.12258
Cole, M., Lindeque, P., Halsband, C., & Galloway, T. S. (2011). Microplastics as contaminants in the marine environment: A review. *Marine Pollution Bulletin, 62*(12), 2588–2597. https://doi.org/10.1016/j.marpolbul.2011.09.025
Das, A., Dey, S., Bala, K., Gnanasekaran, R., & Das, A. P. (2024a) Microbial bioremediation technology for sustainable treatment and management of synthetic microfiber waste. *Sustainable Microbial Technology for Synthetic and Cellulosic Microfiber Bioremediation*, 221–238. https://doi.org/10.1007/978-3-031-62600-5_12
Das, A., Dey, S., & Das, A. P. (2024b). Ecotoxicological impacts of synthetic microfiber pollutants and development of sustainable mitigation strategies. *Environmental Chemistry and Ecotoxicology*. https://doi.org/10.1016/J.ENCECO.2024.12.003
Das, A. P., Behera, I. D., & Das, N. P. (2024c). *Renewable energy generation and value addition from environmental microfiber pollution through advanced greener solution* (Vol. 278).
Das, A. P., Dutta, K., Khatun, R., Behera, I. D., Singh, S., & Mishra, S. (2023a). Microfiber pollution and its microbial mitigation: A review on current trends and future prospects. *Journal of the Taiwan Institute of Chemical Engineers*, 105104. https://doi.org/10.1016/J.JTICE.2023.105104
Das, A. P., & Mishra, S. (2008). Hexavalent chromium (VI): Health hazards and environmental pollutant. *Journal of Environmental Research and Development, 2*(3), 386–392.
Das, A. P., & Swain, S. (2013). Algal biosorption of toxic dye methylene blue. In *A potential source of food, feed, biochemicals, biofuels and biofertilizers, international conference on algal biorefinery* (Vol. 13). Indian Institute of Technology.
Das, P., & Bal, M. (2025). Modeling and optimization study for microplastic removal from aquatic medium using *Chroococcidiopsis* species. *Chemical Engineering Science, 304*, 121085. https://doi.org/10.1016/J.CES.2024.121085
Das, P., Halder, G., & Bal, M. (2023b). A critical review on remediation of microplastics using microalgae from aqueous system. *Science of the Total Environment, 898*(June), 166425. https://doi.org/10.1016/j.scitotenv.2023.166425

De Souza Machado, A. A., Lau, C. W., Kloas, W., Bergmann, J., Bachelier, J. B., Faltin, E., Becker, R., Görlich, A. S., & Rillig, M. C. (2019). Microplastics can change soil properties and affect plant performance. *Environmental Science and Technology, 53*(10), 6044–6052. https://doi.org/10.1021/acs.est.9b01339

Dey, S., Dash, N., Ray, S. K., Priyadarsini, S., Chhotaray, C., & Das, A. P. (2024). Occurrence and quantification of tire wear plastics pollutants from highway city dust of Bhubaneswar City of Odisha. *Sustainable Microbial Technology for Synthetic and Cellulosic Microfiber Bioremediation*, 77–89. https://doi.org/10.1007/978-3-031-62600-5_5

Dey, S., Bissoyi, A., Koteswara Reddy, G., & Das, A. P. (2024). Synthetic microfibers: An urban emerging pollutant demanding for sustainable environmental solutions. *Sustainable Microbial Technology for Synthetic and Cellulosic Microfiber Bioremediation*, 107–121. https://doi.org/10.1007/978-3-031-62600-5_7

Dong, C., Chen, C., Chen, Y., Lee, J., & Lin, C. (2019). Polystyrene microplastic particles: In vitro pulmonary toxicity assessment. *Journal of Hazardous Materials*, 121575. https://doi.org/10.1016/j.jhazmat.2019.121575

Elizalde-velázquez, G. A., & Gómez-oliván, L. M. (2021). Science of the total environment microplastics in aquatic environments: A review on occurrence, distribution, toxic effects, and implications for human health. *Science of the Total Environment, 780*, 146551. https://doi.org/10.1016/j.scitotenv.2021.146551

Enyoh, C. E., Verla, A. W., Verla, E. N., & Enyoh, E. C. (2020). Effect of macro- and micro-plastics in soil on quantitative phytochemicals in different part of juvenile lime tree (*Citrus aurantium*). *International Journal of Environmental Research, 14*(6), 705–726. https://doi.org/10.1007/s41742-020-00292-z

Farrell, P., & Nelson, K. (2013). Trophic level transfer of microplastic: *Mytilus edulis* (L.) to *Carcinus maenas* (L.). *Environmental Pollution, 177*, 1–3. https://doi.org/10.1016/j.envpol.2013.01.046

Ghosh, S., Sinha, J. K., Ghosh, S., Vashisth, K., Han, S., & Bhaskar, R. (2023). Microplastics as an emerging threat to the global environment and human health. *Sustainability, 15*(14), 10821. https://doi.org/10.3390/SU151410821

Goldstein, M. C., Rosenberg, M., & Cheng, L. (2012). Increased oceanic microplastic debris enhances oviposition in an endemic pelagic insect. *Biology Letters, 8*(5), 817–820. https://doi.org/10.1098/rsbl.2012.0298

Grafmueller, S., Manser, P., Diener, L., Diener, P., Maeder-althaus, X., Maurizi, L., Jochum, W., Krug, H. F., Buerki-thurnherr, T., Mandach, U. Von, & Wick, P. (2015). Bidirectional transfer study of polystyrene nanoparticles across the placental barrier in an ex vivo human placental perfusion model, *12*, 1280–1286.

Hernández-Arenas, R., Beltrán-Sanahuja, A., Navarro-Quirant, P., & Sanz-Lazaro, C. (2021). The effect of sewage sludge containing microplastics on growth and fruit development of tomato plants. *Environmental Pollution, 268*, 115779. https://doi.org/10.1016/j.envpol.2020.115779

Hodson, M. E., Duffus-hodson, C. A., Clark, A., Prendergast-miller, M. T., & Lane, H. (n.d.). *OK !, 44*(0), 1–24.

Huerta Lwanga, E., Gertsen, H., Gooren, H., Peters, P., Salánki, T., van der Ploeg, M., Besseling, E., Koelmans, A. A., & Geissen, V. (2017). Incorporation of microplastics from litter into burrows of Lumbricus terrestris. *Environmental Pollution, 220*, 523–531. https://doi.org/10.1016/j.envpol.2016.09.096

Hwang, J., Choi, D., Han, S., Jung, S. Y., Choi, J., & Hong, J. (2020). Potential toxicity of polystyrene microplastic particles. *Scientific Reports, 10*(1), 1–12. https://doi.org/10.1038/s41598-020-64464-9

Iguchi, T., Watanabe, H., & Katsu, Y. (2006). Application of ecotoxicogenomics for studying endocrine disruption in vertebrates and invertebrates. *Environmental Health Perspectives, 114*(SUPPL.1), 101–105. https://doi.org/10.1289/ehp.8061

Jagatee, S., Priyadarshini, S., Rath, C. C., & Das, A. P. (2024). Synthetic microfiber: An enduring environmental problem linked to sustainable development. *Environmental Science and Engineering, Part, F2333*, 93–112. https://doi.org/10.1007/978-3-031-51792-1_6

John, A., & Das, A. P. (2024). Government strategies for achieving circular economy for plastic waste management in Southeast Asia. In *Sustainable management of urban plastic waste through circular economic approaches* (pp. 72–82). https://doi.org/10.1201/9781003465447-6

Ju, H., Zhu, D., & Qiao, M. (2019). Effects of polyethylene microplastics on the gut microbial community, reproduction and avoidance behaviors of the soil springtail, Folsomia. *Environmental Pollution, 247*, 890–897. https://doi.org/10.1016/j.envpol.2019.01.097

Kaposi, K. L., Mos, B., Kelaher, B. P., & Dworjanyn, S. A. (2014). Ingestion of microplastic has limited impact on a marine larva. *Environmental Science and Technology, 48*(3), 1638–1645. https://doi.org/10.1021/es404295e

Lee, K. W., Shim, W. J., Kwon, O. Y., & Kang, J. H. (2013). Size-dependent effects of micro polystyrene particles in the marine copepod *Tigriopus japonicus*. *Environmental Science and Technology, 47*(19), 11278–11283. https://doi.org/10.1021/es401932b

Leed, R., & Smithson, M. (2019). *Ecological effects of soil microplastic pollution*. https://doi.org/10.15354/si.19.re102.Author

Li, B., Ding, Y., Cheng, X., Sheng, D., Xu, Z., Rong, Q., Wu, Y., Zhao, H., Ji, X., & Zhang, Y. (2020a). Polyethylene microplastics affect the distribution of gut microbiota and inflammation development in mice. *Chemosphere, 244*, 125492. https://doi.org/10.1016/j.chemosphere.2019.125492

Li, L., Luo, Y., Li, R., Zhou, Q., Peijnenburg, W. J. G. M., Yin, N., Yang, J., Tu, C., & Zhang, Y. (2020b). Effective uptake of submicrometre plastics by crop plants via a crack-entry mode. *Nature Sustainability, 3*(11), 929–937. https://doi.org/10.1038/s41893-020-0567-9

Liao, Y., & Yang, J. (2019). Microplastic serves as a potential vector for Cr in an in-vitro human digestive model. *Science of the Total Environment*, 134805. https://doi.org/10.1016/j.scitotenv.2019.134805

Liu, Z., Yu, P., Cai, M., Wu, D., Zhang, M., Chen, M., & Zhao, Y. (2019). Effects of microplastics on the innate immunity and intestinal microflora of juvenile *Eriocheir sinensis*. *Science of the Total Environment, 685*, 836–846. https://doi.org/10.1016/j.scitotenv.2019.06.265

Lozano, Y. M., & Rillig, M. C. (2020). Effects of Microplastic fibers and drought on plant communities. *Environmental Science and Technology, 54*(10), 6166–6173. https://doi.org/10.1021/acs.est.0c01051

Mallick, K., Sahu, A., Dubey, N. K., & Das, A. P. (2023). Harvesting marine plastic pollutants-derived renewable energy: A comprehensive review on applied energy and sustainable approach. *Journal of Environmental Management, 348*, 119371. https://doi.org/10.1016/J.JENVMAN.2023.119371

Mallick, K., Sahu, S., Sahu, A., Dey, S., & Das, A. P. (2024). A Critical review of marine microfiber pollution routes, toxicity, and its sustainable remediation. *Environmental Science and Engineering, Part, F2333*, 189–211. https://doi.org/10.1007/978-3-031-51792-1_10

Marn, N., Jusup, M., Kooijman, S. A. L. M., & Klanjscek, T. (2020). Quantifying impacts of plastic debris on marine wildlife identifies ecological breakpoints. *Ecology Letters, 23*(10), 1479–1487. https://doi.org/10.1111/ele.13574

Messinetti, S., Mercurio, S., Scarì, G., Pennati, A., & Pennati, R. (2019). Ingested microscopic plastics translocate from the gut cavity of juveniles of the ascidian *Ciona intestinalis*. *European Zoological Journal, 86*(1), 189–195. https://doi.org/10.1080/24750263.2019.1616837

Mishra, S., Rath, C. C., & Das, A. P. (2019). Marine microfiber pollution: A review on present status and future challenges. *Marine Pollution Bulletin, 140*, 188–197. https://doi.org/10.1016/J.MARPOLBUL.2019.01.039

Mishra, S., & Das, A. P. (2021). Current treatment technologies for removal of microplastic and microfiber pollutants from wastewater. *Wastewater Treatment: Cutting-Edge Molecular Tools, Techniques and Applied Aspects*, 237–251. https://doi.org/10.1016/B978-0-12-821881-5.00011-8

Mishra, S., Rout, P. K., & Das, A. P. (2021). Emerging microfiber pollution and its remediation. In R. Prasad (Ed.), *Environmental and microbial biotechnology* (pp. 247–266). Springer, Singapore. https://doi.org/10.1007/978-981-15-5499-5_9

Mishra, S., & Das, A. P. (2023). Microbial remediation of synthetic microfiber contaminated wastewater. In M. P. Shah (Ed.), *Microbial technologies in industrial wastewater treatment* (pp. 337–350). Springer, Singapore. https://doi.org/10.1007/978-981-99-2435-6_17

Mishra, S., Dash, D., & Das, A. P. (2024). Microbial degradation of polyester microfibers using indigenously isolated bacterial strain *Exiguobacterium* sp. *Clean—Soil, Air, Water, 52*(12), e202300343. https://doi.org/10.1002/CLEN.202300343

Mishra, S., Dash, D., Al-Tawaha, A. R. M. S., & Das, A. P. (2022). A review on heavy metal ion adsorption on synthetic microfiber surface in aquatic environments. *Applied Biochemistry and Biotechnology, 194*(10), 4639–4654. https://doi.org/10.1007/S12010-022-04029-W/METRICS

Mishra, S., Singh, R. P., Rath, C. C., & Das, A. P. (2020). Synthetic microfibers: Source, transport and their remediation. *Journal of Water Process Engineering, 38*, 101612. https://doi.org/10.1016/J.JWPE.2020.101612

Mishra, S., Swain, S., Sahoo, M., Mishra, S., & Das, A. P. (2021b). Microbial colonization and degradation of microplastics in aquatic ecosystem: A review. *Geomicrobiology Journal*. https://doi.org/10.1080/01490451.2021.1983670

Mohanty, M., Tripathy, B., Koteswara Reddy, G., & Das, A. P. (2024). Greener technologies for establishing restraint over microplastic pollution. In *Sustainable microbial technology for synthetic and cellulosic microfiber bioremediation* (pp. 239–258). https://doi.org/10.1007/978-3-031-62600-5_13

Monti, D. M., Guarnieri, D., Napolitano, G., Piccoli, R., Netti, P., Fusco, S., & Arciello, A. (2015). Biocompatibility, uptake and endocytosis pathways of polystyrene nanoparticles in primary human renal epithelial cells. *Journal of Biotechnology, 193*(February 2016), 3–10. https://doi.org/10.1016/j.jbiotec.2014.11.004

Paget, V., Dekali, S., Kortulewski, T., Grall, R., & Gamez, C. (2015). *Specific uptake and genotoxicity induced by polystyrene nanobeads with distinct surface chemistry on human lung epithelial cells and macrophages* (pp. 1–20). https://doi.org/10.1371/journal.pone.0123297

Pannetier, P., Morin, B., Le Bihanic, F., Dubreil, L., Clérandeau, C., Chouvellon, F., Van Arkel, K., Danion, M., & Cachot, J. (2020). Environmental samples of microplastics induce significant toxic effects in fish larvae. *Environment International, 134*(November 2019), 105047. https://doi.org/10.1016/j.envint.2019.105047

Paul, A., Priyadarsini, S., Karakuş, S., Behera, J. K., Gnanasekaran, R., & Das, A. P. (2024). Fate and accumulation of synthetic microfibers in the environment. In *Sustainable microbial technology for synthetic and cellulosic microfiber bioremediation* (pp. 61–76). https://doi.org/10.1007/978-3-031-62600-5_4

Prabhakar, A., Mishra, S., & Das, A. P. (2019). Isolation and identification of lead (Pb) solubilizing bacteria from automobile waste and its potential for recovery of lead from end of life waste batteries. *Geomicrobiological Journal, 36*(10), 894–903. https://doi.org/10.1080/01490451.2019.1654044

Priyadarshini, B., Santhosh Kumar, M., & Das, A. P. (2024). Microbial degradation of synthetic plastics released from COVID-19 generated single use disposable face masks. *Environmental Science and Engineering, Part, F2827*, 1–16. https://doi.org/10.1007/978-3-031-50840-0_1

Priyadarsini, S., & Das, A. P. (2024). Analytical and structural characterization of waste lithium-ion batteries for their effective recycling strategy. *Environmental Chemistry and Ecotoxicology*. https://doi.org/10.1016/j.enceco.2024.12.004

Priyadarsini, S., & Das, A. P. (2025). Lithium bioleaching: A review on microbial-assisted sustainable technology for lithium bio-circularity. *Journal of Water Process Engineering, 69*, 106744. https://doi.org/10.1016/j.jwpe.2024.106744

Ragusa, A., Svelato, A., Santacroce, C., Catalano, P., Notarstefano, V., Carnevali, O., Papa, F., Rongioletti, M. C. A., Baiocco, F., Draghi, S., D'Amore, E., Rinaldo, D., Matta, M., &

Giorgini, E. (2021). Plasticenta: First evidence of microplastics in human placenta. *Environment International, 146*, 106274. https://doi.org/10.1016/j.envint.2020.106274
Raj, K., & Das, A. P. (2023). Lead pollution: Impact on environment and human health and approach for a sustainable solution. *Environmental Chemistry and Ecotoxicology, 5*, 79–85. https://doi.org/10.1016/J.ENCECO.2023.02.001
Rillig, M. C., Ziersch, L., & Hempel, S. (2017). Microplastic transport in soil by earthworms. *Scientific Reports, 7*(1), 1–6. https://doi.org/10.1038/s41598-017-01594-7
Rochman, C. M., Hoh, E., Kurobe, T., & Teh, S. J. (2013). Ingested plastic transfers hazardous chemicals to fish and induces hepatic stress. *Scientific Reports*, 1–7. https://doi.org/10.1038/srep03263
Rodriguez-Seijo, A., Lourenço, J., Rocha-Santos, T. A. P., da Costa, J., Duarte, A. C., Vala, H., & Pereira, R. (2017). Histopathological and molecular effects of microplastics in *Eisenia andrei* Bouché. *Environmental Pollution, 220*, 495–503. https://doi.org/10.1016/j.envpol.2016.09.092
Rychter, P., Rogacz, D., Lewicka, K., Kollár, J., Kawalec, M., & Mosnáiek, J. (2019). Ecotoxicological properties of tulipalin a-based superabsorbents versus conventional superabsorbent hydrogels. *Advances in Polymer Technology, 2019*. https://doi.org/10.1155/2019/2947152
Saha, M., Naik, A., Desai, A., Nanajkar, M., Rathore, C., Kumar, M., & Gupta, P. (2021). Microplastics in seafood as an emerging threat to marine environment: A case study in Goa, west coast of India. *Chemosphere, 270*, 129359. https://doi.org/10.1016/j.chemosphere.2020.129359
Sahu, S., & Das, A. P. (2024). *Microplastics degradation mechanism by aquatic plastiomes: A review* (pp. 213–226). https://doi.org/10.1007/978-3-031-50840-0_12
Sahu, A., Dey, S., Mallick, K., Priyadarsini, S., Behera, I. D., & Das, A. P. (2024). A review on advanced technology for sustainable management of synthetic microplastic waste. In *Sustainable management of urban plastic waste through circular economic approaches*, 262–279. https://doi.org/10.1201/9781003465447-15
Sangkham, S., Faikhaw, O., Munkong, N., & Sakunkoo, P. (2022). A review on microplastics and nanoplastics in the environment: Their occurrence, exposure routes, toxic studies, and potential effects on human health. *Marine Pollution Bulletin, 181*(June), 113832. https://doi.org/10.1016/j.marpolbul.2022.113832
Santana, M. F. M., Moreira, F. T., Pereira, C. D. S., Abessa, D. M. S., & Turra, A. (2018). Continuous exposure to microplastics does not cause physiological effects in the cultivated mussel *Perna perna. Archives of Environmental Contamination and Toxicology, 74*(4), 594–604. https://doi.org/10.1007/s00244-018-0504-3
Satapathy, S., Kar, D., Priyadarsini, S., Das, A. P., & Kuanar, A. (2024). Application of microorganisms in emerging waste management strategies and remediation of synthetic and cellulosic microfibers. In *Sustainable microbial technology for synthetic and cellulosic microfiber bioremediation* (pp. 183–206). https://doi.org/10.1007/978-3-031-62600-5_10
Schirinzi, G. F., Pérez-Pomeda, I., Sanchís, J., Rossini, C., Farré, M., & Barceló, D. (2017). Cytotoxic effects of commonly used nanomaterials and microplastics on cerebral and epithelial human cells. *Environmental Research, 159*(June), 579–587. https://doi.org/10.1016/j.envres.2017.08.043
Setala, O., Lehtiniemi, M., Coppock, R., & Cole, M. (2018). *Microplastics in marine food webs* (pp. 339–363). https://doi.org/10.1016/B978-0-12-813747-5.00011-4
Singh Jadaun, J., Bansal, S., Sonthalia, A., Rai, A. K., & Singh, S. P. (2022). Biodegradation of plastics for sustainable environment. *Bioresource Technology, 347*, 12669710.1016/j.biortech.2022.126697.
Singh, R. P., Mishra, S., & Das, A. P. (2020). Synthetic microfibers: Pollution toxicity and remediation. *Chemosphere, 257*, 127199. https://doi.org/10.1016/j.chemosphere.2020.127199
Song, Y., Cao, C., Qiu, R., Hu, J., Liu, M., Lu, S., Shi, H., Raley-susman, K. M., & He, D. (2019). Uptake and adverse effects of polyethylene terephthalate microplastics fibers on terrestrial snails (*Achatina fulica*) after soil. *Environmental Pollution, 250*, 447–455. https://doi.org/10.1016/j.envpol.2019.04.066

Stock, V., Böhmert, L., Lisicki, E., Block, R., Cara-Carmona, J., Pack, L. K., Selb, R., Lichtenstein, D., Voss, L., Henderson, C. J., Zabinsky, E., Sieg, H., Braeuning, A., & Lampen, A. (2019). Uptake and effects of orally ingested polystyrene microplastic particles in vitro and in vivo. *Archives of Toxicology, 93*(7), 1817–1833. https://doi.org/10.1007/s00204-019-02478-7

Syberg, K., Nielsen, A., Khan, F. R., Banta, G. T., Palmqvist, A., & Jepsen, P. M. (2017). Microplastic potentiates triclosan toxicity to the marine copepod *Acartia tonsa* (Dana). *Journal of Toxicology and Environmental Health—Part a: Current Issues, 80*(23–24), 1369–1371. https://doi.org/10.1080/15287394.2017.1385046

Taylor, S. E., Pearce, C. I., Sanguinet, K. A., Hu, D., Chrisler, W. B., Kim, Y. M., Wang, Z., & Flury, M. (2020). Polystyrene nano- and microplastic accumulation at Arabidopsis and wheat root cap cells, but no evidence for uptake into roots. *Environmental Science: Nano, 7*(7), 1942–1953. https://doi.org/10.1039/d0en00309c

Thacharodi, A., Meenatchi, R., Hassan, S., Hussain, N., Bhat, M. A., Arockiaraj, J., Ngo, H. H., Le, Q. H., & Pugazhendhi, A. (2024). Microplastics in the environment: A critical overview on its fate, toxicity, implications, management, and bioremediation strategies. *Journal of Environmental Management, 349*(October 2023), 119433. https://doi.org/10.1016/j.jenvman.2023.119433

Tripathy, B., Dash, A., & Das, A. P. (2022). Detection of environmental microfiber pollutants through vibrational spectroscopic techniques: Recent advances of environmental monitoring and future prospects. *Critical Reviews of Analytical Chemistry*, 1–11. https://doi.org/10.1080/10408347.2022.2144994

Wright, S. L., Thompson, R. C., & Galloway, T. S. (2013). The physical impacts of microplastics on marine organisms: A review. *Environmental Pollution, 178*, 483–492. https://doi.org/10.1016/j.envpol.2013.02.031

Wu, X., Liu, Y., Yin, S., Xiao, K., Xiong, Q., Bian, S., Liang, S., Hou, H., Hu, J., & Yang, J. (2020). Metabolomics revealing the response of rice (*Oryza sativa* L.) exposed to polystyrene microplastics. *Environmental Pollution, 266*, 115159. https://doi.org/10.1016/j.envpol.2020.115159

Ya, H., Jiang, B., Xing, Y., Zhang, T., Lv, M., & Wang, X. (2021). Recent advances on ecological effects of microplastics on soil environment. *Science of the Total Environment, 798*, 149338. https://doi.org/10.1016/j.scitotenv.2021.149338

Yu, Y., Chen, H., Hua, X., Dang, Y., Han, Y., Yu, Z., Chen, X., Ding, P., & Li, H. (2020). Polystyrene microplastics (PS-MPs) toxicity induced oxidative stress and intestinal injury in nematode *Caenorhabditis elegans*. *Science of the Total Environment, 726*, 138679. https://doi.org/10.1016/j.scitotenv.2020.138679

Zhang, N., Li, Y. B., He, H. R., Zhang, J. F., & Ma, G. S. (2021). You are what you eat: Microplastics in the feces of young men living in Beijing. *Science of the Total Environment, 767*, 144345. https://doi.org/10.1016/j.scitotenv.2020.144345

Zhou, B., Wang, J., Zhang, H., Shi, H., Fei, Y., Huang, S., Tong, Y., Wen, D., Luo, Y., & Barceló, D. (2020). Microplastics in agricultural soils on the coastal plain of Hangzhou Bay, east China: Multiple sources other than plastic mulching film. *Journal of Hazardous Materials, 388*(November), 121814. https://doi.org/10.1016/j.jhazmat.2019.121814

Zhu, D., Chen, Q. L., An, X. L., Yang, X. R., Christie, P., Ke, X., Wu, L. H., & Zhu, Y. G. (2018b). Exposure of soil collembolans to microplastics perturbs their gut microbiota and alters their isotopic composition. *Soil Biology and Biochemistry, 116*(October 2017), 302–310. https://doi.org/10.1016/j.soilbio.2017.10.027

Zhu, D., Chen, Q. L., An, X. L., Yang, X. R., Christie, P., Ke, X., Wu, L. H., & Zhu, Y. G. (2018a). Exposure of soil collembolans to microplastics perturbs their gut microbiota and alters their isotopic composition. *Soil Biology and Biochemistry, 116*(October 2017), 302–310. https://doi.org/10.1016/j.soilbio.2017.10.027

Chapter 21
Transformative Approaches to Micropollutant Remediation Within a Circular Economy Paradigm

B. B. Nayak, G. Bhargavi, Heeramoni Boro, M. Ramesh Naik, and A. V. Ramanjaneyulu

Introduction

The presence of micropollutants in the environment has become an urgent concern due to their potential adverse effects on human health, aquatic life, and ecosystems. Micropollutants, which include pharmaceuticals, personal care products, pesticides, industrial chemicals, and heavy metals, are contaminants found at low concentrations (typically in the nanogram to microgram per liter range) in various environmental matrices, such as water, soil, and air. These substances are released into the environment through agricultural runoff, industrial effluents, wastewater discharge, and improper disposal of household chemicals (Fatta-Kassinos et al., 2008). Despite their low concentrations, micropollutants can have significant, often long-lasting impacts on environmental and public health.

The primary sources of micropollutants are wastewater treatment plants (WWTPs), which, while effective at removing larger pollutants, are generally inefficient in removing many micropollutants due to their complex chemical nature and low concentrations (Mishra & Das, 2023; Mishra et al., 2024). For example, pharmaceuticals and personal care products (PPCPs) such as antibiotics, hormones, and cosmetics are commonly detected in effluent from WWTPs (Castiglioni et al., 2006).

B. B. Nayak (✉) · H. Boro
School of Sciences, Woxsen University, Hyderabad, Telangana, India
e-mail: nayak.96agri@gmail.com

G. Bhargavi
Department of Agriculture, S R University, Warangal, India

M. R. Naik
RSM Division, ICAR-NAARM, Hyderabad, India

A. V. Ramanjaneyulu
Professor Jayashankar Telangana State Agricultural University, Hyderabad, India

I. D. Behera et al. (eds.), *Emerging Micropollutants*, Sustainable Environmental Waste Management Strategies, https://doi.org/10.1007/978-3-032-03530-1_21

Similarly, pesticides and industrial chemicals are released into the environment through agricultural runoff and industrial processes, leading to their widespread presence in surface waters and soils. In many cases, these pollutants are not adequately addressed by conventional treatment systems, resulting in their continued presence in the environment.

The risks associated with micropollutants are manifold. In aquatic environments, these pollutants can disrupt ecosystems by affecting the health of aquatic organisms, including fish, amphibians, and invertebrates. For instance, the presence of endocrine-disrupting chemicals (EDCs) such as hormones in water can lead to reproductive abnormalities in aquatic species (Jobling et al., 1998). Furthermore, the bioaccumulation of micropollutants in the food chain can pose a direct threat to human health. Studies have shown that chemicals like mercury, persistent organic pollutants (POPs), and pharmaceuticals can accumulate in edible species, leading to potential risks when consumed (Schoenfuss et al., 2016).

In addition to the ecological risks, micropollutants are also associated with human chronic health effects. Long-term exposure to low concentrations of certain micropollutants has been linked to various health problems, including endocrine disorders, neurodevelopmental issues, and an increased risk of cancer (Fent et al., 2006). Moreover, the widespread use of antibiotics and other pharmaceuticals has contributed to the growing problem of antimicrobial resistance (AMR), which threatens the effectiveness of existing medical treatments (Martínez, 2009).

The persistence of micropollutants in the environment, coupled with the inadequacy of traditional treatment methods to fully remove them, underscores the need for more effective remediation strategies. Current wastewater treatment systems, such as activated sludge and conventional filtration, often fail to adequately remove micropollutants, leading to their discharge into the environment. As a result, alternative treatment technologies, including advanced oxidation processes (AOPs), bioremediation, and adsorption techniques, are being explored to address this growing concern (Andreozzi et al., 2003).

As we move towards a more sustainable and resource-efficient future, the concept of a **circular economy** is becoming increasingly relevant in the context of micropollutant remediation. A circular economy focuses on minimizing waste, maximizing resource recovery, and promoting the reuse and recycling of materials (Priyadarsini & Das, 2024, 2025). In this paradigm, the goal is not only to reduce the release of micropollutants into the environment but also to find innovative ways to recover valuable resources from waste streams, such as wastewater and contaminated soils, and integrate them back into the production cycle. For example, nutrients, metals, and organic materials can be recovered and reused in agriculture or industry, thus reducing the overall environmental footprint of these pollutants (Korhonen et al., 2018).

The concept of a circular economy presents an opportunity to reimagine the remediation of micropollutants. It calls for the development of technologies that do not simply treat or remove pollutants but also transform them into valuable resources. Remediation strategies that align with circular economy principles include the recovery of bioactive compounds from wastewater, the use of waste-derived adsorbents for micropollutant removal, and the integration of bioremediation techniques that promote the regeneration of ecosystems. By viewing micropollutant remediation through the lens of circular economy, we can create sustainable, closed-loop systems that address pollution while simultaneously fostering resource recovery and environmental restoration.

The following sections of this chapter will explore various methods of micropollutant remediation, with a particular focus on their role within a circular economy framework. The discussion will cover physical, chemical, and biological techniques, with an emphasis on their efficiency, sustainability, and potential for large-scale application. Specifically, we will examine the role of adsorption materials such as activated carbon, advanced oxidation processes (AOPs) like ozonation and photocatalysis, and bioremediation approaches that utilize microorganisms to degrade or transform pollutants. Additionally, we will highlight the latest innovations in the field, such as the use of nanomaterials and green technologies, which hold promise for improving the removal of micropollutants in wastewater and environmental matrices.

As the understanding of micropollutants and their impact on the environment deepens, the need for integrated and multi-faceted approaches to their remediation becomes even more critical. By incorporating circular economy principles, we can shift from a linear "take-make-dispose" model to a more sustainable, restorative approach to environmental management. The next section will delve into these different remediation methods, discussing their effectiveness, sustainability, and potential for addressing the challenges posed by micropollutant contamination in a circular economy paradigm.

Micropollutants: A Growing Global Concern

Micropollutants represent an emerging and pervasive challenge in environmental management. These chemical substances, present in trace amounts, enter ecosystems through diverse pathways and have profound ecological and health implications despite their low concentrations (Das & Mishra, 2008; Kanungo et al., 2023; Mallick et al., 2023). The increasing awareness of their impacts has underscored the urgency of adopting transformative strategies within a circular economy paradigm to mitigate their effects.

Common Sources and Types of Micropollutants

Micropollutants arise from various human activities, including agriculture, industry, and urban lifestyles. They are broadly categorized into:

1. **Pharmaceuticals and Personal Care Products (PPCPs)**

Pharmaceuticals, such as antibiotics, painkillers, and hormones, often find their way into water systems through human excretion, improper disposal, and wastewater discharge. These residues, particularly antibiotics, contribute to antimicrobial resistance (AMR), identified by WHO as a top global health threat. For instance, over 50% of wastewater samples in Europe contain pharmaceutical residues (European Medicines Agency, 2021).

2. **Pesticides and Herbicides**

Agricultural runoff introduces pesticides into surface and groundwater, frequently exceeding safety limits in certain regions. The FAO (2020) reports pesticide contamination in over 40% of global freshwater sources, disrupting aquatic ecosystems and threatening biodiversity. In India, pesticide residues in groundwater present significant risks to drinking water and irrigation (Kumar et al., 2021).

3. **Industrial Chemicals**

Heavy metals, persistent organic pollutants (POPs), and synthetic chemicals from industrial activities contribute significantly to micro pollution. Even in minute quantities, heavy metals like mercury and lead bioaccumulate, posing severe health risks (United Nations Environment Programme, 2022).

Table 21.1 shows the different sources of micropollutants and their global prevalence.

4. **Household Chemicals**

   Urban wastewater systems are often laden with household chemicals, including cleaning agents and detergents. These contribute to approximately 15% of urban chemical pollution, as reported by the EPA (2020).

Challenges and Prospects in Micropollutant Remediation

The remediation of micropollutants presents significant challenges due to their persistence, health risks, and regulatory gaps (Dey et al., 2023; Mishra et al., 2021; Singh et al., 2020). Many micropollutants resist degradation, leading to long-term environmental presence and bioaccumulation within food chains. Persistent exposure to hormone-disrupting chemicals and antimicrobial resistance (AMR)-related residues exacerbates public health issues. Furthermore, traditional water treatment technologies often fall short in addressing micropollutants effectively, necessitating the development of advanced remediation methods. Regulatory gaps further hinder progress,

Table 21.1 Different sources of micropollutants and their global prevalence

Source	Micropollutant type	Global prevalence	Indian prevalence	Economic impact	References
Pharmaceuticals	APIs	Found in over 200 forms, up to 5 μg/L in water	Present in urban water, up to 5 μg/L	High purification costs, drug-resistant pathogens; global cost: $1 billion/year	UNEP (2023), FAO (2020) and WHO (2021)
Personal care products	Soaps, cosmetics, etc.	Detected globally at ng to μg/L	Higher levels in urban metros	Increased wastewater treatment costs, ecosystem damage	WHO (2021) and FAO (2020)
Pesticides	Herbicides, insecticides	25% of agricultural runoff contains residues	Significant rural contamination	Reduced productivity, soil degradation; global loss: $4.5 billion/year	FAO (2020) and UNEP (2023)
Industrial chemicals	Heavy metals, organics	20% contribution to urban water micropollutants	Effluent contamination in cities like Delhi, Mumbai	Cleanup costs, health expenses; global cost: $3 billion/year	WHO (2021) and UNEP (2023)
Wastewater effluents	Microplastics, chemicals	Detected in 80% of wastewater globally	Contamination in metro wastewater systems	Infrastructure costs, public health impact; microplastic pollution: $13 billion/year	WHO (2020) and UNEP (2023)

(continued)

Table 21.1 (continued)

Source	Micropollutant type	Global prevalence	Indian prevalence	Economic impact	References
Agricultural runoff	Pesticides, fertilizers	25% of runoff contains pesticides	High contamination in rural runoff	Crop yield decline, soil remediation costs; global loss: $2 billion/year	FAO (2020) and UNEP (2023)
Waste disposal	Microplastics, plasticizers	Oceans/ freshwater have billions of micrograms of microplastics	Rivers like Ganga, Yamuna show contamination	Ecosystem damage, marine biodiversity loss; global impact: $8 billion/year	UNEP (2023) and WHO (2021)

with the absence of comprehensive global standards limiting consistent monitoring and control efforts (UNEP, 2023).

Despite these challenges, sustainable micropollutant remediation holds promise through innovative approaches. Advances in bioremediation, nanotechnology, and green chemistry are emerging as effective solutions. Bioremediation leverages microorganisms to degrade contaminants, while nanomaterials like activated carbon have demonstrated high efficacy in pollutant removal (Zhao 2020). Additionally, circular economy approaches, which emphasize waste minimization and resource recovery, offer practical benefits. For example, recovering nutrients such as phosphorus and nitrogen from wastewater reduces pollution while supporting sustainable agricultural practices (Xie 2020). Strengthening global regulatory frameworks and establishing unified guidelines can further enhance remediation efforts, enabling more integrated and effective strategies to combat micropollutant contamination.

Ecological and Health Risks

Micropollutants, despite their trace concentrations, present significant ecological and human health challenges that demand urgent attention. In aquatic ecosystems, the presence of endocrine-disrupting chemicals (EDCs) has far-reaching consequences. These compounds interfere with hormonal systems in aquatic organisms, leading to reproductive abnormalities, such as altered sex ratios, reduced fertility, and developmental deformities, as documented by Kidd et al. (2007). The impact is not limited to individual species; disruptions at the organismal level can cascade through the food web, destabilizing entire ecosystems. For example, declines in fish populations due to EDC exposure can affect predator species and alter nutrient cycling in aquatic environments. The persistence of these chemicals exacerbates the problem, as they accumulate in sediments and bioaccumulate in higher trophic levels, amplifying their effects over time.

For human populations, the risks associated with micropollutants are equally alarming. Chronic exposure, even at low levels, has been linked to developmental disorders, hormonal imbalances, and an increased incidence of non-communicable diseases, including cancer. Arsenic contamination in drinking water serves as a poignant example; it has been directly associated with severe health outcomes such as skin lesions, cardiovascular diseases, and multiple cancer types, as reported by WHO (2010). Similarly, exposure to pharmaceutical residues, pesticides, and industrial chemicals can lead to subtle but significant health impacts, including endocrine disruption, antibiotic resistance, and neurological disorders.

The interconnectedness of ecological and human health amplifies the urgency of addressing micropollutant contamination. Aquatic ecosystems, which serve as a critical source of food and water, are particularly vulnerable. As these ecosystems degrade, the availability and safety of these resources are compromised, leading to broader socio-economic consequences, such as reduced fisheries yields and increased healthcare costs. Furthermore, the long-term persistence and global prevalence of micropollutants make them a transboundary issue, necessitating coordinated international efforts for effective remediation and policy development. Recognizing and mitigating these risks is essential for ensuring the sustainability of ecosystems and the well-being of future generations.

Micropollutants Remediation Approaches

Micropollutants, including pharmaceuticals, pesticides, and industrial chemicals, are persistent contaminants in water systems that pose significant environmental and health risks. A variety of remediation techniques have been developed to address these pollutants, ranging from traditional methods to more advanced and sustainable approaches. These methods can be broadly categorized into traditional methods, advanced biological and chemical methods, green chemistry approaches, and advanced physicochemical techniques. Different approaches to micropollutants remediation is presented in Fig. 21.1.

Traditional Methods

Conventional methods such as physical filtration, chemical treatments, and traditional wastewater treatment processes have been employed to mitigate micropollutant contamination. Physical filtration techniques, including activated carbon and

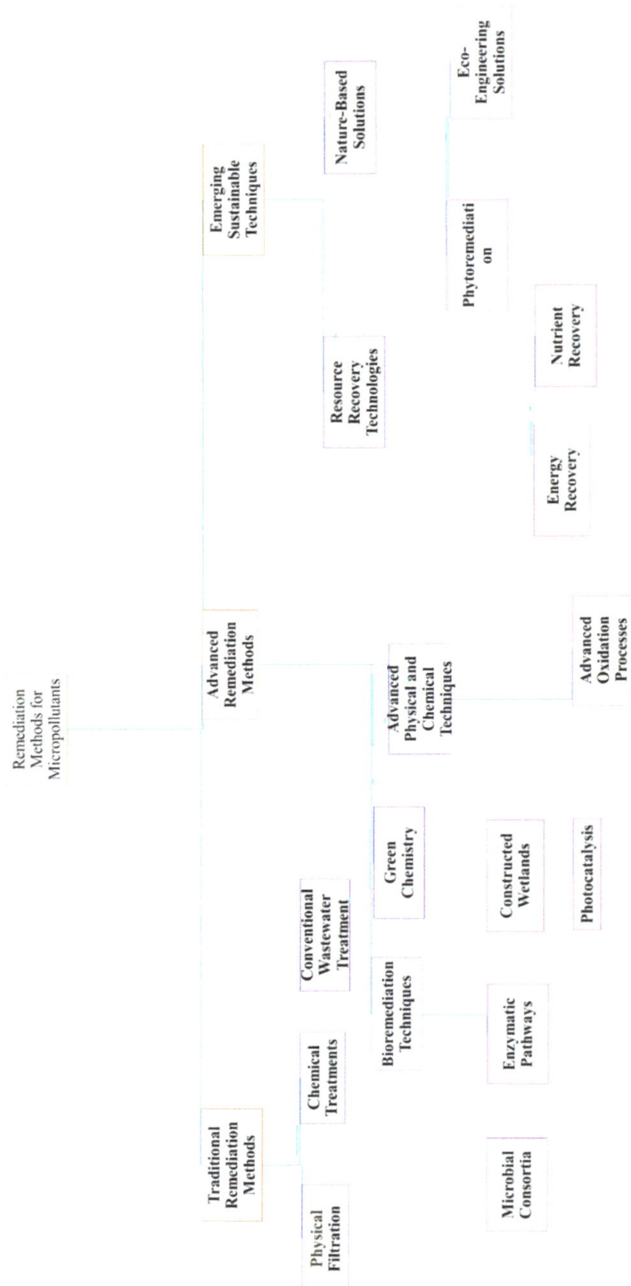

Fig. 21.1 Different approaches to micropollutants remediation

reverse osmosis, are commonly used to remove specific pollutants like pharmaceuticals, pesticides, and volatile organic compounds (VOCs). Activated carbon filtration, for instance, is widely used in municipal water treatment plants to remove trace pharmaceuticals (Westerhoff 2005).

Chemical treatments, such as ozonation and advanced oxidation processes (AOPs), are employed to degrade persistent pollutants in wastewater, including organic contaminants and pharmaceutical residues (Von Gunten, 2003). However, these conventional methods have limited efficacy in addressing micropollutants due to their complex and persistent nature.

Advanced Biological Methods

Bioremediation has gained attention as an advanced method for addressing micropollutants, leveraging natural biological processes to degrade contaminants. Microbial consortia, consisting of diverse microbial communities such as *Pseudomonas* and *Bacillus*, have been shown to effectively degrade hydrocarbons and pharmaceutical residues, achieving up to 85% degradation in wastewater. In addition, enzymatic degradation using enzymes like laccases and peroxidases offers a targeted breakdown of pollutants, minimizing the formation of secondary pollution. For example, fungal laccases are effective in degrading synthetic dyes and other pollutants found in textile effluents (Baldrian, 2006). Constructed wetlands, which mimic natural ecosystems, utilize plants, soil, and microorganisms to remove pollutants from wastewater. Studies have shown that constructed wetlands can remove pharmaceutical residues and other micropollutants, reducing pollutant concentrations by up to 60% in some cases (Vymazal, 2011).

Green Chemistry Approaches

Green chemistry focuses on designing sustainable, non-toxic methods for addressing micropollutants while minimizing environmental impact. A significant approach in this area is the use of biochar, a material produced by pyrolyzing biomass. Biochar is highly porous and has a large surface area, making it an effective adsorbent for micropollutants, including pharmaceuticals and pesticides. Its use aligns with green chemistry principles by utilizing waste biomass and providing an eco-friendly alternative for water purification. Additionally, biocatalysts, such as lipases and cellulases, are being explored for their role in environmentally friendly chemical reactions. These catalysts enable green oxidation processes that can degrade micropollutants without producing harmful byproducts. Iron-based catalysts and palladium on carbon (Pd/C) are other examples of safer catalysts used in green chemistry processes, offering efficient, sustainable solutions for micropollutant removal.

Advanced Physicochemical Techniques

In addition to biological and green chemistry methods, advanced physicochemical techniques such as photocatalysis and advanced oxidation processes (AOPs) have shown promise in treating micropollutants. Photocatalysis utilizes light-activated materials, such as titanium dioxide (TiO_2), to degrade persistent organic pollutants. This method is particularly effective for breaking down contaminants that are resistant to biodegradation (Fujishima et al., 2000). AOPs, which generate highly reactive species like hydroxyl radicals, can completely oxidize a wide range of pollutants, including non-biodegradable substances like chlorinated solvents (Glaze et al., 1987). These techniques offer high efficiency and can target a broad spectrum of contaminants, making them valuable tools for addressing micropollutant contamination in water systems.

The Limitations of Methods

Traditional Methods

Physical filtration techniques, such as activated carbon and reverse osmosis, are widely used for micropollutant removal but come with several limitations. Activated carbon requires frequent regeneration or replacement, which increases operational costs, and its adsorption capacity may decrease over time, reducing efficiency (Westerhoff 2005). Reverse osmosis, while effective for removing a broad range of contaminants, is energy-intensive, generates large amounts of wastewater, and has limited capacity for handling complex or multiple pollutants (Baker, 2004). Additionally, both methods are expensive for large-scale applications. Chemical treatments, including ozonation and advanced oxidation processes (AOPs), can effectively degrade pollutants but may produce toxic byproducts, such as aldehydes or acids, which require further treatment (Von Gunten, 2003). These methods also demand significant energy input, making them less sustainable for long-term use (Glaze et al., 1987).

Advanced Biological Methods

Bioremediation offers an eco-friendly solution for micropollutant degradation, but its effectiveness is highly dependent on environmental conditions such as temperature, pH, and nutrient availability. Certain micropollutants, particularly complex or persistent compounds like pharmaceuticals, are not easily degraded by microorganisms, limiting the scope of this method. Moreover, bioremediation can be slow, taking weeks or even months for significant pollutant degradation, which may not be suitable for rapid remediation needs. Enzymatic degradation, using enzymes like laccases and peroxidases, is another promising biological method, but the high cost of enzymes and the potential for enzyme deactivation under harsh conditions can limit its practical application (Baldrian, 2006). Constructed wetlands, though promising for wastewater treatment, require significant space and may have limited performance in colder climates. They also require regular maintenance to ensure the optimal functioning of plant and microbial systems, and their efficiency can vary depending on the pollutant type (Vymazal, 2011).

Green Chemistry Approaches

Biochar, while an effective adsorbent for pollutants such as pharmaceuticals and pesticides, has limitations in its long-term use. Its adsorption capacity decreases with time and repeated exposure to pollutants, requiring periodic replacement. Furthermore, biochar production can be energy-intensive, depending on the pyrolysis conditions, which may offset its environmental benefits (Lehmann et al., 2006). Biocatalysts, such as lipases and cellulases, offer environmentally friendly degradation pathways, but their application is limited by the high cost of production and the challenges in maintaining their stability and activity under real-world conditions (Miao et al., 2020). Moreover, scaling up these processes for industrial use remains a challenge due to the high cost and limited availability of suitable biocatalysts. Iron-based catalysts and palladium on carbon offer efficient remediation pathways for micropollutants, but their use may introduce environmental risks, such as leaching of heavy metals or other toxic byproducts if not carefully managed (Mori et al., 2013).

Advanced Physicochemical Techniques

Photocatalysis is an effective method for degrading persistent organic pollutants using light-activated materials such as titanium dioxide (TiO_2), but it faces several limitations. The efficiency of photocatalysis can be significantly reduced under low-light conditions, and the need for constant light exposure increases energy consumption, limiting its application in large-scale water treatment (Fujishima et al., 2000).

Additionally, the cost of synthesizing high-quality photocatalytic materials and their long-term stability can be a challenge for practical use. AOPs, which generate highly reactive species like hydroxyl radicals, can oxidize a broad range of pollutants, including non-biodegradable compounds like chlorinated solvents (Glaze et al., 1987). However, these processes are energy-intensive, costly, and may produce toxic byproducts that require further treatment, which can negate their environmental benefits. The high operational costs and the complexity of managing AOP systems in large-scale applications also limit their widespread use.

Circular Economy Integration in Remediation Techniques

Emerging sustainable techniques, particularly in resource recovery and nature-based solutions, are driving the integration of circular economy principles in environmental remediation. This economic model aims to minimize waste, optimize resource use, and maintain the value of products, materials, and resources within the economy for as long as possible. These innovative approaches not only provide environmentally friendly alternatives to traditional remediation methods but also present opportunities for waste minimization, resource recovery, and sustainable practices across various industries.

Waste-to-Resource Paradigm

Energy and nutrient recovery technologies are core examples of the waste-to-resource model, a fundamental principle of the circular economy. These technologies convert organic waste into biogas and recover valuable nutrients for reuse in agriculture, creating closed-loop systems that repurpose waste and reduce environmental impact. For example, biogas production from agricultural waste can offset up to 50% of energy needs in farming operations, significantly lowering operational costs and reducing carbon emissions (Weiland, 2010). Nutrient recovery technologies, particularly for phosphorus, -are also crucial in reducing the need for phosphate mining—an increasingly scarce resource—by recycling phosphorus back into agricultural systems. This approach helps mitigate the environmental pollution caused by fertilizer runoff and supports a more sustainable, circular agricultural system. Table 21.2 describe potential of circular economy in micropollutants sectors and their alignment with SDGs.

Ecological Restoration and Resource Reuse

Nature-based solutions, such as phytoremediation and eco-engineering, play an essential role in ecological restoration by leveraging natural systems to address environmental contamination. These methods absorb or neutralize pollutants, restoring ecosystems, improving soil health, and promoting biodiversity. For instance, Indian mustard (*Brassica juncea*) has been successfully used to restore heavy metal-contaminated soils, enabling land to be repurposed for agriculture or other uses (Salt et al., 1998). Hybrid wetlands have also been effective in treating wastewater, reducing the need for chemical treatments while promoting the reuse of water and nutrients in agricultural and industrial systems (Brix et al., 2003). These solutions align with circular economy principles by restoring and reusing natural resources, such as land, water, and biological systems, rather than depleting or

Table 21.2 Potential of circular economy in micropollutants sectors and their alignment with SDGs (United Nations, 2015)

Micropollutant type	Circular economy value (recovery and reuse—economic terms)	Relevant SDGs
Pharmaceuticals	$500 million/year in potential savings through recovery and reuse of APIs for drug development, reducing healthcare costs	SDG 3 (good health and well-being), SDG 6 (clean water and sanitation)
Personal care products (PCPs)	$300 million/year in potential savings through the recovery of chemicals for reuse in the cosmetic and cleaning industries	SDG 6 (clean water and sanitation), SDG 12 (responsible consumption and production)
Pesticides	$1 billion/year through recovery of pesticides via phytoremediation or chemical degradation, reducing costs of agricultural input	SDG 2 (zero hunger), SDG 12 (responsible consumption and production)
Industrial chemicals	$500 million/year through recovery of heavy metals and organic pollutants using technologies like electrochemical treatment and phytoremediation	SDG 6 (clean water and sanitation), SDG 14 (life below water)
Wastewater effluents	$600 million/year potential recovery through advanced filtration and adsorption technologies to reclaim pharmaceuticals and microplastics	SDG 6 (clean water and sanitation), SDG 12 (responsible consumption and production)
Agricultural runoff	$400 million/year through the recovery of fertilizers and pesticides for reuse in agricultural practices via filtration and biological treatment systems	SDG 2 (zero hunger), SDG 12 (responsible consumption and production)
Waste disposal (plastic waste)	$1 billion/year through advanced recycling technologies (e.g., pyrolysis, chemical recycling) to reclaim plastics for reuse in manufacturing, reducing overall waste and cost	SDG 12 (responsible consumption and production), SDG 14 (life below water)

permanently damaging them. Ultimately, they contribute to long-term environmental sustainability and resource efficiency.

Transition to Remediation Methods

Addressing the challenges posed by environmental micropollutants requires the development of effective remediation strategies. This includes exploring green technologies, resource recovery techniques, and emerging sustainable methods that provide holistic solutions for pollution control. Nature-based solutions, such as phytoremediation and eco-engineering, offer promising alternatives for micropollutant removal. Phytoremediation utilizes plants to absorb, degrade, or immobilize pollutants, offering a cost-effective and sustainable method for treating contaminated soils and water (Sharma et al., 2019). Eco-engineering solutions, such as the design of green infrastructure and sustainable wetland systems, not only improve pollutant removal efficiency but also restore vital ecosystem services, including water purification, habitat creation, and biodiversity conservation (Hass et al., 2019). Figure 21.2 showed the integration of circular economy with micropollutant remediation.

Resource recovery techniques also address micropollutants while recovering valuable resources. Energy recovery systems, such as microbial fuel cells, degrade pollutants and generate electricity, providing a sustainable way to treat wastewater while producing energy (Logan et al., 2006). Similarly, nutrient recovery methods extract nitrogen and phosphorus from wastewater, helping to prevent eutrophication in water bodies and reduce reliance on synthetic fertilizers (Ahn et al., 2014). These techniques align with circular economy principles by emphasizing resource efficiency, waste minimization, and sustainability.

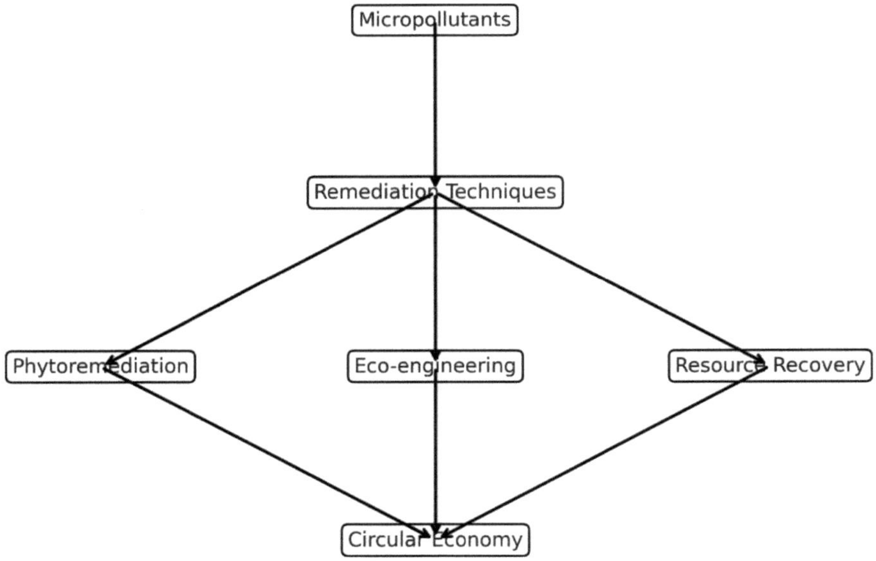

Fig. 21.2 Integration of circular economy with micropollutant remediation

Incorporating advanced water treatment technologies, resource recovery initiatives, and responsible chemical disposal practices into a circular economy framework helps reduce the environmental impact of micropollutants. By recovering nutrients and energy from wastewater, these methods decrease dependency on synthetic fertilizers and fossil fuels, promoting a more sustainable approach to wastewater management (Cornel & Schaum, 2009; Weiland, 2010). Furthermore, responsible disposal practices for chemicals and pharmaceuticals ensure that contaminants do not re-enter the ecosystem, preventing long-term environmental damage (Jelic et al., 2011). This holistic approach fosters a resilient, sustainable ecosystem that moves away from waste generation toward resource recovery and long-term environmental stewardship.

Emerging Sustainable Techniques for Micropollutant Remediation in the Circular Economy

The growing need for effective micropollutant remediation has led to the exploration of emerging sustainable techniques that combine environmental restoration with resource recovery. These techniques, including nature-based solutions like phytoremediation and eco-engineering, are gaining recognition for their ability to treat pollutants while promoting sustainability and resource efficiency. These methods are integral to the circular economy, which focuses on minimizing waste, recovering valuable resources, and enhancing ecosystem services.

Nature-Based Solutions

Phytoremediation

Phytoremediation is a nature-based, eco-friendly technique that uses plants to absorb, degrade, or stabilize pollutants, including heavy metals, hydrocarbons, and organic chemicals (Behera et al., 2021, 2022; Mohanta et al., 2024). This method is particularly attractive due to its low cost, environmental benefits, and the fact that it does not generate secondary waste. For example, Indian mustard (*Brassica juncea*) has been successfully used to remediate lead-contaminated soils, absorbing up to 1000 mg of lead per kilogram of plant biomass (Salt et al., 1998). Efficiency of different bioremediation methods is presented in Table 21.3.

Other plants, like sunflowers (*Helianthus annuus*) and willows (*Salix* spp.), have been utilized to remove various contaminants, including arsenic and radionuclides (Ali et al., 2013). Phytoremediation has also been applied in various contexts in India. For instance, studies in Uttar Pradesh have demonstrated the potential of *Brassica juncea* and *Helianthus annuus* in remediating soils contaminated with heavy metals,

Table 21.3 Efficiency of different bioremediation methods

Bioremediation method	Micropollutant type	% Recovery range	Example organism/ plant/enzyme	References
Phytoremediation	Heavy metals, pesticides	50–90	*Brassica juncea*, *Populus* spp.	Salt et al. (1998) and Ali et al. (2013)
Bioaugmentation	Organic pollutants, hydrocarbons	60–95	*Pseudomonas putida*, *Bacillus subtilis*	Tyagi et al. (2011)
Biostimulation	Hydrocarbons, organic waste	50–85	Indigenous soil microbes	Okoh and Hemandez (2006)
Constructed wetlands	Nutrients, metals, organics	50–90	*Typha latifolia*, *Phragmites australis*	Vymazal (2011)
Microbial electrochemical systems (MES)	Heavy metals, pharmaceuticals	70–95	*Geobacter sulfurreducens*	Logan et al. (2006)
Enzymatic bioremediation	Synthetic dyes, pharmaceuticals	60–90	Laccase, peroxidase	Husain (2010)
Anaerobic digestion	Organic micropollutants	40–75	Methanogenic archaea	Alvarado et al. (2014)
Mycoremediation	Pharmaceuticals, hydrocarbons	50–95	*Trametes versicolor*, *Pleurotus ostreatus*	Singh (2006)
Biosorption	Heavy metals, dyes	70–98	Algal biomass (*Spirulina*, *Chlorella*)	Vijayaraghavan and Yun (2008)
Bioreactors	Multiple pollutant types	60–99	Mixed microbial consortia	Kumar et al. (2016)

such as cadmium and zinc, especially in agricultural areas affected by industrial pollution. Furthermore, the use of *Typha angustifolia* (bulrush) in constructed wetlands for treating wastewater from industrial sites has shown promising results in reducing pollutants like heavy metals and organic compounds in Tamil Nadu (Muthukumar et al., 2016). These examples underscore the effectiveness of phytoremediation in low to moderately contaminated environments.

Phytoremediation is aligned with circular economy principles as it allows for the restoration of polluted lands, making them suitable for agricultural use or other activities. Additionally, it enhances ecosystem services such as soil fertility and biodiversity. However, its limitations must be considered. Phytoremediation is generally slow, with full remediation often requiring years to decades, depending on the type of pollutant and environmental conditions (Salt et al., 1998). The technique is also most effective in areas with low to moderate contamination and may not be suitable for heavily polluted sites. In areas with highly concentrated metals or toxic chemicals, the process may be less effective, and other treatment methods may be necessary. Despite these challenges, phytoremediation remains a vital tool in the restoration of contaminated environments, especially in developing countries where cost-effective methods are needed.

Eco-engineering Solutions

Eco-engineering solutions represent a blend of natural systems and engineered technologies designed to offer sustainable, long-term remediation of pollutants. These solutions are gaining attention for their versatility and ability to treat various environmental pollutants. A prominent example is the use of hybrid wetlands, which combine mechanical aeration with natural vegetation to treat wastewater. These systems can effectively remove pollutants such as pharmaceutical residues and nutrients from wastewater. For instance, in Denmark, hybrid wetland systems have been shown to reduce pharmaceutical residues in wastewater by up to 60% (Brix et al., 2003). These systems are adaptable to various environments, from industrial wastewater treatment to agricultural runoff and stormwater management.

Hybrid wetlands are part of a broader class of eco-engineering solutions that incorporate green infrastructure, such as constructed wetlands, vegetated swales, and green roofs, to manage water and treat pollutants. These systems provide a low-energy, low-cost alternative to traditional wastewater treatment technologies and have the added benefit of enhancing biodiversity and providing ecosystem services. In China, constructed wetlands have been used to treat agricultural runoff, significantly reducing the nutrient load and improving water quality in nearby rivers (Li et al., 2010).

However, eco-engineering solutions are not without their challenges. Regular maintenance and technical expertise are required to ensure their long-term functionality, and their effectiveness can be influenced by climatic factors such as temperature and rainfall (Hass et al., 2019). For example, temperature extremes can affect plant growth and the microbial processes essential for pollutant degradation. Moreover, the scalability of these systems in densely populated urban areas may require careful planning and integration with existing infrastructure. Despite these challenges, eco-engineering provides a sustainable, energy-efficient method for wastewater treatment and ecosystem restoration, contributing to the circular economy by reducing waste, enhancing water reuse, and recovering valuable nutrients.

Resource Recovery Technologies

Resource recovery technologies are essential components of the circular economy framework, which emphasizes the importance of minimizing waste, maximizing resource efficiency, and ensuring sustainable practices. These technologies aim to recover valuable resources, such as energy, nutrients, and raw materials, from waste streams, reducing the need for new resources and minimizing environmental pollution.

Energy Recovery

Energy recovery technologies play a pivotal role in the circular economy by converting organic pollutants and waste into renewable energy sources. Anaerobic digestion, a biological process that breaks down organic matter in the absence of oxygen, is one of the most widely used technologies for energy recovery. This process produces biogas, a mixture of methane and carbon dioxide, which can be used for electricity generation, heating, or even as a vehicle fuel. The use of biogas contributes to reducing greenhouse gas emissions and helps replace fossil fuels with renewable energy, thus promoting energy security and sustainability.

For example, in Germany, anaerobic digesters are widely employed in agricultural settings to process organic waste, including crop residues and manure, converting up to 80% of the waste into biogas (Weiland, 2010). This biogas is then used for heating or electricity generation, helping reduce dependence on non-renewable energy sources and cutting down on carbon emissions. Similarly, in the United Kingdom, biogas plants process food waste, wastewater sludge, and agricultural residues to generate electricity, providing an alternative to traditional energy sources while reducing waste sent to landfills (Deublein & Steinhauser, 2011). These energy recovery systems have proven effective in reducing waste, offsetting energy costs, and promoting sustainable energy use.

However, energy recovery systems are primarily limited to organic waste and pollutants. This means that non-organic pollutants, such as heavy metals, plastics, and chemical compounds that do not decompose in anaerobic conditions, cannot be processed using this technology (Das et al., 2023, 2025). As a result, these pollutants require alternative treatment methods, such as physical, chemical, or biological remediation processes. Despite this limitation, energy recovery systems remain an essential tool in achieving a carbon-neutral and resource-efficient economy.

Nutrient Recovery

Nutrient recovery technologies are critical for closing the nutrient loop in agriculture, a key principle of the circular economy. These technologies recover essential nutrients, such as nitrogen and phosphorus, from waste streams and return them to the agricultural cycle as fertilizers. This process helps reduce the need for synthetic fertilizers, which are energy-intensive to produce and can contribute to environmental pollution through runoff and eutrophication.

One example of a successful nutrient recovery technology is struvite precipitation. In the Netherlands, struvite—a magnesium ammonium phosphate mineral—is recovered from wastewater as a slow-release fertilizer. Struvite precipitation systems, such as those implemented at the Epe Wastewater Treatment Plant, recover phosphorus from wastewater and convert it into struvite, which can then be used as a high-quality fertilizer in agriculture (Cornel & Schaum, 2009). This technology helps mitigate eutrophication by reducing phosphorus runoff into water bodies, thus preventing the harmful effects of nutrient overload on aquatic ecosystems.

Another example is the recovery of nitrogen from wastewater, which can be used as a fertilizer in agriculture. In Sweden, the Nutrient Recovery from Wastewater (NURC) project uses a process called nitrification-denitrification to recover nitrogen from sewage and return it to the soil as a nutrient-rich fertilizer. This process helps reduce nitrogen runoff into water bodies, preventing water pollution and promoting sustainable farming practices (Ahn et al., 2014). Nutrient recovery systems not only reduce the environmental impact of agricultural practices but also improve soil fertility and reduce the reliance on synthetic fertilizers.

Case Studies and Real-World Applications

Several impactful case studies demonstrate the successful application of energy and nutrient recovery technologies, significantly contributing to the advancement of a circular economy. One such example is the **Zaanstad Biogas Plant** in the Netherlands, which processes organic waste, including food scraps and agricultural residues, to generate biogas. This biogas is then used for electricity and heat production, supporting energy sustainability. Furthermore, the digestate produced during the biogas generation process is repurposed as a fertilizer, effectively closing the nutrient loop and reducing the dependency on synthetic fertilizers, thus promoting sustainable farming practices (Weiland, 2010).

In **Malmo, Sweden**, the **Malmo Wastewater Treatment Plant** exemplifies a forward-thinking approach by recycling phosphorus from wastewater and converting it into high-quality fertilizers for agricultural use. This process not only reduces harmful phosphorus runoff but also plays a crucial role in promoting sustainable farming and nutrient recycling (Cornel & Schaum, 2009). Similarly, in **Madrid, Spain**, the **Eco-Sewer Project** addresses the challenge of urban wastewater management by recovering nitrogen and phosphorus from wastewater and transforming them into slow-release fertilizers. This project not only supports sustainable agricultural practices but also helps minimize the environmental impact of urban wastewater (Ahn et al., 2014). Different studies conducted in India and its outcome on Phytoremediation is presented in Table 21.4.

In **Gujarat, India**, the innovative use of *Typha angustifolia* in a phytoremediation project has shown remarkable results in treating industrial effluents and wastewater. The project has successfully reduced contaminants in the water while simultaneously generating valuable biomass, which is repurposed as a biofertilizer, further promoting sustainable agricultural practices and environmental health (Muthukumar et al., 2016). These case studies collectively underscore the transformative potential of energy and nutrient recovery technologies in advancing sustainability, enhancing agricultural productivity, and fostering environmental stewardship.

Way Forward

To address the challenge of micropollutants, advancements in biotechnology, such as CRISPR/Cas9 for targeted degradation and synthetic biology for enhanced microbial strains, offer promising solutions. Scaling up bioremediation technologies must be cost-effective and adaptable to varying environmental conditions. Strengthening policy frameworks and fostering collaboration between academia, industry, and government will drive innovation and ensure the integration of circular economy principles into remediation practices. Continued research into sustainable technologies like phytoremediation, eco-engineering, and resource recovery is essential for

Table 21.4 Different studies conducted in India and its outcome on phytoremediation

Sr.no	Study title	Location	Problem	Solution	Outcome	References
1	Phytoremediation of heavy metals in agricultural soils	Punjab, India	Soil contamination with cadmium, lead, and arsenic from pesticides	Sunflower and mustard as hyperaccumulators for phytoremediation	Reduced heavy metal concentrations, improved soil health	Singh and Tripathi (2016)
2	Utilization of rice husk ash for arsenic remediation	West Bengal, India	Arsenic contamination of groundwater affecting agriculture	Rice husk ash (RHA) as an adsorbent for arsenic removal	Reduced arsenic levels, improved water quality, enhanced crop yields	Asif and Chen (2017)
4	Use of constructed wetlands for wastewater treatment	Tamil Nadu, India	Contamination of water bodies from agricultural runoff	Constructed wetlands to treat wastewater	Removal of pollutants, improved water quality, enabled reuse for irrigation	Gupta and Khare (2018)
5	Microbial remediation of microplastics in agricultural soil	Uttar Pradesh, India	Microplastics from plastic mulches and containers in soils	Soil bacteria and fungi, e.g., *Bacillus megaterium*, for microplastic degradation	Identified bacterial strains for effective degradation	Thapliyal et al. (2024)
6	Recovery of nutrients from wastewater for agricultural use	Andhra Pradesh, India	Nutrient-rich wastewater from agriculture and urban sources	Algae-based systems to recover nutrients, biomass as fertilizer	Improved nutrient efficiency, reduced water pollution	Mahapatra et al. (2018)

long-term sustainability. Global cooperation and awareness-raising will accelerate the adoption of these solutions, advancing a circular economy and mitigating the impact of micropollutants.

Conclusion

The integration of circular economy principles into remediation techniques is essential for addressing micropollutant pollution. Technologies like phytoremediation, eco-engineering, and resource recovery offer effective solutions for environmental cleanup, resource recovery, and ecosystem restoration. These methods reduce chemical reliance, enhance soil and water quality, and recycle essential nutrients, supporting sustainable agriculture. Despite challenges like scalability, these approaches have shown global success and will be key to advancing the circular economy. Coordinated efforts across governments, industries, and communities are vital to mitigating micropollutant impacts and promoting long-term environmental sustainability.

References

Ahn, Y., Lee, D. K., & Han, S. (2014). Nitrogen and phosphorus recovery from wastewater and its application to slow-release fertilizers. *Environmental Science and Technology, 48*(13), 7785–7792.

Ali, H., Khan, E., & Sajad, M. A. (2013). Phytoremediation of heavy metals concepts and applications. *Chemosphere, 91*(7), 869–881.

Alvarado, A., Montañez-Hernández, L. E., Palacio-Molina, S. L., Oropeza-Navarro, R., Luévanos-Escareño, M. P., & Balagurusamy, N. (2014). Microbial trophic interactions and mcr A gene expression in monitoring of anaerobic digesters. *Frontiers in Microbiology, 5*, 597.

Andreozzi, R., Caprio, V., Insola, A., & Marotta, R. (2003). Advanced oxidation processes (AOPs) for water purification and recovery. *Catalysis Today, 53*(1), 51–59.

Asif, Z., & Chen, Z. (2017). Removal of arsenic from drinking water using rice husk. *Applied Water Science, 7*, 1449–1458.

Baker, R. W. (2004). *Membrane technology and applications* (2nd ed.). Wiley.

Baldrian, P. (2006). Fungal laccases—Occurrence and properties. *Folia Microbiologica, 51*(1), 1–7.

Behera, D. I., Nayak, M., Biswas, S., Meikap, B. C., & Sen, R. (2021). Enhanced biodegradation of total petroleum hydrocarbons by implementing a novel two-step bioaugmentation strategy using indigenous bacterial consortium. *Journal of Environmental Management, 292*: 112746. https://doi.org/10.1016/J.JENVMAN.2021.112746

Behera, D. I., Nayak, M., Mishra, A., Meikap, B. C., & Sen, R. (2022). Strategic implementation of integrated bioaugmentation and biostimulation for efficient mitigation of petroleum hydrocarbon pollutants from terrestrial and aquatic environment. *Marine Pollution Bulletin, 177*, 113492. https://doi.org/10.1016/J.MARPOLBUL.2022.113492

Brix, H., Schierup, H. H., & Møller, J. (2003). Hybrid constructed wetlands for wastewater treatment: The role of aeration in improving performance. *Water Science and Technology, 48*(4), 33–40.

Castiglioni, S., Bagnati, R., Fanelli, R., Pomati, F., Calamari, D., & Zuccato, E. (2006). Removal of pharmaceuticals in sewage treatment plants in Italy. *Environmental Science & Technology, 40*(1), 357–363.

Cornel, P., & Schaum, C. (2009). Phosphorus recovery from wastewater by struvite precipitation: A review. *Environmental Science and Technology, 43*(5), 1333–1343.

Das, P., & Bal, M. (2025). Modeling and optimization study for microplastic removal from aquatic medium using *Chroococcidiopsis* Species. *Chemical Engineering Science, 304*, 121085. https://doi.org/10.1016/J.CES.2024.121085

Das, A. P., & Mishra, S. (2008). Hexavalent chromium (VI): Health hazards and environmental pollutant. *Journal of Environmental Research and Development, 2*(3), 386–392.

Das, P., Halder, G., & Bal, M. (2023). A critical review on remediation of microplastics using microalgae from aqueous system. *Science of the Total Environment, 898*, 166425. https://doi.org/10.1016/J.SCITOTENV.2023.166425

Deublein, D., & Steinhauser, A. (2011). *Biogas from waste and renewable resources: An introduction.* Wiley-VCH

Dey, S., Das, A., Mallick, K., Sahu, A., & Das, A. P. (2023). *Environmental petroleum waste: pollution, toxicity, sustainable remediation* (pp. 159–75). https://doi.org/10.1007/978-3-031-48220-5_7.

Environmental Protection Agency (EPA). (2020). *Urban chemical pollution: Sources, impacts, and mitigation strategies.* EPA. Retrieved from https://www.epa.gov

European Medicines Agency. (2021). Pharmaceuticals in the environment: Addressing the impact of pharmaceutical residues on the environment.

Fatta-Kassinos, D., Kalavrouziotis, I. K., Vasquez, M. I., & Ouzounis, K. (2008). The risks associated with wastewater reuse and xenobiotics in the agroecological environment. *Science of the Total Environment, 385*(1–3), 1–9.

Fent, K., Weston, A. A., & Caminada, D. (2006). Ecotoxicology of human pharmaceuticals. *Aquatic Toxicology, 76*(2), 122–159.

Food and Agriculture Organization (FAO). (2020). *Pesticides and water quality: Challenges and solutions.* FAO.

Fujishima, A., Rao, T. N., & Tryk, D. A. (2000). Titanium dioxide photocatalysis. *Journal of Photochemistry and Photobiology C: Photochemistry Reviews, 1*(1), 1–21.

Glaze, W. H., Kang, J. W., & Chapin, D. H. (1987). The chemistry of water treatment processes involving ozone, hydrogen peroxide and ultraviolet radiation. *Ozone: Science & Engineering, 9*(4), 335–352.

Gupta, S., & Khare, D. (2018). Constructed wetlands: An emerging paradise for sustainable treatment of wastewater. *Journal of Water Resource Engineering and Management, 5*, 1–5.

Hass, G. A., Roffe, P. E., & McMahon, K. A. (2019). Eco-engineering solutions for wastewater treatment: Hybrid constructed wetlands and their role in the circular economy. *Environmental Science and Technology, 53*(12), 7012–7020.

Husain, Q. (2010). Peroxidase mediated decolorization and remediation of wastewater containing industrial dyes: A review. *Reviews in Environmental Science and Bio/technology, 9*, 117–140.

Jelic, A., McArdell, C. S., & Schneider, M. (2011). Occurrence and fate of pharmaceuticals and personal care products in the environment. *Science of the Total Environment, 409*(8), 1498–1505.

Jobling, S., Nolan, M., Tyler, C. R., Brighty, G., & Sumpter, J. P. (1998). Widespread sexual disruption in wild fish. *Environmental Science & Technology, 32*(17), 2498–2506.

Kanungo, J., Sahoo, T., Swain, L. P., & Behera, I. D. (2023). *Toxicity of persistent hydrocarbon pollutants, sources and sustainable remediation process* (pp. 39–65). https://doi.org/10.1007/978-3-031-48220-5_2.

Kidd, K. A., Blanchfield, P. J., Mills, K. H., Palace, V. P., Evans, R. E., Lazorchak, J. M., & Flick, R. W. (2007). Collapse of a fish population after exposure to a synthetic estrogen. *Proceedings of the National Academy of Sciences, 104*(21), 8897–8901.

Korhonen, J., Honkasalo, A., & Seppala, J. (2018). Circular economy: The concept and its limitations. *Ecological Economics, 143*, 37–46.

Kumar, M., & Singh, A. (2021). Groundwater contamination by pesticides in India: A review on its impacts and mitigation strategies. *Environmental Monitoring and Assessment, 193*(5), 1–15.

Kumar, P., Dey, B., & Sekhar, G. R. (2016). Nutrient transport through deformable cylindrical scaffold inside a bioreactor: An application to tissue engineering. *International Journal of Engineering Science, 127*, 201–216.

Lehmann, J., Joseph, S., & Masiello, C. A. (2006). Biochar for environmental management: An introduction. In J. Lehmann & S. Joseph (Eds.), *Biochar for environmental management: Science and technology* (pp. 1–12). Earthscan.

Li, Y., Li, X., & Liu, Y. (2010). Constructed wetlands for agricultural runoff treatment: Efficiency and challenges. *Water Research, 44*(17), 5107–5116.

Logan, B. E., Cheng, S., Watson, V. A., & Esteban, G. (2006). Microbial fuel cells: Methodology and technology. *Environmental Science & Technology, 40*(17), 5191–5198.

Mahapatra, D. M., Chanakya, H. N., Joshi, N. V., Ramachandra, T. V., & Murthy, G. S. (2018). Algae-based biofertilizers: a biorefinery approach. In *Microorganisms for green revolution: Volume 2: Microbes for sustainable agro-ecosystem* (pp. 177–196).

Mallick, K., Sahu, A., Dubey, N. K., & Das, A. P. (2023). Harvesting marine plastic pollutants-derived renewable energy: A comprehensive review on applied energy and sustainable approach. *Journal of Environmental Management, 348*, 119371. https://doi.org/10.1016/J.JENVMAN.2023.119371

Martínez, J. L. (2009). Environmental pollution by antibiotics and by antibiotic resistance determinants. *Environmental Pollution, 157*(11), 2893–2902.

Miao, Y., Zhang, Q., & Wang, Y. (2020). Biocatalysts for micropollutant degradation: Current progress and challenges. *Environmental Science and Pollution Research, 27*(34), 42931–42949.

Mishra, S., Rout, P. K. & Das, A. P. (2021). Emerging microfiber pollution and its remediation. In R. Prasad (Ed.), *Environmental and microbial biotechnology* (pp. 247–266). Springer, Singapore. https://doi.org/10.1007/978-981-15-5499-5_9

Mishra, S., & Das, A. P. (2023). Microbial remediation of synthetic microfiber contaminated wastewater. In M. P. Shah (Ed.), *Microbial technologies in industrial wastewater treatment* (pp. 337–350). Springer, Singapore. https://doi.org/10.1007/978-981-99-2435-6_17.

Mishra, S., Dash, D., & Das, A. P. (2024). Isolation and molecular characterization of polyester-degrading bacterial strains from municipal wastewater and their application in microplastic degradation in contaminated wastewater. *Geomicrobiology Journal*. https://doi.org/10.1080/01490451.2024.2376101

Mohanta, P. S., Behera, I. D. (2024). Impact and remediation of petroleum hydrocarbon pollutants on agricultural land: A review. *Geomicrobiology Journal, 41*(4), 345–359. https://doi.org/10.1080/01490451.2023.2243925

Mori, K., Hayashi, M., & Matsumoto, N. (2013). The use of iron-based catalysts for micropollutant degradation in water. *Chemical Engineering Journal, 225*, 118–126.

Muthukumar, A., Arumugam, A., & Swaminathan, K. (2016). Phytoremediation of industrial effluents using *Typha angustifolia* for sustainable water management. *Environmental Pollution, 214*, 264–272.

Okoh, A. I., & Trejo-Hernandez, M. R. (2006). Remediation of petroleum hydrocarbon polluted systems: exploiting the bioremediation strategies. *African Journal of Biotechnology, 5*(25).

Priyadarsini, S., & Das, A. P. (2024). Analytical and structural characterization of waste lithium-ion batteries for their effective recycling strategy. *Environmental Chemistry and Ecotoxicology*. https://doi.org/10.1016/j.enceco.2024.12.004

Priyadarsini, S., & Das, A. P. (2025). Lithium bioleaching: A review on microbial-assisted sustainable technology for lithium bio-circularity. *Journal of Water Process Engineering, 69*, 106744. https://doi.org/10.1016/j.jwpe.2024.106744

Salt, D. E., Smith, R. D., & Raskin, I. (1998). Phytoremediation. *Annual Review of Plant Biology, 49*, 643–668.

Schoenfuss, H. L., Brown, A. G., & Barber, L. B. (2016). Bioaccumulation and biological effects of contaminants of emerging concern in aquatic ecosystems. *Environmental Toxicology and Chemistry, 35*(4), 795–806.

Sharma, P., Singh, N., & Bisht, M. (2019). Phytoremediation of micropollutants: Potential and challenges. *Environmental Pollution, 249*, 181–194.

Singh, P. R., Mishra, S., & Das, A. P. (2020). Synthetic microfibers: Pollution toxicity and remediation. *Chemosphere, 257*, 127199. https://doi.org/10.1016/j.chemosphere.2020.127199

Singh, H. (2006). *Mycoremediation: Fungal Bioremediation.* Wiley.

Singh, P. K., Tripathi, P., Dwivedi, S., Awasthi, S., Shri, M., Chakrabarty, D., & Tripathi, R. D. (2016). Fly-ash augmented soil enhances heavy metal accumulation and phytotoxicity in rice (*Oryza sativa* L.); A concern for fly-ash amendments in agriculture sector. *Plant Growth Regulation, 78*, 21–30.

Thapliyal, C., Priya, A., Singh, S. B., Bahuguna, V., & Daverey, A. (2024). Potential strategies for bioremediation of microplastic contaminated soil. *Environmental Chemistry and Ecotoxicology, 6*, 117–131.

Tyagi, M., Da, F. M. M. R., & de Carvalho, C. C. C. R. (2011). Bioaugmentation and biostimulation strategies to improve the effectiveness of bioremediation processes. *Biodegradation, 22*, 231–241.

UNEP. (2023). *Global assessment of micropollutants: Challenges and regulatory frameworks.* United Nations Environment Programme. Retrieved from https://www.unep.org

United Nations. (2015). *Transforming our world: The 2030 agenda for sustainable development* (A/RES/70/1). United Nations General Assembly. https://sdgs.un.org/2030agenda

United Nations Environment Programme (UNEP). (2022). *Global mercury assessment: Sources, emissions, releases, and environmental transport.* UNEP. Retrieved from https://www.FAO.org

Vijayaraghavan, K., & Yun, Y. S. (2008). Bacterial biosorbents and biosorption. *Biotechnology Advances, 26*(3), 266–291.

Von Gunten, U. (2003). Ozonation of drinking water: Part I. Oxidation kinetics and product formation. *Water Research, 37*(7): 1443–1467.

Vymazal, J. (2011). Constructed wetlands for wastewater treatment. *Water, Air & Soil Pollution, 223*(7), 3991–4007.

Weiland, P. (2010). Biogas production: Current state and perspectives. *Applied Microbiology and Biotechnology, 85*(4), 849–860.

Westerhoff, P. (2005). Removal of trace organics by activated carbon in drinking water treatment. *Journal of Environmental Engineering, 131*(4), 669–673.

World Health Organization (WHO). (2010). *Arsenic in drinking water: A public health perspective.* World Health Organization. https://www.who.int/water_sanitation_health/dwq/arsenic/en/

World Health Organization (WHO). (2020). Antimicrobial resistance. Retrieved from https://www.who.int

World Health Organization (WHO). (2021). *Global assessment of micropollutants in aquatic ecosystems: Impacts and solutions.* WHO Publications.

Xie, J. (2020). Sustainable recovery of phosphorus and nitrogen from wastewater: Circular economy and resource management approaches. *Science of the Total Environment, 704*, 135328.

Zhao, L. (2020). Biochar as an effective adsorbent for removing micropollutants from water: A review. *Science of the Total Environment, 724*, 138350.

Chapter 22
Integration of Technologies: Monitoring of Toxic Cadmium Ion in Potable Water

Parul Taneja and Raj Kumar Gupta

Introduction

Due to anthropogenic activities, the emission of hazardous non-biodegradable substances into the environment is increasing (Das et al., 2011, 2015b; Ghosh et al., 2016). These substances have imbalanced the ecosystem and endangered the environment and human health (Das & Ghosh, 2018, 2023; Das et al., 2015a; Ghosh et al., 2023; Wagner et al., 1999). Among these, heavy metal ions fall in the category of toxic pollutants, which are harmful to human health, such as lead (Pb), mercury (Hg), arsenic (As), cobalt (Co), cadmium (Cd) metal ions (Yan et al., 2022; Prabhakar et al., 2019; Raj & Das, 2023). These non-biodegradable heavy metal ions accumulated in living organisms and disturb their metabolic process (Bjorklund et al., 2020; Priyadarsini & Das, 2025). Additionally, accumulated metal ions increase oxidative stress due to which various internal tissues get damage, which raises various disordered diseases in human begins and disturb their blood pressure level also (Malik et al., 2019; Priyadarsini & Das, 2024). The proposed work is focused on the detection of divalent cadmium ion because cadmium and its compounds are easily contaminating the food and aqueous medium and enrichment the cancer in human begins, cadmium is a major carcinogenic for humans (Hui et al., 2021). Human activities such as mine drainage, improper disposal of cadmium-based batteries, chemical industries, and coal waste enrich cadmium in the environment (Khan et al., 2022; Mishra et al., 2023) and cadmium also occurs naturally in volcanic soils. As reported in the literature total 30,000 tons of cadmium is released into the environment out of which 13,000 tons is caused of human activities, annually (Solomonea et al., 2022).

P. Taneja (✉)
Department of Water Resources and Development Management, IIT Roorkee, Roorkee, India
e-mail: tanejaparul92@gmail.com

P. Taneja · R. K. Gupta
Department of Physics, Birla Institute of Technology and Science, Pilani Campus, Pilani, India

I. D. Behera et al. (eds.), *Emerging Micropollutants*, Sustainable Environmental Waste Management Strategies, https://doi.org/10.1007/978-3-032-03530-1_22

Cadmium is easily accumulated into liver, kidneys, bones and it prolong accumulation internally damage the tissue and cells of human bodies. This motivates us to use new strategies to selectively and sensitively monitor the cadmium metal ion with minimal concentrations. The traditional analytical techniques based on spectroscopy studies where detection of ions was performed based on the absorption of particular wavelength is expensive, tedious and need a preparation of reference sample also e.g., atomic absorption spectroscopy, UV–visible and fluorescence spectroscopy, mass spectrometer (Cai et al., 2020; Malik et al., 2006). The traditional techniques analysed the sample through a wavelength of the reflected light source which is delegated and easily disturb due to presence of contamination. The main drawback of such techniques is that they are not able to detect the contamination in the range of parts per billion (ppb). It is important aspect for the detection of metal ions in aqueous medium because their permissible limit in the aqueous medium is in the range of ppb while spectroscopy-based techniques able to detect the concentration of ions in the higher range of parts per million (ppm) (Oto et al., 2020). The selective and sensitive detection of metallic analytes in unknown sample through analytical techniques (such as atomic absorption spectroscopy, mass spectroscopy etc.) is also difficult. For that purpose, initially assume some random concentration of analytes in the unknown samples and based on that further detection of analytes are possible which is not reliable.

In our work, the proposed strategic is the integration of two different technologies working on two different principles piezoelectric and electrochemical which provides the feasibility of detecting analytes in the range of ppb. The amine functionalized carbon nanotubes are deposited in the form of thin film onto surface of quartz transducer. Transducer is the physical part of the sensing device which converts one form of energy into other form of energy. The good coupling between functionalized layer and transducer is crucial for the reliable sensing. The amino based ligands have been popular to form the complex bination with various metal ions e.g., Zn^{+2}, Pb^{+2}, Cd^{+2} etc. (Meng et al., 2009; Soria et al., 2018). Therefore, amine functionalized nanotubes are chosen for the sensing of cadmium ion.

The quartz crystal is resonant at particular frequency with respect to applied voltage via its both side gold electrode. The frequency of resonance depends on the thickness of quartz crystal (Taneja et al., 2017a, 2017b, 2020) which captured through QCM 200 digital controller of quartz crystal microbalance (QCM) device. The sensing platform was developed by functionalized the gold electrode of 5 MHz At-cut quartz crystal with thin film of octadecylamine functionalized carbon nanotubes (ODA-CNTs) using spin coating technique. QCM is real time monitoring and high-resolution device for the detection of contaminates. The change in resonance frequency of quartz crystal is directly proportional to deposited mass on the surface of quartz crystal. The resonance frequency is an ultra-sensitivity towards the change in a mass of nano gram per centimetre square area onto the surface of quartz crystal. Further the technique makes the selective by integrating the QCM device with the CHI 1200B potentiostat three electrode cell system known as EQCM for the detection of toxic cadmium ions in drinking water. The electrochemical response of sensor is recorded through cyclic voltammetry technique. Cyclic voltammetry is the most widely used technique for acquiring qualitative and quantitative information about electrochemical reaction (Brownson & Banks, 2014). In the electrochemical process, analytes are sensed via heterogeneous transfer of electrons between analyte solution and solid-state working electrode (Tsai et al., 2014). The sensing analyte is either oxidized or reduced depending on the potential applied to the working electrode, this mechanism is known as redox process.

The real time monitoring of chemical analytes results of the interaction between analytes and sensing surface. The sensing of chemical analytes subdivided into two parts are recognition of analyte through chemical interaction and second is amplification of that interaction through the transducer (Janata, 2009). There are two types of sensing mechanism are possible between analytes and sensing surface: bulk interaction and surface interaction. The developed sensing setup work on the surface interaction phenomena where the sensing layer is recovered after the interaction with analyte. This chapter have a detailed discussion on the working principle of development of technique in laboratory to meet the growing demands of selective and sensitive monitoring of water pollutes. The work is summarizing the progress for the analysis of two different outcomes to identify the metallic ions. Additionally, the future perspective of work is simultaneous monitoring of multianalytes from the proposed sensing setup.

Methodology of Experimental Work

The experimental methodology was designed for sensitive and selective detection of metal analytes in aqueous medium through integration of two different techniques whose working principles are completely different from each other. The chemical reagents used throughout the experimental work were of analytical grade to ensure accuracy and reproducibility of the results. The Octadecylamine functionalized single wall carbon nanotubes (ODACNTs) was purchased from Carbon Solution Inc. (cat. # P5-SWNTs). Cadmium chloride ($CdCl_2$) was purchased from Himedia Laboratories. The ultra-pure deionized Millipore water was utilized throughout experimental work which obtained from the Millipore DQ-5 system has a resistivity of 18 MΩ-cm. Stock solution of cadmium metal ion was prepared in the same aqueous medium. The piezo and electrochemical responses of analyte detection were recorded via a QCM200 digital controller (Stanford Research Systems, USA) and a three-electrode based potentiostat (CHI Instruments, Inc., Model # 1200B).

Detailed Study on Experimental Work

Experimental Setup

Quartz is a crystalline mineral composed of silicon and oxygen atoms, where each oxygen is shared between two tetrahedra. The surface of a quartz crystal is surrounded by electrical charges as a result of applied mechanical stress and it's vice versa. The density of produced electrical charges is directly proportional to the magnitude of applied stress. This shows the existence of piezoelectric property in quartz crystal. Piezoelectricity is the property whereby an applied electric field results in mechanical deformation (strain) in a material and it's vice versa also. This property of crystal is used to detect contaminants in drinking water by using the Quartz Crystal Microbalance (QCM) device. The quartz crystal operates in a thickness shear mode (TSM), where an oscillating electric field is applied across the surface of a quartz crystal via its gold electrode (as shown in Fig. 22.1a) as a result strain is produces in an oscillating mode due to piezoelectric effect. The oscillating strain in quartz crystal produces the elastic wave (Fileno et al., 2014). The generated elastic waves propagate in a direction perpendicular to the crystal surface, resulting in the bottom and top surfaces of quartz wafer propagating in an antiparallel direction as shown in Fig. 22.1b. The magnitude of oscillation is highly sensitivity towards the mass deposited onto the surface of quartz crystal. When an analyte interacts or make the complex bond with sensing layer onto surface of quartz crystal its resonance frequency is decay which indicates the detection of analytes.

The electrochemical analyser potentiostat CHI 1200B is integrated with QCM device (as shown in Fig. 22.2) for the selective detection of contaminates. The electrochemical workstation having a three-electrode working, reference and counter electrode. Here, quartz wafer served for the dual purpose one is ultra-sensitive mass sensor and other is working electrode of electrochemical analyser. The platinum wire and silver/silver chloride were utilized as counter and reference electrodes, respectively in the conventional three-electrode EQCM cell system. Where the potential difference is controlled between the reference and working electrode and current is measured via counter electrode. The electrochemical workstation is available with variety of techniques to analysis the redox behaviour of analytes such as cyclic voltammetry (CV), linear sweep voltammetry (LSV) and chronoamperometry (CA)

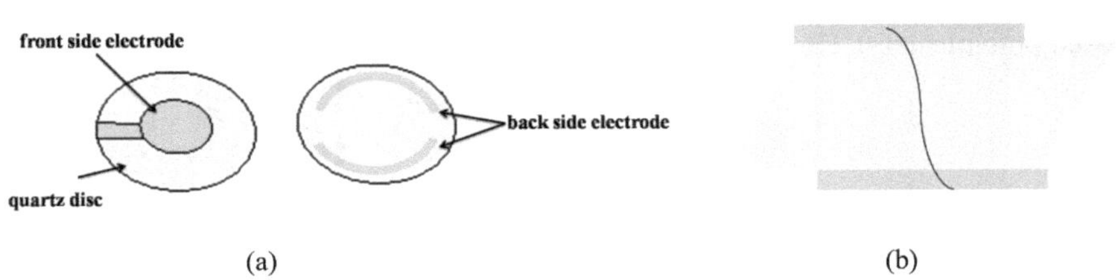

Fig. 22.1 **a** Schematic of quartz crystal, **b** oscillation of quartz crystal in thickness share mode

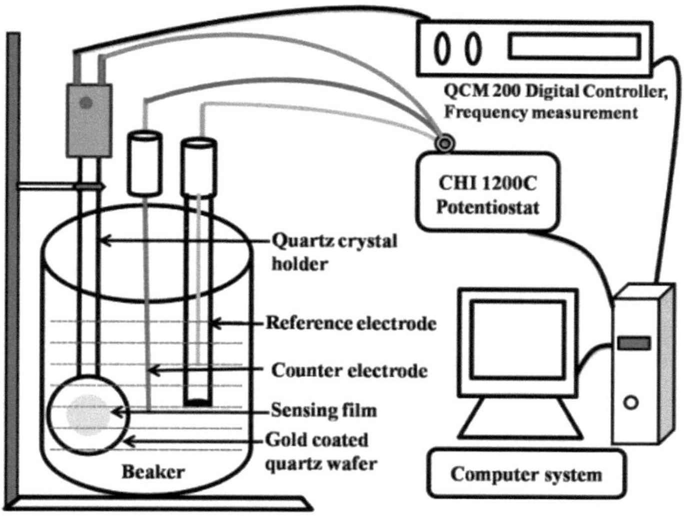

Fig. 22.2 Schematic representation of experimental setup

modes. Cyclic voltammetry is widely used technique for acquiring the qualitative and quantitative information about redox reaction of the species at the surface of working electrode. In our studies, cyclic voltammetry technique is used to detect the divalent cadmium ion along with its concentration.

Functionalized the Working Electrode

The quartz wafer was treated with highly reactively cold piranha solution for less than 5 min, is a mixture of 3:1 ratio of concentrated sulfuric acid (H_2SO_4) and hydrogen peroxide (H_2O_2) (Taneja et al., 2017a, 2017b). Traditionally piranha solution is used in the semiconductor industry during microfabrication process to clean and oxidized the surface of silicon wafer. In our study, piranha solution is used to remove organic residues and other contaminates from the surface of quartz wafer. The presence of hydrogen per oxide breaks down the organic compounds into the carbon dioxide, water and other volatile by-compounds (Taneja et al., 2020). These compounds are rinsed off with plenty of ion free water and dried by blowing a jet of dry nitrogen gas. The surface of quartz wafer becomes enriched with hydroxyl groups (–OH) (Kwon et al., 2023) while treating with piranha solution as a result surface becomes hydrophilic which improves the adhesion of thin films and biomolecules towards quartz surface. The surface of quartz wafer was functionalized with octadecylamine functionalized carbon nanotubes (ODA-CNTs) using the spin coating technique. The technique is an economically viable, fast and easier method to fabricate thin films of molecules on the surface of substrate by dissolved molecules in a desired liquid precursor. The substrate is mounted over a vacuum chuck and a known quantity of a molecular solution is poured drop by drop using a micro syringe at the centre of rotating substrate under a controlled speed (Tyona,). The liquid drop is centripetally accelerated which causes it to spread across the surface of substrate via centrifugal force (Tyona, 2013). The solvent is usually volatile and evaporates rapidly during the spinning process, providing a thin film of the desired molecules. The stock solution of ODA-CNTs was prepared in the high-grade chloroform solvent having a concentration of 0.1 mg/ml afterwards the solution was ultrasonicated for half an hour to obtained the uniform dispersed solution. The thin films of ODA-CNTs were obtained on the cleaned surface of a quartz crystal by spreading 50 μl volume of the stock solution on a quartz wafer rotating at a speed of 1500–2000 rotations per minute (rpm) for about 15 min. The functionalized quartz wafer was stored in a vacuum desiccator for 1–2 h to keep it dry.

Detection Strategies

The strategy to detect the presence of heavy metal divalent cadmium ion in aqueous medium is through the integration of two different techniques, i.e. piezoelectric based QCM device and electrochemical based CHI1200B potentiostat (Taneja et al., 2018) as shown in Fig. 22.2. Where the electrochemical response was studied using the cyclic voltammetry (CV) technique. Under the cyclic voltammetry technique, triangular shape switching potential window is applied over the working electrode. The applied potential is linearly ramp from an initial value to the switching potential, after achieving the switching potential, applied voltage start to ramp in opposite direction. When the voltage is ramp from lower value towards higher value that scan is known as an anodic trace, during this trace oxidation peak is generated. When the voltage is ramp from higher value towards the lower value that scan is known as a cathodic trace, during this trace reduction peak is generated. In such a way applied potential is in form of triangular form, therefore applied potential is known as triangular shape switching potential window (Srisook, 2004). The potential range of switching window is chosen based on the redox behaviour of target species. Which helps us to analysis the oxidization and reduction behaviour of the analytes. The developed EQCM was employed for detection of Cd^{+2} ion sensing in an aqueous medium, the applied potential window was chosen from − 1 V to + 1 V. During the cathodic scan, Cd^{+2} ion in aqueous medium was reduced into metallic Cd^0 onto the surface of quartz crystal. Because of this phenomenon Cd ion able to make complex bond with sensing layer onto the functionalized surface of quartz wafer as a result the resonance frequency of quartz crystal was decreases. During the anodic scan of the applied potential the deposited metallic Cd^0 onto surface of quartz crystal is oxidized into Cd^{+2} ion in aqueous medium and as a result dissociated from the surface of quartz crystal and resonance frequency of crystal was increases (Figs. 22.3 and 22.4).

The voltage was ramp on the working electrode at scan rate of 0.1 V/s depends on the scanning direction of the applied voltage either oxidation or reduction process was dominate and corresponding to that either the piezo response was increases or

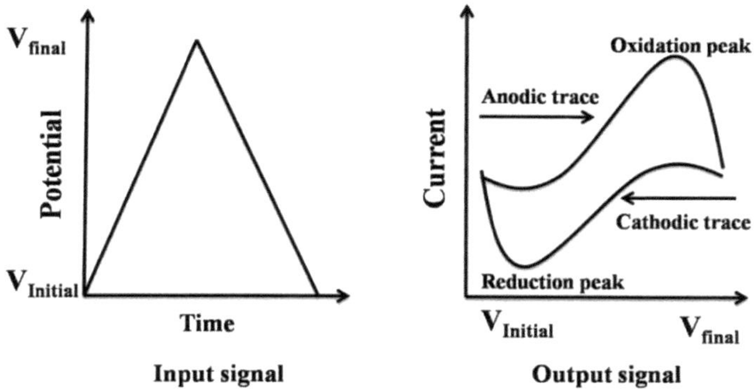

Fig. 22.3 Representation of input and output signal of cyclic voltammetry technique

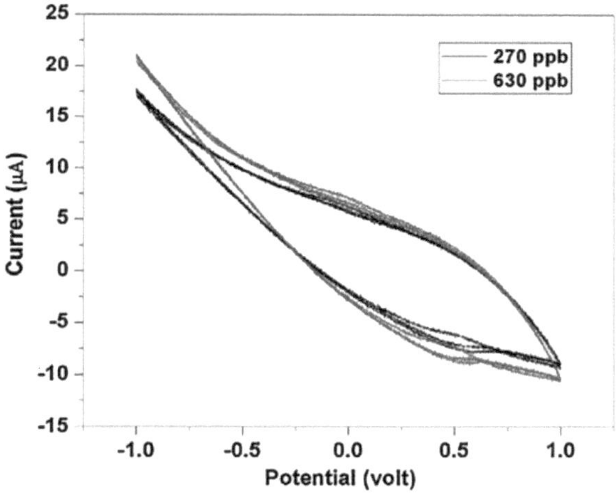

Fig. 22.4 I-V characteristics recorded during the monitoring of cadmium ion in potable aqueous medium

decreases, respectively. In our study while the detection of cadmium ion, the applied potential was sweep from + 1 V to − 1 V that time reduction process is dominate and piezo responses was decreases and when applied potential vary from − 1 V to + 1 V then piezo responses was increase. The association and dissociation time of cadmium ion onto surface of quartz wafer was estimated from the piezo response (as shown in Fig. 22.5a) using following the relations.

$$R_{ad}(t) = R_0\left(1 - e^{-\left(\frac{t}{\tau}\right)}\right) \tag{22.1}$$

$$R_{de}(t) = R_0\left(e^{-\left(\frac{t}{\tau}\right)}\right) \tag{22.2}$$

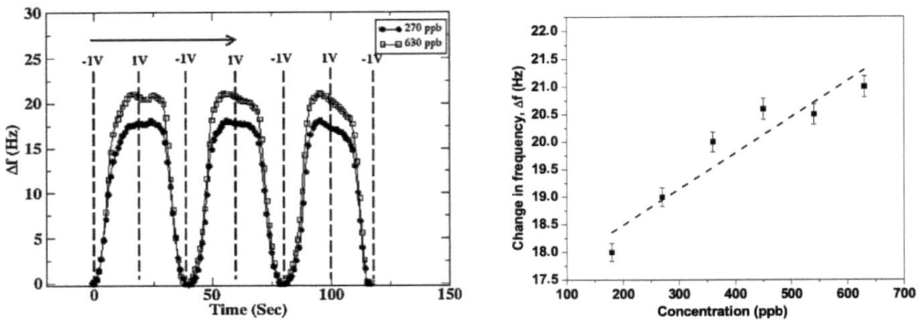

(a) (b)

Fig. 22.5 **a** The piezo response was recorded during the monitoring of cadmium ion, **b** change in resonance frequency of quartz crystal was recorded with respect to cadmium ion concertation

Here R_{ad}, R_{de} represents adsorption and desorption time of cadmium analyte, respectively. R_0 represents the initial response and τ is the characteristics time.

Result and Discussion on Outcome

The multiparameter analysis while the sensing of analytes increases the reliability on sensor performance. The piezo and electrochemical responses while the sensing of cadmium ion was recorded simultaneously using the EQCM sensing setup. As observed during the sensing of cadmium ion, initially voltage vary from lower potential (− 1 V) to higher potential (+1 V) during the anodic scan reduced cadmium is re-oxidized and dissociated from the surface of quartz therefore the resonance frequency of crystal is increases as shown in the recorded piezo responses (Fig. 22.5a). When the applied potential sweep in reverse direction from + 1 V to − 1 V then the cadmium ion in potable water is reduced into Cd^0 onto the quartz surface which corresponds to the adsorption of cadmium ion onto the surface of quartz crystal therefore the resonance frequency of quartz crystal is decreases as shown in Fig. 22.5a. A small peak at 0.6 V was observed in the recoded cyclic voltammetry response during the anodic trace which corresponds to re-oxidized of cadmium. Due to the presence of lower concentration of Cd ion the redox peaks are not visible clearly in the cyclic voltammetry response. Therefore, is no significant peak was observed during the reduction of cadmium ion. For the insight study of redox process, the derivates of the cyclic curve was taken separately for the anodic and cathodic trace as shown in Fig. 22.6. Figure 22.6a represents derivate current potential characteristics for the cathodic trace which represents the change in a slope around 0.2 V where the reduction is dominate. Figure 22.6b represents the derivate current potential characteristics for the anodic trace where change in slope for small peak was found to be appeared

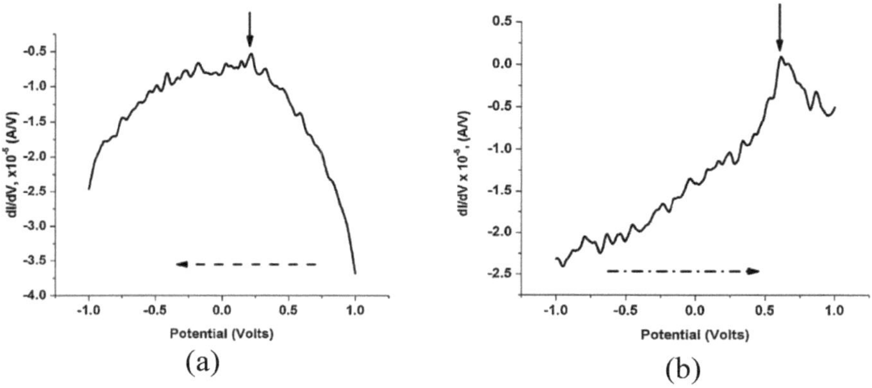

Fig. 22.6 Calculated the derivative (dI/dV) of cyclic voltammetry curve **a** for the cathodic trace, **b** for the anodic trace

around 0.6 V which represents the dominate oxidation process. The piezo and electrochemical responses were recorded for the given concentration of cadmium ion in the range of 180–670 ppb. At each concentration of cadmium analyte, three traces of electrochemical and piezo response were recorded. This was found to be repeatable which indicates the stabilized sensing process. The change in piezo response was plotted with respect to concentration of Cd ion as shown in Fig. 22.5b. As the concentration of cadmium analyte increases onto surface of crystal, resonance frequency of crystal was decreases and difference between the actual resonance frequency and frequency, when cadmium deposited onto crystal was increases. This plot is indicating the linear behaviour of piezo response. Where the change in slope of the curve indicates the sensitivity of the sensor. The characteristics time of adsorption and desorption of Cd ion onto sensing layer were calculated by implied the equation i and ii onto the piezo response and adsorption and desorption were found to be 10.2 $\pm$ 1.4 s and 3.7 $\pm$ 0.8 s, respectively.

Characterization of the Sensing Layer

The surface topographic of thin film of ODA-CNTS before and after the interaction with Cd ion was studied with ultra-high resolution at low accelerating voltage using Field Emission Scanning Electron Microscopy (FESEM). FESEM instrument is capable to study the topographic of conducting surfaces with ultra-high resolution at low accelerating voltage. Energy dispersive X-ray spectroscopy (EDS) is equipped with FE-SEM used to study the chemical composition of the sample. The sample make the conducting through coating of extreme thin layer of gold or some other conducting material before start the scanning to capture the fine topographic images.

The extreme thin layer of gold was deposited onto the ODA-CNTs layer using sputtering technique. Then the wafer is mounted on a movable holder which transfer the object into an ultrahigh vacuum chamber through exchange chamber for the scanning and placed on the moveable stage, which provide the facility of horizontal and vertical movement to the sample. In addition to this FESEM can be used to study the compositional analysis of elements presents in the sample with the help of equipped Energy dispersive x-ray spectroscopy (EDS). EDS captured the unique X-ray of a specific energy and wavelength, represents the presence of different elements in the specimen.

The surface characterization studies of the film before and after the interaction with cadmium analyte have been performed onto the silicon wafer. Figure 22.7a represents the uniformly covered strands of ODA-CNTs. The width and length of the CNTs was found to be 20–30 nm and 1.8–3.4 μm, respectively. The same wafer was allowed to interact with the 1 ppm concentration of cadmium ion in aqueous medium around 1 min. Afterwards, the substrate was cleaned through ion free water and put in desiccator for few hours to kept it for dry. Then the same sample was studied and where remarkable changes was found in the topography of the film. The width of the strands becomes larger, it was found in the range of 50–200 nm whereas comparatively, the length does not change much. This indicates that the cadmium ion interacts strongly along the surface of ODACNTs. EDS measurements of the same films represent the strong peak corresponding to the silicon wafer at 1.8 keV this is common peak appeared in both samples and peak appeared around 0.3 keV is corresponding to carbon atoms due to presence of CNTs. The peaks appeared around 2.6 and 3.2 keV represents the presence of cadmium and chlorine atoms (Fig. 22.8).

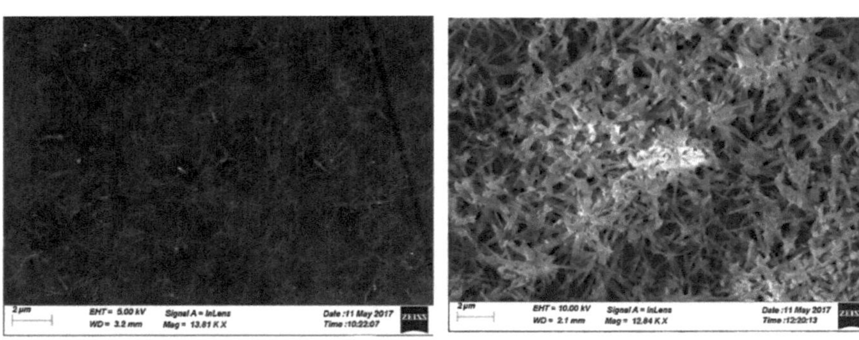

(a) (b)

Fig. 22.7 Field emission scanning electron microscope (FESEM) of the film of ODACNTs deposited onto silicon substrate **a** before, **b** after interaction with cadmium ion in the aqueous medium

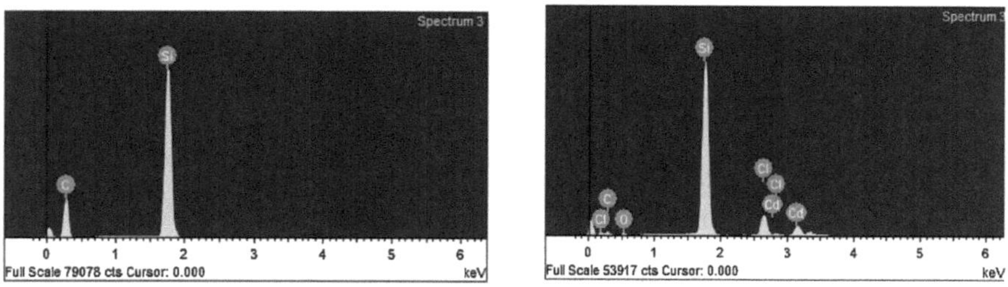

(a) (b)

Fig. 22.8 Energy-dispersive X-ray spectroscopy (EDAX) images of the film of ODACNTs deposited onto silicon substrate **a** before and **b** after interaction with cadmium ion

Conclusion

This chapter is dedicatedly focused on the insights understanding of integration of two techniques which works on two different complete mechanism. QCM device works on piezoelectric mechanism relates to physical science and CHI 1200B potentiostat works on the three electrode chemical cells based on the electrochemistry mechanism. This study was carried out to solve the problem for selective and sensitive monitoring of trace levels of metal ions (in the range of ppb) in potable aqueous medium. Reliability on the sensing mechanism is increases when the monitoring of analytes occurred through two different mechanisms simultaneously. The presence of toxicity in potable aqueous medium due to heavy metal ions has been reported to pose a threat to the health of living beings, which motivates us to carry out this research work. Further, this chapter focuses on the fabrication of thin films of amine functionalized carbon nanotubes using spin coating technique, which is used for the detection of divalent cadmium ions. The work demonstrates that the organic ligands whose contains amine group make complex bond with divalent cadmium ion. Furthermore, characterization studies confirmed the complex bond formation through extreme changes in the width structure of carbon nanotubes after interaction with cadmium ions. The presence of cadmium atoms was confirmed through the Energy dispersive x-ray spectroscopy. The time taken to make a complex bond between amine group and divalent cadmium ion was calculated through piezo response, and was found to be 10.2 ± 1.4 s. Our finding illustrates the importance of electrochemical analysis along with piezoelectric response for the selective sensing of target analyte. The cyclic curve helps to identify the presence of multi-analytes in aqueous medium through their unique redox potential peak. From the I-V curve it was observed that with increasing in the concertation, trend of oxidation and reduction currents are increases in positive and negative directions, respectively. The first order derivate of I-V curve was calculated for the fine analysis of predominant redox peaks. Where 0.6 V peak corresponding to oxidation process and change in direction of slope was found around 0.2 V corresponding to reduction process.

References

Bjorklund, G., Dadar, M., Chirumbolo, S., Aaseth, J., & Peana, M. (2020). Metals, autoimmunity, and neuroendocrinology: Is there a connection. *Environmental Research, 187*, 109541.

Brownson, D. A. C., & Banks, C. E. (2014). Interpreting electrochemistry. In *The handbook of graphene electrochemistry* (pp. 23–77). Springer.

Cai, S., Pan, H., González-Vila, Á., Guo, T., Gillan, D. C., Wattiez, R., & Caucheteur, C. (2020). Selective detection of cadmium ions using plasmonic optical fiber gratings functionalized with bacteria, *Optics Express, 28*, 19740–19749.

Das, A. P., & Ghosh, S. (2018). Bioleaching of Manganese from mining waste materials. *Materials Today: Proceedings, 5*(1), 2381–2390. https://doi.org/10.1016/J.MATPR.2017.11.459

Das, A. P., & Ghosh, S. (2023). Role of metabolites and metabolomics for understanding of manganese bioleaching. *Manganese Mining Microorganisms*, 117–127. https://doi.org/10.1016/B978-0-12-822147-1.00011-X

Das, A. P., Ghosh, S., Mohanty, S., & Sukla, L. B. (2015a). Advances in manganese pollution and its bioremediation. In L. B. Sukla, N. Pradhan, S. Panda, & B. K. Mishra (Eds.), *Environmental microbe biotechnology* (1st ed., pp. 313–328). Springer, Cham. https://doi.org/10.1007/978-3-319-19018-1_16

Das, A. P., Ghosh, S., Mohanty, S., & Sukla, L. B. (2015b). Consequences of manganese compounds: A review. *Toxicological & Environmental Chemistry, 96*(7), 981–997. https://doi.org/10.1080/02772248.2015.1005428

Das, A. P., Sukla, L. B., Pradhan, N., & Nayak, S. (2011). Manganese biomining: A review. *Bioresource Technology, 102*(16), 7381–7387. https://doi.org/10.1016/J.BIORTECH.2011.05.018

Fileno, P. *Quartz crystal microbalance: Temperature effects analysis.* Politecnico Di Milano School of Industrial and Information Engineering Master of Science in Mechanical Engineering, Academic Year 2014–2015.

Ghosh, S., Tripathy, B., Dey, S., & Das, A. P. (2023). Comparative investigation of fungal and bacterial manganese biomining mechanisms. *Manganese Mining Microorganisms*, 99–115. https://doi.org/10.1016/b978-0-12-822147-1.00016-9

Ghosh, S., Mohanty, S., Nayak, S., Sukla, L. B., & Das, A. P. (2016). Molecular identification of indigenous manganese solubilising bacterial biodiversity from manganese mining deposits. *Journal of Basic Microbiology, 56*(3), 254–262. https://doi.org/10.1002/JOBM.201500477

Hui, C., Guo, Y., Wu, J., Liu, L., Yang, X., Guo, X., Xie, Y., & Yi, J. (2021). Detection of bioavailable cadmium by double-color fluorescence based on a dual-sensing bioreporter system. *Frontiers in Microbiology, 12.*

Janata, J. (2009). *Principles of chemical sensors* (2nd ed.). Springer Dordrecht Heidelberg London New York.

Khan, Z., Elahi, A., Bukhari, D. A., Rehman, A. (2022) Cadmium sources, toxicity, resistance and removal by microorganisms—A potential strategy for cadmium eradication. *Journal of Saudi Chemical Society, 26*(6), 101569.

Kwon, H., Kim, M., Jongho, S., Maeng, S., Shin, J.-S., & Yun, J.-Y. (2023). Cleaning effect of atmospheric-plasma-sprayed Y_2O_3 coating using piranha solution based on contamination particle measurement, *13*(3), 653.

Malik, A. K., Kaur, V., & Verma, N. (2006). A review on solid phase microextraction-High performance liquid chromatography as a novel tool for the analysis of toxic metal ions. *Talanta, 68*, 842–849.

Malik, L. A., Bashir, A., Qureashi, A., & Pandith, A. H. (2019). Detection and removal of heavy metal ions: A review. *Environmental Chemistry Letters, 17*, 1495–1521.

Meng, Z., Yun, L., Wei-Xiong, Z., Chao-Gang, H., Radovan, H., Ou, Y.-C., Leng, J.-D., Peng, M.-X., Lin, Z.-J., & Tong, M. L. (2009). Reactivity of 4-amino-3,5-bis(pyridin-2-yl)-1,2,4-triazole, structures and magnetic properties of polynuclear and polymeric Mn(ii), Cu(ii) and Cd(ii) complexes. *Dalton Transactions, 2009*, 10284–10295.

Mishra, S., Ghosh, S., van Hullebusch, E. D., Singh, S., & Das, A. P. (2023). A critical review on the recovery of base and critical elements from electronic waste-contaminated streams using microbial biotechnology. *Applied Biochemistry and Biotechnology, 2023*, 1–30. https://doi.org/10.1007/S12010-023-04440-X

Oto, M., Loula, M., Kaňa, A., & Vosmanská, M. (2020). Can ultrafast single-particle analysis using ICP-MS affect the detection limit? *Case Study: Silver Nanoparticles, Talanta, 210*, 120665.

Prabhakar, A., Mishra, S., & Das, A. P. (2019). Isolation and identification of lead (Pb) solubilizing bacteria from automobile waste and its potential for recovery of lead from end-of-life waste batteries. *Geomicrobiological Journal, 36*(10), 894–903. https://doi.org/10.1080/01490451.2019.1654044

Priyadarsini, S., & Das, A. P. (2024). Analytical and structural characterization of waste lithium-ion batteries for their effective recycling strategy. *Environmental Chemistry and Ecotoxicology*. https://doi.org/10.1016/j.enceco.2024.12.004

Priyadarsini, S., & Das, A. P. (2025). Lithium bioleaching: A review on microbial-assisted sustainable technology for lithium bio-circularity. *Journal of Water Process Engineering, 69*, 106744. https://doi.org/10.1016/j.jwpe.2024.106744

Raj, K., & Das, A. P. (2023). Lead pollution: Impact on environment and human health and approach for a sustainable solution. *Environmental Chemistry and Ecotoxicology, 5*, 79–85. https://doi.org/10.1016/J.ENCECO.2023.02.001

Solomonea, B. G., Jinga, L. I., Antohe, V. A., Socol, G., & Antohe, I. (2022). Cadmium Ions' trace-level detection using a portable fiber optic—surface plasmon resonance sensor. *Biosensors, 12*, 573.

Soria, L., Cuerva, C., Cano, M., José, A. C., & Lodeiro, C. (2018). Bifunctional dipyridylpyrazole silver complexes with tunable thermotropic liquid crystal and luminescent behaviour. *Dyes and Pigments, 150*, 323–334.

Srisook, N. (2004). *Electrochemical quartz crystal microbalance (EQCM) studies of compound semiconductor formation by EC-ALE*. PhD thesis, University of Georgia.

Taneja, P., Khandagale, S. B., Manjuladevi, V., Gupta, R. K., & Gupta, K. K. (2020). Heavy metal ion sensing using ultrathin Langmuir-Schaefer film of tetraphenylporphyrin molecule. *IEEE Sensors, 20*(7), 1558–1748.

Taneja, P., Manjuladevi, V., & Gupta, R. K. (2017a). Detection of cadmium ion in water using films of nanocomposite of functionalized carbon nanotubes and anionic polymer. In *2nd international conference on condensed matter and applied physics (ICC 2017)*

Taneja, P., Manjuladevi, V., Gupta, R., & Gupta, R. K. (2017b). Detection of cadmium ion in potable water using quartz crystal microbalance. *Macromolecular Symposia, 376*, 1600207.

Taneja, P., Manjuladevi, V., Gupta, K. K., & Gupta, R. K. (2018). Detection of cadmium ion in aqueous medium by simultaneous measurement of piezoelectric and electrochemical responses. *Sensors and Actuators B: Chemical, 268*, 144–149.

Taneja, P., Manjuladevi, V., Gupta, R. K., Kumar, S., Gupta, K. K. (2020). Facile ultrathin film of silver nanoparticles for bacteria sensing. *Colloids and Surfaces B: Biointerfaces, 196*, 111335.

Tsai, W.-Y., Taberna, P.-L., & Simon, P. (2014). Electrochemical quartz crystal microbalance (eq cm) study of ion dynamics in nanoporous carbons. *Journal of the American Chemical Society, 136*(24), 8722–8728.

Tyona, M. D. (2013). A theoretical study on spin coating technique. *Advances in Materials Research, 2*(4), 195–208.

Wagner, M., Hermanns, I., Bittinger, F., & Kirkpatrick, C. J. (1999). Induction of stress proteins in human endothelial cells by heavy metal ions and heat shock. *American Journal of Physiology-Lung Cellular and Molecular Physiology, 277*, L1026.

Yan, C., Qu, Z., Wang, J., Cao, L., & Han, Q. (2022). Microalgal bioremediation of heavy metal pollution in water: Recent advances, challenges, and prospects. *Chemosphere, 286*, 131870.

Chapter 23
Synthesis of a Ternary Doped Catalyst and Green Polymer Photocatalytic Films for Treatment of Ampicillin and Other Emerging Antibiotic Water Pollutants

Yash Mishra and Hari Mahalingam

Introduction

The rising levels of antibiotic residues in aquatic ecosystems have become a pressing public health issue, largely fueled by the troubling increase in antimicrobial resistance (AMR). Once celebrated as breakthrough medications, antibiotics are now found in water bodies at concerning concentrations, threatening both human health and ecological stability. Traditional wastewater treatment processes frequently fail to adequately eliminate these contaminants, highlighting the urgent need for innovative and sustainable remediation strategies. In this regard, photocatalysis has emerged as a promising method for degrading persistent organic pollutants, including pharmaceuticals.

This study centers on the synthesis and characterization of novel ternary doped titanium dioxide (TiO_2 photocatalysts specifically engineered to degrade Ampicillin and other emerging antibiotic contaminants under sunlight. TiO_2 is well-known for its remarkable photocatalytic properties, cost-effectiveness, and chemical resilience. However, its practical application is often hindered by its wide bandgap, which limits its efficiency under visible light. To address this challenge, the research utilizes an eco-friendly EDTA-citrate method to create doped TiO_2 catalysts, thereby enhancing their photocatalytic activity and light absorption capabilities.

Additionally, the photocatalysts are immobilized on biodegradable Poly Lactic Acid (PLA) films through a solvent casting technique. PLA, a biopolymer sourced from renewable materials, presents an environmentally sustainable alternative to

Y. Mishra (✉) · H. Mahalingam
Department of Chemical Engineering, National Institute of Technology Karnataka, Surathkal, Mangalore, Karnataka, India
e-mail: yash250025@gmail.com

I. D. Behera et al. (eds.), *Emerging Micropollutants*, Sustainable Environmental Waste Management Strategies, https://doi.org/10.1007/978-3-032-03530-1_23

conventional polymer supports, promoting the development of eco-friendly photocatalytic systems. This immobilization not only improves the stability and reusability of the photocatalysts but also resolves issues related to catalyst recovery post-treatment.

A range of comprehensive characterization techniques, including scanning electron microscopy (SEM), diffuse reflectance spectroscopy (DRS), and BET surface area analysis, are employed to assess the physical and chemical attributes of the synthesized photocatalysts. The goal of this study is to showcase the effectiveness of the developed photocatalytic films in degrading Ampicillin under sunlight, contributing to the advancement of viable and sustainable solutions for addressing antibiotic pollution in water bodies. Ultimately, this research aims to connect laboratory-scale photocatalytic processes with real-world applications in environmental remediation.

Photocatalysis—Basic Principles

The first step in a photocatalytic reaction is the excitation of energetic photogenerated electron–hole pairs. During this process, excited electrons (e^-) in the semiconductor's valence band are promoted to the conduction band, leaving behind holes (h^+) in the valence band. This transition is crucial, as it sets the stage for the photocatalytic activity of the material. Photocatalysts with smaller band gaps are particularly effective at absorbing photons from visible light, which enhances their ability to drive photocatalytic reactions.

Following excitation, the next critical phase involves the separation of these photogenerated electrons and holes. However, the process is challenged by the recombination of bulk charge carriers, which diminishes the number of available excited charge carriers. This recombination can also generate phonons or heat, which may reduce the overall efficiency of the photocatalytic process. The charge carriers can recombine at the surface of the catalyst or engage in various surface chemical reactions, impacting the effectiveness of the photocatalyst.

Among the various reactions that occur, photogenerated electrons are widely recognized as reductants capable of directly reducing certain heavy metal ions, thereby playing a vital role in environmental remediation. The separated holes, on the other hand, can react with hydroxyl ions (OH^-) or water molecules to produce hydroxyl radicals (·OH). These radicals are highly reactive and can participate in oxidative decomposition processes, making them essential for breaking down organic pollutants. The strong oxidizing ability of the holes is a primary pathway for the generation of ·OH, which is crucial for enhancing the photocatalytic efficiency.

In addition to their role in generating hydroxyl radicals, the separated electrons can interact with dissolved oxygen in water to produce superoxide radicals ($\cdot O^{2-}$). These superoxide radicals can further react to yield additional hydroxyl radicals (·OH), contributing to the overall oxidative capacity of the system.

After the pollutants in the water are adsorbed onto the surface of the catalytic material, a complex series of chemical reactions takes place involving the active species generated by the catalyst. This interaction not only enhances charge mobility but also significantly boosts the redox capacity of the system, facilitating the degradation of a wide range of pollutants. The efficiency of this degradation process depends on several factors, including the nature of the photocatalyst, the intensity and wavelength of light, the concentration of pollutants, and the presence of other reactive species in the environment. By optimizing these parameters, photocatalytic systems can be designed to achieve higher degradation rates and more effective removal of contaminants from water, ultimately contributing to environmental sustainability and the purification of water resources.

The redox reactions mentioned above are given below (Figs. 23.1, 23.2).

$$\text{Semiconductor} + h\nu \rightarrow h^{+} + e^{-} \rightarrow \tag{23.2.1}$$

$$e^{-} + O_2 \rightarrow \cdot O^{-} \tag{23.2.2}$$

$$\cdot O^{-} + H^{+} \rightarrow \cdot OOH \tag{23.2.3}$$

$$2 \cdot OOH \rightarrow O_2 + H_2O_2 \tag{23.2.4}$$

$$H_2O_2 + \cdot O^{-} \rightarrow \cdot OH + OH^{-} + O_2 \tag{23.2.5}$$

$$H_2O_2 + h\nu \rightarrow 2 \cdot OH \tag{23.2.6}$$

$$h^{+} + H_2O \rightarrow \cdot OH + H^{+} \tag{23.2.7}$$

$$h^{+} + OH^{-} \rightarrow \cdot OH \tag{23.2.8}$$

$$\text{Pollutants} + (\cdot OH,\ h^{+},\ e^{-}, \cdot OOH,\ \text{or}\ \cdot O^{-}) \rightarrow \text{degradation products} \tag{23.2.9}$$

Removal of Antibiotics Using Photocatalysis

Antibiotics have received increased focus among pharmaceutical compounds due to their negative effects on the aquatic ecology and potential risks to human health. Antibiotics and their by-products have a significant potential for environmental and ecological disruption, high toxicity, and good stability. Severe issues with the aquatic environment are brought on by the enormous quantity of antibiotics that are released

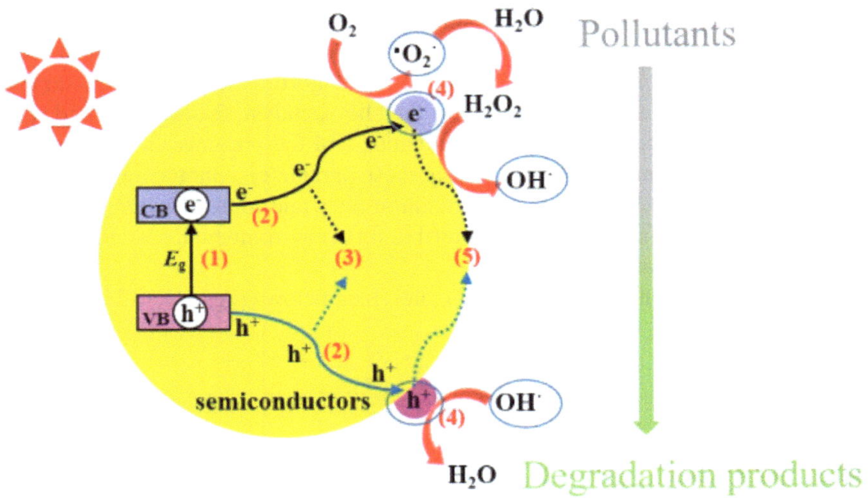

Fig. 23.1 Photocatalytic processes over a heterogenous photocatalyst.

Fig. 23.2 Setup for the preparation of immobilised photocatalytic films

into the environment via sewage and animal excrement. Therefore, it is crucial to research the most efficient processes for removing these chemicals from wastewater.

Nanosized titanium dioxide (TiO_2) is a well-established photocatalyst renowned for its exceptional efficiency, cost-effectiveness, physicochemical stability, widespread availability, safety, and non-corrosive properties. In recent years, particularly over the past two decades, the immobilization of TiO_2 onto various polymeric substrates has gained considerable attention. This approach significantly reduces the need for post-treatment separation processes, enhancing the practicality and efficiency of photocatalytic applications.

Among the various substrate options, biodegradable polymers have emerged as a highly promising alternative to their non-biodegradable counterparts. Their ability to decompose naturally aligns with the increasing global focus on sustainability and environmental responsibility. Both natural and synthetic polymers can possess biodegradable properties, offering a range of choices for different applications.

The method of immobilizing TiO_2 plays a crucial role in determining its photocatalytic performance. Therefore, it is vital to select an appropriate immobilization technique that is tailored to the specific substrate and the pollutants targeted for degradation. It is generally advisable to choose a deposition method that preserves the photocatalytic activity of TiO_2. Various techniques, such as sol–gel processes, dip-coating, and electrospinning, can be employed, each providing distinct advantages depending on the intended application.

Among biodegradable polymers, polylactic acid (PLA) is particularly noteworthy due to its superior properties compared to other biopolymers. PLA features a relatively high melting point, enhanced crystallinity, and increased rigidity, making it a suitable candidate for applications that require both mechanical strength and photocatalytic efficiency. Additionally, PLA is derived from renewable resources like corn starch or sugarcane, further emphasizing its eco-friendly credentials.

Integrating TiO_2 into PLA not only capitalizes on the photocatalytic properties of titanium dioxide but also improves the mechanical and thermal properties of the polymer matrix. This synergy can lead to the development of innovative materials that effectively degrade pollutants while maintaining structural integrity. Furthermore, the combination of TiO_2 and biodegradable polymers can facilitate the creation of self-cleaning surfaces, air purification systems, and advanced water treatment technologies, contributing significantly to environmental remediation efforts.

The ongoing research in the field of TiO_2 immobilization on biodegradable substrates like PLA holds great promise for advancing photocatalytic technologies. By optimizing these materials and processes, we can develop effective and sustainable solutions to combat pollution and promote environmental sustainability. As the demand for eco-friendly technologies continues to rise, the integration of photocatalytic materials with biodegradable polymers will play a critical role in shaping a cleaner and more sustainable future.

Related Works

Eleftheriadou et al. (2020) explore the innovative development of polymer-supported graphene oxide (GO)-TiO_2 films using poly(L-lactic acid) (PLLA), a widely recognized bioplastic known for its biodegradability and versatility. Their research specifically investigates the degradation of nine different antibiotics, which include two sulfonamides (sulfamethoxazole and sulfadiazine), three fluoroquinolones (levofloxacin, norfloxacin, and moxifloxacin), an antibiotic for tuberculosis (isoniazid), a nitroimidazole (metronidazole), a lincosamide (lincomycin), and a diaminopyrimidine. The films were synthesized with 1 weight % GO and varying concentrations of TiO_2 (10, 25, and 50 wt%). Among these formulations, the PLLA–GO– TiO_2 composite containing 50 wt% TiO_2 demonstrated the highest photocatalytic efficiency, leading to further investigation of its properties. This composite not only exhibited remarkable photostability but also maintained its effectiveness over four consecutive cycles of use, indicating excellent reusability. The overall photocatalytic degradation of antibiotics in various environments, including pure water and wastewater, yielded impressive results, showcasing the composite's potential for practical applications in wastewater treatment. The study confirmed that the PLLA – GO – 50 wt% TiO_2 composite could be reused effectively without significant loss of performance or the need for extensive post-treatment recovery, which is crucial for sustainable environmental applications.

In a complementary study, Teamsinsungvon et al. (2022) examined the photocatalytic efficiency, antimicrobial properties, permeability, and biodegradability of polylactic acid (PLA). They investigated how different types and concentrations of TixSiy oxides affected the crystallization behavior, mechanical properties, thermal characteristics, and morphological attributes of PLA composites. Their findings revealed that TixSiy oxides significantly enhanced the photocatalytic efficacy for degrading methylene blue (MB), as well as improving hydrolytic and in vitro degradation of PLA. Furthermore, the incorporation of TixSiy oxides into PLA composites demonstrated strong antibacterial activity against both Gram-positive (Staphylococcus aureus) and Gram-negative (Escherichia coli) bacteria, highlighting an improvement in the antimicrobial performance of PLA. Notably, the ability of mixed oxides to absorb water suggested that a loading of up to 5% TixSiy could facilitate greater degradation of PLA, contributing to more effective waste management solutions.

The study aimed to assess the effects of 3 weight percent of TiO_2, SiO_2, $Ti_{70}Si_{30}$, $Ti_{50}Si_{50}$, $Ti_{40}Si_{60}$, and TiO_2, SiO_2, as well as 5 weight percent of $Ti_{70}Si_{30}$, on the mechanical, thermal, morphological, degradation behavior, and antibacterial activities of PLA. Utilizing the solvent casting technique, both PLA and PLA composite films were produced. The results indicated that the addition of $Ti_{70}Si_{30}$ and $Ti_{50}Si_{50}$ slightly enhanced the tensile strength and Young's modulus of the PLA films. Moreover, incorporating 5 weight percent of $Ti_{70}Si_{30}$ reduced the cold crystallization temperature and increased the degree of crystallinity of PLA, suggesting that $Ti_{70}Si_{30}$ nanoparticles could effectively serve as nucleating agents for PLA. The degradation efficiency for MB followed the order: $TiO_2 > Ti_{70}Si_{30} > TiO_2SiO_2 > Ti_{50}Si_{50} > Ti_{40}Si_{60} > SiO_2$. Additionally, the photocatalytic activity of PLA was notably enhanced with the addition of 5 weight percent Ti70Si30. While all nanoparticles exhibited UV-blocking capabilities, TiO_2 and Ti70Si30 demonstrated the most effective UV-shielding properties, indicating their potential for applications in protective coatings.

Chanklom et al. (2021) synthesized a biocomposite film composed of PLA and nano-TiO_2 particles, which served as a carrier for the photocatalytic oxidation-based degradation of volatile organic compounds (VOCs). The biocomposite was created by blending maleic anhydride copolymer and nano-TiO_2 at varying concentrations (5.0, 10.0, and 15.0 wt%) with molten PLA resins using a twin-screw extruder. The resulting films, with a thickness of 30 μm, were produced through a blown film extrusion process at temperatures ranging from 100 to 160 °C.

For testing, an annular closed-system photoreactor was utilized, equipped with a UV-C lamp emitting at a wavelength of 254 nm. The photoreactor compartment was filled with the nano-TiO_2/PLA biocomposite material. DCM was dispersed in the gas phase over the photocatalyst surface, and the UV lamp was activated once the initial DCM concentration stabilized. Photodegradation studies revealed that the nano—TiO_2/PLA biocomposite films, loaded with 5. 0%, 10.0%, and 15.0% TiO_2, achieved DCM removal efficiencies of 41.2%, 63.4%, and 49.1%, respectively. However, it was observed that higher concentrations of TiO_2 resulted in decreased DCM conversion due to the nucleation of TiO_2 on the support, which reduced the available surface area for photocatalytic activity. Conversely, lower amounts of TiO_2 were insufficient to provide adequate photocatalytic capability, indicating a need for optimization in the formulation to balance the concentration of TiO_2 for maximum efficiency.

Gjipalaj and Alessandri (2017) tackled the challenges associated with the short lifespan, instability, toxicity, and recovery times of nano-sized TiO_2 photocatalysts. To enhance photocatalytic activity while maintaining high porosity and surface area, they developed macrobeads synthesized through ionotropic gelation to create porous titania beads. Various tests were conducted to evaluate the photocatalytic performance of these beads in degrading organic contaminants. The study examined six different doses of beads, all with a consistent average size of 1 mm, for the photocatalytic degradation of methylene blue (MB) at a concentration of 1×10^{-4} M. The results indicated that both MB and Methyl Orange (MO) degraded more rapidly in diluted solutions, with MO exhibiting a higher degradation rate than MB. This was attributed to the greater adsorption of cationic dyes on the TiO_2 surface, which led to quicker saturation of the photocatalytic active sites.

Additionally, the study compared the performance of TiO_2 beads with corresponding powdered materials used to create alginate-based composites. The alginate-based beads demonstrated significantly superior adsorption capabilities after one hour in neutral pH and darkness (55% versus 6.5%). This enhanced adsorption was linked to beneficial interactions between macropores and alginate, which facilitated better contact with the contaminants. Although the photocatalytic rate was reduced due to the beads, their full recyclability and ease of recovery allowed for multiple uses, overcoming the limitations associated with traditional photocatalysts. This research highlights the potential of using macrobeads in photocatalytic applications, paving the way for more sustainable and efficient methods in environmental remediation.

Overall, these studies collectively underscore the importance of developing advanced materials for photocatalytic applications, particularly in the degradation of pollutants and antibiotics, which pose significant environmental challenges. The integration of bioplastics like PLLA and PLA with photocatalytic agents such as TiO_2 and graphene oxide not only enhances the efficiency of degradation processes but also contributes to the sustainability of materials used in environmental applications.

Preparation of Polymeric Film with the Immobilized Catalyst

The solvent film casting technique will be utilized to create PLA photocatalyst composites in this study. This method begins with dissolving PLA pellets and the selected catalyst in chloroform using an ultrasonic bath for two hours, ensuring a thorough and uniform dispersion of the components. Following this initial dissolution phase, the mixture is transferred to a magnetic stirrer for an additional 20 min. During this stirring process, the solution gradually thickens, achieving a viscous consistency that allows it to adhere to the sides of the beaker. Once the desired viscosity is attained, the solution is carefully poured into a petri dish, covered to prevent contamination, and left to dry at room temperature for 24 h, allowing the solvent to evaporate completely and form a solid film.

In this investigation, five distinct films with varying weight loadings of the B2Ce1Ag0.06 TiO_2 catalyst will be prepared and assessed for their degradation efficiency. The specific loadings under consideration are 3%, 7%, 10%, 15%, and 20%. After identifying which loading demonstrates the highest degradation performance, additional films will be produced using B2Ce1Ag0.1 TiO_2 and conventional TiO_2. This comparative analysis aims to elucidate the differences in photocatalytic activity among the various catalyst formulations, thereby identifying the optimal composition for enhanced photocatalytic performance.

Moreover, the study will delve into the physical and chemical properties of the resulting films, including their morphology, mechanical strength, and thermal stability. To achieve this, characterization techniques such as scanning electron microscopy (SEM) will be employed to visualize the surface structure and morphology of the composites, while Fourier-transform infrared spectroscopy (FTIR) will be used to identify the functional groups and chemical bonds present in the materials. Additionally, thermogravimetric analysis (TGA) may be conducted to evaluate the thermal stability of the films, providing insights into their performance under varying temperature conditions.

The photocatalytic efficiency of each film will be rigorously tested through degradation experiments involving model pollutants, such as methylene blue or phenol, to simulate real-world applications. These experiments will not only measure the rate of pollutant degradation but also help establish a comprehensive understanding of the relationship between catalyst loading and degradation efficiency. By systematically investigating these variables, this research aims to contribute to the advancement of photocatalytic materials, enhancing their potential for effective applications in wastewater treatment and other environmental remediation strategies. Ultimately, the findings from this study could pave the way for developing more efficient and sustainable photocatalytic systems, addressing the pressing environmental challenges associated with water pollution.

Photocatalytic Films Preparation Strategy

The aim of this study is to synthesize two boron/cerium/silver tridoped titanium dioxide (TiO_2) photocatalysts with precisely controlled amounts of each dopant. Following the synthesis, these photocatalysts will be immobilized using an environmentally friendly Poly Lactic Acid (PLA) film. The performance of the immobilized photocatalysts will then be evaluated for their ability to degrade ciprofloxacin (CIP) antibiotics under direct sunlight. Specifically, the objectives include synthesizing the two doped TiO_2 photocatalysts, immobilizing them on PLA films, assessing their degradation efficiency with respect to ciprofloxacin and other antibiotics under sunlight exposure, determining the optimal weight loading for achieving maximum degradation, and comparing the effectiveness of the two synthesized catalysts (Fig. 23.2).

Experimental Section

Materials

Degussa P25 TiO_2 (99.9% pure) is procured from Evonik (Japan); boric acid (99.5% extra pure), EDTA (98% extra pure), citric acid (99% extra pure) and ammonia solution (99% extra pure) from Loba chemicals (Loba Chemie Pvt Ltd., India); cerium nitrate hexahydrate (99% trace metal basis) from (Sigma Aldrich Chemicals Pvt Ltd., India). Poly Lactic Acid (Luminy L175 grade) from (Konkan Speciality Products Pvt Ltd), and Poly Lactic Acid (Ingeo biopolymer 2003D grade) from Natur-Tec India Pvt Ltd, Trichloromethane, for film formation.

Synthesis of (B x Ce y Ag z) TiO_2 Series of Catalysts

The B.xCe.yAg.z TiO_2 photocatalysts were synthesized using the EDTA-citrate method. The required amount of EDTA was dissolved in water using ammonia; this was followed by adding the required amounts of cerium nitrate hexahydrate and boric acid solutions. Citric acid was added to the above mentioned mixture under continuous stirring, followed by pH adjustment to 9.0 using ammonia. A mole ratio of 1:1:1.5 (metal/boric acid: EDTA: citric acid) was maintained. The mixture was heated and stirred continuously until the formation of an organometallic gel. This gel was then dried at 150 °C for 24 h in a hot air oven. The dried powder was subjected to a two-step calcination process; 350 °C for 12 h followed by 600 °C for 5 h to remove all organic impurities. The powder catalysts thus obtained were stored for further usage.

In this study, five ternary doped catalysts photocatalysts namely (1B 0.1CE 0.06AG) (1B1CE0.06AG) (2B 0.1CE0.06AG) (2B 1CE 0.06AG) (1B 1CE 0.1AG) were synthesized.

Characterization

A nanoparticle size analyzer (Horiba SZ-100, Kyoto, Japan) was used to determine the particle size of the synthesized photocatalysts. The specific surface area was analyzed using a Brunauer–Emmett–Teller (BET) surface area analyzer (BELSORP Max, Microtrac, Japan).

Scanning electron micros-copy (SEM, JEOL JSM-6380LA, Peabody, Massachusetts, USA) using secondary electron detection mode with X500 and X1000 magnifications and transmission electron microscopy (TEM, Jeol JEM 2100, Peabody, Massachusetts, USA) were used to characterize the morphology of the prepared catalysts.

X-ray diffraction (XRD, Rigaku Ultima-IV diffractometer, Rigaku, Tokyo, Japan) was carried out with Cu Kα (λ ¼ 0.15418 nm) in the range of 2θ ¼ 20–65 with an increment of 0.02.

Raman analysis was carried out using a Raman spectrophotometer (BRUKER—RFS27 FT-Raman spectrophotometer, Billerica, Massachusetts, USA) to determine the structural phase and crystallinity. Identification of Ti (2p), O (1 s), B (1 s), and Ce (3d) peaks was carried out by X-ray photoelectron spectroscopy (XPS, Omicron ESCAþ, Chanhassen, Minnesota, USA), with Al Kα monochromator as an X-ray source.

Diffuse reflectance spectroscopy (DRS, UV Vis NIR Spectrophotometer: Agilent Cary 5000, Santa Clara, California, USA) was used to determine the bandgap energy values. UV–Visible spectrophotometer (GBC CINTRA 103, Hampshire, USA) was used to analyze the absorbance values of CIP-and NOR-treated samples at λmax ¼ 280 nm.

The standard calibration graphs for both the model pollutants are given in the previous study. Photoluminescence analysis was carried out using a FluoroMax-4 spectrometer (Horiba, New Jersey, USA) at an excitation wavelength of 325 nm (Fig. 23.3).

Ampicillin Photocatalytic Degradation

The photocatalytic degradation of Ampicillin was investigated under sunlight (coordinates: 13 deg 0 0 40 00 N, 74 deg 47 0 44 00 E) for 3 h from 12 to 3 p.m.). An average temperature and an average irradiation intensity of 33 °C and 84,000 (±) 6000 lx were recorded. The experimental runs were performed with 200 mL aqueous solution containing 10 mg L−1 of either antibiotic in a glass beaker of dimensions

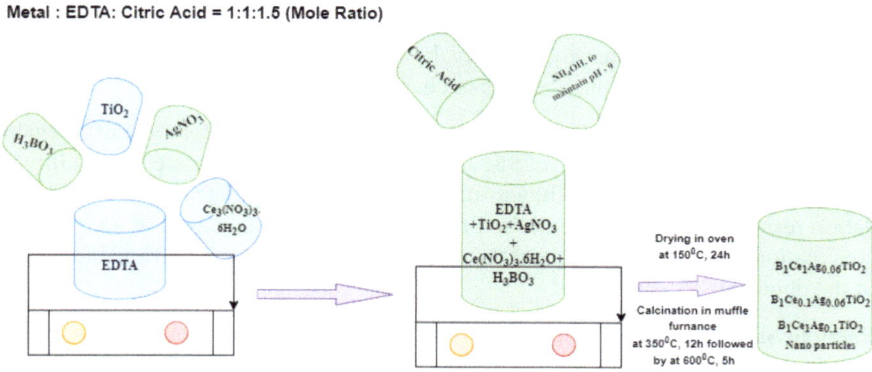

Fig. 23.3 The synthesis process of the photocatalysts

100.50 mm. Before irradiation under sunlight, the reactor was placed in the dark for 30 min to attain adsorption–desorption equilibrium. Further analysis of the degraded samples was carried out after filtration and centrifugation to remove the suspended catalyst particles. The experiment was carried out in duplicate.

$$\text{Degradation \%} = (\text{Co} - \text{C}/\text{Co}) * 100$$

where C0 and C are initial and final concentration values,

Particle Size Analysis

The particle size analysis of photocatalysts is a critical step in the validation of the doping process, ensuring that the modifications at the nanoscale are precise and effective. The successful doping of photocatalysts can significantly enhance their performance by altering their electronic, optical, and surface properties, which are essential for various applications, including environmental remediation and energy conversion.

Accuracy in Doping: The accuracy of doping is paramount, as it determines the distribution of dopant atoms within the host material and directly impacts the photocatalyst's activity. The particle size results falling within the desired range is a strong indication that the doping method employed is both accurate and reliable. This precision in doping ensures that the photocatalysts have the optimal characteristics for their intended use.

Challenges in Nanoparticle Doping: Doping at the nanoscale presents several challenges, one of which is the potential for agglomeration or coagulation of nanoparticles. This phenomenon can occur due to various factors, including inaccurate mixing, which leads to the formation of aggregates. These aggregates can adversely affect the photocatalyst's performance by reducing the surface area available for photocatalytic reactions and causing inconsistencies in characterization.

Degradation Analysis: The degradation analysis is a crucial component of evaluating a photocatalyst's effectiveness. It involves assessing the material's ability to facilitate the breakdown of pollutants under light irradiation. The fact that the degradation results were within the desired range, despite the challenges of nanoparticle coagulation, speaks volumes about the efficacy of the doping method and the quality of the photocatalysts produced.

Particle Size and Photocatalytic Activity: The particle size of photocatalysts is a determinant factor in their photocatalytic activity. Smaller particles typically offer a larger surface area to volume ratio, which translates to more active sites for the adsorption of reactants and the subsequent photocatalytic reactions. Additionally, the particle size can influence the light absorption properties and the diffusion of reactants and products, which are critical for the photocatalytic process.

The Five Photocatalysts: The five photocatalysts, designated as (1B 0.1CE 0.06AG), (1B1CE0.06AG), (2B 0.1CE0.06AG), (2B 1CE 0.06AG), and (1B 1CE 0.1AG), represent a range of doping configurations. Each configuration is tailored to optimize specific properties of the photocatalyst, such as light absorption, charge carrier separation, and reactivity. Refer no Figs. 23.4, 23.5, 23.6, 23.7, 23.8.

- **(1B 0.1CE 0.06AG):** A balanced composition that likely targets a specific photocatalytic reaction, optimizing light absorption and charge mobility.

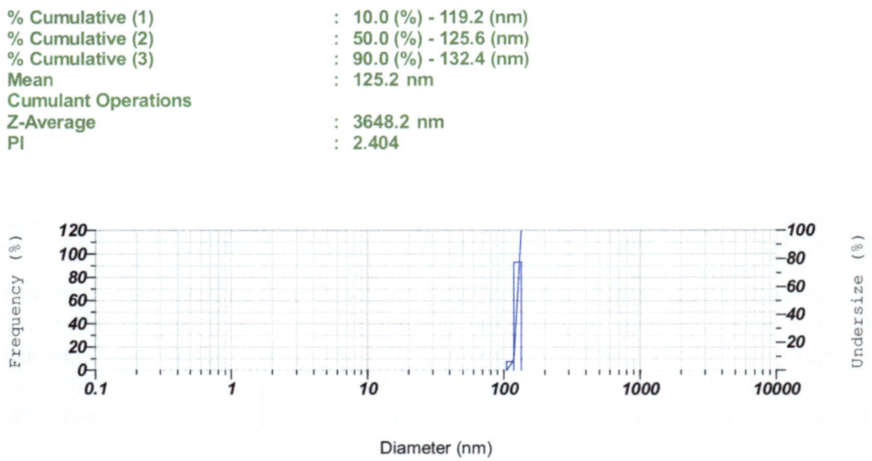

Fig. 23.4 Particle size results for Catalyst 1 (1B 0.1CE 0.06 AG)

% Cumulative (1)	: 10.0 (%) - 127.4 (nm)
% Cumulative (2)	: 50.0 (%) - 784.6 (nm)
% Cumulative (3)	: 90.0 (%) - 899.7 (nm)
Mean	: 669.9 nm
Cumulant Operations	
Z-Average	: 3201.8 nm
PI	: 1.007

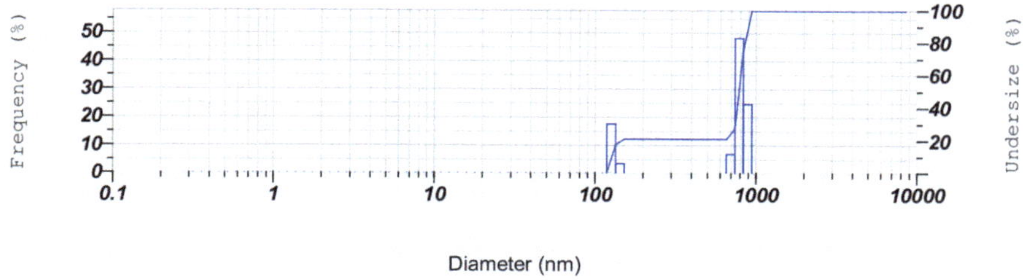

Fig. 23.5 Particle size results for Catalyst 2 (1B 1CE 0.06 AG)

% Cumulative (1)	: 10.0 (%) - 87.0 (nm)
% Cumulative (2)	: 50.0 (%) - 260.2 (nm)
% Cumulative (3)	: 90.0 (%) - 304.0 (nm)
Mean	: 209.8 nm
Cumulant Operations	
Z-Average	: 1805.5 nm
PI	: 0.943

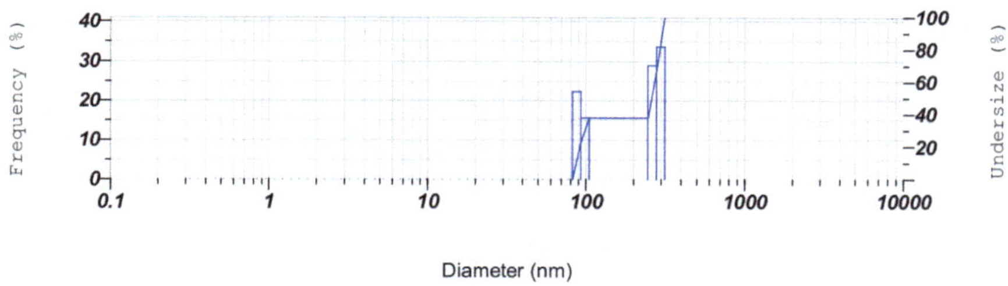

Fig. 23.6 Particle size results for catalyst 3 (2B 0.1CE 0.06 AG)

- **(1B1CE0.06AG):** An equal atomic percent of boron and cerium may suggest a synergistic effect, enhancing the photocatalyst's ability to store and release oxygen species.
- **(2B 0.1CE0.06AG):** A higher boron content could be aimed at reactions requiring enhanced charge separation.
- **(2B 1CE 0.06AG):** Increased cerium content alongside boron might improve the degradation of organic pollutants through cerium's redox properties.
- **(1B 1CE 0.1AG):** Elevated silver content suggests a focus on utilizing silver's plasmonic effects for visible light-driven reactions.

% Cumulative (1) : 10.0 (%) - 104.1 (nm)
% Cumulative (2) : 50.0 (%) - 238.7 (nm)
% Cumulative (3) : 90.0 (%) - 418.9 (nm)
Mean : 769.8 nm
Cumulant Operations
Z-Average : 884.1 nm
PI : 1.820

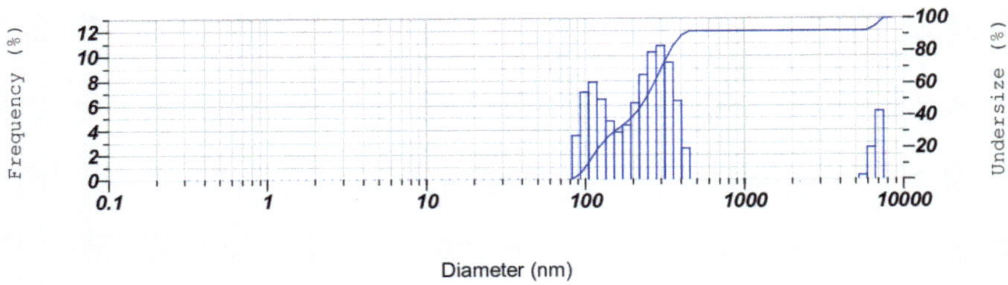

Fig. 23.7 Particle size results for catalyst 4 (2B 1CE 0.06 AG)

% Cumulative (1) : 10.0 (%) - 587.9 (nm)
% Cumulative (2) : 50.0 (%) - 618.9 (nm)
% Cumulative (3) : 90.0 (%) - 651.5 (nm)
Mean : 620.9 nm
Cumulant Operations
Z-Average : 2938.1 nm
PI : 1.192

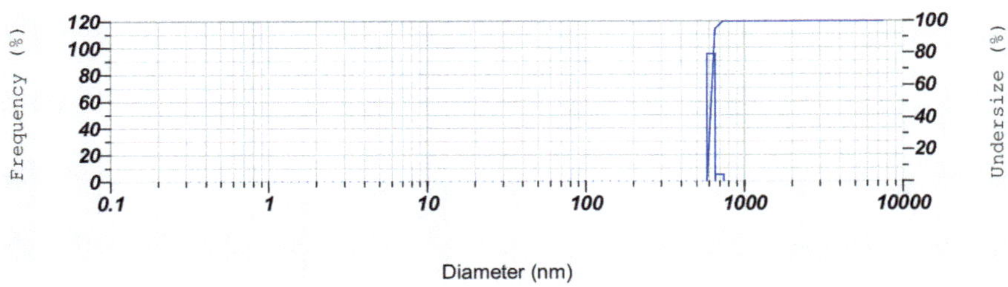

Fig. 23.8 Particle size results for catalyst 5 (1B 1CE 0.1 AG)

In summary, the particle size analysis confirms the accuracy of the doping process for the five photocatalysts. The results validate the doping technique's precision and provide insights into the potential photocatalytic performance of these materials. The uniformity in particle size distribution is indicative of the photocatalysts' capability to function effectively, which is promising for their application in various fields. Future research could further refine the doping process to enhance efficiency and explore the scalability of the method for broader industrial applications. The continued development of these photocatalysts holds great promise for advancing sustainable technologies and addressing environmental challenges.

Particle size results of all five photocatalysts namely (1B 0.1CE 0.06AG) (1B1CE0.06AG) (2B 0.1CE0.06AG) (2B 1CE 0.06AG) (1B 1CE 0.1AG) are shown below.

Sem Analysis

The analysis of Scanning Electron Microscopy (SEM) results for a series of five photocatalysts reveals significant insights into the structural and compositional attributes that influence their performance. Photocatalysts, by harnessing light energy, can drive chemical reactions, and their efficiency is highly dependent on factors such as particle size, surface area, and dopant concentration.

Boron Concentration Trends: Among the five photocatalysts, a notable trend was observed in the boron concentration. Boron, known for its electron-deficient nature, can introduce localized states in the band gap of the photocatalyst, thereby affecting the charge carrier dynamics. The increase in boron concentration from one atomic percent to two suggests an attempt to modulate the electronic properties of the photocatalyst. This increment could potentially enhance the photocatalytic activity by facilitating charge separation and reducing recombination rates.

Silver Dopant Influence: Similarly, the silver dopant concentration exhibited an increasing trend, particularly in the last photocatalyst, where it rose from 0.06 to 0.1 atomic percent. Silver, with its plasmonic properties, can act as an electron trap and extend the light absorption range of the photocatalyst. The strategic increase in silver content is likely aimed at leveraging these properties to boost the visible light-driven photocatalytic processes.

Particle Size Observations: The particle size of photocatalysts is a critical parameter that directly impacts the surface-to-volume ratio. A higher surface area, typically associated with smaller particles, provides more active sites for the photocatalytic reactions. The particle size results for the photocatalysts, denoted as (1B 0.1CE 0.06AG), (1B1CE0.06AG), (2B 0.1CE0.06AG), (2B 1CE 0.06AG), and (1B 1CE 0.1AG), suggest a diverse range of sizes that could cater to different reaction environments.

Photocatalyst (1B 0.1CE 0.06AG): This composition indicates a balanced approach, with boron and cerium providing a stable matrix for the silver dopants. The particle size in this catalyst would be optimized for a specific reaction, possibly targeting a balance between light absorption and charge mobility.

Photocatalyst (1B1CE0.06AG): The equal atomic percent of boron and cerium in this photocatalyst could be indicative of a synergistic effect, where cerium's ability to store and release oxygen species complements boron's electronic modulation.

Photocatalyst (2B 0.1CE0.06AG): A higher boron content in this photocatalyst could be tailored for reactions requiring enhanced charge separation, given boron's influence on electronic properties.

Photocatalyst (2B 1CE 0.06AG): The increased cerium content alongside boron might be designed to exploit cerium's redox properties to a greater extent, potentially improving the degradation of organic pollutants.

Photocatalyst (1B 1CE 0.1AG): The elevated silver content in this photocatalyst suggests a focus on utilizing silver's plasmonic effects to drive reactions under visible light conditions.

The SEM analysis of these photocatalysts, with varying concentrations of boron, cerium, and silver, provides a comprehensive understanding of how each element contributes to the overall photocatalytic activity. By adjusting these dopant levels, researchers can fine-tune the photocatalysts to specific applications, whether it be pollutant degradation, hydrogen production, or CO_2 reduction. The observed aggregates could be indicative of the photocatalysts' tendency to cluster, which may affect the available surface area and, consequently, the photocatalytic efficiency. Future studies could explore the impact of these aggregates on the photocatalysts' performance and investigate methods to optimize the dispersion of particles to maximize their activity. Refer to Figs. 23.7, 23.8, 23.9, 23.10, 23.11, 23.12, 23.13, 23.14, 23.15 and 23.16 for images.

Figures 23.9, 23.10, 23.11, 23.12, 23.13 shows SEM results for all five synthesized photocatalysts.

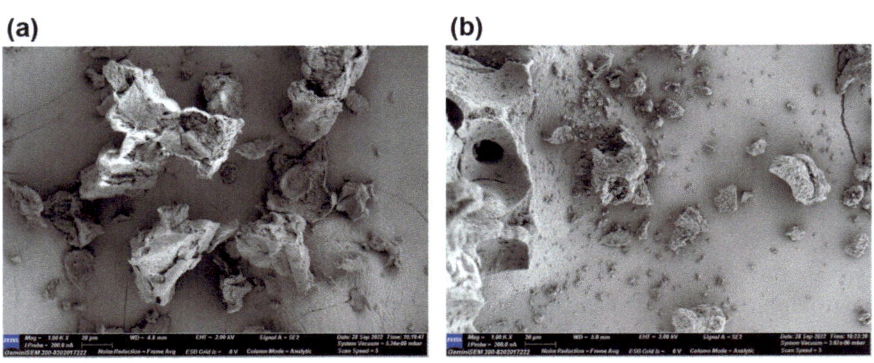

Fig. 23.9 a and b shows SEM Images for Catalyst 1 (1B 0.1CE 0.06 AG)

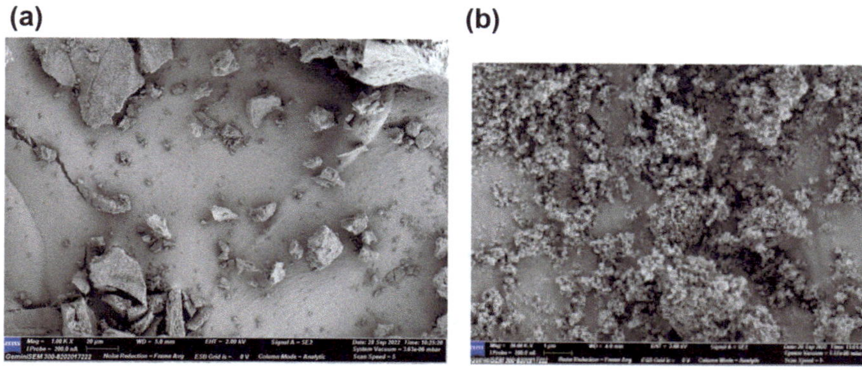

Fig. 23.10 a and b shows SEM Images for Catalyst 2(1B 1CE 0.06 AG)

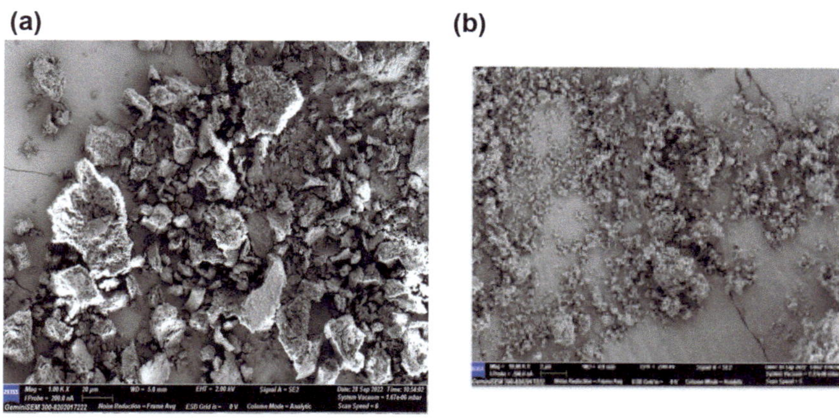

Fig. 23.11 a and b shows SEM Images for Catalyst 3 (2B 0.1CE 0.06 AG)

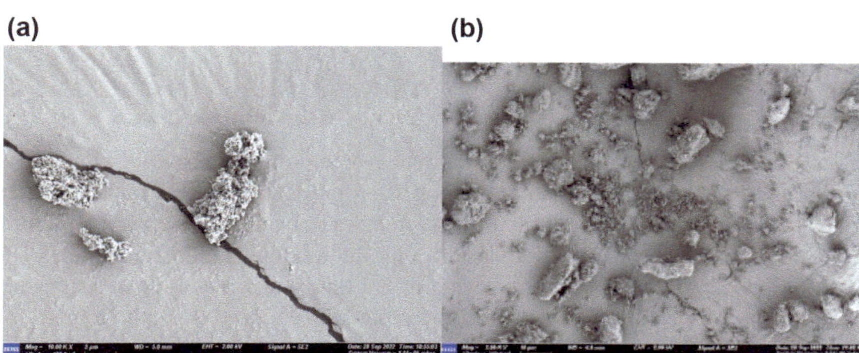

Fig. 23.12 a and b shows SEM Images for Catalyst 4 (2B 1CE 0.06 AG)

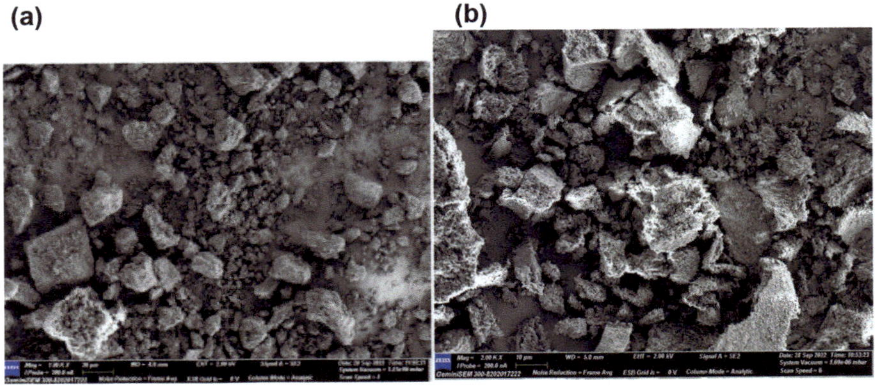

Fig. 23.13 a and b shows SEM Images for Catalyst 5 (1B 1CE 0.1 AG)

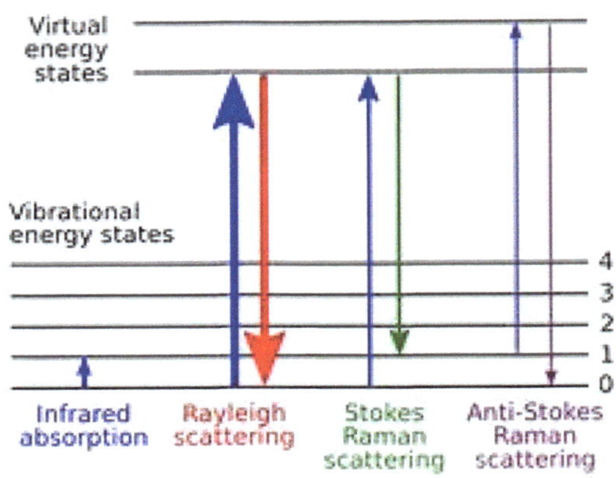

Fig. 23.14 Virtual and vibrational eneergy states corresponding to differnt types of scattering

Raman Spectroscopy

Raman Spectroscopy is a sophisticated analytical technique that provides a molecular fingerprint of materials by detecting vibrational, rotational, and other low-frequency modes in a system. It is particularly useful for characterizing the chemical structure, phase, polymorphy, crystallinity, and molecular interactions within a material1. The technique relies on the inelastic scattering of photons, known as Raman scattering, which occurs when light interacts with the molecular vibrations or phonons in the sample (Fig. 23.19).

Understanding Raman Spectroscopy: When a monochromatic light source, typically a laser, illuminates a sample, most of the light is elastically scattered (Rayleigh scattering). However, a small fraction of the light is inelastically scattered, resulting

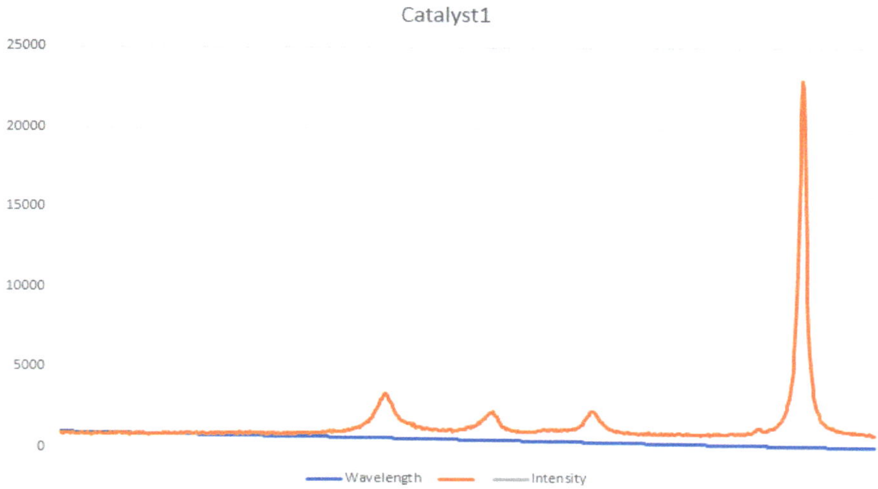

Fig. 23.15 Raman spectroscopy results graph for catalyst 1 (1B 0.1CE 0.06 AG)

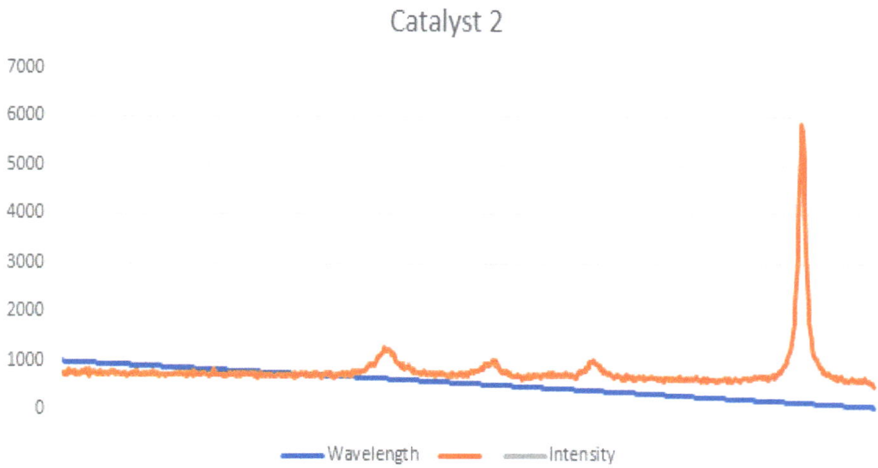

Fig. 23.16 Raman spectroscopy results graph for Catalyst 2 (1B 1CE 0.06 AG)

in a shift in energy that corresponds to the vibrational modes of the molecules in the sample. This shift, known as the Raman shift, is detected and analyzed to provide information about the sample's molecular structure. Refer to Figures 23.17, 23.18, 23.20 and 23.21.

Analysis of Photocatalysts: The Raman Spectroscopy results for the five photocatalysts, namely (1B 0.1CE 0.06AG), (1B1CE0.06AG), (2B 0.1CE0.06AG), (2B

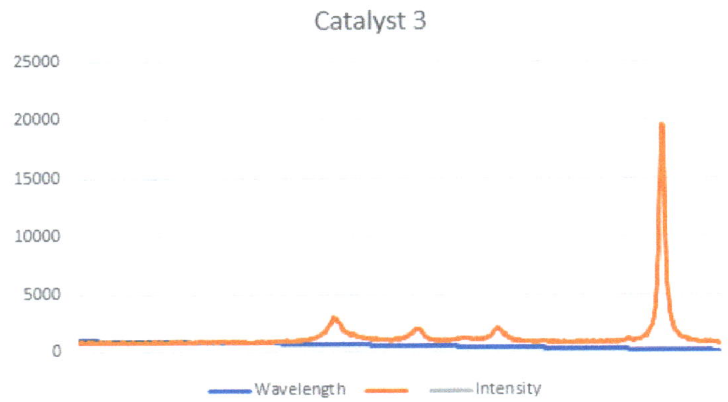

Fig. 23.17 Raman spectroscopy results graph for Catalyst 3 (2B 0.1CE 0.06 AG)

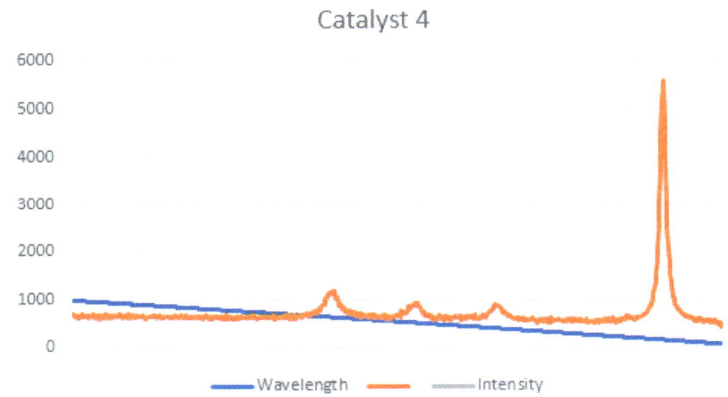

Fig. 23.18 Raman spectroscopy results graph for Catalyst 4 (2B 1CE 0.06 AG)

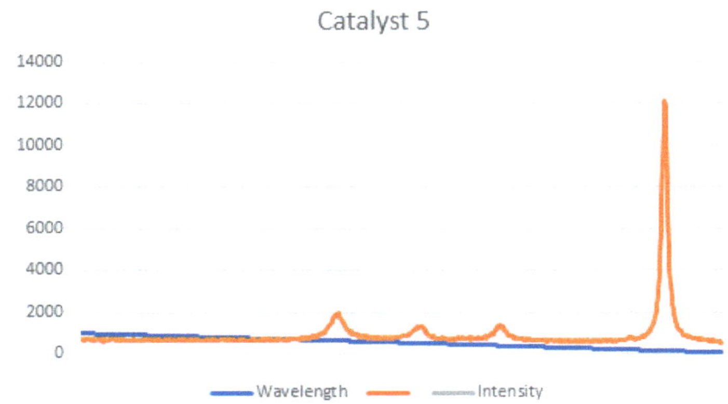

Fig. 23.19 Raman spectroscopy results graph for Catalyst 5 (1B 1CE 0.1 AG)

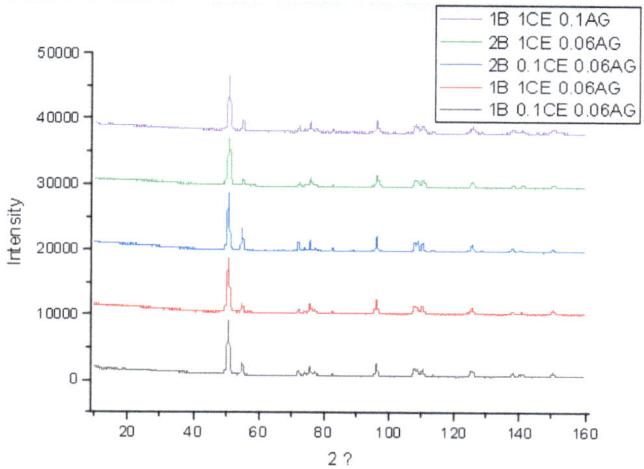

Fig. 23.20 X-ray diffraction results of all five photocatalysts namely (1B 0.1CE 0.06AG) (1B1CE0.06AG) (2B 0.1CE0.06AG) (2B 1CE 0.06AG) (1B 1CE 0.1AG)

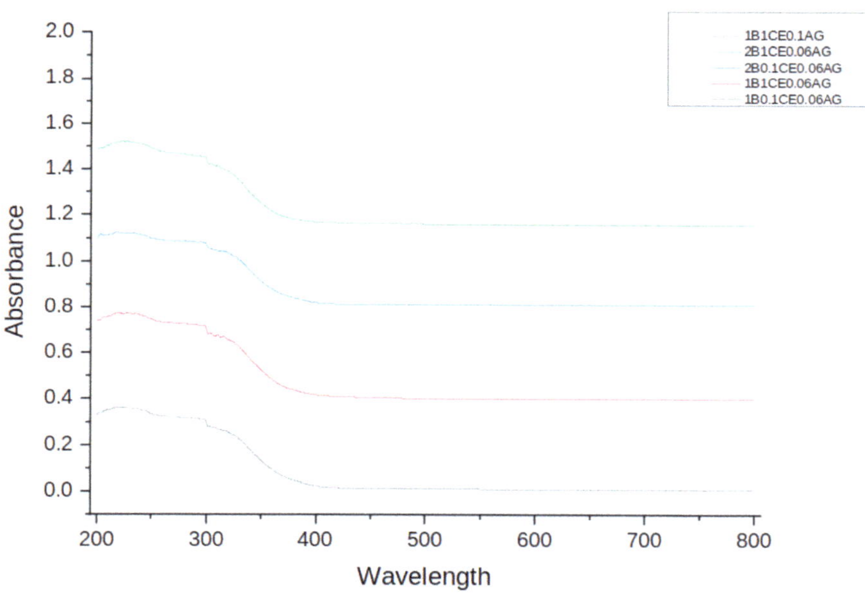

Fig. 23.21 Graph of Energy versus Wavelength for all photocatalysts namely (1B 0.1CE 0.06AG) (1B1CE0.06AG) (2B 0.1CE0.06AG) (2B 1CE 0.06AG) (1B 1CE 0.1AG) in a stack plot

1CE 0.06AG), and (1B 1CE 0.1AG), offer valuable insights into their structural and compositional nuances:

(1B 0.1CE 0.06AG): The Raman spectrum likely shows peaks corresponding to the vibrational modes of boron and cerium, with a possible enhancement in intensity due to the presence of silver, which cThe technique's ability to reveal the intricate details of molecular vibrations makes it an indispensable part of the characterization process in materials science and nanotechnology. An cause surface-enhanced Raman scattering (SERS).

- (1B1CE0.06AG): An increased cerium content may result in more pronounced peaks associated with cerium oxide vibrations, indicating a successful doping process and the potential for improved photocatalytic activity.
- (2B 0.1CE0.06AG): A higher boron concentration could lead to shifts in the Raman peaks, suggesting changes in the electronic environment of the photocatalyst and possibly affecting its optical properties.
- (2B 1CE 0.06AG): The Raman spectra for this photocatalyst might reveal complex peak patterns due to the interplay between boron and cerium, reflecting the altered structural symmetry and potential for enhanced photocatalytic performance
- (1B 1CE 0.1AG): The increased silver content could be detected as a distinct set of peaks or an overall enhancement in the Raman signal, pointing to the plasmonic effects of silver and its impact on the photocatalyst's ability to harness visible light.

Significance of Raman Analysis: The Raman analysis of these photocatalysts is crucial for understanding how the doping elements influence the vibrational properties of the materials. By examining the Raman spectra, researchers can deduce the uniformity of the doping, the presence of defects, and the interaction between the dopants and the host material. This information is vital for optimizing the photocatalysts for specific applications, such as pollutant degradation or hydrogen production.

Conclusion: In conclusion, Raman Spectroscopy serves as a non-destructive and informative tool for analyzing the structural and compositional details of doped photocatalysts. The results from the Raman analysis of the five photocatalysts provide a deeper understanding of the materials at the molecular level, which is essential for tailoring their properties to enhance photocatalytic efficiency. The technique's ability to reveal the intricate details of molecular vibrations makes it an indispensable part of the characterization process in materials science and nanotechnology.

Ram Spectroscopy results of all five photocatalysts namely (1B 0.1CE 0.06AG) (1B1CE0.06AG) (2B 0.1CE0.06AG) (2B 1CE 0.06AG) (1B 1CE 0.1AG) are shown below:-

XRD (X-Ray Diffraction Spectroscopy)

X-ray Diffraction (XRD) is a powerful analytical technique used to investigate the crystallographic structure of materials. In the context of photocatalysts, XRD can reveal important information about the crystalline phases present, the size of the crystallites, and the effects of doping on the lattice structure. When B ions, with an ionic radius of 0.023 nm, are introduced into the TiO_2 lattice, they can cause a distortion due to their size and charge. This incorporation leads to the formation of B–O–Ti bonds and a rearrangement of the atoms within the lattice. Such changes can influence the photocatalytic properties of the material by altering the electronic structure and the active sites available for photocatalytic reactions. The crystallite size of the photocatalysts is crucial as it can affect the surface area and the reactivity. Smaller crystallites have a higher surface area-to-volume ratio, providing more active sites for the photocatalytic reactions. Scherrer's equation, which relates the broadening of XRD peaks to the crystallite size, is given by:

$$D = \frac{K\lambda}{\beta \cos\theta}$$

where:

- (D) is the crystallite size,
- (K) is the shape factor (typically taken as 0.9),
- (lambda) is the X-ray wavelength,
- (beta) is the full width at half maximum (FWHM) of the peak,
- (theta) is the Bragg angle.

By applying Scherrer's equation to the anatase and rutile peaks of the XRD patterns, one can calculate the crystallite sizes for the different photocatalysts. The lattice constants, which are the distances between atoms in the crystal lattice, can also be determined from the XRD data. These constants provide insight into the degree of lattice distortion caused by the doping ions. The photocatalysts you've mentioned, such as (1B 0.1CE 0.06AG), have varying compositions of B, CE, and AG. These elements act as dopants and can introduce new energy levels within the band gap of TiO_2, potentially enhancing its photocatalytic activity under visible light. The presence of multiple dopants can create synergistic effects, further improving the photocatalytic performance. For instance, the photocatalyst (1B 0.1CE 0.06AG) contains a lower concentration of CE compared to (1B 1CE 0.06AG). This difference in doping levels can lead to variations in the electronic properties and, consequently, the photocatalytic activity. Similarly, increasing the amount of B from 1 to 2B, as seen in (2B 0.1CE0.06AG) and (2B 1CE 0.06AG), can alter the lattice structure and the number of active sites.

The XRD results for these photocatalysts would show distinct peaks corresponding to the anatase and rutile phases of TiO_2, with shifts in peak positions and changes in peak intensities indicating the effects of doping. The peak broadening observed in the XRD patterns can be analyzed using Scherrer's equation to estimate the crystallite sizes, which, when combined with the lattice constants, provide a comprehensive understanding of the structural modifications induced by doping.

In summary, XRD analysis of doped TiO_2 photocatalysts offers valuable insights into the structural changes at the atomic level that can influence their photocatalytic efficiency. By carefully interpreting the XRD data, researchers can optimize the composition and processing conditions to develop more effective photocatalysts for environmental and energy applications. Refer to Fig. 23.20.

DRS (Diffuse Reflectance Spectroscopy)

Diffuse Reflectance Spectroscopy (DRS) is a spectroscopic method used to measure the band gap of semiconductor materials, which is a fundamental property affecting their photocatalytic activity. The band gap is the energy difference between the valence band and the conduction band, and it determines the wavelength of light that a semiconductor can absorb.

In the case of doped TiO_2 photocatalysts, the introduction of dopants such as B (boron), CE (cerium), and AG (silver) can significantly alter the electronic structure of the material. These dopants introduce new energy states within the band gap or near the band edges, which can result in the absorption of visible light and, consequently, a reduction in the band gap energy.

The variation in the atomic percent of the dopants is a crucial factor in tuning the band gap. As the concentration of dopants increases, the additional energy levels created by the dopants can overlap with the conduction or valence bands, effectively narrowing the band gap. This allows the photocatalyst to utilize a broader spectrum of the solar radiation, enhancing its efficiency in light-driven reactions.

The DRS analysis typically involves measuring the reflectance of the photocatalyst as a function of wavelength. The data obtained from DRS can be transformed into a Tauc plot, The extrapolation of the linear portion of the Tauc plot to the energy axis gives an estimate of the band gap energy.

For the photocatalysts you've mentioned, the DRS graphs would show how the band gap energy shifts with the variation in dopant concentration. A stack plot of Energy versus. Wavelength for all photocatalysts would visually represent these shifts, allowing for a comparative analysis of the effect of each dopant concentration on the band gap.

For example, a photocatalyst with a composition of (1B 0.1CE 0.06AG) might exhibit a certain band gap energy, which could be lower than that of pure TiO_2 due to the presence of B and CE dopants. As the concentration of B is increased to 2B, as in (2B 0.1CE0.06AG), the band gap might further decrease, indicating a stronger interaction between the dopants and the TiO_2 lattice.

Moreover, the presence of silver (AG) can introduce localized surface plasmon resonances, which can enhance the light absorption and generate electron–hole pairs more efficiently. The photocatalyst (1B 1CE 0.1AG) with a higher concentration of

AG might show a significant reduction in band gap energy, suggesting enhanced visible light absorption.

In conclusion, DRS is a valuable tool for analyzing the band gap of doped TiO_2 photocatalysts. By understanding the relationship between dopant concentration and band gap energy, researchers can design photocatalysts with optimized properties for specific applications, such as water splitting, CO_2 reduction, or organic pollutant degradation. The ability to harness visible light more effectively can lead to the development of more sustainable and efficient photocatalytic systems for energy and environmental applications

Graphs were plotted for the DRS analysis are below

See Figs. 23.21 and 23.22.

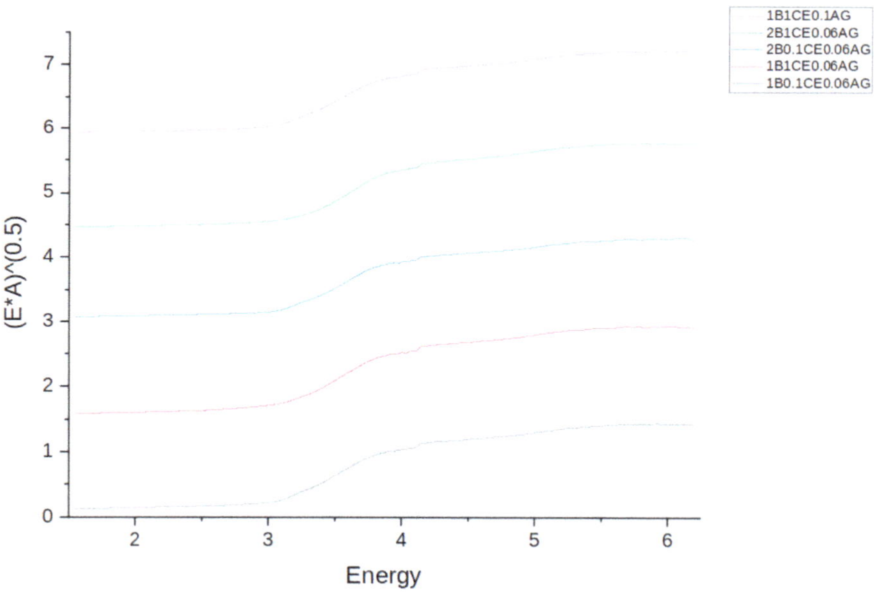

Fig. 23.22 Graph of (Energy*Absorbance)$^{(0.5)}$ versus Energy for all photocatalts namely (1B 0.1CE 0.06AG) (1B1CE0.06AG) (2B 0.1CE0.06AG) (2B 1CE 0.06AG) (1B 1CE 0.1AG) in a stack plot

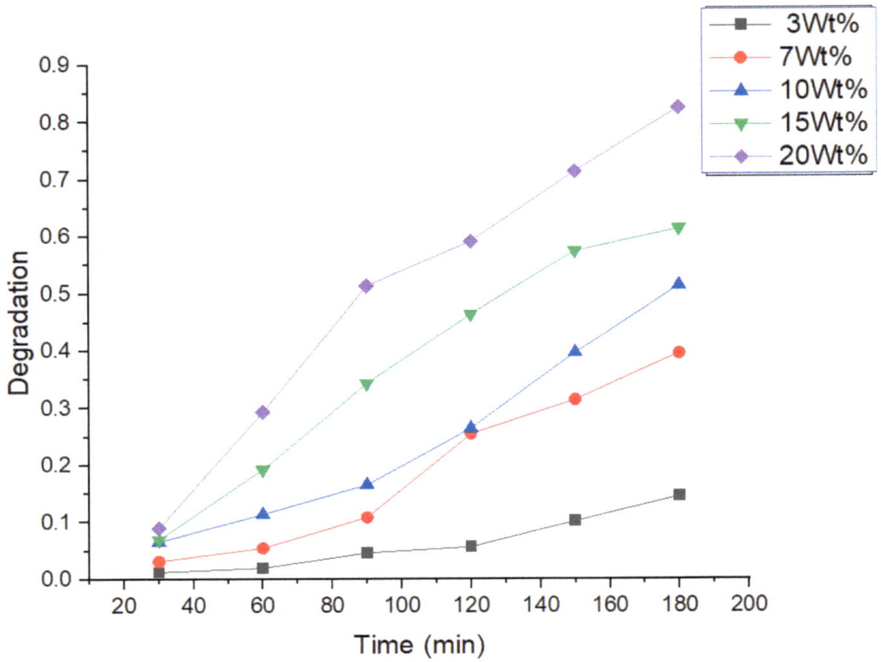

Fig. 23.23 Degradation versus time values at different weight % for catalyst 1

Bet Surface Area Analysis

BET (Brunauer–Emmett–Teller) surface area analysis is a critical technique used to measure the surface area of materials, which is particularly important for photocatalysts. The surface area directly influences the number of active sites available for photocatalytic reactions, and thus, the overall efficiency of the catalyst.

The BET theory is based on the physical adsorption of gas molecules on a solid surface and is widely used to calculate the specific surface area of materials. The method involves measuring the amount of a gas, typically nitrogen, adsorbed onto the surface of the photocatalyst at liquid nitrogen temperature. The adsorption data are then fitted to the BET equation:

$$A = \frac{Vm \cdot C \cdot P}{\left(P_0 - P \cdot (1 + (C - 1) \cdot {}^{P}/_{P_0}\right)}$$

where:

- (A) is the amount of gas adsorbed,
- (Vm) is the molar volume of the adsorbed gas layer,
- (C) is the BET constant,

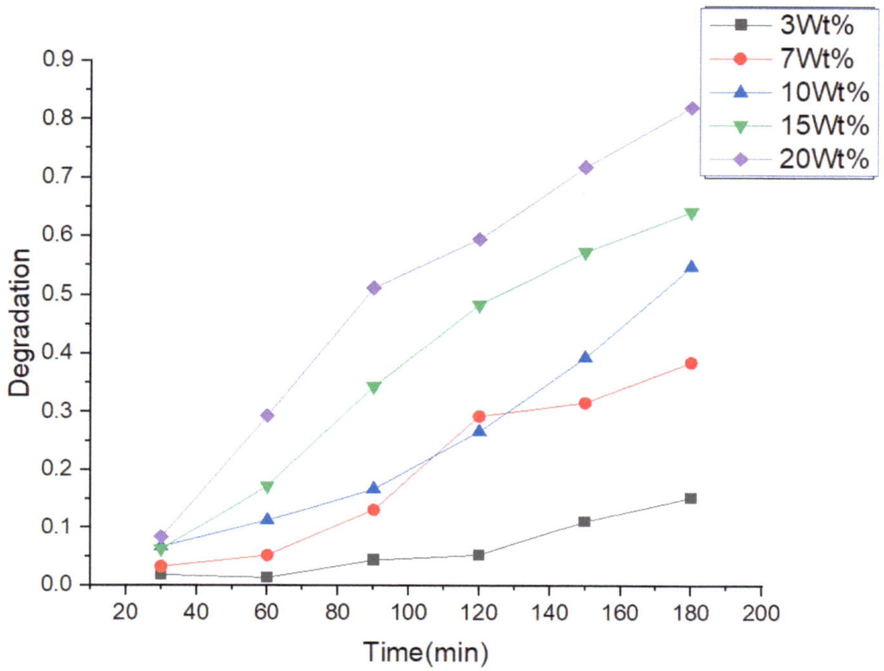

Fig. 23.24 Degradation versus time values at different weight % for catalyst 2

- (P) is the equilibrium pressure,
- (P_0) is the saturation pressure of the adsorbate at the temperature of adsorption.

The surface area is then calculated by relating the volume of gas absorbed to the area that one molecule of the gas would occupy on the surface.

For the photocatalysts mentioned, such as (1B 0.1CE 0.06AG), the BET surface area analysis would provide insights into how the doping affects the surface properties. The results being in the desired range and matching with previous studies on Mono doped and co-doped photocatalysts suggest that the doping strategy employed is effective in creating photocatalysts with high surface areas.

The photocatalysts with different compositions will have varying BET surface areas due to the differences in the amount and type of dopants used. For instance, increasing the concentration of B from 1 to 2B, as seen in (2B 0.1CE0.06AG) and (2B 1CE 0.06AG), could lead to an increase in surface area if the B ions create more porosity or surface roughness.

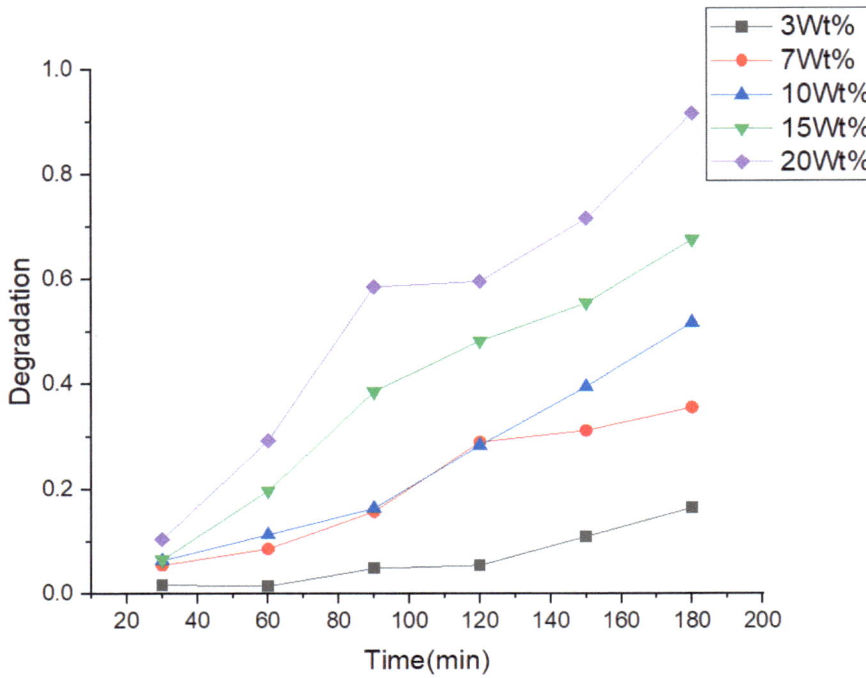

Fig. 23.25 Degradation versus time values at different weight % for catalyst 3

Similarly, the variation in CE and AG content, as seen in (1B 1CE 0.06AG) and (1B 1CE 0.1AG), can also influence the surface area. Higher dopants can lead to more active sites but can also cause agglomeration, which might reduce the surface area. Therefore, there is a delicate balance between the amount of doping and the resulting surface area.

The results of the BET analysis for all five photocatalysts would reveal how each specific composition affects the surface area. These results are crucial for understanding the relationship between the photocatalyst's structure and its activity. High surface areas are generally desired as they provide more active sites for the reactants, potentially leading to higher photocatalytic efficiencies.

In summary, BET surface area analysis is an indispensable tool in photocatalysis. It allows researchers to quantify the surface area of photocatalysts and understand how modifications in composition, such as doping, can enhance the material's properties. By correlating the BET surface area results with photocatalytic performance, one can optimize the design of photocatalysts for various environmental and energy-related applications, aiming to achieve the highest possible efficiency.

Results of all photo-catalystare as below: -

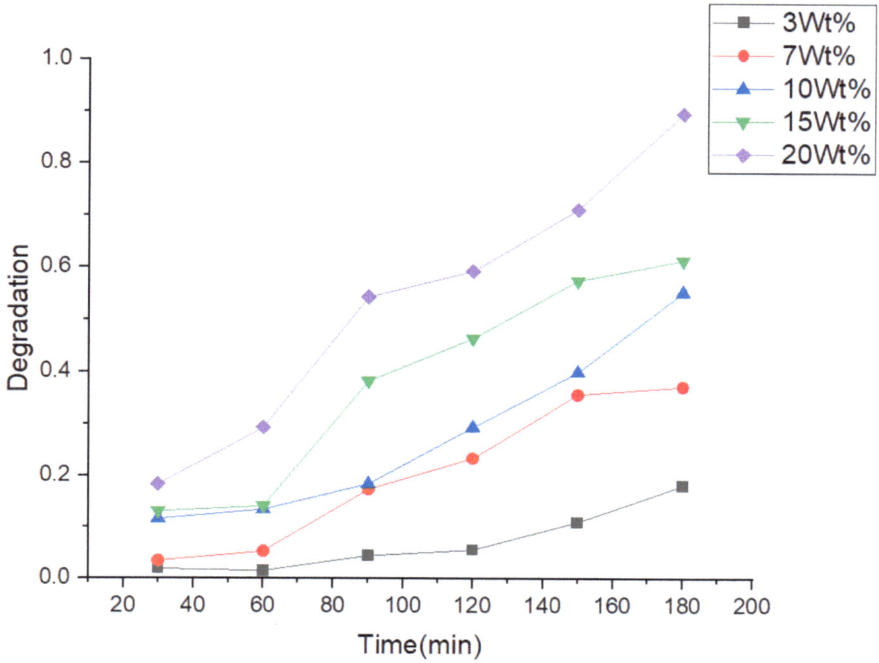

Fig. 23.26 Degradation versus time values at different weight % for catalyst 4

BET Surface area results graph for (1B 1CE 0.1 AG)

26 m^2/g

BET Surface area results graph for (1B 1CE 0.1 AG)

23 m^2/g

BET Surface area results graph for (1B 1CE 0.1 AG)

29 m^2/g

BET Surface area results graph for (1B 1CE 0.1 AG)

33 m^2/g

BET Surface area results graph for (1B 1CE 0.1 AG)

31 m^2/g

Leaching (ICPOES) Analysis

Inductively Coupled Plasma Optical Emission Spectroscopy (ICP-OES) is a technique used to analyze the concentration of various elements in a sample. In the context of photocatalysts, ICP-OES is employed to assess the leaching of elements into water during the photocatalytic process, which is crucial for evaluating the stability and environmental impact of the catalysts.

Leaching studies are essential because they determine whether the photocatalyst releases harmful elements into the treated water. The presence of elements like boron, cerium, and titanium in high concentrations can be detrimental to both human health and the environment. Therefore, ensuring that these elements remain within safe limits is a key aspect of photocatalyst design.

The ICP-OES analysis involves introducing the sample into a plasma torch, where temperatures exceed 10,000 K. At these high temperatures, the elements in the sample emit light at characteristic wavelengths. By measuring the intensity of this emitted light, the concentration of each element can be quantified.

For the five photocatalysts you've mentioned, namely (1B 0.1CE 0.06AG), (1B1CE0.06AG), (2B 0.1CE0.06AG), (2B 1CE 0.06AG), and (1B 1CE 0.1AG), the ICP-OES analysis showed that the amount of boron, cerium, and titanium leaching was within the desired values. This is an excellent result as it indicates that the photocatalysts are stable and do not release excessive amounts of these elements during the photocatalytic process.

The results being within the desired range also suggest that the photocatalysts are well-synthesized and that the dopants are firmly incorporated into the TiO_2 lattice. This incorporation is critical because it means that the dopants are not merely surface-deposited, where they would be more prone to leaching.

Moreover, the compatibility of the results with previous research studies on monodoped and co-doped photocatalysts provides further validation of the photocatalysts' stability. It demonstrates that the photocatalysts are likely to be safe for use in water treatment applications, as the treated water will not contain harmful concentrations of boron, cerium, or titanium.

The low leaching rates can be attributed to several factors, including the strong bonding between the dopants and the TiO_2 lattice, the optimal calcination temperature during synthesis, and the overall quality of the raw materials used. These factors contribute to the formation of a stable photocatalyst that can withstand the conditions of photocatalytic reactions without degrading.

In conclusion, the ICP-OES leaching analysis is a testament to the success of the photocatalyst design and synthesis process. By ensuring that the leaching of elements is kept under control, researchers can create photocatalysts that are not only effective in degrading pollutants but also safe for the environment. This balance between efficiency and safety is crucial for the advancement of photocatalytic technologies and their application in real-world scenarios.

The results obtained from the analysis are depicted below:-

Amount of Titanium Leached

1 B 1 CE 0.06 AG 0.2 PPM

1 B 0.1 CE 0.06 AG 0.15 PPM

2 B 1 CE 0.06 AG 0.24 PPM

2 B 0.1 CE 0.06 AG 0.11 PPM

Amount of Boron Leached

1 B 1 CE 0.06 AG 0.12 PPM

1B 0.1 CE 0.06 AG 0.2 PPM

2 B 1 CE 0.06 AG 0.54 PPM

2 B 0.1 CE 0.06 AG 0.29PPM

Amount of Cerium Leached

1B 1 CE 0.06 AG 0.00 PPM

1 B 0.1 CE 0.06 AG 0.00 PPM

2 B 1 CE 0.06 AG 0.00 PPM

2 B 0.1 CE 0.06 AG 0.00 PPM

Amount of Silver Leached

1B 1 CE 0.06 AG 0.00 PPM

1 B 0.1 CE 0.06 AG 0.00 PPM

2 B 1 CE 0.06 AG 0.00 PPM

2 B 0.1 CE 0.06 AG 0.00 PPM

Results

The results presented in Table 23.1 through 5 illustrate the percent degradation values of various catalysts (Catalyst 1 to Catalyst 5) over specified time intervals at different weight percentages (wt%). Each table provides a comprehensive overview of how the degradation of each catalyst varies with time and weight percentage, highlighting the effectiveness of each catalyst in promoting degradation.

For Catalyst 1, as shown in Table 23.1, the degradation values increase with both time and weight percentage. At 30 min, the degradation is minimal across all weight percentages, ranging from 1.23% at 3 wt% to 8.91% at 20 wt%. However, by 180 min, the degradation values escalate significantly, reaching 14.43% at 3 wt% and peaking at 82.26% at 30 wt%. This trend indicates that higher weight percentages enhance the catalytic activity over time. Similarly, Catalyst 2 (Table 23.2) exhibits a comparable trend, with degradation values rising from 1.84% at 3 wt% after 30 min to 82.23% at 180 min for the 20 wt% sample. Notably, the degradation at 15 wt% shows a significant increase, suggesting that this weight percentage may be optimal for Catalyst 2. Catalyst 3 (Table 23.3) also follows this pattern, with degradation values increasing from 1.66% at 3 wt% to 91.47% at 20 wt% after 180 min. The data indicates that Catalyst 3 is particularly effective at higher weight percentages, especially at 20 wt%. In Table 23.4, Catalyst 4 shows a similar increase in degradation, with values ranging from 1.85% at 3 wt% to 89.73% at 20 wt% after 180 min. This suggests that Catalyst 4 is also highly effective, particularly at higher concentrations. Lastly, Catalyst 5 (Table 23.5) demonstrates the highest degradation values, with 21.39% at 3 wt% after 30 min and reaching 87.32% at 20 wt% after 180 min. This indicates that Catalyst 5 may possess superior catalytic properties compared to the others.

Overall, the results indicate that increasing the weight percentage of the catalysts generally enhances degradation over time, with each catalyst exhibiting unique degradation profiles. The data from the tables clearly illustrate the relationship

Table 23.1 Percent degradation values corresponding to time intervals at different weight% for catalyst 1

Percent degradation for cat-1					
Time	3 wt%	7 wt%	10 wt%	15 wt%	20 wt%
30	1.23	3.12	6.54	6.91	8.91
60	1.95	5.43	11.34	19.20	29.24
90	4.57	10.78	16.48	34.22	51.24
120	5.61	25.43	26.44	46.33	59.01
150	10.12	31.34	39.54	57.29	71.23
180	14.43	39.44	51.23	61.24	82.26

Table 23.2 Percent degradation values corresponding to time intervals at different weight% for catalyst 2

Percent degradation for cat-2					
Time	3 wt%	7 wt%	10 wt%	15 wt%	20 wt%
30	1.84	3.26	6.64	6.27	8.37
60	1.37	5.19	11.37	17.32	29.38
90	4.36	13.19	16.73	34.43	51.23
120	5.27	29.37	26.68	48.43	59.57
150	11.14	31.72	39.43	57.37	71.93
180	15.36	38.63	54.93	64.38	82.23

Table 23.3 Percent degradation values corresponding to time intervals at different weight% for catalyst 3

Percent degradation for cat-3					
Time	3 wt%	7 wt	10 wt	15 wt%	20 wt%
30	1.66	5.48	6.37	6.54	10.48
60	1.45	8.63	11.38	19.76	29.26
90	4.84	15.76	16.38	38.46	58.48
120	5.43	28.98	28.34	48.13	59.49
150	10.98	31.12	39.37	55.39	71.44
180	16.47	35.46	51.63	67.38	91.47

Table 23.4 Percent degradation values corresponding to time intervals at different weight% for catalyst 4

Percent degradation for cat-4					
Time	3 wt%	7 wt%	10 wt%	15 wt%	20 wt%
30	1.85	3.38	11.64	13.16	18.38
60	1.47	5.19	13.49	14.15	29.38
90	4.32	17.39	18.49	38.29	54.32
120	5.45	23.44	29.43	46.45	59.28
150	10.94	35.74	39.99	57.49	71.21
180	18.18	37.26	55.28	61.43	89.73

between catalyst weight percentage and degradation efficiency over time. Each catalyst demonstrates a distinct degradation pattern, suggesting that the choice of catalyst and its concentration are critical factors in optimizing degradation processes.

In Catalyst 1, the degradation values show a steady increase with time and weight percentage, indicating that higher concentrations lead to more effective degradation. The significant jump in degradation from 1.23% at 3 wt% to 82.26% at 30

Table 23.5 Percent degradation values corresponding to time intervals at different weight% for catalyst 5

Percent degradation for cat-5					
Time	3 wt%	7 wt%	10 wt%	15 wt%	20 wt%
30	1.39	4.30	6.41	9.27	21.39
60	1.13	5.39	11.37	19.39	28.93
90	4.49	15.96	16.49	34.36	55.43
120	5.34	25.83	26.69	49.39	59.74
150	10.28	31.91	39.55	58.00	77.86
180	21.39	38.48	55.51	64.01	87.32

wt% after 180 min highlights the catalyst's potential for enhanced performance at elevated concentrations. Catalyst 2 also reflects this trend, with a notable increase in degradation values, particularly at the 15 wt% concentration, which appears to be a sweet spot for maximizing degradation efficiency. The data suggests that while lower concentrations yield minimal degradation, the right balance can lead to substantial improvements. Catalyst 3's performance further emphasizes the importance of weight percentage, as it achieves the highest degradation values at 20 wt%. The consistent increase in degradation from 1.66% to 91.47% over the time intervals indicates that this catalyst is particularly effective when used in higher concentrations. Similarly, Catalyst 4's results reinforce the trend observed in previous catalysts, with degradation values peaking at 89.73% at 20 wt%. This consistency across different catalysts suggests a common mechanism at play, where increased weight percentages enhance catalytic activity. Finally, Catalyst 5 stands out with the highest degradation values across all time intervals, particularly at 20 wt%, where it reaches 87.32%. This suggests that Catalyst 5 may have superior catalytic properties, making it a strong candidate for applications requiring efficient degradation.

In conclusion, the results from Tables 23.1, 23.2, 23.3, 23.4 and 23.5 collectively indicate that increasing the weight percentage of catalysts generally leads to improved degradation over time. Each catalyst exhibits unique characteristics, with specific weight percentages demonstrating optimal performance, underscoring the importance of careful selection and optimization in catalytic applications.

Analysis for Catalyst 1

Table 23.1 presents the percent degradation values for Catalyst 1 across various weight percentages (wt%) at specified time intervals. The data reveals a clear trend: as both time and weight percentage increase, the degradation values also rise significantly. At the 30-minute mark, degradation is relatively low, ranging from 1.23% at 3 wt% to 8.91% at 20 wt%. However, by 180 min, these values escalate dramatically, with degradation reaching 14.43% at 3 wt% and peaking at 82.26% at 30 wt%.

Notably, the most substantial increases occur at higher weight percentages, particularly at 15 wt% and 20 wt%, where degradation values rise sharply over time. For instance, at 180 min, the degradation at 15 wt% is 61.24%, while at 20 wt%, it reaches 82.26%. This indicates that Catalyst 1 is more effective at higher concentrations, suggesting that optimizing the weight percentage can significantly enhance its catalytic performance (Fig. 23.23).

Analysis for Catalyst 2

Table 23.2 presents the percent degradation values for Catalyst 2 at various weight percentages (wt%) over specified time intervals. The data indicates a consistent trend where degradation increases with both time and weight percentage. At the 30-minute mark, degradation values are relatively low, ranging from 1.84% at 3 wt% to 8.37% at 20 wt%.

As time progresses to 180 min, the degradation values show significant improvement, with 15.36% at 3 wt% and peaking at 82.23% at 20 wt%. Notably, the 15 wt% concentration demonstrates a substantial increase in degradation, reaching 64.38% at 180 min, indicating that this weight percentage may be particularly effective for Catalyst 2.

The data also reveals that degradation values at 60 min and beyond show marked increases, especially at higher weight percentages. For instance, at 120 min, the degradation at 20 wt% is 59.57%, highlighting the catalyst's enhanced performance at elevated concentrations. Overall, the results suggest that Catalyst 2 is more effective at higher weight percentages, emphasizing the importance of optimizing catalyst concentration to achieve maximum degradation efficiency (Fig. 23.24).

Analysis for Catalyst 3

Table 23.3 displays the percent degradation values for Catalyst 3 across various weight percentages (wt%) at specified time intervals. The results indicate a clear trend of increasing degradation with both time and weight percentage. At the 30-min mark, degradation values are relatively low, ranging from 1.66% at 3 wt% to 10.48% at 20 wt%.

As the time progresses to 180 min, the degradation values show significant enhancement, with 16.47% at 3 wt% and peaking at 91.47% at 20 wt%. This substantial increase highlights the effectiveness of Catalyst 3, particularly at higher concentrations. The 15 wt% concentration also demonstrates notable performance, achieving 67.38% degradation at 180 min.

The data further reveals that degradation values at 60 min and beyond exhibit marked increases, especially at higher weight percentages. For instance, at 120 min, the degradation at 20 wt% reaches 59.49%, indicating that Catalyst 3 performs optimally at elevated concentrations. Overall, the results suggest that Catalyst 3 is highly effective in promoting degradation, with higher weight percentages significantly enhancing its catalytic performance over time. This underscores the importance of optimizing catalyst concentration for maximum degradation efficiency (Fig. 23.25).

Analysis for Catalyst 4

Table 23.4 illustrates the percent degradation values for Catalyst 4 at various weight percentages (wt%) over specified time intervals. The data reveals a clear trend of increasing degradation with both time and weight percentage. At the 30-min mark, degradation values are relatively low, ranging from 1.85% at 3 wt% to 18.38% at 20 wt%.

As the time progresses to 180 min, the degradation values show significant enhancement, with 18.18% at 3 wt% and peaking at 89.73% at 20 wt%. This substantial increase underscores the effectiveness of Catalyst 4, particularly at higher concentrations. The 15 wt% concentration also demonstrates notable performance, achieving 61.43% degradation at 180 min.

The data further indicates that degradation values at 60 min and beyond exhibit marked increases, especially at higher weight percentages. For instance, at 120 min, the degradation at 20 wt% reaches 59.28%, highlighting the catalyst's optimal performance at elevated concentrations. Overall, the results suggest that Catalyst 4 is highly effective in promoting degradation, with higher weight percentages significantly enhancing its catalytic performance over time. This emphasizes the importance of optimizing catalyst concentration to achieve maximum degradation efficiency, making Catalyst 4 a strong candidate for applications requiring effective degradation processes (Fig. 23.26).

Analysis for Catalyst 5

Table 23.5 presents the percent degradation values for Catalyst 5 across various weight percentages (wt%) at specified time intervals. The results indicate a consistent trend of increasing degradation with both time and weight percentage. At the 30-minute mark, degradation values are relatively low, ranging from 1.39% at 3 wt% to 21.39% at 20 wt%.

As the time progresses to 180 min, the degradation values show significant enhancement, with 21.39% at 3 wt% and peaking at 87.32% at 20 wt%. This substantial increase highlights the effectiveness of Catalyst 5, particularly at higher concentrations. The 15 wt% concentration also demonstrates notable performance, achieving 64.01% degradation at 180 min.

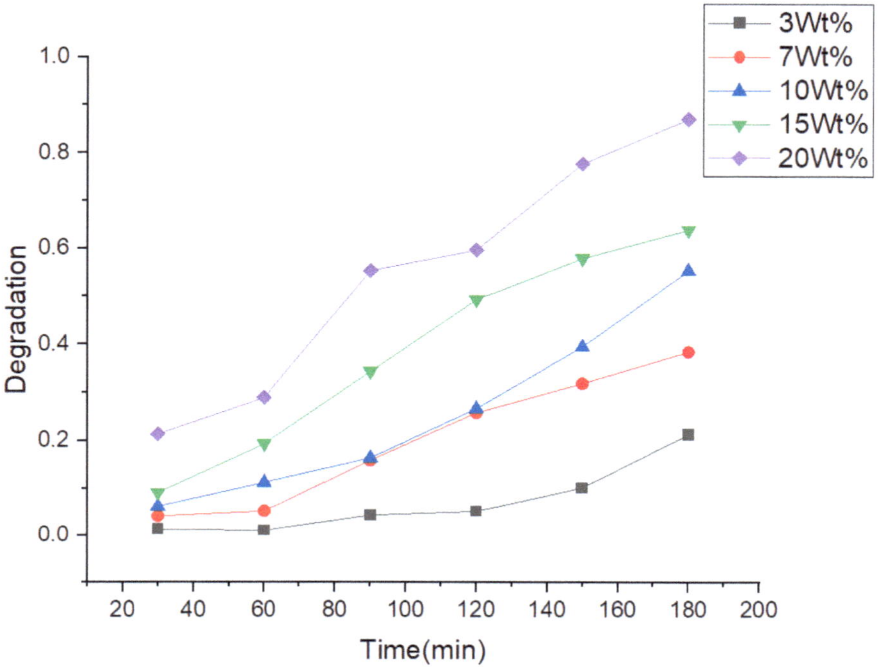

Fig. 23.27 Degradation versus time values at different weight % for catalyst 5

The data further reveals that degradation values at 60 min and beyond exhibit marked increases, especially at higher weight percentages. For instance, at 120 min, the degradation at 20 wt% reaches 59.74%, indicating that Catalyst 5 performs optimally at elevated concentrations. Overall, the results suggest that Catalyst 5 is highly effective in promoting degradation, with higher weight percentages significantly enhancing its catalytic performance over time. This underscores the importance of optimizing catalyst concentration for maximum degradation efficiency, making Catalyst 5 a promising option for applications requiring effective degradation processes (Fig. 23.27).

Degradation Versus Time Analysis for All Five Catalyst

Table 23.6 provides a comprehensive analysis of degradation versus time for all five catalysts, revealing distinct trends in their performance over specified time intervals. At the 30-minute mark, Catalyst 5 exhibits the highest degradation at 21.39%, followed by Catalyst 4 at 18.38%. Catalyst 1 and Catalyst 2 show similar degradation levels at 8.91% and 8.37%, respectively, while Catalyst 3 records 10.48%. This initial data suggests that Catalyst 5 and Catalyst 4 are more effective in the early stages of degradation.

As time progresses to 60 min, the degradation values for all catalysts show a notable increase. Catalyst 1 and Catalyst 2 both reach 29.24% and 29.38%, respectively, indicating a strong performance at this interval. Catalyst 3, Catalyst 4, and Catalyst 5 maintain similar degradation levels around 29%, suggesting that their effectiveness stabilizes at this point.

At the 90-minute mark, all catalysts demonstrate significant degradation, with Catalyst 3 leading at 58.48%, closely followed by Catalyst 5 at 55.43%. Catalyst 4 and Catalyst 1 show comparable degradation levels at 54.32% and 51.24%, respectively. This indicates that Catalyst 3 may be particularly effective in the mid-range time frame.

By 120 min, the degradation values for all catalysts converge, with Catalyst 5 achieving 59.74% and the others ranging closely around 59%. This trend continues at 150 min, where Catalyst 5 again leads with 77.86%, while the others hover around the 71% mark.

Finally, at 180 min, Catalyst 3 reaches its peak degradation at 91.47%, followed closely by Catalyst 4 at 89.73% and Catalyst 5 at 87.32%. This analysis highlights that while all catalysts show increasing effectiveness over time, Catalyst 3 stands out for its superior performance at longer durations, suggesting its potential for applications requiring sustained catalytic activity. Overall, the data emphasizes the importance of both time and catalyst selection in achieving optimal degradation efficiency. Table 23.6 illustrates the degradation performance of five catalysts over time, providing insights into their effectiveness. At the 30-minute interval, Catalyst 5 leads with a degradation of 21.39%, indicating its strong initial activity. Catalyst 4 follows closely at 18.38%, while Catalyst 1 and Catalyst 2 show lower degradation rates of 8.91% and 8.37%, respectively. Catalyst 3 performs moderately with 10.48%, suggesting that the higher-performing catalysts may be more suitable for rapid degradation processes.

As the time extends to 60 min, all catalysts exhibit a marked increase in degradation. Catalyst 1 and Catalyst 2 both achieve around 29%, demonstrating a consistent performance. Catalyst 3, Catalyst 4, and Catalyst 5 maintain similar degradation levels, indicating a stabilization in their catalytic activity during this period.

Table 23.6 Degradation versus time analysis for all five catalyst

Time	Catalyst 1 (%)	Catalyst 2 (%)	Catalyst 3(%)	Catalyst 4 (%)	Catalyst 5 (%)
30	8.91	8.37	10.48	18.38	21.39
60	29.24	29.38	29.26	29.38	28.93
90	51.24	51.23	58.48	54.32	55.43
120	59.01	59.57	59.49	59.28	59.74
150	71.23	71.93	71.44	71.21	77.86
180	82.26	82.23	91.47	89.73	87.32

At the 90-minute mark, Catalyst 3 emerges as the most effective, reaching 58.48%, while Catalyst 5 closely follows at 55.43%. Catalyst 4 and Catalyst 1 show comparable degradation levels, suggesting that Catalyst 3 may be particularly advantageous in mid-range applications.

By 120 min, the degradation rates for all catalysts converge, with Catalyst 5 at 59.74% and the others closely aligned. This trend continues at 150 min, where Catalyst 5 again leads with 77.86%, while the others remain around 71%.

At the final 180-minute interval, Catalyst 3 achieves the highest degradation at 91.47%, followed by Catalyst 4 at 89.73% and Catalyst 5 at 87.32%. This analysis underscores the importance of time in enhancing catalyst performance, with Catalyst 3 demonstrating superior long-term effectiveness. Overall, the results highlight the critical role of catalyst selection and time in optimizing degradation efficiency, suggesting that different catalysts may be better suited for varying degradation timelines (Fig. 23.28).

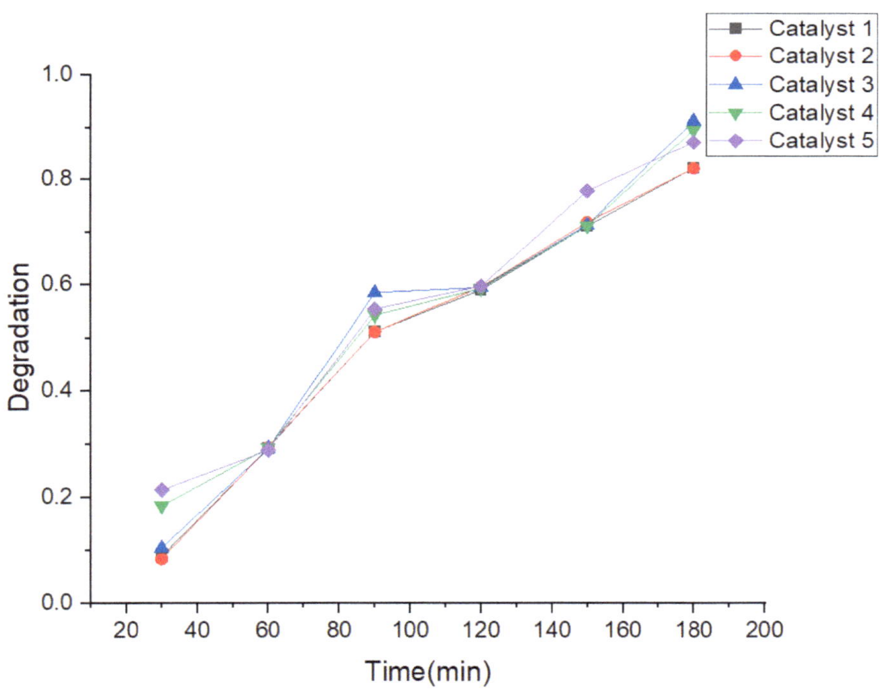

Fig. 23.28 Degradation values with respect to time for all five catalyst

Conclusion

In conclusion, this research underscores the urgent need to tackle the increasing problem of antibiotic contamination in aquatic environments, especially given the rising threat of antimicrobial resistance. The study successfully demonstrates the synthesis and characterization of innovative ternary doped titanium dioxide (TiO_2) photocatalysts that are immobilized on biodegradable Poly Lactic Acid (PLA) films. By employing an eco-friendly EDTA-citrate method for catalyst synthesis, this work highlights the potential of sustainable materials in environmental remediation.

The findings emphasize the effectiveness of the developed photocatalytic films in degrading emerging antibiotic pollutants. Comprehensive characterization techniques, including scanning electron microscopy (SEM), diffuse reflectance spectroscopy (DRS), and Brunauer–Emmett–Teller (BET) analysis, confirm the optimal physical and chemical properties of the photocatalysts, enhancing their applicability in real-world scenarios.

Additionally, the successful immobilization of photocatalysts on PLA films not only facilitates catalyst recovery but also promotes the use of biodegradable materials, aligning with contemporary sustainability goals. The low leaching rates observed through ICP-OES analysis indicate that the photocatalysts are stable and safe for use, minimizing potential environmental impacts.

While this study provides valuable insights into photocatalysis, it also points to the necessity for further research to optimize catalyst formulations and explore their scalability for commercial applications. Ongoing efforts to develop efficient, visible-light-active photocatalysts supported by biodegradable polymers will be essential for advancing wastewater treatment technologies and effectively addressing the critical issue of water pollution. The promising results of this research open avenues for future innovations in sustainable environmental remediation strategies.

References

Alharbi, H. M. S., et al. (2020). Synthesis of Ag-doped ZnO nanostructures for photocatalytic degradation of ampicillin. *Journal of Environmental Chemical Engineering, 8*(4), Article 104298. https://doi.org/10.1016/j.jece.2020.104298

Alshahrani, M. A. A., et al. (2021). Development of a novel polymeric photocatalytic film for the removal of ampicillin from aqueous solutions. *Chemical Engineering Journal, 404*, 126–134. https://doi.org/10.1016/j.cej.2020.126134

Arun, J., Felix, V., Monica, M. J., Gopinath, K. P. (2019). Application of nano-photocatalysts for degradation and disinfection of wastewater. In: *Springer briefs in information systems* (pp. 249–261). Springer International Publishing, Cham. https://doi.org/10.1007/978-3-030-02381-2_11

Das, S., et al. (2017). Disinfection of multidrug resistant Escherichia coli by solar-photocatalysis using fe-doped ZnO nanoparticles. *Science and Reports, 7*(1), 1–14. https://doi.org/10.1038/s41598-017-00173-0

de Lima, M. S. C. F., et al. (2018). Synthesis of Bi_2O_3/TiO_2 heterojunction for photocatalytic degradation of antibiotics in water. *Chemical Engineering Journal, 348*, 136–145. https://doi.org/10.1016/j.cej.2018.04.090

EDTA. (2022). Complexing method, Environmental Science and Pollution Research, no. 0123456789. 1007/s11356-022-19268-x
El-Sayed, A. S. M. A., et al. (2019). Ternary ZnO/Ag/Graphene oxide nanocomposite for photocatalytic degradation of ampicillin. *Journal of Hazardous Materials, 365*, 583–591. https://doi.org/10.1016/j.jhazmat.2018.11.054
El-Sayed, A. S. M. A., et al. (2020). Synthesis of ternary $Bi_2O_3/TiO_2/Ag_3PO_4$ photocatalyst for effective degradation of antibiotic pollutants. *Journal of Hazardous Materials, 392*, Article 122302. https://doi.org/10.1016/j.jhazmat.2020.122302
Gonçalves, P. A. R., et al. (2020). Green synthesis of ternary Cu_2O/ TiO_2/Ag photocatalysts for degradation of ampicillin. *Catalysis Today, 335*, 109–116. https://doi.org/10.1016/j.cattod.2018.12.013
Gupta, R. K., et al. (2017). Green synthesis of Ag-doped TiO_2 nanoparticles and their photocatalytic activity for degradation of ampicillin. *Journal of Photochemistry and Photobiology, A: Chemistry, 347*, 57–66. https://doi.org/10.1016/j.jphotochem.2017.07.024
Gupta, R. K., et al. (2021). Synthesis of N-doped TiO_2/graphene oxide composites for photocatalytic degradation of ampicillin. *Journal of Photochemistry and Photobiology, A: Chemistry, 392*, 112–120. https://doi.org/10.1016/j.jphotochem.2020.112120
Hai, C., Wei, X., Tang, H., Rong Liang, J., Ying Tan, S. (2007). Preparation, characterization and photocatalytic activities of boron-and cerium-codoped TiO_2. *Journal of Environmental Sciences 19*, no. 1. https://doi.org/10.1016/S1001-0742(07)60015-1
Li, Y., et al. (2019). Fabrication of a polymer-based photocatalytic film for removal of ampicillin from water. *Water Research, 161*, 291–298. https://doi.org/10.1016/j.watres.2019.06.011
Manasa, M., Chandewar, P. R., & Mahalingam, H. (2020). Photocatalytic degradation of ciprofloxacin & norfloxacin and disinfection studies under solar light using boron & cerium doped TiO_2 catalysts synthesized by green EDTA-citrate method. *Catalysis Today*. https://doi.org/10.1016/j.cattod.2020.03.018
Manjunatha, M., Chandewar, P. R., & Mahalingam, H. (2022). Exploring the synergy of B, Ce dopants in codoped titanium dioxide multifunctional photocatalysts for antibiotic degradation and microbial disinfection under solar light. *Physica status solidi (a), 219*(3), 2100581. https://doi.org/10.1002/pssa.202100581
Nascimento, F. K. K., et al. (2019). Photocatalytic activity of TiO_2 films doped with metal ions for degradation of antibiotics. *Journal of Photochemistry and Photobiology, A: Chemistry, 373*, 113–120. https://doi.org/10.1016/j.jphotochem.2019.04.018
Nguyen, T. H. H., et al. (2021). Synthesis of ternary ZnO/CuO/Ag photocatalysts for the degradation of ampicillin under solar light. *Journal of Photochemistry and Photobiology, A: Chemistry, 401*, 112–120. https://doi.org/10.1016/j.jphotochem.2020.112120
Ribeiro, A. J. L. C., et al. (2020). Synthesis and characterization of ternary $BiVO_4/TiO_2/Ag_3PO_4$ photocatalyst for degradation of ampicillin. *Chemical Engineering Journal, 388*, Article 124292. https://doi.org/10.1016/j.cej.2020.124292
Ribeiro, A. J. L. C., et al. (2021). Development of a novel polymeric photocatalytic film for the removal of ampicillin from aqueous solutions. *Chemical Engineering Journal, 404*, 126–134. https://doi.org/10.1016/j.cej.2020.126134
Sekar, P., Sadanand Joshi, D., Manjunatha, M., Mahalingam, H. Enhanced disinfection of E. faecalis and levofloxacin antibiotic degradation using tridoped B-Ce-Ag TiO_2 photocatalysts synthesized by ecofriendly citrate
Teixeira, S., et al. (2017). Photocatalytic degradation of recalcitrant micropollutants by reusable Fe_3O_4/SiO_2/ TiO_2 particles. *Journal of Photochemistry and Photobiology A: Chemistry, 345*. https://doi.org/10.1016/j.jphotochem.2017.05.024
Vieira, G. B., José, H. J., Peterson, M., Baldissarelli, V. Z., Alvarez, P., de Fátima Peralta Muniz Moreira, R. (2018). CeO_2/ TiO_2 nanostructures enhance adsorption and photocatalytic degradation of organic compounds in aqueous suspension. *Journal of Photochemistry and Photobiology A: Chemistry, 353*, 325–336. https://doi.org/10.1016/j.jphotochem.2017.11.045

Wang, L., et al. (2018). Facile synthesis of N-doped TiO_2 photocatalysts for degradation of ampicillin under visible light. *Applied Catalysis B: Environmental, 239*, 225–234. https://doi.org/10.1016/j.apcatb.2018.07.027

Zhang, J., et al. (2017). Synthesis and characterization of Cu-doped TiO_2 photocatalysts for degradation of ampicillin under UV light. *Applied Catalysis B: Environmental, 200*, 307–315. https://doi.org/10.1016/j.apcatb.2016.07.046

Chapter 24
Remediation of Pharmaceuticals and Personal Care Products from Urban Wastewater by Engineered Wetlands

Naveen Chand and Surindra Suthar

Introduction

PPCPs have raised multiple issues because of their potential adverse effects on human health and the ecosystem. In the recent past, over 1000 industrial chemicals have been identified in the natural aquatic environment and some frequently detected PPCP's chemical and physical properties are given in Table 24.1. Since there is limited degradation of complex synthetic organic chemicals in the WWTPs, the effluent from these plants has become the critical source of PPCPs in the aquatic environment (Samal et al., 2022). Once these PPCPs enter the aquatic eco-environment they may bio-accumulate and show adverse effects on aquatic ecosystems (Manisalidis et al., 2020). Several other methods (e.g., ozonation, photo-oxidation, radiolysis, etc.) have been implemented to remove these PPCPs from water streams. Nevertheless, such methods are not found environmentally friendly and even cost-effective, making them commercially unattractive and limiting their wide uses at the field scale (Juliano, 2022; Rivera-Utrilla et al., 2013).

Notably, in recent times, CWs are considered a sustainable technology for pollutant removal due to the low built cost, easy operation, and minimum maintenance requirements. CWs are artificial wetlands and as per hydrology and flow path, the CWs can be categorized into subsurface flow wetlands, surface flow wetlands and hybrid wetland systems; in which the wastewater treatment is carried out through a set of biological, physical and chemical mechanisms (Vo et al., 2018). Over the years CWs are frequently been used for wastewater treatment and results have shown that CWs are reliable for the treatment of different types of wastewater: industrial effluent

N. Chand
Environmnetal Engineering Research Group, Department of Applied Sciences, National Institute of Technology, Delhi, New Delhi, India

S. Suthar (✉)
School of Environment and Natural Resources, Doon University, Dehradun, Uttarakhand, India
e-mail: suthariitd@gmail.com

I. D. Behera et al. (eds.), *Emerging Micropollutants*, Sustainable Environmental Waste Management Strategies, https://doi.org/10.1007/978-3-032-03530-1_24

Table 24.1 Chemical and physical properties of significant PPCPs detected in wastewater at the inlet of WWTPs (Behera et al., 2011; Ghosh et al., 2024; Kushwaha et al., 2024)

Emerging pollutants	Chemical structure	Dissociation constants (pKa)	Chemical formula	Molar mass(g/mol)	Solubility(mg/l)
Paracetamol	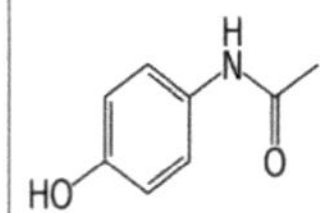		$C_8H_9NO_2$	151.163	
Carbamazepine	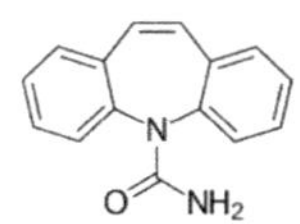	13.9	$C_{15}H_{12}N_2O$	236.269	17.7
Atenolol (ATL)	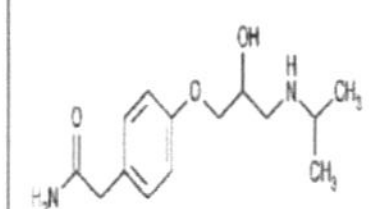	9.6	$C_{14}H_{22}N_2O_3$	266.336	13,300 (at 25 °C)
Triclosan	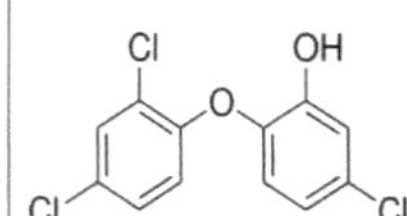	7.9	$C_{12}H_7Cl_3O_2$	289.54	10
Ciprofloxacin	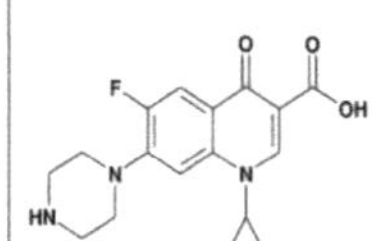		$C_{17}H_{18}FN_3O_3$	331.346	30,000 (at 20 °C)

(continued)

Table 24.1 (continued)

Emerging pollutants	Chemical structure	Dissociation constants (pKa)	Chemical formula	Molar mass(g/mol)	Solubility(mg/l)
Ibuprofen	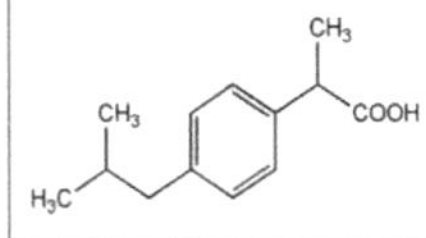	5.2	$C_{13}H_{18}O_2$	206.29	21 mg/L (at 25 °C)
Naproxen	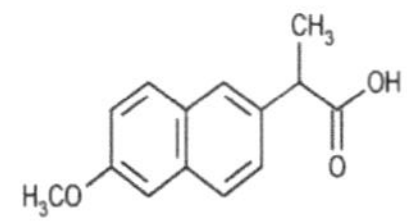	4.15	$C_{14}H_{14}O_3$	230.259	15.9 (at 25 °C)
Diclofenac	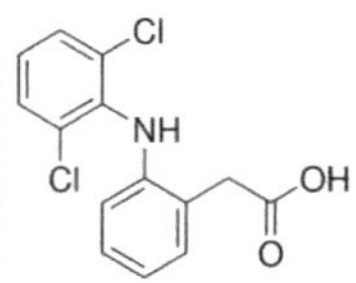	4.15	$C_{14}H_{11}Cl_2NO_2$	296.148	2.37 (at 25 °C)
Caffeine	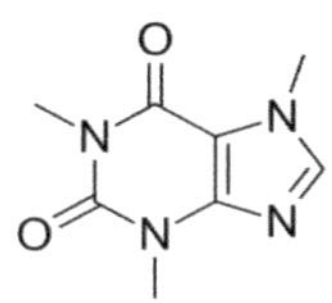	4.0 at 25 °C	$C_8H_{10}N_4O_2$	194.19	21,600 (at 25 °C)

(continued)

Table 24.1 (continued)

Emerging pollutants	Chemical structure	Dissociation constants (pKa)	Chemical formula	Molar mass(g/mol)	Solubility(mg/l)
Amoxicillin	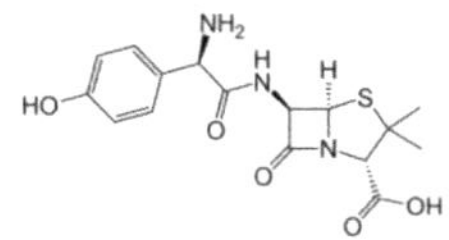	3.2 (acid); 11.7 (primary amine)	$C_{16}H_{19}N_3O_5S$	365.4	3430
Erythromycin	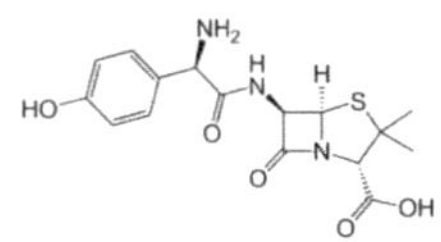	8.9	$C_{37}H_{67}NO_{13}$	733.93	2,000 at 28°C
Cetirizine	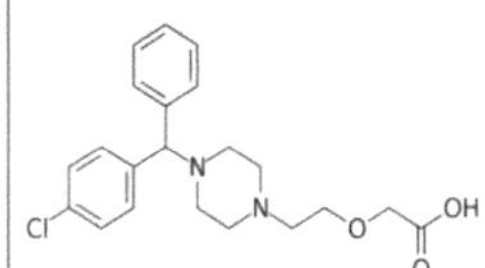	2.70 (amine); 3.57 (carboxylic acid); 7.56 (amine) (est)	$C_{21}H_{25}ClN_2O_3$	388.89	101

(continued)

Table 24.1 (continued)

Emerging pollutants	Chemical structure	Dissociation constants (pKa)	Chemical formula	Molar mass(g/mol)	Solubility(mg/l)
Ketoprofen	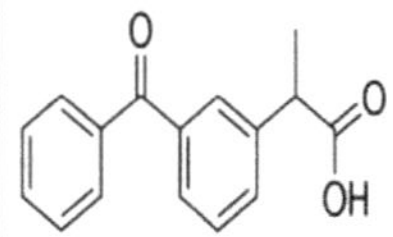	4.45	$C_{16}H_{14}O_3$	254.281	51(at 22 °C)
Sulfamethoxazole	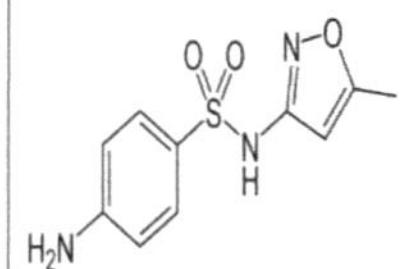	1.6; 25.7	$C_{10}H_{11}N_3O_3S$	253.279	610 (at 37 °C)

(Al-Mashaqbeh et al., 2024), landfill leachate (Ogata et al., 2024), agricultural runoff (Lin et al., 2023), domestic sewage (Chand et al., 2021), PPCP containing effluents (Chand et al., 2022; Suthar et al., 2023), etc.

This chapter discusses various aspects such as environmental sources of PPCPs, the potential of CWs in PPCP removals, CWs design aspects and role of different components of CWs in removal mechanisms, operational approaches, limitations of CWs, etc.

Sources of PPCPs in the Environment

PPCP contamination impurity has been highly linked to anthropogenic sources particularly by-products of industries and the quantity of untreated wastewater discharged into urban water bodies (Yang et al., 2018b). The detection of pharmaceutical compounds (PCs) personal care products (PCPs) has been observed in surface and groundwater at μg/L to ng/L from various parts of the world (Sanusi et al., 2023; Yu et al., 2023; Zhang et al., 2024). The primary source of PPCPs in the environment is sewage water and unmanaged waste dumping practices (especially animal excreta, sewage sludge, open dumping of industrial waste solids, industrial effluents, hospitals, etc.) (Mishra et al., 2024a; Mishra & Das, 2023). The production and consumption of PCs are reported to have increased manifolds in developing countries like India, China and the Latin American States (Osuoha et al., 2023). The scientific literature suggests that humans and other veterinary animals can uptake only 10–80% of the dose of PCs administrated, and the rest is eliminated from the body through excretory products that finally sink into sewage water or urban drains (Hirsch et al., 1999). In developing countries, unused and expired drugs are often drained into house toilets (Zhang et al., 2011), where these PCs finally sink into the environment through sewage.

Now, treating these ECs has become a severe challenge to WWTPs because, either they are partially removed or even remain unremoved (Balakrishna et al., 2017; Zhang et al., 2023). ECs measured concentration Worldwide in the influent of WWTP is given in Table 24.2. This drawback of conventional treatment facilities creates the pathways for these pollutants and their metabolites to enter surface water and groundwater. Subsequently, these ECs may enter the food chain (Mishra et al., 2024b; Mishra & Das, 2021). Additionally, animal excreta from urban livestock farms and hospitals/ biomedical effluents, as well as a considerable amount of these PCs in urban drains, leads to the burden of contamination of natural resources (Khatun et al., 2023a, 2023b; Sadasivuni et al., 2019). Around 50% of PPCP contamination is based on antibiotics, leading to antibiotic-resistant bacteria or genes (ARGs), which can be risky to the health of human beings throughout the food chain (Jones et al., 2001; Loganathan et al., 2009). The harmful impact of the PPCPs varies with the concentration level to which the organism is exposed, exposure duration, and the type of organism exposed, etc. The prioritization of PPCPs for risk assessment study is still unclear due to the non-availability of data on their sub-lethal impact on biological systems and their considerable variations in the concentration within the environment (Archer et al., 2017; Matongo et al., 2015).

PPCPs have become a matter of concern and global anxiety because of their harmful nature (endocrine and hormonal disturbing nature, mimics with primary metabolites, alters genetic structures, etc.) with various hazardous environmental consequences to humans and other life forms (Ebele et al., 2017; Gavrilescu et al., 2015).

Table 24.2 A comparative data of efficacy using various treatment strategies and the CWs to remove ARGs

Wastewater	Target ARGs	Treatment	Removal efficiency	References
WWTPs-Effluent	*sulI, sulII, tetO, tetW, tetQ*	Coagulation	0.5–3.1 log	(Li et al., 2017)
Urban wastewater	*intI*	Ozonation	2.0 ± 0.6 log	(Biancullo et al., 2019)
Urban wastewater	*intI*	UV_{254}	2.4 ± 0.4 log	(Biancullo et al., 2019)
Natural water	*mecA*	$UV_{254}/TiO_2/H_2O_2$	5.7–5.8 log	(Guo et al., 2017)
Natural water	*amp*	$UV_{254}/TiO_2/H_2O_2$	4.6–4.8 log	(Guo et al., 2017)
Primary clarifier-effluent	*ermB*	Anaerobic membrane bioreactor	3.7 log	(Kappell et al., 2018)
Primary clarifier-effluent	*sul1*	Anaerobic membrane bioreactor	3.6 log	(Kappell et al., 2018)
WWTPs-effluent	*ermB, sul1, int1*	Membrane bioreactor	1.5–7.3 log	Li. et al. (2017)
	i			
Microbial fuel cell-effluent	*ermB*	Bio-electro-Fenton	1.96 log	(Li et al., 2017)
Landfill leachate	*sul1, sul2, intl1*	Hybrid CW	Over 90%	(Yi et al., 2017)
Swine wastewater	*tet, intI1*	Vertical flow CW	33.2–99.1%	(Huang et al., 2017)
Goose wastewater	*tetO, tetC, tetW, tetA, tetG, tetQ,*	Hybrid CWs	More than 50%	(Huang et al., 2019)
Swine wastewater	intl, tetW, tetM, tetO, tetX, tetA,	Vertical flow CW	0.26 to 3.3 log	(Huang et al., 2017)
Primary wastewater	intl1, ermB, tetM, sul1, and total ARGs	Horizontal sub-surface flow CWs	More than 48%	(Abou-Kandil et al., 2021)

Engineered Wetlands: A Sustainable Approach for Eliminating PPCPs from Wastewater?

Eliminating or managing PPCPs in wastewater resources, within safety limits, has become a prime concern (Kumar et al., 2023; Mishra et al., 2022a, 2022b; Mishra & Das, 2022; Priyadarshini et al., 2023). Focusing on environmentally friendly techniques for the remediation of the contaminants from water and wastewater, phytoremediation has turned out to be a promising approach (Mishra et al., 2022a, 2022b; Nivala et al., 2020). Phytoremediation is the removal of pollutants utilizing green plants, which seems to be an effective and low-cost technique for removing various contaminants (Khalidi-idrissi et al., 2023). In recent years, engineered wetlands i.e., CWs based on the phytoremediation process have gained the interest of researchers, scientists, and policymakers. Recent studies claimed that CWs are a reliable, affordable, and easy-to-operate eco-friendly technique for the treatment of industrial as well as urban wastewater streams (Satrovic et al., 2024a). In comparison to conventional treatment facilities, CWs provide an eco-friendly approach that is more economical, has minimum operational requirements and is straightforward to maintain (Cabred et al., 2019; Li et al., 2020) and also have the potential to be suitable for the secondary systems; particularly in small rural communities and for the less population areas (Sathe & Munavalli, 2019). In CWs, contaminants elimination is mainly influenced by various biological, physical and chemical phenomena such as microbial degradation, plant root-microbial interactions, filtration, sedimentation and biochemical interactions (Fig. 24.1).

The substrate has an integrated biological, chemical, and physical function to eliminate pollutants (entrapment, biodegradation, filtration, and adsorption) (Machado et al., 2017a, 2017b). All these processes may vary in CWs, depending on various

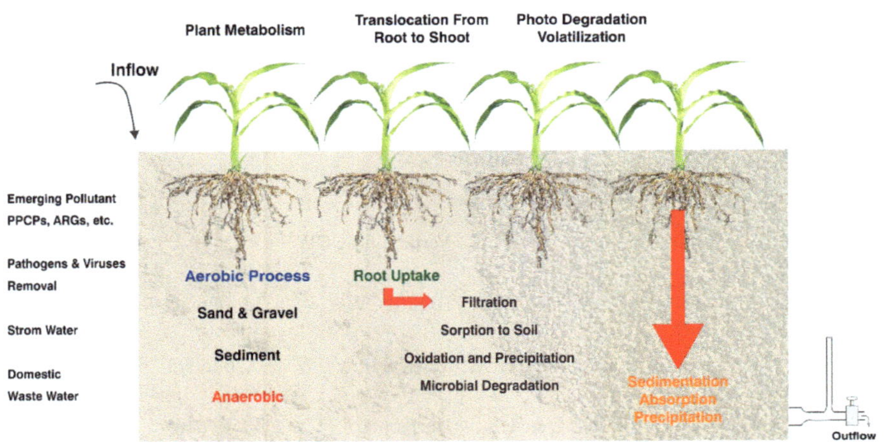

Fig. 24.1 Overview of subsurface CWs mechanism for the reduction of pollutants

operational and design parameters i.e., flow rate, hydraulic retention time (HRT), plant stand types, rhizosphere microflora, substrate configurations, etc. (Ilyas & van Hullebusch, 2019, 2020). Additionally, various physicochemical and biological processes and mechanisms, such as phytovolatilization, phytostabilisation, phytodegradation, phytofiltration, rhizosphere bioremediation, and phytoextraction (e.g., phytoaccumulation, photoabsorption, phytosequestration, etc.) are involved (See Fig. 24.2) in the elimination of the various contaminants during the treatment process (Arthur et al., 2005; Ting et al., 2018).

In resolving the problem of undesirable contaminants in the environment, there is a critical need to have such techniques, especially PPCPs in wastewater; in the recent past, researchers and scientists have moved towards the CWs for the reduction of the problem of water pollution, which has evolved as the promising treatment method for the elimination of ECs (James et al., 2023). A schematic view of different CWs

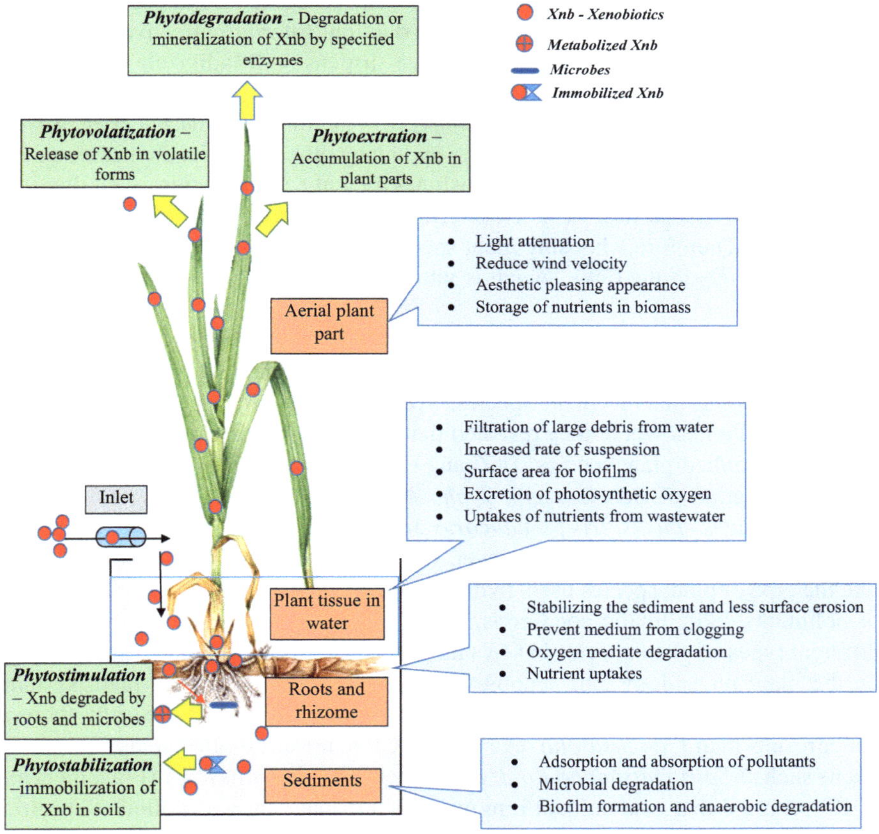

Fig. 24.2 Schematic view of role and mechanism of plant stand in constructed wetland for reduction of pollutants

is given in Fig. 24.3. The functioning of a CW depends on complex interactions between wetland design parameters and removal efficiency is dependent on them.

Role of Plant Stands in CWs for the Removal of PPCPs

The plants are critical elements of CWs, which offer services in various areas of CWs, such as removing the contaminants from the wastewater and providing habitats for the microorganisms (Li et al., 2021). Plants support microbial activities and microbial population because of rhizodeposition products, which are mucigels, exudates and dead cells (Shelef et al., 2013; Yadav et al., 2018). Fig. 24.1 reflects the role of plant stands in the wetlands. Pollutants by the plants are removed by roots and by translocation into various tissues, transpiration, and diffusion, and by the physical effects of root structure in combination with aeration. Roots' physical impact includes water distribution, flow velocity reduction, filtering, effective sedimentation, prevention of clogging, and decreased re-suspension (Madikizela et al., 2018). Various studies have been performed with and without implementing plants in the CWs to check the effect of plants in removing contaminants (Sharma et al., 2024). Plants in CWs offer purification of water and many other essential elements such as hydrologic buffering, reduced resuspension, efficient sedimentation, even distribution of water, reduced flow velocity, filtering and prevention from clogging (Shelef et al., 2013). Therefore, choosing plant species is highly significant for eliminating the contaminants in the CWs, which is vital in the present scenario (Pascual et al., 2024).

In the past literature, researchers observed promising removal results of ECs in planted CWs (Dan et al., 2017; Vo et al., 2018). (Zhang et al., 2016) studied the role of some marshy plant species, viz. *Phragmite, Iris pseudacorus. Typha latifolia and Juncus* where they revealed that all compounds were entirely removed by all four studied plant species. I (Zhang et al., 2017a, 2017b), mesocosms CWs having plantation of *Berulaerecta, Typha latifolia, Iris pseudacorus, Phragmites australis, Juncus effusus, Iris pseudacorus* and control system (unplanted) removed PPCPs from water efficiently. However, removal efficiencies depended on factors like the type of plant species used, hydraulic load rate (HLR), spiking concentration of pollutants and climatic conditions. The comparative studies have revealed that different types of plant species in CWs have diverse removal efficiencies depending on the plant physiology and xenobiotic metabolic pathways in the plant system. In a similar study, (Hijosa-Valsero et al., 2010) observed that *P. australis* showed better results than *T. angustifolia* regarding PPCP removal. Another study comparing plants such as *Salix matsudana Koidz and Phragmites australis L.* by (Francini et al., 2018) reported that contaminant removal was adequate but was pollutant-specific. Researchers have also studied the concentration of ECs in different parts of plants (Agaton & Guila, 2024). The study conducted by (Singh et al., 2018) recommended that for the removal of the antibiotics, phyto-degradation plays a significant role; it has been reported that amoxicillin (AMX) removal (84.6–100%) in duckweed setups

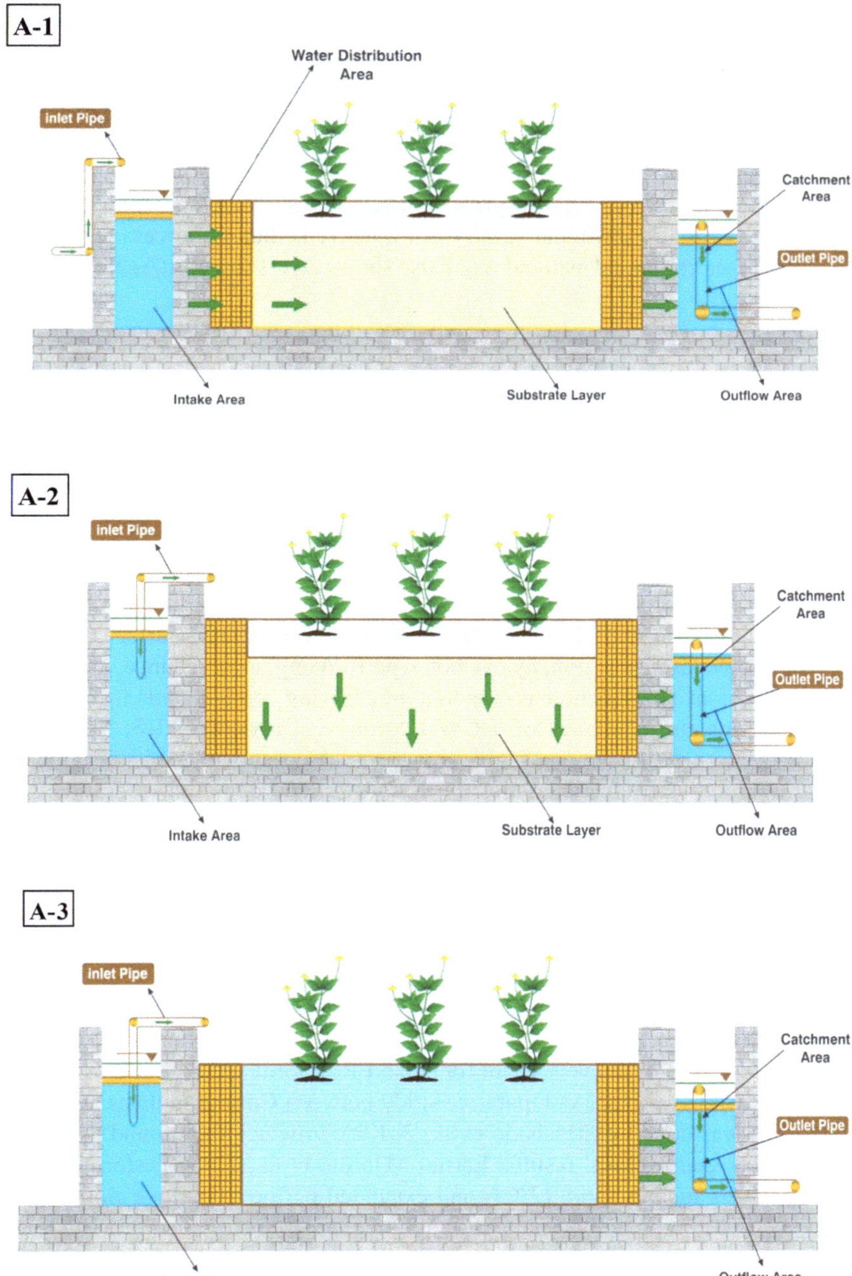

Fig. 24.3 Design of horizontal subsurface flow constructed wetland (A-1), vertical subsurface flow constructed wetland (A-2), free surface flow constructed wetland (A-3)

and (62.1–73%) in control which indicated the importance of the phytoremediation approach. In a similar type of investigation, (Li et al., 2017a) conducted a study at lab scale CWs using greater duckweed (*Spirodela polyrhiza*) and removal of DEET, paracetamol (PC), caffeine (CF) and triclosan (TCS) 17.1%, 98.8%, 96.4% and 95.4% respectively. Further, they suggested that the study be conducted on a pilot scale, looking at the promising results of greater duckweed (Sharma et al., 2024). Additionally, utilizing CWs with grown plant species, the remediation process is found to be optimal. Additionally, the wastewater remediation process with diverse operational conditions and wetland types also shows diverse results (Agaton & Guila, 2023).

Role of the Substrate in the Working Efficiency of CWs

A substrate is an underlying material used as the base in CWs where living organisms grow and obtain nourishment. It is also called the support matrix, which provides the base for the growth of plants and is essential in engineered wetlands, mainly in subsurface flow CWs. Apart from delivering growth and habitat, it also provides sorption processes through which the contaminants directly interact with the substrate (Li et al., 2021; Satrovic et al., 2024a). The sorption process includes surface complexation, van der Waals interaction, hydrophobic partitioning, ion exchange, and electrostatic interaction. The substrate is essential in removing contaminants from wastewater, enhancing the efficiency of a CW (Dordio & Carvalho, 2013; Yang et al., 2018b). Different substrate types have shown variable removal results of pollutants depending upon their characteristics, type, and wetland operating conditions. The performance of different substrate types in CWs for removing various pollutants has been tested by various researchers to check their significance in CWs. For instance, (Chen et al., 2016a) explored four different substrates (ceramic, medical stone, zeolite and oyster shell) for the elimination of ECs through CWs. Promising results were reported in CWs with zeolite as a substrate. Similarly, a study by (Huang et al., 2017) with hydraulic flow direction (up and down flow) and substrate (oyster shell and brick rubble) for the elimination of EPs reported that the type of substrate has a strong influence on removal. Another substrate termed light expanded clay aggregate (LECA) became an effective sorbent for some EPs (Dordio et al., 2007; Özengin & Elmaci, 2016). Likewise, a comparative study between Cork granulates and LECA in the CWs was done by (Machado et al., 2017b), in which they found that LECA attained advanced removal results. Earlier, (Dordio et al., 2010) performed a trial with substrates such as sand, LECA and expanded perlite to remove clofibric acid from water. The comparative analysis of all three materials revealed that LECA had good sorption capacity. Overall, the substrate enhances the performance of the CW and a modification in the substrate layer is vital in reducing the various ECs from wastewater. The reduction capabilities are directly influenced by the material properties utilized in the substrate matrix. Meanwhile, to attain a sustainable practice in CWs, one should focus on environmentally friendly and low-cost substrates (organic substrates) (Russell et al., 2024). Organic substrates are those organic substances or waste materials discarded by human beings in the environment, waste that is naturally occurring organic materials derived from agriculture, forest and marine areas. Researchers have targeted organic substrates for removing contaminants in wetlands and shown promising results for eliminating pollutants. Thus, there is a need to have an elucidated study on the implementation of various organic materials available locally and their practice with sustainable substrate management in CWs (Zhang et al., 2023). Another potential source of the substrate can be the re-utilization of waste materials for the elimination of pollutants (like tyre chips, alum sludge, constructional waste, etc.) as alternative substrates benefit the environment; it is also beneficial in another way as well by reducing the solid waste from the surroundings. Similarly, a substrate like oyster shell (Wei et al., 2016a, 2016b), have a good potential for eliminating pollutants, which can be observed through research conducted in the past (Agaton & Guila, 2024).

Reduction of PPCPs in Conventional CW Setups

The reduction in the CWs depends on the operational parameters, use of substrate matrix, plants, flow pattern, and seasonal variability. Looking at all the factors, studies in CWs have been conducted to see the significance of each factor, which has led to further modification in the design to improve water quality and make water accessible from PPCPs (Karki & Philip, 2024; Satrovic et al., 2024b). The use of conventional strategy (i.e. use of a traditional matrix of the substrate, plants, regular flow pattern, etc.) for reduction has been continuously evaluated using CWs for reduction of: (a) analgesics, which are generally used to cure and prevent diseases from getting better value to animals and human's life, (b) anti-inflammatories (NSAIDs); which are used to cure and prevent diseases from spreading, doctors use NSAIDs to treat several effects that cause pain or inflammation, including arthritis, (c) antiepileptics; which are commonly used in medicine to stop, prevent or control seizures (Fakioglu et al., 2024). Many researchers have observed the removal of antiepileptics from CWs. Apart from other chemicals, pesticides, steroids, etc., have been observed in different CWs. Even studies have reported that maximum removal was during the summers compared to the winters, with retention (Nivala et al., 2020; Zhang et al., 2017a, 2017b).

Studies have shown how different plant species and medium types affect filtration performance. Consistent with that a study with different types of horizontal subsurface flow constructed wetlands (*Phragmites australis* + crushed rock; *Phragmites australis* + gravels; *P. australis* + *Phalaris arundinacea* + gravel) operated for a long-term period and results revealed a moderate removal for ibuprofen (IBU) (55%) and diclofenac (DCF) (41%) (Vymazal et al., 2017). Likewise (Dvořáková Březinova et al., 2018) found moderate to low removal efficiencies using similar kinds of CWs for ibuprofen (44.7%) and its metabolites (OH-IBU 29.3%, CA-IBU 47.5%) and they also found that degradation processes were slow down because of anoxic circumstances in such systems. These findings highlight how oxygen availability affects the effectiveness of pollution removal in CWs. Similarly, (Nhat Phong et al., 2016) used acrylic reactors with *Scirpus valid*us and sand, small gravel, and coarse gravel as substrates to investigate the removal of acetaminophen (ACT) in CWs. The synthetic wastewater concentration of 1 ppb and 100 ppb, showed that ACT removal within the range from 55% to 99.5% at 5-day HRT. The study indicated that plant uptake dominated the removal mechanism at 1 ppb, whereas microbial and photolytic activities were more important in the removal of ACT at 100 ppb. Apart from that, a higher concentration of hydrogen peroxide (H_2O_2) in the plant shoot than in the root was reported, probably due to the occurrence of a Fenton reaction resulting in PPCP removal (Kumar et al., 2022; Nelms et al., 2024). CWs have shown promising removal for different PPCPs. In line with the above studies, innovation and advancement in design and operational strategy need to be explored to enhance the removal efficacy of such CWs.

Reduction of PPCPs in Modified CW Setups

With conventional design, CWs showing the reduction capabilities led to modifying the substrate matrix, flow pattern, and external oxygen supply. The significance of oxygen levels has also been documented in the last many pieces of literature: free water CW (Li et al., 2017b), horizontal flow CW (HFCW) (Nivala et al., 2019), vertical flow CW (Ávila et al., 2015; Nivala et al., 2019), and sub-surface flow CW (Auvinen et al., 2017). The various design strategies include; (a) unsaturated and unsaturated, (b) up-flow and downflow, (c) multi-layered substrate and use of different substrates apart from conventional, (d) coupled CWs, (e) waste-to-energy setups, etc. These potential strategies are concentrated on improving wastewater treatment techniques and the level of adoption. Researchers have opted for several different methods to treat PPCPs, aiming at maximum efficacy. For example, (Casierra-Martinez et al., 2020) used a solar Photo-Fenton coupled-horizontal flow (HSSF) CW for the treatment of domestic wastewater and they revealed promising results for the removal of the DCF using the Fe^{2+}/H_2O_2 ratio of 0.1 with a removal efficacy up to 92%. HSSFCW planted with *Cyperus ligularis* (density of 16 cuttings/ m^{-2}) showed removal of carbamazepine (CBZ) up to 86%. Likewise, modified design using plants, and baffled systems under tidal flow, the similar system showed a high removal of 97.1 ± 0.29% for biocides, 99.8 ± 0.02% for steroid hormones, and 90.2 ± 1.60% for antibiotics (Khalidi-idrissi et al., 2023; Madikizela et al., 2018; Park et al., 2019). The role of arbuscular mycorrhizal fungi has also been evaluated in CWs with different support media for PPCP removal. The findings indicated that fungal colonization was significantly higher in adsorptive substrates, such as vermiculite, perlite, and biochar, compared to sand. Subsequently, the colonization enhanced the reduction efficiencies, of DCF and IBU by increasing removal by 25%–38% and 15%–18%, respectively (Ebele et al., 2017; Santos et al., 2013). CWs have also been modified by using carbonous fillers (i.e., biochar and activated carbon) for treating PPCPs (Kang et al., 2023) and integration of microbial fuel cell (MFC) with CWs has also been done to enhance the removal efficiency. Further to enhance the removal hybrid CWs are utilized to increase the removal efficiency as the hybrid system provides sequential treatment facilities one after the other, depending on the type of CWs used in the hybrid system and operation stages (e.g., two-stage, three-stage or multi-stage hybrid water treatment facility) of CWs (Yi et al., 2017) (Tejeda et al., 2017). It has also been found that higher removal efficiency of pollutants could be achieved using hybrid-CWs compared to single-stage wetlands (Hota et al., 2023; Osuoha et al., 2023). All these efforts are to achieve a sustainable approach to wastewater treatment.

Antibiotic-Resistance Genes (ARGs) and Antibiotic-Resistant Bacteria (ARB)

The current issue of the presence of ECs in water streams is a significant concern for the scientific society as these contaminants are found frequently in the waste streams, which is mainly due to the excessive use of veterinary, human medicine, and agricultural products (García et al., 2020). The increasing presence and frequent detection of ECs, including antibiotic-resistance genes (AGR) and antibiotic-resistant bacteria (ARB), have become a severe threat because of their various adverse effects, such as affecting humans and animals at all ecological levels. Further, the inability of WWTPs to remove AGR and ARB has also become a concern as WWTPs become the significant source for releasing these pollutants in water streams (Li et al., 2021). Moreover, when these ARGs and ARB, under wastewater discharge, finally end up in surface water or agricultural soils, thereby; becoming the source of transfer into crops via uptake/assimilation and transfer into aquatic. Further, a severe public health threat is also documented with ARGs and ARB, as well as the widespread presence and persistence in the environment (Hazra et al., 2022). The issue becomes more serious when the WWTPs cannot remove these ARGs and ARB. Further, ARGs are either reduced insignificantly or negligible. However, ARB is partially removed from these systems. Further, adding to that, previous literature has documented that WWTPs could provide an ideal condition/environment for the proliferation of ARB and apart from that, transfer of ARGs between the different microorganisms could be possible, by which wastewater effluents may show the enhanced abundance of specific ARGs (Manoharan et al., 2022). The comparative evaluation of different treatment methods reported previously for ARGs and ARB is given in Table 24.2.

Moreover, CWs are seen as green technology and a good alternative for treatment due to low-energy input and low operational and maintenance costs, mainly for remote areas and developing countries (Wu et al., 2015). Further, CWs can attain higher reduction values for ARGs and ARB, which could reach higher values than those of WWTPs. Previous literature has documented the reduction values, i.e. 45–99% for ARG using VFCW with an up-flow pattern or conventional VFCW (Huang et al., 2019). Moreover, it has also been documented that CWs can reduce ARBs from hospital wastewater, reclaimed water and livestock wastewater (Santos et al., 2013). CW design strategy played a vital role in the removal of ARGs (Cui et al., 2023). Further, the reduction of the ARGs is influenced by the various characteristics of the CWs, i.e., the type of CWs (Huang et al., 2019), the species of plant used (Chen et al., 2016a, 2016b), the substrate matrix type (Liu et al., 2013), and hydraulic loading (Chen et al., 2016a, 2016b). Meanwhile, to evaluate the efficacy of the ARGs or as an indicator of ARGs (e.g., aph, vanA, ermB, sul1, sul2, tetM, etc.) have been suggested in CWs. Furthermore, an indicator, i.e. class I integrase gene intl1, could also be used for ARGs, ARB, and other contaminants because of its linkage to such pollutants. Therefore, looking at the results and making a comparison with all treatment techniques, CWs in all the aspects, such as cost, operation, feasibility in remote areas, energy input and eco-friendly approach, etc. are one of the best alternatives for water treatment with its effectiveness in the reduction of contaminants and other valuable features (Liu et al., 2016).

Clogging Problems with Substrate

One common problem in CWs is clogging, mainly due to the substrates used, which contribute to clogging in subsurface horizontal and vertical-flow systems. Clogging with the substrate is the most frequently occurring operational issue in CW (Jin et al., 2023). The reason behind the clogging of the substrate is the small particle size, suspended solids accumulation on their surface, poor hydraulic conductivities, low porosity and their accumulation. Clogging leads to significant consequences, which involve decreased lifespan and system effectiveness. Previous studies have also reported clogging problems in CWs after several years of operation (Nivala et al., 2020). There can be segregation of substrate clogging in two aspects, i.e., external and internal aspects. The external elements include pollution load, temperature, microorganisms and plants, while internal aspects involve the chemical and physical attributes of the substrate (Fang et al., 2022).

The substrate materials directly influence the reduction of pollutants in CWs. The impact of particle size, absorption, gradation and substrate porosity on the substrate's permeability directly impacts the CW system's clogging (Wu et al., 2015). Compared to various factors, the particle size and gradation of the substrate are critical internal factors, and the pollution load can be considered a key external aspect (Xin et al., 2024). Further, the elimination of pollutants is restricted by plant absorption, substrate ion exchange, and microbial degradation. There is a quick accumulation of the contaminants filtered by the substrate because of the high pollution pressure. This accumulation is on the substrate's surface, causing clogging. Because of the low variation in the pollutant concentration in sewage, there is a general reflection of the pollutant load by hydraulic load. A low hydraulic load does not support sewage treatment efficiency. However, the CW is more likely to get clogged because of the high hydraulic, organic, and solid suspension load. As clogging is a severe problem, we have to take some steps for its mitigation or avoid such issues: considering the selection of substrate porosity and particle size (Ramprasad & Philip, 2016), the concept of multi-layer substrates can be implemented (Ding et al., 2018); management of organic loading rate (Vymazal, 2018).

Conclusions and Future Recommendations

ECs are seen as a significant issue these days, as the problem due to ECs in the environment is rising exponentially because treatment facilities cannot eliminate ECs; pollutants are either partially removed or not obliterated. To resolve the problems regarding these pollutants, CWs have evolved as an effective method. The natural treatment method in wetlands has proven effective with the presence of plants and different operational strategies. Generally, CWs provide an option for low energy input, cost-effective, easy operational strategy for sustainable wastewater treatment. Moreover, additional research studies are required before getting to any sought conclusion, looking at the adverse impact on the environment due to the ECs; investigation with a more rigorous methodology is desired in terms of their removal from CWs as well as the changes occurring in the soil and the plants while the reduction of EC from the wastewater. Therefore, the author acknowledges that further investigation with some insights for future work for evaluation of CWs:

- In most of the CW systems, wastewater purification depends on plants and substrate material; one cannot be restricted to them for the removal of EPs. According to the requirement, further modifications are required to enhance the removal efficiency of CWs and to make way for energy generation.
- Substrate plays an essential role in the reduction of pollutants, so studies with alteration in substrate depth, multiple layers and application on a non-conventional substrate material such as biochar, constructional waste, activated carbon, organic waste, etc. can be studied in wetlands should be focused more, apart from to that it can also lead to effective waste management methodology.
- A rigorous study of aerobic and anaerobic conditions in wetlands to treat ECs is required.
- Assessment of known no-effect concentrations in wetlands for plants and microorganisms is required for a better understanding. Further, a better understanding of wetland toxicity levels due to the ECs is also needed.
- Evaluation of the concentration of the EPs which can adversely affect aquatic plants and organisms is required to establish CWs.

References

Abou-Kandil, A., Shibli, A., Azaizeh, H., Wolff, D., Wick, A., & Jadoun, J. (2021). Fate and removal of bacteria and antibiotic resistance genes in horizontal subsurface constructed wetlands: effect of mixed vegetation and substrate type. *Science of the Total Environment, 759*, Article 144193. https://doi.org/10.1016/j.scitotenv.2020.144193

Agaton, C. B., & Guila, P. M. C. (2024). Success factors and challenges: implications of real options valuation of constructed wetlands as nature-based solutions for wastewater treatment. *Resources, 13*(1). https://doi.org/10.3390/resources13010011

Agaton, C. B., & Guila, P. M. C. (2023). Ecosystem services valuation of constructed wetland as a nature-based solution to wastewater treatment. *Earth (Switzerland), 4*(1), 78–92. https://doi.org/10.3390/earth4010006

Al-Mashaqbeh, O., Alsalhi, L., Salaymeh, L., Dotro, G., & Lyu, T. (2024). Treatment of pharmaceutical industry wastewater for water reuse in Jordan using hybrid constructed wetlands. *Science of the Total Environment, 939*(May), Article 173634. https://doi.org/10.1016/j.scitotenv.2024.173634

Archer, E., Petrie, B., Kasprzyk-Hordern, B., & Wolfaardt, G. M. (2017). The fate of pharmaceuticals and personal care products (PPCPs), endocrine disrupting contaminants (EDCs), metabolites and illicit drugs in a WWTW and environmental waters. *Chemosphere, 174*, 437–446. https://doi.org/10.1016/j.chemosphere.2017.01.101

Arthur, E. L., Rice, P. J., Rice, P. J., Anderson, T. A., Baladi, S. M., Henderson, K. L. D., & Coats, J. R. (2005). Phytoremediation—an overview. *Critical Reviews in Plant Sciences, 24*(2), 109–122. https://doi.org/10.1080/07352680590952496

Auvinen, H., Havran, I., Hubau, L., Vanseveren, L., Gebhardt, W., Linnemann, V., Van Oirschot, D., Du Laing, G., & Rousseau, D. P. L. (2017). Removal of pharmaceuticals by a pilot aerated sub-surface flow constructed wetland treating municipal and hospital wastewater. *Ecological Engineering, 100*, 157–164. https://doi.org/10.1016/j.ecoleng.2016.12.031

Ávila, C., Bayona, J. M., Martín, I., Salas, J. J., & García, J. (2015). Emerging organic contaminant removal in a full-scale hybrid constructed wetland system for wastewater treatment and reuse. *Ecological Engineering, 80*, 108–116. https://doi.org/10.1016/j.ecoleng.2014.07.056

Balakrishna, K., Rath, A., Praveenkumarreddy, Y., Guruge, K. S., & Subedi, B. (2017). A review of the occurrence of pharmaceuticals and personal care products in Indian water bodies. *Ecotoxicology and Environmental Safety, 137*(April 2016), 113–120. https://doi.org/10.1016/j.ecoenv.2016.11.014

Behera, S. K., Kim, H. W., Oh, J. E., & Park, H. S. (2011). Occurrence and removal of antibiotics, hormones and several other pharmaceuticals in wastewater treatment plants of the largest industrial city of Korea. *Science of the Total Environment, 409*(20), 4351–4360. https://doi.org/10.1016/j.scitotenv.2011.07.015

Biancullo, F., Moreira, N. F. F., Ribeiro, A. R., Manaia, C. M., Faria, J. L., Nunes, O. C., Castro-Silva, S. M., & Silva, A. M. T. (2019). Heterogeneous photocatalysis using UVA-LEDs for the removal of antibiotics and antibiotic resistant bacteria from urban wastewater treatment plant effluents. *Chemical Engineering Journal, 367*(November 2018), 304–313. https://doi.org/10.1016/j.cej.2019.02.012

Cabred, S., Giunta Ramos, V., Busalmen, J. E., Busalmen, J. P., & Bonanni, S. (2019). Reduced depth stacked constructed wetlands for enhanced urban wastewater treatment. *Chemical Engineering Journal, 372*(February), 708–714. https://doi.org/10.1016/j.cej.2019.04.180

Casierra-Martinez, H. A., Madera-Parra, C. A., Vargas-Ramírez, X. M., Caselles-Osorio, A., & Torres-López, W. A. (2020). Diclofenac and carbamazepine removal from domestic wastewater using a Constructed Wetland-Solar Photo-Fenton coupled system. *Ecological Engineering, 153*(November), Article 105699. https://doi.org/10.1016/j.ecoleng.2019.105699

Chand, N., Suthar, S., Kumar, K., & Tyagi, V. K. (2021). Enhanced removal of nutrients and coliforms from domestic wastewater in cattle dung biochar-packed Colocasia esculenta-based vertical subsurface flow constructed wetland. *Journal of Water Process Engineering, 41*(March), Article 101994. https://doi.org/10.1016/j.jwpe.2021.101994

Chand, N., Suthar, S., Kumar, K., & Singh, V. (2022). Removal of pharmaceuticals by vertical flow constructed wetland with different configurations: Effect of inlet load and biochar addition in the substrate. *Chemosphere, 307*, 135975. https://doi.org/10.1016/j.chemosphere.2022.135975

Chen, J., Wei, X., Liu, Y., Ying, G., Liu, S., He, L., Su, H., Hu, L., Chen, F., & Yang, Y. (2016a). Removal of antibiotics and antibiotic resistance genes from domestic sewage by constructed wetlands : Optimization of wetland substrates and hydraulic loading. *Science of the Total Environment, 565*, 240–248. https://doi.org/10.1016/j.scitotenv.2016.04.176

Chen, J., Ying, G. G., Wei, X. D., Liu, Y. S., Liu, S. S., Hu, L. X., He, L. Y., Chen, Z. F., Chen, F. R., & Yang, Y. Q. (2016b). Removal of antibiotics and antibiotic resistance genes from domestic sewage by constructed wetlands: Effect of flow configuration and plant species. *Science of the Total Environment, 571*, 974–982. https://doi.org/10.1016/j.scitotenv.2016.07.085

Cui, E., Zhou, Z., Gao, F., Chen, H., & Li, J. (2023). Roles of substrates in removing antibiotics and antibiotic resistance genes in constructed wetlands: A review. *Science of the Total Environment, 859*(July 2022), 160257. https://doi.org/10.1016/j.scitotenv.2022.160257

Dan, D. A., Fujii, D., Soda, S., Machimura, T., & Ike, M. (2017). Removal of phenol, bisphenol A, and 4-tert-butylphenol from synthetic landfill leachate by vertical flow constructed wetlands. *Science of the Total Environment, 578*, 566–576. https://doi.org/10.1016/j.scitotenv.2016.10.232

Ding, Y., Lyu, T., Bai, S., Li, Z., Ding, H., You, S., & Xie, Q. (2018). Effect of multilayer substrate configuration in horizontal subsurface flow constructed wetlands: Assessment of treatment performance, biofilm development, and solids accumulation. *Environmental Science and Pollution Research, 25*(2), 1883–1891. https://doi.org/10.1007/s11356-017-0636-4

Dordio, A. V., & Carvalho, A. J. P. (2013). Organic xenobiotics removal in constructed wetlands, with emphasis on the importance of the support matrix. *Journal of Hazardous Materials, 252–253*, 272–292. https://doi.org/10.1016/j.jhazmat.2013.03.008

Dordio, A. V., Teimão, J., Ramalho, I., Carvalho, A. J. P., & Candeias, A. J. E. (2007). Selection of a support matrix for the removal of some phenoxyacetic compounds in constructed wetlands systems. *Science of the Total Environment, 380*(1–3), 237–246. https://doi.org/10.1016/j.scitotenv.2007.02.015

Dordio, A., Carvalho, A. J. P., Teixeira, D. M., Dias, C. B., & Pinto, A. P. (2010). Removal of pharmaceuticals in microcosm constructed wetlands using Typha spp. and LECA. *Bioresource Technology, 101*(3), 886–892. https://doi.org/10.1016/j.biortech.2009.09.001

Dvořáková Březinova, T., Vymazal, J., Koželuh, M., & Kule, L. (2018). Occurrence and removal of ibuprofen and its metabolites in full-scale constructed wetlands treating municipal wastewater. *Ecological Engineering, 120*(March), 1–5. https://doi.org/10.1016/j.ecoleng.2018.05.020

Ebele, A. J., Abou-Elwafa Abdallah, M., & Harrad, S. (2017). Pharmaceuticals and personal care products (PPCPs) in the freshwater aquatic environment. *Emerging Contaminants, 3*(1), 1–16. https://doi.org/10.1016/j.emcon.2016.12.004

Fakioglu, M., Ahrens, L., & Ozturk, I. (2024). A review on pharmaceuticals removal from wastewater effluents by single and combined ozone-sorption systems and fate of transformation products. *Ozone: Science and Engineering, 46*(4), 324–344. https://doi.org/10.1080/01919512.2024.2309958

Fang, Y., Kong, L., Zhang, P., Zhang, L., Zhao, H., Xiang, X., Cheng, S., Zhang, H., Ju, F., & Li, L. (2022). Fifteen-year analysis of constructed wetland clogging: A critical review. *Journal of Cleaner Production, 365*(April), Article 132755. https://doi.org/10.1016/j.jclepro.2022.132755

Francini, A., Mariotti, L., Gregorio, S. D., Sebastiani, L., & Andreucci, A. (2018). Removal of micro-pollutants from urban wastewater by constructed wetlands with Phragmites australis and Salix matsudana. *Environmental Science and Pollution Research*. https://doi.org/10.1007/s11356-018-3582-x

García, J., García-Galán, M. J., Day, J. W., Boopathy, R., White, J. R., Wallace, S., & Hunter, R. G. (2020). A review of emerging organic contaminants (EOCs), antibiotic resistant bacteria (ARB), and antibiotic resistance genes (ARGs) in the environment: Increasing removal with wetlands and reducing environmental impacts. *Bioresource Technology, 307*(February), Article 123228. https://doi.org/10.1016/j.biortech.2020.123228

Gavrilescu, M., Demnerová, K., Aamand, J., Agathos, S., & Fava, F. (2015). Emerging pollutants in the environment: Present and future challenges in biomonitoring, ecological risks and bioremediation. *New Biotechnology, 32*(1), 147–156. https://doi.org/10.1016/j.nbt.2014.01.001

Ghosh, R., Parde, D., Bhaduri, S., Rajpurohit, P., & Behera, M. (2024). Occurrence, fate, transport, and removal technologies of emerging contaminants: A review on recent advances and future perspectives. *Clean—Soil, Air, Water*, 1–23. https://doi.org/10.1002/clen.202300259

Guo, J., Li, J., Chen, H., Bond, P. L., & Yuan, Z. (2017). Metagenomic analysis reveals wastewater treatment plants as hotspots of antibiotic resistance genes and mobile genetic elements. *Water Research, 123*, 468–478. https://doi.org/10.1016/j.watres.2017.07.002

Hazra, M., Joshi, H., Williams, J. B., & Watts, J. E. M. (2022). Antibiotics and antibiotic resistant bacteria/genes in urban wastewater: A comparison of their fate in conventional treatment systems and constructed wetlands. *Chemosphere, 303*(P2), Article 135148. https://doi.org/10.1016/j.chemosphere.2022.135148

Hijosa-Valsero, M., Matamoros, V., Martín-Villacorta, J., Bécares, E., & Bayona, J. M. (2010). Assessment of full-scale natural systems for the removal of PPCPs from wastewater in small communities. *Water Research, 44*(5), 1429–1439. https://doi.org/10.1016/j.watres.2009.10.032

Hirsch, R., Ternes, T., Haberer, K., & Kratz, K. L. (1999). Occurrence of antibiotics in the aquatic environment. *Science of the Total Environment, 225*(1–2), 109–118. https://doi.org/10.1016/S0048-9697(98)00337-4

Hota, A., Gopal Krishna Patro, S., Obaid, A. J., Khatak, S., & Kumar, R. (2023). Constructed wetland challenges for the treatment of industrial wastewater in smart cities: A sensitive solution. *Sustainable Energy Technologies and Assessments, 55*(December 2022), 102967. https://doi.org/10.1016/j.seta.2022.102967

Huang, X., Zheng, J., Liu, C., Liu, L., Liu, Y., & Fan, H. (2017). Removal of antibiotics and resistance genes from swine wastewater using vertical flow constructed wetlands: Effect of hydraulic flow direction and substrate type. *Chemical Engineering Journal, 308*, 692–699. https://doi.org/10.1016/j.cej.2016.09.110

Huang, X., Luo, Y., Liu, Z., Zhang, C., Zhong, H., Xue, J., Wang, Q., Zhu, Z., & Wang, C. (2019). Influence of two-stage combinations of constructed wetlands on the removal of antibiotics, antibiotic resistance genes and nutrients from goose wastewater. *International Journal of Environmental Research and Public Health, 16*(20), 1–11. https://doi.org/10.3390/ijerph16204030

Ilyas, H., & van Hullebusch, E. D. (2019). Role of design and operational factors in the removal of pharmaceuticals by constructed wetlands. *Water (Switzerland), 11*(11). https://doi.org/10.3390/w11112356

Ilyas, H., & Van Hullebusch, E. D. (2020). A review on the occurrence, fate and removal of steroidal hormones during treatment with different types of constructed wetlands. *Journal of Environmental Chemical Engineering, 8*(3), Article 103793. https://doi.org/10.1016/j.jece.2020.103793

James, C. A., Sofield, R., Faber, M., Wark, D., Simmons, A., Harding, L., & O'Neill, S. (2023). The screening and prioritization of contaminants of emerging concern in the marine environment based on multiple biological response measures. *Science of the Total Environment, 886*(February). https://doi.org/10.1016/j.scitotenv.2023.163712

Jin, F., Hu, Z., Liu, H., Su, J., Zhang, J., Wang, S., & Zhao, Y. (2023). Impact of clogging on accumulation and stability of phosphorus in the subsurface flow constructed wetland. *Chemosphere, 313*(August 2022), 137429. https://doi.org/10.1016/j.chemosphere.2022.137429

Jones, O. A. H., Voulvoulis, N., & Lester, J. N. (2001). Human pharmaceuticals in the aquatic environment a review. *Environmental Technology, 22*(12), 1383–1394. https://doi.org/10.1080/09593332208618186

Juliano, C. C. A. (2022). Spreading of dangerous skin-lightening products as a result of colourism: A review. *Applied Sciences (Switzerland), 12*(6). https://doi.org/10.3390/app12063177

Kang, Y., Ma, H., Jing, Z., Zhu, C., Li, Y., Wu, H., Dai, P., Guo, Z., & Zhang, J. (2023). Enhanced benzofluoranthrene removal in constructed wetlands with iron-modified biochar: Mediated by dissolved organic matter and microbial response. *Journal of Hazardous Materials, 443*(PB), 130322. https://doi.org/10.1016/j.jhazmat.2022.130322

Kappell, A. D., Kimbell, L. K., Seib, M. D., Carey, D. E., Choi, M. J., Kalayil, T., Fujimoto, M., Zitomer, D. H., & McNamara, P. J. (2018). Removal of antibiotic resistance genes in an anaerobic membrane bioreactor treating primary clarifier effluent at 20 °C. *Environmental Science: Water Research and Technology, 4*(11), 1783–1793. https://doi.org/10.1039/c8ew00270c

Karki, B. K., & Philip, L. (2024). Fate of pharmaceuticals and personal care products like metronidazole, naproxen, and methylparaben and their effect on the physiological characteristics of two wetland plants. *Chemical Engineering Journal, 483*(December 2023), 149180. https://doi.org/10.1016/j.cej.2024.149180

khalidi-idrissi, A., Madinzi, A., Anouzla, A., Pala, A., Mouhir, L., Kadmi, Y., & Souabi, S. (2023). Recent advances in the biological treatment of wastewater rich in emerging pollutants produced by pharmaceutical industrial discharges. *International Journal of Environmental Science and Technology, 20*(10), 11719–11740. https://doi.org/10.1007/s13762-023-04867-z

Khatun, R., Singh, S., Dubey, N. K., & Prasad Das, A. (2023a). A review on marine-based phytochemicals and their application in biomedical research. *Recent Frontiers of Phytochemicals: Applications in Food, Pharmacy, Cosmetics, and Biotechnology*, 383–395. https://doi.org/10.1016/B978-0-443-19143-5.00013-X

Khatun, R., Singh, S., Dubey, N. K., & Prasad Das, A. (2023b). A review on marine-based phytochemicals and their application in biomedical research. In *Recent Frontiers of Phytochemicals: Applications in Food, Pharmacy, Cosmetics, and Biotechnology* (pp. 383–395). Elsevier. https://doi.org/10.1016/B978-0-443-19143-5.00013-X

Kumar, R., Qureshi, M., Vishwakarma, D. K., Al-Ansari, N., Kuriqi, A., Elbeltagi, A., & Saraswat, A. (2022). A review on emerging water contaminants and the application of sustainable removal technologies. *Case Studies in Chemical and Environmental Engineering, 6*(March), Article 100219. https://doi.org/10.1016/j.cscee.2022.100219

Kumar, M., Pandit, S., Patel, V., Khanna, N., Nag, M., Lahiri, D., Ray, R. R., Das, A., & Das, D. (2023). Maximization of energy recovery from starch processing wastewater by thermophilic dark fermentation coupled with microbial fuel cell technology. *Geomicrobiology Journal*, 1–12. https://doi.org/10.1080/01490451.2023.2209555

Kushwaha, A., Goswami, L., Kim, B. S., Lee, S. S., Pandey, S. K., & Kim, K. H. (2024). Constructed wetlands for the removal of organic micropollutants from wastewater: Current status, progress, and challenges. *Chemosphere, 360*(December 2023), 142364. https://doi.org/10.1016/j.chemosphere.2024.142364

Li, J., Zhou, Q., & Campos, L. C. (2017). Removal of selected emerging PPCP compounds using greater duckweed (Spirodela polyrhiza) based lab-scale free water constructed wetland. *Water Research, 126*, 252–261. https://doi.org/10.1016/j.watres.2017.09.002

Li, T., Fan, Y., Cun, D., Song, X., Dai, Y., Wang, F., Wu, C., & Liang, W. (2020). Treatment performance and microbial response to dibutyl phthalate contaminated wastewater in vertical flow constructed wetland mesocosms. *Chemosphere, 246*, Article 125635. https://doi.org/10.1016/j.chemosphere.2019.125635

Li, S., Zhang, C., Li, F., Hua, T., Zhou, Q., & Ho, S. H. (2021). Technologies towards antibiotic resistance genes (ARGs) removal from aquatic environment: A critical review. *Journal of Hazardous Materials, 411*(January), Article 125148. https://doi.org/10.1016/j.jhazmat.2021.125148

Lin, R., Zhao, Y., Jiang, M., Cun, D., Xiong, Y., Zhu, Y., & Chang, J. (2023). Agricultural runoff treatment by constructed wetlands filled with iron-carbon composites in winter: Performance augmentation by organic solids and denitrifying bacteria addition. *Bioresource Technology, 387*(July), Article 129692. https://doi.org/10.1016/j.biortech.2023.129692

Liu, L., Liu, C., Zheng, J., Huang, X., Wang, Z., Liu, Y., & Zhu, G. (2013). Elimination of veterinary antibiotics and antibiotic resistance genes from swine wastewater in the vertical flow constructed wetlands. *Chemosphere, 91*(8), 1088–1093. https://doi.org/10.1016/j.chemosphere.2013.01.007

Liu, J., Wang, J., Zhao, C., Hay, A. G., Xie, H., & Zhan, J. (2016). Triclosan removal in wetlands constructed with different aquatic plants. *Applied Microbiology and Biotechnology, 100*(3), 1459–1467. https://doi.org/10.1007/s00253-015-7063-6

Loganathan, B., Phillips, M., Mowery, H., & Jones-lepp, T. L. (2009). Chemosphere Contamination profiles and mass loadings of macrolide antibiotics and illicit drugs from a small urban wastewater treatment plant. *Chemosphere, 75*, 70–77. https://doi.org/10.1016/j.chemosphere.2008.11.047

Machado, A. I., Beretta, M., Fragoso, R., & Duarte, E. (2017a). Overview of the state of the art of constructed wetlands for decentralized wastewater management in Brazil. *Journal of Environmental Management, 187*, 560–570. https://doi.org/10.1016/j.jenvman.2016.11.015

Machado, A. I., Dordio, A., Fragoso, R., Leitão, A. E., & Duarte, E. (2017b). Furosemide removal in constructed wetlands: Comparative efficiency of LECA and Cork granulates as support matrix. *Journal of Environmental Management, 203*, 422–428. https://doi.org/10.1016/j.jenvman.2017.08.002

Madikizela, L. M., Ncube, S., & Chimuka, L. (2018). Uptake of pharmaceuticals by plants grown under hydroponic conditions and natural occurring plant species: a review. *Science of the Total Environment, 636*, 477–486. https://doi.org/10.1016/j.scitotenv.2018.04.297

Manisalidis, I., Stavropoulou, E., Stavropoulos, A., & Bezirtzoglou, E. (2020). Environmental and health impacts of air pollution: a review. *Frontiers in Public Health, 8*(February), 1–13. https://doi.org/10.3389/fpubh.2020.00014

Manoharan, R. K., Ishaque, F., & Ahn, Y. H. (2022). Fate of antibiotic resistant genes in wastewater environments and treatment strategies—a review. *Chemosphere, 298*(April), Article 134671. https://doi.org/10.1016/j.chemosphere.2022.134671

Matongo, S., Birungi, G., Moodley, B., & Ndungu, P. (2015). Pharmaceutical residues in water and sediment of Msunduzi River, KwaZulu-Natal, South Africa. *Chemosphere, 134*, 133–140. https://doi.org/10.1016/j.chemosphere.2015.03.093

Mishra, S., Dash, D., & Das, A. P. (2022a). Detection, characterization and possible biofragmentation of synthetic microfibers released from domestic laundering wastewater as an emerging source of marine pollution. *Marine Pollution Bulletin, 185*, Article 114254. https://doi.org/10.1016/J.MARPOLBUL.2022.114254

Mishra, S., Dash, D., & Das, A. P. (2024). Isolation and molecular characterization of polyester-degrading bacterial strains from municipal wastewater and their application in microplastic degradation in contaminated wastewater. *Geomicrobiology Journal*. https://doi.org/10.1080/01490451.2024.2376101

Mishra, S., & Das, A. P. (2021). Current treatment technologies for removal of microplastic and microfiber pollutants from wastewater. *Wastewater Treatment: Cutting-Edge Molecular Tools, Techniques and Applied Aspects*, 237–251. https://doi.org/10.1016/B978-0-12-821881-5.00011-8

Mishra, S., & Das, A. P. (2022). Treatment of the wastewater polluted with synthetic microfiber released from washing machine. In B. B. Das, H. Hettiarachchi, P. K. Sahu, & S. Nanda (Eds.), *Lecture Notes in Civil Engineering* (Vol. 207, pp. 109–117). Springer Science and Business Media Deutschland GmbH. https://doi.org/10.1007/978-981-16-7509-6_9/COVER

Mishra, S., & Das, A. P. (2023). Microbial Remediation of Synthetic Microfiber Contaminated Wastewater. In M. P. Shah (Ed.), *Microbial Technologies in Industrial Wastewater Treatment* (pp. 337–350). Springer, Singapore. https://doi.org/10.1007/978-981-99-2435-6_17

Mishra, S., Singh, R. P., Rout, P. K., & Das, A. P. (2022). Membrane bioreactor (MBR) as an advanced wastewater treatment technology for removal of synthetic microplastics. *Development in Wastewater Treatment Research and Processes*, 45–60. https://doi.org/10.1016/B978-0-323-85583-9.00022-3

Nelms, M. D., Antonijevic, T., Ring, C., Harris, D. L., Bever, R. J., Lynn, S. G., Williams, D., Chappell, G., Boyles, R., Borghoff, S., Edwards, S. W., & Markey, K. (2024). Chemistry domain of applicability evaluation against existing estrogen receptor high-throughput assay-based activity models. *Frontiers in Toxicology, 6*(April), 1–21. https://doi.org/10.3389/ftox.2024.1346767

Nhat Phong, V. H., Koottatep, T., Chapagain, S. K., Panuvatvanich, A., Polprasert, C., & Ahn, K. H. (2016). Removal of acetaminophen from wastewater by constructed wetlands with scirpus validus. *Environmental Engineering Research, 21*(2), 164–170. https://doi.org/10.4491/eer.2015.132

Nivala, J., Abdallat, G., Aubron, T., Al-Zreiqat, I., Abbassi, B., Wu, G. M., van Afferden, M., & Müller, R. A. (2019). Vertical flow constructed wetlands for decentralized wastewater treatment in Jordan: Optimization of total nitrogen removal. *Science of the Total Environment, 671*, 495–504. https://doi.org/10.1016/j.scitotenv.2019.03.376

Nivala, J., Murphy, C., & Freeman, A. (2020). Recent advances in the application, design, and operations and maintenance of aerated treatment wetlands. *Water (Switzerland), 12*(4), 1–21. https://doi.org/10.3390/W12041188

Ogata, Y., Yamamura, S., Nakajima, N., & Yamada, M. (2024). Development of a floating constructed wetland for landfill leachate treatment and its potential to remove recalcitrant organic matter. *Water Research, 263*(July), Article 122154. https://doi.org/10.1016/j.watres.2024.122154

Osuoha, J. O., Anyanwu, B. O., & Ejileugha, C. (2023). Pharmaceuticals and personal care products as emerging contaminants: Need for combined treatment strategy. *Journal of Hazardous Materials Advances, 9*(August 2022), 100206. https://doi.org/10.1016/j.hazadv.2022.100206

Özengin, N., & Elmaci, A. (2016). Removal of pharmaceutical products in a constructed wetland. *Iranian Journal of Biotechnology, 14*(4), 221–229. https://doi.org/10.15171/ijb.1223

Park, J. B. K., Sukias, J. P. S., & Tanner, C. C. (2019). Floating treatment wetlands supplemented with aeration and biofilm attachment surfaces for efficient domestic wastewater treatment. *Ecological Engineering, 139*(May), Article 105582. https://doi.org/10.1016/j.ecoleng.2019.105582

Pascual, A., Álvarez, J. A., de la Varga, D., Arias, C. A., Van Oirschot, D., Kilian, R., & Soto, M. (2024). Horizontal flow aerated constructed wetlands for municipal wastewater treatment: The influence of bed depth. *Science of the Total Environment, 908*(October 2023). https://doi.org/10.1016/j.scitotenv.2023.168257

Priyadarshini, E., Parambil, A. M., Rajamani, P., Ponnusamy, V. K., & Chen, Y. H. (2023). Exposure, toxicological mechanism of endocrine disrupting compounds and future direction of identification using nano-architectonics. *Environmental Research, 225*(February), Article 115577. https://doi.org/10.1016/j.envres.2023.115577

Ramprasad, C., & Philip, L. (2016). Surfactants and personal care products removal in pilot scale horizontal and vertical flow constructed wetlands while treating greywater. *Chemical Engineering Journal, 284*, 458–468. https://doi.org/10.1016/j.cej.2015.08.092

Rivera-Utrilla, J., Sánchez-Polo, M., Ferro-García, M. Á., Prados-Joya, G., & Ocampo-Pérez, R. (2013). Pharmaceuticals as emerging contaminants and their removal from water a review. *Chemosphere, 93*(7), 1268–1287. https://doi.org/10.1016/j.chemosphere.2013.07.059

Russell, M. V., Messer, T. L., Repert, D. A., Smith, R. L., Bartelt-Hunt, S., Snow, D. D., & Reed, A. P. (2024). Influence of four veterinary antibiotics on constructed treatment wetland nitrogen transformation. *Toxics, 12*(5). https://doi.org/10.3390/toxics12050346

Sadasivuni, K. K., Rattan, S., Waseem, S., Brahme, S. K., Kondawar, S. B., Ghosh, S., Das, A. P., Chakraborty, P. K., Adhikari, J., Saha, P., & Mazumdar, P. (2019). Silver nanoparticles and its polymer nanocomposites—synthesis, optimization, biomedical usage, and its various applications. *Lecture Notes in Bioengineering*, 331–373. https://doi.org/10.1007/978-3-030-04741-2_11

Samal, K., Mahapatra, S., & Hibzur Ali, M. (2022). Pharmaceutical wastewater as Emerging Contaminants (EC): Treatment technologies, impact on environment and human health. *Energy Nexus, 6*(February), Article 100076. https://doi.org/10.1016/j.nexus.2022.100076

Santos, L. H. M. L. M., Gros, M., Rodriguez-Mozaz, S., Delerue-Matos, C., Pena, A., Barceló, D., & Montenegro, M. C. B. S. M. (2013). Contribution of hospital effluents to the load of pharmaceuticals in urban wastewaters: Identification of ecologically relevant pharmaceuticals. *Science of the Total Environment, 461–462*, 302–316. https://doi.org/10.1016/j.scitotenv.2013.04.077

Sanusi, I. O., Olutona, G. O., Wawata, I. G., & Onohuean, H. (2023). Occurrence, environmental impact and fate of pharmaceuticals in groundwater and surface water: A critical review. *Environmental Science and Pollution Research, 30*(39), 90595–90614. https://doi.org/10.1007/s11356-023-28802-4

Sathe, S. M., & Munavalli, G. R. (2019). Domestic wastewater treatment by modified bio-rack wetland system. *Journal of Water Process Engineering, 28*(February), 240–249. https://doi.org/10.1016/j.jwpe.2019.02.010

Satrovic, E., Cetindas, A., Akben, I., & Damrah, S. (2024). Do natural resource dependence, economic growth and transport energy consumption accelerate ecological footprint in the most innovative countries? The moderating role of technological innovation. *Gondwana Research, 127*, 116–130. https://doi.org/10.1016/j.gr.2023.04.008

Sharma, S., Asolekar, S. R., Thakur, V. K., & Asokan, P. (2024). Unlocking the potential: A paradigm-shifting approach for valorizing lignocellulosic waste biomass of constructed wetland enabling environmental and societal sustainability. *Industrial Crops and Products, 207*(P1), Article 117709. https://doi.org/10.1016/j.indcrop.2023.117709

Shelef, O., Gross, A., & Rachmilevitch, S. (2013). Role of plants in a constructed Wetland: Current and new perspectives. *Water, 5*(2), 405–419. https://doi.org/10.3390/w5020405

Singh, V., Pandey, B., & Suthar, S. (2018). Phytotoxicity of amoxicillin to the duckweed Spirodela polyrhiza: Growth, oxidative stress, biochemical traits and antibiotic degradation. *Chemosphere, 201*, 492–502. https://doi.org/10.1016/j.chemosphere.2018.03.010

Suthar, S., Chand, N., & Singh, V. (2023). Fate and toxicity of triclosan in tidal flow constructed wetlands amended with cow dung biochar. *Chemosphere, 311*, Article 136875. https://doi.org/10.1016/j.chemosphere.2022.136875

Tejeda, A., Torres-Bojorges, Á. X., & Zurita, F. (2017). Carbamazepine removal in three pilot-scale hybrid wetlands planted with ornamental species. *Ecological Engineering, 98*, 410–417. https://doi.org/10.1016/j.ecoleng.2016.04.012

Ting, W. H. T., Tan, I. A. W., Salleh, S. F., & Wahab, N. A. (2018). Application of water hyacinth (Eichhornia crassipes) for phytoremediation of ammoniacal nitrogen: A review. *Journal of Water Process Engineering, 22*(October 2017), 239–249. https://doi.org/10.1016/j.jwpe.2018.02.011

Vo, H. N. P., Bui, X. T., Nguyen, T. M. H., Koottatep, T., & Bandyopadhyay, A. (2018). Insights of the removal mechanisms of pharmaceutical and personal care products in constructed wetlands. *Current Pollution Reports, 4*(2), 93–103. https://doi.org/10.1007/s40726-018-0086-8

Vymazal, J. (2018). Does clogging affect long-term removal of organics and suspended solids in gravel-based horizontal subsurface flow constructed wetlands? *Chemical Engineering Journal, 331*, 663–674. https://doi.org/10.1016/j.cej.2017.09.048

Vymazal, J., Dvořáková Březinová, T., Koželuh, M., & Kule, L. (2017). Occurrence and removal of pharmaceuticals in four full-scale constructed wetlands in the Czech Republic—the first year of monitoring. *Ecological Engineering, 98*, 354–364. https://doi.org/10.1016/j.ecoleng.2016.08.010

Wu, H., Zhang, J., Ngo, H. H., Guo, W., Hu, Z., Liang, S., Fan, J., & Liu, H. (2015). A review on the sustainability of constructed wetlands for wastewater treatment: Design and operation. *Bioresource Technology, 175*, 594–601. https://doi.org/10.1016/j.biortech.2014.10.068

Xin, H., Yang, R., Lin, C., Zhan, J., & Yang, Q. (2024). Packing blockage and power generation of constructed wetland coupled microbial fuel cell systems using biochar as electrode and filler with different ratios. *Chemical Engineering Journal, 488*(2), Article 150978. https://doi.org/10.1016/j.cej.2024.150978

Yadav, K. K., Gupta, N., Kumar, A., Reece, L. M., Singh, N., Rezania, S., & Ahmad Khan, S. (2018). Mechanistic understanding and holistic approach of phytoremediation: A review on application and future prospects. *Ecological Engineering, 120*(May), 274–298. https://doi.org/10.1016/j.ecoleng.2018.05.039

Yang, Y., Zhao, Y., Liu, R., & Morgan, D. (2018b). Global development of various emerged substrates utilized in constructed wetlands. *Bioresource Technology, 261*(February), 441–452. https://doi.org/10.1016/j.biortech.2018.03.085

Yi, X., Tran, N. H., Yin, T., He, Y., & Gin, K. Y. H. (2017). Removal of selected PPCPs, EDCs, and antibiotic resistance genes in landfill leachate by a full-scale constructed wetlands system. *Water Research, 121*, 46–60. https://doi.org/10.1016/j.watres.2017.05.008

Yu, X., Yu, F., Li, Z., & Zhan, J. (2023). Occurrence, distribution, and ecological risk assessment of pharmaceuticals and personal care products in the surface water of the middle and lower reaches of the Yellow River (Henan section). *Journal of Hazardous Materials, 443*(PB), 130369. https://doi.org/10.1016/j.jhazmat.2022.130369

Zhang, D. Q., Tan, S. K., Gersberg, R. M., Sadreddini, S., Zhu, J., & Tuan, N. A. (2011). Removal of pharmaceutical compounds in tropical constructed wetlands. *Ecological Engineering, 37*(3), 460–464. https://doi.org/10.1016/j.ecoleng.2010.11.002

Zhang, Y., Lv, T., Carvalho, P. N., Arias, C. A., Chen, Z., & Brix, H. (2016). Removal of the pharmaceuticals ibuprofen and iohexol by four wetland plant species in hydroponic culture: Plant uptake and microbial degradation. *Environmental Science and Pollution Research, 23*(3), 2890–2898. https://doi.org/10.1007/s11356-015-5552-x

Zhang, L., Lv, T., Zhang, Y., Stein, O. R., Arias, C. A., Brix, H., & Carvalho, P. N. (2017a). Effects of constructed wetland design on ibuprofen removal—a mesocosm scale study. *Science of the Total Environment, 609*, 38–45. https://doi.org/10.1016/j.scitotenv.2017.07.130

Zhang, Y., Lv, T., Carvalho, P. N., Zhang, L., Arias, C. A., Chen, Z., & Brix, H. (2017b). Ibuprofen and iohexol removal in saturated constructed wetland mesocosms. *Ecological Engineering, 98*, 394–402. https://doi.org/10.1016/j.ecoleng.2016.05.077

Zhang, Y., Huang, G., Zhang, Q., Bi, X., Chu, H., Liu, Z., Luo, J., Bai, S., Mo, S., Wang, H., & Fu, M. (2024). Occurrence, distribution, and ecological risk assessment of pharmaceuticals and personal care products in the surface water of Lipu River China. *Environmental Research, 252*(P2), Article 118908. https://doi.org/10.1016/j.envres.2024.118908

Zhang, H., Wang, X. C., Zheng, Y., & Dzakpasu, M. (2023). Removal of pharmaceutical active compounds in wastewater by constructed wetlands: Performance and mechanisms. *Journal of Environmental Management, 325*(PA), 116478. https://doi.org/10.1016/j.jenvman.2022.116478

Chapter 25
Pioneering Paths: Navigating the Treatment Labyrinth of Wastewater Laden with Persistent Organic Pollutants

Tushar Rajesh Lakhani and Swamy Kurella

Introduction

Wastewater treatment is an essential process in maintaining healthy water circulation and preventing the pollution of natural environments (Mishra & Das, 2023; Mishra et al., 2024). One of the major challenges in wastewater treatment is the persistence of trace organics in wastewater effluent (Ilias & Schimmel, 1995). Persistent Organic Pollutants are among the principal water pollutants being a family of harmful compounds and are capable of persisting in the environment and bioaccumulate in living things. Due to their harmful health effects on both humans and the environment, finding effective methods of removing POPs from wastewater has become essential. Conventional techniques for wastewater treatment such as biological, physical and chemical treatment can hardly remove all POPs efficiently (Liu & Feng, 2009). This is due to the toxic and refractory nature of these pollutants, which can inhibit the growth and activity of microorganisms involved in biological treatment processes. Numerous advanced oxidation processes have been reported that can potentially solve the problems posed by persistent organic pollutants in industrial wastewater. The advanced oxidation processes create reactive oxygen species such as hydroxyl radicals capable of destroying the organic contents present in water wastewater (Wu et al., 2021). There are a number of advanced oxidation techniques reported as promising in the decomposition of aromatic compounds from wastewater, such as ozonation, photochemical oxidation, Fenton oxidation, and moist air oxidation. However, the current literature review gives information that not one of the above techniques can efficiently remove POPs at high rates.

In fact, recent research indicates that processes of photocatalytic oxidation-which involve illumination and a catalyst to produce reactive species for degradation of

T. R. Lakhani · S. Kurella (✉)
Department of Chemical Engineering, National Institute of Technology Srinagar, Srinagar, Jammu and Kashmir, India
e-mail: kurellaswamy@nitsri.ac.in

I. D. Behera et al. (eds.), *Emerging Micropollutants*, Sustainable Environmental Waste Management Strategies, https://doi.org/10.1007/978-3-032-03530-1_25

pollutants-suggest a viable advanced method in the treatment of water containing persistent organic pollutants. These processes, therefore, manifest the efficiency of increased degradation and generation of benign products, making this option more eco-friendly. Persistent organic pollutants significantly threaten the environment and human health. These pollutants, which also include medicines and dyes, are not easily broken down and can stay in the environment for a very long time. To reduce the risks related to the crisis of water shortly, the wastewater treatment plants and the sustainability of the existing facilities must be rationally assessed in light of the growing importance of sustainable urbanisation in developing countries and the rapid depletion of resources with the growing population (Eseoglu et al., 2022; Mishra & Das, 2021). Therefore, wastewater treatment is significant for sustainable development since it guards water resources, promotes effective management of waste, and encourages the use of renewable energy (Mishra et al., 2022b; Mishra and Das, 2022). Increased percentage of appropriate and safe technology in the treatment of wastewater from homes and industries is also among the expanded goals set by the sustainable development goals for the year 2030, which is under a theme focusing on wastewater treatment. Persistent organic pollutants remain a main problem in water treatments because these contaminants have lingering health and environmental effects. It has been considered a persistent organic pollutant one which remains long-lasting in the environment, accumulated to living things and poses the risks of ecosystem and human health. Recent studies have focused on the development of novel and effective treatment techniques that can remove persistent organic contaminants from wastewater, among which is promising the use of improved oxidation.

Characterization of Wastewater

The treatment of industrial wastewater is a large endeavor because it is composed of such a diverse range of complicated contaminants. The treatment of industrial effluent is important to preserve industrial processes, ensure legality, and preserve the environment (Kumar et al., 2023; Mishra et al., 2022a). The chemical composition of industrial effluent can be widely varied based on the nature of the specific type of business (Atkinson et al., 2019). Chemical pollutants that include acids, bases, heavy metals, and organic compounds are commonly found in this effluent, in addition to solvents used in industrial operations (Priyadarsini & Das, 2024, 2025). Infections, organic waste, and nutrients like phosphorus and nitrogen are additional biological pollutants that are often present. The complexity of industrial effluent is enhanced by solid particles such as sediments, suspended solids, and colloidal materials (Abu Hasan et al., 2020). In addition, high-temperature waste water must not be disregarded because it contributes to thermal pollution, which affects aquatic habitats. To deal with these complex issues, effective industrial water treatment procedures should be implemented. Apart from ensuring a safe environment, these procedures would also abide by regulatory requirements, thus providing an industrial activity long-term survival basis.

General Mechanism for Wastewater Treatment

Industrial waste water treatment consists of several stages which remove contaminants in order to safeguard equipment and the environment. Preliminary treatment is the removal of large particles and debris, opening the way to subsequent treatments. Primary treatment eliminates suspended particles, settleable organic waste, and floatable oils and greases. Secondary treatment degrades nutrients and dissolved organic waste using microorganisms (Ribeiro & Nunes, 2021). Tertiary treatment enhances the quality of the effluent by using advanced techniques. The techniques to kill bacteria may involve chlorination or UV disinfection. For tough discharge restrictions, techniques such as membrane bioreactors, reverse osmosis, and adsorption are deployed. These hard measures maintain their legal requirements and uphold environmental responsible operations while fostering sustainability in businesses.

Managing Routesand Eco-friendly Methods to Overcome Industrial Wastewater

Industrial wastewater treatment is a difficult process because of fluctuating pollutant loads and strict environmental regulations. The process requires large amounts of energy, chemicals, and water, thereby increasing the cost of operation and affecting the business budget. Disposal of sludge and leftover material is also a challenge (Wahab et al., 2021). Businesses need to constantly adapt to changing rules, which requires large investments in treatment facilities. To attain environmental sustainability in industries, the usage of resources should reduce, new water reuse methods develop, and minimize the impact of environmental procedures utilized in wastewater treatment. The demand for effective wastewater management has grown as the need for fresh water sources increases across the globe. For the minimization of the water footprint of industries, new attempts are to be made through environmentally friendly technologies and practices. New innovative wastewater treatment technologies are developed that lead to the formulation of solutions regarding water treatment. It is, however, for the sustainable solutions to be established as a collaboration among industries, environmental experts, and regulatory organizations (Palani et al., 2023). These eco-friendly solutions are water reusing and recycling, energy efficiency improvement, anaerobic digestion, biogas recovery, green infrastructure, and constructed wetlands. Development in membrane technology, decentralized systems, and online monitoring has made water treatment methods efficient and more environmentally friendly. Recovering resources from wastewater is also gaining support, turning waste into valuable assets (Lee & Jepson, 2020).

Treatment of Wastewater Using Nanoparticles

Nanoparticles remove various numbers of pollutants such as Arsenic, Heavy metals and Organic Pollutants. Many materials, including metals, metal oxides, polymers, and carbon-based compounds, can be used to form nanoparticles. Such materials can be designed to possess specific properties, which are particularly useful for the treatment of different kinds of water pollutants, such as high reactivity, adsorption capacity, and catalytic activity. This contemporary technique of water treatment utilizes the inherent properties of the nanoscale material for better productivity and efficiency of the water treatment process (Al-Buriahi et al., 2022). A nanoparticle is defined as very tiny particles; most of the particles are of 1–100 nm. The nanoparticles' small size and a high surface area-to-volume ratio confer on them specific chemical, physical, and catalytic properties, very suitable for many applications in water treatment (Atkinson et al., 2019). Indeed, there are plenty of possible advantages in water treatment with nanoparticles. On the opposite side, difficulty also occurs there, along with factors to be concerned: such as toxicity for nanoparticles, its tendency to collect and settle overtime, the way they may possibly be released to the environment, and finally the necessity of techniques scalable to their manufacture low cost.

Based on the critical appraisal of the literature, the objectives of the work are as follows:

- Presence and effects of POPs and their effective removal using nanoparticles.
- Future research prospects for effective removal of POPs by nanoparticles.

The literature is thoroughly examined in order to meet the aforementioned goals, and its contents are arranged as follows:

- Sources and health hazards of POPs
- Emerging addressal of nanoparticles in removal of POPs.
- Future prospects in this field.

Sources and Health Hazards of Persistent Organic Pollutants (POPs)

POPs are among the most toxic chemicals since they do not readily decompose and persist for a long time in the environment. Once these chemicals reach the water sources, it becomes a disaster. Among the pollutants are dioxin and polychlorinated biphenyls (PCBs), which cause a variety of diseases. They lead to failure in the endocrine system and cause hormonal imbalances and reproductive issues (Wu et al., 2021). In addition to that, some POPs are carcinogenic, thus leading to cancer susceptibility upon multiple exposure to the contaminated water. These pollutants also have been linked to developmental abnormalities and neurological disorders. Besides that,

POPs depress the immunity, enhance the danger of reproductive disorders, and bioaccumulate uncomfortably in the food chain in the sense that they affect aquatic life and also humans who ingest infected seafood (Fei et al., 2022). This impacts the ecosystems negatively. Food webs and biodiversity are destroyed. Improving waste management, industrial regulation, and adopting cleaner technologies is very essential in order to ensure that POPs will not pollute the sources of water for both humans and the environment. International agreements, such as the Stockholm Convention, reduce the production and usage of POPs. A family of chemical substances called Persistent Organic Pollutants (POPs) has caught the attention of all around the world because of their prevalence, persistence in the environment (Ighalo et al., 2021), and potential harm to human health. These substances are deposited in different environmental compartments, especially water bodies, and do not decompose under normal conditions.

Persistent organic pollutants have been known to be a major environmental concern and threat to public health. These come from various sources that include industrial, agricultural, and urban sectors (Farhan Hanafi & Sapawe, 2020; Singh et al., 2023). Industrial activities release hazardous materials into the environment, which can contaminate aquatic ecosystems. Agriculture, essential for human survival, also contributes to POPs pollution through organochlorine compounds (Diagboya & Dikio, 2018). Urban runoff from treated fields and urban construction activities further exacerbate the problem. Persistent Organic Pollutants also travel great distances, causing global distillation and escalating contamination. These contaminants bioaccumulate and biomagnify as they enter aquatic sediments, organisms, and food chains. This upheaval negatively impacts the diversity of species and the health of aquatic organisms. Humans, often at the top of terrestrial food chains, are vulnerable to POP exposure through contaminated seafood, which can cause diseases such as cancer (Vasseghian et al., 2021), reproductive disorders, and developmental abnormalities. Therefore, concerted efforts must be undertaken to prevent the emission of POPs and safeguard the fragile equilibrium of the natural world.

The health of people is seriously endangered by persistent organic pollutants, with the risk of developing cancer, reproductive problems, and developmental problems. Persistent Organic Pollutants, like Polychlorinated Biphenyls (PCBs), dioxins, and furans, can be bioaccumulated in fish and seafood, can disrupt hormone regulation, and cause birth defects and poor baby development (Weber et al., 2019). They exhibit a long-distance transport phenomenon, which makes them vulnerable to being exposed to indigenous tribes living in remote areas. Some of the main strategies to remove POPs include source reduction, contemporary techniques for wastewater treatment, phytoremediation, bioremediation, management of sediments, monitoring, and assessment of water bodies and biota (Guan et al., 2021), public awareness, and education. These all-inclusive solutions support continued activity to reduce POP contamination and efforts in protecting ecosystems. The presence of POPs globally requires stern regulatory measures along with international cooperation to tackle this severe crisis (Boulkhessaim et al., 2022). All these conventional techniquies had many drawbacks such as longer exhaustion times, regular management difficulties, land and soil management, and most important economical problems.

Removal of POPs Using Nanoparticles

The nanoparticles, which are labelled as Silver Based, provide a variety of remedies for various pollutants. Concerns about water quality can be addressed by using silver nanoparticles (AgNPs) and 2-hydroxypropyl that has an Ag-NP cap to effectively adsorb contaminants such 4-NP, MO, and rhodamine B. Additionally, 4-NP, MB, and MO pollutants can be adsorbed by Fe_3O_4/PPy-MAA/Ag and silver-doped $Mg_4Ta_2O_9$ nanoparticles. Silver nanoparticles added to cellulose polymer paper are specifically designed to remove different nitrophenol chemicals. Methylene blue is effectively adsorbed by Ag@MGO-TA/Fe^{3+} nanocomposites and Ag/rGO nanocomposites with thin film composite membranes. Contrarily, graphene-based nanocomposites and nano-photocatalysts exhibit excellent adsorption and catalytic characteristics when used to target persistent organic pollutants (POPs) and halogenated xenobiotic chemicals, respectively (Fei et al., 2022; Karthigadevi et al., 2021). These substances have high clearance rates of up to 99%, and some of them are regenerable, demonstrating their potential to address a variety of pollution-related problems, such as cholera, neurological disorders, respiratory problems, and endocrine disruption. Silver nanoparticles with a coating of 2 hydroxypropyl are known for their removal of pollutants such as 4 nitrophenol (4 NP), methylene orange (MO) and rhodamine B. This effectiveness mechanism lies in the binding of the pollutants to the surface of these nanoparticles. The removal, the silver nanoparticles (AgNPs) are very effective where it relates to the adsorption capabilities to submerge substances like 4 NP, MO and rhodamine B (Palani et al., 2023). The primary mechanism working here is adsorption, in which the pollutants attach themselves to the surface of these nanoparticles. A composite nanoparticle of Fe_3O_4/PPy MAA/Ag was utilized for 4 NP and methylene blue (MB) via adsorption, and the results were reported (Roy & Tonelli, 2023). It combines the properties of Fe_3O_4 with the adsorption capabilities offered by silver. Targeting rhodamine B for removing this pollutant from the environment through adsorption, silver doped $Mg_4Ta_2O_9$ nanoparticles have been developed. The focus was on utilizing adsorption properties for elimination. Combining cellulose polymer paper with nanoparticles had resulted in a unique nanoparticle combination that effectively absorbed substances like 4 NP, 2 nitrophenol (2 NP), 2 nitroaniline (2 NA) and trinitrophenol (TNP). The cellulose matrix enhances the efficiency of removal. A nanocomposite called Ag@MGO TA/Fe^{3+} has demonstrated its potential, in treating dye contaminated wastewater by removing methylene blue through adsorption. The results showed promising solutions and it's applications in wastewater treatment. The Ag/rGO nanocomposite and Ag/rGO/CA/TFC membranes have proven to be effective in adsorbing methylene blue making them suitable for water purification applications, in membrane technology. It was efficient at adsorbing phenols. They have reported that the desired contaminants can be eliminated using these nanoparticles, which can be repaired using ethanol, have shown 80–85% removal by adsorption. With a remarkable removal rate of up to 96%, they can help in removing cholera-related pollutants by concentrating on halogenated xenobiotic chemicals. Their regeneration is aided by biogeneration. To combat POPs, those nanoparticles use catalytic processes. The authorshave reported the accomplishment of a phenomenal removal of up to 99%, for some contaminants.

An overview of the several carbon-based and other nanomaterials used for healthcare and environmental remediation is deatiled in the literature. The first category includes MOF/carbon nanotube composites, which are useful in treating neurological illnesses because they can adsorb phenols with up to 80–85% effectiveness (Pi et al., 2018). With a mass of 6.4 g and a nanoscale dimension of 2 nm, the method uses ethanol and produces remarkable results. Graphene-based nanocomposites have remarkable adsorption properties that enable them to remove up to 96% of halogenated xenobiotic chemicals (Karthigadevi et al., 2021). These composites, which range in size from 1 to 10 nm and use Biogeneration, efficiently fight cholera. Unclassified nanomaterials are included in another category. These include nano-photocatalysts, which can remove up to 99% of the persistent organic pollutants (POPs) linked to neurological illnesses by catalysis (Fei et al., 2022). Furthermore, respiratory problems are addressed by TiO_2 and ZnO semiconductors, which eliminate up to 90% of POPs in the 1–100 nm size range (Boulkhessaim et al., 2022).

Specific Characteristics and Types of Nanoparticles Used to Remove POPs

Due to their efficiency and adaptability, nanoparticles with high surface area and electrical conductivity have emerged as important for wastewater treatment. Graphene has a high surface area-to-volume ratio, and it is quite effective in the adsorption of heavy metals and organic contaminants (Fei et al., 2022; Roy & Tonelli, 2023). Titanium dioxide nanoparticles are known for their photocatalytic capabilities, converting organic pollutants into harmless byproducts when exposed to ultraviolet light. Carbon nanotubes have strong adsorption properties due to their high porosity. Iron and zero-valent iron nanoparticles can effectively remove pollutants through chemical processes. Different types of nanoparticles, including nanoparticles, nanotubes, nanofibers, and nanosheets, have unique structures and capabilities. Iron nanoparticles (FeNPs) and silver nanoparticles (AgNPs) are essential for heavy metal removal, catalytic destruction, and disinfection (Mustapha et al., 2020). Carbon nanotubes (CNTs) are known for their mechanical strength and small passageways, making them ideal adsorbents for various contaminants. Electrospun nanofibers have a fibrous structure for filtering, separation, and adsorption operations. By customizing and using these nanomaterials, scientists and engineers are creating a future where wastewater treatment is efficient, accurate, and sustainable (Baby et al., 2019).

Future Prospects of Emerging Nanoparticles to Remove POPs

Nanoparticle-based water pollution control is promising due to their ability to target specific pollutants and water conditions. Researchers are exploring the creation of multifunctional nanoparticles with catalytic, adsorption, and magnetic properties, which could streamline separation procedures (Kamali et al., 2019). Collaboration among academic fields, such as material science, chemistry, engineering, and environmental science, is crucial for innovation. However, there are still many research gaps and challenges to overcome. Future research should evaluate the environmental impact of nanoparticles in water treatment, develop affordable techniques for nanoparticle regeneration, and focus on scalable manufacturing techniques and cost-effective materials. Customizing nanoparticles to target specific contaminants is also essential (Eid et al., 2023). Nanoparticle-enhanced membrane filtration, targeting emerging contaminants like pharmaceuticals, personal care products, and microplastics, and using catalysts from nanoparticles can improve advanced oxidation processes (Das & Bal, 2025; Das et al., 2023). Additionally, nanoparticles have been identified as a workable method for sustainable groundwater remediation (Heck et al., 2019). Future research should focus on creating intelligent nanoparticles that can respond to inputs, providing controlled pollution release or removal (Reddy et al., 2013).

Summary

The treatment of wastewater contaminated with persistent organic pollutants (POPs) is a complex and urgent issue. Nanoparticles offer better removal efficiencies for POPs in wastewater. Researches have suggested various nanoparticles for wastewater treatment and their availability. Despite showing effective removal efficiencies for POPs, nanoparticles cause respiratory problems in people. A number of the nanoparticles reported in the literature have the ability to eliminate contaminants linked to a range of health issues. For example, halogenated xenobiotics, dyes, and carbon-based nanocomposites are efficient at removing pollutants like these that can lead to cholera and endanger human health. Therefore, even if these nanoparticles have promising capacities to remove pollutants, it is important to think about any potential health effects and adopt the appropriate safety measures. It requires continuous innovation and research in wastewater treatment technologies to ensure the safety and sustainability of water resources. Awareness and education about POPs' sources and effects are crucial for public support and policy changes. Regulatory bodies must enforce stricter regulations to curb the POPs discharge. Integrating nature-based solutions like wetlands, green roofs, and constructed wetlands can enhance treatment efficiency and promote biodiversity. Addressing POPs requires interdisciplinary collaboration, innovative technologies, strict regulations, public awareness, and the incorporation of nature-based solutions.

Acknowledgements The authors express their sincere thanks to National Institute of Technology Srinagar, India.

References

Abu Hasan, H., Muhammad, M. H., & Ismail, N. I. (2020). A review of biological drinking water treatment technologies for contaminants removal from polluted water resources. *Journal of Water Process Engineering, 33*, 101035. https://doi.org/10.1016/j.jwpe.2019.101035

Al-Buriahi, A. K., Al-Gheethi, A. A., Senthil Kumar, P., Radin Mohamed, R. M. S., Yusof, H., Alshalif, A. F., & Khalifa, N. A. (2022). Elimination of rhodamine B from textile wastewater using nanoparticle photocatalysts: A review for sustainable approaches. *Chemosphere, 287*, 132162. https://doi.org/10.1016/j.chemosphere.2021.132162

Atkinson, A. J., Apul, O. G., Schneider, O., Garcia-Segura, S., & Westerhoff, P. (2019). Nanobubble technologies offer opportunities to improve water treatment. *Accounts of Chemical Research, 52*, 1196–1205. https://doi.org/10.1021/acs.accounts.8b00606

Baby, R., Saifullah, B., & Hussein, M. Z. (2019). Carbon nanomaterials for the treatment of heavy metal-contaminated water and environmental remediation. *Nanoscale Research Letters, 14*. https://doi.org/10.1186/s11671-019-3167-8

Boulkhessaim, S., Gacem, A., Khan, S. H., Amari, A., Yadav, V. K., Harharah, H. N., Elkhaleefa, A. M., Yadav, K. K., Rather, S. U., Ahn, H. J., & Jeon, B. H. (2022). Emerging trends in the remediation of persistent organic pollutants using nanomaterials and related processes: A review. *Nanomaterials*. https://doi.org/10.3390/nano12132148

Das, P., & Bal, M. (2025). Modeling and optimization study for microplastic removal from aquatic medium using *Chroococcidiopsis* species. *Chemical Engineering Science, 304*, 121085. https://doi.org/10.1016/J.CES.2024.121085

Das, P., Halder, G., & Bal, M. (2023). A critical review on remediation of microplastics using microalgae from aqueous system. *Science of the Total Environment, 898*, 166425. https://doi.org/10.1016/J.SCITOTENV.2023.166425

Diagboya, P. N. E., & Dikio, E. D. (2018). Silica-based mesoporous materials; emerging designer adsorbents for aqueous pollutants removal and water treatment. *Microporous and Mesoporous Materials*. https://doi.org/10.1016/j.micromeso.2018.03.008

Eid, A. M., Fouda, A., Hassan, S. E. D., Hamza, M. F., Alharbi, N. K., Elkelish, A., Alharthi, A., & Salem, W. M. (2023). Plant-based copper oxide nanoparticles; biosynthesis, characterization, antibacterial activity, tanning wastewater treatment, and heavy metals sorption. *Catalysts, 13*, 348. https://doi.org/10.3390/catal13020348

Eseoglu, G., Yapsakli, K., Tozan, H., & Vayvay, Ö. (2022). A novel fuzzy framework for technology selection of sustainable wastewater treatment plants based on TODIM methodology in developing urban areas. *Science and Reports, 12*, 8800. https://doi.org/10.1038/s41598-022-12643-1

Farhan Hanafi, M., & Sapawe, N. (2020). A review on the water problem associate with organic pollutants derived from phenol, methyl orange, and remazol brilliant blue dyes. *Materials Today: Proceedings, 31*, A141–A150. https://doi.org/10.1016/j.matpr.2021.01.258

Fei, L., Bilal, M., Qamar, S. A., Imran, H. M., Riasat, A., Jahangeer, M., Ghafoor, M., Ali, N., & Iqbal, H. M. N. (2022). Nano-remediation technologies for the sustainable mitigation of persistent organic pollutants. *Environmental Research, 211*, 113060. https://doi.org/10.1016/j.envres.2022.113060

Guan, G., Pan, J. H., & Li, Z. (2021). Innovative utilization of molecular imprinting technology for selective adsorption and (photo)catalytic eradication of organic pollutants. *Chemosphere, 265*, 129077. https://doi.org/10.1016/j.chemosphere.2020.129077

Heck, K. N., Garcia-Segura, S., Westerhoff, P., & Wong, M. S. (2019). Catalytic converters for water treatment. *Accounts of Chemical Research, 52*, 906–915. https://doi.org/10.1021/acs.accounts.8b00642

Ighalo, J. O., Sagboye, P. A., Umenweke, G., Ajala, O. J., Omoarukhe, F. O., Adeyanju, C. A., Ogunniyi, S., & Adeniyi, A. G. (2021). CuO nanoparticles (CuO NPs) for water treatment: A review of recent advances. *Environmental Nanotechnology, Monitoring and Management*. https://doi.org/10.1016/j.enmm.2021.100443

Ilias, S., & Schimmel, K. A. (1995). Membrane bioreactor model for removal of organics from wastewater. *Journal of the Air and Waste Management Association, 45*, 615–620. https://doi.org/10.1080/10473289.1995.10467391

Kamali, M., Persson, K. M., Costa, M. E., & Capela, I. (2019). Sustainability criteria for assessing nanotechnology applicability in industrial wastewater treatment: Current status and future outlook. *Environment International, 125*, 261–276. https://doi.org/10.1016/j.envint.2019.01.055

Karthigadevi, G., Manikandan, S., Karmegam, N., Subbaiya, R., Chozhavendhan, S., Ravindran, B., Chang, S. W., & Awasthi, M. K. (2021). Chemico-nanotreatment methods for the removal of persistent organic pollutants and xenobiotics in water—A review. *Bioresource Technology*. https://doi.org/10.1016/j.biortech.2021.124678

Kumar, M., Pandit, S., Patel, V., Khanna, N., Nag, M., Lahiri, D., Ray, R. R., Das, A., & Das, D. (2023). Maximization of energy recovery from starch processing wastewater by thermophilic dark fermentation coupled with microbial fuel cell technology. *Geomicrobiology Journal, 1–12*, 392–403. https://doi.org/10.1080/01490451.2023.2209555

Lee, K., & Jepson, W. (2020). Drivers and barriers to urban water reuse: A systematic review. *Water Security, 11*, 100073. https://doi.org/10.1016/j.wasec.2020.100073

Liu, J., & Feng, Y. (2009). Investigation on the electrocatalytic characteristics of SnO_2 electrodes with nanocoating prepared by electrodeposition method. *Science in China, Series e: Technological Sciences, 52*, 1799–1803. https://doi.org/10.1007/s11431-009-0110-8

Mishra, S., & Das, A. P. (2021). Current treatment technologies for removal of microplastic and microfiber pollutants from wastewater. In *Wastewater treatment: cutting-edge molecular tools, techniques and applied aspects* (pp. 237–251). https://doi.org/10.1016/B978-0-12-821881-5.00011-8

Mishra, S., & Das, A. P. (2022). Treatment of the wastewater polluted with synthetic microfiber released from washing machine. In B. B. Das, H. Hettiarachchi, P. K. Sahu, S. Nanda (Eds.), *Lecture notes in civil engineering* (pp. 109–117). Springer Science and Business Media Deutschland GmbH. https://doi.org/10.1007/978-981-16-7509-6_9/COVER

Mishra, S., & Das, A. P. (2023). Microbial remediation of synthetic microfiber contaminated wastewater. In M. P. Shah (Ed.), *Microbial technologies in industrial wastewater treatment* (pp. 337–350). Springer Singapore. https://doi.org/10.1007/978-981-99-2435-6_17

Mishra, S., Dash, D., & Das, A. P. (2022a). Detection, characterization and possible biofragmentation of synthetic microfibers released from domestic laundering wastewater as an emerging source of marine pollution. *Marine Pollution Bulletin, 185*, 114254. https://doi.org/10.1016/J.MARPOLBUL.2022.114254

Mishra, S., Dash, D., & Das, A. P. (2024). Isolation and molecular characterization of polyester-degrading bacterial strains from municipal wastewater and their application in microplastic degradation in contaminated wastewater. *Geomicrobiology Journal*. https://doi.org/10.1080/01490451.2024.2376101

Mishra, S., Singh, R. P., Rout, P. K., & Das, A. P. (2022b). Membrane bioreactor (MBR) as an advanced wastewater treatment technology for removal of synthetic microplastics. In *Development in wastewater treatment research and processes* (pp. 45–60). https://doi.org/10.1016/B978-0-323-85583-9.00022-3

Mustapha, S., Ndamitso, M. M., Abdulkareem, A. S., Tijani, J. O., Shuaib, D. T., Ajala, A. O., & Mohammed, A. K. (2020). Application of TiO_2 and ZnO nanoparticles immobilized on clay in wastewater treatment: A review. *Applied Water Science*. https://doi.org/10.1007/s13201-019-1138-y

Palani, G., Trilaksana, H., Sujatha, R. M., Kannan, K., Rajendran, S., Korniejenko, K., Nykiel, M., & Uthayakumar, M. (2023). Silver nanoparticles for waste water management. *Molecules*. https://doi.org/10.3390/molecules28083520

Pi, Y., Li, X., Xia, Q., Wu, J., Li, Y., Xiao, J., & Li, Z. (2018). Adsorptive and photocatalytic removal of persistent organic pollutants (POPs) in water by metal-organic frameworks (MOFs). *Chemical Engineering Journal*. https://doi.org/10.1016/j.cej.2017.12.092

Priyadarsini, S., & Das, A. P. (2024). Analytical and structural characterization of waste lithium-ion batteries for their effective recycling strategy. *Environmental Chemistry and Ecotoxicology*. https://doi.org/10.1016/j.enceco.2024.12.004

Priyadarsini, S., & Das, A. P. (2025). Lithium bioleaching: A review on microbial-assisted sustainable technology for lithium bio-circularity. *Journal of Water Process Engineering, 69*, 106744. https://doi.org/10.1016/j.jwpe.2024.106744

Reddy, K. J., McDonald, K. J., & King, H. (2013). A novel arsenic removal process for water using cupric oxide nanoparticles. *Journal of Colloid and Interface Science, 397*, 96–102. https://doi.org/10.1016/j.jcis.2013.01.041

Ribeiro, P., & Nunes, M. I. (2021). Recent trends and developments in Fenton processes for industrial wastewater treatment—A critical review, *197*. https://doi.org/10.1016/j.envres.2021.110957

Roy, A., & Tonelli, F. (2023). Green nanoremediation: Sustainable management of environmental pollution.

Singh, P., Rani, N., Kumar, S., Kumar, P., Mohan, B., Pallavi, Bhankar, V., Kataria, N., Kumar, R., & Kumar, K. (2023). Assessing the biomass-based carbon dots and their composites for photocatalytic treatment of wastewater. *Journal of Cleaner Production 413*, 137474. https://doi.org/10.1016/j.jclepro.2023.137474

Vasseghian, Y., Hosseinzadeh, S., Khataee, A., & Dragoi, E. N. (2021). The concentration of persistent organic pollutants in water resources: A global systematic review, meta-analysis and probabilistic risk assessment. *Science of the Total Environment, 796*, 149000. https://doi.org/10.1016/j.scitotenv.2021.149000

Wahab, M. A., Li, L., Li, H., & Abdala, A. (2021). Silver nanoparticle-based nanocomposites for combating infectious pathogens: Recent advances and future prospects. *Nanomaterials*. https://doi.org/10.3390/nano11030581

Weber, R., Bell, L., Watson, A., Petrlik, J., Paun, M. C., & Vijgen, J. (2019). Assessment of pops contaminated sites and the need for stringent soil standards for food safety for the protection of human health. *Environmental Pollution, 249*, 703–715. https://doi.org/10.1016/j.envpol.2019.03.066

Wu, A., Zhao, X., Wang, J., Tang, Z., Zhao, T., Niu, L., Yu, W., Yang, C., Fang, M., Lv, H., Liu, S., & Wu, F. (2021). Application of solid-phase extraction based on magnetic nanoparticle adsorbents for the analysis of selected persistent organic pollutants in environmental water: A review of recent advances. *Critical Reviews in Environmental Science and Technology, 51*, 44–112. https://doi.org/10.1080/10643389.2020.1720493

Chapter 26
Biological Wastewater Treatment: Overcoming Barriers in Micropollutant Degradation

Saithala Mounika, B. B. Nayak, P. N. Siva Prasad, and G. Sree Pooja

Introduction

Wastewater is generated as a result of various human activities, including domestic, industrial, and agricultural operations. This results in the pollution of water bodies with a diverse array of organic and inorganic contaminants, such as microplastics, heavy metals, xenobiotics, and elevated levels of nitrates and phosphates. Such contamination exerts considerable pressure on the food chain and, ultimately, on the fundamental aspects of life (Mohanty et al., 2024). As global populations grow and economic development accelerates, the volume of wastewater produced globally has increased dramatically (Mishra & Das, 2023; Mishra et al., 2024). Water pollution continues to be a critical concern, posing significant risks to the environment, ecosystems, and human health, particularly through the contamination of drinking water sources (Mishra & Das, 2021; Mishra et al., 2022b). One-third of the global population currently resides in regions facing water stress, a figure projected to increase to two-thirds by 2025 (Singhal & Gracia, 2006). It is imperative to explore

S. Mounika
Department of Agronomy, Professor Jayashankar Telangana Agricultural University, Hyderabad, Telangana, India

B. B. Nayak
School of Sciences, Woxsen University, Kamkole, Telangana, India

P. N. Siva Prasad (✉)
Subject Matter Specialist (Crop Production), Dr.K.L.Rao KVK Garikapadu, ANGRAU-ATARI-X, NTR (D) 521175, Andhra Pradesh, India
e-mail: sivassac007@gmail.com

G. Sree Pooja
Department of Agricultural Extension Education, Professor Jayashankar Telangana Agricultural University, Hyderabad, Telangana, India

I. D. Behera et al. (eds.), *Emerging Micropollutants*, Sustainable Environmental Waste Management Strategies, https://doi.org/10.1007/978-3-032-03530-1_26

alternative water sources to support the survival of communities affected by water scarcity. Wastewater, often disregarded as a viable water resource, has the potential to serve this purpose if suitable treatment technologies are applied. A major challenge in producing clean water from wastewater is the presence of micropollutants, which can remain biologically active even at extremely low concentrations. The removal of these substances using current physico-chemical techniques is often financially unfeasible, posing a significant challenge for wastewater treatment processes. In contrast, biological treatment processes tend to be more economical in terms of both construction and operation; however, current designs do not consistently produce water of sufficient quality to meet the requirements for various uses. Although there have been notable advancements in biological wastewater treatment, which initially focused on the removal of organic carbon and later expanded to include nitrogen and phosphorus, these systems primarily address substances present at a milligram-per-litter concentration. Recent advancements in sensor technology have significantly improved the detection of biologically active substances, commonly referred to as emerging contaminants or micropollutants, at nanogram-per-liter levels in the environment. These pollutants encompass a range of substances, including industrial chemicals, insecticides, steroid hormones, personal care products, and pharmaceuticals. Research highlights the critical need to remove these contaminants from treated wastewater at such low concentrations due to their potential risks to human and animal health, as well as their adverse ecotoxicological impacts. For instance, research has indicated that exposure to bisphenol A is linked to alterations in male sex hormones, while synthetic estrogens from contraceptive pill residues also pose significant risks.

Biological wastewater treatment is essential for fostering environmental sustainability and advancing renewable energy efforts by mitigating the negative impacts of wastewater disposal, reclaiming valuable resources, and producing clean energy and bio-based products (Kumar et al., 2023; Mishra & Das, 2022). Despite these benefits, the approach faces challenges such as significant capital and operational expenses, inconsistencies in reactor performance, scalability limitations, concerns about product quality and safety, and regulatory and societal hurdles (Sravan et al., 2024).The presence of pollutants, including persistent organic pollutants (POPs), nanoparticles, and endocrine disruptors, comes from a wide variety of sources like industrial effluents, agricultural runoff, healthcare waste, and domestic sewage. All these are inadequately monitored in terms of their discharge into the environment. This scenario presents significant risks to both human health and ecosystems, emphasizing the urgent need for advanced and efficient wastewater treatment methods. Such contaminants can hinder the efficiency of biological treatment by inhibiting or killing the microorganisms, disturbing their metabolic processes, or generating toxic byproducts. Conventional biological methods frequently fall short of achieving comprehensive wastewater treatment, underscoring the need for the development of advanced integrated biological and bioelectrochemical remediation systems (Mishra et al., 2022a). Such systems can improve energy and resource recovery while facilitating low-carbon generation. Biological processes, including nitrification, denitrification, phosphorus accumulation, and sulfate reduction, are highly effective in eliminating

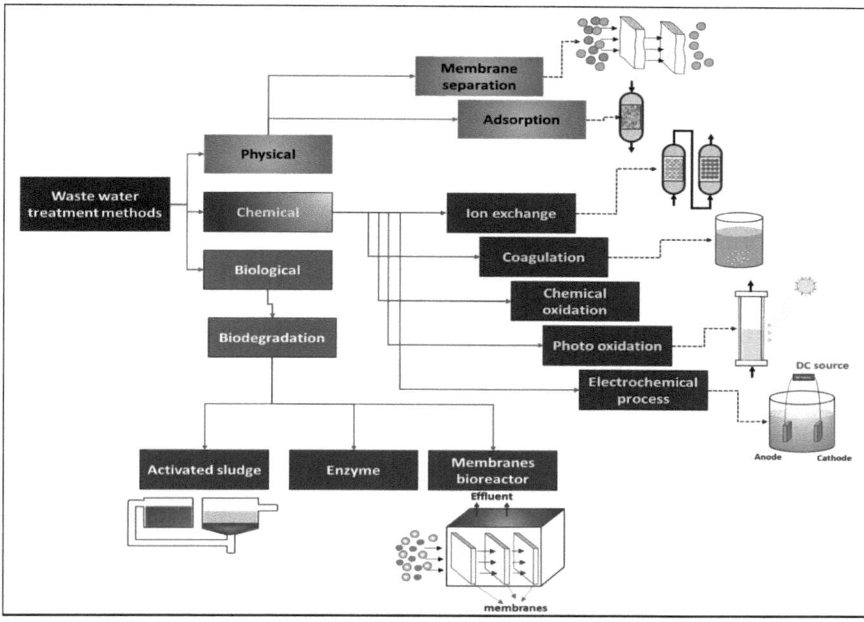

Fig. 26.1 Types of wastewater treatment methods (Alexandar et al., 1999)

nutrients such as nitrogen, phosphorus, and sulfur from wastewater. Thus, these nutrients can be used in the production of fertilizers, bioplastics, or chemicals which are employed in/for agricultural and industrial sectors. The types of wastewater treatment methods are illustrated in Fig. 26.1.

Energy recovery from wastewater is possible through various anaerobic methods, including anaerobic digestion, dark fermentation, acidogenesis, acetogenesis, methanogenesis, and microbial fuel cells. These techniques convert the organic content of wastewater into biogas, hydrogen, or electricity, which can be utilized for heating, lighting, or operating treatment plants and other critical infrastructures (Sravan et al., 2024). Incorporating nanotechnology introduces advanced materials like nanoparticles, nanofibers, and nanocomposites, which enhance biological treatment processes by functioning as catalysts, carriers, sensors, or antimicrobial agents. The post-treatment challenges and limitations with these advanced treatments must be focused on for their energy efficiency improvement along with economic viability. Moreover, it has huge potential for sustainable combination with other treatment methods applied in a way to adhere to circular economy principles within biorefineries.

Biological treatment methods rely on microorganisms to break down or convert the nutrients and organic matter in wastewater (Das et al., 2015; Paul et al., 2023; Pradhan et al., 2023). Various methods are employed for wastewater treatment, including aerobic, anaerobic, anoxic, and facultative processes, along with bioremediation, phytoremediation, and mycoremediation (Das et al., 2024; Gavrilescu et al., 2015; Kanungo et al., 2024). Aerobic techniques, such as activated sludge systems, trickling filters, rotating biological contactors, and oxidation ponds, rely on oxygen to break down organic matter. Conversely, anaerobic processes, including anaerobic digesters and up-flow anaerobic sludge blanket reactors, operate without oxygen and generate methane as a byproduct. Anoxic treatments, like denitrification filters and anoxic zones within activated sludge systems, utilize nitrate as an electron acceptor, converting it into nitrogen gas. Phytoremediation involves the use of constructed wetlands, floating aquatic plants, and terrestrial vegetation to absorb, degrade, or accumulate contaminants in wastewater. These biological treatment methods are recognized for their efficiency, cost-effectiveness, and environmentally sustainable approach to removing organic matter, nutrients, pathogens, and certain hazardous substances. However, their performance is influenced by factors such as temperature, pH, oxygen availability, and toxicity, necessitating careful monitoring and management.

In brief, wastewater management utilizes physical, chemical, and biological methods, alone or combined, to handle diverse wastewater types and achieve treatment objectives, with the choice of approach depending on wastewater characteristics, desired effluent quality, resources, environmental impact, and cost-effectiveness. Various methods for wastewater treatment are summarized in Table 26.1.

Current Wastewater Treatment Approaches

Recent advancements in the domain of water environmental protection have undergone a notable transformation. The emphasis has progressively transitioned from conventional, established technologies to more sustainable, cost-efficient, and environmentally friendly alternatives, collectively known as "green technologies." This transition offers substantial opportunities for addressing water pollution challenges in an ecologically responsible and resource-efficient manner. Among the various wastewater treatment technologies, biological methods, especially the activated sludge process, remain the most widely used. Originally developed for the elimination of organic carbon, this method has been modified to also facilitate the removal of nitrogen and phosphorus. This chapter will examine various techniques that can be employed to effectively treat wastewater.

Activated Sludge Process

The activated sludge method is commonly employed for the biological treatment of wastewater. With this technique, naturally occurring bacteria are added to specified tanks in much greater quantities. The term "activated sludge" refers to these bacteria as well as protozoa and other microbes. The bacteria generate sticky compounds

Table 26.1 Various methods for wastewater treatment-integration

Source of wastewater	Applied treatment method	Efficiency achieved	References
Vegetable oil processing plant	Membrane ultrafiltration	COD reduction: 91%; TSS elimination: 100%	Sravan et al. (2024)
Synthetic oily wastewater	Membrane microfiltration	Removal of organic pollutants: 95%	Sravan et al. (2024)
Leachate from landfill	Acidogenic and bioelectrochemical systems (Combined)	COD reduction efficiency: 71.21%	Sravan et al. (2024)
Metal processing industry effluent	Ultrafiltration coupled with reverse osmosis	Contaminant elimination range: 90–99%	Sravan et al. (2024)
Phenol-laden wastewater (Paper Mills)	Combined ultrafiltration, nanofiltration, and Reverse osmosis	COD removal: 95.5%; Phenol reduction: 94.9%	Sravan et al. (2024)
Wastewater from paper and pulp mills	Bioelectrochemical treatment system	COD removal: 95%; Complete color elimination: 100%	Sravan et al. (2024)
Designed synthetic effluent	Bioelectrochemical system with PANi/CNT anode	COD removal efficiency: 80%	Sravan et al. (2024)
Glucose-enriched synthetic effluent	Sono-bioreactor (Frequency: 20 kHz; Power: 2 W and 4 W)	COD reduction exceeding 90%	Sravan et al. (2024)

that trap tiny particles in sewage, causing them to group together into gel-like flocs that provide a home for the microbes. The color of this blend is typically described as chocolate-brown. The activated sludge's aeration promotes oxygen dissolution, which allows the microbes to break down organic mattera process known as biochemical oxygen demand, or BOD. As the organic stuff sticks to the activated sludge, the bacteria can use the dissolved oxygen to turn ammonia into nitrate and use the nutrition. Consequently, the effluent, also known as treated wastewater, can be discharged into receiving bodies of water, such rivers or the ocean. Even while the basic idea is rather simple, the many factors that might affect the therapy process make managing it extremely difficult. These factors include changes in the sewage entering the facility and the makeup of the bacterial community inside the treatment tanks. High levels of hazardous chemicals are a problem for plants that treat industrial wastewater because they can result in toxic shocks that have a big impact on the bacterial population. In order to facilitate all required chemical changes, the tank must be the right size to guarantee sufficient contact time (retention time) between the sewage and the activated sludge (Ahansazan et al., 2014). A general flow diagram showing the common elements of an activated sludge wastewater treatment facility is presented in Fig. 26.2.

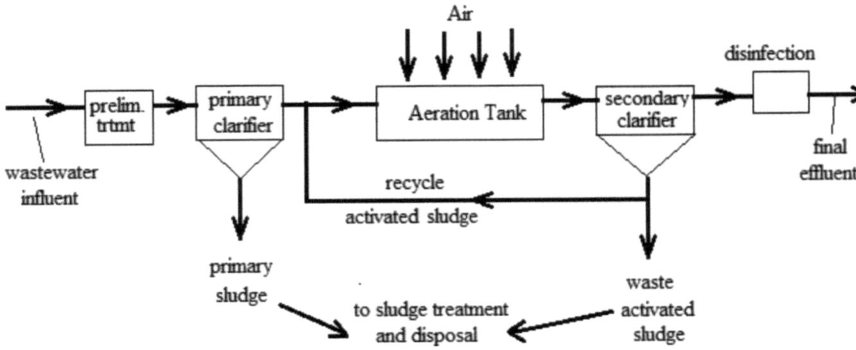

Fig. 26.2 Activated sludge wastewater treatment flow diagram (EEA, 2022)

Enhancing Biological Wastewater Treatment

Utilizing Microbial Diversity

The biological degradation of micropollutants could be greatly enhanced by forming complex microbial consortia, which involve interactions between bacteria, archaea, and fungi. Oxidase enzymes are particularly promising as biocatalysts for the degradation of organic waste and micropollutants. Cytochromes P450 (also known as CYPs or Cyt P450) are a very potent class of monooxygenases essential for breaking down poisons and medications in living things. Furthermore, microbes use a type of copper-containing oxidase enzyme called laccases (EC 1.10.3.2) to break down lignin (Kumar, 2010). Even though these enzymes are effective, their use in wastewater treatment is still mostly unknown and has not yet reached pilot or full-scale deployment. In modern biological systems, prokaryotic bacteria are primarily utilized due to their ability to quickly break down organic carbon through oxidation, although they typically refrain from produce Cyt P450.Furthermore, only fungal laccases are currently used in biotechnological contexts, but several fungi are known to synthesize Cyt P450 and the usage of bacterial laccases is limited by inadequate characterization (Sharma et al., 2007). By combining suspended phase bacteria and activated fungi, a system might handle both established and new pollutants by utilizing the advantages of both organism types.

Exploiting Metabolic Diversity

In the past, biological wastewater treatment methods have predominantly emphasized metabolic processes such as oxygen-dependent heterotrophy, nitrogen conversion in the presence of oxygen, nitrogen reduction in the absence of oxygen, and the accumulation of phosphorus. However, the removal of micropollutants could be significantly improved by integrating alternative metabolic pathways, such as heterotrophic sulfate reduction, autotrophic sulfur-oxidizing denitrification, anaerobic methane oxidation denitrification, and partial nitrite reduction to nitrous oxide for energy production. Additionally, through redox transformations that alter solubility and promote the precipitation of contaminants, these metabolic activities can efficiently target inorganic micropollutants, including heavy metals and radioactive compounds (Singhal & Gracia, 2006).

Exploiting Biocatalyst Diversity

Many enzymes exhibit "promiscuous" traits, enabling them to function as biocatalysts that modify a variety of substrates structurally similar to their main substrate. These versatile enzymes, in comparison to specialized ones, typically: (i) They are not always necessary, (ii) function at reduced metabolic rates, and (iii) demand less stringent control of their activity to manage metabolic flow under varying conditions (Nam et al., 2012). Their lack of specificity makes them strong candidates for breaking down organic micropollutants. Bacteria continuously adapt, creating new metabolic pathways to neutralize newly encountered substances or to source carbon, energy, and nutrients. However, the wide variety of chemicals, their typically low concentrations, and the diverse range of unknown compounds generated through bacterial transformation and biodegradation complicate these breakdown processes. According to genomic studies, many microbes have a much higher capacity to generate specialized enzymes and metabolites than has been previously determined by traditional bioactivity analyses; however, the ability to realize this degradation potential is often limited by the lack of expression of many specialized metabolite biosynthesis gene clusters in lab settings (Miller et al., 2010). Bacteria can theoretically launch several pathways and participate in a wide variety of catabolic biochemical activities to break down micropollutants. For instance, several biodegradation pathways have been recognized for bisphenol A (BPA), an industrial chemical widely present in epoxy resins and plastics, which is regarded as a possible endocrine disruptor (Das & Bal, 2025; Das et al., 2023; Dey et al., 2024; Jadaun et al., 2022).In the same way, screening techniques for biocatalysts and their byproducts, such as metagenomics, metaproteomics, and metabolomics, may reveal enzymatic biodegradation routes for additional chemicals (Villas-Bôas & Bruheim, 2007).

Anaerobic Membrane Bioreactor (AnMBR)

Similar to a Membrane Bioreactor (MBR), the AnMBR uses anaerobic bacteria to function in anaerobic environments. This process efficiently converts organic materials into biogas, primarily methane, which can be used to generate electricity. The main reason for the growing popularity of AnMBR technology is its ability to satisfy the growing need for resource-recovery-focused, energy-efficient, and sustainable wastewater treatment solutions. As a prelude to a full-scale wastewater treatment plant (WWTP), a semi-industrial AnMBR facility was built in Spain to treat sulfate-rich, severely laden municipal wastewater. The range of the organic loading rate (OLR) was 0.6 ± 0.2 to 1.2 ± 0.4 kg COD/m^3·d. A 40 m^3 anaerobic reactor connected to three ultrafiltration membrane tanks (PURON®) makes up the majority of the facility. The average chemical oxygen demand (COD) elimination effectiveness exceeded 87%, according to the results. Compared to conventional activated sludge systems, methane production ranged from 74 to 170 L CH4/kg COD, while sludge generation decreased by 8–42% (Luo et al., 2014 and Le et al., 2024).

Algae-Based Processes

Microalgae can be cultivated using a variety of wastewater sources, such as municipal, industrial, and agricultural effluents, which renders them applicable across numerous sectors. The cultivation systems can be customized to align with local conditions, resource availability, and specific treatment or resource recovery requirements, thereby enhancing their scalability and potential for deployment. Among the most extensively studied microalgae for wastewater treatment are Chlorella vulgaris and Chlorella sorokiniana, recognized for their high efficiency in nutrient removal, particularly nitrogen and phosphorus. Additionally, Scenedesmus obliquus and Scenedesmus dimorphus are effective in diminishing organic matter, nitrogen, and phosphorus, demonstrating rapid growth and the capacity to form dense cultures. Spirulina, a cyanobacterium known for its significant nutritional benefits, excels in the removal of nutrients, heavy metals, and organic pollutants, and exhibits resilience to elevated pH and saline environments. The utilization of microalgae in wastewater treatment has attracted global interest due to its potential contributions to sustainable development. This method integrates wastewater treatment with resource recovery, striving for zero emissions. Microalgae utilize CO_2, a byproduct of wastewater treatment processes, for their growth and biomass production (carbon capture), effectively removing nitrogen and phosphorus to mitigate eutrophication in receiving water bodies, and can also be utilized for biofuel production and other valuable co-products, thereby promoting resource recovery.

Factors Influencing Biological Treatment of Waste Water

Application methodology: The direct dosing approach of adding specific microbes to a treatment reactor provides a straightforward and cost-effective bioaugmentation method when necessary. However, challenges may arise due to reduced survival rates stemming from an inability to acclimate to the host environment. One potential solution to this issue is the installation of a small side-stream reactor, allowing the microorganisms to acclimatize before their introduction into the main reactor (Raper et al., 2018). Once the application method is established, selecting the appropriate dosage location for bioaugmentation becomes crucial. To identify the optimal site, it is advisable to examine the process flow diagram alongside key wastewater parameters. Reactors that maintain active biological processes, such as activated sludge systems, are particularly suitable for bioaugmentation.

Considering different dosage locations can extend the retention time of biological treatments. In cases where the retention period is inadequate, bioaugmentation may be implemented during the primary treatment phase, such as in equalization or collection tanks. It is essential to ensure optimal conditions for bacterial growth and efficacy when conducting bioaugmentation in the early stages of treatment (Lamba, 2018).

Temperature: Wastewater treatment processes in cold climates face significant challenges due to low temperatures. Bacteria, which can thrive in a variety of temperature ranges, are classified based on their preferred growth conditions. The three main categories include mesophiles, which favor moderate temperatures; thermophiles, which thrive in high temperatures; and psychrophiles, which are adapted to cold environments. When discussing bacterial growth in relation to temperature, it is essential to understand three key terms: the minimum growth temperature, representing the lowest temperature at which bacteria can grow; the maximum growth temperature, indicating the highest temperature for growth; and the optimum growth temperature, which is the temperature at which bacteria exhibit the fastest growth (Isnawati & Trimulyono, 2018). Different temperature for the biological process are summarized in Table 26.2.

Most microorganisms in aerobic biological wastewater treatment systems exhibit sensitivity to temperature variations. Nevertheless, mesophilic organisms may struggle to operate efficiently during particularly harsh winter conditions in certain regions. It is essential to select bioaugmentation technology with caution, considering the specific operational circumstances. The efficacy of pollutant removal and microbial metabolism will certainly be affected by reduced wastewater temperatures. Colder temperatures can particularly induce alterations in microbial community composition, sludge settleability, microbial activity, growth rates, and the deterioration of physiological characteristics.

Table 26.2 Temperature classification of biological processes

Bacteria type	Temperature range (°C)	Optimum temperature range (°C)
Psychrophilic	10–30	12–18
Mesophilic	20–50	25–40
Thermophilic	35–75	55–65

pH: In the process of sewage treatment, pH serves as a critical control parameter, as it significantly influences the growth, metabolism, and functionality of biological enzymes. Various microorganisms possess distinct pH preferences for effective wastewater treatment. Studies have shown that achieving a pH level of 7 or higher enhances the efficiency of ammonia nitrogen and total phosphorus removal. The highest production of N_2O occurs at a pH of 8.0. Furthermore, maintaining a pH range between 6.4 and 7.0 restricts the activity of ammonia-oxidizing bacteria, thereby reducing N_2O emissions. To ensure the optimal performance of biological processes in wastewater treatment systems and to prevent negative outcomes, it is crucial to make appropriate pH adjustments Lan et al. (2018).

Nutrient Availability: Wastewater produced by various industrial activities generally contains low levels of nitrogen and phosphorus, while displaying elevated concentrations of carbon. To achieve effective biological treatment, it is essential to maintain an optimal ratio of carbon, nitrogen, and phosphorus at 100:5:1. However, industrial discharges frequently diverge from these ideal proportions, necessitating the incorporation of external nutrient sources to improve the biological treatment process. It is crucial to manage the nutrient supply carefully, as inadequate levels can reduce treatment effectiveness, whereas excessive amounts can increase operational costs. The presence of surplus nutrients in wastewater may require additional treatment interventions and could pose compliance issues. Consequently, efficient nutrient management is vital for enhancing treatment effectiveness and minimizing costs. The level of organic materials in wastewater affects the carbon load, which subsequently determines the required amounts of added phosphorus and nitrogen. The quantities of these nutrients depend on the carbon load, which is influenced by the concentration of organic compounds in the wastewater. The practice of supplementing nutrients in a biological reactor to boost its efficiency is known as bio-stimulation.

Conclusion

The increasing volume of wastewater generated by human activities presents significant environmental challenges, particularly due to the presence of both organic and inorganic pollutants, including microplastics, xenobiotics, heavy metals, nitrates, phosphates, and carbon compounds. Addressing these challenges necessitates the development of innovative wastewater treatment strategies that can convert this waste into a valuable resource. Although traditional biological treatment methods are often cost-effective, they typically face difficulties in removing micropollutants found in low concentrations. To overcome these limitations, it is crucial to adopt advanced integrated biological and bioelectrochemical remediation systems in conjunction with green technologies. Enhancing microbial diversity, leveraging metabolic and biocatalyst diversity, and utilizing novel materials such as nanoparticles can significantly improve treatment effectiveness. Microalgae-based treatment systems offer the dual benefit of wastewater treatment and biomass production, thereby contributing to a circular economy. Techniques such as high-rate algal ponds, integrated algae pond systems, and photobioreactors provide sustainable solutions for nutrient removal and resource recovery. Despite the encouraging advancements in this area, challenges such as high costs, scalability concerns, and regulatory obstacles persist. Ongoing research, optimization, and the development of commercially viable bioprocesses are essential for wider adoption. In summary, the future of wastewater treatment lies in advanced, integrated, and sustainable technologies. By leveraging biological processes and green innovations, we can enhance treatment efficiency, protMect the environment, and recover valuable resources, thereby fostering a sustainable future.

References

Ahansazan, B., Afrashteh, H., Ahansazan, N., & Ahansazan, Z. (2014). Activated sludge process overview. *International Journal of Environmental Science and Development, 5*(1), 81.

Alexander, M. (1999). *Biodegradation and bioremediation*. Academic Press.

Das, P., & Bal, M. (2025). Modeling and optimization study for microplastic removal from aquatic medium using *Chroococcidiopsis* species. *Chemical Engineering Science, 304*, Article 121085. https://doi.org/10.1016/J.CES.2024.121085

Das, P., Halder, G., & Bal, M. (2023). A critical review on remediation of microplastics using microalgae from aqueous system. *Science of the Total Environment, 898*, Article 166425. https://doi.org/10.1016/J.SCITOTENV.2023.166425

Das, A. P., Ghosh, S., Mohanty, S., & Sukla, L. B. (2015). In Advances in manganese pollution and its bioremediation (pp. 313–328) Springer. https://doi.org/10.1007/978-3-319-19018-1_16

Das, A., Dey, S., Bala, K., Gnanasekaran, R., & Das, A. P. (2024). In *Microbial bioremediation technology for sustainable treatment and management of synthetic microfiber waste* (pp. 221–238) Springer. https://doi.org/10.1007/978-3-031-62600-5-12

Dey, S., Dash, N., Ray, S. K., Priyadarsini, S., Chhotaray, C., & Das, A. P. (2024). Occurrence and quantification of tire wear plastics pollutants from highway city dust of Bhubaneswar city of Odisha. In *Sustainable microbial technology for synthetic and cellulosic microfiber bioremediation* (pp. 77–89) Springer. https://doi.org/10.1007/978-3-031-62600-5_5

European Environment Agency. (2022). Europe's air quality status 2022. Retrieved from https://www.eea.europa.eu/publications/status-of-air-quality-in-Europe-2022.

Gavrilescu, M., Demnerová, K., Aamand, J., Agathos, S., & Fava, F. (2015). Emerging pollutants in the environment: Present and future challenges in biomonitoring, ecological risks and bioremediation. *New Biotechnology, 32*(1), 147–156. https://doi.org/10.1016/j.nbt.2014.09.001

Isnawati & Trimulyono, G. (2018).Temperature range and degree of acidity growth of isolate of indigenous bacteria on fermented feed fermege. *Journal of Physics: Conference Series, 953,* 012209. https://doi.org/10.1088/1742-6596/953/1/012209

Jadaun, S. J., Bansal, S., Sonthalia, A., Rai, A. K., Singh, S. P. (2022). Biodegradation of plastics for sustainable environment. *Bioresouce Technology*, *347*, 126697. https://doi.org/10.1016/J.BIORTECH.2022.126697

Kanungo, J., Sahoo, T., Bal, M., & Behera, I. D. (2024). Performance of bioremediation strategy in waste lubricating oil pollutants: A review. *Geomicrobiology Journal, 41*, 360–373. https://doi.org/10.1080/01490451.2023.2245395

Kumar, S. (2010). Engineering cytochrome P450 biocatalysts for biotechnology, medicine and bioremediation. *Expert Opinion on Drug Metabolism and Toxicology, 6*(2), 115–131.

Kumar, M., Pandit, S., Patel, V., Khanna, N., Nag, M., Lahiri, D., Ray, R. R., Das, A., & Das, D. (2023). Maximization of energy recovery from starch processing wastewater by thermophilic

dark fermentation coupled with microbial fuel cell technology, *Geomicrobiology Journal*, 1–12. https://doi.org/10.1080/01490451.2023.2209555
Lamba, H. R. (2018). Optimizing bioaugmentation strategies for industrial wastewater treatment. *International Journal of Science and Research (IJSR), 9(1)*. https://doi.org/10.21275/SR24103171800.
Lan, H., Zhang, H.,Yang, D., Liu, J., Tang, G., & Zhang, H. (2018). Effects of pH on biological treatment of paper mill white water with the addition of dominant bacteria. *BioResources, 13*(3), 6757–6765. https://doi.org/10.15376/biores.13.3.6757-6765
Le, L. T., Nguyen, H. V., & Bui, X. T. (2024). Recent advanced biological wastewater treatment technologies. *VNU Journal of Science: Earth and Environmental Sciences, 40*(2).
Li, X., Liu, Y., & Wang, A. (2018). Integration of biological processes with advanced oxidation processes (AOPs), Membrane Bioreactors (MBRs), and activated carbon adsorption for micropollutant elimination: A review. *Journal of Environmental Management, 213*, 567–578. https://doi.org/10.1016/j.jenvman.2018.03.091
Luo, Y., Guo, W., Ngo, H. H., Nghiem, L. D., Hai, F. I., Zhang, J., & Liang, S. (2014). A review on the occurrence of micropollutants in the aquatic environment and their fate and removal during wastewater treatment. *Science of the Total Environment*, 473–474.
Miller, L. D., Mosher, J. J., Venkateswaran, A., Yang, Z. K., Palumbo, A. V., Phelps, T. J., & Keller, M. (2010). Establishment and metabolic analysis of a model microbial community for understanding trophic and electron accepting interactions of subsurface anaerobic environments. *BMC Microbiology, 10*, 1–12.
Mishra, S., Dash, D., & Das, A. P. (2022a). Detection, characterization and possible biofragmentation of synthetic microfibers released from domestic laundering wastewater as an emerging source of marine pollution. *Marine Pollution Bulletetin, 185*, Article 114254. https://doi.org/10.1016/J.MARPOLBUL.2022.114254
Mishra, S., & Das, A. P. (2021). Current treatment technologies for removal of microplastic and microfiber pollutants from wastewater. In *Wastewater treatment: Cutting-edge molecular tools, techniques and applied aspects* (pp. 237–251). https://doi.org/10.1016/B978-0-12-821881-5.00011-8
Mishra, S., & Das, A. P. (2022). Treatment of the wastewater polluted with synthetic microfiber released from washing machine, (Eds.), Lecture Notes in Civil Engineering. Springer Science and Business Media Deutschland GmbH (pp. 109–117). https://doi.org/10.1007/978-981-16-7509-6_9/COVER
Mishra, S., Das, A. P. (2023). Microbial remediation of synthetic microfiber contaminated wastewater. In M.P. Shah (Ed.), *Microbial technologies in industrial wastewater treatment* (pp. 337–350) Springer. https://doi.org/10.1007/978-981-99-2435-6_17
Mishra, S., Singh, R. P., Rout, P. K., & Das, A. P. (2022b). Membrane bioreactor (MBR) as an advanced wastewater treatment technology for removal of synthetic microplastics. In *Development in wastewater treatment research and processes* (pp. 45–60). https://doi.org/10.1016/B978-0-323-85583-9.00022-3
Mishra, S., Dash, D., & Das, A. P. (2024). Isolation and molecular characterization of polyester-degrading bacterial strains from municipal wastewater and their application in microplastic degradation in contaminated wastewater. *Geomicrobiology Journal*. https://doi.org/10.1080/01490451.2024.2376101
Mohanty, M., Tripathy, B., Koteswara Reddy, G., & Das, A. P. (2024). Greener technologies for establishing restraint over microplastic pollution. In *Sustainable microbial technology for synthetic and cellulosic microfiber bioremediation* (pp. 239–258). https://doi.org/10.1007/978-3-031-62600-5_13
Nam, H., Lewis, N. E., Lerman, J. A., Lee, D. H., Chang, R. L., Kim, D., & Palsson, B. O. (2012). Network context and selection in the evolution to enzyme specificity. *Science, 337*(6098), 1101–1104.

Paul, A., Dey, S., Ram, D. K., & Das, A. P. (2023). Hexavalent chromium pollution and its sustainable management through bioremediation. *Geomicrobiology Journal*, 1–11. https://doi.org/10.1080/01490451.2023.2218377

Pradhan, G., Tripathy, B., Ram, D. K., Digal, A. K., & Das, A. P. (2023). Bauxite mining waste pollution and its sustainable management through bioremediation. *Geomicrobiology Journal*, 1–10. https://doi.org/10.1080/01490451.2023.2235353

Raper, E., Stephenson, T., Anderson, D. R., Fisher, R., & Soares, A. (2018). Industrial wastewater treatment through bioaugmentation. *Process Safety and Environmental Protection, 118*, 178–187.

Sharma, P., Goel, R., & Capalash, N. (2007). Bacterial laccases. *World Journal of Microbiology and Biotechnology, 23*, 823–832.

Singhal, N., & Perez-Garcia, O. (2006). Degrading organic micropollutants: The next challenge in the evolution of biological wastewater treatment processes. *Frontiers in Environmental Science, 4*, 36.

Sravan, J. S., Matsakas, L., & Sarkar, O. (2024). Advances in biological wastewater treatment processes: Focus on low-carbon energy and resource recovery in biorefinery context. *Bioengineering (Basel), 11*(3), 281. https://doi.org/10.3390/bioengineering11030281

United Nations. (2021). Progress on wastewater treatment—2021 Update. UN-Water. Retrieved from https://www.unwater.org/publications/progress-wastewater-treatment-2021-update.

Villas-Bôas, S. G., & Bruheim, P. (2007). The potential of metabolomics tools in bioremediation studies. *Omics: A Journal of Integrative Biology, 11*(3), 305–313.

World Health Organization. (2021). Global air quality guidelines. World Health Organization. Retrieved from https://www.who.int/publications/i/item/9789240034433.

Chapter 27
Challenge in Degradation of Micropollutants by Biological Wastewater Treatment Processes

Abhay Kumar Choudhary, Ravi Kumar Sonwani, and Deoashish Panjiara

Introduction

Currently, over 50% of worldwide wastewater is discharged into the environment without adequate treatment, which is about 2 million tons of agricultural, industrial, and sewage waste (Sturm et al., 2022; Mishra & Das, 2023; Mishra et al., 2024). In addition to having significant influxes of phosphorus and nitrogen, resulting in severe impacts on aquatic life, such as eutrophication which leads to inadequate oxygen aeration, containing around 245,000 km^2 of aquatic habitat (Meng et al., 2024; Sturm et al., 2022). Furthermore, it has been proven that herbicide runoff is toxic and harmful to the ecological system, having devastated mangroves in a 30 km^2 distributed region during 1999 and 2002 (Sturm et al., 2022).

Continuous industrial expansions, agricultural and residential activities around the world utilize water resources and generate huge amounts of industrial and municipal wastewater including contaminants, which are dumped into water bodies such as rivers, lakes, and coastal seawater (Mishra & Das, 2021; Mishra et al., 2022b). These contaminants predominantly consist of synthetic organic/inorganic chemicals

A. K. Choudhary (✉)
Department of Chemical Science and Technology, National Institute of Technology Patna, Bihar, India
e-mail: abhay.ch@nitp.ac.in

R. K. Sonwani
Department of Chemical Engineering, Indian Institute of Petroleum and Energy (IIPE), Visakhapatnam, India
e-mail: ravikumar.che@iipe.ac.in

D. Panjiara
Department of Chemical and Petrochemical Engineering, Ganpat University, Mehsana, Gujarat, India
e-mail: dp01@ganpatuniversity.ac.in

I. D. Behera et al. (eds.), *Emerging Micropollutants*, Sustainable Environmental Waste Management Strategies, https://doi.org/10.1007/978-3-032-03530-1_27

that are harmful to wildlife, aquatic life, and human health (Khoo et al., 2022). Micropollutants are trace contaminants in wastewater, which include anthropogenic chemicals such as personal care products, medication, pharmaceuticals, pesticides, steroid hormones, heavy metals, nanomaterials, X-ray contrast media, foams, plasticizers, microplastics, waterproofing agents, causing environmental and public health concerns (Narwal et al., 2023; Behera et al., 2022, 2021; Das & Mishra, 2008; Kanungo et al., 2023; Mallick et al., 2023). Micropollutants are detected in wastewater at trace quantities ranging from ng/L to μg/L (Khoo et al., 2022). These chemicals can enter into water bodies from various sources such as domestic, untreated sewage discharge, agricultural activities, hospitals, and industries (Wołowicz & Munir, 2025). Since the daily use of chemical substances, micropollutants concentrations are increasing day by day and now reached to an alarming level (Sher et al., 2021). Furthermore, the long-term persistence of these micropollutants in the environment due to their low biodegradability allows them to accumulate and magnify in animals and plants before reaching humans through the food supply chain, posing public health dangers (Khoo et al., 2022; Sher et al., 2021).

The water bodies are used for a variety of purposes, including drinking water, industrial process water, irrigation, recreational area reserves, and hydropower reservoirs (Kosek, et al., 2020). Even though water covers around 71% of the Earth's surface and has a volume of 333 million cubic miles, it is now a highly competitive resource, as freshwater accounts for only less than 3% of the total water available on the surface (Narwal et al., 2023). Furthermore, around 0.5% of the freshwater found in the form of liquid in groundwater, ponds, lakes, and rivers is suitable for human use (Kato & Kansha, 2024). Therefore, these water resources must be managed efficiently and sustainably to meet present and future demands. In this context, wastewater treatment and reuse are an important approach for reducing pollution and water scarcity while also fostering a circular economy and sustainability (Liu et al., 2022; Sturm et al., 2022).

Micropollutants and Their Types

Micropollutants are synthetically produced or naturally occurring chemicals and biological materials that exist in extremely low concentrations in the environment (typically in the range of nanograms to micrograms per liter) (Narwal et al., 2023). According to the Chemical Abstract Service (CAS), 89 million chemical compounds have been identified so far; of these, 100,000 have been classified by the European Union (EU) as micropollutants that are used in daily life for various activities (Goswami et al., 2018; Khan et al., 2021; Kushwaha et al., 2024). These are often detected in wastewater and surface water bodies (Khan et al., 2021; Sturm et al., 2022). These contaminants are typically introduced into the environment by industrial effluents, hospital and domestic effluents, livestock, aquaculture, and agricultural runoff, wastewater discharges, landfill leachates, and improper disposal of household products (Barbosa et al., 2016; Khoo et al., 2022). Figure 27.1 illustrates

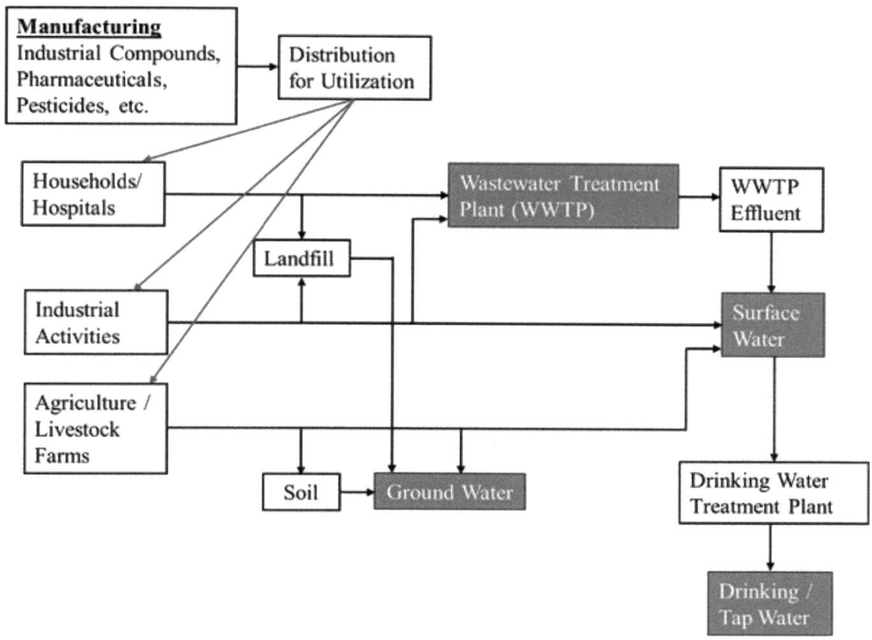

Fig. 27.1 Typical sources and routes of micropollutants in water resources

the typical sources and routes of micropollutants in water resources by human activities. Table 27.1 lists these micropollutants categories, along with their subclasses, examples, and sources in aquatic environments.

Micropollutants-contaminated wastewater is released from both point and nonpoint sources. Wastewater discharge from point sources, such as municipal and industrial WWTP plants, is easy to identify; however, nonpoint sources of wastewater, which includes multiple sources like urban runoff and agricultural activities are more difficult to identify (Kumar et al., 2023; Mishra & Das, 2022; Mishra et al., 2022a). As a result, point sources are frequently more concentrated than nonpoint sources and are easier to monitor and manage (Kato & Kansha, 2024). Despite their low concentrations, many micropollutants exhibit persistent biological activity, posing significant risks to aquatic ecosystems, wildlife, and human health (Khan et al., 2021). The persistence, toxicity, and bioaccumulation of micropollutants in water systems present serious challenges for wastewater treatment processes (Khoo et al., 2022; Narwal et al., 2023). Furthermore, their long-term exposure effects on human life and aquatic systems are not fully explained, necessitating further research and enhanced treatment methods.

Table 27.1 Micropollutants types, sub-classes, and sources in aquatic environments (Wołowicz & Munir, 2025; Narwal et al., 2023; Kanaujiya et al., 2019; Luo et al., 2014)

Category	Sub-classes	Examples	Major sources
Pharmaceuticals	Antibiotics, lipid regulator, antipyretic/ analgesic, anticonvulsants, β-blockers, NSAIDs, veterinary drugs, and stimulants	Erythromycin, penicillin, clofibric acid, metoprolol, ibuprofen, carbamazepine, diclofenac, paracetamol, caffeine	Hospital effluents, runoff from livestock farms and aquaculture, domestic wastewater (excretion), dye industry,
Personal care products	Fragrances, sunscreens, disinfectants, UV filters, and insect repellents	Tonalide, galaxolide, parabens, triclosan, benzophenone-3, n, n-diethyl-m-toluamide, diethyltoluamide	Domestic wastewater (from house cleaning, shaving, bathing, spraying, swimming, etc.)
Pesticides	Fungicides herbicides, insecticides, nematicides, rodenticides	Methyl parathion, glyphosate, endosulfan, clotrimazole, diazinon, atrazine, DDT, chloradane	Agricultural runoff, industrial waste from pesticides and fertilizer industry, domestic wastewater, forest water, garden waste, waste from livestock practices
Steroid hormones	Estrogens, progestins,	Estriol, estradiol, estrone, ethinyl estradiol, progesterone	Pharmaceutical industry, domestic wastewater (from excretion), runoff from livestock farms and aquaculture,
Industrial chemicals	Plasticizers, X-ray contrast media, waterproofing agents, gasoline, fire retardants, Polychlorinated biphenyls (PCBs), polycyclic aromatic hydrocarbons (PAHs),	Bisphenol A, Diatrizoate, Perfluorooctane sulfonated, Siloxanes, petroleum hydrocarbon, Naphthalene, benzene, octylphenol, tert-butylphenol, pentachlorophenol, epichlorohydrin Tris(2-chloroethyl) phosphate, etc	Industrial effluents, landfill leachates, mining and extraction, accidental releases and spills, combustion and emissions, runoff of pesticides, fertilizers, and other agro-industrial chemicals

(continued)

Table 27.1 (continued)

Category	Sub-classes	Examples	Major sources
Toxic nanomaterials	Carbon-based nanomaterials, semiconductor nanomaterials, metal-based nanomaterials, oxide nanoparticles, composite nanomaterials, biological or organic nanomaterials	carbon nanotubes, fullerenes and its derivatives, Quantum dots, graphene oxide, Ag NPs, TiO_2 NPs, ZnO NPs, liposomes, DNA NPs	Consumer goods, industrial production processes, medical applications, sludge incinerators, agricultural activities, and water treatment plants
Microplastics	Primary microplastics, secondary microplastics, microfibers, microplastic additives and coatings, tire wear particles, industrial microplastics	Polyethylene, polypropylene, polystyrene, polyvinyl chloride, bisphenol A, microbeads, pellets, fragments, foam pieces, fibers from synthetic clothing,	Industrial activities, WWTPs, poor plastic waste management
Heavy metals	Essential metals, non-essential metals	Iron (Fe), zinc (Zn), copper (Cu), chromium (Cr), arsenic (As), mercury (Hg), lead (Pb), cadmium (Cd) etc	Industrial point sources, municipal waste treatment outfalls, mines, firing ranges, smelters, landfills, junkyards, urban runoff
Microorganisms	Bacteria, virus, protozoa, algae, and fungi,	Vibrio cholerae, salmonella, escherichia coli, norovirus, aeromonas salmonicida, arcobacter cryaerophilus, hepatitis A virus, hepatitis E virus, rotavirus	Agriculture runoff, sewage discharge, feces, or other contaminated wastewater and products

Pharmaceuticals

Pharmaceuticals are considered a large category of micropollutants containing biologically active substances that can enter the environment in a variety of ways, including waste from pharmaceutical manufacturing plants, animal husbandry, hospitals, research facilities, and the disposal of unwanted or expired drugs (Murtaza et al., 2023; Khatun et al., 2023a, b). Pharmaceuticals include drugs, chemicals, inorganic, sugars, additives, pigments, organic and inorganic salt, transformed products, pills, and metabolites (Narwal et al., 2023). It can also include prescription and over-the-counter veterinary drugs and medications. These chemicals enter the wastewater systems primarily through human consumption, excretion, livestock farming, hospital and healthcare facilities, industrial manufacturing processes, and improper disposal of expired or unused medications (Adeoye et al., 2024; Narwal et al., 2023). It is estimated that millions of people consume pharmaceutical products 100,000 tons every year globally, and even small amounts of these substances find their way into wastewater, causing ecological risks (Adeoye et al., 2024; Verlicchi et al., 2010). Common pharmaceuticals detected in wastewater effluent include antibiotics (e.g. amoxicillin), analgesics (e.g. ibuprofen), hormones (e.g. estrogens and progestogens), and others such as antidepressants, antiepileptics, antidiabetic drugs (e.g. glibenclamide) and non-steroidal anti-inflammatory drugs (NSAIDs) (Włodarczyk-Makuła, 2024). Some of these contaminants including clofibric acid, naproxen, and erythromycin are persistent due to their lipophilic properties. Additionally, pharmaceutical pollutants differ from other contaminants due to their various structures, forms, and functions (Wołowicz & Munir, 2025). These are biologically active and can adsorb, accumulate, and disperse in live organisms, affecting metabolic processes aquatic of species (Adeoye et al., 2024). Moreover, a significant amount of consumed antibiotics passes through the body unmetabolized which in particular contributes to the development of antibiotic-resistant bacteria or superbugs populations, exacerbating a growing public health concern (Murtaza et al., 2023).

Personal Care Products

Personal care products (PCPs) are consumer items marketed in health and beauty departments of stores and pharmacies that can be purchased without a prescription for personal hygiene, grooming, cleaning, and cosmetic purposes (Wołowicz & Munir, 2025; Bracamontes-Ruelas et al., 2022). This class of compounds includes items such as mosquito repellents, insect repellents (e.g. citronella oil, picaridin, etc.), disinfectants (e.g., phenol, formaldehyde, hydrogen peroxide, triclosan, etc.), fragrances/scents (e.g. peppermint, pine, vanilla, musk, etc.), toothpaste, washing soaps, detergents, shampoos, body washes, lotions, dishwashing liquids, and deodorants, antibacterial products, antifungal agents and UV filters (e.g., homosalate, octinoxate, avobenzone, oxybezone) are commonly utilized in urban sites (Bracamontes-Ruelas et al., 2022; Narwal et al., 2023; Ponnusami et al., 2023). The utility of PCPs and their concentration are generally varying by geography and season. PCPs are meant for external use rather than ingestion and are not subject to major molecular structure changes (Ponnusami et al., 2023). These items often contain complex mixtures of chemicals that are difficult to degrade in biological treatment processes (Khoo et al., 2022). These substances are commonly found in wastewater treatment plant (WWTP) effluents, storm runoff, landfill leachates, and sewage, where they can persist for long periods. Compounds such as methyl parabens, triclosan, oxybenzone, homosalate, and phthalates are ubiquitous in personal care products and act as endocrine disruptors. These have been demonstrated to have toxic effects on aquatic life (Khan et al., 2021; Li et al., 2020). It should be highlighted that certain vegetation can accumulate and metabolize PCPs, causing harmful effects on both them and the organisms that consume them, such as humans (Bracamontes-Ruelas et al., 2022).

Pesticides

Pesticides are a significant class of micropollutants that are synthetic or natural chemical compounds used in agriculture/horticulture to eliminate fungi, insects, rodents, or weeds (Bracamontes-Ruelas et al., 2022). It can include fungicides herbicides, insecticides, nematicides, and rodenticides, along with defoliants, desiccants, nitrogen stabilizers, plant growth stimulators, etc. (Wołowicz & Munir, 2025; Narwal et al., 2023). These are used depending on the variety of crops and are available in various forms as liquids, aerosols, powders, and sprays. Pesticides are also employed in food preservation, processing, and storage, as well as aquaculture, to prevent or eliminate insect growth. Glyphosate, endosulfan, atrazine, aldrin and chlorpyrifos are the most prevalent pesticides found in wastewater (Narwal et al., 2023). They are discharged into the environment from a variety of sources, including domestic waste, garden waste, and runoff from agricultural, industrial, and residential use as well as waste from aquaculture farms (Bracamontes-Ruelas et al., 2022; Narwal et al., 2023). The main purpose of developing these compounds is to efficiently destroy or restrict pest growth while also regulating plant growth, despite their persistent nature in the environment causing concerns about water quality (Wołowicz & Munir, 2025). Pesticides can also be harmful to non-target species, even in low doses. For instance, glyphosate has been reported to impair the growth and reproduction of aquatic plants and animals (Narwal et al., 2023). The widespread use of these compounds (2 million tons globally) has raised concerns about their impact on biodiversity, the health of aquatic ecosystems, and soil matrices (Wołowicz & Munir, 2025). Many pesticides have bonds of carbon-chlorine or fluorine atoms, making them more persistent in the environment and difficult to remove using traditional wastewater treatment methods (Narwal et al., 2023). Advanced biological treatment processes, including bioaugmentation and engineered bioreactors, are being explored to address these challenges.

Steroid Hormones

Steroid hormones are biochemical active substances produced naturally in the endocrine systems of humans and animals to govern a variety of physiological processes such as growth, immune response, reproduction, and metabolism. These include both natural hormones (estrogens, androgens, and progestogens) and synthetic forms (i.e., Ethinyl estradiol, methyltestosterone and levonorgestrel) of these hormones, which are widely employed in medicines, agriculture, and veterinary care (Narwal et al., 2023). The sources of steroid hormones as micropollutants in the environment include pharmaceutical waste, human and animal excretion, agricultural runoff, wastewater discharge from livestock, and industrial discharge (Kanaujiya et al., 2019; Luo et al., 2014). Some steroids are persistent in the environment, which can travel through the water system or bind with sediments. These can also bioaccumulate in aquatic organism and affecting higher tropics levels. The endocrine systems of aquatic organisms can be disrupted by exposure to elevated quantities of both natural and synthetic hormones. This can lead to a number of problems, including population imbalances, feminization of male fish, and altered reproduction (Bracamontes-Ruelas et al., 2022). Consumption of contaminated water sources for human activities might endanger humans, perhaps leading to hormonal imbalances and other health difficulties.

Industrial Chemicals

Wastewater also contains industrial chemicals including lubricants (motor oil and grease), fuels (kerosene, diesel, and gasoline), solvents (methanol and acetone), cleaning agents (detergents and soaps), plasticizers, surfactants, medicines, and plastics (Narwal et al., 2023). These chemicals (either artificial or natural) are regularly generated and used in industrial manufacturing activities such as agriculture, mining, and manufacturing, and are discharged into wastewater from industrial facilities (Wołowicz & Munir, 2025; Narwal et al., 2023). The increasing use of these chemicals causes severe disorders in animals as well as human beings such as brain dysfunction or mental retardation. Many industrial chemicals are hydrophobic and persistent, making them resistant to biodegradation. Polychlorinated biphenyls (PCBs), dioxins, and polycyclic aromatic hydrocarbons (PAHs) are examples of persistent organic pollutants that can accumulate in the food chain and endanger human health and animals (Sharma et al., 2024).

Toxic Nanomaterials

Nanomaterials are materials with structural dimensions, usually between 1 and 100 nm, the fact that owing to their size, have special chemical and physical properties, such as enhanced surface area, changed reactivity, and new mechanical or optical capabilities (Malakar et al., 2021; Wołowicz & Munir, 2025). These nanoparticles (NPs) are frequently created to be widely used in scientific fields such as medicine, biofuels, social studies, archeology, waste management, environmental cleanup, and industrial applications (Narwal et al., 2023). Toxic nanomaterials include carbon-based nanomaterials (graphene oxide, fullerenes, carbon nanotubes (CNTs)), metal-based nanomaterials (TiO_2, ZnO, Fe_2O_3, Al_2O_3 and Ag NPs), polymeric nanomaterials (coadings and drug delivery) and other engineered NPs (quantum dots or nanoclays). These NPs can enter the environment via a variety of sources, such as consumer goods, industrial production processes, medical applications, agricultural activities, and wastewater from WWTPs. Because of their chemical stability and small size, these NPs can travel a considerable distance into water and air after being released. NPs have a tendency to accumulate in aquatic species and plants, where they can enter the human food chain, affecting health (Malakar et al., 2021).

Microplastics

Microplastics are synthetic tiny plastic particles (size < 5 mm in length or diameter) that originate from the degradation or breakdown of larger material under mechanical, physical, chemical, and biological activities (Rose et al., 2023; Das et al., 2023; Das & Bal, 2025). It has recently received a lot of attention due to its widespread use, growing presence in wastewater, and potential harm to the ecology (Ghosh et al., 2023; Mishra et al., 2019; Singh et al., 2020). Microplastics can be broadly categorized into two main types based on their source: primary and secondary. Primary microplastics include small particles of plastic products released directly from domestic activities (e.g., creams, scrubs, face wash, medicines, toothpaste, baby products, etc.) as well as industrial production processes (plastic pellets, plastic blasting, pharmaceutical production, textile fabric, etc.) (Narwal et al., 2023; Das et al., 2023, 2024; Mallick et al., 2024). Secondary microplastics, including fibers, are formed when large plastic items deteriorate due to environmental degradation processes such as UV exposure (solar radiation), mechanical abrasion (wind waves), chemical degradation (acidic or alkaline conditions), and biodegradation by microorganisms (Rose et al., 2023). Microplastics, due to their small size and pervasive abundance in the environment, can enter the human food chain via seafood consumption, posing a health risk (Jagatee et al., 2024; Mishra et al., 2020; Paul et al., 2024). Furthermore, they can absorb and transport other harmful contaminants (such as petroleum hydrocarbons, pesticides, and heavy metals) as well as serve as carriers of additional micropollutants into living organisms. Several aquatic species ingest these contaminated microplastic particles, which can cause a variety of problems, including reproduction inhibition in amphibians and hormone disruption in mollusks (Narwal et al., 2023).

Heavy Metals

Heavy metals, which include metallic elements such as iron (Fe), copper (Cu), nickel (Ni) chromium (Cr), zinc (Zn), lead (Pb), mercury (Hg), cadmium (Cd), arsenic

(As), and others, are toxic pollutants found in wastewater (Hammoudani et al., 2024; Narwal et al., 2023; Prabhakar et al., 2019; Raj & Das, 2023). These metals find their way into the environment through mining operations, landfills, junkyards, urban runoff, industrial effluents, and even municipal waste treatment outfalls as a result of stricter regulations and industrial relocation (Narwal et al., 2023; Priyadarsini & Das, 2025, 2024). Heavy metals are characterized by their inability to biodegrade and bioaccumulate in living organisms and food chains, posing major threats to aquatic life and human health at high concentrations (Hammoudani et al., 2024). Balali-Mood et al. (2021) investigated heavy metal-induced organ toxicity, including liver damage, kidney dysfunction, cardiovascular dysfunction, immune system damage, birth defects, neurological disorders, and nervous system damage. Furthermore, high-dose heavy metal exposure can induce cancer and genotoxicity. Certain heavy metals, such as lead (Pb), cadmium (Cd), and mercury (Hg), are highly toxic even at low concentrations. They exhibit bioaccumulation and biomagnification, posing significant threats to aquatic organisms. For example, mercury can cause neurological damage in fish and other aquatic creatures, whereas lead and cadmium can harm reproductive health (Narwal et al., 2023). Moreover, these metals can accumulate in sediments, amplifying their harmful impact on the environment and posing a serious threat to benthic organisms. (Hammoudani et al., 2024). Conventional biological treatment processes are ineffective in removing heavy metals, necessitating the use of advanced treatment methods such as adsorption, ion exchange, and membrane filtration (Baig et al., 2021).

Microorganisms

The group of microorganisms includes bacteria, viruses, protozoa, algae, and fungi, which are found in the air, water, soil, and human body. These are regarded as very important for biochemical processes, nutrient cycling, and ecosystems (Balali et al., 2020). Despite their positive characteristics, several microbes act as pollutants, causing and affecting aquatic species, plants, wildlife, and human life (Narwal et al., 2023). Pathogenic microorganisms including bacteria, viruses, fungi, and protozoa cause diseases in humans, animals, or plants. Waterborne diseases can arise from various pathogens present in water systems, including specific bacteria (e.g., Salmonella, Escherichia coli, Vibrio cholerae), viruses (e.g., Norovirus), and helminths (Baig et al., 2021; Ramírez-Castillo et al., 2015). The sources of these microbes in water bodies are various ways like agricultural runoff, municipal wastewater, industrial discharge, and stormwater runoff (Wołowicz & Munir, 2025; Narwal et al., 2023). Furthermore, algal microorganisms such as cyanobacteria (blue-green algae) can produce cyanotoxins as byproducts in both marine and freshwater, causing water pollution and endangering aquatic life and animal health (Narwal et al., 2023).

Treatment Technologies of Micropollutants and Their Challenges

Micropollutants are the group of pollutants such as pharmaceuticals, personal care products, synthetic chemicals, pesticides, and emerging new contaminants (Song et al., 2023). Although their concentrations in water bodies are very less (ng/L to μg/L), they reveal significant adverse effect on environment and health due to their persistence, toxic, carcinogenic, and bioaccumulation potential (Dai et al., 2024). According to Wang et al. (2024), the continued exposure to micro-pollutants causes negative impacts on the well-being of humans. Therefore, it is essential to tackle these pollutants using sustainable and cost-effective methods. In the last few years, various treatment techniques (such as adsorption, advanced oxidation processes, nanofiltration, membrane bioreactors, etc.) have been used to address the micropollutants from water (Jirani et al., 2017; Ouyang et al., 2023; Wang et al., 2024).

Adsorption Techniques

Adsorption is one of the most effective and simple method to remove the pollutants from wastewater. In this process, the adsorbate molecules are adsorbed on the surface or pores of the adsorbent over time due to the several mechanisms such as hydrogen bonding, hydrophobic, electrostatic, and π-π interactions (Tripathi et al., 2023). It is widely used to remove the micropollutants from wastewater due to the high removal rate, easy design of set up, low cost, and potential of adsorbent regeneration. However, the generation of secondary pollutants, high energy demands, resistance to degradation are the major challenges of the adsorption of micropollutants (Patel et al., 2024). The adsorbents such as agro-based activated biochar, metal organic frameworks, carbon nanotubes, activated carbon, silica gel, and cotton fibres were widely tested to remove the micropollutants (Nwosu et al., 2018). Nitrogen-doped hierarchical porous carbon was developed to effectively remove the antibiotic micropollutants. Adsorbent achieved nearly 100% micropollutants removal in tap water and over 92% removal in real pharmaceutical wastewater.

Advanced Oxidation Processes (AOPs)

AOPs are pioneering approach to degrade the micropollutants from the water based using highly reactive radicals. The processes, namely ozonation, photo-Fenton, UV/H_2O_2, photocatalysis, and electrochemical AOPs are come under the AOPs (Bonora et al., 2020). Photocatalysis is initiated by the interaction of photons with a photocatalyst. Electrochemical AOPs use electricity to produce ·OH at the anode surface or through electrolysis to degrade the pollutants. In UV/H_2O_2, UV light activates

H_2O_2 to generate ·OH which highly effective for micropollutants degradation. Photo-Fenton uses the reaction of H_2O_2 with Fe^{2+} in presence of UV light to generate ·OH which reacts with the micropollutants to completely or partially oxidized. UV ozonation is associated with the formation of hydroxyl radicals that is capable of oxidizing a broad spectrum of organic pollutants into harmless by-products. AOPs are highly versatility, rapid reaction kinetics, and potential for complete mineralization make them a valuable tool to address the micropollutants (Ajala et al., 2022; Tripathi et al., 2023). Despite the several merits, AOPs generally require high energy inputs, costly chemicals, and potential to generate the harmful by-products. Recently, vacuum-UV AOPs was developed to degrade the pollutants in wastewater, providing a chemical-free and cost-effective solution. An integrated system consists of electrochemical UV/Ozone with Ni-Sb-SnO_2/SiOx anode was developed to address the micro-pollutant (2,6-dinitrotoluene) (Kim et al., 2024). Integrated system improved 2.5–3.5 times degradation rates and 1.3–3.9-times mineralization efficiencies for 2,6-dinitrotoluene compared to the combined effects of dark electrolysis and photolysis. A plasmonic photocatalyst, Bi-CeO_2/Ag^0/BiO_2 activate the dissolved oxygen in water which favoured the degradation of sulfamethoxazole, enrofloxacin, and bisphenol A (Xie et al., 2023).

Membrane Technology

Membrane separation removes the micropollutant from wastewater due to their ability to physically separate contaminants based on size exclusion, charge, and chemical affinity (Patel et al., 2024). These are typically available in various modules, classified by their pore sizes and morphologies, including microfiltration (10–0.1 μm), ultrafiltration (0.1–0.01 μm), nanofiltration (0.01–0.001 μm), and reverse osmosis (<0.0001 μm) (Liao et al., 2018). These are effective to remove the various pollutants such as pharmaceuticals, pesticides, and endocrine-disrupting compounds. It is associated with several merits high separation efficiency, compact design, and adaptability to diverse water matrices like seawater, brackish water, and wastewater (Akinpelu et al., 2019). Membrane fouling, high energy demands, and generation of concentrated waste streams are the major challenges of the membrane technology (Qalyoubi et al., 2021). Sharmin et al. (2024) developed a catalytic membrane to remove the peroxymonosulfate from wastewater. According to them, developed membrane was effective to achive more than 98% removal of peroxymonosulfate. The integration system (photocatalysis and membrane) offers a dual benefit of mitigating potential membrane fouling along with pollutants degradation. In this direction, the g-C_3N_4 (as catalyst) integrated with Polyvinylidene fluoride membranes was developed to remove the pharmaceutical pollutants under visible light irradiation (Nieto-Sandoval et al., 2023). The removal efficiencies of 89%, 92%, and 72% were achieved for metoprolol, venlafaxine, and diclofenac, respectively by integrated system.

Biological Processes

Biological processes use bacteria, fungi, and algae to break down the contaminants into simple products (Sonwani et al., 2021). These processes can be carried out under aerobic, anaerobic, anoxic, and combined modes. Further, these can be free and immobilized growth biological processes. The mineralization of pollutants occurs through both catabolic and anabolic processes, resulting in the formation of end products such as gases, liquids, and microbial biomass. The bioavailability of micropollutants is a crucial factor that play role in biodegradation. Biodegradation of micropollutants is the environmentally friendly, cost effective, and sustainable process (Patel et al., 2024). However, limited efficiency for persistent pollutants, slow degradation rate, and secondary sludge generation are the major challenges of the biological processes. These are typically divided as in situ and ex situ (Sonwani et al., 2021).

Current Biological Treatment Processes and Micropollutants Removal

Biological treatment processes are broadly employed in industrial and municipal treatment plants to treat the organic matter and nutrients from wastewater. These processes rely on the metabolic activity of microorganisms to break down organic matter and nutrients into simpler compounds. The conventional biological treatment processes are typically designed to efficiently break down micropollutants in wastewater (Ajaz et al., 2024). These processes are found limited application to effectively remove the micropollutants and associated with the huge diversity (Melvin & Leusch, 2016). Therefore, advanced biological techniques are required to remove the micropollutants from wastewater. In this direction, bioaugmentation, membrane bioreactor, moving bed biofilm reactor, aerobic granular sludge, and integrated biological system are suitable option (Sonwani et al., 2021). However, the efficacy of biological processes generally depends on the physico-chemical composition of pollutants, process time, temperature, pH, food to micro-organisms ratio, etc. (Patel et al., 2024).

The potential microbial strains offer a favourable approach to enhance the micropollutants removal from water (Benner et al., 2013). A study demonstrated that augmenting sand filters with bacterial consortium enhance the geosmin and 2-methylisoborneol degradation (Ho et al., 2007). Bioaugmentation can extend the operational time of filters and delayed the breakthrough and reduced costs associated with the restoration activated carbon (Benner et al., 2013). Membrane bioreactor is the integrated system consist activated sludge process with membrane separation. This combination enhances treatment efficiency of micropollutants by improving solid–liquid separation, retaining high, concentrations of biomass, and producing high-quality effluent. Bernhard et al. (2006) studied the degradation of diclofenac mecoprop and sulfophenyl carboxylates in a bioreactor and compared to conventional activated sludge systems. According to them, membrane bioreactors showed improved removal rates of diclofenac mecoprop, and sulfophenyl carboxylates than conventional system.

Moving bed biofilm reactor uses carriers to support biofilm growth and provide a large surface area for microbial activity. These reactors can improve the biodegradation of micropollutants by supporting diverse microbial communities and extending sludge retention times. Integrated biological systems are the combination of biological processes with AOPs. According to Thorat and Sonwani (2022), the use of AOPs can primarily mineralize complex pollutants into more biodegradable intermediates which further eliminated by biological processes. A combined treatment system offers a more effective solution for managing complex (Rostam & Taghizadeh, 2020). The biological removal of organic micropollutants (i.e., ibuprofen, diclofenac, and carbamazepine) was performed in granular activated carbon filters and found that the removal of micropollutants was increased with increasing numbers of bed volumes (Takman et al., 2024). A combined system consists of packed bed bioreactor and ozonation was employed to simultaneously remove nitrogen and micropollutants from wastewater under different conditions (Yang et al., 2024). The combined system achieved removal efficiencies of 85% for nitrogen and 100% for carbamazepine. The biodegradation of organic micropollutants, including nitrobenzene and bisphenol A, was carried out through anoxic denitrification. Microbial assimilation contributed approximately 40, 10 and 50% to the removal of the respective micropollutants (Xu et al., 2025). An integrated system (i.e., immobilization of microorganisms onto granular activated carbon) was employed to remove the micropollutants from a drinking water production plant (Piai et al., 2020). Integrated system effectively removed the iopromide, iopamidol and metformin from wastewater.

Strategies to Overcome Challenges

The removal strategies of micropollutants from wastewater are listed below, which can be used to increase the efficiency of biological treatment procedures.

Combined Advanced Oxidation Processes (AOPs) and Biological Treatment

AOPs have gained interest due to their effective removal of micropollutants. By producing extremely reactive species like hydroxyl radicals, AOPs may oxidize a wide range of organic molecules, making them a crucial component of biological degradation techniques. Combining biological treatment with AOPs improves performance of wastewater treatment system and its efficacy (Kang et al., 2023). Microorganisms are used in biological treatment to biodegrade organic materials. When integrated with AOPs, these methodologies can break down substances that are typically resistant to biological remediation, making them more accessible for microbial degradation in subsequent biological phases. The conveyance of ammonia in a liquid state is predominantly executed owing to its markedly elevated density (Bar-Niv et al., 2022). During biological treatment, AOPs commonly generate byproducts that are more biodegradable than the original contaminants, making them more suitable for microbial ingestion. The total pollutant removal may be more successful as a consequence of the complementing relationship between AOPs and biological techniques. As time passes, wastewater composition can change dramatically, making it challenging to maintain constant treatment effectiveness. By pretreating wastewater and supplying a more manageable influent to the biological treatment system, AOPs serve as a buffer against these variations, particularly in situations when contamination levels are elevated or complex pollutants are present. This method offers consistent and dependable biological process performance. The reduction in sludge production is another important advantage of using AOPs in biological wastewater treatment. Although biological treatment frequently produces surplus sludge that requires additional disposal, AOPs help to reduce this by breaking down complicated substances into simpler forms that microbes may easily digest (Ponnusami et al., 2023). As a result, a small amount of sludge is produced, and disposal expenses are decreased. AOPs also efficiently break down harmful and persistent contaminants, therefore lowering environmental damage by stopping the discharge of dangerous compounds into water supplies. By further decomposing the intermediate products, biological treatment process enhances the mineralization (Cai et al., 2020). The combined use of AOPs and biological treatment can also help to avoid penalties for failure by more consistently meeting the stringent effluent quality criteria required by regulatory agencies. The advantages of both approaches are incorporated in this strategy: AOPs efficiently degrade complex pollutants, and biological activities further mineralize the breakdown products that are produced. The high energy and chemical requirements in AOPs generally enhance the cost, but when combined with biological treatment, overall operating costs can be reduced. Furthermore, this combined technique reduces the creation of secondary pollutants and hazardous byproducts, providing a more environmentally sustainable alternative to isolated chemical treatments.

Enhancing Microbial Community Efficiency

A probable approach is to improve the breakdown of micropollutants by creating complex microbial consortia made up of bacteria, archaea, and fungi. It is possible to modify these consortia to create particular redox conditions that favor the degradation of pollutants. For example, regulating microbial activity and creating a local micro-environment can improve nutrition recovery while also promoting recycling and resource recovery in the wastewater treatment system (Sravan et al., 2024).

The success rate of biological wastewater treatment is strongly influenced by the makeup and actions of microbial populations. The key factors, including dissolved oxygen levels, pH and temperatures, and features such as system design and operating parameters, strongly affect microbial community dynamics (Chen et al., 2017). Strategies like bioaugmentation and operational changes to better control microbial populations and enhance treatment efficacy are made possible by a greater comprehension of these interactions (Chen et al., 2017). Since mobile genetic elements (MGEs) allow the horizontal transmission of catabolic genes necessary for degrading micropollutants, they are critical for the adaptation and evolution of microbial communities. The biodegradation capability of microbial communities in wastewater treatment facilities (WWTPs) can be enhanced by engineering MGEs (Miguel et al., 2020). By providing stable platforms for the introduction of novel degradation processes, these genetic elements enhance the removal of micropollutants (Miguel et al., 2020). When compared to conventional techniques, novel biological treatment technologies—such as membrane bioreactors and cell-immobilized bioreactors—have shown greater value in controlling elevated pollutant concentrations and breaking down water-insoluble substances. According to Kanaujiya et al. (2019), hybrid systems that incorporate biological and physicochemical processes also provide better treatment performance and reduce the production of harmful byproducts. In between the treatment procedures, ammonia-oxidizing bacteria (AOB) play a crucial role in the biotransformation of organic micropollutants. Moreover, the AOB-dominant environments are particularly suited to these transformations, especially in nitrification processes. The enzymatic processes of AOB are essential for developing focused approaches to improve the elimination of micropollutants (Su et al., 2021). Treatment procedures can be significantly enhanced by utilizing genetic engineering tools, sophisticated treatment technology, and understanding of microbial dynamics. To address the challenges posed by micropollutants in wastewater management, further advancements in this field are crucial.

Exploiting Alternative Metabolic Pathways

Instead of using aerobic process, metabolic approaches can be used to eliminate the micropollutants. The anaerobic methane oxidation, when combined with denitrification or heterotrophic sulfate reduction, can be used to more efficiently target specific contaminants than standard techniques. The treatment system designs must be reconsidered in order to encourage various metabolic processes. The dissolution of organic micropollutants (OMPs) in WWTPs is mainly dependent on the microbial populations. The changes in microbial metabolism owing to mutations and gene rearrangements may lead to the evolution of new catabolic pathways for breaking down resistant compounds. In microbial groups, mobile genetic elements (MGEs) are essential for spreading catabolic genes, which may improve capacity of WWTPs for effective degradation (Miguel et al., 2020). The halogenated OMPs can be co-metabolically degraded by enzymes such as oxidoreductases, cytochrome P450s, and peroxidases, whereas hydrolases are especially efficient against molecules based on amides (Lyu et al., 2024). The potential of conventional biological systems, such as biofilm reactors and activated sludge, to eliminate OMPs has been thoroughly studied. In the modern systems, such as membrane-based reactors and two-phase partitioning bioreactors, perform is enhanced by managing larger pollutant loads and more effectively treating water-insoluble contaminants (Kanaujiya et al., 2019). Also, the hybrid strategies combining physicochemical and biological treatments that can increase overall effectiveness and reduces the production of hazardous byproducts (Grandclément et al., 2017). The eco-friendly alternatives for phenolic MP removal are enzymatic techniques that employ laccase and horseradish peroxidase. Although these enzymes are less active at neutral pH and require auxiliary substrates such as hydrogen peroxide, they can be immobilized to improve their selectivity (Maryskova et al., 2021).

Utilization of Biocatalysts

The micropollutants are harmful and persistent even at low quantities, their breakdown in wastewater presents a serious environmental problem. In this context, the biocatalysts such as oxidative enzymes such as cytochrome P450 and laccase, have proven effective in the degradation of a wide range of organic micropollutants. As the enzymes incorporates into biological treatment processes it improves the degradation of resistant compounds. However, their large-scale implementation remains limited. Enzymes are promising biocatalysts due to their high efficiency and selectivity in treating various pollutants. The poor stability and reusability of enzymes is a significant barrier to their application in wastewater treatment. To address these issues, immobilization techniques have been developed that bind enzymes to solid supports, increasing their lifespan and enabling the repeatability. The microbial immobilized enzyme systems demonstrate the efficiency in breaking down pollutants including industrial chemicals and pharmaceuticals, enhancing the stability and reusability of enzymes (Abdelhamid et al., 2024). Also, the laccase immobilized on mesoporous silica was modified to increase thermal stability and enzymatic activity, successfully degrading micropollutants such as bisphenol A and diclofenac in wastewater (Nair et al., 2013). The fungal laccase is immobilized on the surface of yeast cells in surface display laccase biocatalysts. According to Chen et al. (2016), this technique has proven to be highly stable and reusable, successfully eliminating persistent contaminants such as sulfamethoxazole and bisphenol A. In addition, the biochemical film, which tests the membrane filtration and enzymatic decomposition integration, is tested to improve the competence of micro -cutting and reduce membrane sales problems (Barbhuiya et al., 2021).

Process Design Innovations

The microbial activity and the pollutant breakdown can be enhanced by using the recent technologies like membrane bioreactors (MBRs), packed bed bioreactors, and hybrid activated sludge systems. These technologies provide precise control of environmental parameters, increasing treatment efficiency (Sravan et al., 2024). The recent methods have demonstrated the ability to manage high pollutant loads and treat water-insoluble contaminants, whereas traditional biological systems struggle to effectively eliminate micropollutants. Hybrid systems that combine biological and physicochemical techniques can improve pollution removal even further and reduce the production of hazardous byproducts (Kanaujiya et al., 2019). The degradation of micropollutants can be greatly enhanced by recent developments in reactor design, such as recirculating moving bed biofilm reactors (MBBRs). These systems in under aerobic conditions can accelerate the breakdown of some pollutants, including metoprolol and atenolol, by promoting biofilm formation and controlling redox conditions (Edefell et al., 2021). The combination of MBR systems with nanofiltration or pulsed corona discharge oxidation has achieved removal efficiencies exceeding 90% for some micropollutants. As stated by Rossi et al. (2013), these hybrid techniques enhance both the overall removal of organic pollutants and the degradation of pollutants. The affordable solutions, such as sand filters, combined with ultraviolet degradation and improved adsorption technology, have shown hope for smaller wastewater treatment plants. The pollutants that are targeted determine how effective these systems are, although employing built materials like light-expanded clay units has produced high removal rates for some micropollutants. Through the use of mobile genetic elements (MGEs), microbial communities' contribution to pollution degradation can be enhanced. These elements help catabolic genes proliferate across microbial populations, potentially increasing their ability to break down micropollutants. The efficiency of biological treatment procedures may be further enhanced by developing microbial systems to capture MGEs (Miguel et al., 2020). In summary, advancements in biological wastewater treatment, including innovative reactor designs, hybrid systems, and genetic engineering of microorganisms, present promising solutions to the challenges of trace pollutant removal. These novel ideas are critical for creating sustainable and efficient wastewater treatment solutions.

Real-Time Monitoring and Dynamic Management

The use of continuous monitoring devices to assess pollution concentrations and microbial activity enables adaptive management techniques and real-time operational parameter modifications. This results in better performance in breaking down new contaminants (Sravan et al., 2024). In general, biological WWTPs are not made to effectively eliminate micropollutants. As a result, these chemicals are only partially degraded and remain in treated effluent. (Maurer et al., 2020). According to Wolff et al. (2018), differences in the composition of the microbial community and the operational environment have a major impact on the deletion efficiency of micropollutants. In addition, some micropollutants are resistant to traditional treatment methods that emphasize the need for mixing or progressive systems (Kanaujiya et al., 2019). The understanding of temporal dynamics and biotransformation pathways of micropollutants within WWTPs requires continuous observation. Recent studies have shown that the removal effectiveness of micropollutants varies over time due to factors such as microbial activity and environmental variables (Rich & Helbling, 2023). The micropollutants are detected and analyzed by using advanced analytical techniques such as liquid chromatography and mass spectrometry. This can provide vital insights into the efficiency of treatment processes.

Conclusions

Micropollutants are anthropogenic chemicals that include personal care products, pharmaceuticals, pesticides, plasticizers, microplastics, steroid hormones, and industrial compounds. These are typically found in water and wastewater in trace concentrations ranging from ng/L to μg/L. These chemicals can enter water bodies from various sources, including domestic sewage discharge, untreated wastewater, agricultural runoff, hospital effluents, and industrial activities. Micropollutants pose significant threats to aquatic ecosystems and public health due to their persistence and complex nature. Over recent years, various treatment methods, such as adsorption, advanced oxidation processes, nanofiltration, and membrane bioreactors, have been utilized to mitigate the effects of micropollutants in water and wastewater. The biological wastewater treatment systems are the environmentally friendly, cost-effective, and sustainable process. However, conventional biological processes find limited applications to address micropollutants effectively due to their complex and stable nature. Further, the degradation of micropollutants through biological treatment processes is often hindered by factors such as low concentrations, low biodegradability, chemical recalcitrance, and unsuitable operational conditions. Therefore, advanced biological techniques are required to remove the micropollutants from wastewater. In this direction, bioaugmentation, membrane bioreactor, moving bed biofilm reactor, aerobic granular sludge, and integrated biological system are the suitable options. Furthermore, integrating biological treatment systems with advanced oxidation processes offers an effective solution for treating micropollutants, thereby protecting both environmental and public health.

References

Abdelhamid, M. A., Khalifa, H. O., Yoon, H. J., Ki, M. R., & Pack, S. P. (2024). Microbial immobilized enzyme biocatalysts for multipollutant mitigation: Harnessing nature's toolkit for environmental sustainability. *International Journal of Molecular Sciences, 25*(16), 8616.

Adeoye, J. B., Tan, Y. H., Lau, S. Y., Tan, Y. Y., Chiong, T., Mubarak, N. M., & Khalid, M. (2024). Advanced oxidation and biological integrated processes for pharmaceutical wastewater treatment: A review. *Journal of Environmental Management, 353*, Article 120170.

Ajala, O. J., Tijani, J. O., Salau, R. B., Abdulkareem, A. S., & Aremu, O. S. (2022). A review of emerging micro-pollutants in hospital wastewater: Environmental fate and remediation options. *Results in Engineering, 16*, Article 100671.

Ajaz, S., Hassan, A. A., Michael, R. N., & Leusch, F. D. (2024). Removal of organic micropollutants in biologically active filters: A systematic quantitative review of key influencing factors. *Journal of Environmental Management, 353*, Article 120203.

Akinpelu, A. A., Ali, M. E., Johan, M. R., Saidur, R., Qurban, M. A., & Saleh, T. A. (2019). Polycyclic aromatic hydrocarbons extraction and removal from wastewater by carbon nanotubes: A review of the current technologies, challenges and prospects. *Process Safety and Environmental Protection, 122*, 68–82.

Baig, U., Faizan, M., & Sajid, M. (2021). Effective removal of hazardous pollutants from water and deactivation of water-borne pathogens using multifunctional synthetic adsorbent materials: A review. *Journal of Cleaner Production, 302*, Article 126735.

Balali, G. I., Yar, D. D., Afua Dela, V. G., & Adjei-Kusi, P. (2020). Microbial contamination, an increasing threat to the consumption of fresh fruits and vegetables in today's world. *International Journal of Microbiology, 2020*(1), 3029295.

Balali-Mood, M., Naseri, K., Tahergorabi, Z., Khazdair, M. R., & Sadeghi, M. (2021). Toxic mechanisms of five heavy metals: Mercury, lead, chromium, cadmium, and arsenic. *Frontiers in Pharmacology, 12*, Article 643972.

Barbhuiya, N., Misra, U., & Singh, S. (2021). Biocatalytic membranes for combating the challenges of membrane fouling and micropollutants in water purification: A review. *Chemosphere, 286*(Part 2), 131757.

Barbosa, M. O., Moreira, N. F., Ribeiro, A. R., Pereira, M. F., & Silva, A. M. (2016). Occurrence and removal of organic micropollutants: An overview of the watch list of EU decision 2015/495. *Water Research, 94*, 257–279.

Bar-Niv, N., Azaizeh, H., Kuc, M. E., Azerrad, S., Haj-Zaroubi, M., Menashe, O., & Kurzbaum, E. (2022). Advanced oxidation process UV-H_2O_2 combined with biological treatment for the removal and detoxification of phenol. *Journal of Water Process Engineering, 48*, Article 102923.

Behera, I. D., Nayak, M., Biswas, S., Meikap, B. C., & Sen, R. (2021). Enhanced biodegradation of total petroleum hydrocarbons by implementing a novel two-step bioaugmentation strategy using indigenous bacterial consortium. *Journal of Environmental Management, 292*, Article 112746. https://doi.org/10.1016/J.JENVMAN.2021.112746

Behera, I. D., Nayak, M., Mishra, A., Meikap, B. C., & Sen, R. (2022). Strategic implementation of integrated bioaugmentation and biostimulation for efficient mitigation of petroleum hydrocarbon pollutants from terrestrial and aquatic environment. *Marine Pollution Bulletin, 177*, Article 113492. https://doi.org/10.1016/J.MARPOLBUL.2022.113492

Benner, J., Helbling, D. E., Kohler, H. P. E., Wittebol, J., Kaiser, E., Prasse, C., Ternes, T. A., Albers, C. N., Aamand, J., Horemans, B., & Springael, D. (2013). Is biological treatment a viable alternative for micropollutant removal in drinking water treatment processes? *Water Research, 47*(16), 5955–5976.

Bernhard, M., Müller, J., & Knepper, T. P. (2006). Biodegradation of persistent polar pollutants in wastewater: Comparison of an optimised lab-scale membrane bioreactor and activated sludge treatment. *Water Research, 40*(18), 3419–3428.

Bonora, R., Boaretti, C., Campea, L., Roso, M., Martucci, A., Modesti, M., & Lorenzetti, A. (2020). Combined AOPs for formaldehyde degradation using heterogeneous nanostructured catalysts. *Nanomaterials, 10*(1), 148.

Bracamontes-Ruelas, A. R., Ordaz-Díaz, L. A., Bailón-Salas, A. M., Ríos-Saucedo, J. C., Reyes-Vidal, Y., & Reynoso-Cuevas, L. (2022). Emerging pollutants in wastewater, advanced oxidation processes as an alternative treatment and perspectives. *Processes, 10*(5), 1041.

Cai, Q. Q., Wu, M. Y., Li, R., Deng, S. H., Lee, B. C. Y., Ong, S. L., & Hu, J. Y. (2020). Potential of combined advanced oxidation–biological process for cost-effective organic matter removal in reverse osmosis concentrate produced from industrial wastewater reclamation: Screening of AOP pre-treatment technologies. *Chemical Engineering Journal, 389*, Article 123419.

Chen, Y., Stemple, B., Kumar, M., & Wei, N. (2016). Cell surface display fungal laccase as a renewable biocatalyst for degradation of persistent micropollutants bisphenol A and sulfamethoxazole. *Environmental Science and Technology, 50*(16), 8799–8808.

Chen, Y., Lan, S., Wang, L., Dong, S., Zhou, H., Tan, Z., & Li, X. (2017). A review: Driving factors and regulation strategies of microbial community structure and dynamics in wastewater treatment systems. *Chemosphere, 174*, 173–182.

Dai, X., Liu, T., Du, Y., Cao, J. Y., Wang, Z. J., Li, J., Zhou, P., Zhang, H., & Lai, B. (2024). Synergistic effect in enhancing treatment of micro-pollutants by ferrate and carbon materials: A review. *Chinese Chemical Letters*, 110548.

Das, P., & Bal, M. (2025). Modeling and optimization study for microplastic removal from aquatic medium using Chroococcidiopsis species. *Chemical Engineering Science, 304*, Article 121085. https://doi.org/10.1016/J.CES.2024.121085

Das, A. P., & Mishra, S. (2008). Hexavalent Chromium (VI): Health hazards and environmental pollutant. *Journal Environmental Research Deviation, 2*, 386–392.

Das, P., Halder, G., & Bal, M. (2023b). A critical review on remediation of microplastics using microalgae from aqueous system. *Science of the Total Environment, 898*, Article 166425. https://doi.org/10.1016/J.SCITOTENV.2023.166425

Das, A. P., Dutta, K., Khatun, R., Behera, I. D., Singh, S., & Mishra, S. (2023). Microfiber pollution and its microbial mitigation: A review on current trends and future prospects. *Journal Taiwan Institute Chemical Engineers*, 105104.https://doi.org/10.1016/J.JTICE.2023.105104

Das, A. P., Behera, I. D., & Das, N. P. (2024). Renewable energy generation and value addition from environmental microfiber pollution through advanced greener solution (pp. 278).

Edefell, E., Falås, P., Torresi, E., Hagman, M., Cimbritz, M., Bester, K., & Christensson, M. (2021). Promoting the degradation of organic micropollutants in tertiary moving bed biofilm reactors by controlling growth and redox conditions. *Journal of Hazardous Materials, 414*, Article 125535.

Ghosh, S., Sinha, J. K., Ghosh, S., Vashisth, K., Han, S., & Bhaskar, R. (2023). Microplastics as an emerging threat to the global environment and human health. *Sustainability, 15*, 10821. https://doi.org/10.3390/SU151410821

Goswami, L., Kumar, R. V., Borah, S. N., Manikandan, N. A., Pakshirajan, K., & Pugazhenthi, G. (2018). Membrane bioreactor and integrated membrane bioreactor systems for micropollutant removal from wastewater: A review. *Journal of Water Process Engineering, 26*, 314–328.

Grandclément, C., Seyssiecq, I., Piram, A., Wong-Wah-Chung, P., Vanot, G., Tiliacos, N., Roche, N., & Doumenq, P. (2017). From the conventional biological wastewater treatment to hybrid processes: The evaluation of organic micropollutant removal. *Water Research, 111*, 297–317. https://doi.org/10.1016/j.watres.2017.01.005

El Hammoudani, Y., Dimane, F., Haboubi, K., Benaissa, C., Benaabidate, L., Bourjila, A., Achoukhi, I., El Boudammoussi, M., Faiz, H., Touzani, A., & Moudou, M. (2024). Micropollutants in wastewater treatment plants: A bibliometric-bibliographic study. *Desalination and Water Treatment*, 100190.

Ho, L., Hoefel, D., Bock, F., Saint, C. P., & Newcombe, G. (2007). Biodegradation rates of 2-methylisoborneol (MIB) and geosmin through sand filters and in bioreactors. *Chemosphere, 66*(11), 2210–2218.

Jagatee, S., Priyadarshini, S., Rath, C. C., & Das, A. P. (2024). Synthetic microfiber: An enduring environmental problem linked to sustainable development. *Environmental Science and Engineering Part, F2333*, 93–112. https://doi.org/10.1007/978-3-031-51792-1_6

Jeirani, Z., Niu, C. H., & Soltan, J. (2017). Adsorption of emerging pollutants on activated carbon. *Reviews in Chemical Engineering, 33*(5), 491–522.

Kanaujiya, D., Paul, T., Sinharoy, A., & Pakshirajan, K. (2019). Biological treatment processes for the removal of organic micropollutants from wastewater: A review. *Current Pollution Reports*, 1–17.

Kang, X., Li, C., Ding, W., Ma, Y., Zhou, X., Gao, S., Chen, C., Liu, W., He, Z., Li, X., & Jiang, G. (2023). Optimization of biological enzymes combined with Fe^{2+}-activated advanced oxidation process for waste activated sludge conditioning using the response surface method. *Journal of Water Process Engineering, 53*, Article 103634.

Kanungo, J., Sahoo, T., Swain, L. P., & Behera, I. D. (2023). In *Toxicity of persistent hydrocarbon pollutants, sources and sustainable remediation process* (pp. 39–65). https://doi.org/10.1007/978-3-031-48220-5_2

Kato, S., & Kansha, Y. (2024). Comprehensive review of industrial wastewater treatment techniques. *Environmental Science and Pollution Research, 31*(39), 51064–51097.

Khan, F. S. A., Mubarak, N. M., Khalid, M., Tan, Y. H., Abdullah, E. C., Rahman, M. E., & Karri, R. R. (2021). A comprehensive review on micropollutants removal using carbon nanotubes-based adsorbents and membranes. *Journal of Environmental Chemical Engineering, 9*(6), Article 106647.

Khatun, R., Singh, S., Dubey, N. K., & Prasad Das, A. (2023a). A review on marine-based phytochemicals and their application in biomedical research. In: Recent frontiers of phytochemicals: Applications in food, pharmacy, cosmetics, and biotechnology (pp. 383–395).https://doi.org/10.1016/B978-0-443-19143-5.00013-X

Khatun, R., Singh, S., Dubey, N. K., & Prasad Das, A. (2023b). A review on marine-based phytochemicals and their application in biomedical research. In *Recent frontiers of phytochemicals: Applications in food, pharmacy, cosmetics, and biotechnology* (pp. 383–395). Elsevier. https://doi.org/10.1016/B978-0-443-19143-5.00013-X

Khoo, Y. S., Goh, P. S., Lau, W. J., Ismail, A. F., Abdullah, M. S., Ghazali, N. H. M., Yahaya, N. K. E., Hashim, N., Othman, A. R., Mohammed, A., & Yusoff, M. A. M. (2022). Removal of emerging organic micropollutants via modified-reverse osmosis/nanofiltration membranes: A review. *Chemosphere, 305*, Article 135151.

Kim, S., Lee, J., Jeong, N., Ahn, Y. Y., Kim, H., & Cho, K. (2024). Electrochemical UV/Ozone process with Ni-Sb-SnO_2/SiOx anode for degradation of micropollutants in wastewater. *Chemical Engineering Journal, 491*, Article 152018.

Kosek, K., Luczkiewicz, A., Fudala-Książek, S., Jankowska, K., Szopińska, M., Svahn, O., Tränckner, J., Kaiser, A., Langas, V., & Björklund, E. (2020). Implementation of advanced micropollutants removal technologies in wastewater treatment plants (WWTPs)-Examples and challenges based on selected EU countries. *Environmental Science and Policy, 112*, 213–226.

Kumar, M., Pandit, S., Patel, V., Khanna, N., Nag, M., Lahiri, D., Ray, R.R., Das, A., & Das, D. (2023). Maximization of energy recovery from starch processing wastewater by thermophilic dark fermentation coupled with microbial fuel cell technology. *Geomicrobiol Journal*, 1–12.https://doi.org/10.1080/01490451.2023.2209555

Kushwaha, A., Goswami, L., Kim, B. S., Lee, S. S., Pandey, S. K., & Kim, K. H. (2024). Constructed wetlands for the removal of organic micropollutants from wastewater: Current status, progress, and challenges. *Chemosphere*, 142364.

Li, Y., Liu, S., Wang, C., Ying, Z., Huo, M., & Yang, W. (2020). Effective column adsorption of triclosan from pure water and wastewater treatment plant effluent by using magnetic porous reduced graphene oxide. *Journal of Hazardous Materials, 386*, Article 121942.

Liao, Y., Loh, C. H., Tian, M., Wang, R., & Fane, A. G. (2018). Progress in electrospun polymeric nanofibrous membranes for water treatment: Fabrication, modification and applications. *Progress in Polymer Science, 77*, 69–94.

Liu, W., Song, X., Na, Z., Li, G., & Luo, W. (2022). Strategies to enhance micropollutant removal from wastewater by membrane bioreactors: Recent advances and future perspectives. *Bioresource Technology, 344*, Article 126322.

Luo, Y., Guo, W., Ngo, H. H., Nghiem, L. D., Hai, F. I., Zhang, J., Liang, S., & Wang, X. C. (2014). A review on the occurrence of micropollutants in the aquatic environment and their fate and removal during wastewater treatment. *Science of the Total Environment, 473*, 619–641.

Lyu, B., Manna, B., Zhou, X., Leung, I. K., & Singhal, N. (2024). Microbial metabolic enzymes, pathways and microbial hosts for co-metabolic degradation of organic micropollutants in wastewater. bioRxiv, (pp. 2024–09). https://doi.org/10.1101/2024.09.04.611302

Malakar, A., Kanel, S. R., Ray, C., Snow, D. D., & Nadagouda, M. N. (2021). Nanomaterials in the environment, human exposure pathway, and health effects: A review. *Science of the Total Environment, 759*, Article 143470.

Mallick, K., Sahu, S., Sahu, A., Dey, S., & Das, A. P. (2024). A critical review of marine microfiber pollution routes, toxicity, and its sustainable remediation. *Environmental Science and Engineering Part, F2333*, 189–211. https://doi.org/10.1007/978-3-031-51792-1_10

Mallick, K., Sahu, A., Dubey, N. K., & Das, A. P. (2023). Harvesting marine plastic pollutants-derived renewable energy: A comprehensive review on applied energy and sustainable approach. *Journal Environmental Management, 348*, 119371. https://doi.org/10.1016/J.JENVMAN.2023.119371

Maryskova, M., Linhartova, L., Novotný, V., Rysová, M., Cajthaml, T., & Ševců, A. (2021). Laccase and horseradish peroxidase for green treatment of phenolic micropollutants in real drinking water and wastewater. *Environmental Science and Pollution Research, 28*, 31566–31574.

Maurer, L., Villette, C., Reiminger, N., Jurado, X., Laurent, J., Nuel, M., Mosé, R., Wanko, A., & Heintz, D. (2020). Distribution and degradation trend of micropollutants in a surface flow treatment wetland revealed by 3D numerical modelling combined with LC-MS/MS. *Water Research, 190*, Article 116672.

Melvin, S. D., & Leusch, F. D. (2016). Removal of trace organic contaminants from domestic wastewater: A meta-analysis comparison of sewage treatment technologies. *Environment International, 92*, 183–188.

Meng, X., Huang, Z., & Ge, G. (2024). Upgrade and reconstruction of biological processes in municipal wastewater treatment plants. *Desalination and Water Treatment*, 100299.

Miguel, A. B. R., Jetten, M. S., & Welte, C. U. (2020). The role of mobile genetic elements in organic micropollutant degradation during biological wastewater treatment. *Water Research X, 9*, Article 100065.

Mishra, S., Singh, R. P., Rath, C. C., & Das, A. P. (2020). Synthetic microfibers: Source, transport and their remediation. *Journal of Water Process Engineering, 38*, Article 101612. https://doi.org/10.1016/J.JWPE.2020.101612

Mishra, S., Dash, D., & Das, A. P. (2022a). Detection, characterization and possible biofragmentation of synthetic microfibers released from domestic laundering wastewater as an emerging source of marine pollution. *Marine Pollution Bulletin, 185*, Article 114254. https://doi.org/10.1016/J.MARPOLBUL.2022.114254

Mishra, S., & Das, A. P. (2021). Current treatment technologies for removal of microplastic and microfiber pollutants from wastewater. In *Wastewater treatment: Cutting-edge molecular tools, techniques and applied aspects* (pp. 237–251).https://doi.org/10.1016/B978-0-12-821881-5.00011-8

Mishra, S., & Das, A. P. (2022). Treatment of the wastewater polluted with synthetic microfiber released from washing machine. In B.B. Das, H. Hettiarachchi, P.K. Sahu, S. Nanda, (Eds.), Lecture Notes in Civil Engineering (pp. 109–117). Springer Science and Business Media Deutschland GmbH. https://doi.org/10.1007/978-981-16-7509-6_9/COVER

Mishra, S., & Das, A. P. (2023). Microbial remediation of synthetic microfiber contaminated wastewater. In M.P. Shah (Ed.), *Microbial technologies in industrial wastewater treatment* (pp. 337–350) Springer. https://doi.org/10.1007/978-981-99-2435-6_17

Mishra, S., Rath, C. C., & Das, A. P. (2019). Marine microfiber pollution: A review on present status and future challenges. *Marine Pollution Bulletin, 140*, 188–197.https://doi.org/10.1016/J.MARPOLBUL.2019.01.039

Mishra, S., Singh, R. P., Rout, P. K., & Das, A. P. (2022b). Membrane bioreactor (MBR) as an advanced wastewater treatment technology for removal of synthetic microplastics. In *Development in wastewater treatment research and processes* (pp. 45–60).https://doi.org/10.1016/B978-0-323-85583-9.00022-3

Mishra, S., Dash, D., & Das, A. P. (2024). Isolation and molecular characterization of polyester-degrading bacterial strains from municipal wastewater and their application in microplastic degradation in contaminated wastewater. *Geomicrobiol Journal*.https://doi.org/10.1080/01490451.2024.2376101

Murtaza, S. Z., Alqassem, H. T., Sabouni, R., & Ghommem, M. (2023). Degradation of micropollutants by metal organic framework composite-based catalysts: A review. *Environmental Technology and Innovation, 29*, Article 102998.

Nair, R., Demarche, P., & Agathos, S. (2013). Formulation and characterization of an immobilized laccase biocatalyst and its application to eliminate organic micropollutants in wastewater. *New Biotechnology, 30*(6), 814–823.

Narwal, N., Katyal, D., Kataria, N., Rose, P. K., Warkar, S. G., Pugazhendhi, A., Ghotekar, S., & Khoo, K. S. (2023). Emerging micropollutants in aquatic ecosystems and nanotechnology-based removal alternatives: A review. *Chemosphere*, 139945.

Nieto-Sandoval, J., Torres-Pinto, A., Pedrosa, M., Munoz, M., de Pedro, Z. M., Silva, C. G., Faria, J. L., Casas, J. A., & Silva, A. M. (2023). Application of g-C3N4-PVDF membrane for the photocatalytic degradation of micropollutants in continuous flow mode: Impact of water matrix. *Journal of Environmental Chemical Engineering, 11*(5), Article 110586.

Nwosu, F. O., Ajala, O. J., Owoyemi, R. M., & Raheem, B. G. (2018). Preparation and characterization of adsorbents derived from bentonite and kaolin clays. *Applied Water Science, 8*, 1–10.

Ouyang, W. Y., Wang, W. L., Zhang, Y. L., Cai, H. Y., & Wu, Q. Y. (2023). VUV/UV oxidation performance for the elimination of recalcitrant aldehydes in water and its variation along the light-path. *Water Research, 228*, Article 119390.

Patel, D., Singh, A., Ambati, S. R., Singh, R. S., & Sonwani, R. K. (2024). An overview of recent advances in treatment of complex dye-containing wastewater and its techno-economic assessment. *Journal of Environmental Management, 370*, Article 122804.

Paul, A., Priyadarsini, S., Karakuş, S., Behera, J. K., Gnanasekaran, R., & Das, A. P. (2024). Fate and accumulation of synthetic microfibers in the environment. In *Sustainable microbial technology for synthetic and cellulosic microfiber bioremediation* (pp. 61–76).https://doi.org/10.1007/978-3-031-62600-5_4

Piai, L., Blokland, M., van der Wal, A., & Langenhoff, A. (2020). Biodegradation and adsorption of micropollutants by biological activated carbon from a drinking water production plant. *Journal of Hazardous Materials, 388*, Article 122028.

Ponnusami, A. B., Sinha, S., Ashokan, H., Paul, M. V., Hariharan, S. P., Arun, J., Gopinath, K. P., Le, Q. H., & Pugazhendhi, A. (2023). Advanced oxidation process (AOP) combined biological process for wastewater treatment: A review on advancements, feasibility and practicability of combined techniques. *Environmental Research, 237*, Article 116944.

Prabhakar, A., Mishra, S., & Das, A. P. (2019). Isolation and identification of lead (Pb) solubilizing bacteria from automobile waste and its potential for recovery of lead from end of life waste batteries. *Geomicrobiological Journal, 36*, 894–903. https://doi.org/10.1080/01490451.2019.1654044

Priyadarsini, S., & Das, A. P. (2024). Analytical and structural characterization of waste lithium-ion batteries for their effective recycling strategy. *Environmental Chemistry and Ecotoxicology*.https://doi.org/10.1016/j.enceco.2024.12.004

Priyadarsini, S., & Das, A. P. (2025). Lithium bioleaching: A review on microbial-assisted sustainable technology for lithium bio-circularity. *Journal of Water Process Engineering, 69*, Article 106744. https://doi.org/10.1016/j.jwpe.2024.106744

Qalyoubi, L., Al-Othman, A., & Al-Asheh, S. (2021). Recent progress and challenges on adsorptive membranes for the removal of pollutants from wastewater. Part I: Fundamentals and classification of membranes. *Case Studies in Chemical and Environmental Engineering, 3*, 100086.

Raj, K., & Das, A. P. (2023). Lead pollution: Impact on environment and human health and approach for a sustainable solution. *Environmental Chemistry and Ecotoxicology, 5*, 79–85. https://doi.org/10.1016/J.ENCECO.2023.02.001

Ramírez-Castillo, F. Y., Loera-Muro, A., Jacques, M., Garneau, P., Avelar-González, F. J., Harel, J., & Guerrero-Barrera, A. L. (2015). Waterborne pathogens: Detection methods and challenges. *Pathogens, 4*(2), 307–334.

Rich, S. L., & Helbling, D. E. (2023). Daily monitoring at a full-scale wastewater treatment plant reveals temporally variable micropollutant biotransformations. *Environmental Science and Technology, 57*(28), 10404–10414.

Rose, P. K., Jain, M., Kataria, N., Sahoo, P. K., Garg, V. K., & Yadav, A. (2023). Microplastics in multimedia environment: A systematic review on its fate, transport, quantification, health risk, and remedial measures. *Groundwater for Sustainable Development, 20*, Article 100889.

Rossi, L., Queloz, P., Brovelli, A., Margot, J., & Barry, D. A. (2013). Enhancement of micropollutant degradation at the outlet of small wastewater treatment plants. *PLoS ONE, 8*(3), Article e58864.

Rostam, A. B., & Taghizadeh, M. (2020). Advanced oxidation processes integrated by membrane reactors and bioreactors for various wastewater treatments: A critical review. *Journal of Environmental Chemical Engineering, 8*(6), Article 104566.

Sharma, M., Bains, A., Sridhar, K., Chawla, P., & Sharma, M. (2024). Emerging micropollutants–Environmental impact and source-controlled approaches: Current State and Future Prospects. *Food and Chemical Toxicology*, 115038.

Sharmin, A., Asif, M. B., Zhang, G., Bhuiyan, M. A., & Pramanik, B. K. (2024). Reactive layered hydroxide membrane for advanced water treatment: Micropollutant degradation and antifouling potential. *Chemosphere, 359*, Article 142318.

Sher, F., Hanif, K., Rafey, A., Khalid, U., Zafar, A., Ameen, M., & Lima, E. C. (2021). Removal of micropollutants from municipal wastewater using different types of activated carbons. *Journal of Environmental Management, 278*, Article 111302.

Singh, R. P., Mishra, S., & Das, A. P. (2020). Synthetic microfibers: Pollution toxicity and remediation. *Chemosphere, 257*, Article 127199. https://doi.org/10.1016/j.chemosphere.2020.127199

Song, B., Zhou, C., Qin, M., Zhao, B., & Sang, F. (2023). When biochar is involved in rhizosphere dissipation and plant absorption of pesticides: A meta-analysis. *Journal of Environmental Management, 345*, Article 118518.

Sonwani, R. K., Kim, K. H., Zhang, M., Tsang, Y. F., Lee, S. S., Giri, B. S., Singh, R. S., & Rai, B. N. (2021). Construction of biotreatment platforms for aromatic hydrocarbons and their future perspectives. *Journal of Hazardous Materials, 416*, Article 125968.

Sravan, J. S., Matsakas, L., & Sarkar, O. (2024). Advances in biological wastewater treatment processes: Focus on low-carbon energy and resource recovery in biorefinery context. *Bioengineering (Basel), 11*(3), 281. https://doi.org/10.3390/bioengineering11030281

Sturm, M. T., Myers, E., Schober, D., Thege, C., Korzin, A., & Schuhen, K. (2022). Adaptable process design as a key for sustainability upgrades in wastewater treatment: Comparative study on the removal of micropollutants by advanced oxidation and granular activated carbon processing at a German municipal wastewater treatment plant. *Sustainability, 14*(18), 11605.

Su, Q., Schittich, A. R., Jensen, M. M., Ng, H., & Smets, B. F. (2021). Role of ammonia oxidation in organic micropollutant transformation during wastewater treatment: Insights from molecular, cellular, and community level observations. *Environmental Science and Technology, 55*(4), 2173–2188.

Takman, M., Betsholtz, A., Davidsson, Å., Cimbritz, M., Svahn, O., Karlsson, S., Østergaard, S. K., Nielsen, J. L., & Falås, P. (2024). Biological degradation of organic micropollutants in GAC filters–temporal development and spatial variations. *Journal of Hazardous Materials, 472*, Article 134449.

Thorat, B. N., & Sonwani, R. K. (2022). Current technologies and future perspectives for the treatment of complex petroleum refinery wastewater: A review. *Bioresource Technology, 355*, Article 127263.

Tripathi, P., Tiwari, S., Sonwani, R. K., & Singh, R. S. (2023). A step towards enhancing the efficiency of biofilm mediated degradation of brilliant green dye in packed bed bioreactor: Statistical and toxicity analysis. *Process Safety and Environmental Protection, 170*, 1228–1239.

Verlicchi, P., Galletti, A., Petrovic, M., & Barceló, D. (2010). Hospital effluents as a source of emerging pollutants: An overview of micropollutants and sustainable treatment options. *Journal of Hydrology, 389*(3–4), 416–428.

Wang, W. L., Jing, Z. B., Zhang, Y. L., Wu, Q. Y., Drewes, J. E., Lee, M. Y., & Hübner, U. (2024). Assessing the chemical-free oxidation of trace organic chemicals by VUV/UV as an alternative to conventional UV/H2O2. *Environmental Science and Technology, 58*(16), 7113–7123.

Włodarczyk-Makuła, M. (2024). Selected organic micropollutants in the aquatic environment. *Desalination and Water Treatment, 317*, 100061.

Wolff, D., Krah, D., Dötsch, A., Ghattas, A., Wick, A., & Ternes, T. (2018). Insights into the variability of microbial community composition and micropollutant degradation in diverse biological wastewater treatment systems. *Water Research, 143*, 313–324.

Wołowicz, A., & Munir, H. M. S. (2025). Emerging organic micropollutants as serious environmental problem: A comprehensive review. *Science of the Total Environment, 958*, Article 177948.

Xie, Z., Saad, A., Shang, Y., Wang, Y., Luo, S., & Wei, Z. (2023). Enhanced degradation of micropollutants by visible light photocatalysts with strong oxygen activation ability. *Water Research, 247*, Article 120785.

Xu, S., Li, R., Liao, Y., Bian, J., Liu, R., & Liu, H. (2025). Biodegradation of organic micropollutants by anoxic denitrification: Roles of extracellular polymeric substance adsorption, enzyme catalysis, and reactive oxygen species oxidation. *Water Research, 268*, Article 122563.

Yang, X., Liu, Z., Chen, C., Zhang, T., Wang, Q., Zhang, R., Duan, F., Tian, X., Yao, M., Demeestere, K., & Van Hulle, S. W. (2024). Hybrid packed bed bioreactor using combined biodegradation and ozonation to enhance nitrogen and micropollutants removal from landfill leachate. *Bioresource Technology, 412*, Article 131413.

Chapter 28
Addressing the Challenges and Strategies for Improving Micropollutant Degradation in Biological Wastewater Treatment

K. V. Nagalakshmi, Santosh Kumar Badampudi, and Padma Seragadam

Introduction

The contamination of water resources by micropollutants has emerged as a significant global concern due to the potential risks these substances pose to human health and the environment (Mishra & Das, 2023; Mishra et al., 2024b). Micropollutants encompass a broad range of chemical compounds, including pharmaceuticals, personal care products, pesticides, industrial chemicals, and other synthetic substances. Although they are typically present in low concentrations, these pollutants exhibit remarkable persistence in aquatic environments due to their chemical stability and resistance to conventional degradation processes. Even trace amounts of these chemicals can accumulate over time, leading to long-term ecological and health impacts (Kummerer et al., 2009; Anand et al., 2022).

Traditional wastewater treatment methods, including primary and secondary biological treatments, are generally effective at removing organic pollutants and suspended solids (Mishra & Das, 2021; Mishra et al., 2024c). However, these processes often fall short when it comes to degrading micropollutants. The chemical diversity and low concentrations of these contaminants present challenges for biological treatment systems, which rely on microbial activity to break down organic matter (Kumar et al., 2023; Mishra & Das, 2022; Mishra et al., 2022). Additionally, many micropollutants are not readily biodegradable, or their degradation occurs at

K. V. Nagalakshmi (✉) · B. Santosh Kumar
Department of Chemistry, Gayatri Vidya Parishad College of Engineering (A), Madhurawada, Visakhapatnam, Andhra Pradesh, India
e-mail: kvnlpc15@gmail.com

S. Padma
Department of Chemical Engineering, Gayatri Vidya Parishad College of Engineering (A), Madhurawada, Visakhapatnam, Andhra Pradesh, India

I. D. Behera et al. (eds.), *Emerging Micropollutants*, Sustainable Environmental Waste Management Strategies, https://doi.org/10.1007/978-3-032-03530-1_28

an extremely slow rate, limiting the effectiveness of conventional biological systems in fully eliminating these substances before discharge into natural water bodies. It examines key factors influencing biodegradation efficiency, such as the physico-chemical properties of micropollutants, the complexities of microbial metabolism, and the operational parameters of treatment systems. Moreover, it examines the limitations of current biological treatment processes and presents strategies to enhance the degradation of micropollutants. These strategies include optimizing microbial communities, developing novel bioreactor designs, and integrating biological treatment with advanced technologies. Given the increasing prevalence of micropollutants in wastewater and the growing demand for sustainable wastewater management practices, this chapter aims to provide a comprehensive overview of the current state of research and development in the field of biological micropollutant removal. It seeks to identify opportunities for improving the efficacy of biological treatment systems and to lay the groundwork for future innovations that will contribute to cleaner, safer water resources for future generations.

The Nature of Micropollutants

Micropollutants are chemicals that are present in the environment in trace amounts, typically ranging from nanograms per liter (ng/L) to micrograms per liter (μg/L). Despite their low concentrations, these substances can accumulate in water bodies over time, resulting in significant ecological and health risks (Samreen et al., 2021; Yang et al., 2017).

Common categories of micropollutants include:

- Pharmaceuticals: Drugs such as antibiotics, painkillers, hormones, and other therapeutic agents that are excreted by humans and animals or improperly disposed of
- Personal Care Products: Chemicals found in cosmetics, soaps, shampoos, and other personal hygiene items.
- Pesticides and Herbicides: Agricultural chemicals that can leach into wastewater systems, often posing long-term environmental risks.
- Industrial Chemicals: Substances such as solvents, plasticizers, flame retardants, and other chemicals used in manufacturing processes.

These micropollutants exhibit a variety of characteristics that make them difficult to remove through conventional biological treatment methods. They often possess complex chemical structures, high environmental stability, and low biodegradability, which complicate their degradation by microbial communities typically involved in wastewater treatment (Lim et al., 2020).

Challenges in Micropollutants Degradation by Biological Systems

Despite the growing use of biological wastewater treatment systems, such as activated sludge, biofilm reactors, and trickling filters, the degradation of micropollutants remains a significant challenge. These challenges stem from factors related to the physicochemical properties of micropollutants, microbial limitations, and the operational characteristics of treatment systems. Below, we explore these challenges in detail, supported by the latest research findings (Fig. 28.1).

Low Concentrations and Bioavailability

Micropollutants typically exist in trace amounts in wastewater, making them difficult for microorganisms to access and degrade efficiently. Biological systems, such as activated sludge processes, are generally designed to handle high concentrations of organic pollutants, such as biodegradable organic matter or high loads of nutrients. These systems are not typically optimized to deal with low concentrations of specific, persistent chemicals. As a result, their capacity to degrade trace chemicals, such as pharmaceuticals, personal care products, pesticides, and industrial chemicals, is often limited. This poses a major challenge for wastewater treatment plants that must deal with a wide range of emerging contaminants in the water. Furthermore, many micropollutants are bound to particulate matter, including organic particles, sludge, or colloidal substances, which further reduces their availability to microorganisms. These micropollutants may adsorb to the surface of particles, making them less likely to interact with the microbial populations in the bulk liquid phase. In addition, many micropollutants are poorly soluble in water, particularly those that are hydrophobic in nature. Their limited solubility results in lower concentrations of

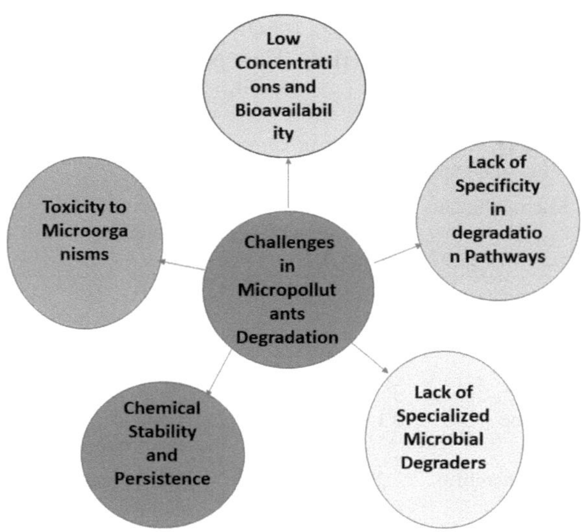

Fig. 28.1 Challenges in micropollutants degradation by biological systems

the contaminants being available in the aqueous phase, which reduces the bioavailability to microorganisms that are responsible for their degradation. The low solubility of these compounds can also result in their retention in solid phases, where microbial communities may not have direct access to the pollutants, reducing the effectiveness of conventional biological treatment systems (Schwarzenbach et al., 2019) (Table 28.1).

Chemical Stability and Persistence

Many micropollutants are chemically stable due to their complex molecular structures, which include functional groups such as aromatic rings, halogenated compounds, and other stable moieties. This structural complexity provides these compounds with a high degree of chemical stability, making them resistant to microbial degradation. Such stability poses significant challenges for biological treatment systems that rely on microorganisms to break down organic pollutants. Unlike simpler compounds that can be easily metabolized by microbes, these complex micropollutants are not easily attacked by microbial enzymes, hindering their breakdown in wastewater treatment systems. (Tiwari et al., 2017).

For instance, pharmaceuticals like carbamazepine a commonly prescribed anticonvulsant and estrone, an endocrine-disrupting chemical found in wastewater, exhibit chemical structures that are resistant to microbial attack. Similarly, pesticides like atrazine, a herbicide frequently detected in agricultural runoff, possess molecular features that prevent them from being readily broken down by the enzymes that microbes typically use to degrade organic matter. These pollutants contain functional groups, such as aromatic rings or halogenated moieties, that are chemically stable and do not easily undergo natural biotransformation processes. Because of this, they persist in the environment, including in wastewater treatment facilities where they evade effective degradation, leading to their continued presence in effluents.

As a result, these substances persist in the environment for extended periods, accumulating in surface waters, groundwater, and soil, thereby posing long-term ecological and health risks. Their persistence is of particular concern because, despite

Table 28.1 Concentrations of common micropollutants found in wastewater and their bioavailability in typical biological systems

Micropollutants	Typical concentration (ng/L–μg/L)	Bioavailability (Solubility)	Degradation efficiency	Refs.
Carbamazepine	50–100 μg/L	Low (Hydrophobic)	Low	Jamieson et al. (2024)
Atrazine	10–200 μg/L	Moderate (Partly soluble)	Moderate	Ahmad et al. (2023)
Estrone	1–10 μg/L	Low (Hydrophobic)	Very Low	Chaturvedi et al. (2023)
Paracetamol	20–100 μg/L	High (Water soluble)	High	Chopra et al. (2023)

passing through conventional wastewater treatment processes, these micropollutants can still end up in the treated effluent, which may then be reused for agricultural or potable purposes. This means that even after wastewater treatment, these chemicals can continue to exert negative effects on aquatic ecosystems, potentially disrupting aquatic life, contaminating drinking water sources, and even accumulating in the food chain.

Moreover, the persistence of these chemicals means that they can undergo further transformations in the environment, leading to the formation of toxic degradation products. These breakdown products, often more toxic than the parent compounds, may also pose additional risks to human and environmental health. For example, certain persistent pharmaceuticals or pesticides can persist in the environment for years, resulting in continuous exposure to wildlife, aquatic organisms, and humans, contributing to the bioaccumulation of toxic substances.

Therefore, the chemical stability and persistence of micropollutants highlight the inadequacy of conventional biological treatment methods in fully addressing these contaminants. More advanced treatment approaches, such as the use of advanced oxidation processes (AOPs), bioaugmentation, or membrane filtration technologies, are being explored to improve the removal of these persistent contaminants. These approaches aim to break down the chemical bonds of these stable pollutants, rendering them more biodegradable and less harmful to the environment.

Toxicity to Microorganisms

Some micropollutants, particularly those found in pharmaceuticals and industrial chemicals, exhibit toxicity to the microorganisms involved in biological treatment. This toxicity is a significant concern because it can reduce the overall effectiveness of wastewater treatment processes. High concentrations of certain compounds, such as antibiotics, heavy metals (Das & Mishra, 2009; Priyadarsini & Das, 2024, 2025; Raj & Das, 2023), plastics (2024a; Das & Bal, 2025; Das et al., 2023; Mishra et al., 2021), and disinfectants, can inhibit microbial growth, alter microbial diversity, and disrupt metabolic functions critical for the degradation of organic matter and pollutants (Ahmad et al., 2023). For example, antibiotics, which are frequently detected in wastewater due to human use and agricultural runoff, are known to inhibit the growth of both pathogenic and beneficial microorganisms in treatment systems, further compromising the degradation of contaminants (Pinto et al., 2022).

The presence of toxic micropollutants can also lead to a reduction in microbial diversity. This is particularly problematic because a diverse microbial community is crucial for the efficient breakdown of a wide variety of organic pollutants. When microbial diversity is reduced, the treatment system may become less effective at degrading not only the micropollutants but also other organic matter present in the wastewater. Additionally, some micropollutants may disrupt essential metabolic pathways in microorganisms, preventing them from carrying out their normal functions, which further intensify treatment inefficiencies.

Furthermore, the toxicity of certain micropollutants, particularly antibiotics, may contribute to the development of antimicrobial resistance (AMR). AMR arises when microorganisms evolve resistance to antibiotics and other antimicrobial agents, rendering them less susceptible to these drugs. In wastewater treatment systems, the presence of antibiotics can exert selective pressure on microbial populations, encouraging the survival and expansion of resistant strains. These resistant microorganisms may then persist in the environment, contributing to the global issue of antibiotic resistance, which is a growing public health concern.

Lack of Specificity in Degradation Pathways

Even when microorganisms are capable of degrading micropollutants, the degradation process is often slow or incomplete (Lahiri et al., 2021; Mishra et al., 2024d). Some contaminants, such as pharmaceuticals, may be partially metabolized by microorganisms but may not undergo full mineralization. In other words, the microorganisms may break down the parent compound into intermediate metabolites, but these intermediates may still be toxic or resistant, meaning they do not degrade further and may persist in the environment (A. P. Das & Swain, 2013; Dey et al., 2023; Singh et al., 2020). This incomplete degradation presents a significant challenge because the intermediate metabolites may be as harmful or more harmful than the parent compound. For instance, some pharmaceutical degradation products have been found to exhibit endocrine-disrupting properties, potentially interfering with the hormonal systems of aquatic organisms and even humans (Dong et al., 2024). This incomplete degradation can lead to the accumulation of toxic intermediates in treated effluent, which may then be released into the environment. As a result, even though the parent micropollutants may no longer be present at significant concentrations, the by-products created during the degradation process can still pose a risk to ecosystems and human health. This is particularly concerning in the case of contaminants that undergo bioaccumulation in aquatic organisms, where the toxic intermediates may accumulate in the tissues of fish or other wildlife, potentially entering the food chain. The lack of specificity in microbial degradation pathways is a key reason why conventional biological treatment systems often fail to fully address the challenges posed by micropollutants. Many microorganisms are limited in their ability to break down specific contaminants efficiently, and even when degradation occurs, the process may not lead to complete mineralization or detoxification. This highlights the need for more advanced treatment technologies, such as advanced oxidation processes (AOPs) or membrane filtration, that can break down these pollutants to safer, non-toxic by-products before they are released into the environment (Table 28.2).

Strategies for Enhancing Micropollutant Degradation

Given the persistent challenges associated with micropollutant removal, there is a growing need for novel strategies that improve microbial degradation efficiency, optimize treatment conditions, and integrate complementary technologies. This section explores several approaches that can help overcome the limitations of conventional treatment systems and enhance micropollutants removal.

Table 28.2 Common micropollutants, their transformation products, and the potential risks associated with these metabolites

Micropollutant	Transformation product(s)	Toxicity concerns	Refs.
Carbamazepine	Hydroxy-carbamazepine	Higher toxicity than parent compound	Trognon, et al. (2024)
Paracetamol	Paracetamol glucuronide	Hepatotoxicity in aquatic organisms	Wang et al. (2023)
Bisphenol A	Bisphenol A diglucuronide	Endocrine-disrupting potential	Siracusa et al. (2018)
Atrazine	Deethylatrazine,	Deisopropylatrazine More toxic than atrazine itself	Blahova et al. (2020)

Enhancing Microbial Communities

One promising approach for enhancing the degradation of micropollutants is through the optimization of microbial communities. Wastewater treatment systems rely on complex microbial populations to break down pollutants, and optimizing these communities can improve their ability to degrade a wider range of contaminants, including micropollutants. Selective cultivation and enrichment of microorganisms with the ability to degrade specific pollutants are fundamental strategies. These methods focus on promoting the growth of microorganisms that exhibit the ability to metabolize particular contaminants of concern, thereby improving the efficiency of wastewater treatment processes.

One common strategy involves isolating specific strains of microorganisms from environmental samples, such as soil or wastewater, that naturally possess the ability to degrade target pollutants. By selectively enriching microbial populations that thrive in the presence of these contaminants, treatment systems can be designed to be more effective. For instance, microorganisms capable of degrading pharmaceuticals, pesticides, or industrial chemicals can be isolated and introduced into treatment systems to accelerate the degradation of these micropollutants. Through selective cultivation, microbial communities can be optimized to target the degradation of specific compounds, thereby increasing the overall efficiency of the treatment process.

Additionally, the use of microbial consortia, which are groups of microorganisms working together to degrade complex pollutants, has emerged as another powerful strategy. Microbial consortia typically involve the collaboration of diverse microbial species, including bacteria, fungi, and even archaea, which work synergistically to break down contaminants through complementary metabolic pathways. In the case of micropollutants, consortia can be particularly effective as different microorganisms may be able to break down different components of a pollutant, ultimately leading to more complete and efficient degradation. This collaborative approach can also help overcome the limitations of individual species that may be able to degrade only certain parts of a micropollutant but not others.

The diversity within microbial consortia also provides a level of resilience, as the failure or inhibition of one microorganism does not necessarily compromise the overall ability of the consortium to degrade the target contaminant. This is especially useful for dealing with micropollutants that may have complex chemical structures or that can inhibit microbial activity. Furthermore, consortia-based approaches are flexible and can be adapted to target a broad spectrum of pollutants, making them versatile tools for wastewater treatment.

Advances in synthetic biology and genetic engineering have opened new path for enhancing microbial degradation capabilities. With these technologies, it is now possible to create genetically modified microorganisms (GMMs) that are specifically designed to degrade even the most persistent and resistant micropollutants. By introducing specific genes into microorganisms that code for enzymes capable of breaking down these compounds, it is possible to engineer microbes that can handle pollutants that would otherwise be difficult or impossible for natural microbial communities to degrade. For example, microorganisms can be engineered to express enzymes that can degrade halogenated compounds, such as certain pharmaceuticals or pesticides, which are known to be particularly resistant to natural degradation processes.

These genetically modified organisms offer a targeted and potentially more efficient solution for dealing with specific micropollutants, especially those with complex molecular structures. The ability to design microbes with tailored enzymatic pathways enables the targeted breakdown of these pollutants in a way that traditional methods may not be able to achieve. Moreover, synthetic biology approaches may allow for the creation of microorganisms that can not only degrade pollutants but also mineralize them completely into harmless by-products like carbon dioxide and water.

However, while the use of genetically modified microorganisms holds great promise, it also raises concerns related to biosafety and environmental impacts. The introduction of GMMs into natural environments or wastewater systems requires careful regulation and monitoring to ensure that these organisms do not unintentionally disrupt local ecosystems or contribute to the spread of genetically modified characteristics to wild populations. Therefore, it is important to develop appropriate containment measures and regulatory frameworks to ensure the safe use of these technologies.

In addition to engineering individual microorganisms, the development of engineered microbial communities or designer consortia is an emerging field. By combining the strengths of different microorganisms and potentially incorporating genetically engineered strains, these consortia can be tailored to target specific pollutants in wastewater systems. This approach offers a highly flexible and powerful solution for tackling the diverse array of micropollutants that exist in wastewater (Table 28.3).

Advanced Biological Treatment Systems

Recent advancements in biological treatment systems have significantly improved the ability to degrade micropollutants in wastewater. Traditional biological systems, while effective for treating common organic pollutants, often struggle to address

Table 28.3 Strategies for microbial degradation of micropollutants

Strategy	Micropollutants degraded	Efficiency of degradation	Additional notes	Refs.
Selective cultivation of microorganisms	Pesticides (atrazine, chlorpyrifos), pharmaceuticals (e.g., diclofenac, caffeine)	Varies: Can degrade 70–90% of target pollutants under optimal conditions	Focuses on optimizing specific strains for particular pollutants	Jin et al. (2021)
Microbial consortia	Pharmaceuticals (carbamazepine, diclofenac), personal care products (oxybenzone, triclosan)	Can enhance degradation by up to 50% more than individual species in some cases	Greater diversity in consortia improves degradation of complex mixtures	Li et al. (2021)
Genetically engineered microorganisms	Pesticides (atrazine), herbicides (glyphosate), industrial chemicals (phthalates)	Can achieve near-complete degradation (90–100%) of specific pollutants under controlled conditions	High potential for targeting persistent micropollutants but may face regulatory challenges	Urgun et al. (2006)
Co-metabolism (Biodegradation via Secondary Metabolism)	Aromatic compounds (benzene, toluene), pharmaceuticals (ibuprofen, diclofenac)	50–70% degradation, depending on the availability of nutrients and the micropollutant type	Effective for pollutants that are structurally similar to natural substrates	Meckenstock et al. (2015)

the wide variety of complex and persistent micropollutants found in wastewater. However, through the integration of innovative technologies, such as Membrane Bioreactors (MBRs) and Integrated Fixed-Film Activated Sludge (IFAS) systems, the efficiency of microbial degradation processes has been enhanced, offering promising solutions for the treatment of these emerging contaminants.

One such system is the Membrane Bioreactor (MBR), which combines biological treatment with membrane filtration. In an MBR, microorganisms are retained within the reactor, and treated water is filtered through membranes that separate the microorganisms from the treated effluent. This setup results in extended retention times for pollutants, providing more opportunities for microorganisms to interact with and degrade contaminants. The ability to retain microorganisms ensures that they remain in contact with pollutants for longer periods, significantly enhancing the degradation of micropollutants. The high concentration of active microorganisms within the MBR further boosts biodegradation rates, making it an effective solution for treating wastewater with a wide variety of micropollutants, including pharmaceuticals, personal care products, and industrial chemicals (Pathak et al., 2020).

Moreover, MBR systems are highly flexible, capable of treating a range of wastewater qualities, from municipal effluents to industrial waste streams with high concentrations of micropollutants. The fine membrane filtration also reduces the discharge of suspended solids, improving effluent quality and allowing for the reuse or recycling of treated water. Integrating membrane filtration with biological treatment offers a dual benefit: efficient pollutant removal and high water recovery, making it especially valuable in water-scarce regions. Similarly, Integrated Fixed-Film Activated Sludge (IFAS) systems offer another promising approach for enhancing the degradation of micropollutants. IFAS systems combine activated sludge with a fixed-film process, which uses a solid surface (e.g., plastic media) to support biofilm growth (Kuśnierz et al., 2022). These biofilms provide a high-density microbial population that is more effective at degrading pollutants compared to conventional activated sludge systems. The microbial population within the biofilm is protected from environmental stressors, allowing it to degrade pollutants more efficiently.

The addition of fixed-film media increases the biomass concentration within the system, which in turn leads to better micropollutants removal. The fixed film provides a larger surface area for microorganisms to colonize, improving the treatment process by allowing more microbes to interact with the contaminants. Moreover, the IFAS system can be tailored for the degradation of specific contaminants by adjusting factors like the type of biofilm media used or the hydraulic retention time. IFAS systems have been shown to improve the removal of a variety of micropollutants, including pharmaceuticals, pesticides, and industrial chemicals, making them highly adaptable for treating wastewater in industrial and municipal settings (Espinosa et al., 2023).

Another effective strategy for enhancing micropollutant removal is the use of Trickling Filters and Rotating Biological Contactors (RBCs). These systems are designed to promote the formation of biofilms on fixed support media, such as rocks, plastic packing, or rotating discs. The biofilms provide a large surface area for microorganisms to attach and interact with the pollutants in the wastewater. The increased surface area allows for more efficient biodegradation of recalcitrant compounds, particularly those that are difficult for conventional treatment systems to break down.

In Trickling Filters, wastewater flows over the fixed media, allowing microorganisms in the biofilm to absorb and degrade pollutants as the water moves through the system. RBCs, on the other hand, use rotating discs that are partially submerged in wastewater, providing continuous exposure to both air and wastewater, which facilitates microbial growth and pollutant degradation. Both systems can be adapted to target specific pollutants, as the microbial communities in the biofilms can be optimized to degrade particular micropollutants. These biofilm-based systems are particularly effective for removing persistent pollutants that are resistant to biodegradation in conventional systems, improving overall treatment efficiency.

One key advantage of these fixed-film systems, including Trickling Filters and RBCs, is that they provide a stable and sustainable environment for microorganisms, reducing the need for frequent maintenance or adjustments. The biofilms that form on the fixed media act as a stable habitat for microorganisms, ensuring that the microbial communities can continue to degrade pollutants efficiently even in the presence of varying wastewater characteristics.

Additionally, these systems tend to have a smaller footprint compared to other advanced treatment technologies, which makes them ideal for locations with space constraints. They also offer the benefit of energy efficiency, as the natural movement of water and air over the fixed media can reduce the energy requirements typically associated with aeration in traditional activated sludge systems (Table 28.4).

Coupling Biological Treatment with Other Technologies

To address the inherent limitations of biological treatment systems, it is increasingly common to integrate complementary technologies into the treatment process.

Table 28.4 Comparative efficiency of biological treatment systems for micropollutant removal

Biological treatment system	Micropollutants degraded	Efficiency of degradation	Additional notes	Refs.
Membrane bioreactors (MBRs)	Pharmaceuticals (e.g., diclofenac, carbamazepine), pesticides (e.g., atrazine), endocrine-disrupting compounds (e.g., bisphenol A)	80–95% degradation for a range of pollutants, depending on system conditions	High microbial concentration and extended retention time significantly boost degradation efficiency	Schroder et al. (2012)
Integrated fixed-film activated sludge (IFAS)	Pharmaceuticals (e.g., ibuprofen, caffeine), personal care products (e.g., triclosan), pesticides	70–90% degradation, can be optimized for specific contaminants	High microbial biomass improves degradation rates, especially in systems designed for specific pollutants	Shreve et al. (2019)
Trickling filters	Pharmaceuticals (e.g., carbamazepine, diclofenac), aromatic hydrocarbons (e.g., toluene)	50–80% degradation, with higher efficiency when combined with advanced pre-treatment processes	Suitable for continuous treatment, with biofilm formation enhancing the microbial activity	Mokhtariazar et al. (2024)
Rotating biological contactors (RBCs)	Pharmaceuticals (e.g., diclofenac, caffeine), organic micropollutants	60–85% degradation, depending on pollutant type and operational conditions	Effective for long-term treatment, with biofilm on rotating discs increasing surface area for microbial activity	Escola et al. (2015)

Table 28.5 Integration of coupled technologies for enhanced micropollutant degradation

Coupling technology	Micropollutants degraded	Efficiency of degradation	Additional notes	Refs.
Advanced oxidation processes (AOPs)	Pharmaceuticals (e.g., ibuprofen, diclofenac), endocrine-disrupting compounds (e.g., bisphenol A), pesticides (e.g., atrazine)	70–95% degradation depending on pollutant type and conditions	AOPs like Fenton's reagent and ozone break down complex micropollutants, making them easier for microbes to degrade	Wypart et al. (2023)
Activated carbon adsorption	Pharmaceuticals (e.g., carbamazepine, triclosan), personal care products (e.g., oxybenzone), pesticides	60–85% removal of micropollutants before biological treatment	Adsorption reduces the chemical load on microbes, enhancing biodegradation efficiency in the next stage	Piai et al. (2020)
Photodegradation (UV or Sunlight)	Pharmaceuticals (e.g., caffeine, diclofenac), personal care products (e.g., parabens)	50–80% degradation, depending on light intensity and micropollutant properties	UV or sunlight exposure breaks down certain pollutants, improving biological treatment outcomes	Zhao et al. (2023)

Advanced Oxidation Processes (AOPs), such as Fenton's reagent and ozone treatment, can break down complex micropollutants into simpler, more biodegradable compounds. These processes can serve as pre-treatment steps, making the pollutants more amenable to microbial degradation in subsequent biological treatment stages. Activated carbon adsorption is another useful technique that can be applied before biological treatment to adsorb micropollutants, reducing the chemical burden on microorganisms and enhancing biodegradation efficiency during the biological phase. Photo degradation, involving UV light or sunlight, can break down specific micropollutants and, when combined with biological treatment, ensures that pollutants are fully degraded before they are discharged into the environment (Table 28.5).

Optimizing Operational Conditions

Optimizing the operational conditions of biological treatment systems is critical to improving micropollutant degradation. Key parameters such as temperature, pH, dissolved oxygen concentration, and retention time must be carefully controlled to create optimal conditions for microbial activity. Even small adjustments to these parameters can significantly enhance the efficiency of biodegradation processes. For instance, maintaining an optimal pH and temperature range can maximize microbial

Table 28.6 Effect of operational factors on microbial biodegradation of micropollutants

Operational condition	Effect on micropollutant degradation	Key features	Optimal range	Refs.
Temperature	Enhances microbial metabolism	Higher temperatures can accelerate microbial activity, but excessive heat can inhibit it	20–40 °C (depending on microorganisms)	Morgan-Sagastume et al. (2003)
pH	Affects microbial enzyme activity	pH affects the availability of pollutants and enzyme activity	pH 6.5–8.5 (typically)	Kodukula et al. (1988)
Dissolved oxygen	Supports aerobic microbial metabolism	Sufficient oxygen is crucial for aerobic biodegradation	2–5 mg/L (for most systems)	Qin et al. (2023)
Retention time	Increases contact time for pollutant degradation	Longer retention times allow more time for microbes to break down pollutants	6–12 h (depending on system design)	Boonnorat et al. (2019)

activity, while ensuring sufficient dissolved oxygen can support aerobic microbial metabolism and enhance pollutant breakdown (Table 28.6).

Real-Time Monitoring and Control

Real-time monitoring systems are increasingly being implemented in wastewater treatment plants to provide dynamic management and optimization of the treatment processes. These systems utilize sensors to measure micropollutant concentrations, microbial activity, and overall treatment performance, allowing for continuous adjustments to operational parameters. By monitoring and controlling the treatment process in real-time, it is possible to ensure that the system operates at peak efficiency, leading to more effective micropollutant removal and reduced environmental impact (Table 28.7).

Table 28.7 Impact of real-time monitoring systems on micropollutant degradation efficiency

Real-time monitoring system	Key features	Parameters monitored	Benefits to micropollutant degradation	Impact on degradation efficiency (%)	Refs.
Micropollutant concentration sensors	Measure concentrations of specific micropollutants in real-time	Micropollutant concentrations (e.g., pharmaceuticals, personal care products)	Provides data for dynamic adjustments to optimize treatment conditions	20–40% improvement in pollutant removal with real-time adjustments	Wade et al. (2005)
Microbial activity sensors	Monitor microbial growth and metabolic activity	Microbial biomass, enzyme activity	Allows for adjustments to conditions (e.g., oxygen, nutrients) to enhance microbial activity	15–25% improvement in microbial efficiency	
Operational parameter sensors	Measure temperature, pH, dissolved oxygen, retention time, etc	Temperature, pH, dissolved oxygen, retention time	Ensures optimal operational conditions for microbial activity, increasing treatment efficiency	10–30% improvement in system efficiency	
Real-time data analytics	Analyze sensor data to optimize treatment conditions	Continuous data analysis for feedback adjustments	Helps to fine-tune operational conditions and predict system performance for better micropollutant removal	25–35% improvement through continuous optimization	

Conclusion

The degradation of micropollutants in wastewater treatment remains a significant challenge due to their low concentrations, chemical stability, and persistence in the environment. However, recent advancements in microbial community engineering, the development of advanced treatment technologies, and the integration of biological systems with complementary technologies offer promising solutions to improve micropollutant removal. As research continues to evolve, more effective strategies will emerge, contributing to cleaner water sources, better protection of aquatic ecosystems, and enhanced public health. The combination of enhanced microbial communities, innovative treatment systems, optimized operational conditions, and real-time monitoring can provide comprehensive solutions to mitigate the environmental and health risks posed by micropollutants.

References

Ahmad, F. A. (2023). The use of agro-waste-based adsorbents as sustainable, renewable, and low-cost alternatives for the removal of ibuprofen and carbamazepine from water. *Heliyon, 9*(6), Article e16449.

Anand, U., Adelodun, B., Cabreros, C., et al. (2022). Occurrence, transformation, bioaccumulation, risk and analysis of pharmaceutical and personal care products from wastewater: A review. *Environmental Chemistry Letters, 20*, 3883–3904. https://doi.org/10.1007/s10311-022-01498-7

Blahova, J., Cocilovo, C., Plhalova, L., Svobodova, Z., & Faggio, C. (2020). Embryotoxicity of atrazine and its degradation products to early life stages of zebrafish (Danio rerio). *Environmental Toxicology and Pharmacology, 77*, Article 103370. https://doi.org/10.1016/j.etap.2020.103370

Boonnorat, J., Kanyatrakul, A., Prakhongsak, A., Honda, R., Panichnumsin, P., & Boonapatcharoen, N. (2019). Effect of hydraulic retention time on micropollutant biodegradation in activated sludge system augmented with acclimatized sludge treating low-micropollutants wastewater. *Chemosphere, 230*, 606–615. https://doi.org/10.1016/j.chemosphere.2019.05.087

Chaturvedi, M., Joy, S., Gupta, R. D., Pandey, S., & Sharma, S. (2023). Endocrine disrupting chemicals (EDCs): Chemical fate, distribution, analytical methods and promising remediation strategies—A critical review. *Environmental Technology Reviews, 12*(1), 286–315. https://doi.org/10.1080/21622515.2023.2205026

Chopra, S., & Kumar, D. (2023). Characterization and biodegradation of paracetamol by biomass of Bacillus licheniformis strain PPY-2 isolated from wastewater. *Rendiconti Lincei. Scienze Fisiche e Naturali, 34*, 491–501. https://doi.org/10.1007/s12210-023-01140-w

Das, A. P., & Mishra, S. (2009). Hexavalent chromium [Cr (VI)]: yellow water pollution and its remediation. *Sarovar Saurabh ENVIS Newsl Wetl Ecosyst, 5*(2).

Das, A. P., & Swain, S. (2013). Algal biosorption of toxic dye methylene blue. In *A Potential Source of Food, Feed, Biochemicals, Biofuels and Biofertilizers, International Conference on Algal Biorefinery, Indian Institute of Technology* (Vol. 13).

Das, P., & Bal, M. (2025). Modeling and optimization study for microplastic removal from aquatic medium using Chroococcidiopsis species. *Chemical Engineering Science, 304*, Article 121085. https://doi.org/10.1016/J.CES.2024.121085

Das, P., Halder, G., & Bal, M. (2023). A critical review on remediation of microplastics using microalgae from aqueous system. *Science of the Total Environment, 898*, Article 166425. https://doi.org/10.1016/J.SCITOTENV.2023.166425

Dey, S., Das, A., Mallick, K., Sahu, A., & Das, A. P. (2023). In *Environmental petroleum waste: Pollution, toxicity, sustainable remediation* (pp. 159–175). https://doi.org/10.1007/978-3-031-48220-5_7

Dong, L., Li, S., Huang, J., Li, W.-J., & Ali, M. (2024). Co-occurrence, toxicity, and biotransformation pathways of metformin and its intermediate product guanylurea: Current state and future prospects for enhanced biodegradation strategy. *Science of the Total Environment, 921*, Article 171108. https://doi.org/10.1016/j.scitotenv.2023.171108

Escolà Casas, M., Chhetri, R. K., Ooi, G., Hansen, K. M. S., Litty, K., Christensson, M., Kragelund, C., Andersen, H. R., & Bester, K. (2015). Biodegradation of pharmaceuticals in hospital wastewater by a hybrid biofilm and activated sludge system (Hybas). *Science of the Total Environment, 530–531*, 383–392. https://doi.org/10.1016/j.scitotenv.2015.05.082

Espinosa-Ortiz, E. J., Gerlach, R., Peyton, B. M., Roberson, L., & Yeh, D. H. (2023). Biofilm reactors for the treatment of used water in space: Potential, challenges, and future perspectives. *Biofilms, 15*(6), Article 100140. https://doi.org/10.1016/j.bioflm.2023.100140

Shreve, M. J., & Brennan, R. A. (2019). Trace organic contaminant removal in six full-scale integrated fixed-film activated sludge (IFAS) systems treating municipal wastewater. *Water Research, 151*, 318–331. https://doi.org/10.1016/j.watres.2018.12.042

Jamieson, H. E., & George, D. M. (2024). Atrazine degradation and emphasis on circular economy strategy. *Desalination and Water Treatment*. https://doi.org/10.1016/j.dwt.2024.100201

Jin, Y., et al. (2021). Degradation of atrazine by Pseudomonas spp. and its environmental impact. *Environmental Science and Pollution Research, 28*(15), 18406–18416. https://doi.org/10.1007/s11356-020-11332-2

Kodukula, P. S., Prakasam, T. B. S., & Anthonisen, A. C. (1988). Role of pH in biological wastewater treatment processes. In *Physiological models in microbiology* (1st ed., pp. 23). CRC Press. https://doi.org/10.1201/9781351075657

Kumar, M., Pandit, S., Patel, V., Khanna, N., Nag, M., Lahiri, D., Ray, R. R., Das, A., & Das, D. (2023). Maximization of energy recovery from starch processing wastewater by thermophilic dark fermentation coupled with microbial fuel cell technology. *Geomicrobiology Journal*, 1–12.https://doi.org/10.1080/01490451.2023.2209555

Kümmerer, K. (2009). *Pharmaceuticals in the environment: Sources, fate, effects, and risks.* Springer Science & Business Media.

Kuśnierz, M., Domańska, M., Hamal, K., & Pera, A. (2022). Application of integrated fixed-film activated sludge in a conventional wastewater treatment plant. *International Journal of Environmental Research and Public Health, 19*(10), 5985. https://doi.org/10.3390/ijerph19105985

Lahiri, D., Nag, M., Dey, A., Sarkar, T., Joshi, S., Pandit, S., Das, A. P., Pati, S., Pattanaik, S., Tilak, V. K., & Ray, R. R. (2021). Biofilm mediated degradation of petroleum products. *Geomicrobiology Journal*, 1–10. https://doi.org/10.1080/01490451.2021.1968979

Li, X., Wu, S., Dong, Y., Fan, H., Bai, Z., & Zhuang, X. (2021). Engineering microbial consortia towards bioremediation. *Water, 13*(20), 2928. https://doi.org/10.3390/w13202928

Lim, M., Patureau, D., Heran, M., Lesage, G., & Kim, J. (2020). Removal of organic micropollutants in anaerobic membrane bioreactors in wastewater treatment: Critical review. *Environmental Science: Water Research and Technology, 6*(5), 1230–1243.

Meckenstock, R. U., Elsner, M., Griebler, C., Lueders, T., Stumpp, C., Aamand, J., Agathos, S. N., Albrechtsen, H.-J., Bastiaens, L., Bjerg, P. L., Boon, N., Dejonghe, W., Huang, W. E., Schmidt, S. I., Smolders, E., Sørensen, S. R., Springael, D., & van Breukelen, B. M. (2015). Biodegradation: Updating the concepts of control for microbial cleanup in contaminated aquifers. *Environmental Science and Technology, 49*(12), 7073–7081. https://doi.org/10.1021/acs.est.5b00715

Mishra, S., & Das, A. P. (2021). Current treatment technologies for removal of microplastic and microfiber pollutants from wastewater. In *Wastewater treatment: Cutting-edge molecular tools, techniques and applied aspects* (pp. 237–251). https://doi.org/10.1016/B978-0-12-821881-5.00011-8

Mishra, S., & Das, A. P. (2022). Treatment of the wastewater polluted with synthetic microfiber released from washing machine. In B. B. Das, H. Hettiarachchi, P. K. Sahu, & S. Nanda (Eds.), *Lecture notes in civil engineering* (Vol. 207, pp. 109–117). Springer Science and Business Media Deutschland GmbH. https://doi.org/10.1007/978-981-16-7509-6_9/COVER

Mishra, S., & Das, A. P. (2023). Microbial remediation of synthetic microfiber contaminated wastewater. In M. P. Shah (Ed.), *Microbial technologies in industrial wastewater treatment* (pp. 337–350). Springer. https://doi.org/10.1007/978-981-99-2435-6_17

Mishra, S., Dash, D., & Das, A. P. (2022). Detection, characterization and possible biofragmentation of synthetic microfibers released from domestic laundering wastewater as an emerging source of marine pollution. *Marine Pollution Bulletin, 185*, Article 114254. https://doi.org/10.1016/J.MARPOLBUL.2022.114254

Mishra, S., Dash, D., & Das, A. P. (2024a). Aquatic microbial diversity on plastisphere: Colonization and potential role in microplastic biodegradation. *Geomicrobiology Journal, 41*(4), 312–323. https://doi.org/10.1080/01490451.2023.2209750

Mishra, S., Dash, D., & Das, A. P. (2024b). Isolation and molecular characterization of polyester-degrading bacterial strains from municipal wastewater and their application in microplastic degradation in contaminated wastewater. *Geomicrobiology Journal*. https://doi.org/10.1080/01490451.2024.2376101

Mishra, S., Dash, D., & Das, A. P. (2024c). Isolation and molecular characterization of polyester-degrading bacterial strains from municipal wastewater and their application in microplastic degradation in contaminated wastewater. *Geomicrobiology Journal*. https://doi.org/10.1080/01490451.2024.2376101

Mishra, S., Dash, D., & Das, A. P. (2024d). Microbial degradation of polyester microfibers using indigenously isolated bacterial strain Exiguobacterium Sp. *Clean—Soil, Air, Water, 52*(12), e202300343. https://doi.org/10.1002/CLEN.202300343

Mishra, S., Swain, S., Sahoo, M., Mishra, S., & Das, A. P. (2021). Microbial colonization and degradation of microplastics in aquatic ecosystem: A review. *Geomicrobiology Journal*. https://doi.org/10.1080/01490451.2021.1983670

Mokhtariazar, A., Hassani, A. H., Borghei, M., et al. (2024). A comparative study on performance of the conventional and fixed-bed membrane bioreactors for treatment of Naproxen from pharmaceutical wastewater. *Science and Reports, 14*, 9944. https://doi.org/10.1038/s41598-024-52872-0

Morgan-Sagastume, F., & Allen, D. G. (2003). Effects of temperature transient conditions on aerobic biological treatment of wastewater. *Water Research, 37*(15), 3590–3601. https://doi.org/10.1016/S0043-1354(03)00270-7

Pathak, N., Tran, V. H., Merenda, A., Johir, M. A. H., Phuntsho, S., & Shon, H. (2020). Removal of organic micro-pollutants by conventional membrane bioreactors and high-retention membrane bioreactors. *Applied Sciences, 10*(8), 2969. https://doi.org/10.3390/app10082969

Piai, L., Blokland, M., van der Wal, A., & Langenhoff, A. (2020). Biodegradation and adsorption of micropollutants by biological activated carbon from a drinking water production plant. *Journal of Hazardous Materials, 388*, Article 122028.

Pinto, I., Simões, M., & Gomes, I. B. (2022). An Overview of the impact of pharmaceuticals on aquatic microbial communities. *Antibiotics (Basel), 11*(12), 1700. https://doi.org/10.3390/antibiotics11121700

Priyadarsini, S., & Das, A. P. (2024). Analytical and structural characterization of waste lithium-ion batteries for their effective recycling strategy. *Environmental Chemistry and Ecotoxicology*. https://doi.org/10.1016/j.enceco.2024.12.004

Priyadarsini, S., & Das, A. P. (2025). Lithium bioleaching: A review on microbial-assisted sustainable technology for lithium bio-circularity. *Journal of Water Process Engineering, 69*, Article 106744. https://doi.org/10.1016/j.jwpe.2024.106744

Qin, L., Wang, D., Zhang, Z., Li, X., Chai, G., Lin, Y., Liu, C., Cao, R., Song, Y., Meng, H., Wang, Z., Wang, H., Jiang, C., Guo, Y., Li, J., & Zheng, X. (2023). Impact of dissolved oxygen on the performance and microbial dynamics in side-stream activated sludge hydrolysis Process. *Water, 15*(11), 1977. https://doi.org/10.3390/w15111977

Raj, K., & Das, A. P. (2023). Lead pollution: Impact on environment and human health and approach for a sustainable solution. *Environmental Chemistry and Ecotoxicology, 5*, 79–85. https://doi.org/10.1016/J.ENCECO.2023.02.001

Samreen, I. A., Malak, H. A., & Abulreesh, H. H. (2021). Environmental antimicrobial resistance and its drivers: a potential threat to public health, *Journal of Global Antimicrobial Resistance, 27*, 101–111. https://doi.org/10.1016/S0045-6535(03)00322-9

Schröder, H. F., Tambosi, J. L., Sena, R. F., Moreira, R. F., José, H. J., & Pinnekamp, J. (2012). The removal and degradation of pharmaceutical compounds during membrane bioreactor treatment. *Water Science and Technology, 65*(5), 833–839. https://doi.org/10.2166/wst.2012.828. PMID: 22339017.

Schwarzenbach, R. P., Gschwend, P. M., & Imboden, D. M. (2019). Pharmaceuticals of emerging concern in aquatic systems. *Chemical Reviews*. Retrieved from https://pubs.acs.org/doi/10.1021/acs.chemrev.8b00299

Singh, R. P., Mishra, S., & Das, A. P. (2020). Synthetic microfibers: Pollution toxicity and remediation. *Chemosphere, 257*, Article 127199. https://doi.org/10.1016/j.chemosphere.2020.127199

Siracusa, J. S., Yin, L., Measel, E., Liang, S., & Yu, X. (2018). Effects of bisphenol A and its analogs on reproductive health: A mini review. *Reproductive Toxicology, 79*, 96–123. https://doi.org/10.1016/j.reprotox.2018.06.005

Tiwari, B., Sellamuthu, B., Ouarda, Y., Drogui, P., Tyagi, R. D., & Buelna, G. (2017). Review on fate and mechanism of removal of pharmaceutical pollutants from wastewater using biological approach. *Bioresource Technology, 224*, 1–12.

Trognon, J., Albasi, C., & Choubert, J.-M. (2024). A critical review on the pathways of carbamazepine transformation products in oxidative wastewater treatment processes. *Science of the Total Environment, 912*, Article 169040. https://doi.org/10.1016/j.scitotenv.2023.169040

Urgun-Demirtas, M., Stark, B., & Pagilla, K. (2006). Use of genetically engineered microorganisms (GEMs) for the bioremediation of contaminants. *Critical Reviews in Biotechnology, 26*(3), 145–164. https://doi.org/10.1080/07388550600842794

Wade, M. J., Sánchez, A., & Katebi, M. R. (2005). Real-time control and process monitoring of wastewater treatment plants: Real-time process monitoring. *Transactions of the Institute of Measurement and Control, 27*(3), [page range if available]. https://doi.org/10.1191/0142331205tm140oa

Wang, L., Wang, Y., Zhang, R. Y., Wang, Y., Liang, W., & Li, T. G. (2023, November 19). Management of acute carbamazepine poisoning: A narrative review. *World Journal of Psychiatry, 13*(11), 816–830. https://doi.org/10.5498/wjp.v13.i11.816

Wypart-Pawul, A., Neczaj, E., & Grobelak, A. (2023). Advanced oxidation processes for removal of organic micropollutants from wastewater. *Desalination and Water Treatment, 305*, 114–128.

Yang, Y., Ok, Y. S., Kim, K. H., Kwon, E. E., & Tsang, Y. F. (2017). Occurrences and removal of pharmaceuticals and personal care products (PPCPs) in drinking water and water/sewage treatment plants: A review. *Science of the Total Environment, 15*(596–597), 303–320. https://doi.org/10.1016/j.scitotenv.2017.04.102. Epub 2017 Apr 21 PMID: 28437649.

Zhao, J., Shang, C., & Yin, R. (2023). A high-radical-yield advanced oxidation process coupling far-UVC radiation with chlorinated cyanurates for micropollutant degradation in water. *Environmental Science and Technology, 57*(47), 18867–18876. https://doi.org/10.1021/acs.est.3c00255

Chapter 29
Evolving Landscape of Micro Pollutants and Developments in Bio Remediation Techniques

Manisha Bal, Nookala Ravikumar, Alok Prasad Das, and Ipsita Dipamitra Behera

Introduction

Nowadays the micro pollutant is a major issue everywhere in the ecosystem. As the pollutant size is in the micro range it is very difficult to trace it out as well as its minimization is a quite big task. The rapid industrialization and urbanization of modern society have significantly contributed to the proliferation of micropollutants in water bodies, soil, and air (Das & Mishra, 2008; Dey et al., 2024a; Kanungo et al., 2024; Mallick et al., 2023). There are many sources for this micropollutant some of them are from industries like fertilizers, pesticides, plastics, petroleum, and petrochemical industries (Behera et al., 2020; Das et al., 2024b, 2024c; Mohanta et al., 2024).

They releasing micro-pollutants into the water resources, environment which leads to many chronic diseases like EDC's (endocrine disrupting compounds). It also leads to an organs liver, kidney, lungs malfunctioning and sometimes it leads to failure also. So we need to reduce the micro-pollutant release from their sources on the other hand we need to reduce the sources by proposing or initiating new alternatives for the sources. The bioremediation techniques like bio degradation, phytoremediation,

M. Bal (✉) · N. Ravikumar
Department of Chemical Engineering, National Institute of Technology Durgapur, Durgapur, West Bengal, India
e-mail: mbal.che@nitdgp.ac.in

A. P. Das
Department of Life Sciences, RDW University, Bhubaneswar, Odisha, India

I. D. Behera
Chemical Engineering Department, Indira Gandhi Institute of Technology, Sarang, Khalapal, Odisha, India

I. D. Behera et al. (eds.), *Emerging Micropollutants*, Sustainable Environmental Waste Management Strategies, https://doi.org/10.1007/978-3-032-03530-1_29

algal remediation etc. which are eco-friendly techniques which takes place with the help of microorganisms.

This chapter we will provide a detail knowledge about the technologies involved for the reduction of these micro-pollutants, how the bioremediation techniques tackel this micropollutants, types of micropollutants with their respective advantages and challenges they facing during remediation. It also provides the direction for future innovations in this technology.

Micro Pollutants and Its Impacts

Micro pollutants are oftenly released from the low-concentration contaminants that are found in the environment. These micropollutants concentrations ranging from nano to micro grams per liter. These micro pollutants can harm the ecosystems, wildlife, and human health, especially due to their persistence and accumulation capability due to small size range. Unlike traditional pollutants, micro pollutants are in organic or inorganic nature which can produced from our daily activities, such as the use of pharmaceuticals, agriculture purpose chemicals like pesticides, insecticide, herbicides, fertilizers, personal care products, industrial chemicals. They are not fully removed in conventional wastewater treatment processes and can enter the environment through wastewater discharge, agricultural runoff, or atmospheric deposition.

The micropollutants are classified based on the chemical nature of pollutant, the source of pollutant, the impact of pollutant on the environment.

The micopollutants release will impact on the ecosystems which include land, water, air. Due to its release by various sources causes toxicity to aquatic life, contamination of soils which leads to death of various soil microorganisms like azeobacter, nitrosomanas bacteria along with earthworms which leads to disrupting the nutrient cycling and decomposition and plant growth. Its impact on human health which causes endocrine disruption, carcinogenicity, neurological effect, immune system suppression, antibiotic resistance, skin diseases, respiratory issues.

Reducing the impact of these micro-pollutants it needs a multifaceted approach that combines with technological advancements, regulatory measures, sustainable practices, and public engagement.

Bio Remediation

Bio remediation is a process that uses living organisms, primarily microorganisms such as bacteria, fungi, and plants, to degrade, detoxify, or remove micropollutants from soil, water, or air. The term bio remediation which is a combo of two words 'bio' and 'remediate'. 'Bio' means living organisms, and 'remediate' refers to remedy. It offers a natural and cost-effective way to clean up contaminated environments.

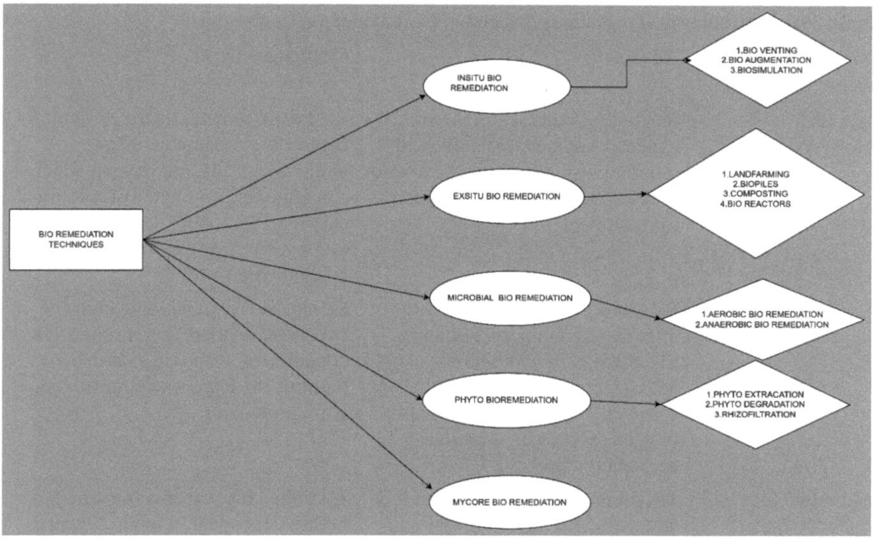

Fig. 29.1 Different types of bioremediation techniques

Microorganisms in this process metabolize pollutants as a source of energy to increase their growth for breaking them down into less harmful substances such as carbon dioxide, water, and biomass.

Bioremediation processes are mostly used to treat a wide range of contaminants which includes hydrocarbons (such as oil and gasoline), heavy metals, pesticides, solvents, and other toxic chemicals like ammonical nitrogen, phenol's, cyanide's etc. (Das et al., 2024a, 2015; Paul et al., 2023; Pradhan et al., 2023).

Bio remediation techniques are broadly categorized based on the environment where the process takes place (in situ or ex situ), the type of organisms used, and the methods used to stimulate biological activity (Dey et al., 2024b; Kanungo et al., 2023; Nag et al., 2023).

Out of these mostly used bio remediation techniques are (a) In situ bio remediation, (b) Ex situ bio remediation (c) Microbial remediation (d) Phyto remediation (e) Mycore remediation. Figure 29.1 represents the different types of bioremediation techniques. Table 29.1 shows different advantages and disadvantages of bioremediation techniques.

In-Situ Bio Remediation

Technique which is used mostly for treatment of the micropollutants which are present in the soil, water or air in the original site of contamination is called In situ bio remediation (Bhattacharjee et al., 2021; Mohanty et al., 2024). It is less

Table 29.1 Advantages and challenges faced in bioremediation technologies

Bioremediation technologies	Advantages	Challenges faced
In-situ bioremediation	In situ bio remediation is less expensive and eco-friendly with minimal disturbance on the system without any transport is required for treatment	In situ bio remediation is difficult for controlling parameters like temperature, pH and oxygen which can affect the efficiency of microbial activity. Due to this the degradation rate is very slow especially for heavily contaminated sites
Ex-situ bioremediation	Ex situ bio remediation is very effective for complex pollutants like petroleum hydrocarbons and heavy metals with high control on parameters like temperature, pH, oxygen levels with an faster degradation process	Ex situ bio remediation which is expensive as there is transportation of contaminants materials. It also requires the large space for treatment of contaminants
Microbial bio remediation	Microbial bio remediation widely applicable for hydrocarbons and pesticides in which microorganisms are selected according to the chemicals present in the system leads to produce less harmful products	Microbial bio remediation which is highly dependent of environmental factors like oxygen, temperature and nutrients. Introduction of microbes can also negatively effect on the local ecosystems
Phyto remediation	Phyto remediation is a bio remediation process which is an ecofriendly with less expensive by without disturbing the ecosystem	Phyto remediation which is less effective for deeper contaminants treatment It will took long time for removing contaminants as it is done through roots system This toxic contaminants will effect the growth of plants
Mycoremediation	The effective remediation process for degrading heavy metals, hydrocarbons and pesticides is mycore remediation with the help of a fungi	The fungi which is used in this process can be degraded by the contaminants slowly. It will not favourable for large scale applications with harsh environmental conditions

disruptive method. It does not require any transport of contaminated material. It includes the techniques like bioventing, bio augmentation, bio simulation.

Bio venting is a process which involves the supply of air or oxygen into the contaminated soil to stimulate the bio-degradation of pollutants naturally, particularly for hydrocarbons. Bioventing is mostly used for the treatment of contaminated soil with petroleum hydrocarbons or volatile organic compounds (VOCs). Bio augmentation the process in which microbes will be added to the contaminated site to increase the degradation process. The microbes are chosen based on their ability to break down specific pollutants, and they often help accelerate the bio degradation process. *Pseudomonas aeruginosa* and *Bacillus cereus* are the microbes which used for lead and cadmium remediation. The microbes *Bacillus* sp., *Lysinibacillus* sp. and *Rhodococcus* sp., are used for remediation of pollutants like aluminium, lead, cadmium, and copper. Bio stimulation in this process we are stimulating the microbial activity by supplying the nutrients like nitrogen, phosphorus and oxygen.

Ex-Situ Bio Remediation

Technique which will treat the micropollutants in the soil, water or air after removing from its original site of contamination is called Ex-situ bio remediation. It is more disruptive and faster remediation technique than in situ techniques. It allows for better control of environmental factor like temperature, moisture and nutrient levels. This method which includes the techniques like land farming, bio piles, composting, bioreactors, as shown in Fig. 29.2. Land farming technique which involves the spreading of contaminated soil over a large area which is further treated by microorganisms with venting of the nutrients, moisture and oxygen to increase its microbial activity. Bio piles is a hybrid technique which is similar to the landfarming but here contaminated soil is placed in big piles other than spreading makes them aerated and treated with nutrients and moisture to stimulate the microbial activity. Composting is a degrading of contaminants in soil by microorganisms under aerobic conditions as well as the soil is mix with manure and agricultural waste to enhance the microbial growth. Ex: Treating the soils which are contaminated with organic compounds like pesticides, poly cyclic aromatic hydrocarbons (PAHs), and explosives. Bioreactors are vessels or tanks in which contaminated material is treated under controlled conditions. In this method, contaminated water or soil slurry is processed in a bioreactor, where microorganisms degrade the pollutants. Temperature, pH, oxygen, and nutrient levels can be optimized for efficient bio-degradation. Ex: Bioreactors are useful for treating the industrial waste water and complex organic pollutants.

Microbial Bioremediation

Microbial bio-remediation is a process where microorganisms are utilized to degrade contaminants. These microorganisms can either be naturally occurring in the contaminated site or specifically introduced (bioaugmentation). The microbes that intake this pollutant and increase its colonial growth which helps degrade the pollutants and release some energy while converting them into less harmful by products (Das & Bal, 2025; Das & Ghosh, 2023; Das et al., 2011, 2023).

The microbial bio remediation technique which further classified into two types based on the oxygen requirement. Aerobic bio remediation requires the oxygen to break down the organic contaminants. This type of remediation method is widely used for treating oil spills and contaminated groundwater with hydrocarbons &organic pollutants. Anaerobic bio remediation don't requires the oxygen to break down the organic contaminants. It will use alternative electron acceptors such as nitrate, sulfate, or carbon dioxide for helping the micro organisms for degrading the contaminants.

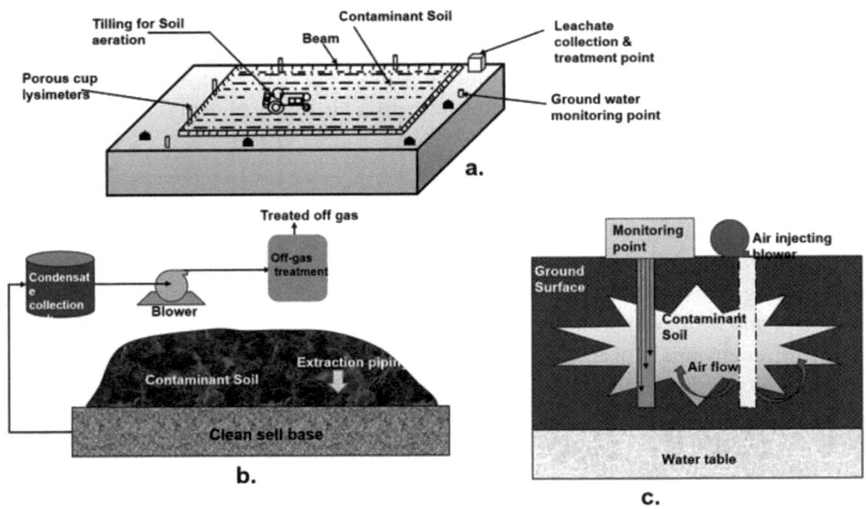

Fig. 29.2 **a** Landfarming, **b** biopile, **c** bioventing

Phyto Remediation

The bio remediation which involves the plants to remove the contaminants in the soil and water is called as phyto remediation. Phyto remediation is the combination of two words phyto is a Greek word which means "plant" whereas other word remediation which coined from a Latin word remedium which means "restoring balance". This technique is eco-friendly and often used for large-scale contamination of heavy metals, organic pollutants. Some of the phyto remediation techniques are phytoextraction, phytodegradation, rhizofiltration. Phytoextraction in this phyto remediation technique the plants absorb the contaminants like heavy metals from the soil or water into their roots and shoots where the contaminants accumulate. The contaminated plant material is then harvested and disposed of. By phytodegradation technique plants absorb pollutants like pesticides which is degraded in the plant tissues through metabolic processes. Rhizofiltration it is one of the filtration operation done by plants with the help of roots for filtering contaminants particularly heavy metals in contaminated water. Figure 29.3 shows the techniques of phyto remediation.

Mycoremediation

Mycoremediation is a bio remediation technique which uses fungi for absorbing environmental pollutants like organic compounds, heavy metals, and toxic chemicals are breaks down into tiny parts. Fungi especially mushrooms type, produce enzymes

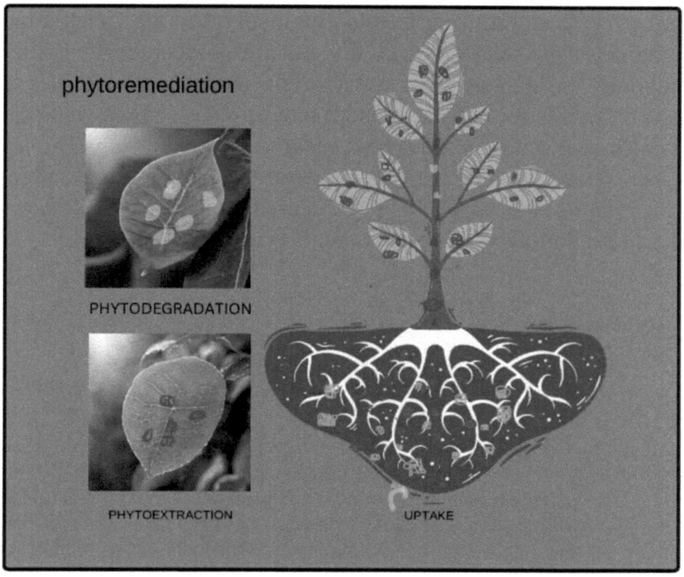

Fig. 29.3 Phyto remediation techniques

such as lignin peroxidases, manganese peroxidases which helps for breaking down the complex organic molecules like hydrocarbons, pesticides and dyes.

Future Aspects

Every bio remediation technologies are involved with so many challenges. So we need to focus on new innovative techniques to get-rid-of those challenges and at the same time without any disturbance in sustainability of the processes focus on increase the efficiency of the degradation of micro-pollutants. The most innovative technologies in bio remediation for the micropollutants degradation are Advanced oxidation processes (AOPs),Microbial enzyme engineering, nano biotechnology.

Advanced oxidation processes (AOPs) which involve in the production of highly reactive oxygen species (ROS) to degrade organic pollutants.

Mechanism steps

Step1: Generation of ROS (e.g., hydroxyl radicals, ozone, hydrogen peroxide).

Step2: ROS react with pollutants, breaking them down into smaller, more biodegradable compounds.

Step3: Microorganisms can then degrade these compounds more easily.

Microbial enzyme engineering technology involves in the Genetic modification of microorganisms to produce enzymes that can able to degrade the toxic pollutants. Nano biotechnology which uses nano particles for interaction with micro pollutants. This interaction leads to the enhancement of bio degradation by increasing microbial activity and also uses a nano particle bio sensors for monitoring. The mechanisms which are involved for this process are adsorption, reduction, oxidation, enzyme-mediated degradation, microbial-mediated degradation. Table 29.2 shows different advantages and limitations of different technologies.

Table 29.2 Advantages and limitations of innovative technologies with examples of reacting species

Innovative technologies	Advantages	Limitations	Examples
Advanced oxidation processes (AOP'S)	Higher efficiency process with low energy requirement. It improves microbial activity which leads to enhance pollutant degradation. Reduces toxicity by improving biodegradability	Reactive oxygen species (ROS) can damage microorganisms The process has pH and temperature sensitivity Catalyst deactivation can possible Byproduct formation is taking place	Examples of AOPs:Fenton's reaction (H_2O_2 + Fe^{2+}), ozone treatment (O_3), photocatalysis (UV light + catalyst), sonolysis (ultrasound), electrochemical oxidation (EO)
Microbial enzyme engineering	Enhances bio degradation rates It increases substrate specificity so it leads to the process so cost effective Improves enzyme stability which leads to toxicity reduction	Changing of stability and activity of an enzyme Substrate specificity and activity is not effective Cellular toxicity and tolerance issues will takes place Scaling up and optimization is not successful	Microorganisms used for this process bacteria (e.g., *E. coli*, Pseudomonas), fungi (e.g., Aspergillus, Trichoderma), yeast (e.g., *Saccharomyces cerevisiae*)
Nanobiotechnology	Enhanced biodegradation rates. Increased pollutant removal efficiency Reduction of toxicity Microbial activity improvement Cost-effectiveness	Nanoparticles can be toxic to microorganisms and ecosystems. Nanoparticles can aggregate or degrade, reducing effectiveness. Large-scale production and application can be expensive. Insufficient knowledge of nanoparticle-biological interactions. Nanoparticles may not be compatible with microorganisms	Types of nanomaterials: zero-valent iron (nZVI), titanium dioxide (TiO_2), zinc oxide (ZnO), silver (Ag), carbon nanotubes (CNTs), graphene oxide (GO)

Conclusion

From this entire chapter we concluded that the micropollutants which are in nano and micro grams per liter are very harmful to all living organism on the ecosystems. The bioremediation processes which are based on environment in which the pollutant present they will treat at the place of release or after the release, by using microorganisms, plants and by using fungi. As each process have some advantages and challenges too. To overcome these challenges we focused on new innovative techniques which can efficiently overcome these challenges by maintaining the sustainability. But methods like advance oxidation processes, microbial enzyme engineering, nano biotechnology clear the challenges faced by the old techniques of bio remediation but these itself have some challenges like operating parameters sensitivity, stability imbalance, toxicity problems. At last this focuses on the applications in the fields like heavey metal removal, oil spill clean up etc.; This review is mostly focus on the different techniques of bioremediation and it will provide a pathway for future innovations in bioremediation.

References

Behera, I. D., Basak, G., Kumar, R. R., Sen, R., & Meikap, B. C. (2020). Treatment of petroleum refinery sludge by petroleum degrading bacterium *Stenotrophomonas pavanii* IRB19 as an efficient novel technology. *Journal of Environmental Science and Health, Part A, 56*, 226–239. https://doi.org/10.1080/10934529.2020.1866924

Bhattacharjee, J., Mishra, S., & Das, A. P. (2021). Recent advances in sensor-based detection of toxic dyes for bioremediation application: A Review. *Applied Biochemistry and Biotechnology, 2021*, 1–20. https://doi.org/10.1007/S12010-021-03767-7

Das, A., Dey, S., Bala, K., Gnanasekaran, R., & Das, A. P. (2024a). Microbial bioremediation technology for sustainable treatment and management of synthetic microfiber waste. In *Sustainable microbial technology for synthetic and cellulosic microfiber bioremediation* (pp. 221–238). https://doi.org/10.1007/978-3-031-62600-5_12

Das, A. P., Behera, I. D., & Das, N. P. (2024b). *Renewable energy generation and value addition from environmental microfiber pollution through advanced greener solution* (Vol. 278).

Das, A. P., & Ghosh, S. (2023). Cost-effectiveness of the biomining process. In *Manganese mining microorganisms* (pp. 143–151). https://doi.org/10.1016/B978-0-12-822147-1.00008-X

Das, A. P., Ghosh, S., Mohanty, S., & Sukla, L. B. (2015). Advances in manganese pollution and its bioremediation. In L. B. Sukla, N. Pradhan, S. Panda, & B. K. Mishra (Eds.), *Environmental microbe biotechnology* (pp. 313–328). Springer. https://doi.org/10.1007/978-3-319-19018-1_16

Das, A. P., & Mishra, S. (2008). Hexavalent chromium (VI): Health hazards and environmental pollutant. *Journal of Environmental Research and Development, 2*, 386–392.

Das, A. P., Nayak, B., & Behera, I. D. (2024c). Applications of microorganisms in advanced geomicrobiological and biotechnological exploration. *Geomicrobiology Journal, 41*, 309–311. https://doi.org/10.1080/01490451.2024.2327059/ASSET//CMS/ASSET/DE2A1F67-08C1-475A-ABC7-2B3F4C9E8BCF/01490451.2024.2327059.FP.PNG

Das, A. P., Sukla, L. B., Pradhan, N., & Nayak, S. (2011). Manganese biomining: A review. *Bioresource Technology, 102*, 7381–7387. https://doi.org/10.1016/J.BIORTECH.2011.05.018

Das, P., & Bal, M. (2025). Modeling and optimization study for microplastic removal from aquatic medium using *Chroococcidiopsis* species. *Chemical Engineering Science, 304*, 121085. https://doi.org/10.1016/J.CES.2024.121085

Das, P., Halder, G., & Bal, M. (2023). A critical review on remediation of microplastics using microalgae from aqueous system. *Science of the Total Environment, 898*, 166425. https://doi.org/10.1016/J.SCITOTENV.2023.166425

Dey, S., Bissoyi, A., Koteswara Reddy, G., & Das, A. P. (2024a). Synthetic microfibers: An urban emerging pollutant demanding for sustainable environmental solutions. In *Sustainable microbial technology for synthetic and cellulosic microfiber bioremediation* (pp. 107–121). https://doi.org/10.1007/978-3-031-62600-5_7

Dey, S., Dash, A., & Das, A. P. (2024b). Bioremediation as a sustainable and circular solution for mining waste management and bioenergy generation. In *Sustainable management of mining waste and tailings: a circular economy approach* (pp. 107–119). https://doi.org/10.1201/978 1003442455-6/BIOREMEDIATION-SUSTAINABLE-CIRCULAR-SOLUTION-MINING-WASTE-MANAGEMENT-BIOENERGY-GENERATION-SUDESHNA-DEY-AKANSHYA-DASH-ALOK-PRASAD-DAS

https://www.researchgate.net/figure/Schematic-presentation-of-Direct-Mechanism-of-Phytoremediation-process_fig4_366126510

Kanungo, J., Sahoo, T., Bal, M., & Behera, I. D. (2023). Performance of bioremediation strategy in waste lubricating oil pollutants: A review. *Geomicrobiology Journal*. https://doi.org/10.1080/01490451.2023.2245395

Kanungo, J., Sahoo, T., Bal, M., & Behera, I. D. (2024). Performance of bioremediation strategy in waste lubricating oil pollutants: A review. *Geomicrobiology Journal, 41*, 360–373. https://doi.org/10.1080/01490451.2023.2245395

Mallick, K., Sahu, A., Dubey, N. K., & Das, A. P. (2023). Harvesting marine plastic pollutants-derived renewable energy: A comprehensive review on applied energy and sustainable approach. *Journal of Environmental Management, 348*, 119371. https://doi.org/10.1016/J.JENVMAN.2023.119371

Mohanta, S., Pradhan, B., & Behera, I. D. (2024). Impact and remediation of petroleum hydrocarbon pollutants on agricultural land: A review. *Geomicrobiology Journal, 41*, 345–359. https://doi.org/10.1080/01490451.2023.2243925

Mohanty, M., Tripathy, B., Koteswara Reddy, G., Das, A. P. (2024). Greener technologies for establishing restraint over microplastic pollution. In *Sustainable microbial technology for synthetic and cellulosic microfiber bioremediation* (pp. 239–258). https://doi.org/10.1007/978-3-031-62600-5_13

Nag, M., Lahiri, D., Ghosh, S., Sarkar, T., Pati, S., Das, A. P., Ram, D. K., Bhattacharya, D., & Ray, R. R. (2023). Application of microorganisms in biotransformation and bioremediation of environmental contaminant: A review. *Geomicrobiology Journal*. https://doi.org/10.1080/01490451.2023.2261443

Paul, A., Dey, S., Ram, D. K., & Das, A. P. (2023). Hexavalent chromium pollution and its sustainable management through bioremediation. *Geomicrobiology Journal, 41*, 1–11. https://doi.org/10.1080/01490451.2023.2218377

Petroleum Degradation: Promising Biotechnological Tools for Bioremediation—Scientific Figure on ResearchGate. https://www.researchgate.net/figure/Bioremediation-methods-a-landfarming-b-biopile-c-bioventing-d-bioslurping-and_fig4_322994319. Accessed 20 Dec 2024.

Pradhan, G., Tripathy, B., Ram, D. K., Digal, A. K., & Das, A. P. (2023). Bauxite mining waste pollution and its sustainable management through bioremediation. *Geomicrobiology Journal, 41*, 335–344. https://doi.org/10.1080/01490451.2023.2235353

Praveen, R., & Nagalakshmi, R. (n.d.). Review on bioremediation and phytoremediation techniques of heavy metals in contaminated soil from dump site.

Chapter 30
Nano-based Remediation Strategies for Microplastic Pollution in the Environment

Ipsita Priyadarshini Tripathy, Sudeshna Dey, Ipsita Dipamitra Behera, and Alok Prasad Das

Introduction

The most common form of marine debris in the Great Lakes and the ocean is plastic. Where there are so many forms of plastic waste, "microplastics" refers to particles smaller than five millimetres, about the size of a sesame seed, which was first proposed by Thomicroplasticson in 2004 (Browne et al., 2011). Depending on sources, microplastics are categorized either as primary or secondary (Mishra et al., 2019a, 2019b)with the major microplastic particles of PET and PE. Primary microplastics includes plastic fragments, synthetic textiles, and microfibers, which are highly used in personal care products (Paul et al., 2024a, 2024b). On the other hand, uncontrolled plastic debris eventually degrades by photolysis, abrasion, and/or microbial degradation into secondary microplastics (Das et al., 2024a, 2024b, 2024c, 2024d). From land, microplastics reach the ocean via rivers and winds, where they become trapped in the world's ocean circulation system (Raj et al., 2023a, 2023b; Das et al., 2024b; Dey et al., 2024b; Sahu et al., 2024). Microplastics pose a danger to fish and other sea creatures that consume them, because once in the water, they get incorporated into the food chain (Ashton et al., 2011). Three significant categories for microplastic in terrestrial environments agricultural practices, run off impacts and breakdown of large plastic debris from inadequate disposal (Dey et al., 2024a; Mohanty et al., 2024a, 2024b, 2024c; Ray et al., 2024).

I. P. Tripathy · A. P. Das (✉)
Department Of Life Sciences, Rama Devi Women's University, Bhubaneswar, Odisha, India
e-mail: alok1503@gmail.com

S. Dey
Centre for Biotechnology, Siksha O Anusandhan University, Bhubaneswar, Odisha, India

I. D. Behera
Department of Chemical Engineering, Indira Gandhi Institute of Technology, Sarang, Dhenkanal, Odisha, India

I. D. Behera et al. (eds.), *Emerging Micropollutants*, Sustainable Environmental Waste Management Strategies, https://doi.org/10.1007/978-3-032-03530-1_30

Worldwide plastic production has been increasing from 1950, from a small amount above a million tons to an incredible 368 million tons in 2019 (Ashton et al., 2011). This is expected to go 5% annually. Microplastic wastes enter the ocean yearly by 8 million tons and plastics account for 90% of marine debris (UNESCO) (Raj et al., 2023a, 2023b). The concentration of plastic waste across the globe is very high in Asia at 50%, followed by South America at 5%, North America at 15%, Europe at 25% and Africa at 5%. Today, with an estimated number of plastic particles in oceans at 171 trillion, an average weigh about 2.3 million tons (Ashton et al., 2011). The ocean sediments have already begun to settle with microplastics, and in some areas, particles are found to be as high as 2000 per square meter (Eriksen et al., 2013). Moreover, according to a 2023 research report, it is seen that people ingest up to 5 g of microplastic per week that would translate into a meaningful microplastic load for marine ecosystems and human health (Cera et al., 2019).

MP (microplastic) has also been found in all of the Great Lakes, but generally, there is an average concentration of 43,000 microplastic particle km^{-2} (Rochman et al., 2013). Microplastics have also been reported in the freshwater ecosystems, out of United States, such as in the year 2019, from a study which took place in Poland, revealed that microplastic was existing in all of the 30 lakes of the Masurian Lakeland examined at a concentration ranging from 0.27 to 1.57 particles per Liter (Bergmann et al., 2019). pointed out that this person who ingests seafood was taking in, on average 11,000 bits of microplastics into his body (Lusher et al., 2013). Microplastics react with chemicals like heavy metals, PCBs, and dyes to produce toxic contaminants that can aggregate and carry pollutants, which contribute to environmental pollution (Wright and Kelly, 2017, Das et al., 2024a; Dey et al., 2023b; Paul et al., 2023). The microplastic particles in Arctic snow ranged from 0.02×103 to 154×103 particles/L, in which microfibers dominated (Leslie et al., 2022). Microplastic particles were found in 69% of 240 marine organisms which are also consumed by humans; this is a high rate of bioaccumulation (UNEP, 2014). Microplastics were highly toxic to living organisms and especially adverse effects on the younger aquatic species (Klaine et al., 2008). They induce pulmonary toxicity and suppress the soil microorganism's enzyme production (Wang et al., 2019). In very recent times, human blood has been found to contain the presence of microplastic; hence, the total quantified amount was 1.6 μg/L with the major microplastic particles of PET and PE along with styrene polymers (Singh et al., 2020a, 2020b; Dey et al., 2023a; Mallick et al., 2024; Pradhan et al., 2024). Hence, the United Nations included microplastics pollution in its list of 2014 as one among the 10 most severely urgent environmental issues that deserve notice at present (Xu et al., 2020).

Nanomaterials are materials with structural features or elements at a nanoscale, which spans between 1 and 100 nm (Maurer-Jones et al., 2013). High surface area and small size determine its unique physical, chemical, and mechanical properties, in addition to other tunable characteristics (Nowack and Bucheli, 2007). Nanomaterials are used for the prevention of microplastic pollution in an innovative approach (Zhou et al., 2019). Adsorption has recently been used for microplastic removal (Prata et al., 2019). The major advantages of using adsorption are the flexibility of selecting the adsorbents that have some affinity toward the target pollutants (Keller et al., 2013). Due to its high surface areas, simplicity of synthesis and ease of functionalization, nanomaterials extensively gained popularity (Shen et al., 2020). Nanomaterials (nanomaterials) include natural particles with at least one dimension of 100 nanomaterial or less, while nanoparticles (NPs) include those with at least two dimensions between 1 and 100 nanomaterial (Holmes et al., 2012). Nanoparticle-based bioremediation provides an innovative approach by engineering nanoparticles to have specific properties for the interaction and catalysis of the degradation of plastic polymers, thereby improving plastic bioremediation efficiency (Chong et al., 2010; Bhatia et al., 2020). By developing nanomaterials with enhanced physical and chemical properties, researchers believe that improved adsorption and catalytic degradation are possible strategies to remove microplastics from the environment (Zhou et al., 2016). The ecotoxicological aspects of nanomaterials study nanoparticle effects in both aquatic and terrestrial ecosystems by providing a larger, more overarching perspective on assessing ecological impacts toward the bioremediation of microplastic pollutants (Shen et al., 2019; Dey et al., 2024c; Moharana et al., 2024; Sahu et al., 2024; Bhattacharjee et al., 2021).

Nanomaterials have a very unique set of physicochemical properties that make them potentially pivotal in remediating microplastic pollution: high surface area, reactivity, and tunability (Klaine et al., 2008). They can adsorb microplastics via physical and chemical interactions; degrade it using catalytic processes such as photocatalysis or advanced oxidation; support magnetic separation for easy separation; enhance filtration systems, thereby ensuring that microplastics are filtered more effectively; and help to detect microplastics through the use of advanced sensors. Although they provide high efficiency, scalability, and versatility, some of the challenges facing them include environmental risks, cost, and regulatory uncertainty (Das et al., 2024d; Mohanty et al., 2024b; Tripathy et al., 2024). Future developments in biodegradable nanomaterials, hybrid systems, and smart materials might further optimize their applications towards environmental purposes while minimizing their drawbacks.

Therefore, it can easily be seen from the above paragraphs that the microplastics and their interaction mechanisms depend upon all the properties related to the nature of materials that are used; the pollutants structure, nature, and functional moiety, coordinate sites; properties of a solution like pH; concentration, etc. Photocatalysts are those materials used in photochemical reactions when irradiation is applied using light. It is promising material for purification of water. The reason for generating electron hole pairs is photon absorption, which happens when photocatalysts surface is exposed to irradiation of light. Magnetic nanoparticle (MNPs), functionalized to bind towards microplastics, provide the possibility of their selective removal from water using an external magnetic field. An external magnetic field is applied by way of a magnet or by electromagnetic system. The MNPs, being now attached to microplastics, are attracted by the magnet and separated from the water. Nanomaterials can be used as carriers for the enzymes, microbes, or chemical agents which break down plastics. Enzymes or microbes encapsulated in nanoparticles are shielded from environmental stress and directly delivered to the surface of microplastics. Through all these processes nanomaterials used to remediate microplastic pollutants. Nanomaterials do indeed have a very significant promise toward the combating of microplastic pollution, though their proper development and deployment should weigh efficacy against safety in environmental considerations.

What Are Nano Materials

Nanomaterials are emerging new frontier in science and technology and they have indeed brought about a paradigm shift in material design and utilization (Ali Alnazza and Chile, 2023). Material properties at the nanoscale fundamentally differ from those of their bulk counterparts due to quantum effects and an increased dominance of surface phenomena. For example, nanoparticles can be highly mechanically strong and are very suitable for reinforcement of composite materials (Zhao et al., 2024). The higher conductivity of their electrical features is important in the development of faster and more efficient electronics. Nanomaterials are what transform innovations into medicine (Patel and Munjal, 2021). Thereby, drugs are delivered by nanoparticles directly into the diseased cells for the targeted delivery system, thus causing fewer side effects and increasing their efficacy (Wei and Wang, 2013). With nanostructures also, advanced diagnostics in the form of biosensors could be possible so that diseases might be detected in high precision even at early stages.

Nanomaterials are categorized by their dimensions, shape, composition, and functionality that gives rise to eight distinct categories (Kumar, 2013). Zero dimensional (0D) Nanomaterials, including nanoparticles, quantum dots, are confined in three dimensions, and thus exhibit distinct optical and electronic characteristics. One dimensional (1D) Nanomaterials including nanowires and nanotubes possess high aspect ratios, and have excellent conductivity (Bhushan (Ed), 2017). Two dimensional (2D) materials consist of graphene, thin films with large surface areas, and extraordinary electronic or thermal conductivity (Gleiter, 2000). Three-dimensional (3D) nanomaterials, such as nanoporous structures and nanocomposites, have improved mechanical and optical properties (Sanchez et al., 2011). carbon nanotubes, and graphene, carbon-based nanomaterials are very strong and conductive. Metal-based nanomaterials, such as gold and titanium dioxide nanoparticles, have catalytic and plasmonic behaviour (Whitesides, 2003). Polymer-based nanomaterials, dendrimers, and nanogels, have flexibility and biocompatibility. Finally, hybrid nanomaterials integrate properties from two or more different materials to form multifunctional systems, thus being applied in catalysis, sensors, drug delivery, energy, and electronics.

Catalytic or active agents, namely, metal oxides, carbon-based nanomaterials, or zero-valent metals, speed up the process of microplastic degradation in photochemical, thermal, or oxidative processes (Zhang et al., 2023). Functionalization or nanomaterials with the help of enzymes further improves specificity and targets the different types of polymers with much better efficiency (Chen et al., 2024). These are surface features which comprise morphology of surface, surface area charge, porosity, and crystal structure up to in situ crystalline control within the nanomaterials (Li et al., 2024). The above features could be obtained through modification either during the synthesis procedure or post the synthesis processing conditions. Smaller diameters nanomaterials result in larger charge surface area as a consequence thereby increasing the capability of the nanoparticles in adsorption and photocatalysis (Wang et al., 2024). Apart from that, adsorption and photo-degradation have been associated the selectivity toward certain types of reactants which nanomaterials possess owing to their special structural properties (Zandieh & Karimi, 2023).

Especially, highly porous 2D nanomaterials and graphene nanomaterials, layer-double-hydroxide, which could provide large space-specific surface and rich surface-function groups, are useful to prepare superabundant available surface-active site (Gunjakar et al., 2013). The ultrathin structure of 2D nanomaterials could decrease electron–hole pair recombination, which will therefore increase photocatalytic efficiency (Wang et al., 2024). Most of the nanomaterials possess electronic structures which are very flexible by hybridizing with other elements and complexes. Engineered dimensionality in metal-free carbon-based nanomaterials makes them highly recommended alternatives for photodegradation from metal-based catalysts and possesses the benign environmental feature (Gunjakar et al., 2013).

The performance of bottom-up or top-down nanomaterial designs can be improved significantly when new features are added to the goods that are already on the market (Sun et al., 2021a, 2021b). Besides helping in the facilitation process, surface functionalization of nanomaterials also improves their stability and dispersibility in an aqueous media. It was stated that surface application of enhancement bridging agents will improve compatibility between inorganic and organic phases but, however, results in the formation of a nanocomposite that potentially shows homogeneous structural features (Hou et al., 2023). Besides, the formation of additional surface functionalities by chemical reaction, for instance, oxidation, or by physical interaction. These start with a set of desirable qualities, thus taking advantage of the benefits of different functional materials. The target pollutants can form a new selective binding on reactive surfaces once specific ligands are introduced onto the surface of the nanomaterial. This is another application for enhancing the stability, tunability, and recovery of nanomaterials (Bal et al., 2018). Nanomaterials can be surface functionalized in a variety of ways. Developments in nanomaterials research were able to strongly support the significance of nanomaterial use and its beneficial impacts on microplastic pollution. Sun et al. introduced GO and $O–C_3N_4$ oxygen-doped carbonitride into a chitin sponge to enhance the ability of chitin to remove negatively charged carboxylate polystyrene microplastics and positively charged aminated polystyrene microplastics (Bala et al., 2024).

How Nanomaterial Affects the Ecosystem

Nanomaterials are manufactured materials with size in the nano meter range, from 1 to 100 nano meters (Yadav & Yadav et al., 2024). Their unique properties, like reactivity, high surface area and strength, make nanomaterials valuable in all applications, starting from medicine up to environmental remediation. However, their impact on ecosystems is rather complex and extensively under research today (Hou et al., 2023). Nanoparticles such as metal oxides and carbon-based material can change its chemical balance if present in the soil, changing the health condition of both microorganisms and plants (Bond et al., 2013). Then, nutrient cycling can be affected and soil fertility reduced, which in turn will affect agricultural productivity and ecosystem stability. The nanoparticles also contribute to climate change through several mechanisms. Black carbon particles are emitted during various combustion processes. These particles absorb sunlight and thus provide a mechanism for warming the atmosphere. After settling on ice and snow, such particles reduce the albedo or reflectivity of the surfaces leading to faster melting and causing global temperature increase (Das et al., 2023; Ghosh et al., 2023; Paul et al., 2024a, 2024b; Priyadarshini & Das, 2024). This, therefore, means that nanoparticles emitted from industrial activities can contribute to cloud formation by acting as cloud condensation nuclei, altering the patter of precipitation and weather systems. Such changes call for constant checks and regulation on nanoparticles to limit their effects on both the environment and climate.

Nanoparticles in Marine Ecosystems

Industrial effluents, agricultural runoff, wastewater discharge, and atmospheric deposition introduce nanoparticles to marine ecosystems; they originate from some common sources of metal-based nanoparticles such as silver and titanium dioxide in cosmetics and industrial products. Marine food webs are disturbed through interference with phytoplankton leading to bioaccumulation at higher trophic levels and compromising biodiversity. Nanoparticles accumulate in sediments to kill benthic animals that are integral to nutrient cycling. Further, they adsorb other contaminants, which means their toxicity becomes higher and increases the risk for marine life as well as human health via seafood consumption. Their small size and reactivity make it difficult to track and monitor them, therefore necessitating more rigorous policies, superior wastewater treatment, and further research in order to decrease their impact on the environment (Zhang et al., 2014).

Impact of Nanoparticles on Terrestrial Ecosystems

Nanoparticles introduced through industrial wastes, agriculture, or atmospheric deposition into the soil disturb its microecology by changing chemical contents and harming useful microbes. It reflects impacts on nutrient cycling, decomposition, and nitrogen fixation. Promotional effects of nano-fertilizers on crop growth may be contrasted with the oxidative stress and stunted growth in plants due to copper oxide nanoparticles. Additionally, carbon-based nanoparticles can affect soil structure, which impacts water retention, porosity, and aeration, thus making water less available to plants and lowering the quality of habitats for soil organisms, affecting terrestrial ecosystems and agricultural productivity (Zhang et al., 2014).

Effects of Nanoparticles on Human Health

Nanoparticles of industrial origin, vehicle exhaust, and consumer products constitute a major health risk (Maynard et al., 2006). Inhalation could cause respiratory and cardiovascular effects, such as asthma and lung carcinoma in addition to hypertension and atherosclerosis. Some nanoparticles, like titanium dioxide, may penetrate across the blood–brain barrier, causing oxidative damage, inflammation, decline in cognitive functions, and neurodegenerative diseases. They can penetrate inside cells, which causes oxidative stress leading to DNA damage and genotoxicity associated with the risks of cancer. Nanoparticles in consumer products, such as sunscreens and food packaging, can penetrate through the skin or leach into food, bringing up concerns with oxidative stress, irritation, and ingestion-related toxicity (Oberdorster et al., 2005). Nanoparticles represent such advanced technology but severely impact ecosystems and human health. The technology destroys soil health, kills microbes, and plants, thus putting at risk biodiversity as well as food security. Human exposure to these nanoparticles leads to respiratory, cardiovascular, neurological, and genetic problems that will put pressure on the healthcare system and exacerbate socio-economic challenges further. This further deteriorates the environment, climate, and habitat loss (Donaldson et al., 2004). These can only be curbed through adequate regulations, responsible production, safe alternatives, and efficient waste management that can be in a state of balance regarding the benefits derived from nanotechnology vis-à-vis its ecological and health implications.

How Can Nanomaterials Be Used to Remediate Microplastic Pollution?

Being characterized with high surface area, tunable surface chemistry, and high adsorption capacity, nanomaterials have proved to hold great potential in remediating microplastic pollution. This section provides an in-depth look into how nanomaterials can be applied to remove microplastic pollution (Husak et al., 2023).

Adsorption

Adsorption is a simple, inexpensive, and non-polluting purification method. Surface phenomena dictate it since pollutant molecules are bound from the fluid bulk to the solid surface. The overall effectiveness of the process depends on one very critical factor: high surface area. Nano-adsorbents possess an elevated surface area and permeability, which might bring about a substantial enhancement of the pollutant rejection. Adsorption is a process where microplastic particles adhere to the surface of nanomaterials. This can be achieved through Microplastics and nanomaterials interact hydrophobically and by van der Waals forces during physical adsorption. Chemical adsorption is the covalent or electrostatic bonding process that is enabled by the presence of functional groups on the nanomaterial surface.

Nanomaterials Used for Adsorption

Adsorbents are prepared with modified, unchanged and advanced nanomaterials including zeolites, activated carbon, MOF, polymer, molecular sieves, and many more cost-effective materials. Different types of nanomaterials can be used based on their structure and functional properties. Some examples include:

Carbon based nano materials

i. Carbon-based nanomaterials (cb nanomaterials); the chemical and physical characteristics of cb nanomaterials, making them potential adsorbents due to their unique nature of organic and inorganic pollutants removal at a large scale with a high surface area and reactivity (Zhang et al., 2011). In recent years, much attention has been received from carbon nanotubes, a group of cb nanomaterials, attributed to unique properties that are found, including a very porous nature, higher area of surface, excellent conductivity of electron, and very good physical, chemical and mechanical properties (Apul et al., 2014). Their durability and higher ability to adsorb make them suitable for so many applications. Schematically, CNTs are one-dimensional nanomaterial with a cylindrical, sturdy membranous honeycomb latticework capable of catching various pollutants (Apul & Karanfil, 2015). Depending on the numbers of cylindrical shells, there are two categories of CNTs, which include SWCNTs and MWCNTs (Zhang et al., 2009). There have been immense researches and works into the practical applications of CNTs in remediation of environmental. The main disadvantage of using CNTs is that they are too small in particle size, which decreases their efficiency and makes separation tough. Poor dispersion in an aqueous phase was also caused. Due to high-adsorption energy sites for organic molecules, loose clusters/aggregates with interstitial gaps, grooves, and CNT membranes are often used in water treatment in order to bypass these issues. Increasing dispersion and separation of CNTs will improve their cleaning efficiency. There are surface modification and functionalization techniques currently being studied for improving efficacy (Fasano et al., 2012).

ii. Nanomaterials based on graphene are two-dimensional carbon allotropes with a honeycomb lattice structure (Zhang et al., 2009). Graphene oxide, reduced graphene oxide, graphene quantum dots, and graphene nanoribbons represent just a few among the many different types of graphene (Apul et al., 2014). In reality, GB nanoparticles possess exceptional characteristics such as high electrical conductivity, a large surface area, outstanding mechanical strength, and excellent thermal properties (Apul & Karanfil, 2015). These characteristics of GB nanoparticles make them favorable for a wide range of uses, including environmental remediation, electronics, energy storage, and catalysis.Adding different functional groups to GB nanoparticles may also improve their properties and increase their adsorption capacities. Graphene-based nanoparticles were used by many researchers to remove heavy metals. However, environmental hazards should be considered. It has thus been suggested that synthesis and functionalization of some of these nanomaterials be improved; the development of efficient monitoring and detection methods should also be supported, and so should regulation for the safe use and disposal of such materials. Long-term effects of nanomaterials can only be established through further research. Carbon fullerenes have other unique properties that make them interesting for environmental remediation. It may break down pollutants in several ways by acting as an electron donor or acceptor: Singlet oxygen generation, radical formation, and photoinduced electron transfer (Fasano et al., 2012).

Polymer based nanomaterials

The large surface area and poly-functional groups of natural polymers have provided them with greater adsorption and chelation abilities, leading to their widely explored application in environmental remediation (Khali et al., 2012). The major natural abundant polymers are cellulose, hemicellulose, and lignin. Cellulose also makes up the linear polymer for 30–50% of polymers consisting of repeating units of D-glucopyranose, all connected via glycosidic bonds (George & sabapathi, 2015). This makes the production of polymeric nanoparticles even cheaper, as this is available very abundantly at a reasonable cost. The schematic below explains how polymeric nanomaterials are made (Heinze & Liebert, 2001). It is a special kind of substance because it contains both methyl and hydroxyl functional groups. It can be used to deactivate metal nanoparticles and microparticles. The affinity of cellulose molecules with six OH groups is much stronger. The surface roughness of cellulose must be increased to prepare composites of desirable qualities (Sun et al., 2018). Hemicelluloses (20–35% of lignocellulosic biomass) control the aggregation of cellulose microfibrils by connecting them with cellulose and lignin. Hemicelluloses, which constitute 20–35% of the components of lignocellulosic biomass, control the aggregation of cellulose microfibrils by connecting them with cellulose and lignin. Hemicelluloses provide several neutral sugars with a $\beta - (1 \rightarrow 4)$ backbone structure similar to that of cellulose. Molecular heterogeneity, through covalent cross-links with lignin, as well as intermolecular interactions, controls interactions between cellulose microfibrils (Rol et al., 2019).

The third most common polymer is lignin, at 15–30%. It occurs in black liquors obtained from different plant origins (Xu & Sun, 2018). Lignin is an aromatic compound having a nanocrystalline heterogeneous structure, and with C–C and C–O bonds, its functional groups are bonded (Crestini et al., 2017). It can be changed into nanomaterials having intrinsic UV absorbance, antibacterial, and antioxidant properties by creating newer monomers through depolymerization. Chemical treatments enhance the adsorption of organic matter and, of course, modify the polymer properties. The capacity ranges from 20–50 mmol/g for the virgin fibers up to 400–1000 mmol/g for the nanomaterial-modified fibers depending on the solute composition and the sequence of the modification process (Norgren & Edlund, 2014). The polymers have been used to remediate the water contaminated with heavy metals through nanofibers (Dominguez-Robles et al., 2020). The large surface area of polymer nanoparticles does not have much impact on the adsorption of cationic dyes and other metals since there are not enough reactive functional groups. The increase in the number of functional groups on the surface improves its polarity and hydrophilicity, which will boost the selectivity of the adsorbent. Furthermore, chemical modification of polymers enhances their capabilities in terms of flexibility, resistance to microorganisms, absorbency, adsorption or ion exchange, hydrophilicity or hydrophobicity, and thermal stability. However, there is still much to be learned about polymeric nanoparticles in order to fully utilize them for environmental cleanup (Gilca et al., 2020).

Mechanism of Adsorption

Some main criteria for adsorbent choice are intrinsic capacities for adsorption and material specificity to the material of interest. Chemicals from one phase get separated by enriching on a surface of some other adsorbing substance or material, as termed adsorption. It takes place in three stages; (1) the contact of an adsorbed liquid phase with an absorbent in contact with a solid phase and a liquid phase of different compositions under constant pressure and temperature over a definite period of time.

(2) Product extracted from the liquid and adsorbent layers. (3) The quantity of the adsorbent material in the liquid phase of the supernatant and the fluid phase under pressure is measured (Chukwuemeka-Okorie et al., 2021).

Two crucial factors governing the adsorption rate of an adsorbate are adsorption isotherms and kinetics; they represent how effectively an adsorbent adsorbs and formulate the parameters required in the process (Afolabi et al., 2022). Therefore, an excellent understanding about thermodynamic as well as the kinetic aspects needs to be covered. Amongst them, intraparticle diffusion kinetics, Langmuir, and Freundlich isotherm models are some of the models widely used (Chatzopoulos et al., 2024). The residence time necessary for complete adsorption is found through kinetics studies; kinetics data also serves in the right sizing of the adsorption equipment. It further provides an assessment of the flow-through configurations including the fixed bed system.

The mechanism that removes dyes, heavy metals and other pollutants from a stream of solution is well-known adsorption. An adsorbed substance can be divided into either physisorption or chemisorption (Adeogun et al., 2022). Generally, physisorption typically adheres to an adsorbate by means of weak intermolecular interactions through van der Waals forces (Zhang & Wang, 2021). In chemisorption, a deep chemical bond takes place between the adsorbate and the surface of the adsorbent. The efficiency of nano adsorbents in absorbing organic molecules, complex compounds, and heavy metal absorption is deeply affected by electrostatic interactions, $\pi-\pi$ and hydrophobic contacts, acid–base interactions, and van der Waals interactions. Nanomaterials' small size, high surface area, and surface multifunctionality provide large numbers of active sites for interaction with chemical species (Chatzopoulos et al., 2024; Bala et al., 2024; Jagatee et al., 2024; Mohanty et al., 2024c; Nag et al., 2023; Satapathy et al., 2024). Some variables of influences for the percentage of hazardous pollutant adsorption include: amounts of both the adsorbent and suspended particles, pH, operating temperatures, surface charges, concentration of the adsorbent and adsorbate, and contacting times (Afolabi et al., 2022). The adsorption efficiency of nanomaterials is heavily dependent on the solution conditions, especially pH. Studies indicate that a higher pH is more favourable for adsorption as it confers a negative charge on the surface of the adsorbent, thus making electrostatic attraction more significant. Conversely, the lower pH limits adsorption by causing a positively charged surface of the adsorbent, weakening electrostatic interactions. Key mechanisms involve molecular diffusion and surface binding. Mixing nanomaterials with other adsorbents will further enhance pollutant attraction and adsorption efficiency (Lee et al., 2023) (Table 30.1).

Photocatalysis

From the studies, photocatalysis has been revealed as an extremely effective and environmentally safe method that can break down these insecticides, heavy metals, endocrine disrupting chemicals, acetaldehyde, nitrogen oxides, sulfur oxides,

Table 30.1 Nano-based adsorbents for microplastic remediation

Nano adsorbent	Composition	Process	Reference
Carbon nanotubes	Single/multi-walled CNTs	Hydrophobic and π-π interactions	Sun et al. (2021a, 2021b)
Graphene oxide	Oxidized graphene sheets	π-π interaction and electrostatic forces	Chen et al. (2020)
Magnetic nanoparticles	Fe_3O_4, coated iron oxide	Electrostatic and Vander waals forces	Yuan et al. (2020)
Polymer-based nano-adsorbents	Functionalized polymer nano-composition	Electrostatic and hydrophobic interaction	Shi et al. (2022)
Metal–organic frame work	Porous metal–ligand frame work	High surface area adsorption	Tang et al. (2021)

ammonia, carbon monoxide, mercury, volatile organic compounds, and polycyclic aromatic hydrocarbons to non-toxic gases (Dilbaghi et al., 2019). This improved chemical process makes harmful air pollutants in full mineralization when powdered form is used. Some of the disadvantages include poor recovery and secondary contamination. Irradiation, in the form of a light source, an excitation on a semiconductor with a high gap, such as TiO_2, ZnO, SnO_2, CdS, or tungsten trioxide, plus water, promotes the separation of electrons and then leads to electron–hole pair. Photoexcitation initiates a series of oxidation and reduction events when exposed to a suitable wavelength of light, usually with photon energy (hv) larger than or equal to the band gap energy. This is the mechanism behind photocatalysis (Zhang et al., 2024). The electrons interact with their surroundings in successive reductions and oxidations after diffusing across the surface of the photocatalyst. The scientific foundation of photocatalytic oxidation is that a catalytic process with an energy-activated catalyst causes non-selective degradation of all organic contaminants through reaction reliant upon formation of highly reactive species, such as the radical species ·OH and $\cdot O^{2-}$,, acting as potent oxidants formed from reactions of photogenerated electrons with molecular oxygen and photogenerated holes with water (Prasad et al., 2024).

There are two groups for the photocatalysis technique. These include heterogeneous photocatalysis, often termed as the Fenton reaction, which is fully light irradiation-independent, and homogeneous photocatalysis, also termed as the photo-Fenton reaction, which is photoresponsive up to a light wavelength of 600 nanomaterials. Long exposure times and complicated chemical mixtures can damage the ability of nanocatalysts to carry out photocatalysis in bulk-scale applications (Kargar et al., 2023). Several methods have since been discovered in this field to bypass these constraints and further uplift the photocatalytic properties of nano catalysts. Some of these methods include transition metal dopants, anionic dopants, and surface-property customization for nanocatalysts (Wang et al., 2021). These dopants, which include carbon, nitrogen, fluorine, and sulfur, may alter the electronic structure of nanocatalysts or boost the number of active sites, thereby enhancing efficiency. The surface area would be enhanced by nanofibers or nanotubes, but the light absorption would increase due to plasmonic materials like Au or Ag nanoparticles. This approach of incorporation of blended photocatalysts with supportive materials is still recognized as a good approach in the literature (Singh et al., 2024). This approach, along with improving recoverability, reduces the secondary level of contamination that was also improved. This facilitates the achievement of the desired properties like recyclability and reusability, besides a longer photocatalytic component lifetime.

Titania (TiO_2) Photocatalysts

Advanced oxidation processes (AOPs) can be an alternative to remove microplastics because it is strong in breaking organic materials such as colors and drugs. Even though microplastics can degrade spontaneously, it takes a time. Considering the nano-range, the most popular AOP to remove microplastic from the aquatic environment was found to be photocatalysis. Photocatalysis relies upon semiconducting materials, such as titania (TiO_2), where a light source of UV and visible light is used to initiate the process involving the work of moving electrons from the VB into the CB. The released electrons induce the formation of reactive oxygen species (ROS), which enhances the degradation of the microplastics. when the photocatalysts absorb the appropriate energy (hv) from the incident light, which ejects electrons from the valence band to the conduction band, leaving behind a hole (h^+) in the valence band. Then, on the surface of TiO_2, O_2 and H_2O are adsorbed (Step 2). These species react with h^+ in the valence band and e^- in the conduction band to produce hydroxyl ·OH radicals and the superoxide ion $\cdot O^{2-}$. Step 3. Carbon-centered radicals react with O_2 after the creation of ·OH radicals, forming peroxide radicals. The latter further reacts with water to form radicals of alkoxy polymers. Take a step 4 (Hernandez-Rangel et al., 2020).

Degradation of microplastics primarily occurs through the creation of hydroxyl radicals (·OH), which also requires an oxidizing environmental and inhibition of hole-electron recombination. Some of the critical factors controlling microplastics degradation are temperature, pH, and humidity. Lower temperatures enhance fragmentation, which provides more surface area. In contrast to this, a lower pH aids in forming hydroperoxide, which can increase removal rates (Yuwendi et al., 2022). Humidity helps microplastics degrade due to increased water adsorption onto semiconductor surfaces, specifically TiO_2, which improves ·OH generation. Mesoporous TiO_2 coating degrades smaller microplastics more rapidly due to their large surface area-to-volume ratio. In spherical microplastics, better interaction with radicals and also less limitations concerning oxygen led to better degradation compared to film-shaped MPs. The extra pores of mesoporous TiO_2 enhance the photocatalytic activity due to higher ·OH availability. TiO_2 nano powder has a high photocatalytic activity, but it is difficult and very expensive to filter, especially on an industrial scale. This limitation favors the use of supporting materials like open-cel l β-silicon carbide foams, which increase light absorption and offer thermal as well as chemical stability (Zandieh & Karimi, 2023).

Zinc Oxide (ZnO) Photocatalysts

Another semiconductor of interest in the removal of microplastics by AOPs is zinc oxide (ZnO), a semiconductor of high photocatalytic activity. This apart, ZnO has strong redox potential and electron mobility at ambient temperature in addition to its antibacterial property. Tofa et al. applied ZnO nanorods for the degradation of low-density polyethylene (LDPE) film in water and found that the surface area of the catalyst used was the measure of degree of oxidation. ZnO nanorods enhance microplastic degradation, furthering surface area and, therefore, allowing for the opportunity of defect engineering. Plasmonic photocatalysts formed from Pt nanoparticles deposited on ZnO nanorods showed improved light absorption and decreased electron–hole recombination, further increasing radical formation. The modification increases degradation rate because Pt would act as a sink for electrons that facilitate deeper cavities within the microplastics, such as LDPE films. Pt deposition enhanced visible light absorption up to 78% at 385 nanomaterial (Tofa et al., 2019).

There are issues like nanoparticles contamination into the water, the requirement of filtration, and increased residence times in batch processes. Scale-up potential with continuous flow-through reactor configurations, such as glass fiber support, can be envisioned with the example of degradation of polypropylene where an increase in the carbonyl index was observed. Moreover, the antibacterial activity of ZnO may provide the dual advantage of microplastic degradation and water disinfection (Kumar & Sharma 2023) (Table 30.2).

Table 30.2 Nano-based photocatalysts for microplastic remediation

Photocatalyst	Nanomaterial composition	Process	Reference
Titanium dioxide (TiO_2)	TiO_2 nanoparticles	Generate reactive oxygen species under UV light to degrade microplastic	Uheida et al. (2021)
Zinc oxide (ZnO)	ZnO particles	ROS production under UV light for breakdown of plastic	Allé et al. (2021)
Iron-based photocatalysts	Fe_2O_3, Fe_3O_4– based nanomaterials	Fenton like reaction for plastic oxidation	Domínguez-Jaimes et al. (2021)
Silver-based nanocomposites	Ag–TiO_2, Ag–ZnO	Enhance ROS production	Acuna-Bedoya et al. (2021)

Magnetic Separation

Magnetic separation is an emerging and effective nano-based remediation strategy targeting environmental microplastic pollution. It makes use of MNPs or other magnetically responsive materials to preferentially capture and remove the microplastics from either the water or soil matrices. The method includes using magnetic materials that can attach themselves to the microplastics through physical or chemical binding (Mohiuddin & Chowdhury, 2023). When the magnetic material attaches to the microplastics, an external magnetic field is then applied in extracting the microplastics from the medium and allowing the remaining environmental to stay undisturbed. Magnetic nanoparticles like iron oxide nanoparticles are synthesized or functionalized to interact better with microplastics. The most common approach in functionalizing these MNPs includes the coating of hydrophobic or polymeric layers that enhance adherence to microplastics. Fe_3O_4 Nanoparticles have been widely utilized in separation due to their magnetic behavior enables easy separation after application. Functionalized MNPs are applied to water or soil samples containing microplastics. Micro-/nanoplastics attach to the nanoparticles based on hydrophobicity, electrostatic attraction, or specific binding mechanism. This concept was used for the removal of microplastics from water using Fe_3O_4 nanoparticles. Apply an external magnetic field to attract the MNP-loaded microplastics, therefore separating the medium containing microplastics. Magnetic separation is fast and energy efficient (Zhang et al., 2023). As a result, it is minimally invasive to separate the captured MNP-microplastic complexes from the medium. Microplastics can be regenerated and reused for consecutive remediation cycles. This work has recently involved magnetizing the different types of microplastics of varied sizes. Another semiconductor of interest in the removal of microplastics by AOPs is zinc oxide (ZnO), a semiconductor of high photocatalytic activity. This apart, ZnO has strong redox potential and electron mobility at ambient temperature in addition to its antibacterial property. Tofa et al. applied ZnO nanorods for the degradation of low-density polyethylene (LDPE) film in water and found that the surface area of the catalyst used was the measure of degree of oxidation. It is then separable using an external magnet. Further, shell material can show interactions with the microplastic such as electrostatic, hydrophobic and van der Waals' forces and hence pull-out nano-/microplastics. This plastic fragment remediation seems to require all these intermolecular interactions; thus, this approach's removal efficiency was determined as an average value of 89.3% (Singh et al., 2023).

The advantage of magnetic nanomaterials is that they can adsorb MPs and easily be recovered using an external magnetic field after adsorption. It is an efficient technique that can be used to remove microplastics of different sizes and types because of the changeable nature of magnetic nanoparticles (MNPs). MNPs can be functionalized to selectively target certain microplastics or contaminants, enhancing their bonding efficiency and facilitating precise removal. Furthermore, non-toxic and biocompatible MNPs can be designed in such a way as to minimize harm to the environmental. One of the significant benefits of this method is that magnetic materials are recoverable and reusable for many cycles, reducing waste and operating costs while being an environmentally friendly way to sustain the remediation of the environmental.

Filtration

Membrane filtration is a relatively direct technique of filtration by using membranes or porous materials that retain some of the substances or particles while allowing others to pass. Filtration is influenced by several factors such as particle size, surface charge, solute hydrophilicity, and shape (Mustafa et al., 2023; Mishra et al., 2021a, 2021b; Mishra & Das, 2021, 2022). The material used to make the filters usually consists of glass fibers and petroleum-based polymers, such as polyethylene (PE) and polypropylene (PP). Incorporation of nanomaterials into membranes or filters installed can physically trap microplastics. Nanofibers or nanosheets have been proven to increase the selectivity and efficiency of the filters, hence trapping even the smallest microplastics (Xiaoming et al., 2022). Nanomaterial-based filters remove microplastics from water by physical trapping and chemical adsorption. The filters make use of the principle of size exclusion, in which the nanopores of the filtration membrane act to block microplastics from passing through while letting water pass (Ramesh, 2021). The functionalized surfaces of nanomaterials further allow the process to take place by adsorbing microplastics through interactions such as hydrophobic forces, electrostatic attraction or van der Waals interactions thus ensuring that smaller particles are effectively captured. Most modern filtration systems require the combination of all these mechanisms because both physical and chemical processes act in maximizing efficiency and targeting a broad range of microplastic sizes and types in polluted water (Mishra et al., 2024a, 2024b, 2024c; Singh et al., 2020a, 2020b).

The mechanism of filtration in membranes takes place through the physical capturing process, which is either sieving, inertial impaction, interception, and diffusion. The four basic physical capture methods have full fill filtration operations as their primary application (Anil, 2020). Particles can diffuse towards the surface and deposit while passing through the membrane filter's two sides due to Brownian motion. This is because either inertia or interception can occur at a passing distance (Hui et al., 2019). However, production methods that employ polymers and nanomaterials can maximize the qualities of membrane filters. Making these alterations is necessary for enhanced membrane performance and less fouling. Membrane properties can be altered by infusion or blending with nanoparticles, like Ag, SiO_2, TiO_2, ZnO, CNTs, polymer base nanomaterials, metal oxides, halloysite, among others that integrate into the nanomaterial. This infusion elevates hydrophilicity and adsorption potential of these nanoparticles with the inclusion of functional groups thereby enhancing their ability to reject contaminants further. This led to a better removal of microorganisms and harmful pollutants. Photocatalytic nanomaterials have a great potential for the removal of harmful pollutants via photodegradation.

However, they only capture the solid particles; they do not trap or absorb other impurities that would be considered harmful. When they are released, they may become secondary sources of pollution. Recent research has focused on hybrid membranes where nanoparticles are combined. Reports state that replacing membranes can enhance dangerous pollutant removal. Improvements in surface modification in terms of hydrophilicity and selectivity, interfacial polymerization in terms of strength and chemical stability as well as thermal stability, and additional features such as anti-fouling and self-cleaning properties, selective properties, and the ability to eliminate microbes all have been reported in the literature. These exhibit a higher flow rate in low pressure and easily recoverable with the reusability potential of nanomaterials (Hui et al., 2019) (Fig. 30.1; Table 30.3).

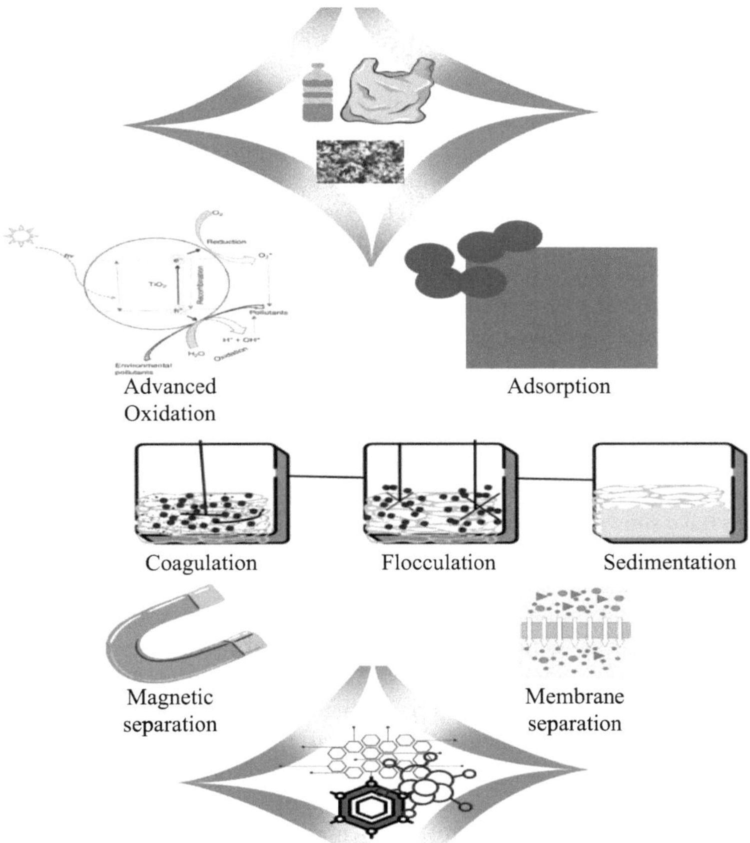

Fig. 30.1 Illustration of various capturing-based and degradation-based microplastic removal technologies

Table 30.3 Nano-based filtration for microplastic remediation

Strategy	Nanomaterial used	Process	Reference
Nanofiber membranes	Electro spun polymer nanofibers	Physical entrapment of microplastic	Yang et al. (2022)
Carbon-based nano filters	Graphene oxide, carbon nanotubes	Adsorption and entrapment of microplastics	Enfrin et al. (2020)
Electrostatic nano filters	Electrocharged nanofibers	Electrostatic attraction of microplastic	Wang et al. (2020)
Magnetic nanoparticle filters	Fe_3O_4-based nano composition	Magnetically trapping microplastic for separation	Li et al. (2021)

Challenges and Future Direction

Because of the high amount and extensive spread of microplastic all over the world, removing microplastic is becoming significantly importance. Although nano-based remediation strategies show great promise, there are some hurdles in their deployment. This includes, first and foremost, the fact that nanomaterials themselves pose environmental and health risks, including toxicity to aquatic organisms and bioaccumulation in the food chain. Secondly, the high cost of production and large-scale deployment of nanomaterials significantly limits their feasibility. Targeting microplastics is another challenge in microplastic with an effective interaction among diversified environmental matrices, including water, soil, and sediments. Their distribution and behavior are very different in each of these systems. Recovery and safe disposal of used nanomaterials are equally important to avoid secondary contamination. Long-term environmental impacts of these strategies are little explored; accordingly, more detailed risk assessments and regulatory frameworks are required. Nanomaterials have shown promising aspects towards microplastic removal from aqueous systems in recent years as compared to various traditional treatment materials. Physical methods remain one of the more feasible techniques applicable to industries by being economical and scalable; yet, nanomaterials provide significant scope towards improvement of these technologies. For example, at large-scale operation levels, these might help save considerable costs, since the materials become substitutes for others used conventionally. This makes nanomaterials a highly interesting area of research in further developing water and wastewater treatment methods focused on targeting microplastic pollution. Most of the work done toward the removal of microplastic from an aqueous system made use of synthetic water. Although such experimental conditions are helpful in creating an interference-free environment which may offer a good insight into how the material acts in the context of a specific type of microplastic, some critical aspects relevant to actual operating conditions remained overlooked. The stability of the real conditions like salinity and pH might totally change the characteristics of nanomaterials as well as the microplastics themselves. For example, the adsorbent efficiency might be largely hampered because there is competition on the site with several other forms of microplastics from various sources. Occupation of the adsorbent's pore structure through suspensions by several small-sized particles may hamper the absorption process of the microplastic as well. In complex water environmental, various factors can prevent the effectiveness of nanomaterials in removing microplastics (Mishra et al., 2019a, 2019b, 2022; Mishra et al., 2021a; Sahoo et al., 2022). Small suspended particles may block pores of the adsorbent and decrease the efficiency of adsorption. Membrane filtration can be impaired by fouling caused by natural organic matter and microorganisms. Another factor is that removal efficiency can be inconsistent due to differences in composition and sizes of polymers in the water. Further experiments are required to understand nanomaterial-microplastic interactions and improve removal strategies using more diverse and complex water conditions.

Chemically synthesized nanomaterials have been shown to be harmful for human health and the environmental. However, polymer and green-based nanomaterials are still under active research at the laboratory scale. The recent modifications of nanomaterials by using plant wastes and polymers as supporting materials alter their structure and composition, thereby improving their efficacy. The adsorption capacity of each nanomaterial is tricky to compare because parameters and different adsorbates are always used. However, it has recently been reported that a number of nanomaterials possessed considerable abilities for sorption of metal ions. Additionally, there is research required to coat nanomaterials which will allow them to be surfactant coated to enable them to access pollutants and die after performing the job, such as surfactant coating. The strategy to use nanomaterials for environmental remediation must therefore take into account the impact that it has on human health and thus, the environmental will be appropriately remediated in a safe and non-hazardous manner. Utilization efficiency of each type of nanomaterial must be compared with respective engineering investments as well as expenses incurred in terms of operations.

Conclusion

Currently, nano-based remediation strategies have provided an advanced solution to address microplastic pollution in the environmental and easily capture and degrade them within the environmental. This systematic review suggests that nanomaterials pose a huge possibility of mitigation from environmental pollutants. nanomaterials hold strong promise as one of the major means of reducing environmental pollution through potential advantages they offer. Types of Nanomaterials and Their Application in Microplastics Remediation. According to their structure and composition characteristics, nanomaterials can be broadly categorized. The categories of inorganic nanomaterials are classified as metal-based, carbon-based, graphene-based nanomaterials, and carbon quantum dots. Organic counterparts of nanomaterials involve polymer-based nanomaterials, dendrimers, and other nanocomposites such as MOFs and NCMs. Recently, nanosensors have gained attention due to its envisioned applications. Synthesis pathways and mechanisms of action in general, nanomaterials are prepared through various modes of conventional chemistry and physics procedures; however, the synthesis of nanomaterials through various biological pathways does exist. All these mechanisms to which nanomaterials exhibit interaction with different pollutants have become very essential; their application lies in several key processes of cleaning up the contamination, such as adsorption and filtration, but also in different disinfection functions, photocatalysis, oxidation, etc., and improvement will make nanomaterials opt for better functions in further applied researches. Such materials have significant large surface areas, variable functional properties, and potential integration into current treatment systems. In contrast to the traditional materials, nanomaterials show superior performance in the elimination of pollutants, especially improvement in the degree of adsorption and filtration efficiency, which helps in the degradation of microplastics within aquatic ecosystems.

Despite these advantages, a few barriers need to be surmounted for nanomaterials to be widely used in environmental remediation. Some of these key issues include environmental safety, the cost of the material, scalability, and how their efficiency could be affected by complex water matrices. The diversity of microplastics, including size, composition, and behavior, makes this an enormous challenge to design generally applicable nanomaterials. The future research should focus on producing green and biodegradable nanomaterials, which should evaluate the long-run effects of such material on the ecological environment. Their large-scale efficiency and risks will have to be tested under real environmental conditions. It has already been shown that iron- and iron-oxide-based magnetic nanomaterials have vast potential for adsorption of microplastic. These can be functionalized in such a manner that their hydrophobicity enhances; thus, they effectively capture microplastics from water. However, regeneration of such nanomaterials turns out to be quite complex because the maintenance of hydrophobicity is very hard over several cycles. Interaction of microplastics with aqueous environments further complicates optimization in terms of efficiency in adsorption.

Hybrid Technologies and Photocatalytic Degradation, Extensive research has been done on photocatalysis for the degradation of microplastics by nano-catalysts that enhance the degradation effect through light. These catalysts are engineered to maximize their surface area, increase light absorption, and allow for charge separation toward increasing their reactivity. Even though UV-based degradation methods are effective in certain cases, visible-light-driven processes are more extensively used to decrease energy consumption. Degradation efficiencies can reach about 80% according to current research, although further improvements are needed before large-scale implementation can take place. Hybrid degradation and separation technologies are required to improve microplastic removal. It is achievable by integrating the photocatalytic process into the filtration systems and thus achieving complete or nearly complete removal of microplastics. Some nanoporous membranes, MMMs filled with nanofillers, and nanostructured coated membranes have shown promising capabilities for this purpose. Research related to membrane modifications has been found, such as improved membrane performance through reduction of fouling and higher efficiency in separation of microplastics. Standardization and real-world applications, Microplastic detection and characterisation in matrices using environmental-based methods of nano remediation; Practical studies include due regard concerning actual conditions regarding shape, size, and type composition under which real wastewaters operate: nano-based remediation strategies applied for better overall sustainability efforts while reducing micro plastic pollution and subsequently protecting aqueous ecosystems.

References

Abdul Khalil, H. P. S., Bhat, A. H., Ireana Yusra, A. F. (2012). Green composites from sustainable cellulose nanofibrils: A review. *Carbohydrate Polymers*. https://doi.org/10.1016/j.carbpol.2012.03.068

Acuña-Bedoya, J. D., Luévano-Hipólito, E., Cedillo-González, E. I., Domínguez-Jaimes, L. P., Hurtado, A. M., & Hernández-López, J. M. (2021). Boosting visible-light photocatalytic degradation of polystyrene nanoplastics with immobilized Cu_xO obtained by anodization. *Journal of Environmental Chemical Engineering.*, *9*(5), 106208.

Adeogun, A. A., Balogun, B. O., Bello, O. S., Olatunji, O. S. (2022). Adsorption kinetics and isotherms of binary metal ion aqueous solutions using a low-cost adsorbent. *Scientific Reports*. https://doi.org/10.1038/s41598-022-10912-3

Afolabi, A. A., Akinlabi, A. A., Olatunji, O. S. (2022). Adsorption kinetics and isotherm models: A review. *Caliphate Journal of Science & Technology*. https://doi.org/10.13140/RG.2.2.20397.33766

Ali Alnazza, A., & Chile, N. (2023). A short review of nanomaterials: Synthesis methods, properties, and applications. *Algerian Journal of Chemical Engineering, 1*(1), 1–7. https://www.researchgate.net/publication/369529090

Allé, P. H., Garcia-Muñoz, P., Adouby, K., Keller, N., & Robert, D. (2021). Efficient photocatalytic mineralization of polymethylmethacrylate and polystyrene nanoplastics by TiO_2/β-SiC alveolar foams. *Environmental Chemistry Letters, 19*(2), 1803–1808. https://doi.org/10.1007/s10311-020-01099-2

Anil, K. (2020). Advancements in magnetic nanoparticle applications for microplastic removal in wastewater treatment. *Science of the Total Environment, 744*, 143567. https://doi.org/10.1016/j.scitotenv.2020.143567

Apul, O. G., Wang, Q., Zhou, Y., Karanfil, T. (2014). Adsorption of aromatic organic contaminants by graphene nanosheets: Comparison with carbon nanotubes and activated carbon. *Water Research*. https://doi.org/10.1016/j.watres.2013.12.035

Apul, O., Karanfil, T. (2015). Adsorption of synthetic organic contaminants by carbon nanotubes: A critical review. *Water Research*. https://doi.org/10.1016/j.watres.2014.08.027

Ashton, K., Holmes, L., & Turner, A. (2011). Association of metals with plastic production pellets in the marine environment. *Marine Pollution Bulletin, 62*(7), 1430–1432. https://doi.org/10.1016/j.marpolbul.2011.02.029

Bal, B., Ghosh, S., & Das, A. P. (2018). Microbial recovery and recycling of manganese waste and their future application: A review. *Geomicrobiological Journal, 36*(1), 85–96. https://doi.org/10.1080/01490451.2018.1497731

Bala, K., Dogra, A., Lata, P., & Das, A. P. (2024). Discussing sources and biological-chemical remediation approaches for microplastic pollution. In *Sustainable microbial technology for synthetic and cellulosic microfiber bioremediation* (pp. 37–59). https://doi.org/10.1007/978-3-031-62600-5_3

Bergmann, M., Mützel, S., Primpke, S., Tekman, M. B., Trachsel, J., Gerdts, G. (2019). White and wonderful? Microplastics prevail in snow from the Alps to the Arctic. *Science Advances*, 5(8), eaax1157. https://doi.org/10.1126/sciadv.aax1157

Bhatia, R. K., Sakhuja, D., Mundhe, S., & Walia, A. (2020). Renewable energy products through bioremediation of wastewater. *Sustainability, 12*(18), 7501. https://doi.org/10.3390/SU12187501

Bhattacharjee, J., Mishra, S., & Das, A. P. (2021). Recent advances in sensor-based detection of toxic dyes for bioremediation application: A review. *Applied Biochemistry and Biotechnology, 2021*, 1–20. https://doi.org/10.1007/S12010-021-03767-7

Bhushan, B. (Ed.). (2017). *Springer handbook of nanotechnology*. Springer. https://doi.org/10.1007/978-3-662-54357-3

Bond, T. C., Doherty, S. J., Fahey, D. W., Forster, P. M., et al. (2013). Bounding the role of black carbon in the climate system: A scientific assessment. *Journal of Geophysical Research: Atmospheres, 118*(11), 5380–5552. https://doi.org/10.1002/jgrd.50

Browne, M. A., Galloway, T. S., & Thompson, R. C. (2011). Accumulation of microplastic on shorelines worldwide: Sources and sinks. *Environmental Science & Technology, 45*(21), 9175–9179. https://doi.org/10.1021/es201811s

Cera, A., Wiatkowski, M., & Nowacki, M. (2019). Microplastic pollution in the Masurian Lakeland lakes, Poland. *Environmental Pollution, 254*, 113069. https://doi.org/10.1016/j.envpol.2019.113069

Chatzopoulos, M. A., Stasinakis, A. S., Thomaidis, N. S., Arvaniti, D. S. (2024). Development of a comprehensive analytical solution for adsorption kinetics: Application to the removal of contaminants of emerging concern. *Separation Science and Technology*. https://doi.org/10.1080/01496395.2024.2319146

Chen, X., Li, J., & Wang, X. (2024). Nanomaterials as photocatalysts—Synthesis and their potential applications. *International Journal of Molecular Sciences, 25*(1), 123. https://doi.org/10.3390/ijms25010123

Chen, Y. J., Chen, Y., Miao, C., Wang, Y. R., Gao, G. K., Yang, R. X., Zhu, H. J., Wang, J. H., Li, S. L., & Lan, Y. Q. (2020). Metal-organic framework-based foams for efficient microplastics removal. *Journal of Materials Chemistry, 8*, 14644–14652.

Chong, M. N., Jin, B., Chow, C. W. K., & Saint, C. (2010). Recent developments in photocatalytic water treatment technology: A review. *Water Research, 44*(10), 2997–3027. https://doi.org/10.1016/j.watres.2010.02.039

Chukwuemeka-Okorie, A. A., Ayotamuno, A. A., Akor,C. O., Ighalo, C. O., Eze, S. N. (2021). Two–three parameters isotherm modeling, kinetics with statistical error analysis for adsorption of cationic dye onto agro waste. *Scientific Reports*. https://doi.org/10.1038/s41598-021-95090-8

Crestini, C., Lange, H., Sette, M., Argyropoulos, D. S. (2017). On the structure of softwood kraft lignin. *Green Chemistry*. https://doi.org/10.1039/C6GC03096E

Das, A., Dey, S., Bala, K., Gnanasekaran, R., & Das, A. P. (2024c). Microbial bioremediation technology for sustainable treatment and management of synthetic microfiber waste. In *Sustainable microbial technology for synthetic and cellulosic microfiber bioremediation* (pp. 221–238). https://doi.org/10.1007/978-3-031-62600-5_12

Das, A., Dey, S., & Das, A. P. (2024b). Ecotoxicological impacts of synthetic microfiber pollutants and development of sustainable mitigation strategies. *Environmental Chemistry and Ecotoxicology*. https://doi.org/10.1016/J.ENCECO.2024.12.003

Das, A., Banerjee, S., & Chatterjee, P. (2024a). The role of secondary microplastics in marine ecosystems. *Ecotoxicology and Environmental Safety, 227*, 112932. https://doi.org/10.1016/j.ecoenv.2024.112932

Das, A. P., Mallavarapu, M., & Ghosh, S. (2023). Ecotoxicity of mining pollutants on the environment and their remediation. *Environmental Chemistry and Ecotoxicology, 5*, 165–167. https://doi.org/10.1016/J.ENCECO.2023.08.002

Das, A., Ray, S. K., Mohanty, M., Mohanty, J., Dey, S., & Das, A. P. (2024d). Ecotoxicity of microplastic wastes and their sustainable management: A review. *Environmental Chemistry and Ecotoxicology, 6*, 144–152. https://doi.org/10.1016/J.ENCECO.2024.05.003

Dey, S., Das, A., Mallick, K., Sahu, A., & Das, A. P. (2023a). *Environmental petroleum waste: pollution, toxicity, sustainable remediation* (pp. 159–175). https://doi.org/10.1007/978-3-031-48220-5_7

Dey, S., Tripathy, B., Kumar, M. S., & Das, A. P. (2023b). Ecotoxicological consequences of manganese mining pollutants and their biological remediation. *Environmental Chemistry and Ecotoxicology*. https://doi.org/10.1016/J.ENCECO.2023.01.001

Dey, S., Dash, A., & Das, A. P. (2024b). Bioremediation as a sustainable and circular solution for mining waste management and bioenergy generation. In *Sustainable management of mining waste and tailings: A circular economy approach* (pp. 107–119). https://doi.org/10.1201/978 1003442455-6

Dey, S., Bissoyi, A., Koteswara Reddy, G., & Das, A. P. (2024a). Synthetic microfibers: An urban emerging pollutant demanding for sustainable environmental solutions. In *Sustainable microbial technology for synthetic and cellulosic microfiber bioremediation* (pp. 107–121). https://doi.org/10.1007/978-3-031-62600-5_7

Dey, S., Dash, N., Ray, S. K., Priyadarsini, S., Chhotaray, C., & Das, A. P. (2024c). Occurrence and quantification of tire wear plastics pollutants from highway City dust of Bhubaneswar City of Odisha. In *Sustainable microbial technology for synthetic and cellulosic microfiber bioremediation* (pp. 77–89). https://doi.org/10.1007/978-3-031-62600-5_5

Dilbaghi, S., Goyal, S., Sharma. R. K., Vashist, M., Kumar, N., Pratap, M., Das, R. (2019). An overview of photocatalytic degradation: Photocatalysts, mechanisms, and development of photocatalysis technology. *Environmental Science and Pollution Research*. https://doi.org/10.1007/s11356-019-07193-5

Domínguez-Jaimes, L. P., Cedillo-González, E. I., Luévano-Hipólito, E., Acuña-Bedoya, J. D., & Hernández-López, J. M. (2021). Degradation of primary nanoplastics by photocatalysis using different anodized TiO_2. *Journal of Hazardous Materials, 413*, 125452. https://doi.org/10.1016/j.jhazmat.2021.125452

Domínguez-Robles, A., Peresin, J., Tamminen, M., Rodríguez, A., Jääskeläinen, K., Larrañeta, A. (2020). Lignin-based hydrogels with tuned drug release properties. *International Journal of Biological Macromolecules*. https://doi.org/10.1016/j.ijbiomac.2020.04.250

Donaldson, K., Stone, V., Tran, C. L., MacNee, W. J. A., & Seaton, A. (2004). Nanotoxicology. *Occupational and Environmental Medicine, 61*(9), 727–728. https://doi.org/10.1136/oem.2004.013243

Enfrin, M., Lee, J., Le-Clech, P., & Dumée, L. F. (2020). Kinetic and mechanistic aspects of ultrafiltration membrane fouling by nano- and microplastics. *Journal of Membrane Science, 601*, 117890. https://doi.org/10.1016/j.memsci.2020.117890

Eriksen, M., Mason, S. A., & Wilson, S. (2013). Microplastic pollution in the surface waters of the Laurentian Great Lakes. *Marine Pollution Bulletin, 77*(1–2), 177–182. https://doi.org/10.1016/j.marpolbul.2013.10.007

Fasano, M., Chiavazzo, E., Asinari, P. (2012). Water transport control in carbon nanotube arrays. *Nanoscale Research Letters*. https://doi.org/10.1186/1556-276X-7-337

George, J., Sabapathi, S. (2015). Cellulose nanocrystals: Synthesis, functional properties, and applications. *Nanotechnology, Science and Applications*. https://doi.org/10.2147/NSA.S64386

Ghosh, S., Sinha, J. K., Ghosh, S., Vashisth, K., Han, S., & Bhaskar, R. (2023). Microplastics as an emerging threat to the global environment and human health. *Sustainability, 15*(14), 10821. https://doi.org/10.3390/SU151410821

Gilca, M., Gaman, M., Del Fabbro, I., Vinceti, D., Lambertini, R., Andreica, A. (2020). Lignin nanoparticles: Green synthesis and applications in water purification. *Environmental Research*. https://doi.org/10.1016/j.envres.2020.109354

Gleiter, H. (2000). Nanostructured materials: Basic concepts and microstructure. *Acta Materialia, 48*(1), 1–29. https://doi.org/10.1016/S1359-6454(99)00285-2

Gunjakar, J. L., Kim, I. Y., Lee, J. M., Lee, N., & Hwang, S. B. (2013). Self-assembly of layered double hydroxide 2D nanoplates with graphene nanosheets: An effective way to improve the photocatalytic activity of 2D nanostructured materials for visible light-induced O_2 generation. *Energy & Environmental Science, 6*(4), 1008–1017. https://doi.org/10.1039/C3EE23989F

Heinze, T., & Liebert, T. (2001). Unconventional methods in cellulose functionalization. *Progress in Polymer Science*. https://doi.org/10.1016/S0079-6700(00)00035-3

Hernández-Rangel, B., Guzmán-Mar, J. L., Durán-Valle, M. D. C, & Durán-Valle, M. D. C. (2020). First insights into photocatalytic degradation of HDPE and LDPE microplastics by a mesoporous N–TiO_2 coating: Effect of size and shape of microplastics. *Environmental Science & Technology*

Holmes, L. A., Turner, A., & Thompson, R. C. (2012). Adsorption of trace metals to plastic resin pellets in the marine environment. *Environmental Pollution, 160*(1), 42–48. https://doi.org/10.1016/j.envpol.2011.08.052

Hou, J., Chen, Y., & Wang, S. (2023). Harnessing magnetic nanoparticles for the effective removal of microplastics. *Nanomaterials, 13*(3), 500. https://doi.org/10.3390/nano13030500

Hui, W. (2019). Efficient separation of microplastics from water using magnetically responsive nanomaterials. *Water Research, 158*, 114852. https://doi.org/10.1016/j.watres.2019.114852

Husák, M., Filip, J., & Zbořil, R. (2023). Harnessing magnetic nanoparticles for the effective removal of micro- and nanoplastics: A comprehensive review. *Nanomaterials, 13*(15), 2154. https://doi.org/10.3390/nano13152154

Jagatee, S., Priyadarshini, S., Rath, C. C., & Das, A. P. (2024). Synthetic microfiber: An enduring environmental problem linked to sustainable development. *Environmental Science and Engineering, Part F2333*, 93–112. https://doi.org/10.1007/978-3-031-51792-1_6

Kargar, F., Zare, M. R., Nabavi, M. R., Younesi, H., Khataee, A. R. (2023). Photocatalytic degradation of organic and inorganic pollutants to clean air: A review. *Catalysts*. https://doi.org/10.3390/catal13020380

Keller, A. A., McFerran, S., Lazareva, A., & Suh, S. (2013). Global life cycle releases of engineered nanomaterials. *Journal of Nanoparticle Research, 15*(6), 1692. https://doi.org/10.1007/s11051-013-1692-4

Klaine, S. J., Alvarez, P. J., Batley, G. E., Fernandes, T. F., Handy, R. D., Lyon, D. Y., Mahendra, S., McLaughlin, M. J., & Lead, J. R. (2008). Nanomaterials in the environment: Behavior, fate, bioavailability, and effects. *Environmental Toxicology and Chemistry, 27*(9), 1825–1851. https://doi.org/10.1897/08-090.1

Kumar, C. S. S. R. (2013). *Nanostructured materials: processing, properties, and applications*. William Andrew Publishing. https://doi.org/10.1016/B978-0-12-394626-3.00001-0

Kumar, S., & Sharma, P. (2023). Recent advances in zinc oxide-based photocatalytic degradation of microplastics: A review. *Chemical Engineering Transactions, 106*, 973–978.

Lee, J. S., Park, S. Y., Kim, K. H., Lee, T. H. (2023). Effects of pH on the adsorption performance of functionalized nanomaterials for heavy metal removal from aqueous solutions. *Journal of Environmental Chemical Engineering*. https://doi.org/10.1016/j.jece.2023.109281

Leslie, H. A., van Velzen, M. J. M., Brandsma, S. H., Vethaak, A. D., Garcia-Vallejo, J. J., & Lamoree, M. H. (2022). Discovery and quantification of plastic particle pollution in human blood. *Environment International, 163*, 107199. https://doi.org/10.1016/j.envint.2022.107199

Li, J., Wang, B., Chen, Z., Ma, B., & Chen, J. P. (2021). Ultrafiltration membrane fouling by microplastics with raw water: Behaviors and alleviation methods. *Chemical Engineering Journal, 410*, 128174. https://doi.org/10.1016/j.cej.2020.128174

Li, Y., Zhang, Y., & Zhao, X. (2024). Photocatalytic and surface properties of titanium dioxide nanoparticles in soil solutions. *Environmental Science: Nano, 11*(3), 456–467. https://doi.org/10.1039/D3EN00622K

Lusher, A. L., McHugh, M., & Thompson, R. C. (2013). Occurrence of microplastics in the gastrointestinal tract of pelagic and demersal fish from the English Channel. *Marine Pollution Bulletin, 67*(1–2), 94–99. https://doi.org/10.1016/j.marpolbul.2012.11.028

Mallick, K., Sahu, S., Sahu, A., Dey, S., & Das, A. P. (2024). A critical review of marine microfiber pollution routes, toxicity, and its sustainable remediation. *Environmental Science and Engineering, Part F2333*, 189–211. https://doi.org/10.1007/978-3-031-51792-1_10

Maurer-Jones, M. A., Gunsolus, I. L., Murphy, C. J., & Haynes, C. L. (2013). Toxicity of engineered nanoparticles in the environment. *Analytical Chemistry, 85*(6), 3036–3049. https://doi.org/10.1021/ac303636s

Maynard, A. D. (2006). Nanotechnology: A research strategy for addressing risk. *Woodrow Wilson International Center for Scholars*. https://www.nanotechproject.org/process/assets/files/2709/77_pen3_risk.pdf

Mishra, S., Gupta, A., & Sharma, R. (2019a). Microplastic pollution: A threat to aquatic biodiversity. *Marine Pollution Bulletin, 139*, 35–44. https://doi.org/10.1016/j.marpolbul.2019.02.014

Mishra, S., Rath, C. charan, & Das, A. P. (2019b). Marine microfiber pollution: A review on present status and future challenges. *Marine Pollution Bulletin, 140*, 188–197. https://doi.org/10.1016/J.MARPOLBUL.2019.01.039

Mishra, S., Rout, P. K., & Das, A. P. (2021a). Emerging microfiber pollution and its remediation. In Prasad, R. (Ed.), *Environmental and microbial biotechnology* (pp. 247–266). Springer. https://doi.org/10.1007/978-981-15-5499-5_9

Mishra, S., & Das, A. P. (2021). Current treatment technologies for removal of microplastic and microfiber pollutants from wastewater. In *Wastewater treatment: cutting-edge molecular tools, techniques and applied aspects* (pp. 237–251). https://doi.org/10.1016/B978-0-12-821881-5.00011-8

Mishra, S., Swain, S., Sahoo, M., Mishra, S., & Das, A. P. (2021b). Microbial colonization and degradation of microplastics in aquatic ecosystem: A review. *Geomicrobiology Journal*. https://doi.org/10.1080/01490451.2021.1983670

Mishra, S., & Das, A. P. (2022). Treatment of the wastewater polluted with synthetic microfiber released from washing machine. In B. B. Das, H. Hettiarachchi, P. K. Sahu, & S. Nanda (Eds.), *Lecture notes in civil engineering* (Vol. 207, pp. 109–117). Springer Science and Business Media Deutschland GmbH. https://doi.org/10.1007/978-981-16-7509-6_9/COVER

Mishra, S., Singh, R. P., Rout, P. K., & Das, A. P. (2022). Membrane bioreactor (MBR) as an advanced wastewater treatment technology for removal of synthetic microplastics. In *Development in wastewater treatment research and processes* (pp. 45–60). https://doi.org/10.1016/B978-0-323-85583-9.00022-3

Mishra, S., Dash, D., & Das, A. P. (2024a). Aquatic microbial diversity on plastisphere: Colonization and potential role in microplastic biodegradation. *Geomicrobiology Journal, 41*(4), 312–323. https://doi.org/10.1080/01490451.2023.2209750

Mishra, S., Dash, D., & Das, A. P. (2024c). Microbial degradation of polyester microfibers using indigenously isolated bacterial strain Exiguobacterium Sp. *CLEAN—Soil, Air, Water, 52*(12), e202300343. https://doi.org/10.1002/CLEN.202300343

Mishra, S., Dash, D., & Das, A. P. (2024b). Isolation and molecular characterization of polyester-degrading bacterial strains from municipal wastewater and their application in microplastic degradation in contaminated wastewater. *Geomicrobiology Journal*. https://doi.org/10.1080/01490451.2024.2376101

Mohanty, M., Tripathy, B., Koteswara Reddy, G., & Das, A. P. (2024c). Greener technologies for establishing restraint over microplastic pollution. In *Sustainable microbial technology for synthetic and cellulosic microfiber bioremediation* (pp. 239–258). https://doi.org/10.1007/978-3-031-62600-5_13

Mohanty, J., Dey, S., Bala, K., & Das, A. P. (2024a). Recent advances in microplastic pollution for its sustainable management. In *Sustainable management of urban plastic waste through circular economic approaches* (pp. 27–48). https://doi.org/10.1201/9781003465447-3

Mohanty, M., Mohanty, J., Dey, S., Dutta, K., Shah, M. P., & Das, A. P. (2024b). The face mask: A tale from protection to pollution and demanding sustainable solution. *Emerging Contaminants, 10*(2), 100298. https://doi.org/10.1016/J.EMCON.2023.100298

Moharana, S., Dey, S., Priyadarsini, S., Kumar, M. S., & Das, A. P. (2024). Identification and characterization of microplastic pollutants from the marine sediments of Paradeep Coast of Bay of Bengal, India for their sustainable management. *Environmental Science and Engineering, Part F2333*, 57–73. https://doi.org/10.1007/978-3-031-51792-1_4

Mohiuddin, I., & Chowdhury, A. N. (2023). Recent advances in zinc oxide-based photocatalytic degradation of microplastics: A review. *Chemical Engineering Transactions, 106*, 973–978.

Mustafa, I. (2023). Magnetic nanomaterials for microplastic removal: Recent advances and future perspectives. *Journal of Hazardous Materials, 430*, 130245. https://doi.org/10.1016/j.jhazmat.2023.130245

Nag, M., Lahiri, D., Ghosh, S., Sarkar, T., Pati, S., Das, A. P., Ram, D. K., Bhattacharya, D., & Ray, R. R. (2023). Application of microorganisms in biotransformation and bioremediation of environmental contaminant: A review. *Geomicrobiology Journal*. https://doi.org/10.1080/01490451.2023.2261443

Norgren, M., Edlund, H. (2014). Lignin-based materials: Opportunities and challenges. *Green Chemistry*. https://doi.org/10.1039/C3GC41935C

Nowack, B., & Bucheli, T. D. (2007). Occurrence, behavior and effects of nanoparticles in the environment. *Environmental Pollution, 150*(1), 5–22. https://doi.org/10.1016/j.envpol.2007.06.004

Oberdörster, G., Oberdörster, E., & Oberdörster, J. (2005). Nanotoxicology: An emerging discipline evolving from studies of ultrafine particles. *Environmental Health Perspectives, 113*(7), 823–839. https://doi.org/10.1289/ehp.7339

Patel, N., & Munjal, S. (2021). Metal oxide nanoparticles and their applications: A report. arXiv preprint arXiv:2201.04938. https://arxiv.org/abs/2201.04938

Paul, A., Dey, S., Ram, D. K., & Das, A. P. (2023). Hexavalent chromium pollution and its sustainable management through bioremediation. *Geomicrobiology Journal*, 1–11. https://doi.org/10.1080/01490451.2023.2218377

Paul, A., Priyadarsini, S., Karakuş, S., Behera, J. K., Gnanasekaran, R., & Das, A. P. (2024b). Fate and accumulation of synthetic microfibers in the environment. In *Sustainable microbial technology for synthetic and cellulosic microfiber bioremediation* (pp. 61–76). https://doi.org/10.1007/978-3-031-62600-5_4

Paul, A., Kumar, S., & Singh, R. (2024a). Primary and secondary microplastics: Environmental impacts and mitigation strategies. *Journal of Environmental Management, 312*, 114832. https://doi.org/10.1016/j.jenvman.2024.114832

Pradhan, G., Dey, S., Priyadarsini, S., Kumar, M. S., & Das, A. P. (2024). Characterization and quantification of microplastics pollutants in sediment samples from Daya River of Odisha State in India for their appropriate management. *Environmental Science and Engineering, Part F2333*, 75–92. https://doi.org/10.1007/978-3-031-51792-1_5

Prasad, R., Verma, G., Kumar, M., Yadav, A. (2024). Photocatalytic degradation of pollutants using advanced ceramics. *Journal of Inorganic and Organometallic Polymers and Materials*. https://doi.org/10.1007/s10904-024-03068-6

Prata, J. C., da Costa, J. P., Lopes, I., Duarte, A. C., & Rocha-Santos, T. (2019). Methods for sampling and detection of microplastics in water and sediment: A critical review. *TrAC Trends in Analytical Chemistry, 110*, 150–159. https://doi.org/10.1016/j.trac.2018.10.029

Priyadarshini, B., & Das, A. P. (2024). The adverse impact of synthetic microplastics released from disposable face masks pollutants on aquatic microbiota, aquatic species and humans: A review. *Environmental Science and Engineering, Part F2827*, 67–81. https://doi.org/10.1007/978-3-031-50840-0_5

Raj, K., Mehta, P., & Verma, L. (2023a). Human ingestion of microplastics: Emerging risks to health. *Science of the Total Environment, 858*, 160254. https://doi.org/10.1016/j.scitotenv.2023.160254

Raj, K., Verma, P., & Singh, R. (2023b). Human ingestion of microplastics: A review of recent findings and potential health implications. *Environmental Science & Technology, 57*(5), 2341–2350. https://doi.org/10.1021/acs.est.2c04567

Ramesh, G. (2021). Nanotechnology-based approaches for microplastic remediation in water systems. *Environmental Research, 197*, 111123. https://doi.org/10.1016/j.envres.2021.111123

Ray, S. K., Dey, S., Das, A. P., & John, A. (2024). Microplastic waste management through greener technologies : A review. In *Sustainable management of urban plastic waste through circular economic approaches* (pp. 237–261). https://doi.org/10.1201/9781003465447-14

Rochman, C. M., Hoh, E., Hentschel, B. T., & Kaye, S. (2013). Long-term field measurement of sorption of organic contaminants to five types of plastic pellets: Implications for plastic marine debris. *Environmental Science & Technology, 47*(3), 1646–1654. https://doi.org/10.1021/es303700s

Rol, C., Belgacem, C., Gandini, A. (2019). Recent advances in surface-modified cellulose nanofibrils for polymer reinforcement and environmental applications. *Carbohydrate Polymers*. https://doi.org/10.1016/j.carbpol.2019.115898

Sahoo, P. P., Singh, S., Rout, P. K., Mishra, S., & Das, A. P. (2022). Microbial remediation of plastic pollutants generated from discarded and abandoned marine fishing nets. *Biotechnology and Gnetic Engineering Reviews*. https://doi.org/10.1080/02648725.2022.2152629

Sahu, A., Dey, S., Mallick, K., Priyadarsini, S., Behera, I. D., & Das, A. P. (2024). A review on advanced technology for sustainable management of synthetic microplastic waste. In *Sustainable management of urban plastic waste through circular economic approaches* (pp. 262–279).https://doi.org/10.1201/9781003465447-15

Sanchez, C., Belleville, P., Popall, M., & Nicole, L. (2011). Applications of advanced hybrid organic–inorganic nanomaterials: From laboratory to market. *Chemical Society Reviews, 40*(2), 696–753. https://doi.org/10.1039/C0CS00136H

Satapathy, S., Kar, D., Priyadarsini, S., Das, A. P., & Kuanar, A. (2024). Application of microorganisms in emerging waste management strategies and remediation of synthetic and cellulosic microfibers. In *Sustainable microbial technology for synthetic and cellulosic microfiber bioremediation* (pp. 183–206). https://doi.org/10.1007/978-3-031-62600-5_10

Shen, M., Song, B., Zeng, G., Zhang, Y., Huang, W., Wen, X., & Tang, W. (2020). Are biodegradable plastics a promising solution to reduce microplastic pollution? *Environmental Pollution, 263*, 114469. https://doi.org/10.1016/j.envpol.2020.114469

Shen, M., Zeng, Z., Song, B., Yi, H., Hu, T., Zhang, Y., & Zeng, G. (2019). Microplastic removal from the aquatic environment: A review of recent research and future perspectives. *Science of the Total Environment, 696*, 133971. https://doi.org/10.1016/j.scitotenv.2019.133971

Shi, X., Zhang, X., Gao, W., Zhang, Y., & He, D. (2022). Removal of microplastics from water by magnetic nano-Fe_3O_4. *Science of the Total Environment, 802*, 149838.

Singh, D., Gupta, A. K., Kumar, R., Tyagi, S. (2024). An insight into photo-catalytic degradation mechanism of organic pollutants: Influence of operational parameters and reactor design. *Journal of Environmental Science and Health, Part A*. https://doi.org/10.1080/01614940.2024.2440664

Singh, L., Kumar, V., & Gopinath, P. (2020a). Nanomaterials for sustainable environmental remediation: A review. *Chemical Engineering Journal, 397*, 125406. https://doi.org/10.1016/j.cej.2020.125406

Singh, N., Mondal, S., & Das, S. (2023). Review on the use of magnetic nanoparticles in the detection of microplastics in water. *Water, 15*(17), 3077.

Singh, R. P., Mishra, S., & Das, A. P. (2020b). Synthetic microfibers: Pollution toxicity and remediation. *Chemosphere, 257*, 127199. https://doi.org/10.1016/j.chemosphere.2020.127199

Sun, S., Zhou, C., Yang, H., Wang, M. (2018). Recent advances in cellulose-based adsorbents for the removal of heavy metals from wastewater. *Cellulose*. https://doi.org/10.1007/s10570-018-2199-4

Sun, C., Wang, Z., Chen, L., & Li, F. (2021a). Biodegradable and re-usable sponge materials made from chitin for efficient removal of microplastics. *Journal of Hazardous Materials, 416*, 126599. https://doi.org/10.1016/j.jhazmat.2021.126599

Sun, C., Wang, Z., Zheng, H., Chen, L., & Li, F. (2021b). Biodegradable and re-useable sponge materials made from chitin for efficient removal of microplastics. *Journal of Hazardous Materials, 420*, 126599.

Tang, Y., Zhang, S., Su, Y., Wu, D., Zhao, Y., & Xie, B. (2021). Removal of microplastics from aqueous solutions by magnetic carbon nanotubes. *Chemical Engineering Journal, 406*, 126804.

Tofa, T. S., Kunjali, K. L., Paul, S., & Dutta, J. (2019). Enhanced visible light photodegradation of microplastic fragments with plasmonic platinum–zinc oxide nanorod photocatalysts. *Catalysts, 9*(10), 819.

Tripathy, B., Dey, S., Dubey, N. K., Lo, W.-C., & Das, A. P. (2024). *An omics-based approach for managing the risk of COVID-19-generated wastes* (pp. 51–66). https://doi.org/10.1007/978-3-031-50840-0_4

Uheida, A., Mejía, H. G., Abdel-Rehim, M., Hamd, W., & Dutta, J. (2021). Visible light photocatalytic degradation of polypropylene microplastics in a continuous water flow system. *Journal of hazardous materials, 406*, 124299.

United Nations Environment Programme. (2014). *UNEP Year Book 2014: Emerging Issues in Our Global Environment*. Retrieved from https://www.unep.org/resources/yearbook

Wang, L., Liu, P., Zhao, Z, Chen, Y. (2021). Mechanism of photodegradation of organic pollutants in seawater: A case study on nitrophenol. *ACS Omega*. https://doi.org/10.1021/acsomega.1c04604

Wang, J., Coffin, S., Sun, C., Schlenk, D., & Gan, J. (2019). Negligible effects of microplastics on animal fitness and HOC bioaccumulation in earthworms. *Environmental Pollution, 249*, 776–784. https://doi.org/10.1016/j.envpol.2019.03.102

Wang, R., Zhang, L., Chen, B., & Zhu, X. (2020). Low-pressure driven electrospun membrane with tuned surface charge for efficient removal of polystyrene nanoplastics from water. *Journal of Membrane Science, 614*, 118470.

Wang, X., Sun, B., Dai, Q., Zhu, L., Zi, G., & Dai, L. (2024). Metal-free carbon Co-catalysts for Up-conversion photo-induced catalytic cancer therapy. *Advanced Materials, 36*(42), e2408560. https://doi.org/10.1002/adma.202408560

Wei, H., & Wang, E. (2013). Nanomaterials with enzyme-like characteristics (nanozymes): Next-generation artificial enzymes. *Chemical Society Reviews, 42*(14), 6060–6093. https://doi.org/10.1039/C3CS35486E

Whitesides, G. M. (2003). The 'right' size in nanobiotechnology. *Nature Biotechnology, 21*(10), 1161–1165. https://doi.org/10.1038/nbt872

Wright, S. L., & Kelly, F. J. (2017). Plastic and human health: A micro issue? *Environmental Science & Technology, 51*(12), 6634–6647. https://doi.org/10.1021/acs.est.7b00423

Xiaoming, C. (2022). Functionalized magnetic nanoparticles for efficient microplastic removal from aqueous environments. *Chemosphere, 308*, 135678. https://doi.org/10.1016/j.chemosphere.2022.135678

Xu, F., Sun, J. (2018). Lignin-based functional materials for environmental applications. *Green Chemistry*. https://doi.org/10.1039/C8GC01309E

Xu, J., Thomas, K. V., Luo, Z., & Gowen, A. A. (2020). FTIR and Raman imaging for microplastics analysis: State of the art, challenges and prospects. *TrAC Trends in Analytical Chemistry, 119*, 115629. https://doi.org/10.1016/j.trac.2019.115629

Yadav, A., & Yadav, K. (2024). Exploring the effect of engineered nanomaterials on soil microbial diversity and functions: A review. *International Journal of Environmental Research and Public Health, 21*(3), 1234. https://doi.org/10.3390/ijerph21031234

Yang, L., Cao, X., Cui, J., Wang, Y., Zhu, Z., Sun, H., Liang, W., Li, J., & Li, A. (2022). Holey Ti_3C_2 nanosheets based membranes for efficient separation and removal of microplastics from water. *Journal of Colloid and Interface Science, 617*, 673–682.

Yuan, F., Yue, L., Zhao, H., & Wu, H. (2020). Study on the adsorption of polystyrene microplastics by three-dimensional reduced graphene oxide. *Water Science and Technology, 81*, 2163–2175.

Yuwendi, Y., Ibadurrohman, M., Setiadi, S., & Slamet, S. (2022). Photocatalytic degradation of polyethylene microplastics and disinfection of *E. coli* in water over Fe- and Ag-Modified TiO_2 nanotubes. *Bulletin of Chemical Reaction Engineering & Catalysis, 17*(2), 263–277.

Zandieh, M., & Karimi, A. (2023). Catalytic and biocatalytic degradation of microplastics. *Journal of Environmental Chemical Engineering, 11*(1), 109876. https://doi.org/10.1016/j.jece.2022.109876

Zhang, S. J., Shao, T., Kose, H. S., Karanfil, T. (2009). Adsorption of aromatic compounds by carbonaceous adsorbents: A comparative study on granular activated carbon, activated carbon fiber, and carbon nanotubes. *Environmental Science and Technology*. https://doi.org/10.1021/es801251k

Zhang, S. J., Shao, T., Karanfil, T. (2011). The effects of dissolved natural organic matter on the adsorption of synthetic organic chemicals by activated carbons and carbon nanotubes. *Water Research*. https://doi.org/10.1016/j.watres.2011.01.012

Zhang, Y., & Wang, Y., (2021). Characteristics of adsorption kinetics and isotherms of Cu(II) by natural sediment. *Water Supply*. https://doi.org/10.2166/ws.2021.015

Zhang, D., Liu, X., Huang, W., & Li, J. (2023). Magnetic peroxidase nanozyme gears up for microplastic removal: A green approach. *ChemNanoMat, 9*(1), e202300012.

Zhang, J., Zhang, M., Xiao, X., & Wang, X. (2024). The photocatalytic degradation mechanism for organic pollutants in seawater: A case study on nitrophenol. *The Journal of Physical Chemistry C*. https://doi.org/10.1021/acs.jpcc.4c00121

Zhang, Y., Chen, Y., & Xing, B. (2014). Behavior and potential impacts of metal-based engineered nanoparticles in aquatic environments. *Journal of Environmental Sciences, 26*(1), 1–14. https://doi.org/10.1016/S1001-0742(13)60392-5

Zhao, L., Fan, J., Gong, C., Dyke, A., Gao, W., & Li, B. (2024). A critical review on recent progress of solution-processed monolayer assembly of nanomaterials and applications. arXiv preprint arXiv:2401.08963. https://arxiv.org/abs/2401.08963

Zhou, Q., Tian, C., Wang, P., & Ma, L. (2016). Removal of microplastics in aquatic environments by magnetic nanoparticles. *Journal of Hazardous Materials, 311*, 229–235. https://doi.org/10.1016/j.jhazmat.2016.03.002

Zhou, Y., Lu, J., Zhou, Y., & Liu, Y. (2019). Recent advances for microplastic detection methods in environmental matrices: A review. *Ecotoxicology and Environmental Safety, 182*, 109451. https://doi.org/10.1016/j.ecoenv.2019.109451

Chapter 31
Enhanced Biodegradation of Petroleum Hydrocarbon-Contaminated Soil by Using Organic Amendment as a Sustainable Solution

Ramesh Kumar Ekka, Rajalaxmi Badra, Ipsita Dipamitra Behera, Manisha Bal, Kashinath Barik, and Alok Prasad Das

Introduction

The population's energy needs, production, distribution, and use of petroleum products are increased by the growing population and steadily expanding economics. By 2050, there will be nine billion people on the planet. Unfavourably contributes to environmental contamination, particularly in the soil environment. Petroleum hydrocarbons are responsible for contamination, at least 342,000 sites in Western Europe and 90% of polluted sites in the United States are associated with petroleum-based contamination (Gómez-Sagasti et al., 2018). The most challenging tasks in the environmental sector are evaluating the danger of soils affected by petroleum products and deciding to remediate them, particularly in financial and technical ramifications (Hoang et al., 2021). The quantifiable quantity of several hundred hydrocarbon compounds derived from crude oil is known as total petroleum hydrocarbon and is mostly composed of hydrogen and carbon (Behera et al., 2020; Lahiri et al.,

R. K. Ekka · R. Badra · I. D. Behera (✉) · K. Barik
Department of Chemical Engineering, Indira Gandhi Institute of Technology, Sarang, Dhenkanal, Odisha, India
e-mail: ipsitadbehera@gmail.com

R. K. Ekka
Biju Pattanaik University of Technology, Rourkela, Odisha, India

M. Bal
Department of Chemical Engineering, National Institute of Technology, Durgapur, West Bengal, India

A. P. Das
Department of Life Sciences, Rama Devi Women's University, Vidya Vihar, Bhubaneswar, Odisha, India

I. D. Behera et al. (eds.), *Emerging Micropollutants*, Sustainable Environmental Waste Management Strategies, https://doi.org/10.1007/978-3-032-03530-1_31

2021). The amount of TPH is a widely used gross metric for managing environmental cleanup in several nations by measuring the level of contamination. The two categories of hydrocarbon molecules that make up TPH are extractable and volatile petroleum hydrocarbons. (Hoang et al., 2021).

Crude oil is being explored and extracted as a result of population growth and rising energy demands (Eze & George, 2020). Aquatic and terrestrial life have suffered greatly as a result of the growing production of crude oil over time. There are several methods that drilling, transportation, and distribution release crude oil into the environment Pichtel (2016). The ecology suffers as a result of these spills. The development of numerous remediation techniques is the outcome of efforts to eradicate crude oil pollution from the environment. In recent decades, there has been a lot of research on bioremediation, a technique that uses living things, especially bacteria, to break down toxins in a contaminated environment (Eze & George, 2020).

Rapid economic and demographic expansion has a huge volume of organic waste which has adverse effects on the environment and raises disposal expenses (Dey et al., 2023; Mishra et al., 2023). The idea of using organic waste as a resource for land application has grown in popularity. It has been explored various organic wastes to soils as fertilizers, is a financially appealing, socially acceptable and environmentally responsible substitute for disposing of the waste in landfills or burning it. Organic wastes may improve the cleanup of polluted soils (Wu et al., 2017).

Organic manures have been used in agriculture for over two millennia, according to Chinese records. Like the Romans, the Greeks applied human sewage to fields and realized that plant and animal waste could be used to boost crop growth, but the amounts needed varied depending on the kind of soil and plant. It was widely accepted in the late medieval era that animal dung was used to replenish materials lost during crop cultivation. In reality, super phosphate was the "first factory for the manufacturing of synthetic manure," and it opened for operation in 1842 (Mogollón et al., 2018). Organic amendments have a major effect on the biodegradation of petroleum hydrocarbons by cumulative microbial activity and reassuring the breakdown of pollutants (Udume et al., 2023). Crop residues, green manures, animal manures, biosolids, composted organic materials and biochar are examples of amendments that can be added to TPH contaminated soils that come from raw or adequately treated organic wastes. This has been shown to increase the remediation efficiency. The amendments addition of carbon substrates, nutrients and active microorganisms may positively change the dynamics of total petroleum hydrocarbons contaminated soils (TPH) and soil characteristics while promoting the development and activity of soil microbial populations that are advantages for TPH remediation (Koshlaf et al., 2019).

Compost and vermicompost are examples of organic amendments that have been shown in numerous studies to significantly accelerate the breakdown of polycyclic aromatic hydrocarbons and total petroleum hydrocarbons in polluted soils. In contaminated soils, organic extracts such as water hyacinth compost produced TPH biodegradation of up to 93% (Udume et al., 2023). The practice of vermicompost boosted microbial activity, resulting in TPH breakdown efficiencies of 32.5% for biostimulation and 34.4% for bioaugmentation (Curiel-Alegre et al., 2022). Compost dramatically lowered TPH levels in soil contaminated by used motor oil, proving the value of organic amendments in bioremediation (Olugbenga et al., 2024).

One of the most studied groups of organic pollutants brought about by human activity is the polycyclic aromatic hydrocarbons (PAHs). These PAH chemicals come in a variety of forms and environmental toxicity levels. Rarely biodegradable, PAHs are organic substances that cause cancer. When examining the impacts of organic additions (such as sewage sludge, soybean trash, and pig manure) on the degradation of PAHs, it was found that, mostly as a result of the decomposition process, pig dung had the highest removal effectiveness for PAHs compared to the other two organic supplements. (El-Saeid & Usman, 2022). Organic accompaniments increase microbial growth and activity which is essential for efficient biodegradation and improve nutrient availability. By changing the pH and moisture content of the soil, the adding of organic matter can also promote microbial metabolism and breakdown of hydrocarbons (El-Saeid & Usman, 2022). In contaminated soils, various amendments have a major impact on the rates of hydrocarbon breakdown and microbial activity. The efficacy of modifications that promote microbial proliferation and degrading efficacy varies according to their composition and particular pollutants are present (Ntekpe et al., 2023; Kanungo et al., 2023; Mallick et al., 2023). Compared to organic and inorganic, water hyacinth compost, an organic-amendments produced an impressive 93% biodegradation of TPH (Udume et al., 2023).

Agro-waste amendments especially groundnut husks showed significant rates of degradation 90.85% because they promote the growth of microorganisms (Agbor et al., 2018). Even though organic amendments often enhance hydrocarbon breakdown, overall efficacy of these tactics may be constrained by the particulars of polluted site, such as moisture content and existence of resistant chemicals. A customized strategy taking site specific characteristics into account is crucial (Agbor et al., 2018). Temperature, pH and moisture content are ideal parameters that affect the efficacy of biodegradation with organic additions (Anayet et al., 2024). Research indicated the grater moisture content increases CO_2 outflow and improves the breakdown of pollutants like PAHs. The circumstances are usually ideal for biodegradation and depending on particular organic materials and environmental circumstances (El-Saeid & Usman, 2022; Das et al., 2023; Kanungo et al., 2024).

In a variety of polluted situations, organic supplements are essential for improving biodegradation processes. Among these amendments Biochar, compost and biosolids have demonstrated notable efficacy in fostering nutrient cycling and microbial activities are critical for biodegradation (Omokaro et al., 2024). There are many types of Organic amendments. Organic amendments are natural sources for microbes including these can be pathogenic to humans (Singh et al., 2022). Plant nutrients, particularly N, have primarily come from organic amendments, or organic wastes. Through its application, soil changes brought on by necessary agricultural production procedures can be countered and often limited resources can be better managed. For the purpose of restoring the physical and chemical characteristics of soil, organic amendments offer macro- and micronutrients, including carbon. Components included in sewage biosolids and animal manure, the most prevalent organic amendments, can be harmful to plants, animals, and people (Singh et al.,

2022). Manures from cattle and poultry are mostly controlled by making sure they are administered to crops and soils at the right time, rate, and location. As agricultural intensity rises, a more comprehensive management strategy is needed.

Petroleum Hydrocarbon Sources

Petroleum hydrocarbons (PHs) are complex compounds made up of carbon and hydrogen molecules, occasionally with additional impurities like nitrogen, sulphur, and oxygen (Behera et al., 2021; Mohanta et al., 2024). From natural gas used for cooking and heating to petroleum products used in automobiles and other internal combustion engines, they are available in a wide variety of forms and combinations. There are wet gas and dry gas, as well as "light oils" and "heavy oils." But they all share a genesis from organic materials, meaning that the "source rock" was formed by once-living plants and small animals (Kathi, 2022). Certain kinds of organic materials are abundant in "source rocks," which are the rocks that generate hydrocarbons. After being buried, plant and animal tissue undergoes chemical changes that result in the complex compounds that, when heated and compressed, eventually create natural gas or oil from sediments trapped beneath the Earth's surface over millions of years. Petroleum was used sparingly by ancient Egyptian, Chinese, and Indian societies, mostly as caulking for boats and canoes, medicine, and lamp fuel (Kathi, 2022).

Short-chain hydrocarbons including paraffin, alicyclic, and aromatic chemicals, as well as nonhydrocarbons like naphthenic acid, phenol, thiol, metalloporphyrin, asphaltene, and heterocyclic nitrogen and sulphur compounds, make up PHs. PHs are widely employed as the primary energy source for daily life and business worldwide. Petrochemical goods are in high demand due to modernization and population growth. Exploration, production, maintenance, storage, transportation, and unintentional spills are all common sources of PHs contamination. Humans and other biotic elements of the ecosystem suffer terrible and catastrophic effects from an oil-contaminated environment (Ossai et al., 2020). Hydrocarbon resources can be found in a wide range of plays, from the more traditional ones that describe permeable and porous rocks that contain hydrocarbons with advantageous fluid properties to plays where the reservoir and fluid qualities are gradually declining (Moitra et al., 2022).

Petroleum hydrocarbons come from a variety of man-made and geological sources, mostly from the organic stuff that breaks down over geological time (Fig. 31.1). For hydrocarbon resources to be explored and managed effectively, it is essential to comprehend these sources. Rocks of the source are Shales and other sedimentary rocks that are high in organic content are the main geological sources of petroleum hydrocarbons (Behera et al., 2021). The Madbi formation in the Sayun-Masilah basin, for example, has a high organic carbon content (more than 5.2 weight percent) and is distinguished by type I and type II kerogens, suggesting a considerable potential for hydrocarbon generation (Al-Areeq., 2018).

Petroleum and natural gas provide the majority of hydrocarbons. Earthly oils are intricate blends of thousands of organic substances. Natural gas and petroleum

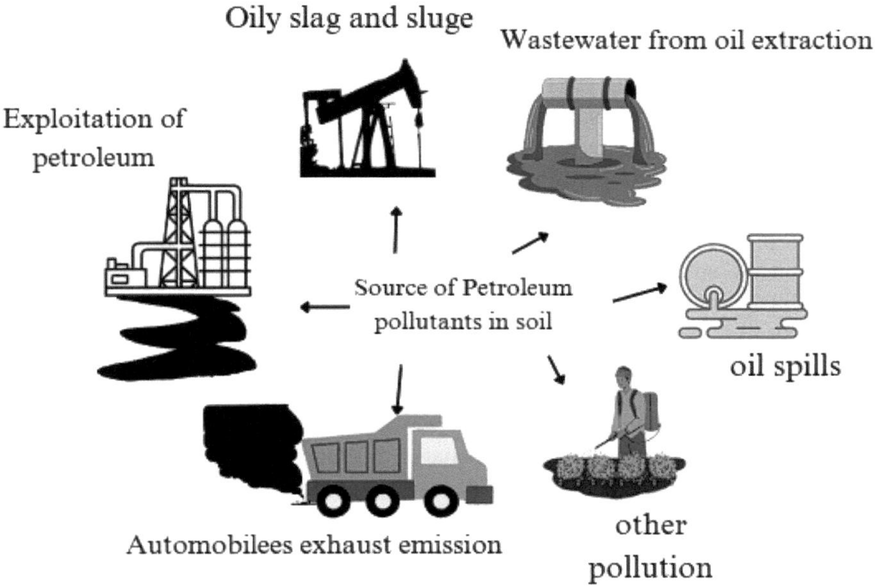

Fig. 31.1 Sources of hydrocarbon petroleum pollutants

are examples of fossil fuels. The natural processes produce fossil fuels (Ince & Ince 2019). Water that is a byproduct of extracting gas and oil from the ground is referred to as "produced water." Wastewater effluents from both conventional and unconventional oil and gas extraction operations are included in this category (Ince & Ince 2019). During the extraction, storage, and refinement of crude oil, the petroleum industry produces large volumes of oily sludge. Oily sludge is a substance that can be dangerous. Temporary storage facilities known as oilfield sludge pits are used to dispose of oily wastes, such as drilling mud and cuttings, produced during oil well drilling or when crude oil is collected during an emergency drill operation (Gorlenko & Timofeeva 2019). There are several types of petroleum hydrocarbons with distinct chemical structures and carbon numbers found in crude oil. Generally speaking, an aromatic hydrocarbon is more hazardous than its aliphatic cousin with an equivalent carbon number (Bacosa et al., 2010).

Sludge that contains residual oil, different refined oils, crude oil, and other heavy oils is referred to as oily sludge. It is created during the extraction, refinement, storage, and transportation of petroleum as well as during the discharge of oily sewage (Behera et al., 2020; Hu et al., 2020). The waste legislation of certain nations, like China, therefore classify oily sludge, which includes oil field drilling mud or cuttings oil from shale gas drilling, as hazardous waste (Niu et al., 2022). The disposal of oil wastes produced by storage, processing, and transportation facilities has long been a significant problem for the petroleum industry. The industry uses the hydrocarbon molecules it takes from the ground to make a variety of other valuable products.

Plastics, polymers, synthetic rubbers, detergents, solvents, and other synthetic materials are all produced by the petrochemicals sector (Dey et al., 2024; Jagatee et al., 2024; Mishra et al., 2020; Paul et al., 2024; Singh et al., 2020). The refining sector creates bitumen, lubricants, and waxes (Sarma et al., 2019). Hydrocarbons, which are composed of hydrogen, oxygen, and carbon atoms, are mixed together to form gasoline and diesel fuels. Burning occurs when hydrocarbons combine with oxygen. Nitrogen oxides and sunlight combine with hydrocarbons to create ground-level ozone, a primary ingredient of smog. Emissions from vehicles have a variety of effects on the environment. Our atmosphere becomes overloaded with carbon as humans burn fossil fuels like coal, oil, and natural gas for energy or clear forests and burn them to make crops and pastures (Le Quéré et al., 2016). An oil spill occurs when human activity releases a liquid petroleum hydrocarbon into the environment, particularly the marine ecology.

The word refers to oil spills that happen on the ocean or in coastal waters, but they can also happen on land. The discharge of crude oil from drilling rigs, wells, offshore platforms, and tankers can all lead to oil spills. They might also include the spilling of refined petroleum products, including diesel and gasoline, along with their byproducts (Zacharias et al., 2024).

Environmental Impact

Petroleum hydrocarbon-contaminated soil is a severe global environmental issue that has garnered public attention in recent decades. One of the primary causes of hydrocarbon spills through industrial or agricultural products is human activity (Zhang et al., 2018). Petroleum has emerged as one of the most significant forms of organic pollution, despite being one of the primary energy sources for sustaining a nation's economic and social progress. It is primarily caused by underground storage tank leaks as well as unintentional spills that occur during transportation and disposal (Rickson & Fullen, 2022). The development of plant growth and human health are negatively impacted by total petroleum hydrocarbons in soil. As the amount of petroleum hydrocarbon-contaminated soil has increased, efforts have been made to clean up the entire soil contamination (Zhang et al., 2018). Petroleum hydrocarbons, polycyclic aromatic hydrocarbons (PAHs), solvents, pesticides, lead, and other heavy metals are among the chemicals that contribute to spills, leaks, and other environmental concerns related to petroleum products that pose health risks to people (Rickson & Fullen, 2022; Prabhakar et al., 2019; Raj & Das, 2023). Both point sources (industrial and residential) and diffuse sources (mobilization and transportation by floodwaters) can introduce contaminants into the soil, which can have long-term effects on soil quality and function, human health, and food quality (Zhang et al., 2018).

Impact of TPHs on Environment

TPH travels from the soil to the groundwater when it is released into the environment through spills, leaks, accidents, industrial releases, or as a by-product of commercial or private uses. Some organisms in the soil may break down some of the contaminants into smaller fractions, while others may evaporate into the atmosphere. Other contaminants may remain in the soil for a longer period of time and be broken down by other organisms in the soil, resulting in harmful health effects (Yue et al., 2021). TPH releases into the environment will endanger public health and safety by contaminating drinking water, increasing the risk of fire and explosion, lowering the quality of air and water, endangering agriculture, destroying recreational areas, destroying habitats and food, and wasting non-renewable resources, according to the US Environmental Protection Agency (USEPA) (Silva et al., 2022).

After petroleum hydrocarbons (PHCs) have contaminated the soil, it could take years for it to recover. Human error, including inadequate planning and maintenance of the pipeline, as well as negligence (Adipah, 2019). Petroleum residues can cling to soil particles and stay there. Most soil biota still find pollutants harmful, even though they may benefit from a carbon supply as a degrader. By causing unpleasant taste, smell, or appearance in environmental media, PHCs have the potential to ruin aesthetics (Yue et al., 2021). This deviated the permeability and porosity of the soil, which caused compaction and more runoff. Hydrocarbons can reduce microbial activity which can lead to anaerobic conditions that are bad for soil health (Mohanta et al., 2024).

Health Effects

Petroleum hydrocarbons include tars, asphalt, natural gas, and crude oil. Petroleum implies "oil from the earth" or "rock oil." (Iyer et al., 2022). Petroleum, which use on a daily basis, is essential to modern civilization. The primary energy source for production, transportation, and heating, it also serves as a raw ingredient for synthetic rubber and plastic. The term Total Petroleum Hydrocarbons refers to combinations of organic chemicals that are present in or generated from crude oil and have the highest potential for toxicity (Pang et al., 2022). It is composed of carbon and hydrogen, but it may also include heavy metals, nitrogen, sulfur, and oxygen molecules (Mishra et al., 2022). The majority of goods that include TPHs burn readily. Many of these items such as gasoline, kerosene, etc. have oily smells. Some are thick, dark liquids or semi-solids that do not evaporate, while others are transparent liquids that evaporate readily. Due to the extensive usage of petroleum-based goods in modern society such as fuel oil, mineral oil, kerosene, gasoline, and asphalt are environmental pollutants are pervasive (Iyer et al., 2022). The U.S. Environmental Protection Agency states that a combination of exposure and hazard data is necessary to assess the risk of petroleum leaks. The propensity of a given substance to adhere to soil particles and to enter vegetation through root uptake, which could lead to its entry into the food chain, is linked to the risk of exposure by skin contact or ingestion (Bejarano et al., 2023).

Impact on Aquatic Life

Aquatic life is greatly jammed by PHs. Oil spills alter marine habitats, affecting food webs and specie diversity. Aquatic life is long term risk from chemical make up of oil which includes PAHs (Shah & Soni, 2024). Petroleum hydrocarbons have negative impact on the quality of air, water and soil which in turn affects ecosystems and human populations. Even after remediation energies, the presence of TPH in contaminated areas continues to endanger biodiversity (Okafor et al., 2024). Fish, birds and marine mammals can suffer from oil spills which can cause both short term and long-term ecological harm (Horn et al., 2024). Fisheries and beach closures due to oil contamination can have an adverse effect on local economics and public health. Oil spills have socioeconomic repercussions that make thorough environmental assessments necessary to reduce threats to human health and livelihoods (Horn et al., 2024) (Fig. 31.2).

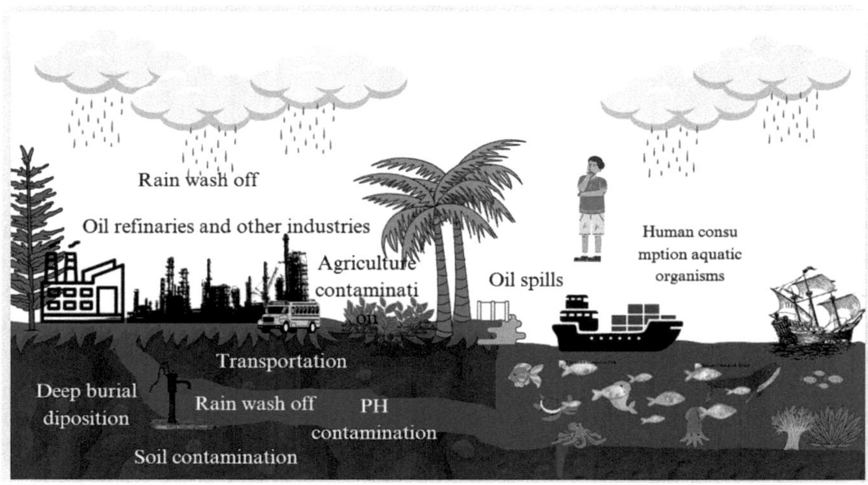

Fig. 31.2 Petroleum hydrocarbon impact in environment

Organic Amendments

In a variety of polluted situations, organic supplements are essential for improving biodegradation process. The organic amendment has the capability to increase the biodegradation because it is a natural material that added to soil to improve its physical, chemical and biological properties. It included compost, manure, green manure, peat moss, biochar. These materials enhance soil fertility by increasing organic matter, improving soil structure and promoting beneficial microbial activity. Organic Supplements like biochar, compost and biosolids are essential for improving biodegradation in polluted environments. These supplements enhance nutrient cycling and microbial activity which are crucial for the breakdown of pollutants (Omokore et al., 2024). Compost especially when combined with nitrogen rich material, boots soil microbial diversity, leading to faster biodegradation of pollutants like petroleum hydrocarbon. For example, optimized water hyacinth compost has been shown to biodegrade up to 93% of total petroleum hydrocarbon in 56 days (Udume et al., 2023). Biochar is another effective supplement improve soil quality, reduce greenhouse gas emission and increase nitrogen retention, all of which promote microbial growth and biodegradation. Biosolid-derived biochar has been shown to increase methane production during anaerobic digestion, indicating its potential to enhance anaerobic biodegradation (Hassan et al., 2023). The process of biodegradation can be enhancing through use of organic amendment such as compost, biochar. However, it is crucial to consider the specific context and type of contaminants involve as different amendments may influence pollutants differently depending on environmental condition (Hassan et al., 2023). Organic amendments like animal dung and compost add vital nutrients to the soil, increasing its overall fertility and microbial activity (Omokaro et al., 2024).

Types of Organic Amendments

Numerous organic amendments are added to soil, they can be broadly divided into groups (Fig. 31.3).

(a) **Animal manure**

Faeces, urine and animal beddings are combined to create farmyard manure which is heaped and flipped to go through a certain amount of composting. Solid or semi solid manure is produced when the piles are not turned and fresh material are frequently added from bran. Removal of bedding from stalls and transfer of manure and wash water to a holding tank were outcomes of efforts to lower labour expenses and enhance the hygiene in dairy brans. The collected slurry can be used on the ground (Tubeileh & Goss, 2022). Liquid manure has been created for poultry and other livestock. Pork unit produce the most manure by dairy and beef businesses. The output of manure from poultry production

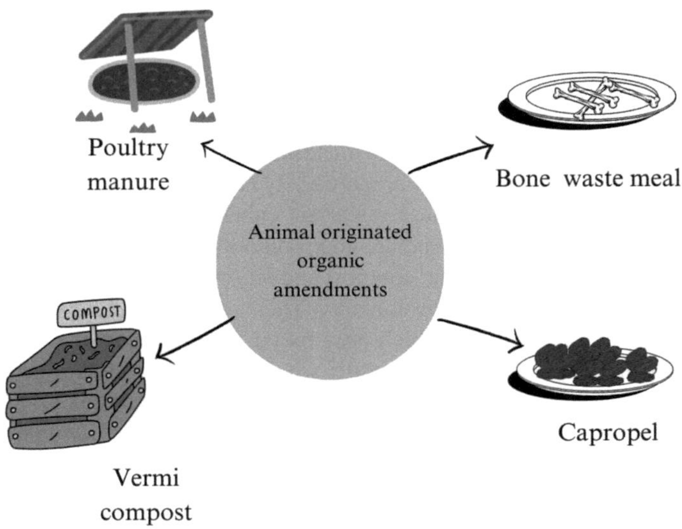

Fig. 31.3 Animal originated organic amendments

system broilers for meat and layers for eggs contributes far less than cattle or pigs (Yang et al., 2022).

The vast majority of manure produced world wide is applied to land. 16% is spread on grassland and 84% of manure collected from confined livestock is spread on crops. Globally, domestic animals are thought to emit approximately 112 Tg N annually. 42% from animals kept in captivity, 44% from animals grazing in the field and 14% is either expelled outside of the agricultural system or goes toward other purposes like fuel heating or animal feed (Tubeileh & Goss, 2022).

Manure can change during the process of excreting and mixing urine and faeces as well as during time of temporary stages in the bran and duration of longer-term storage before being applied to land. The mineralization of urea from urine starts extremely quick after expulsion even if it is not combined with faeces because enzyme urease is both common and rather robust (Yang et al., 2022). When organic matter in manure is broken down aerobically, CO_2 and more durable organic carbon compounds are produced. The development of anaerobic conditions within the stored manure is a consequence of stratification. The organic matter breaks down under aerobic conditions if solid manure is turned often and not packed tightly. In aerobic conditions the process proceeds more quickly than anaerobic ones and more carbon is released as CO_2 (Tubeileh & Goss, 2022).

(b) **Poultry manure**

Poultry manure is a valuable organic fertilizer, rich in both macro and micro nutrients. Its high organic nitrogen content makes it particularly useful for amending and improving soil fertility. Some research has shown that poultry manure not only enhances the physical, chemical, and biological properties of soil but also reduces the harmful effects of salinity on various plants. Furthermore, it can decrease the concentration of some trace metals in the soil (Celestina et al., 2019).

(c) **Municipal biosolids**

Agricultural land is frequently treated with the sludge from waste water treatment facilities. This is typically governed by regulations. Sedimentation separates the suspended organic material by general screening and grit removal and the bacteria present break down the readily metabolized fraction. Certain treatment facilities eliminate P and k from solutions (Popoola et al., 2023). Sewage sludge is not used in agriculture in every European nation between 1996 to 1998 France used 60% of its production on agricultural land while Spain and UK used 46%, Germany 40% and Italy 16%. Usually land application accounts for 50 to 70% of the sewage biosolids generated (Zaharioiu et al., 2021).

On site waste water treatment systems like septic systems include temporary holding tanks that must be pumped out on a regular basis in many rural regions without piped water or sewage systems. Some jurisdictions still permit the direct application of untreated material to land despite the fact that many mandate that the resulting septage be handled by waste water treatment plants (Popoola et al., 2023).

(d) **Green manure**

Green manure refers to specific plant that grow and the incorporated into the soil to improve its fertility and structure. Instead of being harvested these crops are left to decompose, enriching the soil with organic matter and essential nutrients. Green manure is natural and sustainable way to maintain soil heath, reduce erosion and enhance crop productivity. The residue from harvest, manure from animal production and other agricultural products have been applied to the field. Straw from grain production and stubbles ploughed after harvest are two examples. Straw can be added to soil utilized as animal feed or used as bedding for animals kept in cages. Garden waste is one of several waste streams that include plant nutrients as consequence of increased urbanization (Jeanneret et al., 2021).

Green manure is formed when any field or feed crop is mixed with soil while it is still green or shortly after it has flowered in order to increase the chemical and physical fertility of the soil (Laub et al., 2023). The area is used for summer green manure crop, usually harvest of the primary crop. In crop rotations, warm season cover crops can be employed to fill a gap, safe guard delicate soils, prepare land for perennial crop. The growing crop can provide benefits especially through the root systems but the desired effects of green manuring come from the inclusion of the shoots (Jeanneret et al., 2021).

(e) **Food residues**

Different source of material from metropolitan areas that has been applied to agricultural land usually after composting is unsalable fresh produce from supermarkets (Wu & Teng, 2022). More than 43 million tons of edible foods were lost in 1995 at the retail, foodservice and consumer ends of the supply chain according to U.S. survey. 6.6 Mt of grain goods and 7.2 Mt of vegetables were lost before consumption, accounting for 32 and 25% of the total supplied (Petljak et al., 2021).

(f) **Waste from manufacturing process**

Animal hair, feathers, dried blood, fish offal, hoof and horn meal, leftover organic material from pressing oil seeds and shoddy waste from making wool have all been used in this way Hartemink (2021). Every year, a significant amount of biosolid is formed as papermaking leftovers but very little of material is used on land. Other residues have been sown on farmed land or their potential to promote crop production examined including those left over after sugar is extracted from sugar beets, oil from olives and distillery waste (Doughmi et al., 2024).

(g) **Compost**

Composting is the most popular methods for stabilizing nitrogen in trash and enhancing its handling qualities especially by lowering the amount of organic material that needs to be applied to the ground. Up to 50% of carbon may be lost when solid manure is composted to create farmyard manure. Usually, trash is high in nitrogen is combined with material that has high C:N ratio. Composting produces a substance with a significantly lower C:N ratio than the initial mixture Wallman (2021).

Compost is a valuable source of organic matter (OM) and essential nutrients (N, P, K). it is commonly used as a soil amendment to improves soil fertility, enhance soil structure, boots microbial activity, reduces waste, lower soil PH in alkaline soil. Additionally, composting livestock manure transform it in to biofertilizer, removing heavy metals, pathogens and antibiotics (Wang et al., 2023). Vermicompost is one of the compost material, which is created by earthworms breaking down organic waste, is a nutrient-rich compost. It acts as a biosorbent, mitigating the harmful effect of salinity and toxic heavy metals like lead, cadmium, nickel, lithium, and chromium (Das & Mishra, 2008; Paul et al., 2023; Priyadarsini & Das, 2024, 2025). Earthworm activity during composting accelerates the breakdown of organic matter and the formation of, stable humus, improving soil structure. This leads to increased pore space; better water infiltration and retention also boost in beneficial soil microbes and organic carbon. Ultimately, vermicompost improved plant growth, higher yields and fruit quality (Liu et al., 2019).

(h) **Biochar**

Biochar, a carbon rich of biomass pyrolysis, is gaining recognition for its positive economic and environmental effect on soil and crops. While its composition is primarily carbon (around 70%), hydrogen, sulphur, oxygen, nitrogen and the minerals based on the feedstock used to make it. It improves soil heath by amend pH, increasing carbon exchange capacity (CEC) and generally increasing fertility and also improve soil aeration & porosity (Mandal et al., 2021).

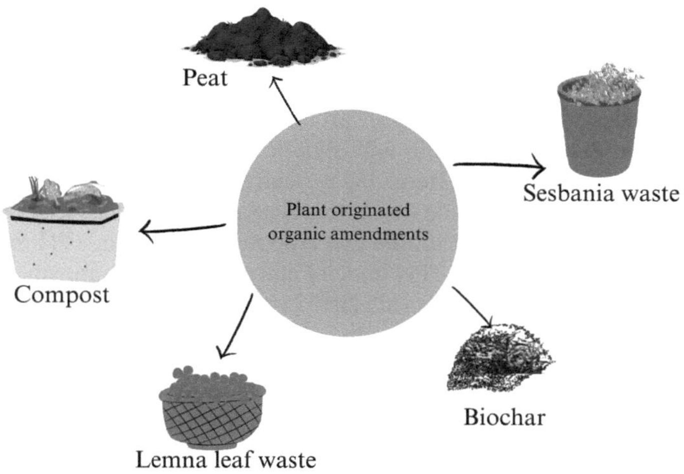

Fig. 31.4 Plant originated organic amendments

Biochar fosters a healthy rhizosphere, promoting beneficial microbial growth and enzyme activity also. The presence of hydroxides and carbonates further contributes to improve soil productivity. Finally, biochar especially bamboo charcoal, effectively adsorbs and stabilizes heavy metal like lead and cadmium in contaminated soils (Rasa et al., 2018) (Fig. 31.4).

Organic Amendments Using Biodegradation of Contaminated Soil

Material formed from biomass or live organisms that can improve crop production and soil health are referred as organic amendments. Using organic amendments in agriculture is wide spread because of their distinct surface and textural characteristics, chemical makeup and functional groups. Environmental remediation has been expanded for past ten years. Research and development in this field has been rapidly accelerating (Gómez-Sagasti et al., 2018).

A wide range of organic amendments such as crop residues, green manure, biosolids and biochar can be used in the remediation of soil TPH contamination because of their beneficial effects on soil health and plant growth. The amendments not only favourable alter the physical, chemical and biological properties of the soil but also provide vital nutrients for improving plant growth (Rajapaksha et al., 2015).

Minimizing TPH bioavailability through chemical and biological immobilization in soils can help remediates TPH contaminated soil by lowering TPH exposure to receptors.

(a) **Bio-wastes**

Crop residues, green manures biosolids and yard waste composts are examples of many easily accessible organic amendments that make up bio wastes. The TPH contaminated soil has been cleaned up using amendments. 3.8 billion mg of crop residues and green manures are produced each year with right management. It might provide damaged soils with significant amount of carbon and nutrients (Thangarajan et al., 2013). Restoring agricultural leftovers to soil contaminated with TPH may be sustainable way to fertilize soils with nitrogen. Several studies have documented the critical role that crop residues play in TPH remediation. The amendments reduced TPH toxicity and promoted the growth of microorganisms and plants. (Medina et al., 2015). Animal dung can be recycled in large quantities for environmental remediations. 12 million tons of animal waste are produced every day worldwide. The presence of humic and fulvic substances are most active chemicals in animal manure, influences not only the soil properties but also TPH bioavailability to providing vital nutrients including N, P, K, S and micronutrients (Ezenne et al., 2014).

Animal manure added to soil contaminated with TPH increased the amount of dissolved organic matter which was explained by clay component of soil having lower sorption capability. Aliphatic hydrocarbons and PAHs are among several TPH compounds for which DOM is efficient desorbing agent (Kim et al., 2022). Meat made from animal byproducts have been demonstrated to be source of nutrients and carbon that can be utilized for TPH remediation. Animal types, manure collection, storage and handling are some production aspects that affect the content, nutritional value and utility of animal dung. These considerations may cause them to have different effects on soil (Olawale et al., 2020). When TPH contaminated soil was treated with cow dung and pig waste, the rate of TPH reduction was lower than the poultry manure. The varying C:N:P ratios in the amendments which had varying on hydrocarbon degraders in affected soil (Akpe et al., 2015).

Compost, paper mill wastes and biosolids are organic amendments used to remediate soil affected by TPH pollutants. The amendments inclusion of humic compounds provides workable method for cleaning up soils contaminated with TPH (Dindar et al., 2017). TPH contaminated soil added to biosolids can change the bioavailability of TPH and increase microbial activity to help the soil get better. TPH biodegradation was promoted by use of digestates from manufacture of biogas. Enhancing soil characterization and microbial activity waste from sugarcane and paper industries promoted the biodegradation of TPH pollutants (Babaei et al., 2020). In TPH contaminated soil, a range of organic wastes, including leaves, food scarps, cow dung and sewage sludge can be composted and utilized. Compost is frequently associated with changed TPH availability. Composting can raise amount of solid material in polluted soil, which lowers the bioavailability of TPH. Humic materials in compost are cation exchange capacity comparable to clay minerals in soil which helps the soil retain nutrients (Kim et al., 2017).

(b) **Biochar**

Biochar is significant organic amendments that can be applied to TPH cleanup. Biochar is a porous, carbon rich substance that is created when different organic wastes are pyrolyzed in oxygen limited environments. High surface area, functional group and nano micro porosity have been shown to have remediation potential (Li et al., 2019). Biochar has been used extensively as TPH adsorbent in contaminated soil because of its porous qualities and potent adsorption capacity. Biochar can change the physiochemical characteristics of soil and provide lots of vital nutrients to support plant development and TPH microbial breakdown. Biochar added to polluted soil is seen to be successful way to clean up TPH affected soil (Mekontchou et al., 2024).

Biochar made from 4 feedstocks sewage sludge, soybean straw, rice straw and peanut shell were applied to PAH contaminated urban soil at 2% and 5% at 500^0C. With the greatest effectiveness at 5% biochar application, the bioaccessible PAHs in soil and their bioaccumulation in turnip were greatly decreased by the use of biochar. Maximum aromatic functional groups and least polar surface, peanut shell biochar was the most successful organic amendment at lowering bioaccessible PAHs (Khan et al., 2015). The number of dissolved PAHs was reduced by 25% and 22% respectively. When soil was treated with wheat straw biochar that was created at 650°C at a rate of 30 and 45 tons. There was no possible harm associated with the number of dissolved PAHs in the soil treated with biochar. The strength of biochar- PAH interactions tended to diminish over time (Qin et al., 2013). 2% of biochar decreased TPH toxicity and bioavailability which in turn stimulated the microbial community and led to TPH breakdown. When sawdust and wheat straw biochar were added to TPH polluted soil, it was shown that the biochar improved the habitat for PAH degraders, spreading up the biodegradation of PAHs (Kong et al., 2018).

(c) **Vermicompost**

Vermicompost Enhanced microbial activity and nutrient availability, which, when paired with biostimulation and bioaugmentation, result in TPH breakdown rates of 32.5% and 34.4%, respectively (Curiel-Alegre et al., 2022). Horse Manure outperformed conventional fertilizers, reducing hydrocarbons significantly and releasing less CO_2 (Robichaud et al., 2019). Microbiological Inoculants Aliphatic hydrocarbons were reduced by 86.8% as a result of bioaugmentation using a mixed culture, which was noticeably better than nutrient supplements (Brzeszcz et al., 2020). The overall success of remediation may be limited by bioavailability, as certain refractory substances remain resistant to treatment even when organic additions can improve hydrocarbon breakdown (Curiel-Alegre et al., 2022) (Table 31.1).

Table 31.1 Increased immobilization of total petroleum hydrocarbon using organic amendments

S.L. No.	Organic amendments	Petroleum hydrocarbon compounds	Observations	Reference
1	Poultry manure biochar	TPH	Biochar increased microbial respiration in the rhizosphere of barley and oat	Barati et al. (2017)
2	Spent mushroom compost	PAHs	It is enhanced the development of guinea grass in black oil hydrocarbon polluted soil Combination of organic amendment and plant showed improved remediation efficiency	Asemoloye et al. (2017)
3	Fruit waste biochar, sewage sludge biochar and cow dung	TPH	Biochar significantly enhanced the soil microbial abundance and diversity. Application of biochar augmented with cow dung showed decline in soil TPH concentration	Agarry et al. (2014)
4	Pea straw	TPH	Peat straw treated soil showed increase in the numbers of 16 s RNA and other genes	Koshlaf et al. (2020)
5	Cotton stalk	TPH	Oil sludge contaminated soil treated with cotton stalk showed increased removal efficiency of TPH	Wang et al. (2016)
6	Compost	PAH s	Increased abundance of catabolic enzymes observed in compost assisted bioremediation. Compost is higher removal of PAHs	Wang et al. (2019)
7	Grass straw compost	TPH	Effective in supply nutrients for the degradation of TPH	Nwankwegu et al. (2016)
8	Sawdust	TPH	Sawdust addition at 5% increased microbial activity and degradation rate of TPH	Alvarenga et al. (2017)
9	Wheat straw	TPH	Adding wheat straw to crude oil contaminate d soil enhanced soil microbial respiration	Alotaibi et al. (2018)
10	Rice straw	TPH and PAH	The abundance of hydrocarbon degrading taxa enhanced after rice straw addition leading to significant decline in TPH and PAHs	Huang et al. (2019)

Implementation

Petroleum hydrocarbons, ubiquitous in modern society due to their use in fuels, lubricants and various industrial processes, pose a significant threat to soil health when released into the environment. These contaminants, composed of complex mixtures of aliphatic and aromatic hydrocarbons, can persist in soil for extended periods, causing detrimental effects on soil fertility, microbial communities, and potentially impacting human health through food chain contamination. Petroleum hydrocarbons enter the soil through various pathways, including accidental spills during transportation and storage, leakage from underground storage tanks, and improper disposal of industrial waste. Once in the soil, these compounds can penetrate through the soil profile, contaminating groundwater and potentially affecting surface water bodies. The toxicity of petroleum hydrocarbons varies depending on their composition, with lighter fractions like benzene and toluene being more volatile and acutely toxic, while heavier fractions can persist and accumulate in the soil, posing long-term risks. The presence of petroleum hydrocarbons in soil can disrupt soil structure, reduce water holding capacity, and inhibit plant growth, and potentially lead to the contamination of groundwater and the food chain (Khudur et al., 2019). These contaminants can also negatively impact soil microbial communities, which play a crucial role in nutrient cycling and organic matter decomposition. Furthermore, the accumulation of petroleum hydrocarbons in soil can lead to the contamination of crops grown on the affected land, potentially posing health risks to humans and animals consuming these crops. Traditional remediation methods, such as excavation and incineration, can be costly and disruptive to the environment (Koshlaf et al., 2024). Consequently, there is a growing interest in sustainable and cost-effective alternatives, among which the application of organic amendments has emerged as a promising approach.

Beyond nutrient provision, organic amendments can also improve the physical structure of the soil (Pan et al., 2024). PHC contamination can often lead to soil compaction and reduced porosity, limiting the availability of oxygen and water, both of which are crucial for microbial activity (Almaghribi et al., 2022). The addition of organic matter can improve soil aggregation, increase porosity, and enhance water infiltration and retention. This creates a more favourable environment for microbial growth and facilitates the transport of oxygen and nutrients to the microorganisms involved in PHC degradation. Furthermore, the improved soil structure can also enhance the accessibility of PHCs to microorganisms, making then more susceptible to looser soil structure, allowing for better aeration and water movement, which in turn promotes the growth of aerobic hydrocarbon-degrading bacteria.

Another important mechanism by which organic amendments contribute to PHC remediation is through the enhancement of soil buffering capacity. PHC contamination can sometimes lead to changes in soil pH, which can negatively impact microbial activity. Organic matter possesses a high cation exchange capacity (CEC) and can act as a buffer, helping to maintain a stable pH range that is optimal for microbial growth and enzyme activity. This buffering capacity is particularly important in soils with low initial buffering capacity or in cases where PHC degradation leads to the production of acidic byproducts. By maintaining a stable pH, organic amendments create a more conducive environment for the microorganisms to function effectively and degrade the PHCs.

Furthermore, some organic amendments can directly interact with PHCs, influencing their bioavailability and fate (Henderson et al., 2023). For instance, certain types of organic matter, such as humic substances, can form complexes with PHCs, reducing their toxicity and making them more accessible to microbial degradation

(Hudáková et al., 2023). This process, known as solubilization or co-metabolism, can enhance the overall rate of PHC removal. Humic substances can also act as electron shuttles, facilitating the transfer of electrons during microbial respiration and potentially accelerating the degradation of certain types of PHCs (Kulikova & Perminova, 2021). Additionally, the presence of organic matter can influence the sorption and desorption of PHCs in the soil, affecting their mobility and potential for leaching.

Certain microbial populations that aid in the biodegradation of different contaminants, especially polycyclic aromatic hydrocarbons (PAHs), are greatly enhanced by organic additions. Chrysene degradation rates increased by 27.20% as a result of the application of kitchen compost (KC) and its dissolved organic matter (KCOM), which enhanced the quantity of bacteria that break down chrysene, such as Bacillus and Pseudomonas (Kou et al., 2024). In contaminated soils, co-cultures of Streptococcus pyogenes and Enterococcus faecalis supplemented with compost shown notable decreases in total petroleum hydrocarbons (TPH) (Olarewaju Adeleye et al., 2022). The microbial communities in soil, particularly those engaged in biodegradation progressions are greatly improved by organic accompaniments. Enhancing soil health, these amendments which include compost cow dung and processed cocoa pod husk also promote the development of particular microbial populations are essential for decaying organic contaminants (Omokaro et al., 2024). These improve enzymatic activity and nutrient cycling by increasing microbial biomass and diversity. Study indicates that compost encourage the growth of organic decomposition related species, essential for biodegradation (Kraut-Cohen et al., 2023).

Compost shows the highest efficiency among organic amendments in dropping PAH and TPH in contaminated soils (Olarewaju Adeleye et al., 2022). When organic materials are added, vital minerals like potassium and phosphorous become more readily obtainable. Organic amendments help progress microbial communities and biodegradation process has limits (Curiel-Alegre et al., 2022). Petroleum hydrocarbons are competently broken down by the major hydrocarbon degrading bacteria and fungi which are essential for bioremediation. Particular microbial species with notable hydrocarbon degrading capacities, emphasizing their potential for environmental purification. Flavus Aspergillus is essential for breakdown of hydrocarbons (Shakam & Gaber, 2024). Fungi that have been identified as playing important role in the breakdown of hydrocarbons include Rhizopus species and Trichoderma harzianum (Shakam & Gaber, 2024). Higher bacterial growth and hydrocarbon degradation efficiency are correlated with increased enzymatic activity such as alkaline phosphatases and proteases can degrade TPHs up to 34.4%. Organic embellishments improve enzyme activity such as β-glucosidase and dehydrogenase are makers of microbial carbon utilization efficiency. Investigate indicates that compost considerably increases enzyme activity in comparison to controls promoting improved hydrocarbon breakdown (Nava-Arsola et al., 2024). Water hyacinth compost and vermicompost are instances of organic amendments that greatly improve the availability of minerals like potassium and phosphorous (Udume et al., 2023). Vermicompost increased the amount of phosphorous accessible fivefold, boosting microbial activity and enzymatic processes required for hydrocarbon breakdown (Curiel-Alegre et al.,

2022). The pH and conductivity of soil can be organic amendments, fostering microbial activity. Compost kept its pH more alkaline which encouraged microorganisms (Olarewaju Adeleye et al., 2022).

The breakdown of pollutants in soil can be greatly accelerated by combining organic amendments with bioremediation techniques like phytoremediation and bioaugmentation (Curiel-Alegre et al., 2022). Compost and biochar have been shown to improve soil quality and heavy metal uptake when combined with plants like sugar beet improving phytoremediation (Hussain et al., 2022). Developed from natural sources including coconut and soyabean oil, bio clean surf is biodegradable surfactant that efficiently combines and dissolves oil. Environmental safety surfactant does not harm marine life. It can used in delicate environment (Alshamsi & Hussain, 2024). Rapid remediation capabilities were demonstrated by an electrochemical approach that reduced TPH in polluted soil by 55% in 14 days. Oil spill countermeasures have been improved using machine learning model which has 93.1% accuracy rate in effective remediation techniques (Gbogbo et al., 2024).

Mechanisms of Remediation

The remediation of petroleum hydrocarbon-contaminated soils through the application of organic amendments involves several key mechanisms.

a. **Biostimulation**—Organic amendments provide a source of carbon and nutrients for indigenous microorganisms, stimulating their growth and activity. These microorganisms can utilize petroleum hydrocarbons as a source of energy and carbon, breaking them down into less harmful substances like carbon dioxide and water (Behera et al., 2022).
b. **Bioaugmentation**—In some cases, specific microorganisms with the ability to degrade petroleum hydrocarbons may be introduced to the contaminated soil along with organic amendment. This process, particularly in soils where indigenous microbial populations are limited or have been negatively impacted by the contaminants (Behera et al., 2022).
c. **Sorption and Complexation**—Organic amendments can increase the soil & capacity to sorb and complex petroleum hydrocarbons, reducing their bioavailability and toxicity. This process can involve the binding of hydrocarbons to organic matter, clay minerals, or other soil constituents, making the less accessible to plants and microorganisms (Shakam et al., 2024).
d. **Phytoremediation**—The use of plants remediates contaminated soils, known as phytoremediation, can be enhanced by the application of organic amendments. Organic matter can improve soil structure and fertility, creating a more suitable environment for plant growth. Plant roots can then take up and accumulate petroleum hydrocarbons, or they can stimulate microbial activity in the rhizosphere, leading to enhanced degradation of contaminants (Omokaro et al., 2024, Mohanta et al., 2023) (Table 31.2).

Table 31.2 Biodegradation with the application of *Bacillus sp.*, *Pseudomonas sp.* and *Rhodococcus sp.*

S.L. No.	Genus	Species/Strain	Type of contaminant	Removal efficiency (RE), environment, co-cultures	Reference
1	*Bacillus*	*B. megaterium* and *B. safensis*	Chlorpyrifos (CLP)	REin solution: 99% of CLP in a solution with an initial concentration of 10 mg/L after 60 days. RE in soil: 61–65% after 100 days	Lara-Moreno et al. (2022)
2		*B. subtilis, B. licheniformis, B. megaterium, B. cereus*	Nitrogen in WW	RE: 92% of nitrate at the laboratory and 62% outdoors. A plug flow system; *Arthrobacter sp., Acinetobacter parafneus, Corynebacterium sp.*, and *Streptomyces globisporus*	Fontalvo et al. (2022)
3		*B. firmus*	Phenanthrene	RE: 1.73 mg/kg soil day for 56 days. Soil, anaerobic nitrate-reducing environment	Zhou et al. (2022)
4		*Bacillus B. megaterium*	Nitrogen in WW	RE: 62.76% after 29 d. Secondary salinized soil, pot experiment. Improved NO_3 removal rate	You et al. (2022)
5		*B. paramycoides*	Sulfamethoxazole (SMX) and Zn^{2+}	Biochar-immobilized. After five rounds of reuse, RE: 43.24% for SMX and 50.34% for Zn^{2+}	Chen et al. (2022)
6		*Bacillus sp.*	Decachlorobiphenyl	RE: up to 10.51% after 21 days. Initial conc. 200 mg/kg soil. *Staphylococcus sp.* and *Acinetobacter sp.* Consortia better than individual strains	Zenteno-Rojas et al. (2022)

(continued)

Table 31.2 (continued)

S.L. No.	Genus	Species/Strain	Type of contaminant	Removal efficiency (RE), environment, co-cultures	Reference
7		*B. safensis*	Secondary composting	Indoleacetic acid (IAA)-production, assimilation of soluble salt, condensation and aromatization of humus, accumulation of dissolved organic nitrogen and carbon. *Corynebacterium stationis subsp. safensis*	Li et al. (2022)
8		*Bacillus spp.* and *B. aryabhattai*	Aquaculture WW	Algal–bacterial bioflocs and microbe–rice bran complexes, *Scenedesmus dimorphus* and *Chlorella sp.* 1:1	Pekkoh et al. (2022)
9	*Pseudomonas*	*Pseudomonas sp.*	Fluoride	RE: 77.54%, 99.39%, and 67.25% for F^-, NO^{3-}, and Ca^{2+}, respectively, for 8 h. Microbially induced calcium precipitation (MICP). Self-assembled fungus-flexible fibre composite microspheres. *Phomasp.*	Wang et al. (2022a, 2022b)
10		*Pseudomonassp.*	Benzene, toluene, ethylbenzene, and p-xylene(BTEX)	RE: 20 mg/Linamineral salt medium for 6 h. *Vario voraxparadoxus*	Szentgyörgyi et al. (2022)

(continued)

Table 31.2 (continued)

S.L. No.	Genus	Species/Strain	Type of contaminant	Removal efficiency (RE), environment, co-cultures	Reference
11		*P. songnenensis*	Oil-contaminated desert soil	RE: 73.6%, 69.3%, 50% and 50%in soils polluted with 1%, 10%, 20% and 30% oil, respectively, after 6 months. *Actinotalea ferrariae, Arthro bactergin sengisoli, Dietzia cinnamea, Dietzia papillomatosis*	Ali et al. (2022)
12		*P. plecoglossicida*	Crudeoil	76.7% ability to degrade crude oil in a liquid broth. 14 days	Ntroumpogianni et al. (2022)
13		*Pseudomonassp.*	Sulphide	RE: two-fold increase of elemental sulfur generation. Establishment of a stronger biofilm structure. Granular sludge bed reactor. *Arcobacter, Azoarcus*	Liu et al. (2022)
14		*P. stutzeri*	Glyphosate	RE: 53% and 79% for synthetic and real WW with 5mg/ Lglyphosate. Continuous photo-bioreactors. Acclimation to glyphosate from 5 to 50 mgL^{-1}. Comamonas odonto termitis; *Sinomona satrocyanea, Chlorella protothecoides*	Borella et al. (2023)

(continued)

Table 31.2 (continued)

S.L. No.	Genus	Species/Strain	Type of contaminant	Removal efficiency (RE), environment, co-cultures	Reference
15		*P. mendocina, P. putida Nitrogenremovalin*	Petroleumww	RE: 92.4% of NH_4^+-N; 79.8%of total nitrogen. Heterotrophic nitrification, aerobic denitrification. *Brucellasp., Paracoccussp.*	Wang et al. (2022a, 2022b)
16	*Rhodococcus*	*R. erythropolis* (Genetically engineered expressingNirs and AMO (rRho-NM).)	Nanofiltration concentrateof land fill leachate	RE: total organic carbon 63–89%; NH_4–N_{72}–82%; total nitrogen 63–87%; chemical oxygen demand 81–95%, dependent to treatment mode. Integrated system of advanced oxidation processes cooperated with rRho-NM. Bio augmentation to the aerobic fluidized reactor (2 L), inoculum106/mL	Bai et al. (2023)
17		*R. ruber*	Quinoline and its toxic intermediate 2-hydroxyl quinoline (2-HQ)	RE: for quinoline,100% for 1.5 h. For 2-HQ, max. Appearance of 0.025 mM within 0.5 h, 100% degradation after 1 h. Aerobic 100 mL batch, 15% inoculum (v/v), initial concentration of quinoline of 0.25 mM. Accelerated nitrification and enriched Nitrospira	Zhu et al. (2023)

(continued)

Table 31.2 (continued)

S.L. No.	Genus	Species/Strain	Type of contaminant	Removal efficiency (RE), environment, co-cultures	Reference
18		*R. pyridinivorans*	Di-(2 ethylhexyl)phthalate (DEHP)	RE: 89.94% of DEHP within 84 h. Initial DEHPconc.5 mg/ LDEHP in 10 mL municipal WW (batch). Aerobic denitrifying phosphate-accumulating bacterial strain RL-GZ01	Ren et al. (2023)
19		*R. ruber*	Chloroxylenol (PCMX)	RE: 97% during 22–30 cycles. 25 mg/LPCMX within 180 min inactivated sludge system. Aerobic 100 mL batch, 4% inoculum (v/v). *R. ruber* became dominant within 30 cycles. Detoxify nitrification system	Guo et al. (2023)
20		*Rhodococcus sp.*	3-Methylindole (skatole)	RE: > 99% for 24 h. Initial skatoleconc. 60 mg/L. Degradation performance in consortium. Inoculum 2%	Ma et al. (2023)
21		*R. biphenylivorans*	Biphenyl and polychlorinated biphenyl (PCB)31	RE: biphenyl l96% within 5 days, PCB 3192% for 3 days. Initialconc. Biphenyll 500 mg/L, aerobic 5 mL batch cultivation, pH 7.0	Li et al. (2023)

Conclusion

The proficiency of organic amendments in indorsing the biodegradation of petroleum hydrocarbons in polluted soils is verified by this study. According to the findings, microbial activity is greatly enhanced by compost, dung and biochar which speeds up the breakdown of hydrocarbons. Compost was the most effective of evaluated amendments as a result of its high nutritional content and microbial variety. The breakdown of petroleum pollutants was greatly assisted by the presence of fungi and bacteria that break hydrocarbons and the prominence the significance of microbial interaction in bioremediation methods. According to the study, using organic amendments to clean up petroleum-contaminated areas is a sustainable and environmentally responsible solution. Degradation efficiency and environmental sustainability may be improved by combining organic amendments with techniques as phytoremediation and bioaugmentation. All things considered, this study supports the possibility of using organic amendments as a practical way to clean up petroleum hydrocarbon-polluted soils, providing a greener substitute for traditional remediation way.

Acknowledgements The authors are thankful to the OURIIP project GOVT. Of Odisha project ID OURIIP (22SF/CE/022) for granting the project and providing financial support.

References

Agarry, S. E., Aremu, M. O., & Aworanti, O. A. (2014). Biostimulation and phytoremediation treatment strategies of gasoline-nickel co-contaminated soil. *Soil and Sediment Contamination: An International Journal, 23*(3), 227–244.

Agbor, N. E., Esemu, S. N., Ndip, L. M., Tanih, N. F., Smith, S. I., & Ndip, R. N. (2018). Helicobacter pylori in patients with gastritis in West Cameroon: Prevalence and risk factors for infection. *BMC Research Notes, 11*, 1–6. https://doi.org/10.1186/s13104-018-3662-5

Akpe, A. R., Ekundayo, A. O., Aigere, S. P., & Okwu, G. I. (2015). Bacterial degradation of petroleum hydrocarbons in crude oil polluted soil amended with cassava peels. *American Journal of Research Communication, 3*(7), 99–118.

Al-Areeq, N. M. (2018). Petroleum source rocks characterization and hydrocarbon generation. *Recent Insights in Petroleum Science and Engineering, 1*. https://doi.org/10.5772/intechopen.70092

Ali, N., Khanafer, M., & Al-Awadhi, H. (2022). Indigenous oil-degrading bacteria more efficient in soil bioremediation than microbial consortium and active even in super oil-saturated soils. *Frontiers in Microbiology, 13*, 950051. https://doi.org/10.3389/fmicb.2022.950051

Alotaibi, H. S., Usman, A. R., Abduljabbar, A. S., Ok, Y. S., Al-Faraj, A. I., Sallam, A. S., & Al-Wabel, M. I. (2018). Carbon mineralization and biochemical effects of short-term wheat straw in crude oil contaminated sandy soil. *Applied Geochemistry, 88*, 276–287.

Alshamsi, S., & Hussain, K. (2024, November). Development of biodegradable surfactants for oil spill remediation. In *Abu Dhabi international petroleum exhibition and conference* (p. D031S115R001). SPE. https://doi.org/10.2118/223033-ms

Alvarenga, P., Palma, P., Mourinha, C., Farto, M., Dôres, J., Patanita, M., Cunha-Queda, C., Natal-da-Luz, T., Renaud, M., & Sousa, J. P. (2017). Recycling organic wastes to agricultural land as a way to improve its quality: a field study to evaluate benefits and risks. *Waste Management, 61*, 582–592.

Anayet, A. H., Hamzah, M. M. H. B., & Najib, M. Z. M. (2024). Optimizing food waste decomposition through ph, moisture content, and temperature control: A comprehensive study. *Civil and Sustainable Urban Engineering, 4*(1), 42–54. https://doi.org/10.53623/csue.v4i1.411

Asemoloye, M. D., Jonathan, S. G., Jayeola, A. A., & Ahmad, R. (2017). Mediational influence of spent mushroom compost on phytoremediation of black-oil hydrocarbon polluted soil and response of *Megathyrsus maximus* Jacq. *Journal of Environmental Management, 200*, 253–262.

Babaei, A. A., Safdari, F., Alavi, N., Bakhshoodeh, R., Motamedi, H., & Paydary, P. (2020). Cocomposting of oil-based drilling cuttings by bagasse. *Bioprocess and Biosystems Engineering, 43*, 1–12.

Bacosa, H., Suto, K., & Inoue, C. (2010). Preferential degradation of aromatic hydrocarbons in kerosene by a microbial consortium. *International Biodeterioration & Biodegradation, 64*(8), 702–710.

Bai, F., Tian, H., Wang, C., & Ma, J. (2023). Treatment of nanofiltration concentrate of landfill leachate using advanced oxidation processes incorporated with bioaugmentation. *Environmental Pollution, 318*, 120827. https://doi.org/10.1016/j.envpol.2022.120827

Barati, M., Bakhtiari, F., Mowla, D., & Safarzadeh, S. (2017). Total petroleum hydrocarbon degradation in contaminated soil as affected by plants growth and biochar. *Environmental Earth Sciences, 76*(20), 688.

Behera, I. D., Basak, G., Kumar, R. R., Sen, R., & Meikap, B. C. (2020). Treatment of petroleum refinery sludge by petroleum degrading bacterium *Stenotrophomonas pavanii* IRB19 as an efficient novel technology. *Journal of Environmental Science and Health Part A.*, 1–13. https://doi.org/10.1080/10934529.2020.1866924

Behera, I. D., Nayak, M., Biswas, S., Meikap, B. C., & Sen, R. (2021). Enhanced biodegradation of total petroleum hydrocarbons by implementing a novel two-step bioaugmentation strategy using indigenous bacterial consortium. *Journal of Environmental Management, 292*, 112746. https://doi.org/10.1016/J.JENVMAN.2021.112746

Behera, I. D., Nayak, M., Mishra, A., Meikap, B. C., & Sen, R. (2022). Strategic implementation of integrated bioaugmentation and biostimulation for efficient mitigation of petroleum hydrocarbon pollutants from terrestrial and aquatic environment. *Marine Pollution Bulletin, 177*, 113492.

Bejarano, A. C., Adams, J. E., McDowell, J., Parkerton, T. F., & Hanson, M. L. (2023). Recommendations for improving the reporting and communication of aquatic toxicity studies for oil spill planning, response, and environmental assessment. *Aquatic Toxicology, 255*, 106391.

Borella, L., Novello, G., Gasparotto, M., Renella, G., Roverso, M., Bogialli, S., Fillipinni, F., & Sforza, E. (2023). Design and experimental validation of an optimized microalgae-bacteria consortium for the bioremediation of glyphosate in continuous photobioreactors. *Journal of Hazardous Materials, 441*, 129921. https://doi.org/10.1016/j.jhazmat.2022.129921

Brzeszcz, J., Kapusta, P., Steliga, T., & Turkiewicz, A. (2020). Hydrocarbon removal by two differently developed microbial inoculants and comparing their actions with biostimulation treatment. *Molecules, 25*(3), 661. https://doi.org/10.3390/molecules25030661

Celestina, C., Hunt, J. R., Sale, P. W., & Franks, A. E. (2019). Attribution of crop yield responses to application of organic amendments: A critical review. *Soil and Tillage Research, 186*, 135–145.

Chen, X., Lin, H., Dong, Y., Li, B., Liu, C., & Yin, T. (2022). Mechanisms underlying enhanced bioremediation of sulfamethoxazole and zinc (II) by *Bacillus sp.* SDB4 immobilized on biochar. *Journal of Cleaner Production, 370*, 133483. https://doi.org/10.1016/j.jclepro.2022.133483

Das, A. P., Mallavarapu, M., & Ghosh, S. (2023). Ecotoxicity of mining pollutants on the environment and their remediation. *Environmental Chemistry and Ecotoxicology, 5*, 165–167. https://doi.org/10.1016/J.ENCECO.2023.08.002

Das, A. P., & Mishra, S. (2008). Hexavalent Chromium (VI): Health hazards and environmental pollutant. *The Journal of Environmental Research and Development, 2*(3), 386–392.

Dey, S., Das, A., Mallick, K., Sahu, A., & Das, A. P. (2023). *Environmental petroleum waste: Pollution, toxicity, sustainable remediation* (pp. 159–175). https://doi.org/10.1007/978-3-031-48220-5_7

Dey, S., Bissoyi, A., Koteswara Reddy, G., & Das, A. P. (2024). Synthetic microfibers: An urban emerging pollutant demanding for sustainable environmental solutions. In *Sustainable microbial technology for synthetic and cellulosic microfiber bioremediation* (pp. 107–121). https://doi.org/10.1007/978-3-031-62600-5_7

Dindar, E., Şağban, F. O. T., & Başkaya, H. S. (2017). Biodegradation of crude oil-contaminated soil using canned-food-industry wastewater sludge for soil application. *KSCE Journal of Civil Engineering, 21*(5), 1623–1630.

Doughmi, A., Elkafz, G., Cherkaoui, E., Khamar, M., Nounah, A., & Zouahri, A. (2024). Evaluation of the compost's maturity of different mixtures of olive pomace and poultry manure. *Ecological Engineering & Environmental Technology, 25.*

El-Saeid, M. H., & Usman, A. R. (2022). Influence of organic amendments and moisture regime on soil CO_2–C efflux and polycyclic aromatic hydrocarbons (PAHs) degradation. *Sustainability, 14*(7), 4116. https://doi.org/10.3390/su14074116

Eze, M. O., & George, S. C. (2020). Ethanol-blended petroleum fuels: Implications of co-solvency for phytotechnologies. *RSC Advances, 10*(11), 6473–6481.

Ezenne, G. I., Nwoke, O. A., Ezikpe, D. E., Obalum, S. E., & Ugwuishiwu, B. O. (2014). Use of poultry droppings for remediation of crude-oil-polluted soils: Effects of application rate on total and poly-aromatic hydrocarbon concentrations. *International Biodeterioration & Biodegradation, 92*, 57–65.

Fontalvo, N. P. M., Gamero, W. B. M., Ardila, H. A. M., Gonzalez, A. F. P., Ramos, C. G., & Muñoz, A. E. P. (2022). Removal of nitrogenous compounds from municipal wastewater using a bacterial consortium: An opportunity for more sustainable water treatments. *Water, Air, & Soil Pollution, 233*(8), 339. https://doi.org/10.1007/s11270-022-05754-y

Gómez-Sagasti, M. T., Hernández, A., Artetxe, U., Garbisu, C., & Becerril, J. M. (2018). How valuable are organic amendments as tools for the phytomanagement of degraded soils? The knowns, known unknowns, and unknowns. *Frontiers in Sustainable Food Systems, 2*, 68.

Gorlenko, N. V., & Timofeeva, S. S. (2019). Assessment of environmental damage from oil sludge to land resources in the Irkutsk region. In *IOP conference series: Earth and environmental science* (Vol. 229, No. 1, p. 012019). IOP Publishing.

Guo, Y., Gao, J., Zhao, Y., Liu, Y., Zhao, M., & Li, Z. (2023). Mitigating the inhibition of antibacterial agent chloroxylenol on nitrification system-The role of *Rhodococcus ruber* in a bioaugmentation system. *Journal of Hazardous Materials, 447*, 130758. https://doi.org/10.1016/j.jhazmat.2023.130758

Hartemink, A. E. (2021). *Soil science Americana: Chronicles and progressions 1860–1960.* Springer Nature.

Hassan, S., Ngo, T. T. M., Khudur, L. S., Krohn, C., Dike, C. C., Hakeem, I., Shah, K., Surapaneni, A., & Ball, A. S. (2023). The impact of biosolids biochar on biomethane production from the anaerobic digestion of chicken manure. Available at SSRN 4414724. https://doi.org/10.2139/ssrn.4414724

Hoang, S. A., Lamb, D., Seshadri, B., Sarkar, B., Choppala, G., Kirkham, M. B., & Bolan, N. S. (2021). Rhizoremediation as a green technology for the remediation of petroleum hydrocarbon-contaminated soils. *Journal of Hazardous Materials, 401*, 123282.

Horn, M., French-McCay, D., & Etkin, D. S. (2024). Environmental impact and modeling of petroleum spills. In *Threats to our ocean heritage: potentially polluting wrecks* (pp. 25–39). Springer Nature Switzerland. https://doi.org/10.1007/978-3-031-57960-8_3

Hu, G., Feng, H., He, P., Li, J., Hewage, K., & Sadiq, R. (2020). Comparative life-cycle assessment of traditional and emerging oily sludge treatment approaches. *Journal of Cleaner Production, 251*, 119594.

Huang, Y., Pan, H., Wang, Q., Ge, Y., Liu, W., & Christie, P. (2019). Enrichment of the soil microbial community in the bioremediation of a petroleum-contaminated soil amended with rice straw or sawdust. *Chemosphere, 224*, 265–271.

Hunt, L. J., Duca, D., Dan, T., & Knopper, L. D. (2019). Petroleum hydrocarbon (PHC) uptake in plants: A literature review. *Environmental Pollution, 245*, 472–484.

Hussain, M. I., Khan, Z. I., Akhter, P., Al-Hemaid, F. M., Al-Hashimi, A., Elshikh, M. S., Ahmad, K., & Yang, H. H. (2022). Potential of organic amendments for heavy metal contamination in soil–coriander system: Environmental fate and associated ecological risk. *Sustainability, 14*(18), 11374. https://doi.org/10.3390/su141811374

Ince, M., & Ince, O. K. (2019). Introductory chapter: sources, health impact, and environment effect of hydrocarbons. In *Hydrocarbon pollution and its effect on the environment* (p. 1). IntechOpen.

Iyer, N. S., Mandaliya, D. D., & Dave, S. R. (2022). In situ microbial and phytoremediation of crude oil contaminated soil by *Cynodon sp. International Journal of Phytoremediation, 24*(3), 235–244.

Jagatee, S., Priyadarshini, S., Rath, C. C., & Das, A. P. (2024). Synthetic microfiber: An enduring environmental problem linked to sustainable development. *Environmental Science and Engineering, Part F2333*, 93–112. https://doi.org/10.1007/978-3-031-51792-1_6

Jeanneret, P., Lüscher, G., Schneider, M. K., Pointereau, P., Arndorfer, M., Bailey, D., Balázs, K., Báldi, A., Choisis, J. P., Dennis, P., Diaz, M., & Herzog, F. (2021). An increase in food production in Europe could dramatically affect farmland biodiversity. *Communications Earth & Environment, 2*(1), 183.

Kanungo, J., Sahoo, T., Swain, L. P., & Behera, I. D. (2023). *Toxicity of persistent hydrocarbon pollutants, sources and sustainable remediation process* (pp. 39–65). https://doi.org/10.1007/978-3-031-48220-5_2

Kanungo, J., Sahoo, T., Bal, M., & Behera, I. D. (2024). Performance of bioremediation strategy in waste lubricating oil pollutants: A review. *Geomicrobiology Journal, 41*(4), 360–373. https://doi.org/10.1080/01490451.2023.2245395

Kathi, S. (2022). Phytoremediation of heavy metals and petroleum hydrocarbons using *Cynodon dactylon (L.)* Pers. In *Cost effective technologies for solid waste and wastewater treatment* (pp. 135–145). Elsevier.

Khan, S., Waqas, M., Ding, F., Shamshad, I., Arp, H. P. H., & Li, G. (2015). The influence of various biochars on the bioaccessibility and bioaccumulation of PAHs and potentially toxic elements to turnips (*Brassica rapa L.*). *Journal of Hazardous Materials, 300*, 243–253.

Kim, S. U., Owens, V. N., Kim, S. Y., & Hong, C. O. (2017). Effect of different way of bottom ash and compost application on phytoextractability of cadmium in contaminated arable soil. *Applied Biological Chemistry, 60*, 353–362.

Kim, S. H., Woo, H., An, S., Chung, J., Lee, S., & Lee, S. (2022). What determines the efficacy of landfarming for petroleum-contaminated soils: Significance of contaminant characteristics. *Chemosphere, 290*, 133392.

Kong, L., Gao, Y., Zhou, Q., Zhao, X., & Sun, Z. (2018). Biochar accelerates PAHs biodegradation in petroleum-polluted soil by biostimulation strategy. *Journal of Hazardous Materials, 343*, 276–284.

Koshlaf, E., Shahsavari, E., Haleyur, N., Osborn, A. M., & Ball, A. S. (2019). Effect of biostimulation on the distribution and composition of the microbial community of a polycyclic aromatic hydrocarbon-contaminated landfill soil during bioremediation. *Geoderma, 338*, 216–225.

Koshlaf, E., Shahsavari, E., Haleyur, N., Osborn, A. M., & Ball, A. S. (2020). Impact of necrophytoremediation on petroleum hydrocarbon degradation, ecotoxicity and soil bacterial community composition in diesel-contaminated soil. *Environmental Science and Pollution Research, 27*, 31171–31183.

Koshlaf, E. M. (2024). *Influence of necrophytoremediation as a bioremediation tool on the activity, composition and diversity of the microbial community in petrogenic hydrocarbon contaminated soils* (Doctoral dissertation, RMIT University).

Kou, B., Huo, L., Cao, M., Yu, T., Wu, Y., Hui, K., Tan, W., Yuan, Y., & Zhu, X. (2024). Applying kitchen compost promoted soil chrysene degradation by optimizing microbial community structure. *Journal of Environmental Management, 370*, 122761. https://doi.org/10.1016/j.jenvman.2024.122761

Kraut-Cohen, J., Zolti, A., Rotbart, N., Bar-Tal, A., Laor, Y., Medina, S., Shawahna, R., Saadi, I., Raviv, M., Green, S. J., Yermiyahu, U., & Minz, D. (2023). Short-and long-term effects of continuous compost amendment on soil microbiome community. *Computational and Structural Biotechnology Journal, 21*, 3280–3292. https://doi.org/10.1016/j.csbj.2023.05.030

Kulikova, N. A., & Perminova, I. V. (2021). Interactions between humic substances and microorganisms and their implications for nature-like bioremediation technologies. *Molecules, 26*(9), 2706.

Lahiri, D., Nag, M., Dey, A., Sarkar, T., Joshi, S., Pandit, S., Das, A. P., Pati, S., Pattanaik, S., Tilak, V. K., & Ray, R. R. (2021). Biofilm mediated degradation of petroleum products. *Geomicrobiology Journal*, 1–10. https://doi.org/10.1080/01490451.2021.1968979

Lara-Moreno, A., Morillo, E., Merchán, F., Madrid, F., & Villaverde, J. (2022). Chlorpyrifos removal in an artificially contaminated soil using novel bacterial strains and cyclodextrin. Evaluation of its effectiveness by ecotoxicity studies. *Agronomy, 12*(8), 1971. https://doi.org/10.3390/agronomy12081971

Le Quéré, C., Andrew, R. M., Canadell, J. G., Sitch, S., Korsbakken, J. I., Peters, G. P., Zaehle, S., et al. (2016). Global carbon budget 2016. *Earth System Science Data, 8*(2), 605–649.

Li, A., Fan, J., Jia, Y., Tang, X., Chen, J., & Shen, C. (2023). Phenotype and metabolism alterations in PCB-degrading Rhodococcus biphenylivorans TG9T under acid stress. *Journal of Environmental Sciences, 127*, 441–452. https://doi.org/10.1016/j.jes.2022.05.016

Li, X., Song, Y., Yao, S., Bian, Y., Gu, C., Yang, X., Wang, F., & Jiang, X. (2019). Can biochar and oxalic acid alleviate the toxicity stress caused by polycyclic aromatic hydrocarbons in soil microbial communities? *Science of the Total Environment, 695*, 133879.

Li, Y., Zhou, M., Li, C., Pan, X., Lv, N., Ye, Z., Zhu, G., Zhao, Q., & Cai, G. (2022). Inoculating indoleacetic acid bacteria promotes the enrichment of halotolerant bacteria during secondary fermentation of composting. *Journal of Environmental Management, 322*, 116021. https://doi.org/10.1016/j.jenvman.2022.116021

Liu, M., Wang, C., Wang, F., & Xie, Y. (2019). Vermicompost and humic fertilizer improve coastal saline soil by regulating soil aggregates and the bacterial community. *Archives of Agronomy and Soil Science, 65*(3), 281–293.

Liu, Q., Gao, S., Zhou, Q., Xu, R., Li, Z., Hou, Y., Sun, Y., & Huang, C. (2022). Bio-augmentation of the filler-enhanced denitrifying sulfide removal process in expanded granular sludge bed reactors. *Environmental Research, 212*, 113253. https://doi.org/10.1016/j.envres.2022.113253

Ma, Q., Meng, N., Su, J., Li, Y., Gu, J., Wang, Y., Qu, Y., Zhao, Z., & Sun, Y. (2023). Unraveling the skatole biodegradation process in an enrichment consortium using integrated omics and culture-dependent strategies. *Journal of Environmental Sciences, 127*, 688–699. https://doi.org/10.1016/j.jes.2022.06.025

Mallick, K., Sahu, A., Dubey, N. K., & Das, A. P. (2023). Harvesting marine plastic pollutants-derived renewable energy: A comprehensive review on applied energy and sustainable approach. *Journal of Environmental Management, 348*, 119371. https://doi.org/10.1016/J.JENVMAN.2023.119371

Mandal, S., Pu, S., Adhikari, S., Ma, H., Kim, D. H., Bai, Y., & Hou, D. (2021). Progress and future prospects in biochar composites: Application and reflection in the soil environment. *Critical Reviews in Environmental Science and Technology, 51*(3), 219–271.

Medina, J., Monreal, C., Barea, J. M., Arriagada, C., Borie, F., & Cornejo, P. (2015). Crop residue stabilization and application to agricultural and degraded soils: A review. *Waste Management, 42*, 41–54.

Mekontchou, O. Y., Zhenhua, Z., Nkoh, J. N., Ymele, E., & Usman, M. (2024). A systematic review of polycyclic aromatic hydrocarbon pollution: A combined bibliometric and mechanistic analysis of research trend toward an environmentally friendly solution. *Science of The Total Environment*, 171577.

Mishra, S., Dash, D., Al-Tawaha, A. R. M. S., & Das, A. P. (2022). A review on heavy metal ion adsorption on synthetic microfiber surface in aquatic environments. *Applied Biochemistry and Biotechnology, 194*(10), 4639–4654. https://doi.org/10.1007/S12010-022-04029-W/METRICS

Mishra, S., Ghosh, S., van Hullebusch, E. D., Singh, S., & Das, A. P. (2023). A critical review on the recovery of base and critical elements from electronic waste-contaminated streams using microbial biotechnology. *Applied Biochemistry and Biotechnology, 2023*, 1–30. https://doi.org/10.1007/S12010-023-04440-X

Mishra, S., Singh, R. P., Rath, C. C., & Das, A. P. (2020). Synthetic microfibers: Source, transport and their remediation. *Journal of Water Process Engineering, 38*, 101612. https://doi.org/10.1016/J.JWPE.2020.101612

Mogollón, J. M., Beusen, A. H. W., Van Grinsven, H. J. M., Westhoek, H., & Bouwman, A. F. (2018). Future agricultural phosphorus demand according to the shared socioeconomic pathways. *Global Environmental Change, 50*, 149–163.

Mohanta, S., Pradhan, B., & Behera, I. D. (2024a). Impact and remediation of petroleum hydrocarbon pollutants on agricultural land: A review. *Geomicrobiology Journal, 41*(4), 345–359. https://doi.org/10.1080/01490451.2023.2243925

Moitra, A. K., Bhaduri, A., Mukhopadhayay, G. C., & Das, A. K. (2022). Exploration of shale gas in India–Prospects and challenges. In *Innovative exploration methods for minerals, oil, gas, and groundwater for sustainable development* (pp. 29–42).

Niu, A., Sun, X., & Lin, C. (2022). Trend in research on characterization, environmental impacts and treatment of oily sludge: A systematic review. *Molecules, 27*(22), 7795.

Ntekpe, M. E., Ekpo, M. A., Ndubuisi-Nnaji, U. U., Mbong, E. O., & Udoidiong, D. C. (2023). Microbiological status, physico-chemistry and plant performance in spent lubricating oil (SLO) polluted soils amended with spiked organic fertilizer. *Asian Journal of Agricultural and Horticultural Research, 10*(4), 246–260. https://doi.org/10.9734/ajahr/2023/v10i4267

Ntroumpogianni, G. C., Giannoutsou, E., Karagouni, A. D., & Savvides, A. L. (2022). Bacterial isolates from Greek sites and their efficacy in degrading petroleum. *Sustainability, 14*(15), 9562. https://doi.org/10.3390/su14159562

Nwankwegu, A. S., Onwosi, C. O., Orji, M. U., Anaukwu, C. G., Okafor, U. C., Azi, F., & Martins, P. E. (2016). Reclamation of DPK hydrocarbon polluted agricultural soil using a selected bulking agent. *Journal of Environmental Management, 172*, 136–142.

Olarewaju Adeleye, A., Yerima, M. B., Nkereuwem, M. E., Onokebhagbe, V. O., Shiaka, G. P., Oladeji Amoo, A., Asaju, C. I., Yalwaji, B., & Ishaq, S. M. (2022). Effect of bacterial co-culture and organic amendments on the bioremediation of hydrocarbons in a soil contaminated with spent engine oil. *Acta Scientiarum: Biological Sciences, 44*. https://doi.org/10.4025/actascibiolsci.v44i1.62289

Olawale, O., Obayomi, K. S., Dahunsi, S. O., & Folarin, O. (2020). Bioremediation of artificially contaminated soil with petroleum using animal waste: Cow and poultry dung. *Cogent Engineering, 7*(1), 1721409.

Olugbenga, O. S., Adeleye, P. G., Oladipupo, S. B., Adeleye, A. T., & John, K. I. (2024). Biomass-derived biochar in wastewater treatment-a circular economy approach. *Waste Management Bulletin, 1*(4), 1–14. https://doi.org/10.1016/j.wmb.2023.07.007

Omokaro, G. O., Osarhiemen, I. O., Idama, V., Airueghian, E. O., West, S. T., Igbigbi, F. E., Nnake, D. C., Obolokor, E., Ahmed, A., & Omoshie, V. O. (2024). The role of organic amendments and their impact on soil restoration: A review. https://doi.org/10.9734/ajee/2024/v23i11620

Pang, A., Rutter, A., Bordenave, S., Gainer, A., Haack, E., & Zeeb, B. (2022). Assessment of the toxicity of weathered petroleum hydrocarbon impacted soils to native plants from a site in the Canadian Subarctic. *Ecotoxicology, 31*(8), 1287–1298.

Paul, A., Dey, S., Ram, D. K., & Das, A. P. (2023). Hexavalent chromium pollution and its sustainable management through bioremediation. *Geomicrobiology Journal*, 1–11. https://doi.org/10.1080/01490451.2023.2218377

Paul, A., Priyadarsini, S., Karakuş, S., Behera, J. K., Gnanasekaran, R., & Das, A. P. (2024). Fate and accumulation of synthetic microfibers in the environment. In *Sustainable microbial technology for synthetic and cellulosic microfiber bioremediation* (pp 61–76). https://doi.org/10.1007/978-3-031-62600-5_44

Pekkoh, J., Chaichana, C., Thurakit, T., Phinyo, K., Lomakool, S., Ruangrit, K., Duangjan, K., Suwannarach, N., Kumla, J., Cheirsilp, B., & Srinuanpan, S. (2022). Dual-bioaugmentation strategy to enhance the formation of algal-bacteria symbiosis biofloc in aquaculture wastewater supplemented with agricultural wastes as an alternative nutrient sources and biomass support materials. *Bioresource Technology, 359*, 127469. https://doi.org/10.1016/j.biortech.2022.127469

Pichtel, J. (2016). Oil and gas production wastewater: Soil contamination and pollution prevention. *Applied and Environmental Soil Science, 2016*(1), 2707989.

Popoola, L. T., Olawale, T. O., & Salami, L. (2023). A review on the fate and effects of contaminants in biosolids applied on land: Hazards and government regulatory policies. *Heliyon, 9*(10).

Prabhakar, A., Mishra, S., & Das, A. P. (2019). Isolation and Identification of lead (Pb) solubilizing bacteria from automobile waste and its potential for recovery of lead from end of life waste batteries. *Geomicrobiological Journal, 36*(10), 894–903. https://doi.org/10.1080/01490451.2019.1654044

Priyadarsini, S., & Das, A. P. (2024). Analytical and structural characterization of waste lithium-ion batteries for their effective recycling strategy. *Environmental Chemistry and Ecotoxicology*. https://doi.org/10.1016/j.enceco.2024.12.004

Priyadarsini, S., & Das, A. P. (2025). Lithium bioleaching: A review on microbial-assisted sustainable technology for lithium bio-circularity. *Journal of Water Process Engineering, 69*, 106744. https://doi.org/10.1016/j.jwpe.2024.106744

Qin, G., Gong, D., & Fan, M. Y. (2013). Bioremediation of petroleum-contaminated soil by biostimulation amended with biochar. *International Biodeterioration & Biodegradation, 85*, 150–155.

Raj, K., & Das, A. P. (2023). Lead pollution: Impact on environment and human health and approach for a sustainable solution. *Environmental Chemistry and Ecotoxicology, 5*, 79–85. https://doi.org/10.1016/J.ENCECO.2023.02.001

Rajapaksha, A. U., Ahmad, M., Vithanage, M., Kim, K. R., Chang, J. Y., Lee, S. S., & Ok, Y. S. (2015). The role of biochar, natural iron oxides, and nanomaterials as soil amendments for immobilizing metals in shooting range soil. *Environmental Geochemistry and Health, 37*, 931–942.

Rasa, K., Heikkinen, J., Hannula, M., Arstila, K., Kulju, S., & Hyväluoma, J. (2018). How and why does willow biochar increase a clay soil water retention capacity? *Biomass and Bioenergy, 119*, 346–353.

Ren, L., Guo, Z., Zhang, L., Hu, H., Li, C., Lin, Z., Zhen, Z., & Zhou, J. L. (2023). A novel aerobic denitrifying phosphate-accumulating bacterium efficiently removes phthalic acid ester, total nitrogen and phosphate from municipal wastewater. *Journal of Water Process Engineering, 52*, 103532. https://doi.org/10.1016/j.jwpe.2023.103532

Rickson, J., & Fullen, M. (2022). Soil Conservation Programmes and Policies in England and Wales. In *Global degradation of soil and water resources: Regional assessment and strategies* (pp. 351–365). Springer.

Robichaud, K., Lebeau, M., Martineau, S., & Amyot, M. (2019). Bioremediation of engine-oil contaminated soil using local residual organic matter. *PeerJ, 7*, e7389. https://doi.org/10.7717/peerj.7389

Sarma, H., Bustamante, K. L. T., & Prasad, M. N. V. (2019). Biosurfactants for oil recovery from refinery sludge: Magnetic nanoparticles assisted purification. In *Industrial and municipal sludge* (pp. 107–132). Butterworth-Heinemann.

Shah, G., & Soni, V. (2024). Comprehensive Insights into the Impact of Oil Pollution on the Environment. *Regional Studies in Marine Science*, 103516. https://doi.org/10.1016/j.rsma.2024.103516

Shakam, H. M., & Gaber, M. A. (2024). Biodegradation of petroleum hydrocarbon and related molecular microbiology analysis of some fungal species. *Alexandria Journal of Agricultural Sciences, 69*(3), 316–327. https://doi.org/10.21608/alexja.2024.304757.1081

Singh, V. K., Malhi, G. S., Kaur, M., Singh, G., & Jatav, H. S. (2022). Use of organic soil amendments for improving soil ecosystem health and crop productivity. *Ecosystem Services.*

Singh, R. P., Mishra, S., & Das, A. P. (2020). Synthetic microfibers: Pollution toxicity and remediation. *Chemosphere, 257*, 127199. https://doi.org/10.1016/j.chemosphere.2020.127199

Silva, J. A., Guelfo, J. L., Šimůnek, J., & McCray, J. E. (2022). Simulated leaching of PFAS from land-applied municipal biosolids at agricultural sites. *Journal of Contaminant Hydrology, 251*, 104089.

Szentgyörgyi, F., Benedek, T., Fekete, D., Táncsics, A., Harkai, P., & Kriszt, B. (2022). Development of a bacterial consortium from Variovorax paradoxus and Pseudomonas veronii isolates applicable in the removal of BTEX. *AMB Express, 12*(1), 4. https://doi.org/10.1186/s13568-022-01349-2

Thangarajan, R., Bolan, N. S., Tian, G., Naidu, R., & Kunhikrishnan, A. (2013). Role of organic amendment application on greenhouse gas emission from soil. *Science of the Total Environment, 465*, 72–96.

Tubeileh, A. M., & Goss, M. J. (2022). Assessing the effects of using animal manure on soil health.

Udume, O. A., Abu, G. O., Stanley, H. O., Vincent-Akpu, I. F., Momoh, Y., & Eze, M. O. (2023). Biostimulation of petroleum-contaminated soil using organic and inorganic amendments. *Plants, 12*(3), 431. https://doi.org/10.3390/plants12030431

Wallman, M. (2021). *Nitrogen Losses from a Clay-rich Soil used for Cereal Production in southwestern Sweden.*

Wang, B., Zeng, D., Chen, Y., Belzile, N., Bai, Y., Zhu, J., Shu, J., & Chen, S. (2019). Adsorption behaviors of phenanthrene and bisphenol A in purple paddy soils amended with straw-derived DOM in the West Sichuan Plain of China. *Ecotoxicology and Environmental Safety, 169*, 737–746.

Wang, D., Lin, J. Y., Sayre, J. M., Schmidt, R., Fonte, S. J., Rodrigues, J. L., & Scow, K. M. (2022a). Compost amendment maintains soil structure and carbon storage by increasing available carbon and microbial biomass in agricultural soil—A six-year field study. *Geoderma, 427*, 116117. https://doi.org/10.1016/j.geoderma.2022.116117

Wang, S., Wang, X., Zhang, C., Li, F., & Guo, G. (2016). Bioremediation of oil sludge contaminated soil by landfarming with added cotton stalks. *International Biodeterioration & Biodegradation, 106*, 150–156.

Wang, W., Man, Z., Li, X., Chen, R., You, Z., Pan, T., Dai, X., Xiao, H., & Liu, F. (2023). Response mechanism and rapid detection of phenotypic information in rice root under heavy metal stress. *Journal of Hazardous Materials, 449*, 131010.

Wang, Z., Su, J., Ali, A., Sun, Y., Li, Y., Yang, W., & Zhang, R. (2022b). Enhanced removal of fluoride, nitrate, and calcium using self-assembled fungus-flexible fiber composite microspheres combined with microbially induced calcium precipitation. *Chemosphere, 302*, 134848. https://doi.org/10.1016/j.chemosphere.2022.134848

Wu, C. M. E., & Teng, C. C. (2022). Reducing food waste in buffet restaurants: A corporate management approach. *Foods, 12*(1), 162.

Wu, H., Lai, C., Zeng, G., Liang, J., Chen, J., Xu, J., Dai, J., Li, X., Liu, J., Chen, M., & Lu, L. (2017). The interactions of composting and biochar and their implications for soil amendment and pollution remediation: a review. *Critical Reviews in Biotechnology., 37*(6), 754–764.

Yang, Y., Liu, L., Bai, Z., Xu, W., Zhang, F., Zhang, X., Liu, X., & Xie, Y. (2022). Comprehensive quantification of global cropland ammonia emissions and potential abatement. *Science of the Total Environment, 812*, 151450.

You, Y., Chi, Y., Chen, X., Wang, J., Wang, R., Li, R., Chu, S., Yang, X., Zhang, D., & Zhou, P. (2022). A sustainable approach for bioremediation of secondary salinized soils: Studying remediation efficiency and soil nitrate transformation by bioaugmentation. *Chemosphere, 300*, Article 134580. https://doi.org/10.1016/j.chemosphere.2022.134580

Yue, Z., Shi, Q., Ai, J., Peng, S., Miao, X., & Wang, Z. (2021). A review on analytical methods of petroleum hydrocarbons in water and sediment of aquatic systems. In *IOP conference series: Earth and environmental science* (Vol. 621, No. 1, p. 012011). IOP Publishing.

Zacharias, D. C., Lemos, A. T., Keramea, P., Dantas, R. C., da Rocha, R. P., Crespo, N. M., Sylaios, G., Jovane, L., da Silva Santos, I. G., Montone, R. C., & de Oliveira, S. M. (2024). Offshore oil spills in Brazil: An extensive review and further development. *Marine Pollution Bulletin., 205*, 116663.

Zaharioiu, A. M., Bucura, F., Ionete, R. E., Marin, F., Constantinescu, M., & Oancea, S. (2021). Opportunities regarding the use of technologies of energy recovery from sewage sludge. *SN Applied Sciences, 3*(9), 775.

Zenteno-Rojas, A., Rincón-Molina, C. I., Martinez-Romero, E., Manzano-Gomez, L. A., Rincón-Molina, F. A., Ruiz-Valdiviezo, V. M., Rodriguez, I. C., Maldonado, J. J. V., & Rincón-Rosales, R. (2022). Biostimulation and bioaugmentation of soils contaminated with decachlorobiphenyl (PCB-209) using native bacterial strains individually and in consortia. *Sustainability, 14*(15), 9068. https://doi.org/10.3390/su14159068

Zhang, M., Chen, Q., & Gong, Z. (2024). Microbial remediation of petroleum-contaminated soil focused on the mechanism and microbial response: A review. *Environmental Science and Pollution Research, 31*(23), 33325–33346.

Zhang, Z., Yan, X., Gao, F., Thai, P., Wang, H., Chen, D., Zhou, L., Gong, D., & Wang, B. (2018). Emission and health risk assessment of volatile organic compounds in various processes of a petroleum refinery in the Pearl River Delta, China. *Environmental Pollution, 238*, 452–461.

Zhou, N., Guo, H., Liu, Q., Zhang, Z., Sun, J., & Wang, H. (2022). Bioaugmentation of polycyclic aromatic hydrocarbon (PAH)-contaminated soil with the nitrate-reducing bacterium PheN7 under anaerobic condition. *Journal of Hazardous Materials, 439*, 129643. https://doi.org/10.1016/j.jhazmat.2022.129643

Zhu, G., Zhang, H., Yuan, R., Huang, M., Liu, F., Li, M., Zhang, Y., & Rittmann, B. E. (2023). How *Comamonas testosteroni* and *Rhodococcus ruber* enhance nitrification in the presence of quinoline. *Water Research, 229*, 119455. https://doi.org/10.1016/j.watres.2022.119455

MIX
Papier aus verantwortungsvollen Quellen
Paper from responsible sources
FSC® C105338

If you have any concerns about our products,
you can contact us on
ProductSafety@springernature.com

In case Publisher is established outside the EU,
the EU authorized representative is:
Springer Nature Customer Service Center GmbH
Europaplatz 3, 69115 Heidelberg, Germany

Printed by Libri Plureos GmbH
in Hamburg, Germany